Environmental Analysis

THE NEPA EXPERIENCE

Edited by

Stephen G. Hildebrand
Johnnie B. Cannon

LEWIS PUBLISHERS
Boca Raton Ann Arbor London Tokyo

Library of Congress Cataloging-in-Publication Data

The Scientific challenges of NEPA: future directions based on 20 years of experience / edited by Stephen G. Hildebrand and Johnnie B. Cannon.
p. cm.
Includes bibliographical references and index.
ISBN 0-87371-908-5
1. Environmental impact analysis—United States—Congresses. 2. United States. National Environmental Policy Act of 1969—Congresses. 3. Environmental policy—United States—Congresses. I. Hildebrand, Stephen G. II. Cannon, J. B. (Johnnie B.)
TD194.65.S35 1993
333.7′14′0973—dc20 92-33227
CIP

Direct all inquiries to CRC Press, Inc., 2000 Corporate Blvd., N.W., Boca Raton, Florida 33431.

PRINTED IN THE UNITED STATES OF AMERICA
1 2 3 4 5 6 7 8 9 0
Printed on acid-free paper

Dedication

This book is dedicated to the memory of H. E. "Bud" Zittel, who was instrumental in forming and managing the National Environmental Policy Act (NEPA) program at Oak Ridge National Laboratory (ORNL), Oak Ridge, Tennessee. Bud had been involved in ORNL's NEPA work since its inception in 1971, shortly after the Calvert Cliffs decision. He served as task group leader and program manager for a number of environmental impact statements for proposed nuclear power plants. As a result of this work, Dr. Zittel became a recognized expert on the environmental impacts of nuclear power production. In the late 1970s, Bud was instrumental in broadening the ORNL NEPA program to include a broad spectrum of energy technologies, including biomass, conservation, geothermal, fusion, and synthetic fuels, and his efforts helped ORNL become a world-recognized center of expertise on assessing the potential environmental impacts of energy technologies. During the 1980s, under Dr. Zittel's leadership, the ORNL NEPA program attracted a variety of other sponsors, including the U.S. Army, U.S. Air Force, and other agencies, further strengthening the depth and experience of the program.

In addition to his role in building a successful NEPA program at ORNL, Bud was perhaps better known for his understanding of the intent of NEPA and the role of the President's Council on Environmental Quality (CEQ) regulations in meeting that intent. The depth of Bud's comprehension of NEPA and the CEQ regulations not only was instrumental in the development of ORNL as a center of NEPA expertise, but also played a key role in training future NEPA project leaders and program managers. Bud, who was retired from ORNL, passed away on April 5, 1990. His knowledge of NEPA, sense of humor, and good nature will be missed. However, many of the NEPA documents, reports, and papers prepared by ORNL staff members reflect Bud's ideas, thoughts, and philosophy, and in this sense, he continues to play an important role in the development and evolution of the NEPA process.

Preface

January 1, 1990, marked the 20th anniversary of the signing of the National Environmental Policy Act (NEPA). Since this law was enacted, numerous institutions have assisted federal agencies in the implementation of NEPA, including the preparation of environmental impact statements and environmental assessments. The Ninth Oak Ridge National Laboratory Life Sciences Symposium was dedicated to the celebration of this anniversary. The symposium was held October 24–27, 1989 in Knoxville, Tennessee. The intent of the symposium was (1) to review what has been learned while performing NEPA assessments, (2) to summarize the state-of-the-art in methods and approaches, and (3) to define future opportunities and new approaches required to link high quality science to the decisionmaking process. The conference consisted of invited papers, contributed papers, and a poster session that maximized participation. Plenary sessions addressed the future of science in environmental policy, environmental impact assessment abroad, the integration of NEPA with other environmental laws, and the future challenges of NEPA. Technical sessions addressed the NEPA process, assessing impacts to ecological resources, assessing social impacts, advanced assessment techniques, quantifying sources and fate of environmental pollutants, regional and global issues, cumulative impacts, state environmental policy act experience, NEPA follow-up studies, the role of public involvement, international experience, and impacts to human health.

Financial sponsors of this symposium included the U.S. Department of Energy, the Oak Ridge National Laboratory and the Hazardous Waste Remedial Actions Program (both managed by Martin Marietta Energy Systems, Inc., for the U.S. Department of Energy under contract DE-AC05-84OR21400); the U.S. Environmental Protection Agency; the U.S. Fish and Wildlife Service; and the Electric Power Research Institute. We are grateful for the resources that provided support for both the conference and the preparation of this book. This is Publication No. 3926, Environmental Sciences Division, ORNL.

We express sincere gratitude to Sandy Bell, Rita Wadlington, Jennifer Seiber, Bonnie Reesor, and Joy Lee for their assistance in preconference arrangements and their logistic support during the conference. Joe Rich led the team that produced the book and was assisted early in the process by Charlie Hagan. Shirley Wright, Felicia Reeves, Lisa Haff, and Lana McDonald spent long hours word processing the book. Their dedication is sincerely appreciated.

Stephen G. Hildebrand
Johnnie B. Cannon
June 1992

Stephen G. Hildebrand is the Associate Director of the Environmental Sciences Division (ESD) at Oak Ridge National Laboratory (ORNL) in Oak Ridge, Tennessee. He also serves as an adjunct faculty member in the Ecology Program at the University of Tennessee, Knoxville, and is a Fellow of the American Association for the Advancement of Science. He joined ORNL in 1973 with an undergraduate degree in zoology from Wabash College, Crawfordsville, Indiana, and M.S. and Ph.D. degrees in Fisheries from the University of Michigan, Ann Arbor. His experience with NEPA includes both technical preparation and management of environmental analyses for nuclear power plants, nuclear fuel cycle facilities, hydroelectric projects, enhanced oil recovery projects, and analyses supporting decisions on Department of Defense operations. His research interests include the distribution of mercury in the environment, biogeochemical cycling of trace elements in the environment, environmental effects of hydroelectric development on aquatic systems, and the basic ecology of stream systems. Prior to becoming Associate Director of ESD at ORNL, he held group leader positions in the Aquatic Ecology Section and the Environmental Impact Program, managed the Environmental Impact Program, was head of the Environmental Analyses Section, and most recently was head of the Ecosystems Studies Section. He was a member of the National Research Council Ad Hoc Panel on Impact of Large Dams, the UNESCO International Hydrological Program Project Team on Energy Policies and Strategies for Water Resources Development, the Council on Environmental Quality/National Science Foundation Panel on Long-Term Ecological Research Needs, the Electric Power Research Institute Technical Review Committee for Hydroelectric Research, the Task Group on Assessments and Policy Analysis of the National Acid Precipitation Assessment Program, and an advisor to the Institute of Ecology for Assistance to the National Commission on Water Quality.

Johnnie B. Cannon is Associate Director of the Energy Division at Oak Ridge National Laboratory (ORNL) in Oak Ridge, Tennessee. He joined ORNL in 1975 with a B.S. degree in Mechanical Engineering from Tuskegee Institute, Alabama, and M.S. and Ph.D. degrees in Mechanical Engineering from the California Institute of Technology, Pasadena. During his 16-year tenure at ORNL, he has contributed to the development of environmental impact statements (EISs) and related research in various capacities: first, as a specialist in contaminant transport, then as team leader of EIS projects, and most recently as manager of the ORNL National Environmental Policy Act (NEPA) program. His NEPA work has involved nuclear power plants, nuclear fuel cycle facilities, magnetic fusion energy, fossil energy projects, low-level waste disposal, disposal of unitary chemical munitions, low-level flying aircraft, hydroelectric projects, and numerous other activities. Prior to becoming Associate Director of the Energy Division at ORNL, he held group leader and section head positions in the division. His current work is focused on energy and resource analysis, environmental impact assessment, and emergency preparedness.

CONTENTS

Chapter 5
Cumulative Impact Assessment

Chapter 6
Regional and Global Analysis

CHAPTER 1

INTRODUCTION

Introduction

S. G. Hildebrand and J. B. Cannon, Oak Ridge National Laboratory, Oak Ridge, TN

A logical outgrowth of the environmental awareness of the 1960s was the passage of the National Environmental Policy Act of 1969 (NEPA), Public Law 91-190, by the 91st Congress on January 1, 1970. The purposes of NEPA were to establish a policy that would encourage harmony between man and the environment, prevent or eliminate damage to the environment, stimulate the health and welfare of man, enrich the understanding of ecological systems and resources important to the nation, and establish the Council on Environmental Quality. One particular segment of NEPA [Section 102(2)(c)] has had a profound impact on the evolution of applied science in the United States and abroad. This section of NEPA established the basis for requiring federal agencies to prepare environmental assessments and environmental impact statements. The challenge to physical scientists, biological scientists, and social scientists presented by NEPA was to predict adverse (and beneficial) impacts on the environment arising from the execution of a wide range of projects, policies, and programs across the federal establishment. The evolution of the NEPA process has been a rocky one, and the development of technical skills, approaches, and methods to provide the necessary predictive capability has been a major focus of many scientists over the last 20 years.

During the 1970s, the early experience with NEPA evoked criticism from the technical community as well as the policy makers. Fairfax suggested that NEPA implementation was a disaster (Fairfax 1978). For many scientists involved in NEPA in the early days, the critical editorial by D. W. Schindler was particularly disturbing (Schindler 1976). He maintained that the idea of NEPA protecting the environment had backfired. He criticized the process for developing a *gray literature* that was nothing more than massive amounts of uninterpreted and incomplete descriptive data. Scientists participating in the process were characterized as a *traveling circus* which produced products that seldom received hard scrutiny similar to peer review in scientific journals. He characterized the methods employed in the NEPA process as *ancient, descriptive textbook techniques*, that did not take advantage of new developments in science. Schindler even suggested the scientific method was in jeopardy. Schindler's viewpoint was refuted (Auerbach et al. 1976), but he certainly got the attention of the research scientists as well as the applied scientific community. Additional criticism of the NEPA process, as practiced in the 1970s, proliferated. Efford (1975) characterized a mid-1970s approach to impact assessment of hydroelectric dams in Canada as *haphazard*. He suggested that old techniques of limnology and ecology were used and that questions to be addressed were not systematically defined. The result was that biologists and ecologists were not capable of providing precise evidence in court on *serious* or *irreparable* injury. Eberhardt (1976) reviewed ecological approaches to making impact assessment quantitative. He concluded that experimental approaches utilizing preproject and postproject studies at affected and control sites suffered from lack of replication in the NEPA process. He questioned the use of basic ecological data on productivity and diversity and recommended, among other things, that improved ecological census methods should be developed.

Two studies in the 1980s further evaluated technical performance in the environmental assessment process. The National Science Foundation commissioned a study of ways to improve the scientific content and methodology of an environmental impact analysis (Indiana University 1982). A major point of this study recognized that impact assessments in the NEPA context are not *scientific* documents, but that *science* is certainly used in the NEPA process. It was emphasized that advancements in science alone will not necessarily strengthen the scientific quality of impact assessments. This policy-oriented analysis began to focus on the social science needs and the way in which science can contribute to policy issues. An excellent study of the role of basic ecology in environmental assessment was completed in Canada (Beanlands and Duinker 1983). This thorough study defined requirements for conducting ecological impact studies and made recommendations pertaining to institutional aspects of impact assessment. The 1980s also witnessed the signing of the "Peace Treaty for the Hudson," a classic case study in the role of science in impact analysis (Christensen et al. 1981). This out-of-court settlement ended 17 years of controversy over the impact of electric power generation on Hudson River fish populations. This case provided one profound lesson for all involved in environmental impact assessment: once technical issues are formally delineated, it is possible to carry out scientific assessments. It is inevitable, however, that these assessments will also be made challenging by the complexity of biological systems, by the limited ability to describe such systems, and by the interests of various parties and institutions whose conflicts provide the basic forces for the assessment process. The case also documented the effectiveness of adjudicatory hearings as a powerful stimulus for defining and resolving issues that are both important to society and amenable to scientific analysis.

Impact assessment today should benefit from more than 20 years experience with the NEPA process. This book presents examples of the state of science in impact assessment based on the learning of two decades. Individual chapters address the process itself; examples of recent experience with ecological impact assessment; evaluation of social impact assessment and the important role the public must play; the difficult challenge of assessing cumulative effects of multiple impacts; the regional and global implications of NEPA; the important role of follow-up studies in the process; and federal, state, and international experience. Contributors of individual papers represent the major sectors that have been key participants in the process from the beginning (academia, national laboratories, federal agencies, state agencies, the private sector, and foreign nations). Both the topical diversity evidenced in this book and the representation of the major players provide a comprehensive snapshot of the NEPA process as we face the NEPA challenges of the 1990s.

The technical challenge of NEPA for the future has several dimensions. There is a continuing need for scientifically defensible analyses in NEPA documents. The NEPA approach to decisionmaking is being utilized for issues of ever-increasing complexity and scope. The difficult technical issues surrounding acidic deposition, global climate change, and management of hazardous wastes are a few examples of the types of problems where scientific uncertainty is great. NEPA assessments must always frame conclusions acknowledging this uncertainty. The NEPA process will serve as an important public forum for evaluating these and other issues, and scientific input to the process continues to be essential for rational decisionmaking. Integrating the complex regulatory requirements for hazardous waste management under the Resource Conservation and Recovery Act, the cleanup of past waste operations under Superfund, and the consideration of environmental consequences of these activities under NEPA are current challenges facing managers of many federal facilities. The technical experience with NEPA documented in this book should be a useful reference for addressing these types of issues.

NEPA is an important statute for both the public and scientific community. Experience dictates that NEPA analyses receive widespread public review and comment and assist federal agencies in giving careful consideration to the environmental effects of their actions. The NEPA process gives other agencies, public interest groups, and the general public the opportunity to express alternative views and debate significant issues. Thus, the NEPA process provides an important public forum for discussing the environmental effects of federal government activities and major environmental issues. The role of scientists in this process should be to continually improve the predictive capability of their respective technical disciplines to address environmental issues and to explicitly state the uncertainty of their predictions. We hope this book stimulates scientific progress in fulfilling the intent of NEPA in the next 20 years equal to the progress realized in the first 20 years.

REFERENCES

Auerbach, S. I., R. W. Brocksen, R. B. Craig, F. O. Hoffman, S. V. Kaye, D. E. Reichle, and E. G. Struxness. 1976. Environmental impact statements (Letter to the Editor). *Science* 193:188.

Beanlands, G. E., and P. M. Duinker. 1983. An Ecological Framework for Environmental Impact Assessment in Canada. ISBX O-7703-0460-S. Dalhousie University, Institute for Environmental Studies, Dalhousie University, Halifax, Nova Scotia, Canada.

Christensen, S. W., W. Van Winkle, L. W. Barnthouse, and D. S. Vaughn. 1981. Science and the law: Confluence and conflict on the Hudson River. *EIA Rev.* 2(1):63–88.

Eberhardt, L. L. 1976. Quantitative ecology and impact assessment. *J. Environ. Manage.* 4:27–40.

Fairfax, S. K. 1978. A disaster in the environmental movement. *Science* 199:743–747.

Indiana University School of Public and Environmental Affairs. 1982. A Study of Ways to Improve the Scientific Content and Methodology of Environmental Impact Analysis. Final Report to the National Science Foundation on Grant PRA-79-10017. Indiana University School of Public and Environmental Affairs, Bloomington, IN.

Schindler, D. W. 1976. The impact statement boondoggle. *Science* 192:50.

CHAPTER 2

THE NEPA PROCESS

Introduction

R. M. Reed and J. B. Cannon, Oak Ridge National Laboratory, Oak Ridge, TN

The process set in motion by the passage of the National Environmental Policy Act (NEPA) at the end of 1969 has evolved over the past 20 years into a well-defined set of procedures based on a concise section in the statute requiring federal agencies to prepare environmental impact statements (EISs) on major federal actions that significantly affect the quality of the human environment. The NEPA process developed in the 1970s as agencies attempted to comply with the Section 102(2)(c) requirement and were challenged in court on the adequacy of the documents they produced. A landmark in the evolution of the process was the publication in 1978 by the President's Council on Environmental Quality (CEQ) of regulations for implementing the procedural provisions of NEPA. These regulations embodied the early experience with the NEPA process and the substantive issues that were addressed by the courts. The regulations continue to provide a strong basis for conduct of the process and have only been modified slightly since their publication. Initial adoption of the regulations was a *pro forma* exercise by many agencies, but as time has gone by, most agencies have been forced to embrace to some degree the spirit as well as the letter of NEPA.

Opinions vary on the success of NEPA over the past 20 years. Although the CEQ regulations attempted to address basic problems with the process by implementing procedures to reduce paperwork, make documents useful to decision makers, and ensure that the environment was considered in the decisionmaking process, the focus was primarily on procedure — substance was not directly addressed. The regulations appear to be based on the assumption that sound, well-defined procedures will result in high quality documents.

This chapter provides several perspectives on the success of the NEPA process over the past 20 years and offers a variety of ideas on future direction that are being or should be pursued. Lynton Caldwell argues that NEPA's implementation has been largely governed by the courts, resulting in an emphasis on procedure rather than substance. NEPA's intent "to draw upon science as an informant and corrective for public policies impacting upon the environment" is a major challenge that has yet to be filled. Phillip Gustafson reviews the implementation of the NEPA process by the Atomic Energy Commission in licensing commercial nuclear power plants, concluding that this experience was a success story providing new challenges to research scientists in dealing with problems without having complete data. Michael Gerrard discusses the judicial review of scientific evidence in EISs and concludes that strict compliance with NEPA procedures is "at least as important as the actual contents of the EISs." He advises that defensible EISs need to describe the scientific basis for all conclusions; explain why certain scientific methodologies were not used; ensure that specific testing methods required by regulatory agencies are used when appropriate; and for projects that are likely to be challenged in court, retain qualified experts to prepare the EIS who will be suitable expert witnesses in any subsequent legal setting. Owen Schmidt states that defining the underlying need for a proposed action is the most important step in defining the scope of an environmental document and determining the range of alternatives that need to be considered. Roger Nelson believes that the NEPA process places too much emphasis on one aspect of planning and has not resulted in a balanced process. He feels that the timing of EISs early in the planning process is inappropriate and suggests that environmental considerations should be incorporated throughout the planning and design process, not tagged on at the front end.

The integration of the NEPA process with requirements of the Comprehensive Environmental Response, Compensation, and Liability Act (CERCLA) and the Resource Conservation and Recovery Act (RCRA) is a current topic of much interest to many federal agencies. A panel with representatives from the CEQ, the Department of Energy (DOE), the Environmental Protection Agency (EPA), and the U.S. Air Force (USAF) discussed overlapping requirements and the concept of "functional equivalence." There is no clear guidance on the applicability of NEPA to cleanup actions. The DOE and USAF have a policy for integrating the decisionmaking process. The CEQ is supportive of integrating the processes and plans to issue guidance at a later date. The EPA is concerned with potential delays in cleanups that could be caused by the NEPA process, based in part on a Department of Justice memorandum.

Future challenges of NEPA were addressed in a panel with representatives from the CEQ, DOE, EPA, and USAF. Challenges of particular interest included the need to strengthen requirements for mitigation and monitoring; the extension of NEPA requirements to areas outside the territorial limits of the United States, particularly in light of increasing concern with global and transboundary environmental issues; the importance of follow-up studies to test the adequacy of predictions made in EISs and the effectiveness of monitoring and mitigation measures, and the need to open up the environmental assessment (EA) process to closer public involvement and scrutiny; improvement of the scientific basis for impact projections; and placement of greater emphasis on broader policy issues and cumulative impacts.

Clear themes arising from these various discussions are the need for federal agencies to address substantive aspects of the statute; that is, the intent or spirit of NEPA. Although the emphasis on procedure is frequently criticized, there is general agreement that NEPA has succeeded in forcing decision makers to consider the environment. Improving the scientific basis of NEPA analyses by evaluating the effectiveness of previous impact predictions and mitigation and monitoring measures is also recognized as important. Some agencies are moving in the direction of implementing follow-up studies, but funding for such activities in the past has rarely been available. Addressing programmatic issues, cumulative impacts, and impacts of global concern will also receive increasing attention as the NEPA process evolves during the next decade.

Achieving the NEPA Intent: New Directions in Politics, Science, and Law

L. K. Caldwell, Indiana University, Bloomington, IN

ABSTRACT

The National Environmental Policy Act (NEPA) has had a significant influence on public policy in the United States and abroad; the procedural reform required by the environmental impact statement (EIS) provision has improved the quality of planning and decisionmaking. But the EIS alone is insufficient to fully achieve the intent declared in NEPA. Title II of NEPA has yet to be fully implemented. The Council on Environmental Quality (CEQ) has done what it could with unduly limited resources, but active presidential support is needed for it to play the role indicated under Title II. In default of positive White House initiatives, the courts have been the principal interpreters and enforcers of NEPA, but their jurisdiction has been largely limited to procedure. If the NEPA intent is to be achieved, consistent with its substantive goals, a basis in constitutional law may be necessary. After two decades of experience, sufficient time has elapsed to permit a fair assessment of what difference the NEPA has made. A new strong upsurge in environmental concern provides an opportune occasion to consider what we have learned from the NEPA experience that might guide environmental policy for the future. This paper develops two general propositions: first, NEPA has had a significant positive impact on policy and administration in the United States and a catalytic effect on policies abroad; second, the substantive goals of NEPA have been only partially attained. To achieve the NEPA intent will require new initiatives — political, institutional, scientific, and legal. To understand why these initiatives are necessary, we need to review the circumstances under which NEPA became law and how it happened that the primary intent of the Act became subordinated to important but secondary considerations.

INTRODUCTION

The broad purpose of the National Environmental Policy Act (NEPA) is stated in its preamble in language appropriate to the declaration of a national policy:

> The purposes of this Act are: To declare a national policy which will encourage productive and enjoyable harmony between man and his environment; to promote efforts which will prevent or eliminate damage to the environment and biosphere and stimulate the health and welfare of man; to enrich the understanding of the ecological systems and natural resources important to the Nation; and to establish a Council on Environmental Quality (CEQ).

NEPA was a legislative response to an upwelling of public concern over the worsening state of the environment. Of at least 35 environmental bills in the 90th and 91st Congresses in the late 1960s, Senate Bill 1075, introduced on February 18, 1969, became the enacted statute intended to declare and implement a national policy for the environment. NEPA underwent extensive revisions in the course of its evolution. It received inputs from many sources, but the enacted statute was basically the work of congressional committees, notably the Senate Committee on Interior and Insular Affairs. NEPA's purpose and precepts reflected a general public anxiety. The statute was not a composite product of environmental pressure groups, few of which were more than vaguely aware of its enactment until its potential for blocking unwanted development projects was discovered.

The legislative history of NEPA explains some of the difficulties in its implementation. The NEPA intent was to redirect federal policy and decisionmaking toward outcomes (stated in Section 101) that took account of environmental quality values. Its substantive precepts were general — seldom reducible under the existing state of law to control of specific acts and decisions — and rarely amenable to judicial review. Unlike most environmental statutes enacted during the 1960s and 1970s, NEPA had no militant organized activists with a possessive concern over its implementation. Moreover, the act and NEPA were thrust upon a reluctant president and a bureaucracy committed to mission policies that traditionally regarded environmental values (if regarded at all) as subordinate to the specific statutory goals of the agency. In effect, NEPA amended all federal agency goals, but to accommodate the new mandate to the old was neither automatic or easy.

PROCESS OVER PURPOSE

On its own accord, and in comparison with other statutes, NEPA has been effective. It has been emulated abroad more than almost any other U.S. statute. "Little NEPAs" have been enacted in roughly half of the states. NEPA provides a model of an analytic procedure for policy development which many municipalities and some private corporations have adopted. Then why should it be regarded by some critics as controversial? There are several reasons, and they help to explain why the NEPA process has overshadowed the NEPA purpose.

Two reasons for controversy relate to the imposition of NEPA procedures and principles upon the customary behavior of the federal bureaucracies. Because NEPA, in effect, amended the basic missions and organic laws of all federal agencies, there was intra-agency resentment against this external imposition of new rules to live by. The environmental impact statement in particular aroused scornful indignation among some older officials who pursued their missions like blindered horses. But many younger agency personnel welcomed NEPA procedures as a more rational and defensible way of program and project planning and decisionmaking. A second bureaucratic objection was the opening to question of agency expertise. Once project authorization was received from Congress, the agencies had customarily felt free to plan dams, highways, airports, drainage projects, and other environmental-shaping activities with little if any public interference. The agencies were seldom obliged to tell the public anything about their plans and were not accustomed to being questioned by persons without official standing. It had been standard operating procedure in many agencies to delay public hearings until the bulldozers were ready to roll and it was too late to alter plans or stop the project. The full-disclosure provision of NEPA, supplementing the Freedom of Information Act, was an affront to bureaucratic autonomy and expertise. NEPA produced a new dimension of public interest which qualified and sometimes contradicted the attitude that the agency knows best.

Some agency personnel, reinforced by clients, saw NEPA as complicating and encumbering their missions. EIS paperwork was alleged to slow progress and drive up costs. Some agencies sought to off-load their project-planning costs on EIS procedures as if environmental impact analysis was an expensive add-on and not a necessary feature of careful planning. Investigations by the Commission on Federal Paperwork and the General Accounting Office did not substantiate these allegations.

Beyond agency orthodoxy there were three sociopsychological traits characteristic of Americans which were incompatible in principle with environmental protection as public policy. Opinion polls show support for environmental protection among most Americans. But what of the negative minority? Three viewpoints may be identified. First, there is the opinion that love of nature (i.e., regard for the environment) is sentimental, impractical, and unmanly. A second belief is that public environmental protection tends to cost more than it is worth — that it burdens the economy and is counterproductive to jobs, earnings, private property, and economic growth. Third is the attitude of individualistic libertarians who argue that aesthetic aspects of environmental quality ought to be bought and paid for by those who feel the need for them; the public should not be asked to bear costs that reflect the values of only some, not all, Americans.

The cumulative effect of these contentions has been to focus attention on NEPA's action-forcing procedure as imposing inequitable economic burdens or obstructing favored developments. Disaffection with EIS procedures tended to be projected against all of NEPA. To some critics the EIS was NEPA — all of it that amounted to anything. But complaints have diminished as experience with impact analysis has accumulated and as the process has become professionalized. To the extent that improved techniques of impact analysis have brought scientific, technical, and nonquantifiable values into the processes of planning and decisionmaking, NEPA has become a more effective instrument of policymaking. Through EIS, the mandate of Section 102(2)(a) to "utilize a systematic, interdisciplinary approach which will ensure the integrated use of the natural and social sciences and the environmental design arts" is put into practice. The risk in an emphasis on impact analyses is that the purpose of NEPA may be lost in the refinement of procedures. To some extent this has, in fact, occurred.

The EIS provides information (at least by implication) about what should not be done where federal action affects the quality of the environment. It does not indicate what should be done, although Section 101(2)(c) requires the agencies to identify alternatives to environmentally impacting proposals. The EIS is essentially cautionary and corrective. But it risks making more acceptable an action that should not be undertaken at all. The EIS is essential to responsible decisions that affect the environment, but it alone is not sufficient to realize the full measures of the NEPA intent.

SCIENCE AND NEPA

The clear intent of NEPA was to draw upon science as an informant and corrective for public policies impacting upon the environment. Science had, of course, previously been invoked and employed in many public programs and projects. These uses of science were often narrowly focused on particular problems of policy implementation. Interdisciplinary approaches were exceptional, as were uses of the social sciences and environmental design arts other than engineering.

In contrast, NEPA required a systematic integrated interdisciplinary use of all sciences that could reveal the probable effects of agency action that significantly impacted the environment. Moreover, the environmental design arts encompassed more than engineering. The informational base of environmental policy was thereby broadened, and new uses of science were introduced into policymaking. As previously noted, this expanded use of science was not welcomed by all administrators and some scientists. Decision processes necessarily became more complex, and administrative procedures were extended to cover issues heretofore neglected. Some mission scientists and disciplinary conservatives objected to what they called "interdisciplinary mish-mash" and argued that science was not adequate to support reliable impact statements.

Even when utilizing science to the fullest extent possible, the EIS is not a scientific document. More than science must be considered in an adequate environmental impact analysis. Under Section 102(2)(b), the federal agencies are authorized and directed to "identify and develop methods and procedures in consultation with the CEQ — which will ensure that presently unquantified environmental amenities and values may be given appropriate consideration in decisionmaking along with economic and technical considerations." By implication, the EIS should reflect these values.

Where science is applicable and adequate to the analysis, it should not be assumed to always yield definitive results. NEPA directs a systematic use of science as a corrective to selective atomistic uses that failed to reveal all of the ascertainable effects of agency proposals. But science today is not holistic; there are many sciences, and among them are gaps in knowledge and unresolved contradictions in their assumptions and interpretations. The inclusion of the social sciences in impact analysis adds to its complexities and uncertainties.

The probability of an environmental impact occurring is generally easier to ascertain than is an evaluation of its significance. When social and economic consequences of environmental impacts are considered, it is not exceptional to find that not all people — or indeed all elements in an ecosystem — are equally affected or affected in the same way. Science is not equipped to resolve many of these anomalies, but it may help to clarify the probable consequences of policy choice.

Perhaps the best use of science in impact analysis is the testing of assumptions underlying the programs or projects being considered. Science seldom is indicative of policy choices to be made, but it contributes to rational decisionmaking when it reveals the extent to which assumptions are supported by verifiable evidence. If the assumptions are falsified, the merits of the proposition are questionable. But the assumptions underlying public policy decisions are not always demonstrably right or wrong. They often express beliefs regarding ethics, equities, or economics that fall beyond the reach of science; these are often beliefs that move political decision makers. Science moves politics when its methods or findings arouse public concern or apprehension — as in the climate change and ozone issues. Indeed, the intended role of science in NEPA went far beyond impact analysis. A secondary purpose of the act, but only partially realized, has been the strengthening of science in relation to environmental issues.

The drafters of NEPA recognized that, at the time, the state of science was inadequate to resolve numerous problems relating to the environment. Therefore, much of the substance of the earlier Ecological Research and Surveys Bill was written into Title II of NEPA. Some of the provisions have been transferred to the Environmental Protection Agency (EPA), but a major role envisioned for the CEQ was to monitor, stimulate, sponsor, and assist advancement in environmental science. It was not to do research, but to become a catalyst with resources to enable it to work with the National Science Foundation, the National Research Council, and the research offices of the executive agencies and nongovernmental organizations in addressing critical environmental problems. To accomplish this task requires appropriate funding and a clear signal of presidential concern.

The primary role of the CEQ is advisory to the president and is properly performed in the Executive Office of the President. It does not belong in a line or regulatory agency such as the EPA. Its role in relation to science could be greatly enhanced by allowing it to pursue seriously and effectively its role under

Title II. One means of ascertaining where better science is needed is through assessing its adequacy in environmental impact analysis.

The role of science in Title I environmental impact analysis is neither clear-cut nor simple. There should be no argument regarding its importance in the analytic process. To the extent that scientific methods and findings become more reliable instruments of prognosis, we may expect science to be of correspondingly greater influence in public policies, but influence does not necessarily translate into decisiveness; it may also extend and intensify political polarization. Among objectors to NEPA and impact analysis are persons whose objectives and interests are called into question in EISs. In our society, a traditional role of science in relation to political ambitions has been to serve not subvert. Thus, the provision of Section 102(2)(c) that requires agencies to show that alternatives to environmentally impacting proposals have been considered may be regarded by sponsors of the proposals as subversive of their right to influence public policies, by irresponsible sources outside the confines of agency expertise or practical politics.

NEPA AS LAW

One of the obstacles to achieving the NEPA intent is the anomalous nature of the act in conventional American practical jurisprudence. Our laws exhibit a tendency, widespread throughout technological society, to become a professionalized specialty — an aspect of governance concerning which only lawyers are entitled to hold credible opinions. The exclusiveness did not characterize public affairs during the founding years of the republic. Thomas Jefferson, in particular, took strong exception to lawyer-dominated government, and he never accepted the doctrine of the judiciary as the sole and exclusive interpreter of the Constitution. But with the growth in volume and complexity of the law, its meaning became less and less accessible to the ordinary citizen. Diversification and heterogeneity in American society have diminished whatever common understandings that may have existed in the earlier years of the republic. Yet not all lawyers claim exclusive competence in law. It should be recorded that William J. Van Ness, Special Counsel to the Senate Interior Committee, probably had more to do with the actual drafting and enactment of NEPA than any other person, and it was he who brought me and many other nonlawyers into the preparation of this exceptional piece of legislation.

Law in its traditional sense reflected the long-standing, widely shared consensus of communities. It was concerned largely with the definition of rights and duties. As of today, it seems fair to say that in the minds of most Americans law is synonymous with some form of regulation or officially required procedure. People may argue over whether there should or should not be a law, but the making of positive law has become the near-exclusive province of lawyers and judges. Laws no longer necessarily conform to popular consensus, and not all legislative acts, including statutes resembling laws, are regarded as real law if their declared intent is unenforceable within the American judicial system.

NEPA is Public Law 91–190. But does calling a statute a law make it one in fact? That NEPA is a declaration of policy by the republic's highest legislative authority is hardly debatable. That the EIS requirement is good law is evidenced by the vast body of litigation and adjudication that it has engendered. But what of the substantive provisions of the act of which the courts take cognizance only in cases of flagrant dereliction and which the executive branch can ignore with impunity? In some other society, the principles declared in Section 101 of NEPA and the mandates of Title II for monitoring, evaluating, forecasting, and researching might be regarded as obligatory on the public executive.

If NEPA is indeed real law, then the constitutional obligation of the president to take care that laws are faithfully executed has been honored indifferently in the White House. It is obvious that there are far more positive laws and regulations in the United States than any president can personally oversee. Consequently, institutional arrangements have been devised to augment presidential action. The Reorganization Act of 1939 (53 Stat 561) established the Executive Office of the President; provided for the White House staff; and enabled the functions of budget, personnel, and planning to be brought under presidential control. The planning function was soon dropped as "socialistic," but the budgetary function evolved to become the Office of Management and Budget (OMB), the most powerful office in the federal government short of the president. In recognition of the power and economistic bias of the OMB, a proposal was made during the drafting of NEPA to place certain responsibilities for compliance on the OMB (and on the General Accounting Office as well). These proposals may have been functionally realistic if the NEPA intent were to be vigorously pursued. They were politically unrealistic, however, since there was little expectation in either the Congress or the executive power that NEPA principles were to be put fully into effect by presidential action, at least not in the near future.

Yet even without legal or institutional innovations beyond the CEQ, there are instruments of executive power sufficient to allow a president far more latitude in advancing NEPA principles than has ever been taken. Over the past two decades, custody and care of the environment has found an occasional place on the list of presidential priorities; but recent rhetoric aside, it has never been near the top and is clearly below where public opinion prefers it. To some persons who place a high value on protection of the environment, Ronald Reagan's deliberate dereliction in the administration of enforceable environmental laws was an impeachable offense. These environmental derelictions were more harmful to the country than the Iran-Contra affair. But some members of Congress found that making an issue of enforcement might not be in their political interest, and that international scandal makes headlines in the news and may be politically advantageous.

ACHIEVING NEPA'S INTENT

NEPA and a large number of environmental protection statutes have brought about measurable improvement in many aspects of the environment. But these achievements look best when compared with past abuse and neglect. They are much less impressive when reviewed in relation to present and future need. To achieve the NEPA intent is now seen to be a task far more formidable and costly than was perceived in 1969. Science is exacerbating apprehensions for the environmental future by announcing new findings regarding the consequences of global warming, stratospheric ozone depletion, acidic precipitation, and tropical deforestation. Thus, how our achievements in environmental law are evaluated depends upon the direction in which one looks — backward at what has been done or forward to what needs to be done.

The goals of NEPA will not be more than partially realized unless the popular will is sufficient to that end. But for popular will to be sufficient, it must be effective politically. Political effectiveness requires more than the preference of political majorities, a condition often elusive in pluralistic societies. In democracies with the social diversity of the United States, pluralities of minorities rather than majorities tend to dominate politics. Minorities that are organized, militant, and directed may successfully impose policies which majorities would not prefer but are disinclined to oppose. If popular preference as indicated by polls is real, environmental values, if held seriously, could be expected to have a more prominent place in American politics than is actually the case.

Budget allocations are one indicator of the relative importance of issues in American politics. For environmental goals, however, public spending is only a partial indicator of political commitment. For many aspects of the natural environment, NEPA goals would be achieved by no public (or private) expenditure but simply by abstaining from development harmful to the environment (an objective of the EIS process). A more revealing test of commitment would be congressional tolerance of politically influential special interests on developments with negative environmental impacts. For example, when Congress funds projects for dams, highways, drainage, or agricultural production, the net effect of which is a degraded environment, public expenditure contradicts the NEPA intent. But projects that have negative impacts on some aspects of the environment may sometimes have positive benefits for others. Value differences are today the principal underlying causes of conflict in environmental politics, although the visible conflicts are often economic.

If NEPA's intent is to be achieved and a declared national policy realized, means must be found to bring the political will closer to what appears to be the nation's popular preference. Governance occurs through institutionalized arrangements. Where these are insufficient for the policies to be pursued, results will fall short of intentions. In reviewing the adequacy of politics and law for achieving the goals of NEPA, two institutional deficiencies are apparent. The first appears in the structured behavior of the major political parties; the second appears in the legal foundation of environmental policy.

Because of the ways in which leadership and control are acquired, legitimized, and retained in political parties, it is often difficult for them to adjust to new political realities. Seniority often equates with resistance to innovation. In recent years, established political parties in both America and Europe have become alienated from substantial numbers of citizens. The rise of "green" parties in

Europe and, to some extent, nonvoting in America reflect disaffection with conventional politics. The "greens" are better adapted to the multiparty systems of Europe than to bipartisan America. But an inclusive, focused, and aggressive environmental political movement might force the two major parties into competition for support. Elements of such an environmental coalition presently exist, but have not yet achieved the scope and coherence needed to offset the influence of campaign contributors and vote-mobilizing minorities in political elections. The intent of NEPA is widely shared among the American people, but it has not yet received the integrated and focused political action needed to move it closer to realization.

In the late 1960s, when the first wave of the environmental movement was cresting, the absence of a comprehensive legal basis for environmental policy caused some concerned citizens and congressmen to consider the possibility of an environmental bill of rights. Representative Richard Ottinger of New York proposed a constitutional amendment to accomplish this purpose. A few observers saw a constitutional basis for environmental policy in the power of the Congress to provide for the common defense on the reasoning that there was no greater threat to the nation's security than impairment of its environment and natural resource base.

How the Americans would have received an environmental amendment to the constitution is conjectural. How most of their political representatives would have received it is not — they were simply not ready for such an innovation. Mindful of their patrons, skeptical of the constancy of the public, and doubtful that a subject of undefined breadth and uncertain implications could be written into basic law, they chose a safer course to satisfy a public demand which they judged to be pressing but possibly ephemeral. The constitutional issue was evaded by basing NEPA on the power of Congress to provide for the administration of the federal government. Congress could thus specify the manner in which the agencies that it created administered the statutes that it enacted. Similarly, other constitutional powers over interstate commerce, public lands, and spending for the general welfare would enable Congress to legislate on environmental issues for which no explicit constitutional basis could be found; hence, an environmental amendment to the constitution was perhaps unnecessary.

After two decades of experience under NEPA and its companion environmental statutes, assurances that the environment needs no constitutional protection are less persuasive. Three reasons for doubt have become apparent. First, most environmental protection legislation is procedural, and on environmental issues the courts have primarily adjudicated questions of legal process. They have been reluctant to review the merits of administrative decisions where (unlike civil rights legislation) the agencies are not bound by law higher than their statutory authorizations. Thus, the substantive goals of NEPA have found little reinforcement in the courts. Second, where the substance of a statute has no clear constitutional base, the president and the bureaucracy are under less pressure to honor its policy intent. Procedural compliance, as with the EIS, evidences conformity to the law. Third, the internationalization of environmental policy places the United States in a unique position of opportunity and obligation. As major agents of change in the world, American government and business have powerful, pervasive, and questionable effects upon the global environment. For example, can the United States retain credibility on international environmental issues if it exports and promotes abroad environmental hazards that it prohibits or curbs at home?

Certainly, so fundamental an area of policy as the environment is more significant for the nation's security and welfare than are many other constitutional amendments that have been adopted or proposed. It is by orders of magnitude more important for the nation's future than, for example, flag burning or prayer in the schools. The environment is the context in which everyone lives — now and in the future. The environment has become one of the major policy areas in international affairs. It is very possible that in the relatively near future the United States may be asking other nations to cooperate in efforts to arrest global climate change and the loss of genetic variety and diversity. Advances in molecular biotechnology are already raising issues of international environmental tension. There is thus a case of growing persuasiveness for reconsideration of a constitutional basis for environmental policy.

Under the Constitution of the United States, ratification of an international treaty confers upon implementing domestic legislation support comparable to a constitutional amendment. Through this device, birds migrating between Canada and the United States were brought under federal protection. Senator Claiborne Pell in 1977 proposed a treaty to internationalize environmental impact analysis. Yet as experience under the Convention on Trade in Endangered Species has shown, a treaty does not carry the force of a formal amendment to the Constitution. If achievement of the NEPA intent is seriously attempted and if international environmental cooperation becomes a major global responsibility of nations, an amendment to the Constitution of the United States may become necessary and feasible. Time would be required to work out the terms of such an amendment and to provide the public debate necessary to its ratification. Such a development is neither utopian nor improbable. It would be best for it to occur with deliberation and forethought rather than by amendment in the urgency and anxiety of some future environmental crisis.

The NEPA Process: Its Beginnings and Its Future

P. F. Gustafson, Argonne National Laboratory, Argonne, IL

ABSTRACT

The National Environmental Policy Act (NEPA) of 1969 was an extraordinary piece of legislation that required evaluation of the potential adverse impacts, and ways in which these impacts might be reduced or eliminated prior to a federal agency taking a major action. This act embodied long-range planning, an action without precedent in the United States, where growth and expansion are viewed as being totally desirable.

The U.S. Atomic Energy Commission (AEC) was the first federal agency to confront and to comply with NEPA as a result of the Calvert Cliffs decision, a decision by a federal court that the AEC had failed to make an *independent* NEPA assessment. The response of Battelle Northwest, Oak Ridge National Laboratory, and Argonne National Laboratory to AEC's request for assistance is described. The adversarial nature of the permitting and licensing processes required the scientists/engineers preparing a given environmental impact assessment (EIS) to defend their conclusions before an Atomic Safety and Licensing Board hearing board. The trauma and triumph of this experience are described. NEPA actions regarding the nuclear power plant licensing process set the tone for EIS development and adequacy and spawned two new professions: environmental assessment and environmental law.

In conclusion, the benefits and the drawbacks of the NEPA process and the need to examine ways in which the process may become more effective, less time consuming, and most importantly, less litigative, are discussed.

DISCUSSION

The late 1960s was a unique time in the history of the United States. We were clearly the richest and most powerful nation on earth, and, aside from being involved in a no-win war in Southeast Asia, life was good. The seeds of an expanded social consciousness developed from the Vietnam War, and the momentum spilled over into environmental matters. It was becoming increasingly apparent that the full-bore productivity of American industry, creating both weapons of war and consumer products, was maintained at a considerable price in terms of environmental quality. In the "good old days" (predepression, postdepression, and through World War II), belching smokestacks were an indicator of food on the table. The generation of the 1960s wanted both food on the table and considerably less belching (of smokestacks).

The proenvironmental movement, which was vocal and well-educated, had the ear of influential congressional leaders. What followed was a legion of environmental protection legislation, which also resulted in the formation of the President's Council on Environmental Quality (CEQ) and the creation of the U.S. Environmental Protection Agency (EPA). The EPA was responsible for enforcing the Clean Air Act, the Clean Water Act, and our beloved National Environmental Policy Act (NEPA).

When NEPA was enacted by Congress in 1969, it was doubtful that 1 person in 100,000 knew what NEPA stood for or really cared. NEPA was an extraordinary piece of legislation in that it required some form of evaluation of the potential adverse impacts and means whereby these impacts might be reduced or eliminated before taking a major federal action. As such, this act embodied what might best be termed balanced long-range planning. Whereas totalitarian governments (for purposes of window dressing) have always indulged in 5-, 10-, 20-year plans setting goals that were seldom met, NEPA was the first attempt in the United States to truly look to the future. Traditionally, big was good, and bigger was better as we ad hoced our expansion to greatness.

In the summer of 1972, the Atomic Energy Commission (AEC) became poignantly aware of NEPA as a result of the Calvert Cliffs court decision; this crucial decision resulted from a case brought in federal court by several environmental groups. They charged that the AEC had not conducted an *independent* analysis of the potential environmental impacts of constructing and operating the Calvert Cliffs Nuclear Power Station on the western shore of Chesapeake Bay and had not investigated ways in which these impacts could be minimized or eliminated. Before the Calvert Cliffs decision (but after NEPA came into effect), the AEC environmental permitting and licensing procedures for nuclear power plants operated as follows. An electric utility wishing to obtain a construction permit or an operating license for a nuclear plant would submit an environmental report (ER) describing in considerable detail the probable or possible environmental impacts of the proposed action (i.e., construction or operation of the reactor) along with a presentation of proposed actions to lessen (mitigate) those adverse environmental effects. The AEC, in turn, made a fairly cursory analysis of the ER and sent the package off to cognizant federal and state agencies for their comments, reaction, and analysis. The process was fairly benign, and Argonne National Laboratory (ANL) provided modest assistance to AEC in this effort. The reactor licensing process involved adjudicatory hearings before an Atomic Safety and Licensing Board (ASLB). The ASLB hearings offered one place where opposition groups could at least be heard. The nuclear power plant licensing process provided a safety or relief valve for those individuals and groups who were essentially fed up with "big brother" doing what he damned well pleased.

The legal impact of Calvert Cliffs on the AEC licensing process was that if the AEC did not provide an independent and substantive review in keeping with the letter of NEPA, there would be no more construction permits or operating licenses granted because of court intervention. This situation put AEC and the nuclear power industry in severe jeopardy. The early 1970s witnessed commercial nuclear power emerge as a boom industry. In 1972, more than 60 construction and licensing applications were on file with AEC for action. However, some segments of the environmental movement were strongly opposed to nuclear power, viewing it as just an A-bomb in a building. Understandably, the antinuclear activists would use any legal means to thwart nuclear power, and NEPA offered them a handy vehicle for a shot at the Achilles' heel of nuclear power development.

The AEC permitting/licensing process was really geared to consideration and resolution of nuclear safety issues. The AEC regulatory staff consisted of nuclear engineers, nuclear safety specialists, and radiation safety experts. The ASLB was also strongly nuclear-safety oriented, although the chairmen of the individual hearing boards were always lawyers. The fact of the matter was that immediately post-Calvert Cliffs, neither the AEC regulatory staff nor the ASLB had any real base of environmental expertise.

However, within the AEC's national laboratory system resided a substantial amount of environmental knowledge and experience gained from the significant research and development programs supported by the agency. In the summer of 1972, the general manager of the AEC sent a polite telegram to the directors of the national laboratories requesting assistance for the Office of Nuclear Regulation in the preparation of suitable, adequate, and independent environment statements (ESs) as they were called in AEC parlance [they were actually environmental impact statements (EISs)]. The assistance was viewed as temporary by the AEC. Funding would come from reprogramming (at the headquarters level) of monies already at the national laboratories for other purposes. Once the AEC had hired its own staff, the laboratories could return to doing their own things. Three laboratories volunteered — ANL, Battelle Northwest, and Oak Ridge National Laboratory (ORNL). All institutions made good staff available, and in recognition that preparation of an EIS would be a multidisciplinary activity, engineers, physical and biological scientists, computer modelers, meteorologists, and other scientific specialists were assembled into loose teams, each team to prepare an EIS for a specific site and a specific action. As the laboratories began to organize, AEC, including the legal staff, also began to organize. This was a truly unique time in terms of a federal agency trying to meet the letter and the spirit of NEPA on a major scale. However, it was a bit like the lame leading the blind. After numerous meetings involving AEC and laboratory staff, a general format and approach was developed between the legal and technical participants. Of utmost importance was the fact that the lawyers deemed the approach legally sufficient. The elements of the approach were to be a detailed treatment of the following key areas and issues:

- description and discussion of the proposed action
- need for the proposed action
- alternatives to the proposed action, including the no-action alternative
- probable environmental impacts of the proposed action and the alternatives thereto
- ways in which to reduce or eliminate the environmental impacts identified above
- an analysis of the economic aspects of the proposed action and the alternatives
- a cost-benefit analysis of the proposed action and the alternatives
- a summary of findings and recommendations

The assistance effort at ANL was organized in the Environmental Statement Project (ESP). The name was chosen partly because it described the activity and partly because it was felt that a fair amount of extrasensory perception would be needed if any degree of success was to be achieved. The ESP was organized in a matrix management system consisting of disciplinary sections (engineering, biology, etc.) containing potential team members. The other side of the matrix consisted of a cadre of team leaders or potential team leaders. When the AEC assigned a project to ANL, a team leader was selected, a budget and schedule were developed, and the team leader negotiated with the disciplinary section heads to assemble the requisite team. They were then off and running, hopefully in the proper direction. Inadvertently, the AEC and the three laboratories were helping to create two new multifaceted disciplines, namely environmental assessment and environmental law.

There were many interesting features to the startup of the ESP. For example, space was scarce, and a nuclear engineer, an aquatic biologist, and a radiochemist might be sharing the same office. Argonne at the time was highly departmentalized. To hear a nuclear engineer expressing interest and concern for fish kills and thermal effects in aquatic biota and to see a biologist learning about basic nuclear plant operations was a broadening and enlightening experience.

Another aspect involved the fact that the ESP staff consisted of research scientists and engineers used to collecting their own data, making their own analyses, and perhaps ultimately subjecting their findings and conclusions to peer review via publication. Now they had to depend solely on data provided by others, namely that presented in the applicants' environmental report and whatever other sources of data they could lay hands on in what often was relatively limited time. No new research was allowed to confirm or negate the available information. Thus, the EIS preparers were in the then-uncomfortable position of analyzing, interpreting, and basing technical conclusions on the work of others. Furthermore, their positions and conclusions would be subject to review by the AEC lawyers and ultimately might have to be testified to before an ASLB Hearing Board and subjected to the adjudicatory process allowed in such hearing process. This situation caused pain and apprehension to some ESP staff; however, in the main, they rose to the occasion.

Most laboratory and field scientists and engineers seemed to welcome the chance to apply their experience and expertise to the resolution of national problems. Most of us have never had nor will have a substantive role in determining the go or no-go decision on a billion dollar project. Something not to be taken lightly. Moreover, the change from laboratory and field research provided a rejuvenating aspect for many of the staff, and this rebirth resulted in a new dimension of talent and innovation.

The ES or EIS preparation process involved three basic phases:

1. draft environmental statement (DES)
2. final environmental statement (FES)
3. public hearing phases

The published DES represented the mutually agreed upon position of the AEC and laboratory staff members involved, plus negotiated changes agreed to by the applicant. Here, it should be stressed that the agreement and negotiation phases did not constitute a NEPA whitewash of the proposed action. In fact, in a number of instances, construction permit applications were withdrawn at this phase. NEPA had teeth. Once finalized, the DES was published and issued for review and comment by cognizant federal and state agencies, concerned environmental groups, and the general public. After an established review period of 30, 60, or 90 days, comments were submitted, and the AEC and laboratory teams had to respond in one way or another to all comments received. This was a tedious process, but one often relieved by some anecdotal hilarity; for example, the time that an intervenor asked the AEC to supply a list of energy production technologies not yet discovered. The reply by our laboratory respondee was a blank page. The AEC lawyers did not find that adequate, so some nebulous response was prepared.

It should be noted that some construction permit applications were withdrawn after the review of the DES. It was not possible with the time and manpower constraints to obtain all relevant information regarding proposed construction at a specific site. Some of the information provided in the review process from a variety of sources clearly made approval of a construction permit untenable.

Phase 2 of the process was preparation of an FES which included an appropriate response to each comment and might involve a reexamination of conclusions presented in the DES. The information presented in the FES constituted the basis for a public hearing before an ASLB Hearing Board. That hearing was the third phase of the process.

The public hearing was an adjudicatory process wherein persons supporting a given conclusion could be called to testify on behalf of that conclusion and be cross-examined by opposing parties, called intervenors. This was a real eye-opening experience or blood bath for some laboratory staff. But it was a challenge to their integrity and to a person they performed professionally. Before the hearing, staff were advised by legal experts on posture and behavior. They were particularly advised to establish their area of expertise and not to answer questions beyond that area. This latter admonition was particularly difficult when one had a broad, but fairly shallow knowledge base. We tend to think of ourselves as "expert" well beyond our actual documentable (in a legal sense) area of expertise.

The NEPA compliance history of the AEC and its successor, the Nuclear Regulatory Commission (NRC), is basically a success story. The competence and objectivity of the analyses by the agency and laboratory staffs prevailed. With few exceptions, the ASLB Hearing Boards approved the proposed action presented to them by the AEC and NRC. On occasion, court decisions, strong local opposition, and economic factors did cause changes in the approved action. For example, the Palisades nuclear plant originally was approved for direct but offshore discharge of heated cooling water to Lake Michigan, but a court order required the use of mechanical draft cooling towers instead.

By 1977, the permitting/licensing system established to abide by NEPA was functioning smoothly. Many impacts or potential impacts could be treated in an almost generic manner. Unfortunately, however, generic issues were never actually treated as such; that is, a decision by one hearing board was not often used as a precedent by another hearing board. Each case was treated *de novo.*

The NRC issued detailed guidance to applicants regarding the amount of information required in the ER, including the need to consider at least two alternative sites to the preferred one. A requirement imposed by the CEQ was that several alternatives be examined. The EISs originally consisted of 80 –100 pages. However, with new issues raised in early hearings, the treatment of which added to the volume, CEQ set a limit of 250 pages. However, an unlimited volume of appendices was allowed, still providing for the danger of a hernia with a complete case!

By the late 1970s, the NRC system was working smoothly and included a standard review plan (for review of the applicant's ER) and ranking of potential sites. Furthermore, intervention before the hearing boards was confined to substantive, rather than trivial issues. However, by this time, applications for new nuclear plant construction were dwindling because of conservation (leading to a slower growth in demand for electricity) and rising construction costs.

Although the AEC and NRC treated reactor NEPA matters on a strict case-by-case basis, there were some issues treated more generically. Among these were the Generic Environmental Statement for Mixed Oxide Fuel, Spent Fuel Storage Capacity, and an impact statement for uranium milling through the year 2000 called the Uranium Milling Operation Generic Environmental Statement. This latter EIS was the nearest NRC came to issuing a regional analysis of environmental impact. Regional looks were a subject championed by the laboratories, but one that bore little fruit. Another deficiency (in my opinion) was that the agency conducted little follow-up on predicted impacts at nuclear stations. A start was made in 1976, but confirmation of predicted environmental impacts was not considered cost-effective. This was unfortunate because such studies would have provided a more sound scientific basis for subsequent environmental impact predictions (i.e., assessments).

As the nuclear work began to wind down, the laboratories were called upon by the Energy Research and Development Administration, and then by the Department of Energy (DOE), to assist in NEPA matters for activities in which those agencies were involved. Principal among these was a programmatic environmental statement for the Fuel Use Act (FUA). The FUA was a congressional response to the oil crisis and involved a nationwide switch from oil and natural gas to coal, where appropriate, for generating electricity. The programmatic statement (prepared jointly by ANL, ORNL, and Battelle Northwest) examined the environmental as well as socioeconomic costs involved. This effort was followed by a true regional impact statement for the northeast, which involved assessing the impacts of shifting 20 oil- or gas-fired power plants to coal. Three issues were involved: (1) coal availability and available transportation capacity, (2) coal and ash handling capacity, and (3) regional air quality preservation.

All in all, the NEPA experience started under the AEC has on balance proven beneficial to the regulatory and research and development agencies and more importantly to environmental protection and national well-being. NEPA is clearly here to stay, but some modifications for future application seem in order. The initial intent of NEPA, namely to consider the probable consequences of a major federal action before taking that action, has been abused by those using NEPA to prevent taking any action at all or at least delaying action. This growing national obstructionist ethic will prove harmful and counterproductive in the long run. The obstructionists demand that more and more trivial impacts be treated in increasingly greater detail. The courts have supported many of their demands, and, as a result, federal agencies tend to go to great lengths to treat what are essentially nonissues in excessive detail just to avoid going to court. Furthermore, the DOE in an effort to restore or establish public (and congressional) credibility is using the NEPA process in detail before taking actions which when implemented will improve the DOE environmental compliance posture. This would appear to be counterproductive to the DOE's needs in both the short and long term in regard to environmental compliance. Perhaps Congress and/or the courts should reexamine NEPA and state what is adequate in terms of an environmental impact analysis (EIA). Otherwise, a sinful amount of paper, time, and money will continue to be wasted on inaction.

The procedure of using a programmatic or generic EIS as a base for a tiering-off process to treat specific actions should increase. The DOE Environmental Restoration and Waste Management Program is a case in point. Finally, it would seem appropriate for Congress to consider requiring the EPA to use the NEPA process prior to promulgating landmark regulations as well as before taking discrete actions such as Superfund cleanups. The EPA asbestos regulations are an example of overkill without enough forethought of the consequences. Many school districts are going broke unnecessarily because of these regulations.

Individual Superfund actions are an example of the site-specific need for NEPA examination to provide equity and uniformity across the nation.

In summary, NEPA is a unique and farsighted piece of legislation which provides a valuable tool for decision makers and, with modification, will be even more valuable to the nation and could provide a substantive, functional model for the rest of the world.

ACKNOWLEDGMENTS

This work supported by the U.S. Department of Energy under Contract No. W–31–109–Eng–38.

Judicial Review of Scientific Evidence in Environmental Impact Statements

M. B. Gerrard, Berle, Kass & Case, New York

ABSTRACT

The contents of many environmental impact statements (EISs) are shaped by the desire of the project sponsors to avoid judicial invalidation. This paper provides a legal analysis of the circumstances under which the courts will review the contents of an EIS and the standards that are applied. It then focuses on the admissibility of scientific evidence in lawsuits under the National Environmental Policy Act (NEPA). Scientific discussion contained in an EIS itself is part of the record on review and thus is automatically before the court. Information not contained in the EIS, and not otherwise in the administrative record, may be excluded under certain circumstances. Even if the court agrees to admit scientific evidence that is outside the record, it may be excluded if it is not based on methodologies that are well accepted in the particular field. Implications of these legal principles for EIS writers and officials of lead agencies are presented.

INTRODUCTION

Environmental impact statements (EISs) are legal documents. They are prepared to comply with statutory or regulatory requirements, usually either the National Environmental Policy Act (NEPA) or its state counterparts. Their sufficiency is often challenged in court, and the fate of the proposed project is often determined by the courts' view as to whether the EIS meets legal requirements.

Unlike most legal documents which rest on readily verifiable facts, EISs contain heavy doses of scientific opinion. The judiciary has developed modes of dealing with such material — modes which may be unfamiliar to the scientists who write and review EISs.

This paper, written from the perspective of a practicing environmental lawyer who frequently attacks or defends EISs in court, discusses the ways the courts assess scientific evidence in EISs. It begins with a discussion of the contexts in which EISs are reviewed and the different burdens of proof found in each context. It then focuses on the most important of these contexts, the trial-level court, and analyzes how the courts decide whether to admit and how to treat scientific evidence in EISs. It concludes with a discussion of the implications of these evidentiary rules for EIS writers.

CONTEXT OF ADJUDICATION AND BURDENS OF PROOF

There are four sequential contexts in which EISs are reviewed for legal sufficiency: the lead agency's review of an EIS drafted by its staff, by outside consultants, or by the applicant; a permit-granting agency's review of an EIS in the course of adjudicating whether to issue a permit to the applicant; a trial court's review of final agency action in approving an EIS and issuing (or withholding) necessary approvals; and an appellate court's review of a trial court's decision.

Lead Agency Review

Under the NEPA regulations of the U.S. Council on Environmental Quality (CEQ), 40 CFR 1508.16, the lead agency must determine the adequacy of the EIS in the first instance. In so doing, it is governed by the CEQ regulations as well as its own NEPA regulations (if any).

The adequacy of a draft EIS is not, however, ordinarily subject to judicial review. The courts have held that EIS adequacy, and other aspects of NEPA compliance, are not ripe for review until the agency makes a final decision on the proposed action. Before then, the dispute may still be academic; deficiencies in the draft EIS may be corrected in the final EIS, or, after the final EIS, the agency may decide to reject the project [e.g., *Natural Resources Defense Council* v. *Nuclear Regulatory Commission*, 539 F.2d 824 (2d Cir. 1976); *Eastern Connecticut Citizens Action Group* v. *Dole*, 638 F. Supp. 1297 (D. Conn.), *aff'd per curiam*, 804 F.2d 804 (2d Cir. 1986), *cert. den.*, 481 U.S. 1068 (1987)].

An exception may arise if the agency meanwhile is doing things that commit it to taking the action. Otherwise, an agency's compliance with NEPA is ordinarily shielded from judicial intervention during this interim period. That does not mean that the agency is any less obligated to comply with NEPA, only that it is the keeper of its own legal conscience at this stage.

Administrative Review

Upon the completion of a final EIS (FEIS), some agencies undertake formal adjudicatory hearings before deciding what action to take. These hearings are before an administrative law judge or hearing officer and proceed much like a judicial trial with sworn testimony, cross-examination of witnesses, formal receipt of evidence, and the like. Such proceedings typically arise before independent regulatory agencies overseeing private utilities or common carriers; the Federal Energy Regulatory Commission and the Nuclear Regulatory Commission are two examples.

Where the action under review was the subject of a federal EIS, the sufficiency of that EIS is often one of the issues for the hearing, especially when there are intervenors opposing the requested approvals. The hearing will often look at the contents of the EIS in depth, without paying any special deference to its contents. The presiding officer is bound to comply with NEPA, but is usually not subject to interim challenge in court. At the conclusion of the hearing, the presiding officer will typically render a report and recommended decision to the heads of the agency, who will make the final agency decision. Then, and only then, is the regulatory agency's compliance with NEPA usually subject to review in court.

Trial-Level Judicial Review

The level of review that is the focus of this paper occurs after the FEIS has been filed and after the administrative agency has made its final decision. At that point, parties unhappy with the decision may challenge it before a trial-level court. In the federal system, that is usually a U.S. district court.

In reviewing federal agency action, the district courts are governed by the Administrative Procedure Act, 5 U.S.C. Section 706 (APA). The APA does not give the district courts free rein to act as they please in overturning agency action with which they disagree. Instead, the courts may consider only a limited range of issues. Of greatest relevance here are the following:

- Did the agency act in excess of its statutory authority?
- Did the agency follow the proper procedures?
- Was the agency action arbitrary, capricious, or an abuse of discretion?
- If the agency held an adjudicatory hearing on the action, was its decision supported by substantial evidence in the record of that hearing?

The meaning of the terms "arbitrary" and "capricious" in the NEPA context has been described by one court as follows: "Normally, an agency's action is held to be arbitrary and capricious when it relies upon factors Congress did not want considered, or utterly fails to analyze an important aspect of the problem, or offers an explanation contrary to the evidence before it, or its explanation is so implausible that it cannot be ascribed to differing views or agency expertise" [*Sierra Club* v. *U.S. Army Corps of Engineers*, 772 F.2d 1043, 1051 (2d Cir. 1985)].

The successful NEPA challenges to federal action have focused on procedural compliance and on whether the agency has considered the relevant and pertinent information. In the words of one commentator, "it is either the inadequacy of the federal agency's procedures and processes, or the inadequacy of the information considered as measured by the `rule of reason,' that has allowed successful challenges to federal action under NEPA" (Snarr 1985).

The courts grant considerable deference to the technical judgments of the administrative agencies and are loathe to substitute their own judgments. The courts are just as competent to decide legal questions as are the agencies, but not factual questions. This requirement was reiterated by the U.S. Supreme Court in a recent decision. *Marsh* v. *Oregon Natural Resources Department*, 109 S.Ct. 1851 (1989), upheld a decision of the U.S. Army Corps of Engineers not to require a supplemental EIS for a dam project in Oregon. The Supreme Court held that

> Resolution of this dispute involves primarily issues of fact. Because analysis of the relevant documents 'requires a high level of technical expertise,' we must defer to 'the informed discretion of the responsible federal agencies.' Under these circumstances, we cannot accept respondents' supposition that review is of a legal question and that the Corps' decision 'deserves no deference.' Accordingly, as long as the Corps' decision not to supplement the FEIS was not 'arbitrary or capricious,' it should not be set aside.

In further elaborating the basis for its decision, the Supreme Court remarked that

> [T]he Corps had a duty to take a hard look at the proffered evidence. However, having done so and having determined based on careful scientific analysis that the new information was of exaggerated importance, the Corps acted within the dictates of NEPA in concluding that supplementation was unnecessary. Even if another decisionmaker might have reached a contrary result, it was surely not 'a clear error of judgment' for the Corps to have found that the new and accurate information contained in the documents was not significant and that the significant information was not new and accurate the Corps conducted a reasoned evaluation of the relevant information and reached a decision that, although perhaps disputable, was not 'arbitrary and capricious.'

The Supreme Court added that the courts "should not automatically defer to the agency," and they must "carefully review the record and satisfy themselves that the agency has made a reasoned decision." Deference to an agency's decisions is substantial but not automatic or absolute. In a prior decision, the Supreme Court did say, however, that when examining scientific determinations by the expert agencies, the courts must be at their most deferential [*Baltimore Gas & Electric Co.* v. *Natural Resources Defense Council*, 462 U.S. 87, 103 (1983)]. The Supreme Court has cautioned that a lower court may not "interject itself within the area of the executive as to the choice of the action to be taken" [*Kleppe* v. *Sierra Club*, 427 U.S. 390, 410 n.21 (1976)] and may not substitute its judgment for that of the agency [*Vermont Yankee Nuclear Power Corp.* v. *Natural Resources Defense Council*, 435 U.S. 519 (1978)].

The proper scope of a district court's review of an EIS was at issue in the litigation over Westway, a proposed interstate highway project to be built along the west side of Manhattan in New York City. (The author's firm represented the plaintiffs in that case.) Because the project would involve substantial landfilling in the Hudson River, permits were required from the U.S. Army Corps of Engineers. The state of New York and the Federal Highway Administration had prepared an EIS, but the district court found it to be inadequate and directed the Corps to prepare its own EIS. The plaintiffs then challenged the Corps' EIS as inadequately describing the landfill's effects on the Hudson River fishery. The court conducted a lengthy trial in which numerous experts testified about the impact of the project on the fisheries. After the trial, the court found the Corps' EIS to be inadequate and issued a permanent injunction against the project.

On appeal, the Court of Appeals found that the district court had erred in the manner in which it conducted the trial. The Court of Appeals found that the district court had held a *de novo* trial, whereas a plenary trial would have been more appropriate. Plenary review is more limited than *de novo* review. As the Court of Appeals explained, plenary review is "permitted when the agency's record is so sparse as to make judicial review ineffectual. Through plenary review a district court may obtain from the agency, either through affidavits or testimony, such additional explanation of the reasons for the agency decision as may prove necessary" [*Sierra Club* v. *U.S. Army Corps of Engineers*, 772 F.2d 1043, 1052 (2d Cir. 1985)].

"The district court here," the Court of Appeals continued, "undertook *de novo* review. By allowing the plaintiffs to call their own expert witnesses, and substituting those witnesses' interpretations of the data for the views of the experts that the Corps had relied upon, the district court moved from a plenary filling in of holes in the Corps presentation to a *de novo* hearing on the fisheries issue" [*Sierra Club* v. *U.S. Army Corps of Engineers*, 772 F.2d 1043, 1052 (2d Cir. 1985].

However, the Court of Appeals found this was not a "reversible error" because the Corps had violated a prior district court order to keep records of all its communications regarding Westway, therefore diminishing its credibility, and the Corps had provided no explanation of a crucial change between its draft EIS and its final EIS, so that the administrative record did not support the granting of a permit. The Court of Appeals converted the permanent injunction into a preliminary injunction, allowing the Corps to go back and correct the defects in the EIS. As it happened, certain funding options for Westway were about to expire, and the Court of Appeals decision rapidly led to the abandonment of the project.

Appellate Review

Decisions of the district courts may be appealed to the U.S. Court of Appeals. Decisions of the U.S. Court of Appeals may be appealed (with permission) to the U.S. Supreme Court.

The scope of review conducted by the appellate courts is even more limited than that by the district courts. An appellate court reviewing a district court is much like a district court reviewing an administrative agency; great deference must be paid to the lower tribunal's findings of fact, while its conclusions of law are not owed any special deference. Thus, if a district court has conducted a plenary trial, an appellate court will not overturn the district court's factual (as opposed to legal) findings unless they are obviously without sound basis in the record or suffer some other fatal legal defect.

A number of federal environmental statutes provide that certain (but by no means all) decisions of the U.S. Environmental Protection Agency (EPA) and some other agencies are to be appealed directly to the Court of Appeals, bypassing the district courts. These typically arise after the agency has gone through an elaborate fact-finding process in which it creates a record which the courts may review.

The appellate courts are strictly limited to consideration of factual material that was present in the record of the trial court or administrative agency below. If a litigant did not submit the evidence to the lower court, it may not later submit the evidence to the appellate court.

TRIAL COURT REVIEW OF SCIENTIFIC EVIDENCE

Record Review

The starting point for any trial court review of administrative action is the record compiled by the agency. If the agency has conducted formal adjudicatory hearings, then the record is rather neatly circumscribed as the transcript, exhibits, and decision of those hearings together with the pleadings and certain other documents.

If, as is more often the case in federal environmental practice, there has been no such hearing, definition of the record is more difficult. It certainly contains the EIS (if one has been prepared) and the permit application (if any). Beyond that, all manner of correspondence, memoranda, studies, and other material that were before the agency when it made the decision may be deemed included.

Since the trial court will primarily be seeking to determine if the agency's decisions are supported by the record and are rational rather than arbitrary and capricious, it will generally be in the agency's interests to make sure that any documents supporting its decision are included in the record.

EVIDENCE OUTSIDE THE RECORD

Availability of Review

The more difficult evidentiary issues arise when a party seeks to introduce documents or expert opinions that were not part of the administrative record. A threshold question is whether such extra-record evidence may be considered at all.

The U.S. Supreme Court considered the extent to which a district court may supplement the agency record in *Citizens to Preserve Overton Park* v. *Volpe*, 401 U.S. 402 (1971). The Secretary of the Interior had to decide whether alternatives were available for locating a federal highway through a public park. The Supreme Court held that the district court must undertake a "substantial inquiry" and take a "hard look" at the availability of alternatives, and it authorized the district court to admit limited categories of supplemental evidence outside the administrative record to further explain the contents of the record.

The leading NEPA case on this point is *County of Suffolk* v. *Secretary of the Interior*, 562 F.2d 1368 (2d Cir. 1977), *cert. den.* 434 U.S. 1064 (1978), a challenge to the Interior Department's decision to accelerate the leasing of outer continental shelf lands for oil and gas exploration. The Court of Appeals found that "in NEPA cases a primary function of the court is to insure that the information available to the decision-maker includes an adequate discussion of environmental effects and alternatives — which can sometimes be determined only by looking outside the administrative record to see what the agency may have ignored."

The Court of Appeals continued that "allegations that an EIS has neglected to mention a serious environmental consequence, failed adequately to discuss some reasonable alternative, or otherwise swept 'stubborn problems or serious criticism . . . under the rug' . . . raise issues sufficiently important to permit the introduction of new matters, both in challenges to the sufficiency of an environmental impact statement and in suits attacking an agency determination that no such statement is necessary."

When it is clear that the record is complete, however, the courts may refuse to admit any new evidence [e.g., *Van Abbema* v. *Fornell*, 807 F.2d 633 (7th Cir. 1986); *Friends of the Earth* v. *Hintz*, 800 F.2d 822 (9th Cir. 1986)]. They may also refuse to consider alternatives propounded by intervenors that were not raised before the agency acted [*Vermont Yankee Nuclear Power Corp.* v. *Natural Resources Defense Council*, 435 U.S. 519 (1978)].

An agency may be unsuccessful in supporting its decision with documentation that was not considered as part of the NEPA process [*Grazing Fields Farm* v. *Goldschmidt*, 626 F.2d 1068 (1st Cir. 1980)].

Scientific Evidence

If the district court agrees to accept evidence from outside the record, the next question is whether the particular piece of evidence is otherwise admissible. Of special interest for this paper is scientific evidence.

The leading case in the United States on the admission of scientific evidence is *Frye* v. *United States*, 293 Fed. 1013 (D.C. Cir. 1923), the first case rejecting the admissibility of polygraph evidence. A criminal defendant sought to prove his innocence by introducing a polygraph report showing he was telling the truth when denying he committed the crime. The Court of Appeals upheld the trial court's exclusion of this evidence because the defendant had not proven that the technique was reliable. The Court of Appeals held that

> Just when a scientific principle or discovery crosses the line between the experimental and demonstrable stages is difficult to define. Somewhere in this twilight zone the evidential force of the principle must be recognized, and while courts will go a long way in admitting expert testimony deduced from a well-recognized scientific principle or discovery, the thing from which the deduction is made must be a field in which it belongs.

This is known as the Frye test or the general acceptance test. It is the most conservative approach used for deciding the admissibility of scientific evidence (Levy 1988), but it gradually spread throughout the state and federal courts and became the standard test (Moenssens 1984). Most of the cases under Frye arose in the forensic context in determining what scientific tests could be used in criminal prosecutions (Black 1988). The burden of proof in criminal cases is much higher than it is in civil cases, so a judicial reluctance to allow new testing techniques in that context is appropriate. Most criminal cases are tried before a jury, and the courts are reluctant to expose jurors to unsuitable evidence, since jurors, unlike judges, are presumed unable to strike unsuitable information from their minds. However, the Frye test has also been used in many civil cases as well, including many tried before a judge without a jury (Riesel 1989). (Pure NEPA cases are never tried before juries because the Seventh Amendment right to a jury trial extends only to suits of a sort that would have been entitled to a jury in the English courts at the time of the ratification of the Constitution.)

In recent years, a number of courts have moved away from the Frye test (Imwinkelried 1984). There is also a debate about whether the Federal Rules of Evidence, enacted by Congress in 1975 for use by the federal courts, have in effect repealed Frye (Levy 1988). A more liberal view was proposed by McCormick (1954).

> General scientific acceptance is a proper condition upon the court's taking judicial notice of scientific facts, but not a criterion for the admissibility of scientific evidence. Any relevant conclusions which are supported by a qualified expert witness should be received unless there are other reasons for exclusion. Particularly, its probative value may be overborne by the familiar dangers of prejudicing or misleading the jury, unfair surprise and undue consumption of time.

One commentator has proposed 11 factors that should be exercised in assessing whether to admit scientific evidence (McCormick 1982):

1. potential error rate in using the technique
2. the existence and maintenance of standards governing its use
3. the presence of safeguards in the characteristics of the technique
4. analogy to other scientific techniques whose results are admissible

5. the extent to which the technique has been accepted by scientists in the field involved
6. the nature and breadth of the evidence adduced
7. the clarity and simplicity with which the technique can be described and the results explained
8. the extent to which the basic data are verifiable by the court and jury
9. the availability of other experts to test and evaluate the technique
10. the probative significance of the evidence in the circumstances of the case
11. the care with which the technique was employed in the case

In the environmental field, the cutting edge in the development of new standards for the admission of scientific evidence is not NEPA. That is largely because, as noted above, the courts are severely constrained in their review of administrative agency action. Instead, most of the activity is in the area of toxic torts. These cases typically involve persons who have contracted illnesses that they believe were caused by exposure to chemicals, radiation, or other environmental conditions. Because the courts try these cases from scratch rather than reviewing administrative decisions, they have a much broader scope in admitting or excluding evidence, and many courts have used this broad scope to explore actively the probative value of new scientific techniques (Black 1988). For example, epidemiological studies and risk assessments, which are, of necessity, based on long strings of assumptions and innovative methodologies, have been admitted in many toxic tort cases (Dore 1985; Gold 1986; Walker 1989).

Numerous substantive environmental statutes provide guidance into what types of scientific evidence should be admissible in NEPA cases, as well as assist EIS writers in determining what kinds of testing to undertake and report. For example, many statutory schemes establish standards with "not-to-exceed" numerical limits (e.g., the ambient air quality standards and emissions standards under the Clean Air Act, the ambient water quality standards and categorical effluent limitations under the Clean Water Act, the maximum contaminant levels under the Safe Drinking Water Act, and the hazardous waste characteristics under the Resource Conservation and Recovery Act).

To help enforce these standards, the EPA and other agencies have established detailed protocols for testing many of these parameters. These regulatorily established protocols would certainly meet the Frye test and any of the more liberal, newer tests for admissibility of scientific evidence as well. Many questions remain concerning the reliability of the results of tests conducted under these protocols (Koorse 1989), but those go to the weight the results should be accorded by the court rather than to whether the results should be admitted into evidence.

Expert Testimony

Closely related to the issue of what scientific evidence may be admitted is just how it is introduced. The normal method is through expert testimony. The Federal Rules of Evidence (FRE), Rule 702, allows testimony by an expert qualified "by knowledge, skill, experience, training, or education" in his field. Under FRE 703, the facts or data testified to by an expert "need not be admissible in evidence" if they are "of a type reasonably relied upon by experts in the particular field in forming opinions or inferences upon the subject." This avoids some of the great difficulty that the rule against hearsay evidence posed to the admission of scientific opinions.

Moreover, certain specific exceptions to the hearsay rule also facilitate scientific evidence. Under FRE 803(8)(C), "the courts may admit actual findings resulting from an investigation made pursuant to authority granted by law." This allows many government reports to be admitted. FRE 803(18) allows cross-examination of expert witnesses based on "published treatises, periodicals, or pamphlets on a subject of... science" if a witness establishes that the publication is authoritative, although the publications themselves may not be admitted into evidence this way. There is also a catch-all exception, FRE 803(24), whereby a statement not specifically within any of the usual hearsay objections can be admitted if the court determines that it is sufficiently probative and reliable.

These rules somewhat ease the way for scientific evidence, but the door is not wide open. As one commentator has noted,

> Facts or data on which an expert testifies must be 'reasonably relied upon by experts,' presenting a possible problem in NEPA cases. The environmental sciences are sufficiently new so that disagreement still exists on issues such as acceptable methodology. They also are highly interdisciplinary, making it difficult to confine an opponent's expert to the legitimate limits of his expertise. Neither may enough studies have been done on a particular subject to provide reasonably reliable conclusions (Mandelker 1984).

A challenger's opportunity to learn the basis of the opinions of an opposing party's expert witnesses is constrained by Rule 26(b) of the Federal Rules of Civil Procedure, which severely limits the rights to take discovery of expert witnesses.

The difficulty in dealing with novel scientific issues has led to proposals for special courts and special juries in toxic tort and other environmental cases (Drazen 1989). One move in that direction was the establishment in 1982 of the U.S. Court of Appeals for the Federal Circuit, which hears appeals from such specialized tribunals as the Board of Patent Appeals, the U.S. Court of International Trade, the International Trade Commission, and the Merit Systems Protection Board.

Implications for EIS Writers

Persons writing EISs and officials of government agencies preparing or reviewing EISs have an interest in shielding those documents from successful judicial challenge. The legal discussion presented above has the following implications for achieving this purpose:

1. Strict compliance with the procedures established pursuant to NEPA is at least as important as the actual contents of the EIS in avoiding judicial invalidation. The NEPA regulations of CEQ and of the particular lead agency set forth detailed instructions for the publication, circulation, review, scoping, and other aspects of an EIS, and these should be strictly followed.
2. An EIS is most likely to withstand attack if it fully sets forth the scientific basis for all its conclusions. The appendices, and at a minimum the bibliography, should be certain to set forth the methodologies utilized and the sources of data relied upon.
3. While it is often prudent to describe methodologies that were rejected so that full consideration of all factors can be demonstrated, it is perilous to do so without explaining why they were rejected. If the methodologies are presented without this explanation, project opponents will have a significantly easier time of getting these methodologies into evidence and demonstrating to the court the different, and possibly unfavorable, results that may derive from them.
4. Where the EPA or other agencies have established particular test methods for use in administering regulatory programs, those methods should be utilized wherever appropriate.
5. If an agency anticipates a serious legal challenge to an EIS, it would do well to retain, during the EIS preparation stage, experts who will be suitable (in terms of recognized expertise, demeanor, and lack of apparent bias) as trial witnesses so that they can participate in the drafting of the EIS. This will make them more credible and authoritative witnesses at any eventual trial.

REFERENCES

Black, B. 1988. Evolving legal standards for the admissibility of scientific evidence. *Science* 239:1508–1512.

Dore, M. 1985. A proposed standard for evaluating the use of epidemiological evidence in toxic tort and other personal injury cases. *Howard Law J.* 28(3):677–700.

Drazen, D. 1989. Special juries in toxic tort litigation. *Environ. Law Rep.* 19(7):10298–10303.

Gold, S. 1986. Causation in toxic torts: Burdens of proof, standards of persuasion, and statistical evidence. *Yale Law J.* 96(2):376–402.

Imwinkelried, E. J. 1984. Judge versus jury: Who should decide questions of preliminary facts conditioning the admissibility of scientific evidence? *William & Mary Law Rev.* 25(4):577–618.

Koorse, S. J. 1989. False positives, detection limits, and other laboratory imperfections: The regulatory implications. *Environ. Law Rep.* 19(5):10211–10222.

Levy, S. J. 1988. The admissibility of scientific proof in environmental and toxic tort litigation. pp. 611–628. In R. J. Lippes, and B. Wrubel (eds.), *The Sixth Annual Seminar on Hazardous Wastes and Toxic Torts: Regulation-Liability.* Law Journal Seminars Press, New York.

Mandelker, D. R. 1984. *NEPA Law and Litigation.* Callaghan, Wilmette, IL.

McCormick, C. 1954. *Handbook of the Law of Evidence.* West, St. Paul, MN.

McCormick, M. 1982. Scientific evidence: Defining a new approach to admissibility. *Iowa Law Rev.* 67(5):879–916.

Moenssens, A. A. 1984. Admissibility of scientific evidence — an alternative to the Frye Rule. *William & Mary Law Rev.* 25(4):545–575.

Riesel, D. 1989. Pre-Trial Discovery of Experts, Scientific Proof, and Examination of Experts in Environmental Litigation. ALI-ABA Course of Study Materials on Environmental Law.

Snarr, S. W. 1985. Evidentiary issues in National Environmental Policy Act litigation under the APA. *Fed. Bar News J.* 32(8):339–342.

Walker, V. R. 1989. Evidentiary difficulties with quantitative risk assessments. *Columbia J. Environ. Law* 14(2):469–499.

The Statement of Underlying Need Determines the Range of Alternatives in an Environmental Document

O. L. Schmidt,[1] U.S. Department of Agriculture, Office of the General Counsel, Portland, OR.

ABSTRACT

Early models of environmental document preparation called for alternatives to the proposed action. Thus, a railroad might be seen as an alternative to a highway because both are modes of transportation.

Bonneville Power Administration (BPA) has been using a model of environmental document preparation where the statement of underlying need defines the range of alternatives. Thus, if the need is to relieve automobile congestion, a railroad is not an alternative and the railroad alternative can be eliminated from detailed discussion.

This model is based in classic decision theory, where an underlying need is first defined, then alternative ways to meet the underlying need are defined, then the alternatives are compared against decision factors (economics, environment, and technical factors), and finally the decision is made. This model is sanctioned by case law upholding federal agencies that have eliminated alternatives that do not meet the underlying need.

Our experience at BPA is that the statement of underlying need is the single most important element in defining the scope of an environmental document. Any

[1] The author was formerly an attorney with the Bonneville Power Administration. The views expressed are those of the author and not necessarily those of the Office of the General Counsel, U.S. Department of Agriculture, or the federal government.

alternative that does not meet the underlying need is not subjected to detailed analysis. Any alternative that does meet the underlying need is deemed to be a reasonable alternative and is analyzed to at least some degree of detail.

Case anecdotes are presented where BPA successfully limited the scope of an environmental document by carefully defining the statement of underlying need.

INTRODUCTION

The year was 1979. I was reading the first environmental impact statement (EIS) prepared by my agency under the new Council on Environmental Quality (CEQ) regulations implementing National Environmental Policy Act (NEPA) [40 CFR Parts 1500–1508]. It was a preliminary draft circulating inside the agency before publication. My job was in-house legal counsel to the agency, and I was reading this EIS to advise on whether it met the requirements of the CEQ's regulations. I got to the fourth section of the EIS — Purpose of and Need for Action. This follows the (1) cover sheet, (2) summary, and (3) table of contents.

What I read on that page had nothing to do with "purpose of and need for action." It had to do with proposed actions, alternative actions, connected actions, and the *impacts of* proposed actions, alternative actions, and connected actions. There was something on the page, as I recall, about the EIS itself. Why it was being prepared — the fact that federal agencies prepare EISs on proposed actions with significant impacts.

But I did not read anything that sounded like "purpose of and need for action." And so I turned to the CEQ's regulations to find out just exactly what is supposed to be in this section of an EIS. The index to the CEQ's regulations has entries both for the phrase "Need for Action" and the phrase "Purpose for Action." That is all the index has. There is no entry for the phrase "Purpose and Need."

Both phrases in the index refer the reader to the same two sections of the CEQ's regulations — a section dealing with the recommended format of an EIS and the section for purpose and need [40 CFR §1502.10(d) and 1502.13]. The "Recommended Format" section [40 CFR §1502.10(d)] says, simply, that the fourth section of a standard EIS should be titled "Purpose of and Need for Action."

The "Purpose and Need" section [40 CFR §1502.13], titled "Purpose and Need," by the way, not the full title of "Purpose of and Need for Action," is supposed to elaborate on that. This section basically repeats itself. It says, "The statement shall briefly specify the underlying purpose and need to which the agency is responding in proposing the alternatives including the proposed action."

I took this section apart into four pieces.

1. There should be something new. The "Purpose and Need" section should say something other than what is said in the rest of the EIS. What I was reading had already been said in both the cover sheet and the summary. It was entirely redundant of other sections of the EIS. EISs are supposed to be "concise" [40 CFR §1502.2(c)], not redundant.

2. The section must be "brief." What I was reading was nearly a full page — not brief. Certainly not "brief," considering that there was nothing new here.
3. I was looking for an "underlying purpose and need." I focused on the word "underlying." A purpose and need "underlying" the agency's actions, not the actions themselves and not the impacts of the actions. Those are covered elsewhere in the EIS. Certainly I was not looking for anything to do with the EIS itself, the fact that the reader is reading an EIS, or why the agency is preparing an EIS, or what process the agency used in scoping the EIS. That information is either self-evident or unnecessary. I was looking for a statement of what was underlying the agency's proposal for action.
4. I was looking for a direct connection between the agency's statement of their proposed action and their statement of purpose and need *underlying* their proposed action. In other words, if the proposal were to build a dam, I would expect to read about an underlying need for flood control, power, recreation, or some good rationale that would underlie a proposal to build a dam.

Having not found any of these four things, I concluded by the fourth page of the EIS that the EIS was defective. But I am in-house counsel, and it is not my job just to declare EISs defective. It is my job to figure out what it would take to make this EIS complete. Just exactly what is supposed to be in the "Purpose and Need" section of an EIS?

There is no real guidance in the CEQ's regulations themselves. Just the one section that says "Purpose and Need" is a brief statement of the underlying purpose and need. There is a brief mention in the preamble to the regulations published in the *Federal Register* in 1978 [43 *Fed. Reg.* 55977, 55983, November 29, 1978]. There is nothing in the "Forty Most Asked Questions" that was published by the CEQ in March 1981 [46 *Fed. Reg.* 18026, March 23, 1981]. So, we are on our own to figure this out.

Why did CEQ use two words, "purpose" and "need," when one might have been enough? Are the words "purpose" and "need" synonyms, interchangeable? What about "goal" or "objective"? Was this a case of lawyers using two words, purpose and need, when either one would have been enough? No matter what CEQ intended, their use of the two words "purpose" and "need" gives us an opportunity to build a model of EIS construction that is immensely useful in preparing a legally adequate EIS.

Webster's unabridged dictionary defines "need" as "a want of something requisite, desirable, or useful," and "a condition requiring supply or relief" [*Webster's Third New International Dictionary (1971)*]. For short, let us just say that a need is the lack of something wanted or the presence of something unwanted. A need is a problem or an opportunity.

Webster's defines "purpose" as "an object to be attained: an end or aim to be kept in view in any plan, measure, exertion, or operation" [*Webster's Third New International Dictionary (1971)*]. For short, let us just say that a purpose is a goal to be attained. At this point, it is fairly clear that the words "purpose" and "need" are not synonyms. They are different words with different meanings.

Let us skip now to the "proposed action." The NEPA process starts with a proposal for action. What action? Action that is intended to meet an underlying need. Action that supplies something that is lacking or takes away something that is not wanted. Action that fixes a problem or seizes an opportunity.

Imagine a target. The bull's-eye is an underlying need. The arrow is a proposed action. Arrows can strike this target one of two ways, either by hitting the bull's-eye or by missing it. An arrow that hits the bull's-eye is an action that meets the underlying need. The action does not have to completely meet the underlying need, be the only way to meet the underlying need, or even be the best way to meet the underlying need. But there is a connection between the proposal for action and the underlying need to which the agency is responding in proposing the action.

Now let us return to purposes. Purposes were goals to be attained. Let us say that purposes are goals to be attained while meeting an underlying need. In our model, purposes are represented by a ring suspended in front of the target. An arrow can strike this arrangement in any of four ways: (1) through the ring to the bull's-eye, (2) miss the ring but hit the bull's-eye, (3) through the ring but miss the bull's-eye, and (4) miss the ring and the bull's-eye.

Following this model, we can say that there are four kinds of alternatives: (1) those that meet the underlying need and the purposes, (2) those that meet the underlying need but not the purposes, (3) those that meet the purposes but not the underlying need, and (4) those that do not meet the underlying purpose or need.

So now we have a model to define the range of alternatives in an environmental document. An *action* is proposed to meet some *underlying* need. Other reasonable ways to meet the underlying need are also to be considered. These are the *alternatives* within the *scope* of an environmental document. In addition to the underlying need, agencies may have other objectives they are pursuing, called *purposes*. Alternatives that meet both the underlying need and the purposes are the most reasonable of the alternatives and the ones that should be analyzed in greatest detail.

How does this help prepare an EIS? The statement of underlying need defines the range of alternatives in an EIS. The statement of purposes defines the alternatives that are analyzed in the greatest detail.

Alternatives that do not meet the underlying need can be eliminated from the EIS. If there is no relationship between the action and the underlying need to which the agency is responding, the action has no place in agency decisionmaking and it has no place in the EIS. "Actions" here include the proposed action and actions that are alternative to the proposed actions.

Alternatives that meet the underlying need but do not meet other stated purposes must be present in the EIS but can be eliminated from detailed analysis. If an action meets the stated underlying need it is perhaps a "reasonable" alternative, but if it does not also meet other stated purposes it is not as "reasonable" as actions that meet both the stated needs and the stated purposes. Such actions are present in an EIS but are not analyzed to the same level of detail as other actions because they are not as reasonable and, consequently, do not have as great a chance of ultimately being adopted.

We have a model. Does it work?

City of New York v. *United States Department of Transportation*, 715 F.2d 732 (2d Cir. 1983), *cert. denied,* 465 U.S. 1055 (1984). The U.S. Department of Transportation (DOT) proposed to regulate the highway transportation of radioactive materials to meet the underlying need for safe highway transport and published an EIS on their rulemaking. The city of New York sued because the EIS did not include the alternative of barging radioactive materials around New York. Does the barging alternative have to be present in an EIS on safe highway transport? No. Barging is an alternative to *highway transport.* But the underlying need as defined by the DOT was the need for *highway safety.*

> The scope of alternatives to be considered is a function of how narrowly or broadly one views the objective of an agency's proposed action. In this case, for example, if DOT's objective is to improve the safety of highway transportation of radioactive materials, relevant alternatives might include a choice of routes, a choice of equipment, and a choice of driver qualifications. If DOT is concerned more broadly with all transportation of these materials, it might consider alternative modes of transportation. If the objective is viewed still more broadly as reducing the hazards of radiation exposure, the Department might consider alternative sources of power that could reduce the generation of spent nuclear fuel (715 F.2d at 739).

The court used the word "objective" rather than "underlying need," but the concept is the same.

Barging is not an alternative to highway *safety,* although it would be an alternative to highway *transport.* The DOT did not have to consider barging radioactive materials around New York as an alternative in its environmental document because it was responding to the need for safe highway transport, not to the need for transportation of some kind.

Natural Resources Defense Council v. *Morton*, 458 F.2d 827 (D.C. Cir. 1972). The Secretary of the Interior proposed oil and gas leasing in the Gulf of Mexico to meet the nation's energy crisis and prepared an EIS. The Natural Resources Defense Council (NRDC) sued for not including the alternatives of eliminating oil import quotas, developing oil shale, tar sands, geothermal resources, desulfurizing coal, and coal liquefaction and gasification. What was the result? The NRDC won.

The Secretary of the Interior defined the underlying need as the need to meet the nation's energy crisis. All reasonable ways to meet the energy crisis had to go into the EIS. Even though it was beyond the secretary's jurisdiction to change the oil import quota or implement the other alternatives, the court required these alternatives in the EIS, noting that the level of detail needed for some kinds of alternatives might be affected by "the needs to which the underlying proposal is addressed" (458 F.2d at 837–38). Once the Secretary of the Interior had defined the underlying need as the need to meet the nation's energy crisis, all reasonable ways to meet the energy crisis had to go into the EIS. Those alternatives that were less reasonable could perhaps be described in less detail.

Izaak Walton League of America v. *Marsh*, 655 F.2d 346 (D.C. Cir. 1981). The U.S. Army Corps of Engineers proposed a replacement lock and dam and did not include (1) a rehabilitation alternative, (2) a railroad alternative, and (3) the alternative of controlling congestion at the existing lock and dam. The Corps was sued; the Corps won.

(1) The rehabilitation alternative was not necessary because that would not meet the underlying need for expanding the lock and dam. (2) The railroad alternative was not necessary because that would not meet the need for more safety at the lock and dam. (3) The "congestion control" alternative was not necessary because that would not meet the need for expanding the capacity of the waterways system.

The Corps defined the underlying need as the need for more capacity at the lock and dam, more capacity in the waterways system, and more safety at the lock and dam. Those needs could only be met by the replacement lock and dam, and so the scope of the Corps' EIS was properly limited.

Trout Unlimited v. *Morton*, 509 F.2d 1276 (9th Cir. 1974). The Department of the Interior prepared an EIS on a proposed dam project, and the EIS was upheld for its alternatives. For the need for flood control, the alternative of levees was considered. For the need for irrigation water, the alternative of groundwater pumping was considered. "The range of alternatives that must be considered need not extend beyond those reasonably related to the purposes of the project" (509 F.2d at 1286). The court used the term "purposes" of the project rather than "underlying needs," but the concept is the same.

Methow Valley Citizens Council v. *Regional Forester*, 833 F.2d 810 (9th Cir. 1987), reversed on other grounds *Robertson* v. *Methow Valley Citizens Council*, 490 U.S. __, 109 S.Ct. 1835, 104 L.Ed.2d 351 (1989). The U.S. Forest Service prepared an EIS on a proposed ski resort in the North Cascades in Washington State. The EIS only analyzed one site for the ski resort — the Sandy Butte site — but the court said that the underlying need was the need for a "winter sports opportunity." The Forest Service lost. The need for "winter sports opportunity" allows more alternatives than the Sandy Butte site that was in the EIS. "Thus, the Forest Service should more clearly articulate its goal, specifically identifying the market and geographic pool of skiers targeted. This will provide a clear standard by which it can determine which alternatives are appropriate for investigation and consideration in its EIS" (833 F.2d 815–16).

Coalition for Better Veterans Care, Inc. v. *Administrator of the Veterans Administration*, Civil No. 81-365-BE (slip opinion October 5, 1981, D. Ore.). The Veterans Administration (VA) proposed to build a new veterans' hospital in Portland, OR, and prepared an EIS with three alternatives: building at Marquam Hill, building at Emmanuel Hospital, and not building at all. The VA did not include the alternative of private care for VA patients.

Summary judgment for the VA; the case was not appealed. The underlying need was for a replacement hospital, and the alternative of using private care facilities was not reasonable in light of an appropriation by Congress to build a replacement hospital, banning the use of the money for any other purpose.

Sylvester v. *U.S. Army Corps of Engineers*, 884 F.2d 394 (9th Cir. 1989). The Corps prepared an environmental assessment (EA) on a destination resort that needed a dredge and fill permit. Connected to the resort was an 18-hole golf course that displaced wetlands. Did the Corps have to consider alternative locations for the golf course? No. In its statement of underlying need, the Corps said "a quality 18-hole golf course is an essential element for a successful alpine destination resort" (slip opinion at 9387-88), and that offsite locations "did not meet [the] basic purpose and need" (slip opinion at 9388). Again, a precise definition of the underlying need successfully limited the range of alternatives in an environmental document.

City of Angoon v. *Hodel*, 803 F.2d 1016 (9th Cir. 1986). This was a challenge to a complex land exchange, log transfer facility, and timber harvest on Admiralty Island for which there is a long litigation history. For the EIS prepared by the Corps, the question was whether the EIS should have included an alternative land exchange while the Corps was considering granting a permit to build the log transfer facility.

The underlying need the Corps identified was the need for a "safe, cost effective means of transferring timber harvested on [permittee's] land to market." The District Court rejected the Corps' statement of need and substituted a more general "commercial timber harvesting" need. For this larger need, the EIS was defective for failing to consider a particular land exchange.

The Ninth Circuit Court of Appeals reversed the District Court. First, the Circuit Court accepted the Corps' definition of underlying need. "The preparation of [an EIS] necessarily calls for judgment, and that judgment is the agency's" [803 F.2d at 1021, quoting *Lathan* v. *Brinegar*, 506 F.2d at 693]. Second, the Corps' more narrow definition of need makes the land exchange alternative irrelevant. "When the purpose is to accomplish one thing, it makes no sense to consider alternative ways by which another thing might be achieved" [803 F.2d 1021]. Third, the Circuit Court pointed out that no matter what the statement of underlying need, the land exchange alternative was "remote and speculative" because for the Corps to find exchangeable land "would involve almost endless speculation" [803 F.2d 1021]. Fourth, the Circuit Court pointed out that a land exchange alternative would require legislation, which would not eliminate an alternative automatically, but, in this case, such an alternative was not ascertainable or reasonably within reach [803 F.2d 1021–22, n. 2].

Citizens Against Burlington, Inc. v. *Busey*, 938 F.2d 190 (Circuit Judge Clarence Thomas, C.A.D.C., June 14, 1991). The city of Toledo proposed to expand their airport and applied to the FAA for a permit. The FAA considered only two alternatives: grant the permit and deny the permit. "[T]he FAA defined the goal for its action as helping to launch a new cargo hub in Toledo and thereby helping to fuel the Toledo economy. The agency then eliminated from detailed discussion the alternatives that would not accomplish this goal." Thus, the FAA eliminated the development of a cargo hub at Fort Wayne or any other city.

When an agency defines the underlying need very narrowly, it wins. And when it defines the need very broadly (or not at all), it loses. It seems that a narrowly scoped EIS is easier to write and to defend than one that is broadly scoped.

When an agency is careful in defining its underlying need, the agency can limit the range of alternatives to those that could reasonably meet that underlying need. This is another way of saying that an agency can safely eliminate those alternatives from an environmental document that do not meet the underlying need.

This model avoids the pitfalls of the phrase "alternatives to the proposed action" for which there may be no boundary. The phrase "alternative ways to meet the underlying need" is very different from the phrase "alternatives to the proposed action." This model makes it comparatively easy to defend the inclusion or exclusion of a particular alternative in an environmental document.

I could stop here with this model. It is useful in defining the range of alternatives in an environmental document. It is the only model I know of for precisely defining the range of alternatives. But I want to extend this model in two ways: first, to define the "no action" alternative as "not meeting the underlying need"; and second, to define "purposes" as "decision factors."

No-Action Alternative

A "no action" alternative is required in EAs and EISs [40 CFR §1508.9(b), 1508.25(b)(2)]. Many environmental documents treat the no-action alternative literally as "not taking the proposed action." If the proposed action is to build a dam, the no-action alternative may be treated simply as not building a dam. The impacts of not building a dam might be treated as not having the impacts of a dam — not flooding farmland, not relocating displaced landowners, not building access roads, etc.

A more insightful approach would be to treat the no-action alternative as not meeting the need. In our example, there is a need to control floods. What would happen if the need were not met? Continued floods! What is the impact of continued floods? And what is the impact of the proposed dam? With this comparison, decision makers and the public would have before them the true difference between building and not building the dam.

One court has acknowledged this approach. In *Trout Unlimited* v. *Morton* (509 F.2d at 1286), the Ninth Circuit upheld an EIS which "considered the alternatives of (1) no development whatsoever (the alternative of not accomplishing the project's purposes)," but the court did not elaborate. The CEQ has also acknowledged this approach in their "Forty Questions." "'No action'... would mean the proposed activity would not take place, and the resulting environmental effects from taking no action would be compared with the effects of permitting the proposed activity or an alternative activity to go forward" [CEQ, Forty Most Asked Questions Concerning CEQ's National Environmental Policy Act Regulations, 46 *Fed. Reg.* 18026, 18027 (1981)]. Following the CEQ's advice, an agency would analyze the impacts of not meeting the need rather than simply not taking the proposed action.

Decision Factors

Returning to our suspended ring, we said that purposes were goals the agency was attempting to reach while taking action to meet an underlying need.

What sort of goals? We can say that there are basically three kinds of goals for any proposed action: (1) economic, (2) engineering (here all technical and legal factors are lumped into "engineering"), and (3) environmental. These three purposes overlap.

Turning this model from the side view to more of a front view, we see that an arrow can strike the bull's-eye in any of eight ways. Double this number to show the arrows that miss the bull's-eye, and we see that in all the world there are 16 kinds of alternatives, including the proposed action. Those alternatives that hit the bull's-eye and only one, or none, of the rings can probably be eliminated from serious consideration. Those alternatives that hit the bull's-eye and two or more of the rings are more reasonable and probably should be analyzed in some amount of detail. Those that hit dead center — the bull's-eye and all three rings — are the very most reasonable of the alternatives. In other words, in this model, even alternatives that meet the underlying need can be eliminated from detailed study if they do not also meet the stated purposes.

There is no case law on this point, but it has a logical place in this model.

Purposes appear in two places in the EIS process: in the "purpose and needs" section of the EIS and again in the Record of Decision (ROD) (40 CFR §1505.2). The CEQ regulations do not say that purposes appear in a ROD, but the CEQ regulations do say that decision factors must be disclosed in the ROD.

> An agency may discuss preferences among alternatives based on relevant factors including economic and technical considerations and agency statutory missions. An agency shall identify and discuss all such factors including any essential considerations of national policy which were balanced by the agency in making its decision and state how those considerations entered into its decision [40 CFR §1505.2(b)].

There are four arguments for tying "purposes" in an EIS to "decision factors" in the ROD.

- First, it is inevitable that the factors used to compare the alternatives in the EIS will be the same factors used when selecting between the alternatives at the time of decision. For example, each alternative must be analyzed in an EIS for compliance with the law [40 CFR §1508.27(b)(9), (10)]. Compliance with the law is always a "purpose" in the sense that it is a goal the agency seeks to achieve, and compliance with the law is always a factor to consider when choosing between alternatives. Thus, compliance with the law is always relevant in an EIS as a purpose and in a ROD as a decision factor. For another example, when economics are relevant and important to a decision, economic information must be included in an EIS [40 CFR §1502.23], and when economics are relevant and important to a decision, economic information must be disclosed in a ROD [40 CFR §1505.2(b)]. Thus, we must ineluctably conclude that purposes in an EIS and decision factors in a ROD are one and the same.

- Second, the ROD should be based on the record before the agency. In many agency proceedings, the EIS is the only record created prior to the decision, and so the EIS is the only place to put information that is relevant to the decision. Dealing with decision factors in the EIS under the heading of "purposes" puts that information into the record. Even in proceedings where the EIS is only a part of the administrative record, the EIS may be the best place to put decision factors. This is where the public will expect to find them, and this may be the only part of the administrative record where it is logical to put relevant decision factors.
- Third, the public will have a chance to review and comment on the decision factors if they are disclosed in the EIS. This is "full disclosure." This is a great way to involve the public in what is really a very important part of the decisionmaking process — the selection of the factors that will be used when deciding among the alternatives presented in the environmental document.
- Fourth, putting these factors into the EIS paves the way for those who follow to write the ROD. The eleventh decisionmaking hour, when ROD writers are hard at work, is not the best time to start figuring out the relevant decisionmaking factors.

Special Rule for Environmental Assessments

EAs must have a statement of underlying need, but need not have a statement of purposes: "Environmental Assessment"...shall include brief discussions of the need for the proposal...[40 CFR §1508.9(b)].

Why did the CEQ leave "purposes" out of EAs? A simple explanation, consistent with the model described above, is that in the CEQ's regulations EISs are followed by RODs, but EAs are not. "Purposes" in an EIS become "decision factors" in a ROD. Because EAs are not followed by RODs, there is no requirement to put purposes into EAs. Agencies may choose to prepare RODs following EAs and thus should put purposes into EAs as well as EISs.

In any event, at the time of decision, when decision makers are balancing all relevant factors, the record of decision must disclose those decision factors, those underlying purposes — those economic, engineering, and environmental factors — the agency was responding to.

CONCLUSION

Action is proposed to meet an underlying need. Other reasonable ways to meet the underlying need belong within the scope of an environmental document. Alternatives which do not meet the underlying need may be left out of an environmental document.

Purposes are goals to be attained while meeting an underlying need. Purposes are always economic, engineering (or technical), and environmental. Those alternatives that best meet both the agency's purposes and needs are the most reasonable alternatives and should be analyzed in the greatest detail.

The "no action alternative" is defined as not meeting the need. "Purposes" in the EIS — economic, technical, and environmental — appear again in the ROD as decision factors.

What happened to that preliminary draft EIS I was reading in 1979? That particular EIS took over 5 years to write and cost over $3 million. It was never revised to match the underlying need with the range of alternatives and was never used to make an administrative decision.

ADDITIONAL READING

Schmidt, O. L. 1988. The statement of underlying need defines the range of alternatives in environmental documents. *Environ. Law* 18:371–381.

Practice Pointers for the Purpose and Need Section

There is no such thing as an inadequate statement of underlying need. An EIS could not be held to violate NEPA or the CEQ's regulations for the sole reason that the need statement is inaccurate.

But there is the possibility of a mismatch between the need and the range of alternatives. This is where the legal peril lies. If the need statement and range of alternatives do not match, there is a good possibility that the range of alternatives will not be adequate.

Thus, it is not possible to analyze a need statement without also analyzing the range of alternatives. Once these are aligned, an environmental document and the decision process are on firm ground.

It's an Iterative Process ...

Write a need statement and see if it yields a reasonable range of alternatives. Conversely, look at the range of alternatives being considered and see if it matches the statement of underlying need. If there is a mismatch, change either the need statement or the range of alternatives.

Drafting the need statement is reiterative with the alternatives. Does the need statement match the range of alternatives? If not, change the need statement. Does the need statement produce the desired range of alternatives? If not, change the need statement.

Start Early ...

Logically, the need statement should be the first sentence drafted. First, a problem or opportunity is perceived. This is the need. Then a proposal for action is conceived. An agency proposes to fix a problem or seize an opportunity.

Practically, it seems that the need statement is often drafted late in the process. This is the pitfall. The underlying need is the bedrock of the decision process, and the sooner it is worked out the whole process will run better.

"Purpose and Needs" Worksheet

The Statement of Underlying Need Determines the Range of Alternatives in an Environmental Document

Ask This Question	Looking For	Recommendation
What action is proposed?	Proposal	
Why?	Underlying need	
What other action would do the same thing, at least in part?	Alternatives	
What are the effects of the proposed action, and alternative actions — in comparative format?	Impacts, actions	
What factors will be used in making the decision between alternatives?	Purposes	
Are there any ways to mitigate adverse effects?	Mitigation	

Needs Are Nouns ...

A noun is a "thing." Problems and opportunities are things. Therefore, problems and opportunities are nouns. In short, a need is a noun.

We propose to take action, a verb, in order to meet a problem or opportunity, a noun. The structure is the key. We propose action (a verb) to meet a need (a noun). If the need were stated as a verb, we would actually be saying we are proposing action (a verb) because we need action (a verb).

The need can be a transitive verb, the only exception to the rule that needs are always nouns. A transitive verb is a verb that acts like a noun. Management is an example. "The underlying need is the need for (some type of) management." Not the need "to manage." That would be a verb. This is a common need for proposals to continue ongoing action, such as plans for public lands (Figure 1).

Needs Have Alternatives, Purposes Do Not

The purpose and needs model is a double-winnowing model. First, we want to eliminate from an environmental document all the alternatives that do not meet our statement of underlying need. Second, among the alternatives present in the environmental document, we want to eliminate from detailed consideration those alternatives that do not meet our stated purposes.

Only alternatives that would meet the underlying need are present in an environmental document. And only those alternatives that meet the stated purposes are considered in detail. This double-winnowing eliminates from serious consideration all but the most reasonable alternatives.

Once again, the structure is the key. Agencies perceive a problem or opportunity. Agencies propose action to fix the problem or seize the opportunity. All reasonable ways to fix the problem or seize the opportunity should be developed. This is the broadest possible range of alternatives; none are overlooked. Then this broad range of alternatives is compared against the second winnowing factor, the

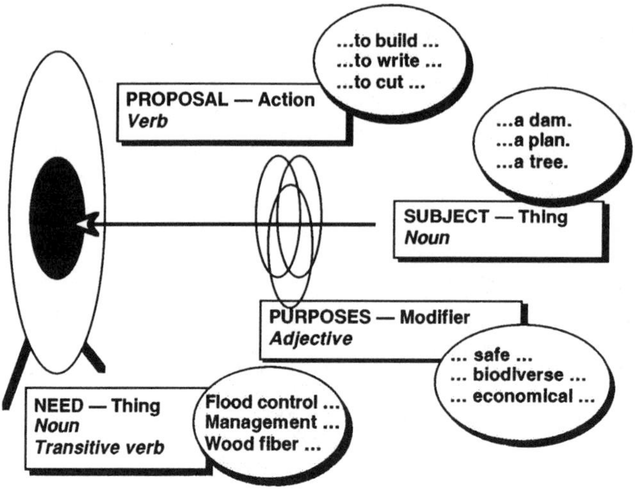

Figure 1. The structure is the key. A need is a thing and is expressed with a noun or transitive verb. A proposal is an action and is expressed with a verb. The subject of the verb is a thing and is expressed with a noun. Purposes modify the subject and are expressed with adjectives.

purposes. Only those alternatives that also meet the stated purposes go through the expense and energy of being analyzed in detail.

Keeping needs as the first winnowing factor guarantees a hard look at all possible ways to deal with a problem or solution. Keeping purposes as the second winnowing factor guarantees that only the more reasonable alternatives get the attention of a detailed analysis.

Purposes Are Adjectives and Decision Factors

At the time of decision, one of the alternatives will be selected over the others. The reasons for picking one over the others will be disclosed in the decision record. The reasons always center on three factors: economic, legal (including technical or engineering), and environmental. These are the qualities that make one alternative more desirable than another or less desirable than another.

These decision factors are, in fact, the purposes listed earlier in the environmental document. These decision factors are adjectives describing the qualities the agency is seeking in an alternative, the purposes the agency is striving to achieve, or the goals the agency is trying to attain.

[1]

The Statement of Underlying Need Determines the Range of Alternatives in an Environmental Document

Owen L. Schmidt

U.S. Department of Agriculture
Office of the General Counsel

[2]

Environmental Impact Statement

40 C.F.R. §1502.10 Recommended format...

(a). Cover sheet.
(b). Summary...
(c). Table of Contents...
(d). Purpose of and Need for Action...
(e). Alternatives Including Proposed Action...
(f). Affected Environment..
(g). Environmental Consequences...
(h). List of Preparers...
(i). List of Agencies, Organizations, and Persons to Whom Copies of the Statement Are Sent...
(j). Index...
(k). Appendices (if any)...

[3]

Need for Action

Purpose and Need

Purpose for Action

Purpose of and Need for Action

[4]

40 C.F.R. §1502.13

"The statement shall briefly specify the underlying purpose and need to which the agency is responding in proposing the alternatives including the proposed action."

FIGURE 1 (continued).

[5]

Need

- "a want of something requisite, desirable, or useful"
- "a condition requiring supply or relief"
- the lack of something wanted
- the presence of something unwanted

[6]

Purpose

- "an object to be attained: an end or aim to be kept in view in any plan, measure, exertion, or operation"
- a goal to be attained

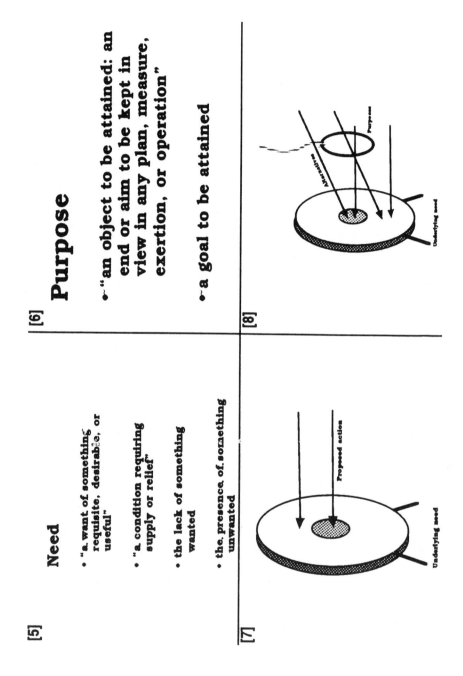

[7]

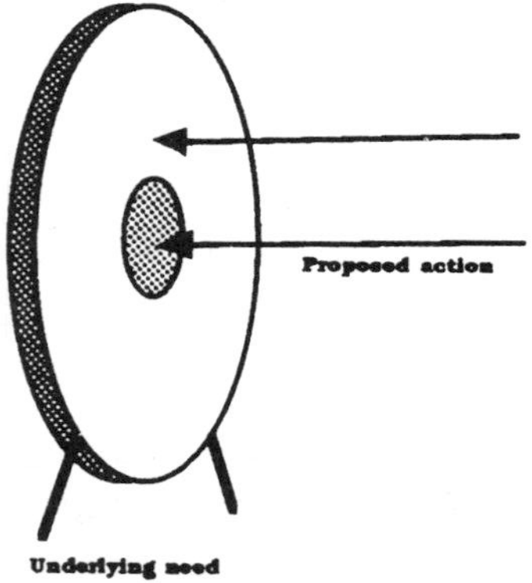

[8]

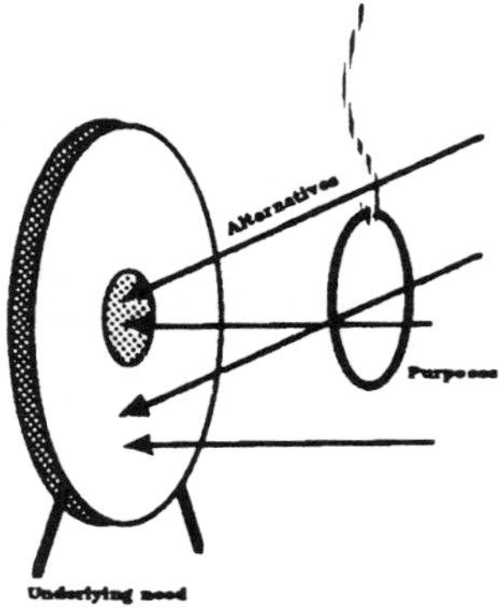

[9]

City of New York v. United States Department of Transportation
715 F.2d 732 (2d Cir. 1983), cert. denied, 465 U.S. 1055 (1984).

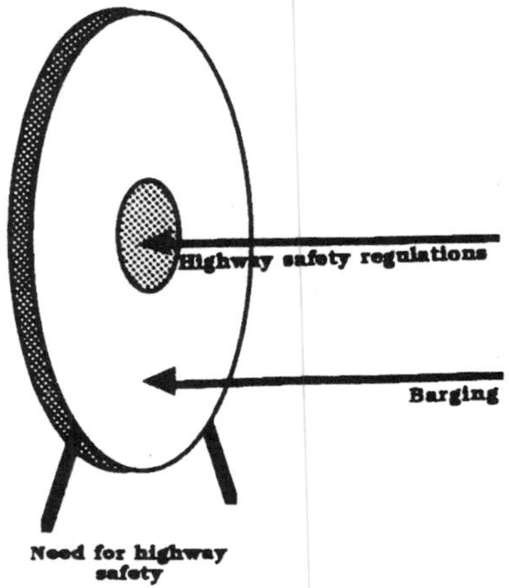

[10]

Natural Resources Defense Council v. Morton
(458 F.2d 827 (D.C. Cir. 1972).

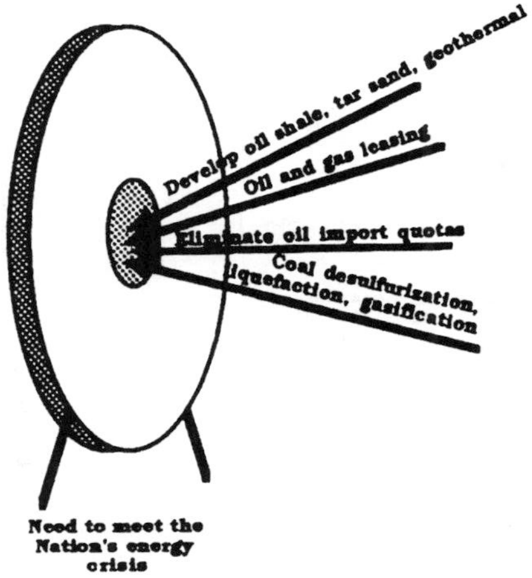

FIGURE 1 (continued).

[11]

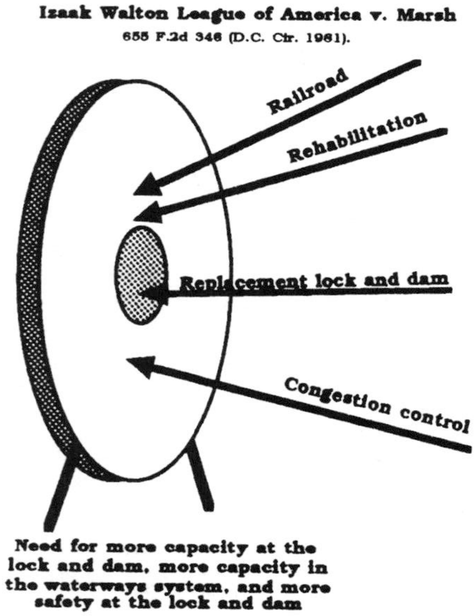
Izaak Walton League of America v. Marsh
655 F.2d 346 (D.C. Cir. 1981).
Railroad
Rehabilitation
Replacement lock and dam
Congestion control
Need for more capacity at the lock and dam, more capacity in the waterways system, and more safety at the lock and dam

[12]

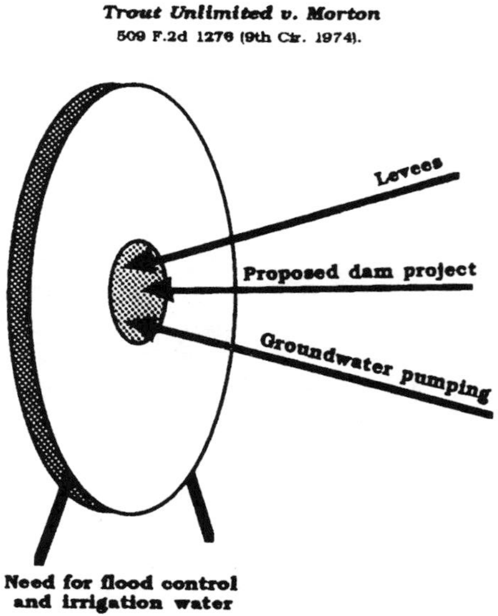
Trout Unlimited v. Morton
509 F.2d 1276 (9th Cir. 1974).
Levees
Proposed dam project
Groundwater pumping
Need for flood control and irrigation water

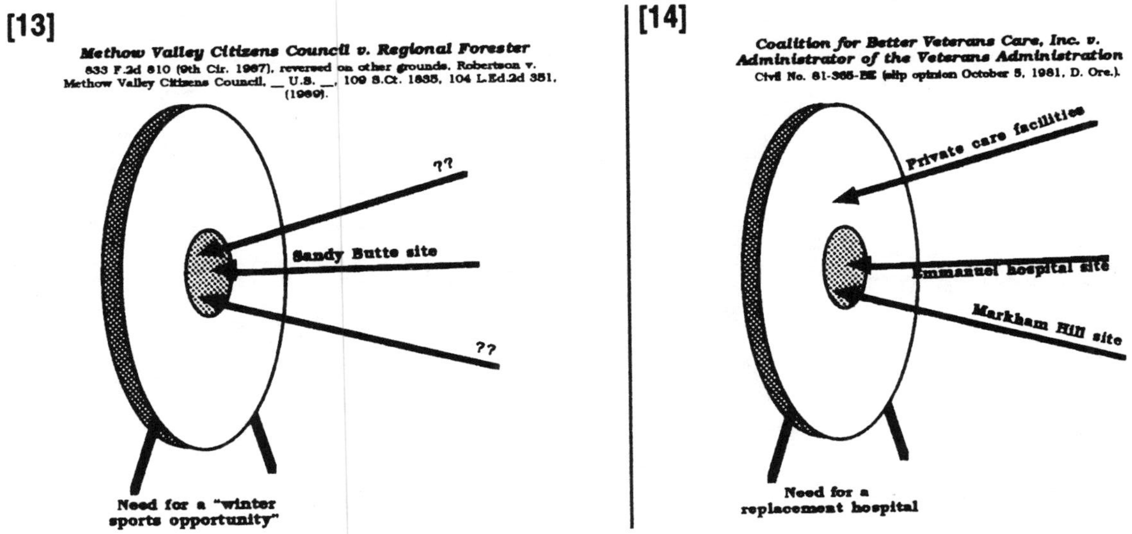

FIGURE 1 (continued).

[15]

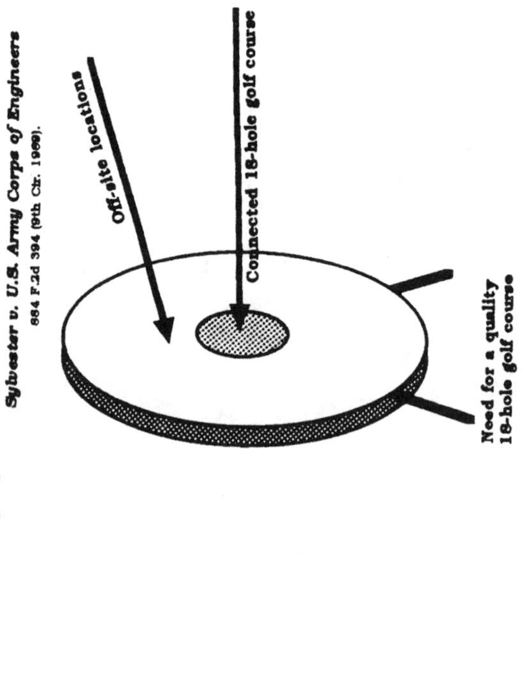
Sylvester v. U.S. Army Corps of Engineers
884 F.2d 394 (9th Cir. 1989).
Off-site locations
Connected 18-hole golf course
Need for a quality 18-hole golf course

[16]

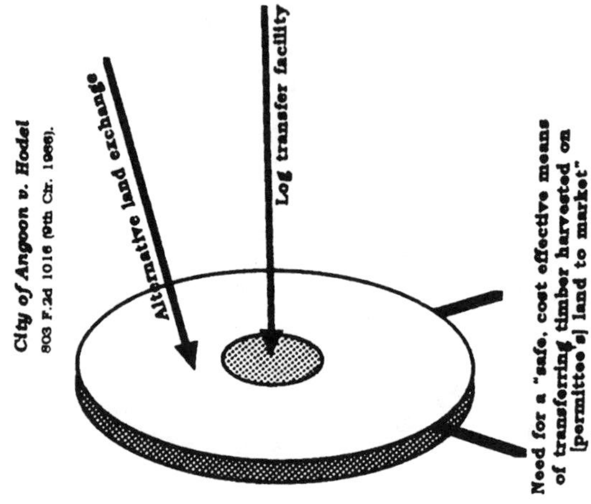
City of Angoon v. Hodel
803 F.2d 1016 (9th Cir. 1986).
Alternative land exchange
Log transfer facility
Need for a "safe, cost effective means of transferring timber harvested on [permittee's] land to market"

[17]

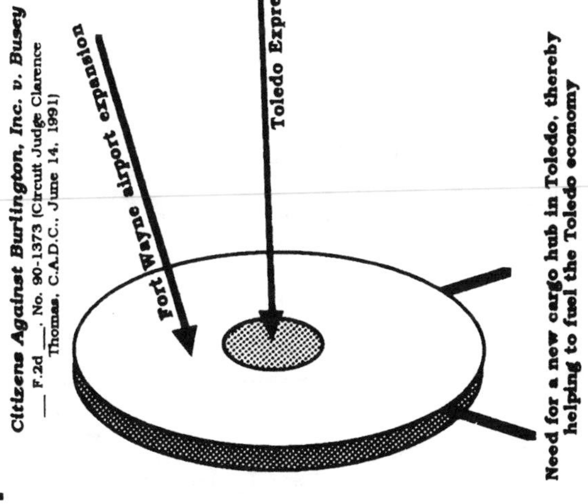
Citizens Against Burlington, Inc. v. Busey
— F.2d ___, No. 90-1373 (Circuit Judge Clarence Thomas, C.A.D.C., June 14, 1991)
Fort Wayne airport expansion
Toledo Express
Need for a new cargo hub in Toledo, thereby helping to fuel the Toledo economy

[18]

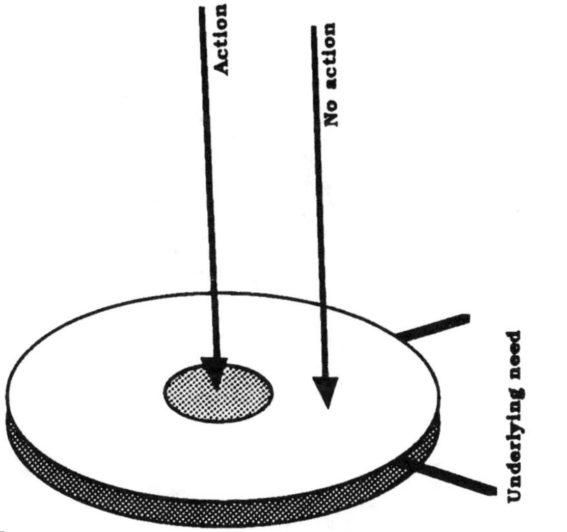
Action
No action
Underlying need

FIGURE 1 (continued).

[19]

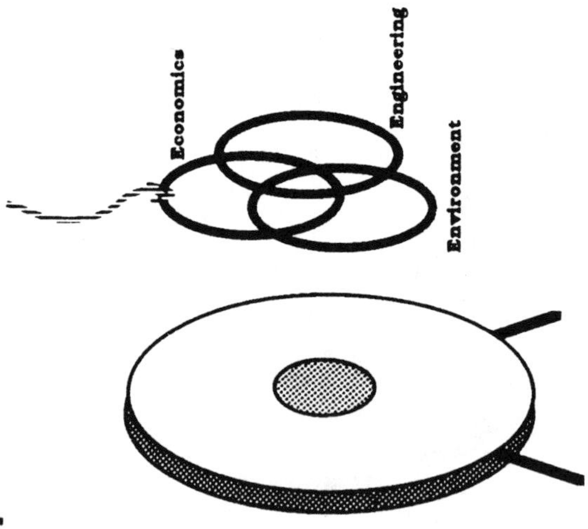

[20]

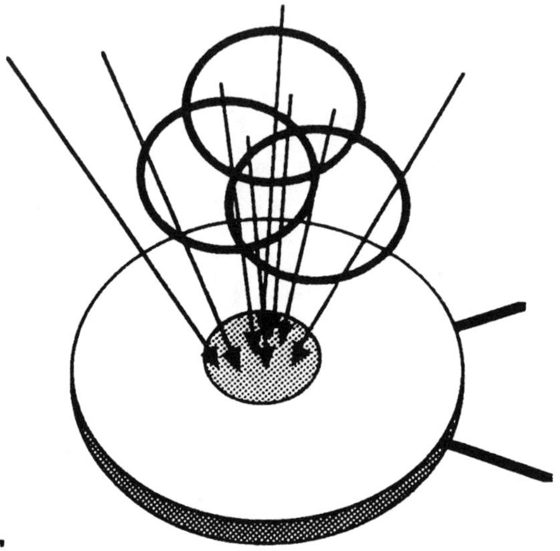

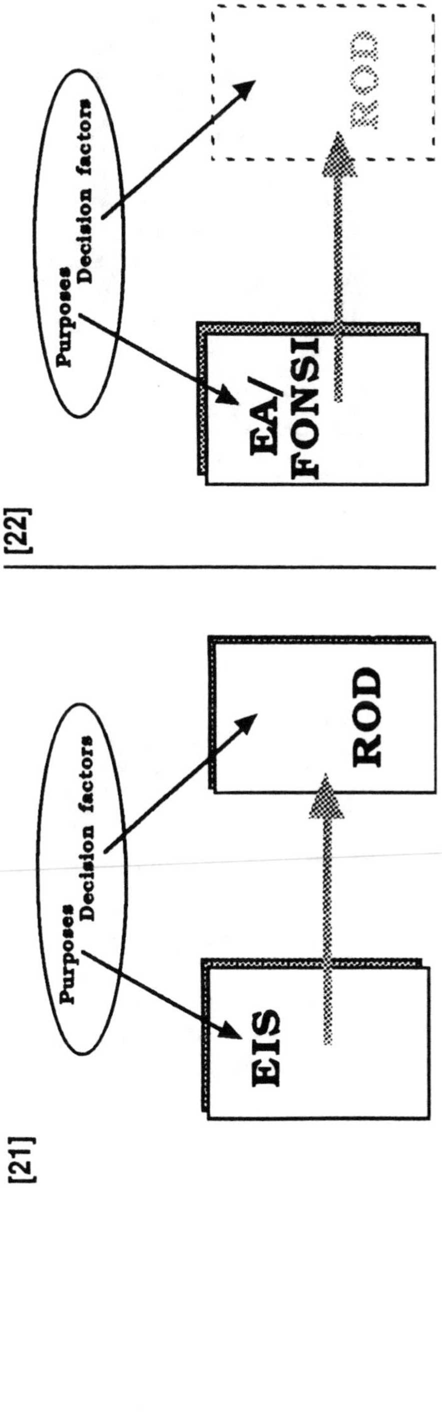

FIGURE 1 (continued).

[23]

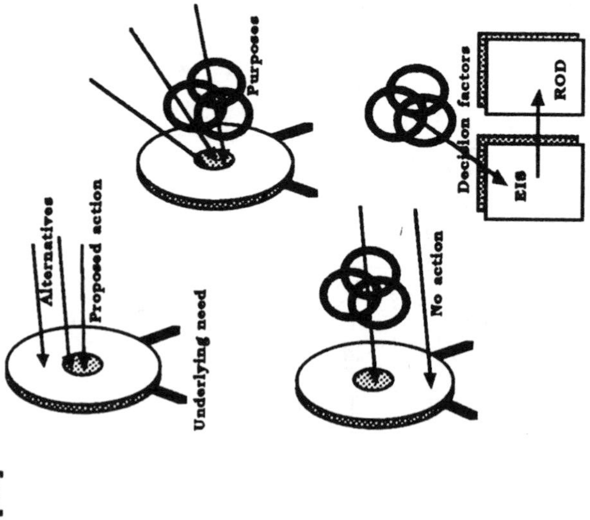
Alternatives
Proposed action
Underlying need
Purposes
No action
Decision factors
EIS
ROD

[24]

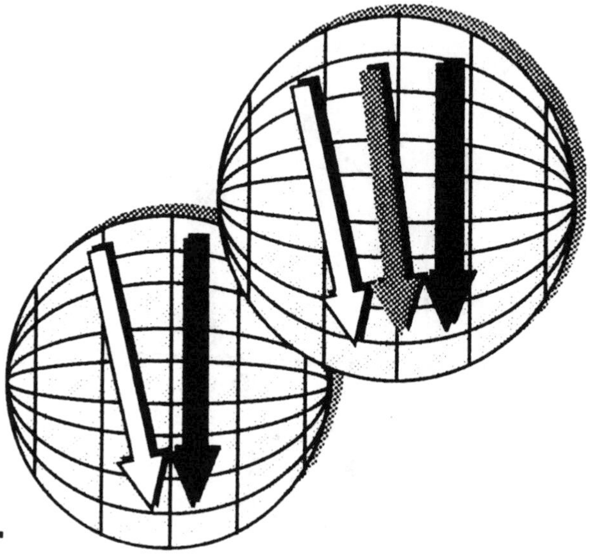

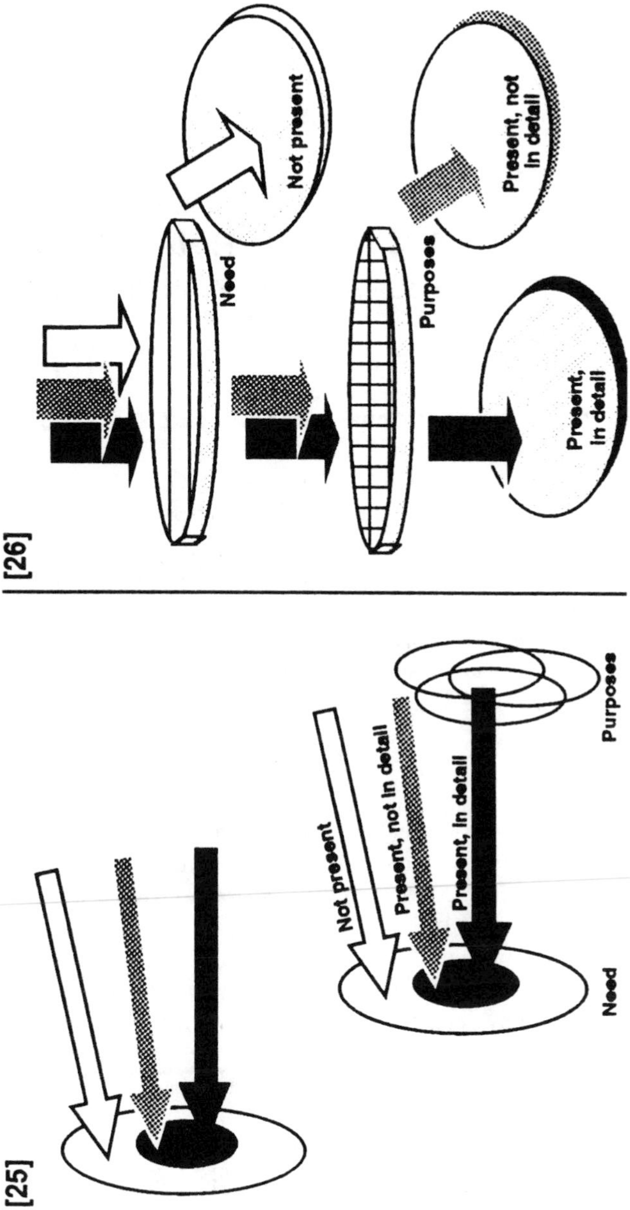

FIGURE 1 (continued).

A Call for a Return to Rational Comprehensive Planning and Design

R. E. Nelson, BHP-Utah International Inc., San Francisco, CA

ABSTRACT

Moving amidst my own people I was never impressed by any of their accomplishments; I never felt the presence of any deep religious urge, or any great aesthetic impulse; there was no sublime architect, no sacred dances, no ritual of any kind. We moved in a swarm, intent on accomplishing one thing — to make life easy. The great bridges, the great dams, the great skyscrapers left me cold. Only nature could instill a sense of awe and we were defacing nature at every turn (Henry Miller, *Nexus*).

Certain people always say we should go back to nature. I notice they never say we should go forward to nature. It seems to me they are more concerned that we should go back, than about nature (Adolph Gottlieb, *The New York School*).

The National Environmental Policy Act (NEPA), passed in 1969 and signed into law January 1, 1970, was the nation's response to the malaise expressed by Henry Miller. Today, however, evidence suggests we are stricken with another malaise. Have we lost the confidence to develop our physical world for the benefit of mankind or is it our vision that six billion human beings should go back to nature?

The basic purpose of NEPA was to heighten attention to environmental considerations in public decisionmaking. At the time of its passage, Senator Jackson, one of its authors, hailed it as "a new concept of planning and management [which] is emerging in the federal government" (Committee Print 1970).

The Declaration of National Policy as stated in Section 101 of NEPA recognizes the validity of natural resource development and economic claims on the environment. It reads in part as follows: "to create and maintain conditions under which man and nature can exist in productive harmony, and fulfill the social, economic, and other requirements of present and future generations of Americans." Yet NEPA has become the single most powerful tool to inhibit development.

Is it time now to set aside the fractured planning approach that results from the environmental impact statement (EIS) process and to integrate environmental considerations with other aspects of planning and design? Does it make sense to continue to design projects, then in tandem run them through a process to highlight their environmental and social implications. Is it logical to continue to pursue a process that by its very nature presumes that change to the physical environment is generally bad and nonchange is generally good? Should we continue to support a process that focuses on documentation rather than on analysis? Should we continue to hold to a process that for the sake of ecology and holistic thinking, in fact, causes designers instead to compartmentalize their analysis into such sectors as engineering analysis, economic analysis, health and safety analysis, social analysis, and environmental analysis?

NEPA was a response to the obvious excesses of the national highway building program by the U.S. Department of Transportation, national dam building programs by the Bureau of Reclamation, flood control programs by the U.S. Army Corps of Engineers, and a host of other federal programs that were rapidly changing the face of the United States. These federal agencies were each pursuing their legislative mandate, and none had the authority to consider environmental values in its planning and design. None was authorized to spend budgeted funds to protect or enhance environmental values.

Given the excesses of development policy that abounded at the time of the passage of NEPA, it is clear that a sharp turn in public policy was required. Although NEPA was not originally perceived to be of any great significance, it became the key policy device to force much needed correction in public policy.

NEPA was one act of many that was part of a federal interventionist outburst in the 1960s and 1970s. This outburst began in 1964 with the Civil Rights Act and continued through the end of the 1970s with a plethora of environmental legislation.

In his book *Crisis and Leviathan* (Higgs 1987), Robert Higgs analyzes the spurts of government growth and intervention in response to public crises throughout the last 100 years of our republic. His unhappy conclusion is that government grows in response to a crisis but fails to atrophy when the crisis has ended. It is my thesis that we are far overdue for a fundamental reexamination of the purposes of NEPA, which I support, and the mechanisms through which it is carried out, which I feel are severely flawed.

The intended purpose of the EIS is to improve the federal planning process through the early consideration of environmental values. Yet the process has failed to substantially improve planning, as Lynton Caldwell noted in his keynote address.

With very rare exceptions, regardless of whether it is an industry or agency proposal, the project objective is to construct something or to modify a management technique to achieve enhancement of a resource or economic objective; in short, to cause change. In this context, environmental objectives become constraints — limiting factors — or criteria to be met while achieving the project objective.

Rational planning, on the other hand, presents a goal — a problem to be solved, a mission to be achieved. Rational planning must compare various possible solutions within a context of a number of criteria. As a minimum, a project must be economically achievable, technically sound, socially acceptable, and ecologically desirable. The EIS deals with only one, or at most two, of these planning criteria.

The problem lies in the fact that the EIS process has become the decision endpoint rather than being a part of the decision process. For example, we do not construct a power plant for the specific objective of enhancing environmental quality; yet the EIS process is used as the policy tool to plan the power plant. The EIS process should establish design criteria and constraints only to protect and enhance environmental quality. It should not masquerade as a comprehensive approach to project planning and design. The EIS process is only one segment of a much larger planning process. In one sense, it is a check on planning to ensure that environmental quality values are adequately considered. If we could achieve a consensus on this basic principle, many of our procedural problems would dissipate.

The EIS has not been very useful as an aid to planning, largely because the EIS process does not mesh with a rational, comprehensive planning and design sequence. This can be illustrated as follows. Let us assume we are designing a large project such as a mine. This project typically requires a 10-year sequence of data collection, analysis, permit application, sales negotiation, and construction. This procedure includes decision points and the consideration of alternatives at a number of junctures during project development. Yet at one point, approximately in the middle of this sequence, an EIS is initiated even though it is recognized that most of the data for this EIS would logically be collected several years later in the project design stage. Worse, the EIS requires analysis of alternatives, presumably the same alternatives that were considered and dismissed in the project planning stage 4 or 5 years earlier. This sequence of events, where the EIS process presumably forces the decision maker to go back and reconsider decisions on alternatives already foreclosed, does not make sense. At the same time, the EIS process forces the development of data years earlier than is rational in a normal planning process.

The one meritorious argument for an EIS is that it provides open review of project development. However, this review has been of greatest use to opponents of projects and not as an aid to design, coordination, and planning as was intended.

The argument can be made that one should merely shift the EIS to a point later in the project development schedule. However, if the project is then rejected or requires substantial modification, the developer has already invested large sums of capital and can ill afford to cancel the project. Other arguments have been made that an EIS serves a useful function by drawing together the results of many studies in disparate fields. In this sense, the EIS can be only a historical account of decisions made. Many agency procedures adopted to conform to environmental legislation enacted later in the 1970s contained such reviews. This raises the question of whether the need for the EIS has been superseded.

The enormously cumbersome EIS process that has grown out of a single sentence in NEPA has ballooned far beyond what the authors envisioned. Clearly, we should not relax our effort to improve the quality of environment that affects each of us every day of our lives. We must, however, come to grips with many other problems that our nation faces. Why is it that a single public policy issue, environmental quality, should alone command such a specialized process as the EIS? In project decisionmaking, we do not have a parallel process for the analysis of health and safety impacts of a proposed project. We do not have a fair employment or nondiscrimination EIS. We do not have a balance of trade impact analysis. We neither have an impact analysis of effects on our cultural heritage, nor do we have an analysis on a host of other possible target institutions that might be worthy of examination such as jobs or taxes.

One of the signs of maturity in an individual is a lack of excesses. We recognize immaturity through acts that are too loud, too colorful, and too fast. Surely our government, our society, and America is approaching a stage of maturity where we can deal appropriately with our desires for a quality environment without the excesses resultant from the leviathan of the EIS process. The EIS process was a useful tool to gain our attention and to force us to change our ways. Its' continued use today, however, provides the basis for too much litigation, too much delay, too much inefficiency, and provides the basis for small groups of citizens to illegitimately hold much needed projects at bay. The not-in-my-backyard syndrome has far too frequently rendered legitimate authorities powerless to achieve needed social objectives. A mature outlook would cause us to conclude that a quality, clean environment comes at a cost to our total society and not just to a company, not just to a project proponent, and not just to the objectives of a specific public agency.

We submit that a "new concept in planning," as Senator Jackson envisioned, has not emerged through the application of the EIS. We feel that if there is to be rational planning, then what is needed is a continuous and flowing system where alternatives are considered early in keeping with industry procedures, where public participation is a part of planning, and where planning and permitting are intertwined. Let us return to a rational comprehensive planning and design approach, one that is sound, efficient, and holistic in its approach to meeting established objectives. This can be accomplished, in part, by incorporating environmental consideration in the planning and design process rather than tagging it on as the NEPA process now requires. We are not opposed to NEPA or the EIS process per se or to environmental standards, but we would like to see NEPA as it was intended, not what it has become.

REFERENCES

Committee Print. 1970. Interior and Insular Affairs Committee. June 1970.
Higgs, R. 1987. *Crisis and Leviathan,* Oxford University Press, New York.

NEPA/CERCLA/RCRA Integration: A Panel Discussion

F. E. Sharples and E. D. Smith, Oak Ridge National Laboratory, Oak Ridge, TN

INTRODUCTION

The Comprehensive Environmental Response, Compensation, and Liability Act (CERCLA) as amended requires that decisions concerning remedial actions at Superfund sites be made through a formal decisionmaking process known as a Remedial Investigation/Feasibility Study (RI/FS). Many of the elements of this process are similar to the steps in the process required to comply with the National Environmental Policy Act (NEPA). Both processes, for example, involve the identification and analysis of alternative courses of action, provide for public disclosure and participation in the processes, and are documented by records of decision (RODs).

Decisions by federal agencies are subject to NEPA, which requires that environmental impact statements (EISs) be prepared for major federal actions that may significantly affect the quality of the human environment. Actions undertaken by the Environmental Protection Agency (EPA), however, are usually treated as exempt from EIS requirements because of the doctrine of "functional equivalence." This doctrine holds that the EPA's organic legislation mandates procedures that are functionally equivalent to those of NEPA, in that they ensure adequate substantive and procedural consideration of environmental issues and afford public participation. The functional equivalence of certain EPA actions has been established by court decisions and by amendments to the Clean Air and Clean Water Acts (Mandelker 1984). As a result of the functional equivalence doctrine, the NEPA process is not followed by the EPA when it undertakes CERCLA remedial actions at nonfederal sites.

Remedial responses at CERCLA National Priorities List (NPL) sites on federal land or at federal facilities are the responsibility of the agency with jurisdiction over the site, subject to review and approval by the EPA. Because the decision–making process for remedial responses at these sites must by law follow the RI/FS process, which is exempt from NEPA when conducted by the EPA, there has been some uncertainty as to whether NEPA applies to CERCLA actions undertaken by agencies other than the EPA. Furthermore, because the two processes are similar but not identical, there has been concern that implementation of both the NEPA and CERCLA processes will result in unnecessary duplication of effort, delay, and possible legal conflicts.

Similar questions are expected to arise as federal facilities are required to conduct corrective actions under the Resource Conservation and Recovery Act (RCRA). Under the RCRA, facilities that manage hazardous wastes may be required to take action to correct past releases from their solid waste management units as a condition of their RCRA hazardous waste permits. The decisionmaking process for RCRA corrective actions is expected to be similar to that for CERCLA remedial actions.

Generally, application of functional equivalence has been viewed as being limited to actions by the EPA alone because the EPA is administering statutes that are environmentally protective. The courts have declined to apply the doctrine of functional equivalence to agencies other than the EPA, including agencies with substantial environmental responsibilities (e.g., the National Marine Fisheries Service). The doctrine can thus be interpreted as relieving only the EPA of the obligation to comply with NEPA's procedural requirements. It has, however, been suggested that functional equivalence is conferred on the RI/FS process when it is conducted by other agencies through the EPA's oversight and involvement. No court has ruled on the validity of this argument.

In the absence of clear legal direction on the applicability of NEPA to CERCLA remedial actions at federal facilities, several federal agencies have adopted policies calling for the environmental planning and review procedures of the RI/FS to be integrated with the NEPA process. Similar policies are likely to apply to RCRA corrective actions. This approach is consistent with the Council on Environmental Quality's (CEQ) directive to "Integrate the requirements of NEPA with other planning and environmental review procedures ... so that all such procedures run concurrently rather than consecutively" [40 CFR 1500.2(c)].

PANEL DISCUSSION

There are sufficient legal ambiguities and practical questions surrounding this subject to result in continuing disagreement over the applicability of NEPA to federal facility remedial actions and the advisability of integrating the NEPA process with the CERCLA and RCRA processes. "NEPA/CERCLA/RCRA Integration" was the subject of a panel discussion conducted during a plenary session of this NEPA symposium. The discussion addressed the questions of (1) whether NEPA applies to CERCLA and RCRA remedial-action decisions and (2) whether and how the two processes should be integrated. The panel members were

- Dinah Bear, General Counsel, President's Council on Environmental Quality (CEQ), Washington, D.C.
- Carol Borgstrom, Director, Office of NEPA Oversight, U.S. Department of Energy (DOE), Washington, D.C.
- David Durham, Special Assistant to the Administrator, U.S. EPA, Washington, D.C.
- Raymond Pelletier, Director, Office of Environmental Guidance, U.S. DOE, Washington, D.C.
- Gary Vest, Deputy Assistant Secretary, Environment, Safety, and Occupational Health, U.S. Air Force, The Pentagon, Washington, D.C.

Frances Sharples of the Oak Ridge National Laboratory (ORNL) served as panel moderator. The main points addressed by each of the panelists are summarized below, followed by a synopsis of the discussion during the question-and-answer period and a brief update on developments since the October 1989 symposium.

David Durham, EPA

Mr. Durham stated that the EPA had no *formal* position on integrating the NEPA and CERCLA processes. In discussing the agency's views on the applicability of NEPA, he was careful to draw a distinction between actions taken by the EPA itself and actions taken by other agencies that are conducting remedial actions under the EPA's oversight. Since 1982, the EPA has treated its own removal actions under CERCLA as not being subject to NEPA and has deemed its own conduct of the RI/FS process to be functionally equivalent to NEPA. This interpretation is consistent with the agency's general approach to its own decisionmaking and with court decisions regarding functional equivalence.

The EPA's views on integration by other agencies are not resolved, but as expressed by Mr. Durham, they appear to be strongly influenced by arguments set forth in a letter from Donald A. Carr, Acting Assistant Attorney General, Land and Natural Resources Division, U.S. Department of Justice (DOJ), to Dinah Bear, General Counsel, Council on Environmental Quality, on March 6, 1989. In this letter, the DOJ stated its position that NEPA does not apply when federal agencies are performing cleanups under CERCLA's authorities. The DOJ believes this view is supported by the language, legislative history, and principles of CERCLA "and does not turn on a 'functional equivalent' analysis." Mr. Durham summarized the major points of the DOJ's analysis as follows:

1. Based on a review of the various versions of the Superfund Amendments and Reauthorization Act of 1990 that were under consideration prior to enactment, the DOJ concluded that Congress intended for SARA to stand apart from other

environmental laws except as expressly provided. In addition, the DOJ believes that the intent for federal agencies to follow the same rules as the EPA in selecting response actions has, in effect, eliminated most additional requirements that might otherwise apply for these agencies.

2. Where Congress wanted federal agencies to be subject to other federal statutory requirements, the DOJ notes, these are explicitly stated in CERCLA. For example, Section 120(i) obligates federal agencies to comply with all substantive requirements of the RCRA and Section 121 requires that CERCLA cleanups meet any standards, requirements, criteria, or limitation under any federal environmental law, including the Toxic Substances Control Act, the Safe Drinking Water Act, the Clean Air and Clean Water Acts, etc. Because NEPA was not specifically enumerated in this list, the DOJ makes the case that the Congress did not intend for NEPA to apply.
3. Congress expressly rejected the application of state NEPA-like procedures to Superfund cleanups.
4. Because CERCLA gives decisionmaking authority to the president and it is "well-settled" that NEPA does not apply to presidential decisionmaking, the DOJ makes the interpretation that Congress could not have intended for CERCLA cleanups to be subject to NEPA.
5. The DOJ also believes that it is significant that Congress established specific public participation requirements under CERCLA "which render compliance with the public participation requirements of NEPA superfluous."
6. Finally, the DOJ notes that Congress has built substantial constraints into the CERCLA remedy selection process, such as limiting the consideration of alternatives to those based on health and environmental cleanup standards and a prohibition on judicial review prior to completion of the remedy. Accordingly, the DOJ views compliance with NEPA as having the potential to violate congressional intent by requiring additional considerations to be made that might interfere with the responsibility of the EPA and other agencies to conduct expeditious cleanups.

The Justice Department's letter concludes by stating, "In sum, ... we conclude that cleanups conducted by EPA and federal agencies under CERCLA are not subject to NEPA." Mr. Durham's conclusion was, nevertheless, that it is not entirely clear whether NEPA applies to CERCLA cleanups. This conclusion suggests that the EPA is not yet convinced that the DOJ's arguments resolve the legal ambiguities.

Dinah Bear, CEQ

Ms. Bear stated that the relationship between NEPA and CERCLA had been the most common subject of questions received by her office in recent months. She pointed out that the DOJ letter represented a theory being submitted to the CEQ for its response and that it had no legal status as guidance. She reiterated that the DOJ letter did not make a functional equivalence argument, but that the CEQ in any case would oppose the extension of functional equivalence to federal agencies other than the EPA. Furthermore, she noted, the courts have held that the CEQ (not the DOJ) is "in charge" with respect to setting NEPA requirements. Much of her subsequent discussion addressed her reasons for disagreeing with the theory of legislative intent expressed in the DOJ letter.

First, Ms. Bear stated the opinion that all other environmental statutes apply under CERCLA unless specifically excluded. Although, for example, the wording of Section 121(d) of the statute does not mention NEPA, it uses the phrase "including, but not limited to" when enumerating other applicable laws, making it clear that the list of laws is not exhaustive. She mentioned that at least one Senate committee report in the legislative history of SARA had indicated that NEPA could apply. In addition, she asserted that the omission of NEPA from the CERCLA list of *substantive* requirements of other laws is not meaningful because NEPA is procedural and *not* substantive. One would not, therefore, expect it to appear on a list of statutes with substantive requirements. Second, she contested the DOJ's assertion that Congress' rejection of state NEPA-like procedures could be construed to mean that Congress also rejected the application of the federal NEPA. Instead, she interpreted this as a limitation on the authority of states over federal actions and asserted that the absence of such a specific rejection indicates that NEPA may apply.

Third, although she agreed that NEPA does not apply to presidential actions, she disagreed with the interpretation that the EPA's CERCLA cleanup activities represent presidential actions. Many presidential responsibilities are delegated to executive agencies, and most such delegated responsibilities are clearly subject to NEPA as federal actions. She also disagreed with the DOJ's interpretation that the establishment of specific public participation requirements under CERCLA makes the NEPA public participation process superfluous, saying that there is no reason CERCLA's requirements cannot be supplemented with those of NEPA.

Finally, on the DOJ argument that preparation of an EIS would delay the remedial action process, she stated that this was based on a wrong assumption by the DOJ (i.e., that an EIS would be prepared *after* completion of the RI/FS). The advice of the CEQ is that the RI/FS and EIS processes not be conducted consecutively, but as a single integrated process. Taking an integrated approach should alleviate the potential problem of delays.

Ms. Bear also addressed the question of whether compliance with NEPA would subject CERCLA remedy selections to citizen lawsuits that are otherwise barred under CERCLA. To prevent CERCLA actions from being delayed by legal action, one of its provisions bars most citizen suits until after the remedial action has been implemented. It has been suggested, however, that integration of NEPA with CERCLA could lead to delays in remedial action by permitting lawsuits under NEPA. Because NEPA is silent on the question of timing of judicial review, it was Ms. Bear's opinion that the CERCLA prohibition on citizen suits would take precedence over NEPA when the two processes are integrated. Thus, citizen suits over an allegedly inadequate EIS would be barred until after remedial action is complete. Ms. Bear concluded by stating that the CEQ would further examine the integration of the NEPA and CERCLA processes, which the CEQ clearly supports, and that the CEQ expected to issue guidance for other federal agencies.

Gary Vest, U.S. Air Force

According to Mr. Vest, the U.S. Air Force believes that NEPA applies to its CERCLA remedial actions and that "functional equivalence" is valid only for the EPA. NEPA compliance is therefore incorporated into Air Force Installation Restoration Program projects. Mr. Vest expressed the view that NEPA fosters informed decisionmaking on remedial action questions. He also suggested that the vast cost of federal facility environmental restoration activities makes it foolish to ignore NEPA. He observed that the NEPA process is necessary for considering remedial-action issues and impacts that involve several sites or geographic regions, such as transportation of cleanup wastes to a different location.

Although the Air Force in the past has conducted some remedial actions with separate CERCLA and NEPA documentation, integration of the processes is preferred because it is seen as saving time, money, and effort. Integrated documents must, of course, be designed to fulfill the requirements of both laws. Mr. Vest noted many similarities between NEPA and CERCLA that facilitate integration of the processes. Both processes call for analysis and comparison of alternatives, including the alternative of no action, public involvement, and issuance of a record of decision. The CERCLA RI/FS process is integrated with the NEPA process at the level of either an EA or an EIS. The Air Force has determined that preliminary assessment/site investigation activities, which precede the RI/FS under CERCLA, qualify as NEPA categorical exclusions.

Mr. Vest mentioned a few practical aspects of integrating NEPA and CERCLA. For example, he noted that the intent to conduct an integrated process should be emphasized at all public and interagency meetings concerned with a remedial-action project.

Mr. Vest stated his belief that NEPA would apply to corrective-action projects under RCRA, although the EPA's rule establishing the requirements for this process had not been issued. The Air Force had little experience to date in integrating the NEPA and RCRA corrective-action processes.

Carol Borgstrom, DOE

The DOE is another agency that has been proactive in establishing a policy calling for the NEPA and CERCLA processes to be integrated. Ms. Borgstrom's presentation reviewed this policy. In August of 1988, the department issued DOE Notice 5400.4, "Integration of Compliance Processes." This notice established that it is the DOE's policy to integrate the requirements of the NEPA and RI/FS processes for remedial actions under CERCLA. Ms. Borgstrom emphasized that the processes are to run concurrently rather than consecutively, thereby reducing the level of resources that would be needed to implement both processes separately. The primary instrument for integration is to be the RI/FS process, "supplemented, as needed, to meet the procedural and documentational requirements of NEPA." She emphasized that the policy is subject to revision pending guidance from the CEQ.

Ms. Borgstrom also stressed that a key element of the integrated process is making early determinations on the level of NEPA documentation needed prior to entering into the RI/FS scoping process or as soon thereafter as possible. She stated that the policy is not entirely mandatory. For example, if a project is already committed to conducting the two processes separately, it is not required to integrate. Also, integration might not be practical where the aggregation of remedial-action and nonremedial-action activities would overly complicate one process or the other.

Ms. Borgstrom then discussed some of the problems the DOE was encountering in implementing this policy. She noted that few people in the department understand both NEPA and CERCLA and that there was a general lack of understanding within the DOE on how to integrate. Internal opposition to the policy was motivated by the fear that NEPA compliance would delay CERCLA actions. The Office of NEPA Oversight recognized that guidance on implementing the policy was needed, and Ms. Borgstrom stated that such guidance would be developed. In addition, the DOE intended to expand its list of categorical exclusions so that some activities associated with the remedial-action process would not require NEPA documentation.

Ray Pelletier, DOE

Mr. Pelletier focused his remarks on some practical problems that may arise in implementing the DOE's integration policy. First, he noted that there is a great deal of public and political pressure on the EPA to produce results from the CERCLA program and that numbers of completed and signed CERCLA RODs are often used as a measure of progress. As a result, he suggested, the EPA has an incentive to divide large remedial-action sites, such as DOE facilities, into many small "operable units," each covered by a separate ROD or separate RODs under CERCLA. For example, the DOE Hanford Reservation in Washington State has 78 CERCLA operable units. This perceived pressure to divide remedial-action projects into many small units would appear to conflict with the NEPA mandates to assess connected actions together and to evaluate cumulative impacts. He also noted that if operable units are grouped together for analysis, NEPA would generally call for grouping by the type or focus of impact, whereas CERCLA would probably require grouping by type of remedial response technology.

Another potential problem identified by Mr. Pelletier is that integrated RI/FS-EIS documents are subject to different and potentially conflicting review requirements. Two completely different parts of the EPA both have review responsibilities: CERCLA program personnel must review and approve all draft RI/FS documents, while the EPA Office of Federal Activities reviews and rates published draft and final EISs under Section 1504 of the CEQ NEPA regulations. This dual review by the EPA could be inefficient and might lead to internal conflicts when one branch of the EPA is called upon to evaluate the work of another branch. Another concern related to the EPA's role under CERCLA is that federal facilities agreements between the EPA and other federal agencies spelling out CERCLA responsibilities do not typically address NEPA integration. As a result, compliance schedules may not allow sufficient time for NEPA document reviews (by the EPA and the public) that are different from required CERCLA reviews.

The conflict of interest provisions of the CEQ regulations pose another potential problem for the DOE in its efforts to integrate NEPA and CERCLA, as they would have the effect of barring the DOE management and operation contractors from preparing RI/FS documents that will also serve as NEPA documents. Mr. Pelletier also noted that the two processes have different expectations and requirements for the length, focus, and readability of documents and that it may be difficult to prepare documents that meet both sets of requirements. EPA guidance on CERCLA calls for reporting of essentially all available information in RI/FS reports, which can be quite lengthy and are not necessarily intended to be readable by the lay public. In contrast, the NEPA regulations call for EISs to be readable documents that focus on significant issues, omit unnecessary detail, and are limited to no more than 300 pages. Another concern Mr. Pelletier expressed is that the significant beneficial impacts of most remedial-action projects might mean that an EIS would be the required level of NEPA documentation for virtually every project.

Mr. Pelletier concluded by stating that the DOE's biggest problem to date in integrating NEPA and CERCLA has arisen from failures to begin NEPA implementation early in the decision process. In most instances, the binding interagency compliance agreements that the DOE is signing for its remedial-action sites fail to allow for NEPA. He stressed that a NEPA strategy should be developed before agencies begin negotiations so that agreements can include any features needed to accommodate NEPA integration. The Hanford agreement dealing with 78 operable units, for example, leaves no room for the preparation of programmatic or other forms of tiered EISs. It is too late to start planning a NEPA compliance strategy after an agreement has already been negotiated and signed.

DISCUSSION

A period of questions, answers, and discussion followed the panelists' initial presentations. One topic of discussion was the policies of other federal agencies with respect to the application of NEPA to remedial actions. The Department of the Army was named as another agency that has issued regulations adopting an integrated approach to NEPA and CERCLA. Ms. Bear noted that the Department of the Interior (DOI) had conducted a remedial action at a wildlife refuge without following NEPA, based on an interpretation that functional equivalence did apply to that particular action. She also knew about instances of federal agencies making remedial-action decisions under NEPA alone. Another meeting participant said that on lands belonging to the DOI's Bureau of Land Management, functional equivalence is deemed to apply only to actions undertaken by the EPA. The EPA's

Region VIII was cited as having supported the successful integration of NEPA with RCRA for actions in the state of Utah.

Several participants addressed themselves to the question of whether the expected environmental benefits of CERCLA and RCRA remedial actions should exempt these actions from NEPA. Ms. Bear remarked that other agencies have argued that highways and dams are entirely beneficial in their impacts and should be exempt from NEPA, but it is now well-accepted that these projects also have adverse impacts and are subject to NEPA. A member of the audience suggested that it is probably invalid to assume that the impacts of a remedial action will always be beneficial. He compared remedial actions to the federal program of financing sewage treatment plant construction, which was also undertaken to improve the environment. According to his statements, the application of NEPA to sewage treatment plant construction has been found to have reduced the adverse impacts of these projects, for example, by reducing impacts on floodplains and wetlands. Mr. Vest observed that the alternative supported by RI/FS analysis is not always the best choice environmentally. At one Air Force CERCLA site, application of the NEPA process led to a decision to select the no-action alternative, which would have been rejected if the decision had been made under CERCLA alone.

Audience member Dan Reicher of the Natural Resources Defense Council questioned whether a decision not to prepare an EIS (i.e., the issuance of a finding of no significant impact) for a CERCLA action would be exempt from judicial review as a result of the CERCLA prohibition of citizen suits. He suggested that some provision for judicial review of such decisions might be necessary to enforce NEPA. In considering the question, Ms. Bear indicated that the CEQ would have to study further the important issue of the effect of the CERCLA prohibition on litigation as it relates to NEPA-CERCLA integration.

RECENT DEVELOPMENTS

Developments since October 1989 have included (1) release of a CEQ memorandum on the applicability of NEPA to CERCLA remedial actions, (2) formalization of the DOE policy calling for integration of NEPA and CERCLA, and (3) publication by the EPA of a proposed rule setting out procedures for conducting corrective actions under RCRA (55 *Fed. Regist.* 30798; July 27, 1990). In addition, ORNL issued a report (Levine et al. 1990) that discusses NEPA-CERCLA integration and presents recommendations on practical aspects of conducting an integrated RI/FS-EIS process and preparing integrated documentation.

The CEQ memorandum (Swartz 1990) was sent to the EPA with a proposal that the CEQ and EPA work together to draft guidance on the legal and practical aspects of integrating NEPA and CERCLA (D. Bear, CEQ; letter to E. D. Elliott and J. M. Strock, U.S. EPA, Aug. 1, 1990). The memorandum presents the CEQ's analysis of the legal arguments concerning the applicability of NEPA to CERCLA actions. The conclusion of this analysis is that the NEPA process applies to federal agency actions under CERCLA "because (1) Congress did not expressly or impliedly repeal the application of NEPA in CERCLA/SARA, (2) the goals of NEPA and CERCLA do not conflict fundamentally, and (3) the functional equivalence doctrine does not apply to actions taken by federal agencies other than EPA." Furthermore, the memorandum states that the "EPA's review and approval of the remedy selected is not a sufficient nexus to [justify a finding of functional equivalence for federal NPL sites or otherwise] allow other federal agencies to disregard the requirements of NEPA." In reaching these conclusions, the memorandum makes, expands upon, and provides legal citations in support of the arguments given by Dinah Bear in the presentation summarized above. Other topics discussed by Swartz (1990) include the theory of functional equivalence and the expectation that implementing NEPA for CERCLA actions will enhance the decision process by providing for earlier and more effective public participation.

The DOE policy on NEPA-CERCLA integration has been formalized in DOE Order 5400.4 (CERCLA Requirements), issued Oct. 6, 1989. In a related action, the DOE listed certain actions taken under CERCLA and RCRA as NEPA categorical exclusions (proposed on April 6, 1990, 55 FR 13064; final publication on Sept. 7, 1990, 55 FR 37174). Actions by the DOE that are now categorically excluded from NEPA documentation (i.e., actions that do not normally require either an EIS or an EA) include certain removal actions under CERCLA and similar actions under RCRA and other authorities, improvements to environmental control systems to comply with environmental permit requirements, and site characterization and environmental monitoring activities under CERCLA and RCRA.

REFERENCES

Levine, M. B., E. D. Smith, F. E. Sharples, and G. K. Eddlemon. 1990. Integrating NEPA and CERCLA Requirements During Remedial Responses at DOE Facilities. ORNL/TM-11564. Oak Ridge National Laboratory, Oak Ridge, TN.

Mandelker, D. R. 1984. *NEPA Law and Litigation.* Callaghan and Company, Wilmette, IL.

Swartz, L. L. 1990. Application of the National Environmental Policy Act to federal agency actions under the Comprehensive Environmental Response, Compensation, and Liability Act. CEQ internal memorandum to D. Bear, General Counsel, July 30, 1990. Council on Environmental Quality, Washington, D.C.

Future Challenges of NEPA: A Panel Discussion

E. D. Smith, Oak Ridge National Laboratory, Oak Ridge, TN

INTRODUCTION

One portion of a plenary session during the conference was a forum on "The Future Challenges of NEPA." The session was a panel discussion, moderated by Dinah Bear, General Counsel on the staff of the President's Council on Environmental Quality (CEQ), Washington, D.C. Other panelists were

- William Dickerson, Deputy Director, Office of Federal Activities, U.S. Environmental Protection Agency (EPA), Washington, D.C.
- Carol Borgstrom, Director, Office of National Environmental Policy Act (NEPA) Oversight, U.S. Department of Energy (DOE), Washington, D.C.
- Gary Vest, Deputy Assistant Secretary, Environment, Safety, and Occupational Health, U.S. Air Force, The Pentagon, Washington, D.C.

Bear identified the panel's focus as "future visions of NEPA, how NEPA might change, and what trends exist." Each of the panelists was to spend 10–15 min talking about their observations on how NEPA is operating, some of the trends they observed, and how they thought NEPA might change in the future. Following the presentations, Bear invited the audience to share questions, comments, and thoughts on the future of NEPA. Topics discussed in this forum included congressional proposals to amend NEPA; possible changes at the CEQ; postdecision monitoring, mitigation, and follow-up studies; applicability of NEPA to international actions of the U.S. government; assessment of global change impacts; and the relationship between NEPA and state "little NEPA" laws. The individual presentations and the subsequent discussion are described and excerpted below.

INDIVIDUAL PRESENTATIONS

Dinah Bear, CEQ

Dinah Bear began with a brief summary of the current status of the CEQ. Michael Deland, previously regional administrator for EPA Region I (New England), had recently become the CEQ chairman. Bear remarked, "His appointment is one indication that President Bush is fulfilling his commitment to reinvigorate CEQ." The CEQ expected a significant budget increase and expected to expand the staff from 8 to 40. Bear said she was often asked whether there would be a full council or just a chairman. She reported that there would not be a full three-member council as in the past; a decision had been made not to fill the two vacant council member slots. Another decision was expected later on whether to reduce the council to a single full-time chairman, institute a part-time council, or replace the full council with part-time advisory committees.

Turning to the subject of the future of NEPA, Bear noted a great deal of recent interest in amending the statute, largely stimulated by concern about specific issues. This interest represented a change from earlier years when Congress paid little or no attention to NEPA and NEPA's supporters feared that any amendments might eviscerate the statute. No changes were expected in the near future in either the overall thrust of the NEPA process or the CEQ regulations. Rather, Bear identified three major areas in which she perceived a potential for change: (1) the international reach of NEPA, (2) the postdecisional aspects of NEPA (i.e., monitoring and mitigation), and (3) the substantive area of NEPA (i.e., additional measures to give substance to the policy goals of the statute). She saw "a great deal of movement" in the first two of these areas and "serious discussion, stimulated in part by recent Supreme Court decisions," in the third area.

A bill (HR 1113) that was passed by the House of Representatives on October 10, 1989, but was not acted upon by the Senate during the 101st Congress would have amended NEPA in several ways. It included provisions related to international issues and to monitoring and mitigation. According to Bear,

> There are several intents in this bill. One is to play "catch up," to amend the statute so that it reflects current practice. NEPA is a unique statute in that it is very general, quite the opposite of Superfund or the Clean Air Act. Over 20 years it has been interpreted by the courts and CEQ in ways that are not actually reflected in the statutory language. For example, the word "public" does not appear anywhere in NEPA §102(2)(C), the section that calls for environmental impact statements (EISs). I get phone calls asking where the statutory requirement is for public hearings, but there is no requirement for public hearings in either NEPA or in the CEQ regulations. It simply developed as a matter of policy and common sense. One amendment would insert the words "and the public" in the part of §102(2)(C) of NEPA that deals with circulating the EIS to federal, state, and local agencies. There are a lot of things like that, elements that are not in the statute but that developed as a result of case law and CEQ regulations.

One of the proposed amendments would both clarify the current status of the law and make additions to address concerns arising from the recent Supreme Court decisions dealing with mitigation and monitoring. It would amend §102(2)(C) to say that the detailed statement must include reasonable alternatives to the proposed action that achieve the same or similar public purposes. That part of the amendment is really addressed to the §404(C) permit area and similar areas where agencies have been struggling with the question, "Do you look at purpose and need from the applicant's point of view or the public point of view?" CEQ has addressed this question in the context of the §404 permit situation and said essentially that you look at both. Obviously, you have to look at the applicant's purpose and need. You always interpret "need" in the light of common sense, but if there were no public purpose and need there would be no federal permit. That is why we have a federal permit in the first place, and the federal agency must look at the proposed action from the standpoint of public purpose and need. There is case law that supports that quite strongly. The second part of the amendment would add "including alternatives that avoid the adverse impacts described in clause ii" (the clause that talks about adverse environmental effects which cannot be avoided) and "alternatives that otherwise mitigate those adverse impacts." I think this provision is a reaction to a concern that the Supreme Court decisions in *Methow Valley Citizens Council* v. *Robertson* and *Oregon Natural Resources Council* v. *Marsh* deemphasize the role of mitigation in the EIS process. Those decisions did say that mitigation must be discussed in EISs, but they drew the line at saying that NEPA imposed mitigation on the agencies or required the agencies to take mitigation. This amendment would not require mitigation either, but it would reinforce the requirement to examine certain alternatives in the context of the EIS.

Another very significant proposed amendment responds to the issue of whether or not mitigation and monitoring provisions are enforceable or must be implemented. This amendment would state that environmental mitigation and monitoring measures and other conditions discussed in a detailed statement or other document prepared under §102(2)(C) and selected by an agency as part of its final decision shall be implemented by the appropriate agency. That is simply a rewording of our regulation, but putting it in the statute will give it greater legal weight.

The proposed amendment in the mitigation area that has received the most attention from agencies, because it would cause the most work even if not necessarily the greatest real-life effect, is one which would require all federal agencies to conduct NEPA follow-up studies. Each agency would select a statistically significant sampling of its EISs, evaluate the accuracy of the predictions in the documents, determine whether or not the mitigation and monitoring called for was actually carried out, and evaluate the effectiveness of the mitigation and monitoring provisions. Originally, this was to be an annual review, but the agencies and CEQ protested that this would be burdensome, expensive, etc. It has now been modified so that CEQ would be required to issue guidelines no later than 6 months after the passage of the act, and it would be up to us to determine the frequency and content of reviews. We would then have to analyze the agency reviews and include our analysis of those reviews in the CEQ annual report.

> The amendments that deal with the extraterritorial reach of NEPA and NEPA's impact on the global commons have caused by far the most internal administration debate and the most interest on Capitol Hill. This is an issue that has been debated for 20 years. I have memos going back to May 20, 1970, in which there are two different positions taken by agencies in the government, and that debate has continued. The reason for not applying NEPA to extraterritorial actions that is cited most often is that application of the environmental impact assessment (EIA) process to actions in other countries would impede foreign policy. I have heard it suggested that we would have to do everything from withdrawing from the United Nations, to shutting down the space shuttle program, to changing our nuclear nonproliferation policy, and a number of other really drastic things. Another summary argument is that it would be environmental imperialism to impose the EIA process overseas. I would suggest that it is environmental imperialism for us to undertake actions in other countries and not use the same degree of care on analyzing the environmental impacts as we do at home.

Other developments that Bear identified as supporting the extraterritorial application of NEPA included (1) the adoption of the EIA procedure by many other countries and multilateral organizations and (2) the experience of the U.S. Agency for International Development, which has for many years successfully conducted an EIA process, though not identical to domestic NEPA, as a result of a settlement with the Natural Resources Defense Council. She suggested that even those countries that do not have EIA processes would have government ministries that would welcome the opportunity to participate with the United States in an EIA process. Some modifications and restrictions on the NEPA process would be appropriate for its extraterritorial application. First, procedures might need to be modified to accommodate local conditions. As Bear remarked, "Issuing scoping meeting notices in the middle of the People's Republic of China without coordination with the government might produce less than desirable results." Second, a provision for exemptions or some flexibility in the process is needed to deal with national security, arms control, and intelligence concerns. Existing Executive Order 12114, which governs the agencies' analysis of environmental effects of major federal actions abroad, contains several such exemptions, and HR 1113 would have directed the CEQ to make its regulations on analysis of extraterritorial impacts consistent with national security and foreign policy. Bear remarked that "Many people would argue that the executive order has too many exemptions and exceptions and is not very effective in today's changing world in terms of the global environment, but I think most people would recognize that some of those exemptions are important."

In the area of global impacts, HR 1113 would have directed the CEQ to issue guidance for federal agencies for assessing the effects, including cumulative effects, of proposed major actions on global climate change, depletion of the ozone layer, loss of biological diversity, transboundary pollution, and other matters of international environmental concern. Bear reported that the CEQ had been studying the question of how to integrate global climate change into the EIS process. Although it was generally accepted that global climate change is an impact that significantly affects the quality of the human environment, there was legitimate concern about how well the processes are understood and how that understanding could be translated into a defensible analysis in an EIS. The CEQ had circulated draft guidance on assessing global climate change to federal agencies. Bear reported receiving some 60–70 comments on the draft guidance. Although most were supportive of the notion of integrating global impacts into the EIS process, many raised questions about how this would be done.

Bear concluded,

> The areas where I foresee important changes in NEPA are (1) international, including extraterritorial impacts or global commons, and global climate change, and (2) monitoring and mitigation. It is interesting to note that I first started hearing about a lot of interest or concern about mitigation and monitoring not from the United States, but from foreign visitors and foreign government officials who are trying to develop an EIA process. Inevitably, when foreign visitors came by my office or called, they requested all of our studies on mitigation and monitoring. Of course, we did not have much to offer. When I thought about it, I then realized that this was very natural. For the first 20 years of NEPA, the emphasis has been on the predecisional aspects of NEPA. That is good, it was necessary and it should not be deemphasized. Now, however, people are starting to ask questions about the effectiveness of the process. I saw the concern go from the foreign community, as they were developing their EIA procedures, to the academic community and then into the federal agencies.
>
> There are really two issues: Are we going to be able to perform impact assessments, in light of budget problems? If we are doing assessments, are they effective? Paul Culhane and Paul Friesema did the first real study [(Culhane et al. 1987)] directed at the latter question, essentially, "Are the predictions in an EIS valid from a scientific point of view?" They looked at a sample of EISs from the 1973 to 1974 range and compared the EIS predictions to the actual projects to see how they matched. I think that study was an important milestone, and it triggered a lot of other interest. There have been graduate students writing theses on it; there was a lot of interest on Capitol Hill; there were several amendments coming out of this committee looking at mitigation and monitoring; and there has been a lot of concern within the federal agencies. Thus, I think that the whole post-decisional aspect of NEPA will be a major focus for the future.

Bill Dickerson, EPA

Bill Dickerson began his remarks by observing, "During most of this conference, we have been talking about the major impacts, the big problems, and the crisis actions. I would like to talk now about the little problems, problems below the so-called 'significance threshold.' Are we perhaps winning the battles, but losing the war, especially in the environmental assessment (EA) process and in EISs where we mitigate down to the significance threshold and stop? How far below significance should we really be concerned?" He suggested that budget constraints in federal agencies inevitably lead to prioritization, with the result that agencies focus on high-priority actions and critical impacts, while lower-priority problems are not addressed. His concern was that this situation results in "chronic degradation (i.e., small unmitigated impacts, a large number of which cumulate over time and space in an entire country). This approach is not in fact fulfilling the policy promise in NEPA of improving the environment and the human habitat; in fact, we may be incrementally losing the war." In his subsequent remarks, he sought to discuss whether this problem could be handled in the NEPA process in a cost-effective way and to identify some of the steps that might be taken.

He observed that "Most of the focus in NEPA is on the higher levels of the NEPA pyramid" (i.e., on the major actions covered in EISs, Dickerson stated that only some 30–50 programmatic EISs and 250–450 project EISs are issued each year, whereas many more decisions are covered in EAs and NEPA categorical exclusions). His best guess at the annual number of EAs was 30,000–50,000, and "nobody even has an estimate for the numbers of categorical exclusions." The EPA is concerned about the cumulative impact of the myriad actions that fall below the threshold of significance. According to Dickerson, "In examining one year's EISs, EPA found that only 6% of the final EISs had significant problems." However, 37% of the final EISs projected "important impacts that could and should have been mitigated, even though the level of impact was below that magic threshold of 'significant.' The agency is examining about 1200 EAs per year throughout the ten regions, and we find that, on average, about 71% have either no mitigation or inadequate mitigation." In many of these EAs, the EPA determined that no mitigation was needed, but in about 40% of the EAs, the agency concluded "that there were impacts that could and should be reduced further. This indicates to us that there could be a big cumulative problem out there that could and should be handled through the NEPA process."

Asserting that high cost is the main reason for failing to apply mitigation, Dickerson said, "I would like to suggest that there are some very cost-effective things that can be done within the process, and that could reduce the overall level of insult to the environment considerably. NEPA has generally been very successful in integrating the environmental ethic into decision making from the planning phase through the EIS/EA phase, but we have not done much in the phases from the record of decision through project construction and operation." He recommended new emphasis on these later stages of the process, emphasizing the importance of effective communication of the mitigation conditions. In many cases, he observed,

> There may be boilerplate conditions that can be put into environmental permits, but projects are handed off from one group to another, and the information often gets lost before the permitting phase, perhaps because the information is not translated well or the people involved are not sensitive to the issues.

In 1987 we looked at this process in five agencies, examining how well it worked through the system [(Bassin et al. 1986)]. All five agencies thought that their process worked pretty well, but there was little oversight on how it was really working; nobody could really tell in the end. Only the Forest Service had formal guidance. Staff from one of our regions visited a project site where some vegetation was supposed to be left along the riverbank after the project was completed. Unfortunately, the bulldozer operator did not get the word, and they went down, and it was all gone. This is an example of the necessity of making sure that the word gets all the way down.

Dickerson offered five recommendations for improving the effectiveness of the NEPA process in reducing lower-level impacts.

First, we need full disclosure of impacts and potential mitigation, even if the decision maker cannot effect the mitigation. It is critically important that the information be made available. The county, state, and local people need to be aware before these issues surface so that they can plan for them and have a chance to take care of the projected impacts. Second, we need to look for mitigation of lower-level impacts. This requires an change in ethic. In many cases, once the impact has been reduced to a certain level, no more analysis is done. Third, as I said before, we need improved consistency of the total planning in the construction/operation process. I think it is critical to make sure that what we say early on is what we can and will in fact do. Fourth, we need to open up the EA process. We need to reexamine it from the standpoint of public involvement and of making sure that the public knows what is going on. We are talking about a lot of actions whose impacts may be small, but they do add up. Fifth and finally, we need to consider postproject-monitoring adjustment. I think that is a concept that would allow for handling situations where we do not have good predictive techniques. You could call for doing the best projection that you can, but hedging your bets by actually monitoring what is on the ground and allowing for adjustment."

Dickerson concluded by calling on the CEQ to provide some leadership in the area he had defined. First, to answer the questions "Is this a problem?" and "How big of a problem is it?," he recommended a comprehensive examination of the NEPA process, the EAs themselves, and the policy. He suggested that if this examination indicates that this is a fruitful area for improved environmental mitigation, the CEQ should look into why the mitigation is not being done "and establish guidance or an executive order or some other mechanism to revitalize the policy aspects of NEPA."

Dinah Bear

Bear remarked that she had identified three areas of change, but discussed only two: (1) international and (2) monitoring and mitigation. She observed that

Dickerson's remarks had suggested a need for revitalization in the substantive area and said,

> I think that is absolutely correct. There is just starting to be a lot of discussion about that, and I am hearing a lot of it at some of these conferences; there have been some papers written about it, suggesting everything from executive orders to constitutional amendments. CEQ will probably look at this, although it has lower priority than issuing all the guidance that everybody has been waiting for on NEPA-CERCLA-RCRA, global climate change, extraterritorial, and so forth. There is a lot in NEPA besides the EIS process, and there are a lot of very substantive goals that the EIS process is designed to implement. We need to get away from the idea that NEPA is just a paper-using statute, a paper process.

Carol Borgstrom, DOE

Carol Borgstrom prefaced her presentation by stating that although her comments would reflect a DOE perspective, they were her own views and not necessarily those of the DOE except where indicated. She then reported on some changes in NEPA policy and procedure being made at the DOE in response to a June 1989 initiative from the Secretary of Energy, Admiral James D. Watkins, that had directed the department to revise its procedures and improve its compliance with NEPA. According to Borgstrom,

> The Secretary indicated his intent to be personally involved in preliminary NEPA decisions and to fully coordinate such decisions with the governors of the states that host our facilities. In the words of Secretary Watkins, 'In the future, if the Department is to err in its judgment as to the extent of NEPA review required of a new project, it will err on the side of full disclosure and complete assessment of potential environmental impacts.' [DOE is drafting new NEPA policies and procedures that] will provide for greater public and state participation in the DOE NEPA process, eliminate certain abuses that had crept into the process, ensure uniform communication of NEPA procedures, and provide for greater line management accountability for NEPA documentation.[1] There will be a renewed emphasis on early and adequate NEPA planning so that NEPA compliance serves, in fact as well as in theory, as a useful planning tool rather than a nuisance or an impediment to achieving mission goals. These [changes] will require more training in NEPA principles and procedures and more resources devoted to NEPA compliance at all levels of DOE.

Turning to the general topic of the future of NEPA, Borgstrom stated,

> Our greatest challenge, I believe, is to ensure that NEPA survives, to do for future generations what it already has done for us. On this 20th anniversary, I think it is

[1] The new DOE NEPA procedures were promulgated as 10 CFR Part 1021, published April 24, 1992 (57 *Fed. Regist.* 15122).

prudent to look forward and consider how we might improve the NEPA process so that it will better address the technical and public policy issues of the future. I see three general challenges facing those of us in the NEPA business: (1) the scientific basis for impact projections must advance along with the complexity and sophistication of our proposed actions, (2) greater emphasis must be placed on risk assessment and risk communication in the NEPA process, and (3) more attention must be given to broad policy issues and cumulative impacts in the NEPA process.

The first of the challenges I named is that the scientific basis for our impact projections must advance commensurate with the complexity and sophistication of our proposed actions. Conferences such as this one provide important opportunities for us to share information on the state-of-the-art in environmental assessment techniques. The public expects, and if we do not deliver, they will demand, that we utilize the appropriate advanced impact assessment methodologies in our NEPA documentation. More quantitative analysis of impacts in all scientific disciplines will be required. For example, as our energy-related projects become technologically more complex, and in some cases, potentially more hazardous, the public will demand that our capability to predict the resultant environmental impacts advance, consistent with the issues posed by these new technologies. This is true whether the impacts pertain to global climate change associated with the use of fossil fuel, the migration of radioactive or other hazardous materials in groundwater, the development of advanced nuclear technologies, or the effects of electromagnetic fields. To improve our predictive capabilities, we should pursue programs to review and evaluate the adequacy and the accuracy of our past impact projections and investigate new methodologies to overcome identified weaknesses.

Little follow-up work has been done to determine whether the impacts we projected for a project on the drawing board a decade ago did in fact materialize. Was the mitigation we recommended implemented? Was it effective? Did unexpected consequences arise? I believe we should study our past projects in order to improve our capability to make accurate impact projections and ensure a well informed decision process.

My second challenge is that greater emphasis must be placed on risk assessment and risk communication in the NEPA process. Within DOE, risk assessment is becoming the most significant, and potentially the most sensitive, part of the NEPA process. NEPA requires us to assess the impact of what might happen, sometimes in the face of considerable uncertainty. It is what might happen, rather than what ought to happen, that is often of greatest concern to the public and the decision makers. We know that NEPA does not require us to engage in wild speculation and that the CEQ has eliminated the use of so-called worst-case analysis. At DOE, however, we are struggling with the problem of what I call a "worsening worst case" and the definition of what is reasonably foreseeable. Let me give you some examples.

DOE was a cooperating agency with the National Aeronautics and Space Administration (NASA) for the EIS to support the recently launched Galileo mission to the planet Jupiter. The project involved a DOE-supplied plutonium power source aboard the space shuttle Atlantis. NASA's analysis showed that no significant impacts would occur from a normal launch and a normal journey of the Galileo spacecraft to Jupiter. If all goes as planned, this leaves little to write about in an EIS. However, if there were a launch accident or an unexpected reentry on one of Galileo's two Earth fly-bys, then plutonium could be released and significant consequences might result. Thus, an accident with a low probability of occurrence could cause a significant impact. Of course, the decision was made to prepare an EIS. Following completion of the NEPA process, NASA and DOE were sued by a coalition seeking to stop the launch. The plaintiffs argued that the risk of an accidental release of plutonium into the environment at the various stages during the mission was too high. In particular, they argued that the EIS "does not address all relevant risks, underestimates the magnitude of the risks, and lacks complete information and is therefore inadequate." However, the District Court in the District of Columbia, in deciding in favor of the government, found the assessment of risk adequate and stated: "It is not the function of this court to decide whether the government's decision to go forward with the Galileo mission is a good one. Instead, the court's function is only to ensure that the government has complied with NEPA by evaluating and weighing the environmental impact of the proposal when making its decision." The court acknowledged that, if given more time, NASA's knowledge of risk might improve, but noted that "uncertainties are involved in any decision."

Recently, DOE addressed a more difficult NEPA question in the case of a project in which no significant impacts would result from normal operations or from design-basis accidents with a probability of occurrence of about 1 in 1 million. Consequently, DOE prepared a draft EA. However, the safety analysis that was prepared in conjunction with that EA identified an unlikely accident scenario with potentially serious consequences. That accident, which had an estimated probability of about 1 in 10 million, involved a plane crash into a proposed facility, resulting in a release of plutonium and a projected number of cancer fatalities. The issue was whether DOE should prepare an EIS solely on the basis of a very low-probability, high-consequence accident. This was not an easy decision. We reviewed the CEQ regulations on incomplete and unavailable information. The provision regarding low-probability, high-consequence accidents appears to apply only to EISs and not to EAs. Therefore, could DOE legitimately prepare an EA, ignoring the very low-probability accident scenario with significant consequences, and issue a finding of no significant impact (FONSI)? On the other hand, if we assumed that the CEQ provisions regarding the issue of unavailable information did apply, at least in principle, then it appeared that an EA and FONSI would not be appropriate. In this regard, we reviewed the CEQ guidance regarding the use of credible scientific analysis. Traditionally, DOE has held the view that an event with the probability of less than 1 in 1 million ($<10^{-6}$) is incredible. But we are not so sure any more about the dividing line between "credible" and "incredible." In the wake of Three Mile Island, Bhopal, Chernobyl, Challenger, and EXXON Valdez, our current thinking at DOE is that accidents with probabilities in the range of 10^{-6} to 10^{-8}

should be covered in EISs if they contribute significantly to overall risk. In the case I just described, we determined that a FONSI would be inappropriate for a number of reasons, including the Secretary's full disclosure policy, and we decided to prepare an EIS.

As our NEPA documents increasingly focus on risk, we must develop our skills in risk communication. That is, we must improve our ability to exchange meaningful information and opinions about risk, both within the scientific community and with the public. We cannot rest on the excuse that risk assessment is too complicated for the public to understand. After all, the public does continue to buy lottery tickets based on its understanding of probabilities and consequences. Risk communication is not viewed as a one-way street, whereby experts explain complex technical uncertainties to nonexperts. Rather, there is some degree of give and take. When effective risk communication is incorporated in the NEPA process, we facilitate public review and well-informed decisions. In this regard, I recommend a book called *Improving Risk Communication* [(NAS 1989)], which was prepared by the Committee on Risk Perception and Communication of the National Research Council of the National Academy of Sciences. This committee is chaired by John Ahearne, who also serves as chairman of DOE's Advisory Committee on Nuclear Facility Safety. In a September 19, 1989, memo to senior DOE officials, Secretary Watkins recommended the Ahearne book and stated his intent "to make improved risk communication a matter of DOE policy so that we are better informed in our decisions and better communicate the bases of our decisions with the media and with the public."

My third challenge is that more attention must be given to broad policy issues and cumulative impacts in the NEPA process. There is a growing recognition at DOE of the need to address the cumulative impacts of connected actions. Broad policies and programs set the course for future agency actions which have environmentally significant consequences, both beneficial and adverse. As national energy policies are better defined, the need to provide comprehensive environmental analyses of far-reaching energy initiatives will increase. Inevitably, more programmatic EISs will be required. In July, the department issued a draft programmatic EIS on the Clean Coal Technology Program; we expect to issue the final EIS shortly. I think this is a step in the right direction. DOE recognizes the need to take a national and regional look at the consequences of encouraging these new technologies. The direct action considered in the programmatic EIS is the proposed selection for cost-shared federal funding of one or more projects to demonstrate innovative clean coal technologies. However, the indirect result of this program is anticipated to be the widespread private-sector commercialization of the successful technologies. The programmatic EIS compares the projected atmospheric releases in the year 2010, assuming maximum commercialization of the innovative technologies, with projected emissions in the absence of such technologies. The EIS looks at potential impact of this program on acid deposition and the contribution of these technologies to emissions of greenhouse gases. Through this EIS DOE has acknowledged the need to begin to address this very difficult issue despite the scientific limitations of our analyses.

In other policy areas, DOE faces a number of major decisions regarding the restoration of its contaminated sites and its overall program of waste management. The future environmental implications of broad policy decisions made now regarding treatment, storage, and disposal of hazardous waste must be addressed. Similarly, as DOE moves to modernize its nuclear weapons complex, one can also anticipate the need for and the benefit of early comprehensive environmental analyses.

Gary Vest, U.S. Air Force

Some 6 and a half years ago, when I took the job that I have now, I was asked to focus on making sure that the Department of the Air Force was effectively complying with NEPA. That has been a challenge, and I think it is a challenge in all federal agencies. This experience is instructive for the future because I believe that one of the major future aspects of NEPA is the need for continued awareness and commitment to NEPA within the federal establishment and the continual need to educate those who come into the federal establishment about the importance of adhering to the precepts of NEPA and actually implementing them.

I also think that we must look at the future of NEPA in the context of what is going on in the world today. I am a defense person, so I watch a lot of what is going on around the world. I am very aware of military threat and military balance, as I think everyone is today, because there are major changes occurring in the world today. To a large degree this nation's agenda is changing greatly, and in many respects concern for our international military role is being replaced by a global environmental awareness.

In the countries where the Department of the Air Force has installations, the political environment and public perceptions are changing. As I look at what we have done in this country and at what those other countries are going to have to do, I see major challenges. If you think back to the environmental situation in this country in the middle or late 1960s and reflect on what we have actually done in 20 years, we have seen tremendous change. I do not want to speculate how much change there will be in the next 20 years, but if it is anything like the last 20, we are going to go through some very challenging times. What other countries in the world are going to have to do will also affect us, because many countries are going to have to compress into a very short time what we have had the opportunity and the resources to work on for many years. Changing environmental awareness and the changing expectations of people throughout the world will influence the way we [implement] NEPA.

In the Air Force we have made a major commitment to environmental quality and environmental compliance. In considering how to deal with the challenges of the 1990s and the 21st century we asked, 'What are the real foundation elements for the kind of program that the American people expect?' We have settled on three foundation elements. One of them is what we call 'base comprehensive planning.' That can be viewed as a data base, a framework to deal with the built environment. The second element is our version of environmental auditing, the Environmental Compliance and Management Program (ECAMP). The third, and perhaps the most important, element of our environmental compliance program for the future is our commitment to compliance with NEPA, not only at the EIS level, but also at the EA level.

We in the Air Force believe that we must continue to emphasize the integration of NEPA into the decision process. We must meet the challenge of educating our decision makers, the nonenvironmental professionals, because that will largely determine our success or failure. We must deal openly with the public. I share the view expressed earlier that we should look at the environmental assessment process and determine how we should involve the public in it. Today in the Air Force we have a great deal of public involvement in a number of our more interesting assessments.

Mitigation will indeed be one of the challenges of the 1990s, and we really have not touched follow-up analysis yet. I think follow-up is a major future aspect of NEPA. I see a lot of programmatic and generic EISs in the 1990s. This is appropriate because the kinds of issues we face warrant programmatic or generic EISs, but it can be quite a challenge. The Air Force is currently working on a generic EIS on low-level flight activities, and we had great difficulty deciding what the proposed action was. Our purpose, however, was to develop a framework and a data base for making specific decisions over time. We also wanted to be able to involve the public in all regions of the country. This is an appropriate use of NEPA, and we expect to do more of it in the future, driven in large part by the need to confront and effectively address the issues of cumulative impact. We really have not touched cumulative impact enough yet; it is, in fact, a challenge of the future.

I think we are going to see a lot more tiering. And we need to better understand how tiering really works, when it is appropriate, and when it is not. I think we are going to spend a lot of time looking more at the substance of the NEPA process in lieu of the procedural aspects. I think there will be a demand for greater accountability by the public and the political apparatus. [They will be] asking us whether we really made the right decision, rather than merely looking at whether we complied with the procedural aspects of NEPA.

In my opinion NEPA must become, if not the single most important element, at least one of the most important elements in our entire decision-making process in the federal government. Additionally, NEPA can and should become the framework for dealing with the whole myriad of environmental laws and environmental requirements in this country. It is the one thing that we have that has the potential for applying true order to environmental affairs.

In the near term, Vest projected that regulated agencies would receive more oversight in the NEPA arena, largely as a result of President Bush's stated commitment to the environment. He also thought that agencies would need to pay more attention to public involvement and cumulative impacts. For the decade of the 1990s, however, he believes "that the overseas international issue may well overshadow most other concerns. The Air Force is already involved in doing environmental analysis overseas because we have bases and facilities in many other countries. How you execute environmental analysis varies from country to country and involves a whole new set of laws and relationships. It is not impossible, but it is a challenge." In closing, he stated, "I really cannot speak very well for the 21st century, nor do I think that anyone can, but I know the 1990s are going to be very exciting."

DISCUSSION

Dinah Bear was asked for her views on the substantive area of NEPA, which she had identified as one of the three major issues for the future but had not really discussed. She noted that the substantive area of NEPA had just recently begun to be discussed, so that proposals for changes were not as mature as they were for the international and monitoring/mitigation areas. She noted that during the Carter Administration there had been a proposal for an executive order to require federal agencies to select the environmentally preferable alternative unless there was a good reason not to, but that that proposal had been dropped after controversy ensued. She did not foresee any concrete proposals emerging in the near future, but she was aware of "the kind of debate, thought, and papers being written that precede the development" of formal proposals.

A member of the audience stated his support for Carol Borgstrom's idea about the need for continued development of assessment methods, saying, "Often when we are writing an EIS, we find ourselves trying to develop the methods at the same time we are trying to apply them, all in a very short period." Noting that financial support for research and development (R&D) of assessment methodology has fallen short of what is really needed, he asked the panelists for their assessments of the potential for additional funding in the future. Gary Vest responded that the identification of R&D needs is one good reason to do more programmatic or generic EISs, saying "that is exactly what the Air Force is doing" in the generic EIS on low-level flight activities. He noted that his was the only agency in the federal government with a continuing R&D program in noise and sonic booms, "and we are doing that in parallel with the generic EIS, as well as simply laying a data base that we can [use a basis for NEPA tiering] and making intelligent decisions." Carol Borgstrom remarked that more R&D funding might become available if federal agencies are required to undertake NEPA follow-up studies and additional monitoring, saying, "I think a natural outgrowth of that will be a greater emphasis on the R&D aspects."

Bear was asked to explain the rationale for reducing the CEQ from a full council to a single chairman. She responded that there were several reasons and that the decision had received informal support from members of Congress, environmental organizations, past council members, and others. The first reason was that the role of the members in the council has never been clear. The chairman has certain statutory responsibilities under NEPA, but the council members do not. During a congressional oversight hearing in 1973, a member of the Senate suggested that either the statute should be amended to give the members some responsibilities or the positions should be eliminated, but Congress did not act. A second concern has to do with the federal government in the Sunshine Act. Although some other presidential boards, such as the Council of Economic Advisors, do not have to comply with this law, the Court of Appeals for the District of Columbia has held that it does apply to the CEQ. Bear said,

> This creates some interesting situations. In the past CEQ has had eight people, three of whom are presidential appointees. Two of those presidential appointees cannot talk to each other about any council business at all without going through staff. This is true even if it is a question of how we should advise the president, for example on what we should do about acid rain. They may not talk without either (1) placing a notice in the Federal Register 10 days in advance and then holding a public meeting or (2) claiming an exemption, and presidential advice does not qualify for an exemption under the Government in the Sunshine Act. We tried our best to comply during the past 8 years, and it meant that council members never discussed issues among themselves; they basically went through me. It is not a satisfactory situation in the context of an agency within the Executive Office of the President [and, consequently,] that is a very good reason to restructure CEQ. A third and very practical reason is the budget. Although we are expecting a big improvement in our budget, it is always going to be smaller than we want. With two council members lacking specific responsibilities and the drain of the Sunshine Act, the thought has been that the budget would be better spent for other matters such as the professional staff. There is a competing concern, however, of wanting to involve people with other expertise besides whatever expertise the chairman possesses. Therefore, consideration is being given to a part-time council or an advisory committee. In the early 1970s, CEQ had several advisory committees attached to it, including a legal advisory committee and a citizens' advisory committee. Some format like that may be revived.

Another member of the audience noted that many states and some localities have enacted their own versions of NEPA. He asked panelists and state representatives to comment on whether the trend of enactment of "little NEPAs" was likely to continue and on how these little NEPAs might be expected to affect the federal NEPA. Bear noted that 27 states have "little NEPA" statutes, only one of which had been enacted within the last few years (the District of Columbia's in 1989).

None of the panelists reported problems with conflicts between federal and state NEPA laws. Usually, when a project is subject to both NEPA and an analogous state law, the two levels of government cooperate in a joint EIA process. Vest commented that the Air Force has "probably had much greater problems dealing with other federal agencies when they are cooperating agencies" than in coordinating with state processes. Bear said that the only problems that had come to her attention had occurred when the state process had been completed before it was discovered that the project required a federal permit or approval and thus would also be subject to federal NEPA. She observed that "some states have provisions that if a document satisfies federal NEPA, it will also satisfy the state NEPA law," but when the need for a federal NEPA process is discovered after the state process is completed, "in many circumstances you have to go back and reinvent the wheel."

Steve Ugoretz of the state of Wisconsin presented a state perspective.

> We have had some experience in working with EPA, especially on wastewater-related projects, and generally, we have worked quite well together. However, we are seeing an increasing trend toward passing authorities for various programs to the states, including construction grants, prevention of significant deterioration, air quality, and possibly the §404 authority, and this trend creates a greater need to look at putting states with equivalent "little NEPAs" into a truly equivalent position with the federal agencies in conducting environmental reviews. I think that provisions that would call for other federal agencies to work more closely with the states would be helpful. Also, provisions to allow federal agencies to adopt state-prepared impact statements, with some degree of appropriate participation in the process, would be very useful.

Another member of the audience asked for clarification on how much mitigation can be included in an action that is covered by a categorical exclusion or by an EA leading to a finding of no significant impact (FONSI). He added, "If we have an EA that requires a large amount of mitigation, I think we are awfully close to needing an EIS."

Dickerson responded that he did not regard mitigation as a problem in the context of categorical exclusions because agency categorical exclusions are conservatively defined and typically limited to those actions "that will not produce, even cumulatively, a level of impact that should be of concern." He identified the topic of mitigated FONSIs, however, as one that was "ripe for discussion," and Bear reported that the CEQ hoped to issue new guidance on their use. She noted that the CEQ's "Forty most asked questions..." (CEQ 1981) had discouraged relying on mitigation as the basis for a FONSI, but said that that particular guidance had been overturned by several judicial circuits, which found that it was inappropriate not to give credit for measures that agencies or applicants are willing to add to a project. Dickerson noted a trend toward "bigger and bigger projects that are subject to EAs and not EISs" and restated his opinion that NEPA practitioners should make an extra effort to identify impacts that might be cost effectively mitigated, even if they are below the threshold of significance or documented in an EA rather than in an EIS. Bear identified public participation in EAs and FONSIs as a subject of current concern, noting that federal agencies differ dramatically in their handling of public participation in these documents.

She suggested that it might be prudent to issue mitigated FONSIs for public review as provided for in 40 CFR 1501.4(e)(2).

A member of the audience observed that the discussion had not yet touched on the social dimension of NEPA. Bear responded that NEPA consideration of social and economic impacts is indeed "a second level of concern, legally speaking," noting that these impacts by themselves do not lead to a requirement for an EIS. Although some agencies now have social scientists on their NEPA staffs, she thought that these issues generally do not receive the full level of discussion that they deserve. Noting that the policy goals in NEPA say a great deal about social and economic impacts, including population impacts and the preservation of cultural resources, she suggested that this "is one of the underdeveloped areas of NEPA."

Vest said that the Air Force frequently addresses socioeconomic concerns in its NEPA assessments. In many instances, he said, "the real issues are socioeconomic rather than environmental, and socioeconomic analysis is included in the EIS because it works and the public expects it." The Air Force often also conducts separate, but related, fiscal impact analyses, which attempt to quantify in terms of dollars the net fiscal impact of their actions. Borgstrom observed that the attention paid to socioeconomic issues in DOE NEPA documents depends largely on whether the proposed action is on one of the department's large existing reservations or is a new activity at a new site. Concern about socioeconomic impacts, and thus the level of socioeconomic analysis, is generally far greater for a new site than for a proposed action at an existing installation.

REFERENCES

Bassin, N. J., and D. R. West. 1986. Analysis of Selected Agencies' Post-Final Environmental Impact Statement Commitments. Environmental Management Support, Silver Spring, MD. Prepared for the Federal Agency Liaison Division, Office of Federal Activities, U.S. Environmental Protection Agency, Washington, D.C.

Council on Environmental Quality. 1981. Forty most asked questions concerning CEQ's National Environmental Policy Act regulations. *Fed. Regist.* 46:18026.

Culhane, P., H. P. Friesema, and J. A. Beecher. 1987. *Forecasts and Environmental Decisionmaking — The Content and Predictive Accuracy of Environmental Impact Statements.* Westview, Boulder, CO.

Methow Valley Citizens Council v. *Robertson*, *Environ. Law Rep.* 19:20743 (U.S. May 1, 1989); *Oregon Natural Resources Council* v. *Marsh*, *Environ. Law Rep.* 19:20749 (U.S. May 1, 1989).

National Academy of Sciences. 1989. *Improving Risk Communication.* National Academy of Sciences, Washington, D.C.

CHAPTER 3

Ecological Impact Assessment

Introduction

M. J. Sale, Oak Ridge National Laboratory, Oak Ridge, TN

In the years since the National Environmental Policy Act (NEPA) established the requirement for conducting environmental impact studies, much effort has gone into refining environmental assessment methodologies. Early environmental impact statements were criticized as a waste of time because these products seemed to be simply paper processing exercises. An encyclopedic approach to discussing potential impacts *can* obscure useful decisionmaking information. Nevertheless, the fact remains that any environmental assessment requires processing a large amount of ecological information, and the presentation and synthesis of this information is usually quite challenging.

Taken as a group, the following papers represent the evolution of the environmental impact statement (EIS) approaches that have been used over the last two decades, from descriptive catalogs of information to modern geographic information systems (GIS). In this chapter, the first two papers by Bruns et al. and by Hlohowskyj et al. are good examples of how to organize and present ecological information in the context of an EIS. Attention to ecological processes, such as the Bruns et al. examples of trophic relationships and plant succession, forces an assessment toward more effective consideration of ecological resources.

The second set of papers in this chapter focuses on new and old methods for specific problems: habitat modification in aquatic systems (Stalnaker), toxic effects in natural ecosystems (Suter), and physiological stresses (Adams et al.). These three problem areas are some of the most commonly confronted in assessments. The habitat evaluation models discussed by Stalnaker have been developed to address common problems of physical changes in surface water systems and subsequent impacts to aquatic ecosystems. The paper by Suter focuses on chemical effects in the environment, including the distinction between more traditional toxicity testing approaches and more progressive ecological risk assessment. The technology of bioindicators, as summarized by Adams et al., promises to be a very useful measure of ecosystem response to both chemical and physical stresses.

The third set of papers in this chapter concentrate on one of the newest and fastest growing areas of impact assessment: GIS and remote sensing technologies. Whereas assessment information has traditionally been presented in terms of lists and matrix elements, GIS technology makes it possible to represent information as pixel elements on a computer screen by combining digital mapping with database management capability. Siegel and Moreno present a clear example of how GIS techniques can be used to overlay many different types of environmental data and identify conflicts between ecological resources and proposed development (e.g., highway routing). The last two papers present additional examples of this exciting technology applied to design of the Yucca Mountain high-level nuclear waste repository in Nevada (Winsor and Rousseau) and to planning for the Big Cypress National Preserve in Florida (Kuykendall et al.).

The threat of "Garbage-In:Garbage-Out" will continue to be present in any ecological assessment, as it is in most other scientific disciplines. The real challenge for improving assessments is to find ways to cull the most relevant information to understanding ecological effects and responses. Hopefully, the examples and insights contained in this chapter will provide some insights that will help improve future assessments.

An Ecosystem Approach to Ecological Characterization in the NEPA Process

D. A. Bruns,[1] C. S. Staley, R. C. Rope, and K. S. Moor, Idaho National Engineering Laboratory, Idaho Falls, ID

ABSTRACT

Recent reviews of environmental impact assessment documents indicate a general lack of incorporation of ecological principles or techniques in environmental assessment studies. Nevertheless, numerous published papers in the peer-reviewed literature recommend the use of an ecosystem-level approach that may be applied to ecological monitoring and assessment to support the National Environmental Policy Act (NEPA) process. Based on a systems approach, we have developed a heuristic conceptual framework and assessment criteria for ecological resources at the Idaho National Engineering Laboratory (INEL). This framework and its associated criteria are being applied in an update of an existing ecological document for NEPA and in planning of future monitoring and characterization studies to support the NEPA process at INEL.

Selected examples of our update efforts from an ecosystem approach are provided for trophic relationships, plant community distributions, and succession. The use of ecological principles in the NEPA process is intended to result in (1) an emphasis on quantitative data, (2) application of ecological parameters (e.g., photosynthetic rates), and (3) examination of aspects of both ecosystem structure (e.g., species richness) and function (nutrient limitation). This approach should facilitate the role of NEPA in protecting and preserving ecological resources.

---

[1]Present address: GeoEnvironmental Sciences and Engineering Department, Wilkes University, Wilkes-Barre, PA, 18766.

INTRODUCTION

The National Environmental Policy Act (NEPA) requires that existing ecological conditions be documented and established in an area before a major development activity is undertaken. This requirement includes the need for follow-up studies during and after these activities in order to assess potential biological/ecological effects (Ausmus 1984). Application of the best available conceptual and analytical tools in environmental assessment for baseline ecological studies is necessary to satisfy the demands of the NEPA process. These assessment tools may include design of an ecosystem conceptual framework for planning purposes (e.g., Beanlands and Duinker 1984), incorporation of recent scientific advances based on the peer-reviewed literature (e.g., Schindler 1976, 1987), better integration of existing data to determine ecological health (Schaeffer et al. 1988) of a site, and use of a systems approach in the design and implementation of the baseline monitoring plan (Wiersma et al. 1984; Ausmus 1984; Beanlands and Duinker 1984).

The primary purpose of this paper is to review the ecological literature on environmental assessment for NEPA and to provide a general conceptual framework for use in updating the Idaho National Engineering Laboratory (INEL) ecological characterization document (EG&G 1984) as part of the NEPA process. Based on the review and conceptual framework, a major goal is to identify a set of assessment criteria to facilitate (1) review of the current INEL ecological characterization document for specific areas to be updated, (2) evaluation of existing INEL ecological data for use in the characterization document, (3) identification of data gaps that will require additional field studies, and (4) design and implementation of a field sampling plan. Selected examples are given to demonstrate ecological characterization based on the conceptual framework and criteria.

APPROACH

A general ecosystem approach to environmental assessment has been used in this report. This approach is based on the peer-reviewed literature for both the NEPA assessment process (e.g., Beanlands and Duinker 1984) and general measurement of ecosystem response to potential anthropogenic stress (e.g., Odum 1985; Schindler 1987). The intent is to provide first a general ecological framework for environmental assessment and then a specific conceptual model for the INEL ecosystem. This will allow better evaluation and integration of existing data for the characterization document and a more cost-effective implementation of additional field studies.

LITERATURE REVIEW: ECOLOGICAL ASSESSMENT

Recent reviews of environmental impact assessment documents (e.g., Rosenberg et al. 1981; Schindler 1976, 1987; Beanlands and Duinker 1984; Beanlands 1987;

Larkin 1984) indicate a general lack of incorporation of ecological principles or techniques in environmental assessment studies. Environmental impact assessments, not published as refereed papers (primary publications and symposia proceedings) and not subjected to the peer-reviewed literature, consistently scored lower in ability to identify main impacts, predict effects, and formulate useable recommendations (Rosenberg et al. 1981). Therefore, it is intended that the use of the peer-reviewed literature will facilitate and improve ecological characterization in the NEPA process.

A literature review was conducted on ecological principles and methods relevant to environmental impact assessment (Table 1). The review is not intended to be comprehensive, but is based on a wide range of representative papers focused on an ecological perspective in environmental assessment. In particular, peer-reviewed, refereed journals were emphasized (14 of 21 papers reviewed in Table 1). Except for one monograph on ecosystem concepts (O'Neill et al. 1986), the reviewed papers that were not in journals came from published proceedings of expert panels. These panels were formed under the National Research Council (NRC 1981), the Scientific Committee on Problems of the Environment (Sheehan et al. 1984), and the Council on Environmental Quality (Draggan et al. 1987).

The objectives of the literature review are to (1) identify an ecological conceptual framework to support ecological characterization in the NEPA process and (2) determine from the literature those ecological parameters sensitive to environmental impact. The literature review is intended to help identify a set of assessment criteria for evaluating NEPA ecological characterization documents. Also, the literature review will provide a basis for using a systems approach (e.g., Wiersma et al. 1984) in the design and implementation of any additional field studies require for the NEPA environmental assessments.

Assessment

All papers reviewed in Table 1 focused on various aspects of ecology and environmental assessment. Eight of these specifically identified NEPA and supported the relevance of ecological principles and techniques to the environmental impact assessment process. The other papers dealt with related aspects of environmental assessment, such as response and measurement of ecosystem stress (Rapport et al. 1985; Odum 1985), ecological health (Schaeffer et al. 1988), air pollution impacts on terrestrial vegetation (Sigal and Suter 1987), and ecotoxicology at community and ecosystem levels (Sheehan 1984a,b). All of these have direct relevance to selection of baseline ecological measurements for environmental impact assessment.

Conceptual Framework

The use of a general conceptual framework for environmental impact assessment and ecosystem response to disturbance is a critical component in all the papers reviewed in Table 1. In a scientific approach to environmental impact assessment, Rosenberg et al. (1981; see also Larkin 1984; Beanlands and Duinker 1984)

Table 1. Literature Review on Ecological Principles Relevant to Environmental Impact Assessment.

Authors	Assessment	Conceptual	Quantitative	Levels[a]	Ecosystem[b]	Parameters[c]
Rosenberg et al. 1981	NEPA	Yes(i)[d]	Yes	All	NA	NA
Schindler 1987	NEPA	Yes(i)	Yes	All	B(S:A) (T:F)	A:composition T:productivity
Karr 1987	General	Yes(e)	Yes	All	S	Guild[e]
Larkin 1984	NEPA	Yes(i)	Yes	NA	NA	NA
Beanlands and Duinker 1984	NEPA	Yes(e)	Yes	All	B	Trophic
Schaeffer et al. 1988	General	Yes(e)	Yes	All	B(S)	Many
Ausmus 1984	NEPA	Yes(e)	Yes	Ecosystem	F	Nutrient (pools) cycling
Rapport et al. 1985	General	Yes(e)	Yes	Ecosystem	B	Several
Hinds 1984	General	Yes(i)	Yes	All	B	Several
Hildebrand et al. 1987	NEPA	Yes(e)	Yes	All	NA	NA
Taub 1987	General	Yes(e)	Yes	All	B	Many
Odum 1985	General	Yes(e)	Yes	Ecosystem	B	Many
Cairns 1986	General	NA	Yes	All	NA	Multispecies
Skalski and McKenzie 1982	NEPA	Yes(e)	Yes	Populations	NA	Abundance
NRC 1981	General	Yes(e)	Yes	All	B	Many
O'Neill et al. 1986	General	Yes(e)	Yes	All	B	Several

Beanlands 1987	NEPA	Yes(e)	Yes	All	B	NA
Sigal and Suter 1987	General	Yes(i)	Yes	All	B	Many
Sheehan 1984a	General	Yes	Yes	Ecosystem	F	Many
Sheehan 1984b	General	Yes(e)	Yes	Individuals/ populations	B	Several
Sheehan 1984c	General	Yes	Yes	Community	S	Many

[a] Levels of organization include populations, communities, and ecosystem.

[b] Ecosystem structure and function: NA = not analyzed, B = both structure and function, S = structure, F = function, A = aquatic, F = forest.

[c] NA = not analyzed, A = aquatic, T = terrestrial; category gives idea of number and/or kind of ecosystem parameters reviewed.

[d] (i) = implicit conceptualization, (e) = conceptual framework explicit in paper.

[e] A guild represents a group of species that are functionally similar in their ecology (this often includes aspects of feeding habits).

recommend (1) definition of scientific objectives, (2) preparation of background material (literature reviews, field reconnaissance, case history studies), (3) identification of main impacts, (4) prediction of effects (surveys, experiments, models), (5) recommendations for project design, and (6) ecological monitoring and assessment (before and after development). Beanlands and Duinker (1984) argue for a related set of requirements for environmental impact assessment within an ecological framework.

A number of the papers provide a conceptualization of ecosystem change relative to baseline conditions, environmental impact, or environmental stress. Ausmus (1984) (see Figure 1) focused on nutrient pools and nutrient cycling at the ecosystem level for environmental monitoring and assessment. Beanlands and Duinker (1984) (Figures 2 and 3) emphasized both structural and functional relationships of the ecosystem in assessing the chain of ecological impact. Other papers (e.g., Odum 1985; Rapport et al. 1985) develop a series of predictions in the response of key ecosystem parameters to environmental stress. These conceptual frameworks are synthesized and applied below in an example to develop an ecosystem approach to facilitate and improve ecological characterization at INEL.

Quantitative Approach

All the papers reviewed (Table 1) require a quantitative approach to measurement of ecosystem parameters for ecological characterization and environmental impact assessment. The sole reliance on qualitative data (e.g., species lists, distributions, and habitats) would be viewed as a serious shortcoming in environmental assessment (Rosenberg et al. 1981; Beanlands and Duinker 1984; Beanlands 1987; Schaeffer et al. 1988). Consideration of statistical design is an important component of a quantitative approach (e.g., Rosenberg et al. 1981; Hinds 1984; Sigal and Suter 1987; Green 1979).

Levels of Ecological Organization

More than one half of the papers reviewed in Table 1 cover all levels of ecological organization from populations [and individuals (e.g., Sigal and Suter 1987)] through communities and ecosystems. This broad coverage reflects several factors. First, consideration of multiple levels of organization is important from a conceptual framework in ecosystem theory (O'Neill et al. 1986), environmental monitoring (Hinds 1984), and ecological assessment (NRC 1981; Beanlands and Duinker 1984; Schaeffer et al. 1988). Second, general literature reviews of ecosystem response to impacts require coverage of various levels of organization to better assess potential effects on ecological resources (e.g., Sheehan 1984a, b, c; Taub 1987; Schindler 1987; Sigal and Suter 1987). Third, it currently appears that an optimal strategy in environmental impact assessment will necessitate the use of multiple measures of ecological response over more than one level of organization (NRC 1981; Hinds 1984; Beanlands and Duinker 1984; Sheehan

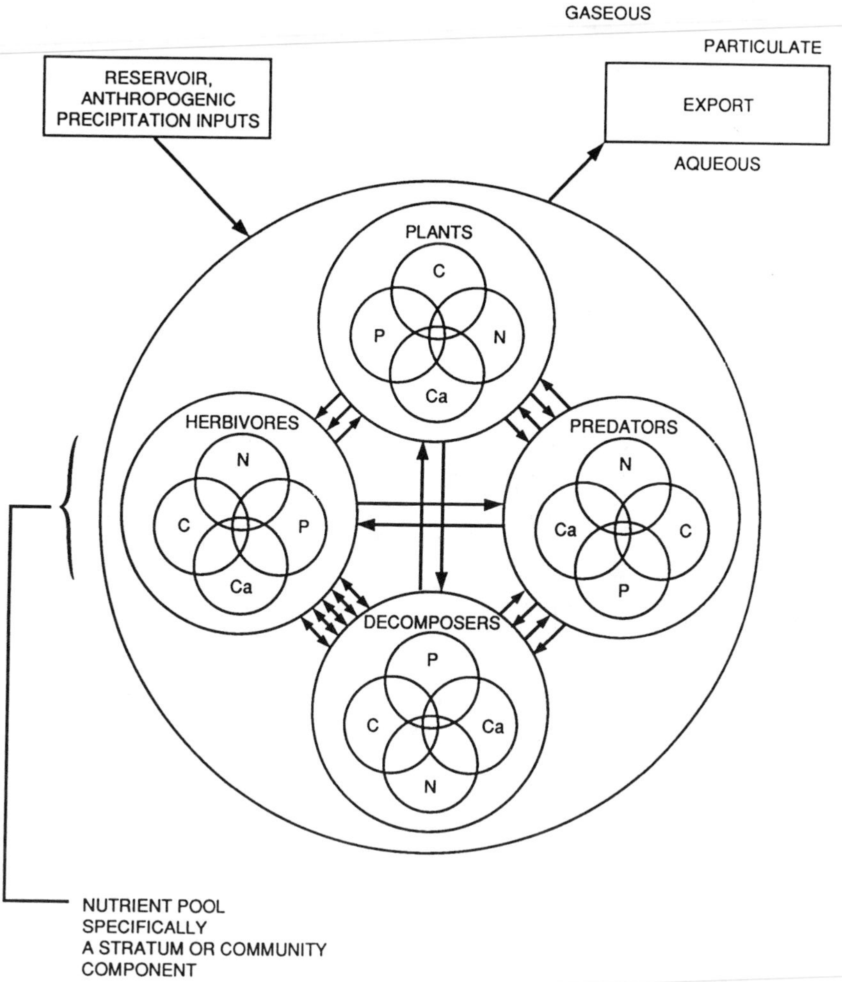

Figure 1. Conceptualization of a soil showing nutrient pools and exports. (From Ausmus, B. S. 1984. *Environ. Monit. Assess.* 4:275–293. With permission.)

1984a, b, c; Rapport et al. 1985; Odum 1985; Schindler 1987; Sigal and Suter 1987; Schaeffer et al. 1988).

It should not be concluded from the above reviews that a long list of parameters over all levels of organization (e.g., the list of more than 50 parameters in Schaeffer et al. 1988) is necessarily required for ecological monitoring and environmental assessment. For example, Odum (1985) and Rapport et al. (1985) developed a framework for evaluating ecosystem response to stress based on a subset of key integrative measures; a similar approach is recommended by Schaeffer

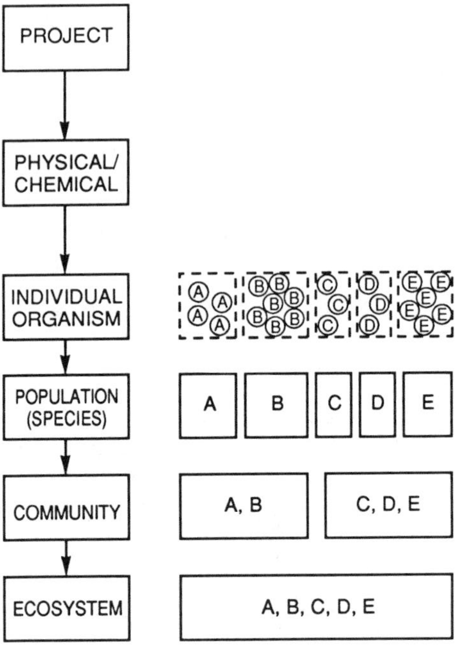

Figure 2. Chain of impact and the structural relationships of biota. (From Beanlands, G. E., and P. N. Duinker. 1984. *J. Environ. Manage.* 18:267–277. With permission.)

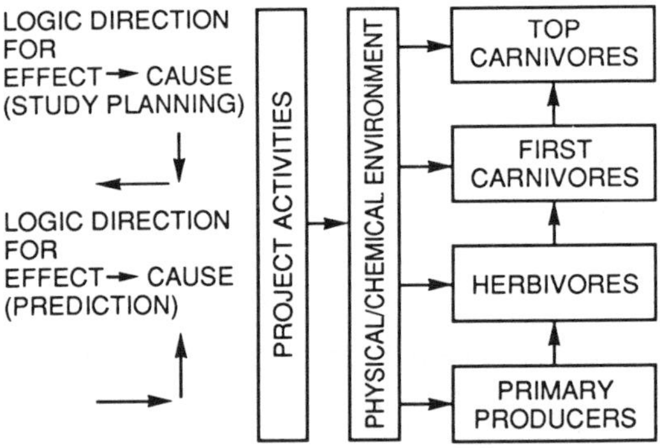

Figure 3. Chain of impact and the functional relationships of biota. (From Beanlands, G. E., and P. N. Duinker. 1984. *J. Environ. Manage.* 18:267–277. With permission.)

et al. (1988) for measuring ecological health. Beanlands and Duinker (1984) recommend early selection of an initial set of valued ecosystem components for an environmental impact assessment of a particular development project. Also, profuse amounts of undigested data may overwhelm decision makers in the environmental impact assessment process (e.g., Rosenberg et al. 1981).

Ecosystem Structure and Function

Except for two papers that focus on guilds (Karr 1987) and nutrient cycling (Ausmus 1984), all papers that explicitly dealt with an ecosystem conceptual framework included aspects of both structure and function in their review, synthesis, or evaluation. This is due largely to the same three reasons outlined above for coverage of various levels of organization. Thus, aspects of both the structure and function of ecosystems need to be considered from a conceptual standpoint (e.g., Hinds 1984; O'Neill et al. 1986), for review and determination of potential impacts to ecological resources (e.g., Sheehan 1984a, b, c; Taub 1987), and as an optimal strategy in the design and implementation of an ecosystem monitoring and assessment program (e.g., based on Rapport et al. 1985; Schaeffer et al. 1988).

Monitoring and Assessment Parameters

Many of the papers reviewed in Table 1 provide lists and discussions of various parameters for potential use in ecological measurements to support environmental assessments of human impacts. For example, Schaeffer et al. (1988) tabulate over 50 parameters for reference in determining ecological "illness," while others discuss a wide range of similar ecological measures in related contexts such as ecotoxicology and damage to terrestrial vegetation (e.g., NRC 1981; Sheehan 1984a, b, c; Taub 1987; Sigal and Suter 1987). However, measuring many different parameters is not necessarily an optimal strategy for monitoring and impact assessment (e.g., see Rosenberg et al. 1981). In many cases, a subset of key ecosystem integrators of environmental impact can be identified (as in Table 2). Table 2 summarizes key ecological parameters over three levels of organization that have been identified in various studies for use in measuring and detecting environmental impact.

Trophic relationships, species diversity, succession (temporal changes in composition), energy flow, and nutrient cycling were the most common parameters that were applied, discussed, and/or recommended. In the context of trophic levels, the use of keystone predators (i.e., those with critical regulatory role) as integrators of ecological impact was often encouraged (Beanlands and Duinker 1984; Odum 1985; O'Neill et al. 1986; Schindler 1987; Schaeffer et al. 1988). Retrogressive succession was a common diagnostic feature where ecosystem stress or illness was being defined (Rapport et al. 1985; Odum 1985; Schaeffer et al. 1988). This refers to temporal changes in community composition that reflect an earlier successional stage where smaller, short-lived, opportunistic (i.e., r-selected) species may predominate.

A number of authors (Odum 1985; Sigal and Suter 1987; Schaeffer et al. 1988) suggest that functional attributes such as energy flow and nutrient cycling are more robust (homeostatic) than structural measures because of functional redundancy and variation in pollutant sensitivity among species. Therefore, functional measures may not be as useful for early detection of ecological stress from human activities, but may better reflect a breakdown of homeostasis in the

Table 2. Ecological Parameters: Recommendations for Monitoring and Assessment of Baseline Conditions and Human Impacts.

	Author(s)[a]												
Parameter	**1**	**2**	**3**	**4**	**5**	**6**	**7**	**8**	**9**	**10**	**11**	**12**	**13**
Population													
Abundance (biomass)					+	+	+	+	+(A)	+			
Reproduction				+	+	+	+		+(A)				
Behavior				+	+	+							
Community Structure													
Trophic relationships		+	+	+	+	+	−	+	+(A)	+	+	+	+
Species diversity	+	+	+	+	+	+	+	+				+	
Succession or change in composition	+	+	+		+		+	+	+(A)	+	+	+	
Size relationships	+	+											
Ecosystem Function													
Energy flow	+	+	+	+	+	+	−	+	−(A) +(T)	+	+		
Nutrient cycling		+	+	+	+	+	+	+(A)	−(A)	+	+		+
Decomposition/respiration		+	+	+	+	+		+(A)	−(A)	+	+		+
Biomass/nutrient pools	+	+	+			+							+

[a] (1) Rapport et al. 1985; (2) Odum 1985; (3) Schaeffer et al. 1988; (4) National Research Council 1981; (5) Sheehan 1984 a,b,c; (6) Taub 1987; (7) Sigal and Suter 1987; (8) O'Neill et al. 1986; (9) Schindler 1987; (10) Hinds 1984; (11) Beanlands and Duinker 1984; (12) Karr 1987; (13) Ausmus 1984.

[b] A = aquatic ecosystem; T = terrestrial ecosystem; + = good potential for monitoring and assessment; and – = robust; not indicative of *early* impacts or stresses. Plus signs reflect a generally positive view of a particular parameter as a key indicator of impact or at least its potential utility to detect anthropogenic perturbations. A negative sign means that an author has found this parameter to be a poor indicator of ecological impact; these parameters were found to be too robust and were not very sensitive to impacts.

later stages of severe ecosystem damage (Odum 1985). Also, when some authors hesitate to recommend certain functional parameters (like system respiration for Odum and total production for Sigal and Suter), indirect or surrogate measures may be encouraged (e.g., microbial activity and litter production for Odum and Sigal and Suter, respectively) because of greater sensitivity, ease of operation, and/or low cost.

A second set of commonly used parameters includes decomposition (respiration), biomass (nutrient pools), and population abundance (biomass). Sheehan (1984c) suggests that measuring changes in population abundance might be most applicable to point-source pollution; in general, he argues for more applications of an ecosystem approach to assessing environmental impacts from pollutants (Sheehan 1984b). Size relationships (i.e., changes in average size of organisms, usually smaller with impacts), behavior, and reproduction were also used for determining human impacts, but to a lesser degree than those indicated above.

In general, almost all of the authors reviewed in Table 2 supported a systems-level approach to environmental monitoring and impact assessment. Although exceptions and certain problems for particular parameters have been identified and discussed by these authors (also, see points below), there is a general view that a number of important ecological impacts can be measured and assessed only at the ecosystem level. This view is emphasized by Sigal and Suter (1987), even though they generally favor the use of structural responses of individual organisms.

Spatial and Temporal Scale and Conceptual Framework

Definition of spatial and temporal scales is necessary in order to select the proper level of ecological organization for ecosystem measurements (O'Neill et al. 1986). On a conceptual basis, a watershed/drainage basin systems approach (e.g., Likens and Bormann 1974; Likens 1985; Minshall et al. 1985) provides a scale that can integrate the various parameters and levels of organization reviewed in Table 2. This system's perspective emphasizes functional linkages between terrestrial and aquatic ecosystems (Figure 4). Within this heuristic context, environmental monitoring and ecosystem assessment programs can be designed and conducted by identifying key ecological compartments (or parameters, see Table 2) of primary concern, delineating potential pollutant (or impact) pathways through the system, and identifying critical receptors. This allows one to view an environmental impact problem as one of pollutant or impact sources and pathways to critical receptor components of the ecosystem (e.g., Beanlands and Duinker 1984) (see Figures 2 and 3). A systems approach has been applied successfully to both pollutant monitoring studies (Wiersma et al. 1984; Wiersma and Otis 1986) and baseline ecosystem studies (e.g., Bruns et al. 1984; Bruns and Minshall 1985).

A drainage basin conceptual framework also facilitates measurement and assessment of key ecological linkages between groundwater and terrestrial and surface water components of the ecosystem. Groundwater is now viewed as one of four major dimensions in running water ecosystems (Ward 1989), and studies in the Rocky Mountain region have indicated significant groundwater communities of invertebrates as integral parts of detrital food chains (Stanford and Gaufin 1974; Pennak and Ward 1986; Stanford and Ward 1988). The relationship between groundwater and the fate of organic matter in streams is discussed by Hynes (1983). Holsinger (1988) provides a recent general review of aquatic faunas in subterranean groundwaters.

SYNTHESIS: CONCEPTUAL FRAMEWORK AND CRITERIA FOR ECOLOGICAL CHARACTERIZATION AND ENVIRONMENTAL IMPACT ASSESSMENT AT THE IDAHO NATIONAL ENGINEERING LABORATORY

Objectives

The objectives of this synthesis section are twofold: to develop, as an example, an integrated conceptual framework applicable to the INEL ecosystem and to propose a set of working criteria to be used in the environmental impact assessments.

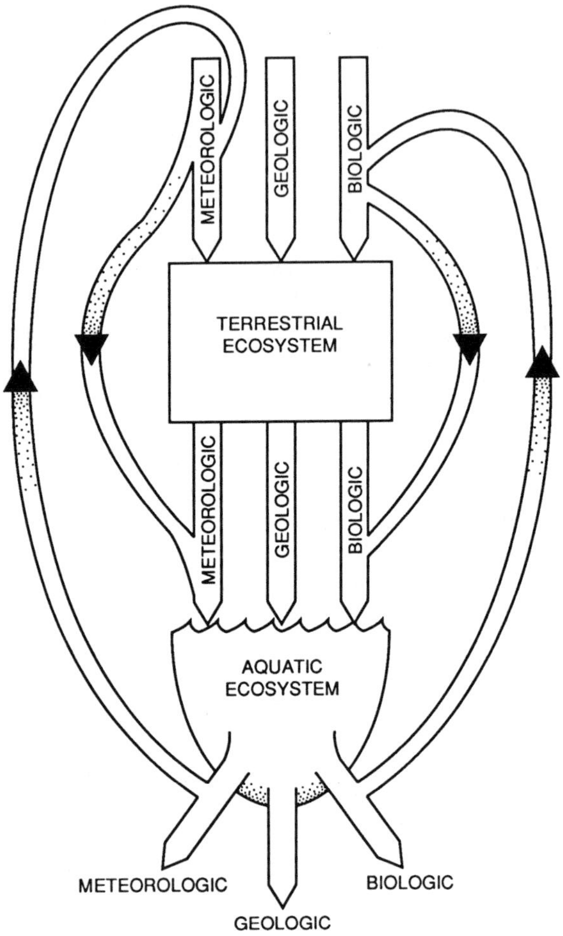

Figure 4. Diagrammatic model of the functional linkages between terrestrial and aquatic ecosystems. Vectors may be meteorologic, geologic, or biologic components moving nutrients or energy along the pathway shown. (From Likens, G. E. 1985. *An Ecosystem Approach to Aquatic Ecology, Mirror Lake and Its Environment.* Springer-Verlag, New York. With permission.)

Both objectives are based on the synthesis and overall integration of the literature review in the previous section. Selected examples on the application of the conceptual framework and environmental impact assessment criteria are provided based on existing INEL ecological data.

Conceptual Framework

Figure 5 represents an integrated conceptual framework for the INEL ecosystem based on a watershed basis (e.g., Likens and Bormann 1974; Likens 1985) that

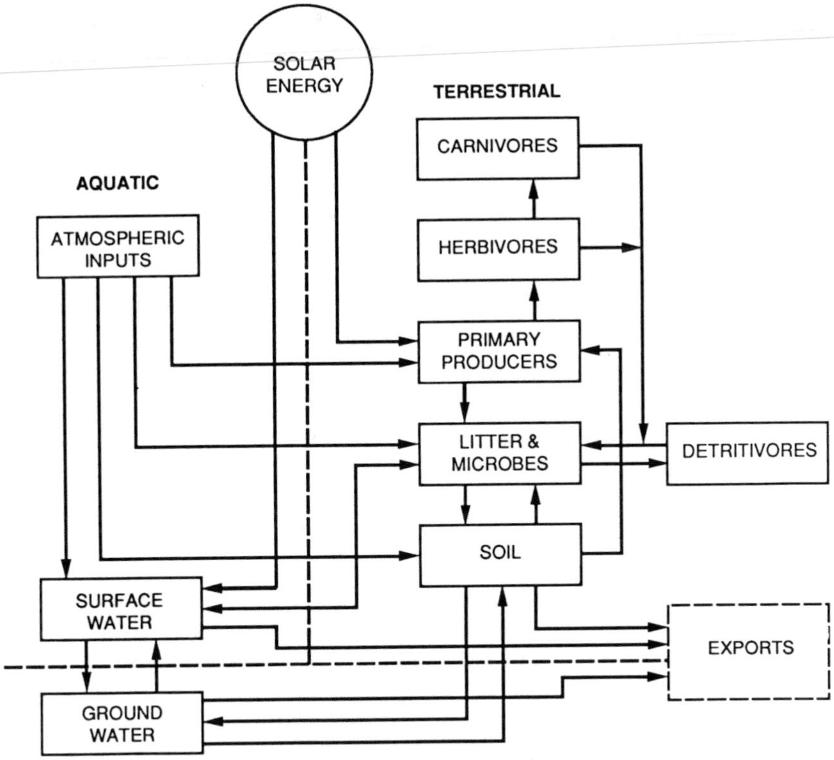

Figure 5. Integrated conceptual framework for the INEL ecosystem. Surface water in this high desert ecosystem is ephemeral and unpredictable.

emphasizes functional linkages between terrestrial, surface water, and groundwater components (see also Figure 4). Within each of these components, relationships among primary producers, consumers, litter (microbes), and detritivores are highlighted. These individual compartments of the model are built upon the diagrams developed by Ausmus (1984) (see Figure 1) and Beanlands and Duinker (1984) (see Figure 3) for environmental assessments in regard to nutrient cycling and energy flow, respectively. As such, the boxes (compartments) can represent biomass or nutrient pools, while the arrows indicate energy flow (one-way direction only) or nutrient flux. Similarly, boxes and arrows may reflect the storage and movement, respectively, of pollutants along key pathways of the ecosystem. The chain of potential environmental impact from a proposed development project is not shown but would be essentially the same as that indicated by Beanlands and Duinker (1984) and presented earlier in Figure 3.

Figure 5 also serves as a first-step conceptual outline for ecosystem structure, even though specific factors like competition, habitat size, spatial heterogeneity, and temporal variability that are known to influence populations and communities (e.g., Southwood 1977; May 1976) are not explicitly shown. Relative to structural

Table 3. Ecological Characterization and Assessment Criteria.

I.	General:	
	A. Conceptual framework	B. Quantitative approach
	C. Levels of organization	D. Ecosystem structure and function
II.	Specific:	
	A. Populations:	(1) abundance, (2) reproduction, (3) behavior
	B. Communities:	(1) tophic relations, (2) species diversity, (3) succession/community composition
	C. Ecosystem:	(1) energy flow, (2) nutrient cycling, (3) decomposition/respiration, (4) biomass/nutrient pools

parameters shown in Table 2, the conceptual model does demonstrate trophic relationships in a straightforward manner, and species diversity, compositional changes (e.g., succession), and size relationships can readily be measured relative to any of the biotic compartments. Also, key population parameters could be evaluated for selected species associated with any of the boxes for trophic levels.

Criteria for Ecological Measurements

Table 3 summarizes a checklist of criteria to be used in the review and update of ecological characterization documents and for planning and implementing any additional field studies as part of the NEPA process. Four of the criteria are general in nature. (1) Has a conceptual framework been used in the overall organization? (2) Are quantitative data summarized for the ecosystem? (3) Are key ecological aspects covered at the level of the population, community, and ecosystem? (4) Are aspects of both ecosystem structure and function represented? These criteria represent consistent themes that were emphasized in various review papers for environmental assessment and ecosystem measurements relative to human impacts (i.e., see literature review in Table 1).

Specific criteria also are listed in Table 3 and are based on the above literature review as summarized in Table 2. The basic question being addressed by these criteria is whether an existing ecological characterization document provides baseline data and information on these parameters covering a range of levels from populations to ecosystem. Data for all of these parameters are not necessarily required to adequately characterize an ecosystem for purposes of environmental impact assessment. In addition, it is not expected that all the criteria would be met by any one document and the criteria are intended as an initial checklist.

Example Applications

Several examples based on an INEL ecological characterization are provided to illustrate use of the ecosystem conceptual framework and environmental impact assessment criteria. Table 4 compares the present characterization document (EG&G 1984) with aspects of the ongoing update relative to selected assessment criteria (from Table 3). The existing document is not based on an ecological

Table 4. Comparison of Present INEL Ecological Characterization Document with Proposed Update Relative to Selected Assessment Criteria.

Criteria	Present Document	Ongoing Update
Conceptual framework	No	Yes[a]
Quantitative data	No	Yes[b]
Food chain relationships	Qualitative	Quantitative[b]
Plant community distributions	Coarse scale	Fine scale[c]
Succession	No	Yes[b]

[a] See Figure 5 and related text earlier in this report.
[b] See examples for food chains and succession in this section of report.
[c] Present vegetation map of plant community distributions based on aerial photography and qualitative ground truthing (i.e., coarse scale); update will be fine-scale mapping based on quantitative data in the field and remote sensing imagery.

conceptual framework and does not present or review quantitative data. Although information is given on selected aspects of food relationships and a basic vegetation map is provided, neither of these are characterized in terms of quantitative data. In contrast, the ongoing update of the INEL ecological characterization document is intended to reflect a basic conceptual framework as shown in Figure 5 and to emphasize quantitative patterns where appropriate. Selected examples are provided below for trophic relationships, plant community distributions, and succession.

Trophic Relationships

The use of keystone predators (i.e., those with critical regulatory role) as integrators of ecological impact was recommended by most authors in this review (Table 2). Other than reporting the presence of such predators, the present INEL ecological characterization document provides no information to aid the selection of a predator for monitoring purposes.

The most widely studied predator on the INEL is the coyote. MacCracken and Hansen (1987) have summarized food habits and relative abundance data over a discontinuous time period, 1977–1979 and 1982–1983. Coyote abundance was shown to increase approximately threefold (Figure 6) during a jackrabbit population explosion (tenfold increase), with a corresponding dominance of jackrabbits in the coyote diet. During periods of low jackrabbit abundance, coyote diet was dominated by Nuttall's cottontail rabbits, with montane voles secondary in importance. These data fit the predictions of optimal-diet models, the jackrabbit being the most profitable (largest) from an energetic standpoint and the vole the least profitable of the three.

From this example, we see that the coyote diet is dominated by three species of smaller mammals and that coyote populations respond to prey abundance (for jackrabbits, at least). Monitoring coyote abundance and food habits, therefore, can provide an integration of impacts on lower trophic level organisms which make up the coyote diet. Once such impacts are detected, then one may need to investigate at a lower trophic level to determine causes for the impacts. Studies

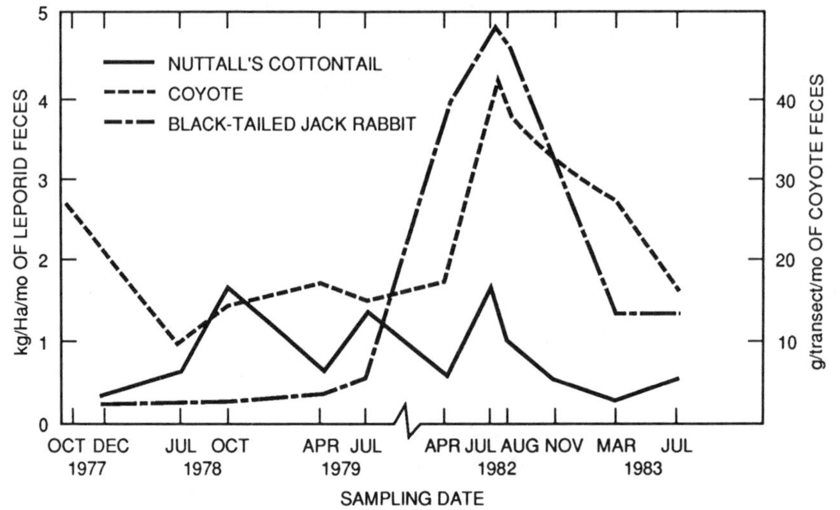

Figure 6. Relative abundance of black-tailed jackrabbits, Nuttall's cottontails, and coyotes, as determined by fecal accumulation indices, at 12 sampling dates on the INEL. (From MacCracken, J. G., and R. M. Hansen. 1987. *J. Wildl. Manage.* 51:278–285. With permission.)

such as this will be incorporated into the updated INEL ecological characterization to enhance our understanding and conceptualization of the INEL ecosystem and to aid in assessment and monitoring of anthropogenic impacts.

Plant Community Distributions

Quantitative mapping of plant community distributions provides relevant data for the NEPA process (Table 3) regarding important species, population distributions and abundances, aspects of species diversity on a spatial scale, and community composition. On a practical basis, such a map allows one to examine the potential degree of impact from construction and operation of a new facility within the larger context of resources within the ecological landscape. Also, it provides a basis for assessing similar concerns about animal populations and communities if animal and habitat (vegetation type) associations have been established.

The existing vegetation map for the INEL is based on aerial photographs and some general qualitative ground truthing in the field. Only coarse-scale relationships can be identified without quantitative data (Table 4), and it is difficult to obtain full benefit of the parameters identified above from an ecosystem approach.

Work has been started on a recent false-color satellite (LANDSAT) image for the INEL as part of our ecosystem characterization of the INEL for the NEPA process. Before this study, only a preliminary classification (e.g., sagebrush, junipers, agriculture, grasslands) had been conducted. Since then, quantitative ground truthing (intensive plot measurements) for each vegetative class has been conducted for the summers of 1990 and 1991. Even with only a preliminary classification, relative to the existing map, the satellite image provides considerably more information on different vegetation types and the degree of spatial heterogeneity within and among these types on a landscape scale. Quantitative ground truthing in the future should enhance our ability to make specific NEPA assessments from a stronger ecological conceptual framework by providing better data on important species, their abundance and distribution, community composition, and species richness over spatial gradients.

Plant Succession

A number of models of succession have been developed to account for vegetation change over time. Many of these have been reviewed by Miles (1979) who suggests that no single model can account for all successions. In general, the early models of Clements (1916) and Gleason (1926) represent two basic schools of thought, and the empirical evidence in more recent studies appears to support the latter (see Miles 1979). Clements' model shows a directional change in vegetation types with time based on successive site modification by each stage of vegetation development; the sequence ends with a stable climax community that is self-maintaining under current environmental conditions. Gleason's concept of succession is based on differences among individual species in their dispersal efficiency; in their ability to persist as seeds; and in their ability to establish, grow, compete, and reproduce under different environmental conditions (see Miles 1979). It is these differences among individual species that account for patterns of vegetation change through time.

Data and information on successional changes in plant populations (and communities) over time also may support NEPA assessments (Table 3). The same parameters identified above in regard to spatial mapping are also applicable but within the context of a temporal scale. Temporal changes in vegetation have not been described in the current INEL ecological characterization document (Table 4). However, a recent report by Anderson and Inouye (1988) provides considerable information on long-term vegetation dynamics at the INEL. Percent cover for each plant species has been measured at 35 study plots at about 5-year intervals from 1950 to 1983.

Figure 7 shows mean cover of shrubs and perennial grasses at the 35 INEL study plots from 1950 to 1983. Anderson and Inouye (1988) suggest that the composition and structure of vegetation in 1950 most likely reflected a history of heavy grazing by domestic livestock plus prolonged drought effects of the 1930s and 1940s. After 1957, total cover of shrubs and perennial grasses increased presumably as a result of more favorable periods of precipitation. However, cover of most dominant species decreased after 1975 except for *Chrysothamnus viscidiflorus* (rabbitbrush) which increased steadily over the study period.

Figure 8 demonstrates mean cover of perennial grasses plotted against mean cover of shrubs for selected study years at the INEL permanent vegetation plots. These authors note several trends. First, perennial grass cover was negatively correlated with shrub cover in sampling years when perennial grass cover was highest (e.g., 1978); water availability was greater in those years and suggests

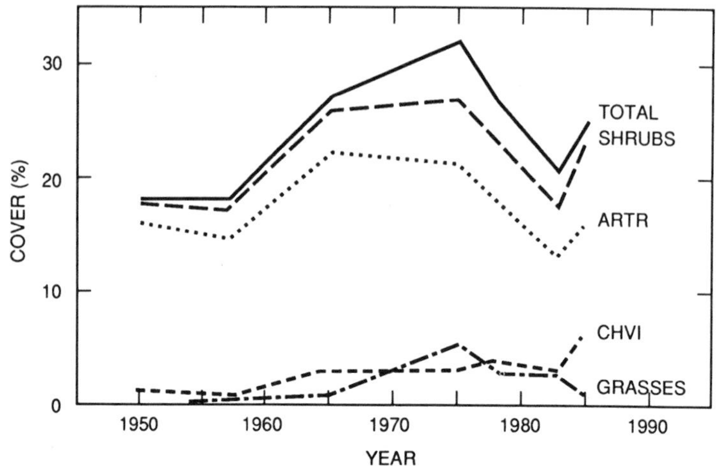

Figure 7. Mean cover of shrubs and perennial grasses for 35 permanent plots at the INEL. ARTR = *Artemesia tridentata*; CHVI = *Chrysothamnus viscidiflorus*. (From Anderson, J. E., and R. Inouye. 1988. Final Report to Radiological And Environmental Services Laboratory, Idaho National Engineering Laboratory, U. S. Department of Energy, Idaho Operations Office, Idaho Falls, ID.)

that competition may influence the abundance of these species under those conditions. Second, the study plots became much more heterogeneous through time as evidenced by the increasing spread of data points along both the x- and y-axes (Figure 8). And third, in general, the data in Figure 8 (and Figure 7) show that both shrub and perennial grass cover may fluctuate considerably over the time span of even a decade in the absence of any major disturbance.

Anderson and Inouye (1988) suggest that the above patterns in plant community variability are probably attributable to the stochastic nature of water availability (year-to-year and longer time scales) and the relatively slow response time of vegetation to these changes in precipitation. The authors indicate that these conditions may preclude factors that would result in predictable successional patterns at the scale of the individual plot.

The results of this vegetation study (Anderson and Inouye 1988) at INEL are relevant to the NEPA process on a number of points. First, it provides an excellent longer-term, baseline data set for use as an historic context from which to evaluate potential environmental impacts. Second, it enhances basic understanding of plant ecology at INEL relative to key factors such as precipitation variability, water availability, and competition. This basic understanding will facilitate an ecosystem approach to environmental impact assessment as developed and reviewed above in this report. Third, the use of a "climax community" concept to extrapolate data or make impact assessments should be avoided given the dynamic nature of vegetation patterns at the INEL, its degree of variability, and the lack of an

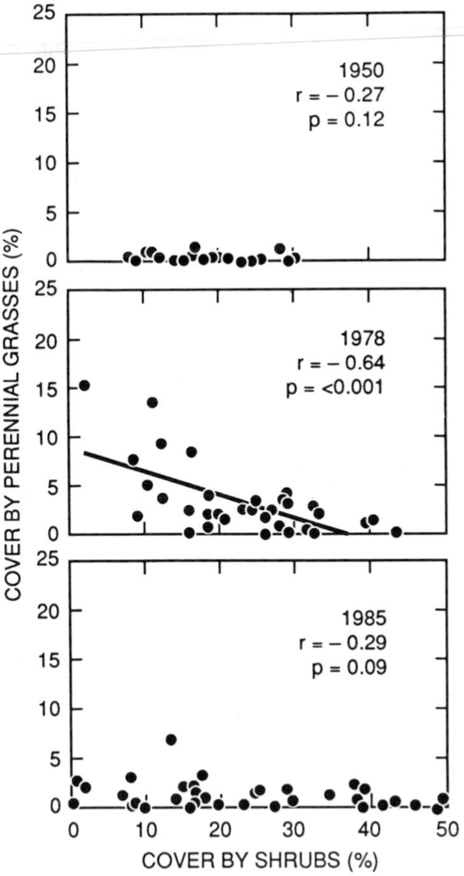

Figure 8. Mean cover of perennial grasses plotted against mean cover of shrubs for 35 permanent plots at the INEL. A trend line is shown for that year (1978) in which there was a significant negative correlation between perennial grasses and shrub cover. (Modified from Anderson and Inouye 1988.)

identified stable plant community. And fourth, the results of this study reinforce the need for preoperational field studies and postoperational monitoring to document and evaluate potential environmental impacts because of the high degree of natural environmental and ecological variation.

CONCLUSIONS

This paper has focused on an ecosystem approach to environmental impact assessment and ecological characterization to support the NEPA process. NEPA regulations require preconstruction studies and postoperational monitoring to assess potential impacts to ecological resources. The ecosystem conceptual framework and assessment criteria developed here are intended to facilitate this process at the INEL. A number of conclusions follow from this work.

1. An increasing number of peer-reviewed papers have been published on the use of ecological principles for environmental impact assessment. The application of a basic conceptual framework and a quantitative approach are typically recommended.
2. Although a wide range of different ecological parameters have been identified for potential use in impact assessment, our literature review indicates that most papers focus on a smaller number of key parameters that encompass both ecosystem structure and function.
3. The available published literature on ecological approaches to environmental impact assessment provided the necessary basis to develop a general conceptual framework for the INEL ecosystem and a set of ecological criteria for assessment.
4. Initial application of the conceptual framework and assessment criteria resulted in identification of important ecological databases and provided insight to the degree of natural variation for both animal and plant populations.
5. Overall, the use of ecological principles for environmental impact assessment should enhance the use of quantitative data and ecosystem parameters, including both structure and function. This approach facilitates the role of NEPA in protecting and preserving ecological resources.

REFERENCES

Anderson, J. E., and R. Inouye. 1988. Long-Term Dynamics of Vegetation in a Sagebrush Steppe of Southeastern Idaho. Final Report to Radiological and Environmental Sciences Laboratory, Idaho National Engineering Laboratory, U.S. Department of Energy, Idaho Operations Office, Idaho Falls, ID.

Ausmus, B. S. 1984. An argument for ecosystem level monitoring. *Environ. Monit. Assess.* 4:275–293.

Beanlands, G. E. 1987. *In situ* contaminants and environmental assessment — an ecological summary. *Hydrobiologia* 149:113–118.

Beanlands, G. E., and P. N. Duinker. 1984. An ecological framework for environmental impact assessment. *J. Environ. Manage.* 18:267–277.

Bruns, D. A., and G. W. Minshall. 1985. River continuum relationships in an 8th-order river reach: Analysis of polar ordination, functional groups, and organic matter parameters. *Hydrobiologia* 127:277–285.

Bruns, D. A., G. W. Minshall, C. E. Cushing, K. W. Cummins, J. T. Brock, and R. L. Vannote. 1984. Tributaries as modifiers of the river continuum concept: Analysis by polar ordination and regression models. *Arch. Hydrobiol.* 99:208–220.

Cairns, J., Jr. 1986. The myth of the most sensitive species. *Bioscience* 36:670–672.

Clements, F. E. 1916. Plant Succession. Carnegie Inst. Wash. Publ. No. 242.

Draggan, S., J. J. Cohrssen, and R. E. Morrison (eds.). 1987. *Preserving Ecological Systems: The Agenda for Long-Term Research and Development.* Praeger, New York.

EG&G Idaho, Inc. 1984. An Ecological Characterization of the Idaho National Engineering Laboratory. Appendix B in INEL Environmental Characterization Report. Volume 2: Appendices A–C. EGG-NPR-6688.

Gleason, H. A. 1926. The individualistic concept of plant association. *Bull. Torrey Bot. Club* 53:7–26.

Green, R. H. 1979. *Sampling Design and Statistical Methods for Environmental Biologists.* Wiley, New York.

Hildebrand, S. G., L. W. Barnthouse, and G. W. Suter. 1987. The role of basic ecological knowledge in environmental assessment. pp. 51–69. In S. Draggan, J. J. Cohrssen, and R. E. Morrison (eds.), *Preserving Ecological Systems: The Agenda for Long-Term Research and Development.* Praeger, New York.

Hinds, W. T. 1984. Towards monitoring of long-term trends in terrestrial ecosystems. *Environ. Conserv.* 11(1):11–17.

Holsinger, J. R. 1988. Troglobites: The evolution of cave-dwelling organisms. *Am. Sci.* 76:147–153.

Hynes, H. B. N. 1983. Groundwater and stream ecology. *Hydrobiologia* 100:93–99.

Karr, J. R. 1987. Biological monitoring and environmental assessment: A conceptual framework. *Environ. Manage.* 11(2):249–256.

Larkin, P. A. 1984. A commentary on environmental impact assessment for large projects affecting lakes and streams. *Can. J. Fish. Aquat. Sci.* 41:1121–1127.

Likens, G. E. (ed.). 1985. The aquatic ecosystem and air-land-water interactions. pp. 430–435. In *An Ecosystem Approach to Aquatic Ecology, Mirror Lake and Its Environment.* Springer-Verlag, New York.

Likens, G. E., and F. H. Bormann. 1974. Linkages between terrestrial and aquatic ecosystems. *Bioscience* 24(8):447–456.

MacCracken, J. G., and R. M. Hansen. 1987. Coyote feeding strategies in southeastern Idaho: Optimal foraging by an opportunistic predator. *J. Wildl. Manage.* 51:278–285.

May, R. M. (ed.). 1976. *Theoretical Ecology: Principles and Applications.* W. B. Saunders, Philadelphia, PA.

McBride, R., N. R. French, A. H. Dahl, and J. E. Detmer. 1978. Vegetation Types and Surface Soils of the Idaho National Engineering Laboratory Site. IDO-12084. U.S. Department of Energy, Idaho Operations Office, Idaho Falls, ID.

Miles, J. 1979. *Vegetation Dynamics.* Chapman and Hall, London.

Minshall, G. W., K. W. Cummins, R. C. Peterson, C. E. Cushing, D. A. Bruns, J. R. Sedell, and R. L. Vannote. 1985. Developments in stream ecosystem theory. *Can. J. Fish. Aquat. Sci.* 42:1045–1055.

National Research Council (NRC). 1981. *Testing for the Effects of Chemicals on Ecosystems.* National Academy Press, Washington, D.C.

Odum, E. P. 1985. Trends expected in stressed ecosystems. *Bioscience* 35(7):419–422.

O'Neill, R. V., D. L. DeAngelis, J. B. Waide, and T. F. H. Allen. 1986. *A Hierarchial Concept of Ecosystems.* Princeton University Press, Princeton, NJ.

Pennak, R. W., and J. V. Ward. 1986. Interstitial faunal communities of the hyporheic and adjacent groundwater biotopes of a Colorado mountain stream. *Arch. Hydrobiol. Suppl.* 74:356–396.

Rapport, D. J., H. A. Regier, and T. C. Hutchinson. 1985. Ecosystem behavior under stress. *Am. Nat.* 125(5):617–640.

Rosenberg, D. M., V. H. Resh, S. S. Balling, M. A. Barnby, J. N. Collins, D. V. Durbin, T. S. Flynn, D. D. Hart, G. A. Lamberti, E. P. McElravy, J. R. Wood, T. E. Blank, D. M. Schultz, D. L. Marrin, and D. G. Price. 1981. Recent trends in environmental impact assessment. *Can. J. Fish. Aquat. Sci.* 38:591–624.

Schaeffer, D. J., E. E. Herricks, and H. W. Kerster. 1988. Ecosystem health: I. Measuring ecosystem health. *Environ. Manage.* 12(4):445–455.

Schindler, D. W. 1976. The impact statement boondoggle. *Science* 192:509.

Schindler, D. W. 1987. Detecting ecosystem responses to anthropogenic stress. *Can. J. Fish. Aquat. Sci.* 44(1):6–25.

Sheehan, P. J. 1984a. Effects on community and ecosystem structure and dynamics. pp. 51–99. In P. J. Sheehan, D. R. Miller, G. C. Butler, and P. Bourdeau (eds.), *Effects of Pollutants at the Ecosystem Level. SCOPE 22.* Wiley, New York.

Sheehan, P. J. 1984b. Functional changes in ecosystems. pp. 101–145. In P. J. Sheehan, D. R. Miller, G. C. Butler, and P. Bourdeau (eds.), *Effects of Pollutants at the Ecosystem Level. SCOPE 22.* Wiley, New York.

Sheehan, P. J. 1984c. Effects on individuals and populations. pp. 23–50. In P. J. Sheehan, D. R. Miller, G. C. Butler, and P. Bourdeau (eds.), *Effects of Pollutants at the Ecosystem Level. SCOPE 22.* Wiley, New York.

Sheehan, P. J., D. R. Miller, G. C. Butler, and P. Bourdeau (eds.). 1984. *Effects of Pollutants at the Ecosystem Level. SCOPE 22.* Wiley, New York.

Sigal, L. L., and G. W. Suter, II. 1987. Evaluation of methods for determining adverse impacts of air pollution on terrestrial ecosystems. *Environ. Manage.* 11(5):675–694.

Skalski, J. R., and D. H. McKenzie. 1982. A design for aquatic monitoring programs. *J. Environ. Manage.* 14:237–251.

Southwood, T. R. E. 1977. Habitat, the templet for ecological strategies. *J. Anim. Ecol.* 46:337–365.

Stanford, J. A., and A. R. Gaufin. 1974. Hyporheic communities of two Montana rivers. *Science* 185:700–702.

Stanford, J. A., and J. V. Ward. 1988. The hyporheic habitat of river ecosystems. *Nature* 335:64–66.

Taub, F. B. 1987. Indicators of change in natural and human-impacted ecosystems: Status. pp. 115–144. In S. Draggan, J. J. Cohrssen, and R. E. Morrison (eds.), *Preserving Ecological Systems: The Agenda for Long-Term Research and Development.* Praeger, New York.

Ward, J. V. 1989. The four-dimensional nature of lotic ecosystems. *J. North Am. Benthol. Soc.* 8(1):2–8.

Wiersma, G. B., C. W. Frank, M. J. Case, and A. B. Crockett. 1984. The use of simple kinetic models to help design environmental monitoring systems. *Environ. Monit. Assess.* 4:233–255.

Wiersma, G. B., and M. D. Otis. 1986. Multimedia design principles applied to the development of the global integrated monitoring network. pp. 317–332. In Y. Cohen (ed.), *Pollutants in a Multimedia Environment.* Plenum Press, New York.

Ecological Assessments at DOE Hazardous Waste Sites: Current Procedures and Problems

I. Hlohowskyj, J. R. Krummel, J. S. Irving, and W. S. Vinikour, Argonne National Laboratory, Argonne, IL

ABSTRACT

Major actions at U.S. Department of Energy (DOE) hazardous waste sites require Comprehensive Environmental Response, Compensation, and Liability Act of 1980 (CERCLA) compliance that meets National Environmental Policy Act (NEPA) of 1969 considerations. Although NEPA compliance includes ecological considerations, neither the Council on Environmental Quality (CEQ) nor DOE provide detailed guidance for conducting ecological assessments under NEPA. However, the identification of the form and magnitude of potential ecological impacts associated with a proposed action is directly dependent on the quality of the baseline data available for a particular site. Using the Surplus Facilities Management Program Weldon Spring, Missouri, site as an example, we discuss the collection of baseline ecological data for the site. This site is surrounded by approximately 17,000 ac of wildlife area. Available wildlife data consists of qualitative, county-level species lists, and vegetation data is in the form of a regional qualitative narrative. Detailed site-specific occurrence data for listed species and high quality natural communities was provided by the Missouri Department of Conservation Heritage database.

INTRODUCTION

The Department of Energy (DOE) currently maintains responsibility for a number of hazardous waste sites throughout the United States, including several sites that are on or proposed for inclusion on the National Priorities List. The remedial actions carried out at these sites are subject to U.S. Environmental Protection Agency (EPA) oversight under Comprehensive Environmental Response, Compensation, and Liability Act (CERCLA) of 1980, as amended by the Superfund Amendments and Reauthorization Act of 1986. Current regulatory compliance guidelines under CERCLA emphasize remediation for human health impacts and development and selection of cleanup technologies.

The DOE is also responsible for meeting the requirements of the National Environmental Policy Act (NEPA) of 1969, which requires federal agencies to consider environmental consequences of a proposed action as part of the decisionmaking process. Although NEPA considerations include ecological assessments, neither the Presidsent's Council on Environmental Quality (CEQ) regulations nor the DOE NEPA Compliance Guide provide detailed guidance for conducting ecological investigations, and ecological assessments at DOE sites are often afforded secondary considerations relative to human health and remediation technology development. Detailed guidance for conducting ecological assessments under CERCLA at these sites has only recently been provided (U.S. EPA 1987, 1989a).

The combination of a relative lack of ecological assessment guidance and a primary emphasis on human health and site cleanup, along with budgetary and time constraints, often preclude quantitative investigations that can forecast potential impacts to ecological systems at these sites. Using the Surplus Facilities Management Program (SFMP) Weldon Spring site in Missouri as an example, we discuss the sources and methods used for the collection of baseline ecological data for this site. We chose the Weldon Spring site because it is surrounded by approximately 17,000 ac of actively managed wildlife area and is thus ecologically diverse. The identification of potential impacts associated with a proposed action and the estimation of the magnitude of these impacts on ecological resources are directly dependent on the quality of the baseline data available for a particular site.

SITE DESCRIPTION

The Weldon Spring site is located in St. Charles County, MO, approximately 48 km west of the city of St. Louis. The site consists of two noncontiguous areas: (1) a raffinate pits and chemical plant area and (2) a quarry (Figure 1). The raffinate pits and chemical plant area occupies approximately 217 ac, and the quarry covers approximately 9 ac. The chemical plant was originally used by the U.S. Army for production of trinitrotoluene (TNT) and dinitrotoluene (DNT) and, subsequently, by the Atomic Energy Commission (AEC) for processing of uranium and thorium ores. The plant has been unused and in caretaker status since 1969. The quarry area had been used by the U.S. Army for the disposal of chemically contaminated explosive material from the early 1940s through the mid-1950s. Between 1957 and 1966, the quarry was used by the AEC for the disposal of uranium and thorium residues and radioactively contaminated building rubble and process equipment. Waste disposal in the quarry ended in 1969.

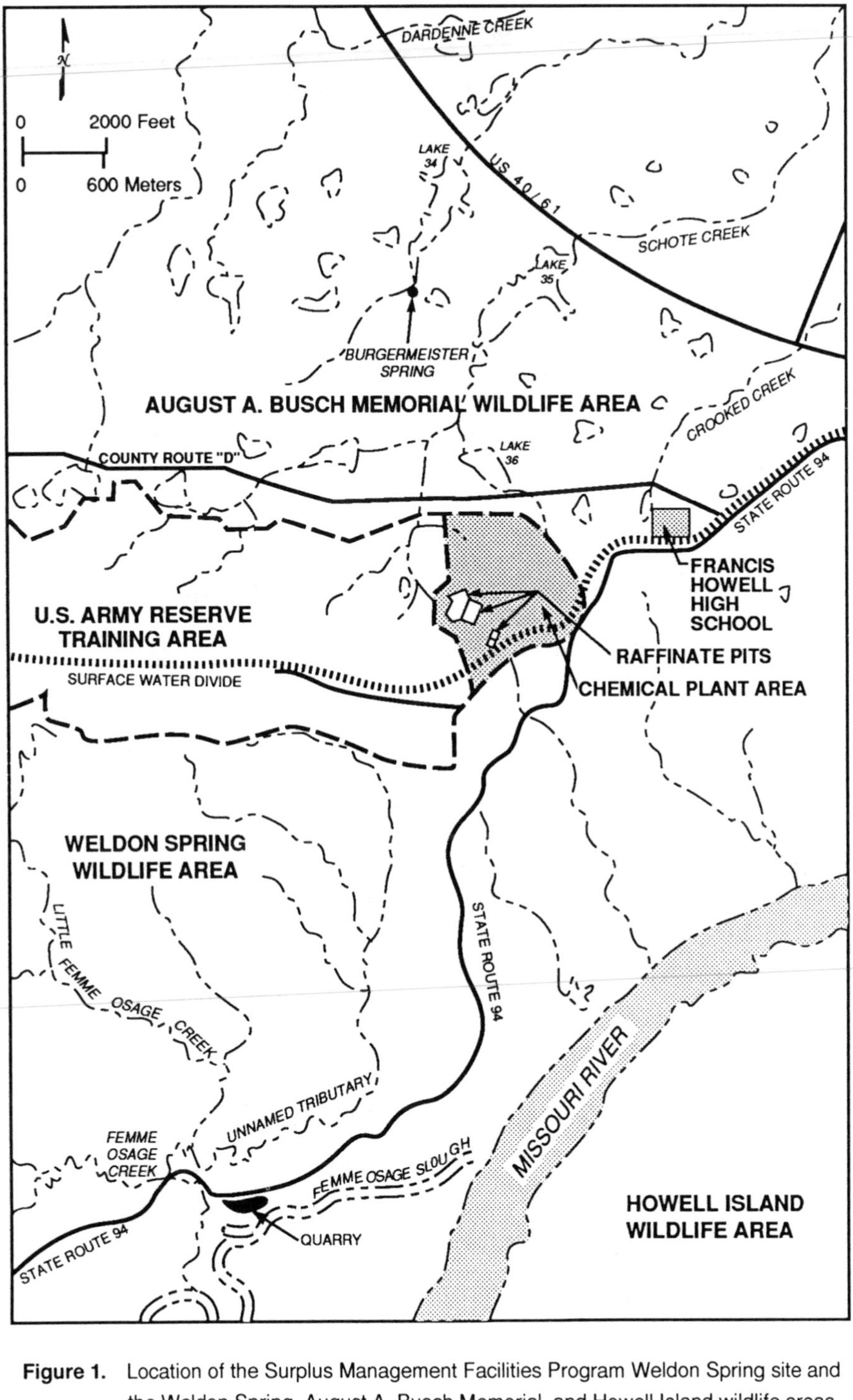

Figure 1. Location of the Surplus Management Facilities Program Weldon Spring site and the Weldon Spring, August A. Busch Memorial, and Howell Island wildlife areas, St. Charles County, MO.

The chemical plant area is bordered to the north by the August A. Busch Memorial Wildlife Area, by the Weldon Spring Wildlife Area to the south and east, and the U.S. Army and National Guard Training Area to the west. The quarry was excavated into a limestone bluff that overlooks a portion of the Missouri River alluvial floodplain and is surrounded by the Weldon Spring Wildlife Area.

August A. Busch Memorial Wildlife Area

The August A. Busch Memorial Wildlife Area (Figure 1) encompasses approximately 7000 ac, including 35 lakes totaling 500 ac, and is under the jurisdiction of the Missouri Department of Conservation (MDOC). The area is used as a demonstration site for wildlife management, integrating the development and use of forestry, water, agricultural, and wildlife resources. A variety of habitats occur within the wildlife area, including open fields and pasture, forests, and cultivated cropland.

In addition to the hunting and fishing provided by the management programs, the wildlife area also supports and provides for a variety of research, education, and nature study programs. The Busch Wildlife Area receives approximately 800,000 visitors each year from St. Louis and surrounding areas.

Weldon Spring Wildlife Area

The Weldon Spring Wildlife Area (Figure 1) is a 7230-ac Special Management Area managed for wildlife by the MDOC. The area is extensively wooded with some areas of old field, grassland, and agriculture. The wildlife management program includes a share cropping program with local farmers that provides area wildlife with high energy cereal grains and winter green browse.

The southern boundary of the wildlife area is delineated by the Missouri River (Figure 1) and includes portions of the 100-year floodplain for the Missouri River and also Little Femme Osage Creek. Some of the floodplain area is used for crop production. Aquatic habitats include numerous small, unnamed creeks, drainages, springs, and ponds located throughout the Weldon Spring Wildlife Area. The estimated 250,000 yearly visitors use the wildlife area for a variety of recreational activities, including hunting, fishing, hiking, and boating (in the Missouri River).

Howell Island Wildlife Area

The Howell Island Wildlife Area is located in the Missouri River across from the Weldon Spring Wildlife Area (Figure 1). This 2575-ac wildlife area is within 2.5 km of both the chemical plant and quarry areas and contains a known night roost for bald eagles overwintering in the area (Gaines 1988).

CONSTRAINTS

The goal of the ecological assessment is to provide sufficient information and analyses on the ecosystems of an area so that an evaluation of potential impacts from the proposed and alternative actions can be performed. Biological resources for which baseline data were deemed necessary included terrestrial and aquatic ecosystems, listed species, unique or high quality natural areas and communities, and wetlands. At the Weldon Spring site, these data were collected to support a feasibility study (FS) under CERCLA. Although not a NEPA document, the FS was to include a NEPA environmental assessment (EA) level of analysis to support a remedial action at the site.

Several problems were encountered during the collection of baseline ecological data for the SFMP Weldon Spring site and surrounding areas. Little site-specific ecological information was available for the Weldon Spring site. Previous site-specific investigations discussed the ecological resources at and in the vicinity of the site only in very general, qualitative terms (U.S. DOE 1987) or consisted primarily of radiological and chemical uptake and bioaccumulation studies on selected biota (RETA 1978; MK-Ferguson and Jacobs Engineering Group 1988).

The collection of baseline ecological data for the Weldon Spring site was further complicated for a variety of reasons. The completion of the EA-level analysis and documentation for the site was scheduled at approximately 7 months, thus limiting the amount of time available for the collection and analysis of baseline data for the area. The CEQ (1981) suggests that the NEPA process for an EA should take no more than 3 months. In addition, the presence of the extensive wildlife areas, along with the time constraints, precluded the field collection of ecological data for the site. The Weldon Spring area includes several actively managed wildlife areas totalling approximately 17,000 ac and contains a large number of habitats.

METHODS

The MDOC Heritage database was the primary source of information on the wildlife of the area. The Heritage database provided species lists of wildlife known to occur in St. Charles County and included information on the status (i.e., rare, transient, common, winter resident only) of the bird and mammal species known to occur in the county. Discussions with naturalists at the Busch and Weldon Spring Wildlife areas provided additional information on the ecological resources in the immediate area of the site.

Life history and additional distribution information on the biota of the Weldon Spring site area was obtained from a variety of regional literature sources, such as *The Fishes of Missouri* (Pflieger 1975) and *Amphibians and Reptiles of Missouri* (Johnson 1987). Quantitative information on the flora of the area was obtained from Bailey's (1976) *Ecoregions of the United States*, the Missouri Botanical Garden (1975), and MDOC information pamphlets for the Busch (MDOC 1978) and Weldon Spring (MDOC 1985) wildlife areas. Additional information on the topography, wildlife habitats, and vegetation of the SFMP Weldon Spring site and surrounding wildlife areas was obtained during a visit to the area. The site visit consisted of a walking inspection of the chemical plant and quarry areas and also included driving and walking through portions of the Busch and Weldon Spring wildlife areas.

The Heritage database provided detailed information on state and federally listed rare and endangered species and on high quality natural communities, occurring in St. Charles County, adjacent St. Louis County, and in the Howell Island, Busch, and Weldon Spring Wildlife areas. Information on the status of federally listed threatened or endangered species was obtained through consultation with the U.S. Fish and Wildlife Service (USFWS) under Section 7(c) of the Endangered Species Act. The USFWS was also contacted for National Wetlands Inventory maps of St. Charles County. However, no base maps were available at the time for this portion of Missouri.

RESULTS

Terrestrial Ecology

The SFMP Weldon Spring site is located along the boundary between two physiographic provinces (Johnson 1987; Thom and Wilson 1980). Both portions of the site occur in areas that are biologically rich and contain significant ecological resources.

The quarry area is situated in the northern portion of the Ozark Border physiographic province. This region occurs in a band along the lower Missouri River and the eastern edge of the state of Missouri along the Mississippi River and is characterized by hills and bluffs, deciduous forests, and wide river valleys. The quarry is surrounded by the Weldon Spring Wildlife Area.

The chemical plant area occurs within the southern portion of the Glaciated Plains physiographic province. This area is characterized by rolling hills and broad flat valleys, and limestone bluffs and steep hills occur at the eastern edge of the province along the Missouri River (Johnson 1987).

Vegetation

The Quarry Area

Although extensively affected by past human activities, little human disturbance currently occurs in this area and vegetation has become reestablished. The quarry floor is old-field habitat with a variety of grasses, herbs, and shrubs. The rim and upper portions of the quarry consist primarily of slope and upland forest; tree species present include cottonwood, sycamore, and oak (personal observation). No list of plant species occurring at the quarry was available.

The Chemical Plant Area

The chemical plant area is essentially grassland and old-field habitat with numerous buildings, roads, parking areas, and other plant facilities. The area has a gently rolling topography and receives active mowing. Little undisturbed vegetation exists at the site, and vegetation is limited to a variety of grasses with scattered small shrubs (personal observation). No list of plant species occurring at the chemical plant area was available.

The Wildlife Areas

Habitats identified during the site visit included open fields and pastures; slope, upland, and bottomland forests; and cultivated farmlands. Plant species common to the open fields and pastures of the area include a variety of grasses, annuals, perennials, and shrubs (U.S. DOE 1987; MDOC 1978, 1985; Missouri Botanical Garden 1975). Forested habitats in the area contain a variety of tree species, such as shagbark hickory, pawpaw, black walnut, eastern cottonwood, and a variety of oaks.

The area south of the quarry is within the 100-year floodplain of Little Femme Osage Creek and the Missouri River. Vegetation in this area consists primarily of crops or grasses and herbaceous species (personal observation). Trees, when present, are restricted to numerous levees throughout the area and to the banks of Little Femme Osage Creek and Femme Osage Slough; these are primarily willow and cottonwood. No listing of plant species occurring in the wildlife areas was found, and none was provided by the MDOC.

Wildlife

Little site-specific information on the wildlife of the chemical plant and quarry areas was found. The MDOC Heritage database identified 25 species of amphibians, 47 species of reptiles, and 29 species of mammals occurring in St. Charles County. Many of these species could occur in the wildlife areas near the site. Although no list of species occurring at the quarry was available, the natural condition of the quarry and the proximity of the Weldon Spring Wildlife Area suggest that a number of the species listed for St. Charles County may inhabit or use habitats at the quarry. In contrast, the chemical plant area probably contains relatively depauperate amphibian, reptile, and mammal communities characterized by species commonly associated with developed urban and residential areas.

Few reptile species would be expected to occur at the chemical plant because of the relative absence of suitable habitat. Similarly, amphibian species, if present, would be restricted to the few aquatic habitats present at the plant, and none of the more terrestrial forms such as the eastern tiger salamander would be expected to occur at the plant. Some amphibians and reptiles, such as the bullfrog and snapping turtle, have been reported from the aquatic habitats at the raffinate pit and chemical plant area (MK-Ferguson Company and Jacobs Engineering Group 1988).

The SFMP Weldon Spring site is situated along the Mississippi Flyway, one of the major north-south routes for migratory North American birds. As a consequence, the Weldon Spring area possesses a very diverse avifauna. More than 295 avian species have been reported from St. Charles County and could occur at the Weldon Spring site. More than 100 species are known to breed in the habitats present at the wildlife areas, and many are common year-round residents. The many ponds and small lakes in the area, and also Little Femme Osage Creek and the Missouri River, provide important habitat for nesting and migratory waterfowl and shorebirds.

Aquatic Ecology

Aquatic habitats at the quarry are limited to a 0.5-ac pond located on the quarry floor. No data are available on the presence or absence of vertebrates in this pond or on the use of this pond by terrestrial species. However, collection of fish from the quarry pond for radiological and chemical uptake studies was unsuccessful (MK-Ferguson Company and Jacobs Engineering Group 1988). Aquatic habitats at the chemical plant include the raffinate pits, two small ponds, and several small intermittent streams and drainages. Collection of fishes at the raffinate pits has been unsuccessful, and only an unidentified species of sunfish (*Lepomis* sp.) has been collected from one of the ponds at the plant (MK-Ferguson Company and Jacobs Engineering Group 1988).

Principal aquatic habitats in the Weldon Spring Wildlife Area are the Missouri River (approximately 1.6 km southeast of the quarry), Little Femme Osage Creek (150 m west of the quarry), Femme Osage Creek (610 m south-southwest of the quarry), and Femme Osage Slough (150 m south of the quarry). Other aquatic habitats in the area include numerous small, unnamed creeks, drainages, springs, and ponds. Several intermittent and perennial streams also occur at the Busch Wildlife Area. In addition, this wildlife area contains 35 ponds and lakes ranging in size from approximately 1 to 182 ac.

The MDOC reports 105 species of fish from St. Charles County, including the Missouri River, and some of these species could occur in the aquatic habitats of the wildlife areas. However, little information was found with regards to the species known to occur in these areas. Many of the aquatic habitats in the area, and especially at the Busch Wildlife Area, are actively managed for recreational fishing activities, and some of the ponds and lakes are stocked with channel catfish, bass, crappie, and other species.

Threatened and Endangered Species and High Quality Communities

Consultation with the USFWS indicated that the only federally listed threatened or endangered species known to occur in the Weldon Spring area is the bald eagle *Haliaeetus leucocephalus*. Overwintering bald eagles roost at night on the Howell Island Wildlife Area. However, no critical habitat for this species exists at either the quarry or the chemical plant. The range of the federally endangered peregrine falcon *Falco perigrinus* includes the Weldon Spring area. Although this species was not identified by the USFWS as occurring in the area, the MDOC Heritage database includes this species as a migrating transient in the area.

The MDOC Heritage database included records of the sturgeon chub *Hybopsis gelida* and the sicklefin chub *Hybopsis meeki* from the Missouri River at the Howell Island and Weldon Spring Wildlife areas. These chubs are classified as Category 2 (C2) species, candidates for federal listing as threatened or endangered species. These species, however, are restricted to open channels of large turbid rivers such as the Missouri River and do not enter tributary streams (Pflieger 1975). Thus, with the exception of the Missouri River proper, these species will not be found in the aquatic habitats of the Weldon Spring area.

An additional three C2 species (Table 1) have been reported from St. Charles County and thus could possibly occur in the Weldon Spring vicinity. One of these species, the pallid sturgeon *Scaphirhynchus albus*, has recently been proposed for endangered status under the Endangered Species Act. This species inhabits the Missouri River. Two C3 (former C2 candidate species) have also been reported from St. Charles County (Table 1) and could potentially be found in the Weldon Spring area. None of these species, however, are known to occur in the immediate vicinity of the SFMP Weldon Spring site.

The MDOC Heritage database contains detailed information on high quality natural areas and on state-listed rare, threatened, and endangered species, including detailed location and status information. The MDOC has identified 17 state endangered and 17 state rare species from St. Charles County. Eight additional species considered by the state to be of special concern are also reported from the county (Table 1). However, except for the bald eagle and the sicklefin and sturgeon chubs, only two state-listed rare or endangered species and one state species of concern are known to occur in the immediate vicinity of the Weldon Spring site.

The Cooper's hawk *Accipiter cooperii*, a state endangered species, is reported to nest in the Weldon Spring Wildlife Area. This species nests in large trees greater than 7 m in height (Bent 1937; Bull 1974), and trees of this size are found in the quarry area. The wood frog *Rana sylvatica* is classified by the state as rare and is known to occur and breed at the Weldon Spring Wildlife Area (Saladin 1989). In Missouri, this species is generally associated with wooded hillsides and breeds in small fishless woodland ponds and pools (Johnson 1987). The quarry pond may provide suitable breeding habitat for this species.

The sedge wren *Cistothorus platensis*, a species on the state's watch list, has been reported at the Weldon Spring Wildlife Area from old-field habitat in the vicinity of the chemical plant. No legal status is associated with this listing; watch list status is given to species of possible concern for which the MDOC is seeking further information.

A search of the MDOC Heritage database identified several high quality natural communities in the area of the Weldon Spring site. A mesic forest/dry-mesic chert forest of approximately 125 ac and containing good, old growth occurs in the Weldon Spring Wildlife Area southeast of the chemical plant. Approximately 81 ac of this forest community lie within the Weldon Spring Natural Area, which is an old-growth mesic forest. In addition, high quality dry chert forest and chert savannah are located in the Weldon Spring Wildlife Area

Table 1. Threatened, Endangered, or Special Concern Species Reported from St. Charles County, MO and Potentially Occurring in the SFMP Weldon Spring Site Area.

	Status	
Species	**Federal[a]**	**State[b]**
Plants		
Starwort (variety)	C2	Endangered
Forbes saxifrage	C3	Watch list
Rose turtlehead	C3	Endangered
Arrow arum		Rare
Star duckweed		Rare
Fish		
Pallid sturgeon	C2	Endangered
Pugnose minnow		Endangered
Sturgeon chub	C2	Rare
Sicklefin chub	C2	Rare
Alligator gar		Rare
Brown bullhead		Rare
Alabama shad		Rare
Starhead topminnow		Watch list
Western sand darter		Watch list
Reptiles and Amphibians		
Western fox snake		Endangered
Eastern massasauga rattlesnake		Endangered
Western smooth green snake		Endangered
Wood frog		Rare
Northern crawfish frog		Watch list
Birds		
Bald eagle	Endangered	Endangered
Peregrine falcon	Endangered	Endangered
Least tern	C2	Endangered
Cooper's hawk		Endangered
Northern harrier		Endangered
Sharp-shinned hawk		Endangered
Osprey		Endangered
Barn owl		Endangered
Double-crested cormorant		Endangered
Snowy egret		Endangered
Bachman's sparrow		Endangered
American bittern		Rare
Yellow-headed blackbird		Rare
Red-shouldered hawk		Rare
Black-crowned night heron		Rare
Little blue heron		Rare
Mississippi kite		Rare
Upland sandpiper		Rare
Henslow's sparrow		Rare
Sedge wren		Watch list
Mammals		
Long-tailed weasel		Rare

Sources: Personal communication from D. F. Dickneite, Environmental Administrator, Missouri Department of Conservation, to I. Hlohowskyj, Argonne National Laboratory, with enclosure, August 24, 1988. Personal communication from E. P. Gaines, Data Manager, Missouri Department of Conservation, to I. Hlohowskyj, Argonne National Laboratory, with enclosure, September 8, 1988.

Table 1 (cont'd.)

---

[a] C2 = federal candidate for listing as a threatened or endangered species. C3 = former federal candidate species.
[b] Special concern species include those classified by the state as rare or on the watch list. Watch list = species of possible concern for which the MDOC is seeking further information; this listing does not imply that these species are imperiled. Undetermined = possibly rare or endangered, but insufficient information is available to determine the proper status.

northwest of the quarry area. These communities contain old-growth vegetation, and the dominant trees (primarily oak) often exceed 50-cm diameter breast height. The chert savannah community, which contains old-growth black and post oak and some unusual plants, is essentially undisturbed and has been classified as rare by the MDOC.

DISCUSSION

Actions and activities undertaken by federal agencies, including remedial activities at hazardous waste sites, can result in a wide range of environmental effects, and CERCLA, NEPA, and the CEQ regulations require considerations of potential impacts that could result from the implementation of a specific action. To estimate the type and magnitude of (ecological) impact that could result from the implementation of a proposed or alternative action, sufficient baseline data of good quality must be available for the impact analyses.

With the exception of listed species and unique natural areas, available information on "typical" species, communities, and ecosystems is generally lacking. However, these are the ecological resources that could be most affected by the implementation of a proposed action, and it is in this regard that the adequacy of the available site-specific baseline ecological data used in many ecological assessments fall short. It should be noted that these shortcomings are not unique to DOE hazardous waste sites and are evident in many NEPA documents, particularly in those for which proposed actions are sited in or near relatively natural areas.

In the present study, relatively detailed data were available for threatened, endangered, and rare species and on high quality natural communities in the vicinity of the quarry and chemical plant areas. These data were provided by the USFWS and the MDOC Heritage database. We were fortunate that the quality of the MDOC Heritage database was so high; comparable databases for other portions of the country may be lacking. The Heritage database has only recently been completed, and if the assessment at the Weldon Spring site had been conducted 2 years earlier, much of the information on listed species and high quality natural areas would have been unavailable.

The MDOC Heritage database was particularly detailed, providing specific location data for the listed species and high quality communities in the area. Life history information for many of these species, obtained from literature sources such as Pflieger (1975) and Johnson (1987), provided additional material useful in determining the potential for some of these species, such as the sicklefin and sturgeon chubs, to occur in the project area. Because these species and communities are uncommon and have been provided some degree of protective status, numerical abundance information is not as critical for estimating potential impacts to these resources as are presence/absence and distribution data. If these resources were found to occur in the immediate area of the site and potential impacts to these resources were identified, then additional, detailed quantitative ecological data would be collected and analyzed.

No National Wetlands Inventory maps for the area were available from the USFWS, and no wetlands were identified from the Weldon Spring location by the MDOC Heritage database. Given the level of detail that exists within the Heritage database for high quality natural areas, important wetlands, if present in the area, probably would have been identified by the MDOC. The absence of wetlands information from this database thus suggests that no major wetlands occur in the area of the Weldon Spring site.

In contrast to the information on listed species and high quality natural areas, very little quantitative data were available for the nonendangered species and nonunique ecosystems at the SFMP Weldon Spring site. Available data for these resources consisted almost exclusively of qualitative presence/absence species lists for a very broad geographical area, St. Charles County. Little or no data on habitation or use of the quarry and chemical plant areas by individual species were available, and basic measures of ecosystem condition or quality are lacking for the site.

CONCLUSIONS

The collection of ecological data should be afforded the same level of importance and effort as occurs with the collection of baseline data on groundwater and surface water resources, air quality, noise, and chemical and radiological contaminants. However, several factors have acted to limit the quality of ecological assessments at hazardous waste sites. The primary concerns at these sites, driven by CERCLA, are for human health (i.e., potential exposure to contaminants through inhalation, ingestion, or other direct contact). As a consequence, ecological concerns are often assigned secondary importance. Recent guidance by the EPA (U.S. EPA 1989a) indicates that ecological assessments at Superfund sites under CERCLA will be expected to include specific, detailed, quantitative data. This information would then be available for use during the preparation of NEPA documentation.

A variety of sources exist that describe techniques and methods for the collection and analysis of baseline ecological data. Among the more commonly used metrics for estimating the condition of populations and ecosystems are measures of species diversity (Margalef 1958; Wilhm and Dorris 1968; Wilson and Bossert

1971; Krebs 1978), measures of faunal similarity (Pinkham and Pearson 1976; Matthews 1986), and measures of community structure and persistence (Simpson 1949; Krebs 1978; Winner et al. 1980; Matthews 1986). In addition, the ecological literature contains a variety of metrics that have been developed for estimating the relative condition of ecological communities with respect to anticipated, observed, or perceived environmental impacts. Examples of these measures include the index of biotic integrity (Karr 1981; Karr et al. 1986), the Hilsenhoff biotic index (Hilsenhoff 1987), and the community degradation index developed by Ramm (1988). Several ecological assessment guidance documents have recently been published by the EPA (U.S. EPA 1987, 1989a, 1989b) and the Bureau of Land Management (U.S. DOI 1986). A more quantitative approach to ecological assessments incorporating some of these ecological metrics and assessment protocols will aid in the characterization of site-specific biotic components and in the identification of the form and severity of potential impacts and thus strengthen the environmental documentation and decisionmaking under CERCLA and NEPA.

ACKNOWLEDGMENTS

Work supported by the U.S. Department of Energy, Assistant Secretary for Nuclear Energy, under Contract W-31-109-Eng-38.

REFERENCES

Bailey, R. G. 1976. Ecoregions of the United States (Map), U.S. Forest Service, Intermountain Region, Ogden, UT.

Bent, A. C. 1937. Life Histories of North American Birds of Prey, Part 1, Smithsonian Institution, U.S. National Museum Bulletin 167.

Bull, J. 1974. *Birds of New York State.* Comstock Publishing Associates, Cornell University Press, Ithaca, NY.

Council on Environmental Quality (CEQ). 1981. Forty most asked questions concerning CEQ's National Environmental Policy Act Regulations. *Fed. Regist.* 46(55):18026–18038.

Gaines, E. P. 1988. Personal communication from E. P. Gaines, Data Manager, Missouri Department of Conservation, to I. Hlohowskyj, Argonne National Laboratory with enclosure (Sept. 8, 1988).

Hilsenhoff, W. L. 1987. An improved biotic index of organic stream pollution. *Great Lakes Entomol.* 20:31–39.

Johnson, T. R. 1987. *The Amphibians and Reptiles of Missouri.* Missouri Department of Conservation, Jefferson City, MO.

Karr, J. R. 1981. Assessment of biotic integrity using fish communities. *Fisheries* 6(6):21–27.

Karr, J. R., K. D. Fausch, P. L. Angermeier, P. R. Yant, and I. J. Schlosser. 1986. Assessing Biological Integrity in Running Waters: A Method and Its Rationale. Special Publ. 5, Illinois Natural History Survey, Springfield, IL.

Krebs, C. J. 1978. *Ecology: The Experimental Analysis of Distribution and Abundance.* 2nd ed. Harper and Row, New York.

Margalef, D. R. 1958. Information theory in ecology. *Gen. Syst.* 3:36–47.

Matthews, W. J. 1986. Fish faunal structure in an Ozark stream: Stability, persistence, and a catastrophic flood. *Copeia* 1986(2):388–397.

Missouri Botanical Garden. 1975. An Introduction to the Biological Systems of the St. Louis Area. Prepared for the East-West Gateway Coordination Council under contract from St. Louis District, Corps of Engineers, the Metro Study, Vol. 1.

Missouri Department of Conservation (MDOC). 1978. August A. Busch Memorial Wildlife Area Auto Tour. Missouri Department of Conservation, Jefferson City, MO.

Missouri Department of Conservation (MDOC). 1985. Weldon Spring Wildlife Area (Map and Information Pamphlet). LDM/8-85 FD. Missouri Department of Conservation, Jefferson City, MO.

MK-Ferguson Company and Jacobs Engineering Group, Inc. 1988. Radiological and Chemical Uptake by Edible Portions of Selected Biota at the Weldon Spring Site. Prepared for the U.S. Department of Energy, Oak Ridge Operations, Weldon Spring Site Remedial Action Project Office, St. Charles, MO.

Pflieger, W. L. 1975. *The Fishes of Missouri.* Missouri Department of Conservation, Columbia, MO.

Pinkham, C. F. A., and J. B. Pearson. 1976. Applications of a new coefficient of similarity to pollution surveys. *J. Water Pollut. Control Fed.* 48:717–723.

Ramm, A. E. 1988. The community degradation index: A new method for assessing the deterioration of aquatic habitats. *Water Res.* 22(3):293–301.

RETA. 1978. Phase III Report: Assessment of Disposition Alternatives, Weldon Spring Chemical Plant, St. Charles County, St. Louis, MO. Ryckman/Edgerley/Tomlinson and Associates, Data Requirement A00B, Contract No. DAAK 11-77-L-0026. Prepared for the Department of the Army, Aberdeen Proving Ground, MD.

Saladin, T. 1989. Personal communication from T. Saladin, Naturalist, Weldon Spring Wildlife Area, to I. Hlohowskyj, Argonne National Laboratory (Jan. 12, 1989).

Simpson, E. H. 1949. Measurement of diversity. *Nature* 163:688–690.

Thom, R. H., and J. H. Wilson. 1980. The natural divisions of Missouri. *Trans. Mo. Acad. Sci.* 14:9–23.

U.S. Department of Energy (U.S. DOE). 1987. Draft Environmental Impact Statement, Remedial Action at the Weldon Spring Site. DOE/EIS-0117D. U.S. Department of Energy, Office of Remedial Action and Waste Technology, Washington, D.C.

U.S. Department of the Interior (U.S. DOI). 1986. Inventory and Monitoring of Wildlife Habitat. A. Y. Cooperrider, R. J. Boyd, and H. R. Stuart (eds.), U.S. Department of the Interior, Bureau of Land Management, Service Center, Denver, CO.

U.S. Environmental Protection Agency (U.S. EPA). 1987. A Compendium of Superfund Field Operations Methods. EPA/540/P-87/100. U.S. Environmental Protection Agency, Office of Emergency and Remedial Response, Washington, D.C.

U.S. Environmental Protection Agency (U.S. EPA). 1989a. Risk Assessment Guidance for Superfund — Environmental Evaluation Manual, Interim Final. EPA/540/1-89/001A. U.S. Environmental Protection Agency, Office of Emergency and Remedial Response, Washington, D.C.

U.S. Environmental Protection Agency (U.S. EPA). 1989b. Rapid Bioassessment Protocols for Use in Streams and Rivers. Benthic Macroinvertebrates and Fish. EPA/444/4-89-001. U.S. Environmental Protection Agency, Office of Water (WH-553), Assessment and Watershed Protection Division, Washington, D.C.

Wilhm, J. L., and T. C. Dorris. 1968. Biological parameters for water quality criteria. *Bioscience* 18:472–481.

Wilson, E. O., and W. H. Bossert. 1971. *A Primer of Population Biology*. Sinauer Associates, Inc., Publishers, Sunderland, MA.

Winner, R. W., M. W. Boesel, and M. P. Farrell. 1980. Insect community structure as an index of heavy-metal pollution in lotic ecosystems. *Can. J. Fish. Aquat. Sci.* 37:647–655.

Fish Habitat Evaluation Models in Environmental Assessments

C. B. Stalnaker, U.S. Fish and Wildlife Service, Fort Collins, CO

INTRODUCTION

Instream water uses include such traditional uses of streamflow as hydroelectric power generation, navigation, and wasteload assimilation (water quality). During the last 20 years, streamflow requirements for protecting fish, waterfowl, riparian habitat, recreation, and for channel integrity have become legally recognized instream uses of water. Legislation in several states provides similar legal recognition for instream-flow uses as for offstream uses (McDonnell et al. 1989).

An instream-flow requirement is that volume of streamflow necessary to maintain a specific instream use or group of instream uses at an acceptable level. Instream-flow requirements are stream specific, often tailored to accommodate differences in the hydrologic and morphologic characteristics of individual stream reaches. The most desirable instream-flow requirement is one which simultaneously satisfies several instream uses and still delivers water for offstream uses downstream.

INSTREAM FLOW FOR FISH HABITAT

Water administrators in virtually every state are faced with instream-flow issues. Many states are developing policies and legislation for resolving instream vs out-of-stream conflicts. Should the total flow of every stream be reported as instream use? Should the lowest amount required to sustain a minimal resident fish population be reported? Should only the officially protected minimum streamflow be reported as instream water use? It is important to recognize that minimum streamflow standards established by policy or by water rights are not necessarily the same thing as the flow necessary to maintain a healthy and highly productive fishery. The problem of quantifying the amount of water flowing in a stream channel is basic to any discussion of instream water use (Stalnaker 1990).

There are several instream-flow methods that, to varying degrees, are capable of either (1) determining a base streamflow necessary to maintain fish habitat at some acceptable level or (2) predicting the response of fish habitat to naturally occurring or human-induced changes in streamflow, stream temperature, sediment transport, or water chemistry. These methods have been referred to as standard setting methods and incremental methods, respectively, by Trihey and Stalnaker (1985). Only incremental methods are appropriate for environmental assessment under the concepts of the National Environmental Policy Act (NEPA) because they quantify the gains and loss of habitat as a function of flow.

Instream-flow assessment methods were developed to resolve conflicts resulting from excessive withdrawal of streamflow for offstream uses. Major stimuli in the evolution of instream-flow methods have been western water law, national water development policy, and, more recently, federal environmental policy. A discussion of the evolution of institutional awareness and the quantification of instream-flow needs is presented by Stalnaker (1982) and McKinney and Taylor (1989).

The following three questions posed by water planners and administrators are just as pertinent to the selection of instream assessment techniques today as they were in the early 1970s:

1. How much water is needed to maintain the fishery?
2. What happens if that much water (or a particular release schedule) cannot be provided?
3. How many fishes are gained or lost with different system operation criteria for the river releases?

PRE-NEPA STREAM ISSUES

During the 1950s and 1960s, a principal concern of federal and state agencies responsible for protecting and maintaining fisheries was determining some way to measure the effects of large irrigation and hydroelectric power projects on fish populations in streams and rivers throughout the western United States. Direct diversion of streamflow into large irrigation canals markedly decreased streamflow and, at times, completely dewatered entire reaches of streams. The widespread occurrence of such perturbations resulted in a search for methods to protect streamflow as a flow rate that would ensure fish survival throughout the year. Thus, the phrase *minimum flow* was coined (Orsborn and Allman 1976). As applied, the minimum flow was a constraint on reservoir operation and storage during runoff and on irrigation diversion during the low-flow season. Various methods were developed during the 1960s and 1970s to identify this minimum flow (Wesche and Rechard 1980; Loar and Sale 1981). These methods were meant to address the question: How much water is needed to maintain the fishery?

Within the institutional framework of the appropriation doctrine and philosophy of western water administrators, minimum flows were reserved (exempted from diversion) by the state to maintain fish populations at some desired level. A variety of minimum-flow methods were developed during the early 1970s for reserving streamflow based on analysis of streamflow records or limited onsite evaluations (Stalnaker and Arnette 1976; Stalnaker 1982).

One of the first methods was the use of hydrological statistics. Although widely touted by water planners because of their ease of use, stream-flow statistics are not well-suited to fish habitat analysis. Such statistics commonly refer to frequency of low-flow events. Although originally introduced to protect and improve water quality, the 7-day Q10 streamflow statistic (7-day, 10-year low flow) became one method used to identify the minimum flow. Several states have used hydrological statistics as a means for setting streamflow standards. Tennessee, for example, has considered the 3-day, 20-year low-flow statistic as a minimum flow. In most circumstances, the proposed minimum-flow statistics result in very low flows indeed.

In practice, the degree of protection afforded fish populations by application of minimum-flow standards varies considerably both within and among the states (Lamb and Meshorer 1983). Minimum streamflow standards should be thought of as policy choices rather than fish population or habitat assessment procedures. As western water projects were operated to maintain an artificially small minimum streamflow, it soon became apparent that downstream fish populations were being adversely affected (Raley et al. 1988). Nevertheless, substantial progress for protecting instream habitat for fish and wildlife was achieved during the last two decades due to two primary reasons:

1. New environmental legislation emerged as a result of a heightened awareness by the public of the growing reduction of our ecosystems and the realization that only through legal protection would future generations be able to enjoy these instream values.
2. Stimulated by demands from the water planning community for quantitative documentation of instream-flow requirements, new techniques were devised that produced persuasive support for the aquatic biologist's recommendations.

The passage of NEPA and emphasis on multiple-use planning coupled with the general inability of minimum-flow standards to actually protect fish populations necessitated that trade-offs between offstream and instream uses be clearly identified and much better quantified. Allred (1976) clearly stated the viewpoint of the state water administrator when he wrote "If instream flow interests expect to compete with other uses for limited water supplies, they must be able to demonstrate with the same type of analysis and approach as other uses, the need for instream flows and the effect of not obtaining those flows." Such emphasis forced the fishery managers to start addressing these questions: What happens if the needed water is not provided and how many fish are gained or lost?

Table 1. Definitions of Methods Adapted from Trihey and Stalnaker.[a]

Instream-flow methods	Techniques to measure, describe, or predict the value of some variable assumed to be important to the general well-being of some instream use or user
Instream-flow methodology	A collection and integration of several methods (techniques) to assess instream flow arranged in an organized process for the purpose of (1) developing flow regimes for stated management objectives, (2) quantifying the effects of potential water management alternatives on instream resources, (3) developing mitigation for specific water management schemes, or (4) negotiating operating rules and flow releases for a reservoir
Standard setting methodology	Measurement and interpretive techniques designed to generate a flow value(s) that is intended to maintain the fishery at some acceptable level.
Incremental methodology	Measurement and interpretive techniques organized into a repeatable process by which (1) a fishery habitat/streamflow relationship and the hydrology of the stream are transformed into a baseline habitat time series, (2) proposed water management alternatives are quantified and compared with the baseline, and (3) project operating rules are negotiated; Items 2 and 3 are often iterative processes involving trade-offs among instream and out-of-stream uses

[a] Adapted from Trihey, E. W., and C. B. Stalnaker. 1985. Evolution and application of instream flow methodologies to small hydropower development: An overview of the issues. pp. 176–183. In F. W. Olson, R. G. White, and R. H. Hamre (eds.), *Proceedings of the Symposium on Small Hydropower and Fisheries.* American Fisheries Society, Bethesda, MD.

Since the mid-1970s, considerable emphasis has been placed on the development of instream-flow methods that describe the relation between streamflow and fish populations. With few exceptions, fish habitat is used as a surrogate for fish populations when expressing relations between fish populations and the hydrologic, hydraulic, structural, or water quality aspects of a stream. The common product of all incremental methods is a discrete relation between a fish habitat index or a fish population index and streamflow that can be used to evaluate the availability of fish habitat or well-being of fish populations at different instream discharges (Trihey and Stalnaker 1985). See Table 1 for definitions.

Important pioneers of the incremental-methods approach included Collings (1972), who used stream habitat mapping techniques and binary-depth, velocity, and substrate criteria to evaluate the effects of incremental changes in streamflow on salmon-spawning habitat in Washington. Waters (1976) introduced computer simulation and weighted-depth, velocity, and rearing-substrate/cover criteria to evaluate the response of rainbow trout habitat to streamflow variations in California. Single-transect methods, such as the U.S. Forest Service's R2 CROSS procedure, introduced the concept of hydraulic modeling and were soon followed by multiple-transect, water-surface profile (WSP) simulation models developed by the U.S. Army Corps of Engineers and the U.S. Bureau of Reclamation (Bovee and Milhous 1978). More recently, hydraulic models and physical-process models for stream temperature, river ice, and euphoric surface area have been applied as integral components of instream-flow studies (Wilson 1981; Harza-Ebasco Susitna Joint Venture 1985; Trihey and Associates, and Entrix, Inc. 1985; Trihey and Baldrige 1985).

INSTREAM-FLOW METHODS

Several excellent reviews and descriptions of instream-flow methods have been published and may help in selecting and applying an instream-flow method (Fraser 1975; Stalnaker and Arnette 1976; Wesche and Rechard 1980; Loar and Sale 1981; Estes 1984; Leonard et al. 1985; Camp, Dresser, and McKee 1986; Morhardt 1986; Scott and Shirvell 1987).

A variety of methods can be applied to assist with the identification of instream-flow requirements. In general, all commonly used instream-flow methods can be classified as either standard setting or incremental.

STANDARD SETTING METHODS

Because these methods are not suitable for environmental assessment under NEPA, only a few of the more commonly used methods are mentioned here for comparative purposes. See Stalnaker (1982) and Trihey and Stalnaker (1985) for further discussion of instream-flow standard setting methods.

As mentioned above, a variety of streamflow statistics have been used by various authors as criteria for identifying acceptable streamflow conditions for fish. Standard setting methods based on streamflow statistics, such as the New England aquatic base flow (ABF) (Larsen 1980) and the Tennant methods (Tennant 1976), are in common use today. In spite of the absence of corroborating biologic studies supporting the choice of the statistics (Morhardt 1986), they are generally accepted as reasonable and reproducible estimates of minimum-flow requirements for fish in stream systems similar to those for which these methods were originally developed.

Aquatic Base Flow

The ABF is considered adequate for the protection of fish during all periods of the year, unless additional releases are necessary for spawning and incubation (Larsen 1980; U.S. Fish and Wildlife Service 1981).

The ABF generally is applied as the minimum instantaneous discharge during normal runoff conditions. During low flows when streamflow into a reservoir may be less than the ABF, minimum releases from the reservoir equal to inflow are generally requested (Knapp 1980).

Streamflow Duration

Hydrologic statistics that describe probability of occurrence have been used in the western states for deriving instream-flow recommendations. The concept of streamflow duration in descriptive hydrology refers to the streamflow that is equaled or exceeded in a certain percentage of all streamflow measurements. When discussing low flows, the 90th percentile streamflow value (calculated from a data set of mean daily streamflow values) would be the streamflow expected to be equaled or exceeded 9 out of 10 days during the year.

Hoppe and Finnell (1970) used flow-duration analysis to determine streamflows for trout spawning, rearing, and gravel flushing, which correspond to the 80th, 40th, and 17th percentiles, respectively, on the flow-duration curve. The state of Iowa has used flow-duration analysis and the 84th percentile annual low-flow statistic as the basis for defining minimum streamflows in its instream-flow policy (Dougal 1979).

Fixed Percentage of Annual Flows

The best known of the fixed-percentage methods and perhaps of all standard setting methods is the Tennant method (Tennant 1975, 1976). The method involves calculating the mean annual flow rate for the stream reach of interest (Tennant 1976). Tennant's (1976) "recommended base flow regimen" includes flushing flows (200% of the mean annual flow), optimum flows for all instream uses (100–60% of the mean annual flow), and a gradation of lesser conditions for fish habitat ranging from excellent (60–40% of the mean annual flow) to severe degradation (less than 10% of the mean annual flow). Additional examples of fixed percentage rule-of-thumb approaches summarized from Stalnaker (1980b) are

- Median monthly flow values equal to 79–100% of the average flow for each month of record.
- Monthly minimum flows equal to the mean monthly flow (MMF) if the MMF <40% of the mean annual flow (MAF). If the MMF >40% MAF, monthly minimum flows equal 40% MAF. If 40% MAF ≤ MMF, monthly minimum flows equal 40% MAF.
- Single values of 60–100% of mean annual flow or 70–140% of the natural characteristic low or base flow.

The applicability of fixed percentages of the mean annual flow in the range of 30–60% for maintaining viable fisheries generally is supported by the results of other investigators (Wood and Whelan 1962; Nelson 1980; Orth and Maughn 1981).

A review of hydrologic methods and the development of flow equations specific to the James River basin in Virginia are presented in Leonard et al. (1985).

Channel Maintenance Flows

Tennant (1976) recommended that 200% of the mean annual discharge would provide flushing flows on regulated salmonid streams. The U.S. Forest Service (1986) developed an analytical approach for determining instream-flow requirements needed to maintain the integrity of the stream channel by flushing it of fine sediments and preventing the encroachment of vegetation into the active stream channel. This method is based on the premise that short-term bankfull discharge is a requisite for channel maintenance.

HYDRAULIC APPROACHES

Another class of approaches for determining the instream-flow recommendation for protecting fish habitat is to evaluate hydraulic and habitat conditions within the stream channel. Loar and Sale (1981) provide a detailed discussion of hydraulic-based instream-flow methods.

The U.S. Forest Service developed several reconnaissance and planning level hydraulic methods for determining instream-flow requirements for fish habitat (Herrington and Dunham 1967; Russell and Mulvaney 1973; Collotzi 1975; Dunham and Collotzi 1975; Swank and Phillips 1976; Bartschi 1976; Parsons et al. 1981). These methods were developed for wadable trout streams in the mountainous western states where ungaged streams are much more numerous than gaged streams. In all the Forest Service's hydraulic methods, habitat types (riffles, runs, pools, spawning areas, or passage barriers) are described by establishing one or more transects across the stream then obtaining channel-geometry and hydraulic measurements along each transect for one or more stream discharges.

Flushing-flow methods have been introduced by Milhous et al. (1986), Wilson (1981), and Trihey (1983) in which site-specific hydraulic and channel-geometry data are used to identify the range of streamflows capable of flushing sand and finer particles from streambeds consisting primarily of gravel. These methods place more emphasis on ensuring the permanence (nontransport) of gravel for spawning than on preventing encroachment of vegetation or otherwise maintaining the flood capacity of the active channel. A review of flushing-flow methods is provided by Reiser et al. (1985).

Staff-Gage Analysis

The method for using a staff gage to determine instream-flow requirements is similar to that for developing a stage-discharge curve. However, a biologic rather than hydrologic emphasis is placed on the staff-gage location, and a habitat-discharge curve rather than stage-discharge curve is derived from the onsite data.

Critical habitat conditions are determined by application of minimum-depth criterion or identifying the "inflection point" in the wetted-perimeter or top-width gage-height curve. The instream-flow requirement most commonly chosen is that which corresponds to the inflection point (Nelson 1984; Spear 1985). Normally, the critical areas are assumed to be the shallowest riffle in the shallowest reach of the stream or stream segment being investigated.

The criteria used in Colorado to select the critical habitat condition for fisheries in streams less than 20 ft wide are the streamflow that wets 50% of the stream bottom and that maintains a mean velocity of 1–1.5 ft/sec and minimum depths between 0.2–0.4 ft across riffles (Wesche and Rechard 1980). In applications elsewhere, the instream-flow requirement for fisheries may be chosen using habitat reduction or habitat retention criteria, such as 75% retention as measured from an optimum or some arbitrary reference habitat condition.

Hydraulic Modeling

This method has replaced repetitive measurements in recent years. Starting in the mid-1970s, models developed by the U.S. Army Corps of Engineers' Hydrologic Engineering Center (the HEC-2 model) and the U.S. Bureau of Reclamation (the PSEUDO model) have been applied in many instream-flow studies (White and Cochnauer 1975; Dooley 1976; Elser 1976; White 1976; Workman 1976; Brusven and Trihey 1978). The manner in which protection for an instream-flow habitat is determined with the hydraulic simulation models is basically the same as with the critical area and wetted perimeter methods.

The advantages of using hydraulic simulation models rather than the critical area method for predicting hydraulic characteristics at unmeasured streamflows is that hydraulic simulation models provide a more theoretical basis for analyzing energy loss, are readily applied to large streams, and can quickly perform the interpretive computations associated with multiple-transect study sites. However, steady flow is still required throughout the range of streamflow to be modeled and the same average hydraulic characteristics (depth, velocity, top width, hydraulic radius, wetted perimeter, and flow area) are computed. Considerable work has been done since the late 1970s to refine hydraulic simulation models that are more applicable to fisheries interests (Bovee and Milhous 1978; Milhous et al. 1989; Trihey and Baldrige 1985). Stalnaker et al. (1989) provide a review of hydrology and hydraulics as applied to fish habitat analyses in large rivers.

INCREMENTAL METHODS

Whereas standard setting policies are intended to provide a single streamflow value for habitat retention or protection, incremental methods are intended to determine habitat-discharge relations of varying degrees of complexity. Incremental methods range in complexity from those that define the relation between usable width or wetted perimeter and streamflow to those that use multiple physical habitat conditions as a surrogate for fish populations. Incremental methods can also describe the availability and quality of physical habitat conditions in terms of stream-channel structure, hydraulics, and water temperature.

Instream-flow methods that simply provide a habitat-discharge relation provide a transition from single-value, standard setting problems to impact analyses based on physical-process and habitat modeling. Habitat modeling methods capable of displaying the time-series response of fish habitats to natural or human-induced changes in streamflow, water temperature, and/or channel geometry represent the state of the art for instream-flow-related habitat assessments. The following is a description of the incremental instream-flow methods more commonly used for environmental assessments over the last 20 years.

USABLE WIDTH

The terms *usable width* and *weighted usable width* originated from instream-flow methods developed for cold water, salmonid streams in Oregon (Sams and Pearson 1963; Thompson 1972, 1974). A streamflow-dependent habitat index is calculated as the percentage of the total stream width that is *usable*, as determined by species/life-function criteria, and site-specific channel-geometry and hydraulic measurements. The usable-width method initially employed single transects placed at critical habitat areas (e.g., spawning beds), and the top-width discharge relation was obtained from repetitive onsite measurements at several discharges. More recently, hydraulic simulation techniques have been used to calculate top-width values, which greatly decreased the number of required site visits while increasing the number of transects that may be used to describe the critical habitat.

The U.S. Forest Service can be credited with pioneering the development of a method for determining optimum streamflow for fish habitat from evaluation of top-width values in a variety of habitat types (Swank and Phillips 1976). The method uses multiple-transect study sites in representative spawning, rearing, and food-producing areas at which depths and velocities are measured across the stream for three or more stream stages. Usable top width for each life function is determined at each stream stage using species-specific criteria and then plotted as a streamflow-dependent variable. The optimum streamflow is defined as the flow that simultaneously provides the most usable top width for all three life functions. The instream-flow recommendation usually is selected through visual evaluation of the curve sets aimed toward comparison of percent optimum habitat from a proposed water development scheme. However, less subjective, arithmetic methods also are used to identify the streamflow that provides the most usable top width.

Thompson (1972) applied the usable top-width concept when determining instream-flow requirements for migrating adult salmonids. Potential migration barriers (shallow riffles or velocity chutes) are identified through onsite reconnaissance. Transects are then established across the stream at the location of the most critical passages. Onsite measurements are made at each transect to determine depths and velocities across the stream for a variety of stream stages. Usable top-width values for fish passages are then determined using species-specific criteria.

The instream-flow requirement for fish passage was determined as the flow that simultaneously provides a minimum of 25% of the channel meeting the species criteria for usable depth and velocity at each transect and along a continuous portion of a longitudinal line across the passage reach accounting for at least 10% of the top width.

The usable-width method also has been used in Oregon and Washington to determine instream-flow requirements for optimal spawning habitats for salmon. The optimum spawning flow is defined as the flow that provides suitable hydraulic conditions (with respect to the selected depth and velocity criteria) over the largest area of gravel at critical spawning areas. The streamflow that provides suitable hydraulic conditions over 80% of the available gravel at the optimum flow is considered to be the minimum streamflow for spawning. Orsborn (1981) has used watershed and channel characteristics to determine optimum flow estimates for spawning salmon.

The only difference between the usable-width and weighted-usable-width methods is the criteria for the usable top width. The usable-width method employs binary criteria, such as from Thompson (1972) or Swank and Phillips (1976), to define usable and nonusable subsegments of the channel width. Weighted usable width is computed from depth and velocity criteria similar to those introduced by Waters (1976) that express the relative suitability of individual depths and velocities within a range of usable hydraulic conditions. If the depth or velocity conditions are considered optimum for the species/life function being evaluated, the weighting factors for these hydraulic variables are given values of 1.0 (optimum). If one or both variables are considered nonusable, their weighting factors are equal to 0. Techniques used to determine weighting factors for depth, velocity, substrate, and cover have ranged from quantitative expressions based on the opinion of fisheries experts to statistical analyses of fish habitat use observations (Baldrige and Amos 1981; Bovee 1986; Bovee and Zuboy 1988).

WYOMING TROUT COVER METHOD

Wesche (1972, 1974, 1980) developed a method for assessing the availability of cover for adult trout at different instream discharges. A cover-rating equation was developed from analysis of site-specific cover conditions used by about 1100 trout in small Wyoming streams. Onsite data collection emphasized quantification of overhanging streambank cover and instream cover using multiple-channel cross-sections and longitudinal streambank transects.

A cover-rating index for the investigated stream reach is computed for three or more stream stages within the range of the evaluated stream stages (discharges). These indices are then plotted against average daily streamflow or percentage of the average monthly or average annual flow to identify the relation between cover availability and streamflow. The instream-flow assessment index is then selected by identifying the optimum point on the cover vs streamflow curve or by applying some predetermined cover-retention criteria, such as 75% of the cover index rating or the percentage that corresponds with bankfull discharge. Impacts are assessed by comparing deviations from the optimum with and without a particular water development.

Preferred Area

The Washington Department of Fisheries in cooperation with the U.S. Geological Survey used onsite surface area mapping techniques to quantify the potentially available streambed area for spawning salmon at different stream discharges (Collings 1972, 1974; Swift 1976, 1979). This approach commonly is referred to as the Washington method and provides the basis for derivation of the Washington toe-width method. Biologic criteria by Collings (1972) consisted of a range of usable depths, velocities, and substrate particle sizes. Onsite activities involved obtaining depth and velocity measurements over suitable gravels along four or more transects at a potential spawning site and preparing an isopleth map of the stream reach to show the distribution of depths and velocities with respect to spawning gravels at a monitored discharge. Areas with suitable substrate that satisfy the velocity and depth criteria are designated as potentially spawnable parts of the stream, and their surface area is determined. This procedure was repeated for five to ten different stream stages to develop a spawnable area vs discharge curve for the range of the evaluated stream stage. Hydraulic modeling was not originally used. All depth and velocity information was obtained directly from onsite measurements. The preferred spawning flow is defined as the flow that provides the maximum spawnable area for the range of the investigated stream stages. The instream-flow recommendation maintains 75% of the maximum spawning area.

A similar method was developed by the Pacific Gas and Electric Co. for quantifying the availability of trout habitat in California streams (Waters 1976). Like the Washington method, two-dimensional analysis of the wetted stream surface is performed, but weighted rather than binary criteria are used to define the surface area of the stream that has acceptable depths and velocities as a function of streamflow. The habitat index introduced by Waters (1976) is referred to as net preferred habitat (NPH) and is calculated as:

$$NPH = \sum_{i=1}^{n} v_i \times d_i \times a_i$$

where v_i = velocity weighting factor between 0 and 1
d_i = depth weighting factor between 0 and 1
a_i = bottom surface area of ith element of the stream reach
n = number of elements in the stream reach

The NPH index is calculated for several instream stages within the range of instream stage being evaluated and is plotted against stream stage. As with the Washington method, hydraulic parameters were not originally modeled. They were determined by direct measurement at each stream stage. However, Waters (1976) introduced computer calculations of the NPH index. With the addition of a weighting factor for substrate or cover and hydraulic simulation modeling to predict depths and velocities of unmeasured streamflows, the NPH method became the California Department of Fish and Game method for assessing effects to potential salmonid habitat (Smith and Aceituno 1987).

INSTREAM-FLOW INCREMENTAL METHODOLOGY

During the latter part of the 1970s and early 1980s, the U.S. Fish and Wildlife Service developed a series of computerized models for calculating usable habitat area as a function of streamflow, stream-channel structure, and water temperature quality, which, in conjunction with hydrologic time series analysis, is known as the Instream-Flow Incremental Methodology (IFIM) (Stalnaker 1978, 1980a; Bovee 1982). The IFIM provides a documented state-of-the-art computational procedure for determining the effects of streamflow or stream-channel alterations on the availability or quality of fish habitat (Orth 1987; Gore and Nestler 1988). The IFIM incorporates many hydrologic, hydraulic, and habitat-discharge aspects of earlier developed methods, but attempts to quantify the temporal and spatial aspects of fish habitats for all life stages. *With* and *without* water management scenarios are compared for identifying habitat impacts.

A library of computer programs, SNTEMP (Stream Network Temperature model), PHABSIM (Physical Habitat Simulation system), and TSLIB (Time Series Library), has been developed to facilitate the analysis (Theurer et al. 1984; Milhous et al. 1989). Guidelines are available for application of the IFIM (Bovee 1982) and onsite data collection (Bovee and Milhous 1978; Trihey and Wegner 1980). The approach combines several factors: (1) multiple-transect data for a representative and/or critical river reach, (2) channel-geometry analysis, (3) reach-specific hydrologic time series, (4) hydraulic simulation modeling to predict mean column and focal point velocity and flow depth at unmeasured streamflows, (5) water-temperature modeling, and (6) species-specific habitat suitability criteria (Bovee 1986). A streamflow-dependent microhabitat index called weighted usable area (WUA) and a linear macrohabitat measure are computed for each life stage of the evaluated species. In conjunction with time-series streamflow data, a plot of total usable area reflects habitat availability and quality by seasonal or annual variations in streamflow. The instream-flow regime is the monthly or weekly series of maximum and minimum streamflows that provide aquatic habitat conditions that are most compatible with the fishery management goals of the user. This methodology is most suitable for designing reservoir operation criteria (rule curves and addressing water management issues) (Stalnaker 1980a, 1982).

Pre-NEPA methods for evaluating instream habitat were developed to set streamflow standards and are still widely used to set water allocation policy at the state level. Post-NEPA environmental assessments of streamflows focus largely on fish habitat, but simulation techniques improved the ability to compare habitat suitability at many unmeasured discharges. Thus, comparisons of alter-

Table 2. Comparison of Selected Methods for Identifying Instream-Flow Requirements for Fish Habitat.

Method	Application Level	Precision Reproducibility	Effort Cost	Data
		Standard Setting (Policy Statements)		
Base flow	Setting broad minimums	Moderate	Low	Minimal
Streamflow duration	Setting broad minimums	Moderate	Low	Minimal/ moderate
Fixed percentage	Setting stream-reach minimums	Moderate/ high	Low	Minimal/ moderate
Channel maintenance and flushing flow	Setting stream-reach high flows	Moderate	Low/ moderate	Minimal/ moderate
Staff-gage analysis	Setting stream-reach optimums and minimums	Low/ moderate	Low/ moderate	Low/ moderate
Critical area	Setting stream-reach optimums and minimums	Low/ moderate	Low/ moderate	Low/ moderate
Hydraulic modeling	Setting stream-reach optimums and minimums	Moderate/ high	Moderate	Low/ moderate
		Incremental (Impact Analyses)		
Usable width	Setting habitat optimum	Moderate	Moderate	Low/ moderate
Wyoming trout cover	Setting habitat optimum	Moderate	Moderate	Moderate
Preferred area	Setting habitat optimum	Moderate	Moderate	Moderate
Instream flow incremental methodology	Setting stream-reach flow regimes	Moderate/ high	Moderate/ high	Moderate/ high

native water management scenarios with incremental habitat methods and time-series simulations have become the common practice for impact analyses during the 1980s. Table 2 briefly compares the more common standard setting and incremental methods based on their reproducibility, cost, and data requirements.

In summary, *standard setting* is most appropriate for:

- setting policy flows for protecting the existing instream resource
- identifying potential water use conflicts in state water plans
- state water allocation permits or reservations
- identifying target flow for use during project feasibility studies

Incremental methods are most appropriate for:

- analyzing temporal and spatial habitat to identify limiting flow conditions
- fine tuning a resource maintenance objective (maximum utilization of available water)

- comparing water management alternatives
- avoiding or minimizing flow-related impacts
- comparing mitigation alternatives

Instream-flow assessment methods that attempt to evaluate fish habitat by the hydraulic, structural, and water quality aspects of the stream environment were not known before the 1980s. Streamflow time series, project operating rules, and species habitat response curves are necessary intermediate products of this phase.

Perceptions in Habitat Modeling for Instream-Flow Negotiation

The use of fish habitat models for negotiating instream-flow regimes has gradually shifted from habitat protection to addressing the many questions that accompany water developments (i.e., responses by populations of aquatic organisms to perturbations). Habitat models were initially developed after the era of large federal reservoir construction in the western United States in the 1950s and 1960s. The fishery management goal to protect and maintain existing riverine resources led to the application of standard setting methods. With the passage of NEPA, more emphasis was placed on the protection of flowing stream resources. Particularly in the northwest and the northeast, standard setting methods were widely applied and approved. In the 1970s and early 1980s, standard setting methods were successfully applied as the design criterion for new small hydro development. Many projects were not built simply because they were not deemed economically feasible after considering the constraints imposed by the instream-flow standards required by the agencies.

Questions were raised quite often by hydro development interests and water administrators who were charged with balancing the competing uses of water. When significant alterations to streamflow, stream temperature, water chemistry, or channel realignment activities were proposed, it became obvious that the flow standards would be violated in many cases. Incremental methods were developed as part of environmental impact studies to analyze trade-offs and alternatives in an attempt to quantify the habitat gains and losses resulting from such perturbations. Incremental methods are still most commonly used to evaluate dredge and fill permits and water development projects with endangered species present.

Starting in the late 1980s and projected through the end of the 1990s, a new era in water resource decisionmaking is forcing fishery agencies to address gains and losses in fish numbers. Nearly 300 hydroelectric facilities require license renewal in the next decade. Such licensing endeavors carry the goal of the applicant to continue to operate the project as in the past and, in many cases, to propose peaking operations, if not already present, whereas the goal of resource agencies is to not only maintain the existing fishery but, where possible, to enhance the quality of the habitat through negotiating new streamflow release schedules. During the 1990s, operating procedures at several existing federal projects are also being reevaluated. The present social emphasis on multiple uses of flowing waters requires a move from the simpler era of protecting instream-flow environments, identifying potential harm, and eliminating undesirable water development alternatives to the more complex situation in U.S. water management requiring the balancing of competing uses. Single-purpose water projects are no longer in vogue, and multiple use in conjunction with river system management is leading the administrators to ask for quantitative analysis of gains and losses to the various water use sectors resulting from various water development and management scenarios. Recent emphasis on fish population modeling along with incremental habitat analysis are being used to project changes in important fish species numbers as a consequence of streamflow management. Such studies have predominately been used for trout and salmon streams in the West and most notably in California (Bio Systems Analysis, Inc. 1988; E. A. Engineering, Science and Technology, Inc. 1989).

Reviewing progress in instream-flow method development during the decade of the 1970s, I identified three areas needing further research effort during the 1980s (Stalnaker 1982). These were

1. Continued development of physical/chemical simulation techniques to be coupled with accelerated research efforts to establish biological criteria for interpreting the results of these simulations for a wider variety of target fish species.
2. New conceptual models were needed before additional elements could be factored into the instream assessment process and water planning. New conceptual models were called for to address freshwater inflow to estuaries, sediment transport and channel change, water requirements for riparian vegetation, riparian wetlands, and the response of aquatic organisms (fish and invertebrates) to rapid fluctuations resulting from hydropeaking and pumpback storage projects.
3. Increased attention to the legal and institutional arena at the state level was needed to provide specific recognition of instream uses of water and develop administrative procedures for quantifying and administering such water uses.

Most progress in the 1980s was in addressing Items 1 and 3. Considerable improvement in hydraulic simulation techniques have come about: emphasis on the computations of velocities at specified points in the water column, near bottom velocities, in areas adjacent to fish holding locations, and scouring velocities which move streambed particles in spawning nests (i.e., redds).

Accuracy and precision of hydraulic simulation has greatly improved, and computer models were transferred from large mainframe computers to desktop microcomputers with more emphasis on easy user interface. Water quality models likewise have improved in their ease of use and are widely available for desktop microcomputers. Development and refinement of physical habitat criteria for fish species received much attention during the 1980s. Many independent data sets were developed (predominately for trout and Pacific salmon) by various agencies, private consultants, and water development interests. This research has focused on the role of fish behavior in describing habitat suitability. The recognition of a need for describing the range of habitat conditions available to the target species and means of factoring out environmental bias has led to several recent studies in different streams and regions (Bovee 1986; Bovee and Zuboy 1988). Many studies are designed with the intent of developing criteria that are transferable across stream types within a larger geographic region for particular species.

Spawning, incubation, and rearing habitats for salmonid fish are very well-described. Suitability criteria for adult salmonid habitat have been studied in small headwater streams and in large coastal streams. The need to stratify habitat suitability criteria by fish size has become obvious.

The legal protection for instream flow expanded in the 1980s to several eastern states. Many states have passed legislation to recognize instream use. McKinney and Taylor (1989) reviewed the status for the western states and Lamb and Meshorer (1988) provide a nationwide overview of state instream-flow programs. Suffice it to say that instream-flow protection and administration at the state level is no longer just a western issue, and in the latter half of the 1980s, much attention has been in the eastern states.

The least amount of progress has been made in the development of new conceptual models. Most ecological research in the 1980s was in multivariate regression analysis that considerably enhanced understanding of important habitat variables (Morhardt 1986; Parsons et al. 1981; Fauch et al. 1988). However, it has not yet led to new conceptual models suitable for simulation of fish community response and comparison of water management schemes (Bain and Boltz 1989). What has apparently caught on is the integration of more traditional single species population dynamic models with habitat suitability models into time-series simulations for comparing alternative stream-flow scenarios (Mattice 1989; Cheslak 1989). Such work has led to the recognition that flow-related habitat limitations (bottlenecks) may occur during the life history of obligate riverine fishes (Burns 1971). This has been demonstrated for western salmonids and is a product of the stream channel geometry and the temporal distribution of flows. Year class strength was shown to be determined by constraints in the physical habitat during the first year of life (Bovee 1988; Nehring and Miller 1987). Constraints may be high flows that scour eggs during the incubation period or completely swamp shallow edge environments important for rearing. Extremely low flows may reduce year class strength by eliminating backwater rearing habitats.

Most progress seems to have been in quantitative descriptions of flushing flows that move fine materials (sand of 2 mm diameter and less) from the interstitial spaces in gravel bed streams (Reiser et al. 1985). Debate over techniques for determining channel maintenance flows continues. The dominant discharge, which is approximated by the bankfull discharge, is still widely touted as the channel-forming flow for alluvial channels and is often proposed as necessary for river management. Physical-process models for predicting scour and fill at specified points in the stream did not receive much attention and testing during the 1980s. Such work is essential to river basin management (Stalnaker et al. 1989).

Several studies have been conducted on responses of fishes and invertebrates to lateral shifts in habitat during peaking hydroelectric facility operations. Attention needs to be focused on this concern (Bain et al. 1988). Dewatering and stranding are important concerns in high-gradient streams, and high flows in low-gradient streams are perhaps more detrimental to aquatic organisms than low flows. High flows simply render much of the wetted area unsuitable from extremely high velocities. This is particularly true for young-of-the-year and small fishes. Progress was insignificant toward understanding and quantifying effects of depletions of freshwater inflows to estuarine environments, which continues to be of great concern (Martin 1987). Sediment transport was given much attention during the 1980s (Chapman 1988).

In the past few years, considerable emphasis has been directed toward riparian wetland habitat protection (Stromberg and Patten 1988). Research was initiated to develop a better understanding of the relationships among soil moisture, vegetation, germination, and growth of woody riparian plants. Only when such information becomes available will water resource analysts have a means of projecting riparian vegetation response to flow manipulations. This area will receive considerable research emphasis during the 1990s.

In conclusion, habitat suitability models are very much a part of the water resources decisionmaking process in the United States and Canada and will continue to be a vital part of the management and regulatory agencies standards for comparison among streamflow alternatives. However, the next 5–10 years will see many of the biological models improved or replaced.

New models:

1. single species fish population models driven by physical/chemical simulations of habitat limitations (bottlenecks)
2. fish species and benthic invertebrate habitat using guilds, particularly for warm water streams
3. development of sophisticated compensatory mortality and growth models for species or species groups
4. more emphasis upon temperature simulations to forecast changes in fish egg incubation and growth rates, as well as effects on migration and egg maturation
5. secondary production models driven by physical/chemical simulations with emphasis on forecasting changes in species diversity, as well as seasonal production and drift
6. herbaceous and woody plant models responsive to flooding and changes in soil moisture brought about by dewatering

Physical models will see marked improvements during the same period.

1. improvement in the precision of simulating velocities at specified points in the water column (near bottom, near surface, and at specified distances around a point)
2. models for developing flow management schemes for *flushing* fine material deposits from within and upon gravel and cobble bars in natural stream channels
3. sediment transport via physical-process models for scour and fill computation at points along a stream course (tested and applied in sand, gravel, and cobble bed streams)
4. stochastic river hydrology models, which allow for risk analyses in water supply and biological response simulations

During the next decade, basic research will improve our understanding of biological community dynamics (species interactions, competition, predation) in riverine and estuarine environments, but developed and tested models that incorporate these concepts are not very likely for widespread application until after the turn of the century.

ACKNOWLEDGMENTS

Portions of this paper have drawn heavily from an earlier draft prepared with E. Woody Trihey for a chapter on water use in a national handbook of recommended procedures for water data acquisition, which is being prepared by an interagency committee for the U.S. Geological Survey.

REFERENCES

Allred, C. S. 1976. Data needs for decisionmaking. In J. F. Orsborn, and C. H. Allman (eds.), *Proceedings of the Symposium and Specialty Conference on Instream Flow Needs.* Vol. 1. American Fisheries Society, Bethesda, MD.

Bain, M. B., J. T. Finn, and H. E. Booke. 1988. Streamflow regulation and fish community structure. *Ecology* 69(2)382–392.

Bain, M. B., and J. N. Boltz. 1989. Regulated streamflow and warmwater stream fish: A general hypotheses and research agenda. *U.S. Fish Wildl. Biol. Rep.* 89(18).

Baldrige, J. S., and D. Amos. 1981. A technique for determining fish habitat suitability criteria: A comparison between habitat utilization and availability. In N.B. Armantrout (ed.), *Acquisition and Utilization of Aquatic Habitat Inventory Information.* American Fisheries Society, Bethesda, MD.

Bartschi, D. K. 1976. A habitat-discharge method of determining instream flows for aquatic habitat. pp. 285-294. In J. F. Orsborn, and C. H. Allman (eds.), *Proceedings of the Symposium and Specialty Conference on Instream Flow Needs.* Vol. 2. American Fisheries Society, Bethesda, MD.

Bio Systems Analysis, Inc. 1988. Chinook Salmon Population Model for the Sacramento River Basin. Submitted to National Marine Fisheries Service. Bio Systems Analysis, Inc., Fort Cronkhite, Building 1064, Sausalito, CA.

Bovee, K. D., and R. T. Milhous. 1978. Hydraulic Simulation in Instream Flow Studies: Theory and Techniques. Instream Flow Information Paper No. 5. FWS/OBS-78/33. U.S. Fish Wildlife Service,

Bovee, K. D. 1982. A Guide to Stream Analysis Using the Instream Flow Incremental Methodology. Instream Flow Information Paper No. 12. FWS/OBS 82/26. U.S. Fish Wildlife Service,

Bovee, K. D. 1986. Development and Evaluation of Habitat Suitability Criteria for Use in the Instream Flow Incremental Methodology. Instream Flow Information Paper No. 21. *U.S. Fish Wildl. Serv. Biol. Rep.* 86(7).

Bovee, K. D. 1988. Use of the instream flow incremental methodology to evaluate influences of micro-habitat variability on trout populations in four Colorado streams. In *Proc. 68th Annu. Conf.* Western Assoc. Fish Wildl. Agencies, Albuquerque, NM.

Bovee, K. D., and J. R. Zuboy. 1988. Proceedings of a workshop on development and evaluation of habitat suitability criteria. *U.S. Fish Wildl. Serv. Biol. Rep.* 88(11).

Brusven, M. A., and E. W. Trihey. 1978. Interacting Effects of Minimum Flow and Fluctuating Shorelines on Benthic Stream Insects. Tech. Completion Report. Idaho Water Resources Research Institute, University of Idaho, Moscow.

Burns, J. W. 1971. The carrying capacity for juvenile salmonids in some northern California streams. *Calif. Fish Game* 57(1):44–57.

Camp, Dresser, and McKee. 1986. Final Report: Minimum Instream Flow Study. Prepared for Commonwealth of Virginia State Water Control Board, Annandale, VA (pages numbered in sections).

Chapman, D. W. 1988. Critical review of variables used to define effects of fines in redds of large salmonids. *Trans. Am. Fish. Soc.* 117(1):1–21.

Cheslak, E. F. 1989. Personal communications. EA Engineering, Science and Technology, Inc. Western Regional Operations, 41 Lafayette Circle, Lafayette, CA.

Collings, M. R. 1972. A methodology for determining instream flow requirements for fish. pp. 72–86. In *Proceedings of Instream Flow Methodology Workshop.* Washington Department of Ecology, Olympia.

Collings, M. R. 1974. Generalization of Spawning and Rearing Discharges for Several Pacific Salmon Species in Western Washington. U.S. Geological Survey Open-File Report.

Collotzi, A. 1975. Aquatic Habitat Inventory Computer Storage Program Guidelines. USDA For. Serv. Rep. Bridger-Teton Natl. For.

Dooley, J. M. 1976. Application of the U.S. Bureau of Reclamation Water Surface Profile Program (WSP). In J. F. Orsborn, and C. H. Allman (eds.), *Proceedings of the Symposium and Specialty Conference on Instream Flow Needs.* Vol. 2. American Fisheries Society, Bethesda, MD.

Dougal, M. D. 1979. Roles and Alternatives of State Agencies in Regulating Stream Alteration. Paper presented at the Stream Alterations Symp. Midwest Fish Wild. Conf., Champaign, IL. Iowa State University, Ames (unpublished manuscript).

Dunham, D. K., and A. Collotzi. 1975. The Transect Method of Stream Habitat Inventory—Guidelines and Applications. USDA For. Serv.

E. A. Engineering, Science and Technology, Inc. 1989. San Joaquin River System Chinook Salmon Population Model Documentation. EA Engineering, Science and Technology, Inc., Lafayette, CA.

Elser, A. A. 1976. Use and reliability of a water surface profile program data on a Montana prairie stream. In J. F. Orsborn, and C. H. Allman (eds.), *Proceedings of the Symposium and Specialty Conference on Instream Flow Needs.* Vol. 2. American Fisheries Society, Bethesda, MD.

Estes, C. C. 1984. Evaluation of methods for recommending instream flows to support spawning by salmon. M.S. thesis. Washington State University, Pullman.

Fauch, K. D., C. L. Hawkes, and M. G. Parsons. 1988. Models that Predict Standing Crop of Stream Fish from Habitat Variables: 1950-1985. Gen. Tech. Rep. PNW-GTR-213. U.S. Dept. Agriculture, Forest Service, Pacific N.W. Res. Stn., Portland, OR.

Fraser, J. C. 1975. Determining Discharges for Fluvial Resources: Rome, United Nations, Food and Agriculture Organization, FAO Fisheries Technical Paper No. 142 (FIRS/T143).

Gore, J. A., and J. M. Nestler. 1988. Instream flow studies in perspective. *Regul. Rivers: Res. Manage.* 2:93–101.

Harza-Ebasco Susitna Joint Venture. 1985. Susitna Case E-VI Alternative Flow Regime. Vol. 1: Main Report. Final Report Feb. 1985. FERC/Susitna Hydroelectric Project (No. 7114), Alaska Power Authority, Vol. 1. Feb. 1985. Doc. No. 2600.

Herrington, R. B., and D. K. Dunham. 1967. A Technique for Sampling General Fish Habitat Characteristics of Streams. USDA For. Serv. Res. Paper INT-41.

Hoppe, R. A., and L. M. Finnell. 1970. Aquatic Studies on Fryingpan River, Colorado, 1969–1970. Bureau Sport Fishing and Wildlife Division River Basin Studies, Albuquerque, NM.

Knapp, W. E. 1980. Instream Flow Policy: A Stimulus for Attitudinal Change. Paper presented at HCRS Conference on River Conservation and Revitalization: Agenda for the 1980's. October 20, 1980, Lake Mohawk, NY. Mimeo.

Lamb, B. L., and H. Meshorer. 1983. Comparing instream flow programs: A report on current status. pp. 435–443. In *Proc. Conf. Advances in Irrigation and Drainage: Surviving External Pressures, July 20–23, 1983*. American Society of Engineers Irrigation and Drainage Division, Jackson, WY.

Larsen, H. N. 1980. Policy memorandum to area manager, New England Area Office from Regional Director, Region 5. U.S. Fish Wildl. Serv., April 11, 1980.

Leonard, P. M., D. J. Orth, and C. J. Goudreau. 1985. Development of a Method for Recommending Instream Flows for Fishes in the Upper James River, Virginia. Virginia Polytechnical Institute and State University, Blacksburg, VA.

Loar, J. M., and M. J. Sale. 1981. Analysis of Environmental Issues Related to Small-Scale Hydroelectric Development. V. Instream Flow Needs for Fishery Resources. ORNL/TM-7861. Oak Ridge National Laboratory, Oak Ridge, TN.

Martin, Q. W. 1987. Estimating freshwater inflow needs for Texas estuaries by mathematical programming. *Water Resour. Res.* 23(2):230–238.

Mattice, J. S. 1989. Personal communication. Electric Power Research Institute, P.O. Box 10412, Palo Alto, CA.

McDonnell, L., T. Rice, and S. J. Shupe (eds.). 1989. *Instream Flow Protection in the Western United States.* Natural Resources Law Center, University of Colorado, Boulder.

McKinney, M. J., and J. G. Taylor. 1989. Western state instream flow programs: A comparative assessment. *U.S. Fish Wildl. Serv. Biol. Rep.* 89(2).

Milhous, R. T., J. B. Bradley, and C. L. Loeffler. 1986. Sediment transport simulation in an armoured stream. pp. 116–126. *Proceedings 4th Federal Interagency Sedimentation Conf.* Vol. 2. Subcommittee on sedimentation of the Interagency Advisory Committee on Water Data.

Milhous, R. T., M. Updike, and D. Schneider. 1989. Computer reference manual for the Physical Habitat Simulation System (PHABSIM) — Version II. Instream Flow Information Paper No. 26. *U.S. Fish Wildl. Serv. Biol. Rep.* 89(16).

Morhardt, J. E. 1986. Instream Flow Methodologies. Report of Research Project 2194-2. Prepared for Electric Power Research Institute, Palo Alto, CA.

Nehring, R. B., and D. D. Miller. 1987. The influence of spring discharge levels on rainbow and brown trout recruitment and survival. Black Canyon of the Gunnison River, Colorado, as determined by IFIM/PHABSIM models. In *Proc. 67th Annu. Conf.* West. Assoc. Fish Wildl. Agencies, Salt Lake City, UT.

Nelson, F. A. 1980. Evaluation of selected instream flow methods in Montana. pp. 412–432. In *Proc. 60th Annu. Conf.* Western Assoc. Fish-Wildlife Agencies, Kalispell, MT.

Nelson, F. A. 1984. Guidelines for Using the Wetted Perimeter (WETP) Program of the Montana Department of Fish, Wildlife, and Parks (unpublished manuscript).

Orsborn, J. F. 1981. Estimating spawning habitat using watershed and channel characteristics. pp. 154–161. In N. B. Armantrout (ed.), *Acquisition and Utilization of Aquatic Habitat Inventory Information.* American Fisheries Society, Bethesda, MD.

Orsborn, J. F., and C. H. Allman. 1976. *Proceedings of the Symposium and Specialty Conference on Instream Flow Needs.* Vols. 1 and 2. American Fisheries Society, Bethesda, MD.

Orth, D. J. 1987. Ecological considerations in the development and application of flow habitat models. *Regul. Rivers* 1:171–181.

Orth, D. J., and O. E. Maughn. 1981. Evaluation of the "Montana method" for recommending instream flows in Oklahoma streams. *Proc. Okla. Acad. Sci.* 61:62–66.

Parsons, M. G., J. R. Maxwell, and D. Heller. 1981. A predictive fish habitat model using geomorphic parameters. pp. 85–91. In N. B. Armantrout (ed.), *Acquisition and Utilization of Aquatic Habitat Inventory Information.* American Fisheries Society, Bethesda, MD.

Raley, C., W. Hubert, and S. Anderson. 1988. Maintenance of flows downstream from water development projects in Colorado, Montana, and Wyoming. *U.S. Fish Wildl. Serv. Biol. Rep.* 88(27).

Reed, S. E. and J. S. Mead. 1990. Use of multiple methods for instream flow recommendations — A state agency approach. In M. B. Bain (ed.) *Ecology and Assessment of Warmwater Streams: Workshop Synopsis.* U.S. Fish Wild. Serv., Biol. Rep. 90(5):40–42.

Reiser, D. W., M. P. Ramey, and T. R. Lambert. 1985. Review of Flushing Flow Requirements in Regulated Streams. Pacific Gas and Electric Co., Dept. Eng. Res., San Ramon, CA.

Russell, R. L., and R. J. Mulvaney. 1973. Recommending streamflows to meet national forest needs. pp. 35–52. In *Proc. Hydrology and Environ. Seminar.* Water Resources Research Institute, University of Wyoming, Laramie.

Sams, R. E., and S. Pearson. 1963. A Study to Develop Methods for Determining Spawning Flows for Anadromous Salmonids. Oregon Fish Comm., Portland, OR (unpublished manuscript).

Scott, W., and C. S. Shirvell. 1987. A critique of the instream flow incremental methodology and observations on flow determination in New Zealand. pp. 27–43. In J. R. Kemper, and J. Craig (eds.), *International Symposium on Regulated Streams* (3rd: 1985: Edmonton, Alberta, Canada). Plenum Press, New York.

Smith, G. E., and M. E. Aceituno. 1987. Habitat Preference Criteria for Brown, Brook, and Rainbow Trout: Eastern Sierra Nevada Streams. Calif. Dept. Fish Game, Stream Evaluation Rep. 87-2 (unpublished manuscript).

Spear, P. W. 1985. A simple method of determining fish maintenance flows downstream of existing New England Hydro Facilities. pp. 195–200. In F. W. Olsen, R. G. White, and R. H. Harme (eds.), *Proc. Symp. on Small Hydropower and Fish,* May 1–3, Aurora, CO.

Stalnaker, C. B. 1978. The IFG incremental methodology for physical instream habitat evaluation. pp. 126–135. In D. E. Samuel, J. R. Stauffer, C. H. Hocutt, and T. W. Mason (eds.), Surface Mining and Fish/Wildlife Needs in the Eastern United States. FWS/OBS-78/812. Proceedings of a Symposium, U.S. Fish and Wildlife Service, December 3–6, 1978, Morgantown, WV.

Stalnaker, C. B. 1980a. The use of habitat structure preferenda for establishing flow regimes necessary for maintenance of fish habitat. In J. V. Ward, and J. A. Stanford (eds.), *The Ecology of Regulated Streams.* Plenum Press, New York.

Stalnaker, C. B. 1980b. Effects on fisheries of abstractions and perturbations in streamflow. pp. 366–383. In J. H. Grover (ed.), *Allocation of Fishery Resources: Proceeding Technical Consultation on Allocation of Fishery Resources.* 20–23 April 1980, Vichy, France. Published by U.N. Food and Agriculture Organization in cooperation with the American Fisheries Society.

Stalnaker, C. B. 1982. Instream flow assessments come of age in the decade of the 1970's. pp. 119–141. In W. T. Mason, Jr., and S. Iker (eds.), Research on Fish and Wildlife Habitat. EPA-600/8-82.022. U.S. Environmental Protection Agency.

Stalnaker, C. B. 1990. Minimum flow is a myth. In M. B. Bain (ed.), Ecology and Assessment of Warmwater Streams: Workshop Synopsis. *U.S. Fish Wild. Ser., Biol. Rep.* 90(5):31–33.

Stalnaker, C. B., and J. L. Arnette. 1976. Methodologies for the determination of stream resource flow requirements — an assessment (Logan, Utah State University). U.S. Fish Wildl. Serv., Office Biol. Serv., Western Water Allocation Project.

Stalnaker, C. B., R. T. Milhous, and K. D. Bovee. 1989. Hydrology and hydraulics applied to fishery management in large rivers. In D. D. Dodge (ed.), Proceedings of the International Large River Symposium. *Can. Spec. Publ. Fish. Aquat. Sci.* 108:13–30.

Stromberg, J. C., and D. T. Patten. 1988. Instream flow requirements for riparian vegetation. pp. 123–130. In G. R. Baumli (ed.), *Legal, Institutional, Financial and Environmental Aspects of Water Issues.* American Society of Civil Engineering, New York.

Swank, G. W., and R. W. Phillips. 1976. Instream flow methodology for the Forest Service in the Pacific Northwest Region. In J. F. Orsborn, and C. H. Allman (eds.), *Proceedings of the Symposium and Specialty Conference on Instream Flow Needs.* Vol. 2. American Fisheries Society, Bethesda, MD.

Swift, C. H. 1976. Estimates of Stream Discharges Preferred by Steelhead Trout for Spawning and Rearing in Western Washington. U.S. Geological Survey Open-File Report 75-155.

Swift, C. H. 1979. Preferred Stream Discharges for Salmon Spawning and Rearing in Washington. U.S. Geological Survey Open-File Report 77-422.

Tennant, D. L. 1975. Instream Flow Regimens for Fish, Wildlife, Recreation and Related Environmental Resources. U.S. Fish Wildlife Service. Mimeo.

Tennant, D. L. 1976. Instream Flow Regimes for Fish, Wildlife, Recreation, and Related Environmental Resources. pp. 359–373. In J. F. Orsborn, and C. H. Allman (eds.), *Proceedings of the Symposium and Specialty Conference on Instream Flow Needs.* Vol. 2. American Fisheries Society, Bethesda, MD.

Theurer, F. T., K. S. Voos, and W. J. Miller. 1984. Instream water temperature model. *U.S. Fish Wildl. Serv. Biol. Rep.* 84(15).

Thompson, K. E. 1972. Determining streamflows for fish life. pp. 31–50. In *Proc. Instream Flow Requirement Workshop.* Pacific NW River Basins Comm., Portland, OR.

Thompson, K. E. 1974. Salmonids — Chapter 7. pp. 85–103. In K. Bayha (ed.), *The Anatomy of a River. Report of the Hells Canyon Task Force.* Pacific NW River Basins Comm., Portland, OR.

Trihey, E. W. 1983. Preliminary Assessment of Access by Adult Salmon into Portage Creek and Indian River. Report for Alaska Power Authority, Anchorage, AK.

Trihey and Associates, and Entrix, Inc. 1985. Instream Flow Relationships Report. Vol. 1. Final Report, May 1985. FERC/Susitna Hydroelectric Project (No. 7114). Alaska Power Authority. Vol. 1., Dec. 1985. Doc. No. 3060.

Trihey, E. W., and J. E. Baldrige. 1985. An empirical approach for evaluating microhabitat response to streamflow in steep-gradient, large bed-element streams. In F. W. Olson, R. G. White, and R. H. Hamre (eds.), *Proceedings of the Symposium on Small Hydropower and Fisheries.* American Fisheries Society, Bethesda, MD.

Trihey, E. W., and C. B Stalnaker. 1985. Evolution and application of instream flow methodologies to small hydropower development: An overview of the issues. pp. 176–183. In F. W. Olson, R. G. White, and R. H. Hamre (eds.), *Proceedings of the Symposium on Small Hydropower and Fisheries.* American Fisheries Society, Bethesda, MD.

Trihey, E. W., and D. L. Wegner. 1980. Field Data Collection Procedures for Use with the IFG-2 and IFG-4 Hydraulic Simulation Models. Coop. Instream Flow Serv. Group. U.S. Fish Wildl. Serv., Fort Collins, CO.

U.S. Fish and Wildlife Service. 1981. Interim Regional Policy for New England Streamflow Recommendations Memorandum from H. N. Larsen, Director, Region 5, Newton Corner, MA. February 13, 1981. Mimeo.

U.S. Forest Service. 1986. Procedure for Quantifying Channel Maintenance Flows. Chapter 30. In Forest Service Handbook FSH2509. 17, Water Information System Handbook.

Waters, B. F. 1976. A methodology for evaluating the effects of different streamflows on salmonid habitat. pp. 254–266. In J. F. Orsborn, and C. H. Allman (eds.), *Proceedings of the Symposium and Specialty Conference on Instream Flow Needs.* Vol. 2. American Fisheries Society, Bethesda, MD.

Wesche, T. A. 1972. Parametric Determination of Minimum Streamflow for Trout. Water Resources Research Institute Report, University of Wyoming, Laramie.

Wesche, T. A. 1974. Relationship of Discharge Reductions to Available Trout Habitat for Recommending Suitable Streamflows. Water Resources Series No. 54, Water Resources Research Institute Report, University of Wyoming, Laramie.

Wesche, T. A. 1980. The WRRI Trout Cover Rating Method: Development and Application. Water Resources Service 78, Water Resources Research Institute, University of Wyoming, Laramie.

Wesche, T. A., and P. A. Rechard. 1980. A Summary of Instream Flow Methods for Fisheries and Related Research Needs. Eisenhower Consortium Bulletin No. 9, Water Resources Research Institute, University of Wyoming, Laramie.

White, R. G. 1976. A methodology for recommending stream resource maintenance flows for large rivers. pp. 376–399. In J. F. Orsborn, and C. H. Allman (eds.), *Proceedings of the Symposium and Specialty Conference on Instream Flow Needs.* Vol. 2. American Fisheries Society, Bethesda, MD.

White, R., and T. Cochnauer. 1975. Stream Resource Maintenance Flow Studies. Idaho Dept. Fish Game and Idaho Coop. Fish Res. Unit Rep.

Wilson, W. J. 1981. An interagency streamflow recommendation analysis for a proposed Alaskan hydroelectric project. pp. 241–250. In N. E. Armantrout (ed.), *Acquisition and Utilization of Aquatic Habitat Inventory Information.* Proceedings of a Symposium, October 1981, Portland, OR. American Fisheries Society, Western Division.

Wood, R. K., and D. E. Whelan. 1962. Low-Flow Regulation as a Means of Improving Stream Fishing. Proc. 16th Annual Conf. SE Assoc. Game and Fish Commissioners, Charleston, SC.

Workman, D. L. 1976. Use of the water surface profile program in determining instream flow needs in 16 Mile Creek, Montana. In J. F. Orsborn, and C. H. Alman (eds.), *Proceedings of the Symposium and Specialty Conference on Instream Flow Needs.* Vol. 2. American Fisheries Society, Bethesda, MD.

Impact, Hazard, and Risk Assessment of Toxic Effects on Nonhuman Organisms

G. W. Suter, II, Oak Ridge National Laboratory, Oak Ridge, TN

ABSTRACT

When the National Environmental Policy Act (NEPA) was enacted, environmental toxicology was nearing the end of a long infancy. Toxicity testing procedures were not standardized, testing was almost entirely limited to determining lethal effects of acute exposures of fully developed organisms, and there was no assessment paradigm for effective use of test data. The requirement of NEPA that all effects of major federal actions be predicted and disclosed presented a major challenge. However, environmental toxicity assessments for NEPA did not generate improvements in either toxicology or assessment techniques. Instead, assessments dealt with chemical emissions largely in terms of their compliance with standards and permits. This paper presents an argument that this approach is not adequate and that ecological risk assessment is the best approach for assessing toxicological effects under NEPA.

INTRODUCTION

When NEPA was enacted in 1969, the field of ecological toxicology was nearing the end of a long infancy. The writings of Rachel Carson and others had drawn the attention of the public to the occurrence of toxic effects on nonhuman organisms and that issue was a major impetus behind the enactment of NEPA. However, the assessment of ecotoxic effects, in general, has been poorly handled in NEPA documents. To this day, NEPA documents routinely attempt to estimate the effects of habitat loss, entrainment and impingement of fish, and collisions of birds with structures, but seldom attempt to estimate ecotoxic effects. Effects of air pollutants, water pollutants, waste materials, and pesticides, to the extent that they are considered at all, are treated as regulatory compliance issues. If the proposed action meets all applicable regulatory requirements, then it is assumed that there are no ecological impacts.

The assumption that regulatory compliance prevents ecological effects is inappropriate for at least six reasons. First, many regulatory requirements are based on technology (e.g., best available control technology) rather than on preventing effects. Second, most chemicals are not subject to regulatory standards. Although there are thousands of chemicals in use and 128 priority pollutants, there are only 33 chronic national water quality criteria for freshwater aquatic life and 28 for marine life (EPA 1986). Third, there are no regulations for quality of soil or sediments. Fourth, many types of effects are not considered in setting criteria and standards. For example, behavioral effects are not included in the calculation of national water quality criteria, and acidification effects were not included in the derivation of the national ambient air quality standard for sulfur dioxide. Fifth, criteria and standards are designed to be broadly applicable. Many characteristics of a site can accentuate or diminish exposure of organisms to pollutants or can modify the sensitivity of organisms to pollutants. Finally, state and federal environmental laws and regulations do not require that all effects be prevented. Instead, they nearly always call for balancing costs and benefits or include some wording about reasonableness. As a result, standards and criteria are set at levels that are expected to have some effect that is considered reasonable.

The following examples illustrate the last point. The chronic national water quality criteria for protection of aquatic life are designed to protect 95% of species from statistically significant chronic effects (Stephan et al. 1985). As a result, biologically significant effects can occur at concentrations below the criteria (Suter et al. 1987a; Barnthouse et al. 1987). For example, the hardness-corrected chronic criterion concentration for zinc caused an approximately 90% decline in fecundity of fathead minnows (*Pimephales promelas*) (Brungs 1969). Similarly, concentrations below the national air quality criteria for sulfur dioxide and ozone can cause visible injury and reduced production in sensitive plants including major crops and forest trees (EPA 1982; Winner et al. 1985; Heck et al. 1988).

If ecotoxic effects can occur despite regulatory compliance, then it is incumbent on the authors of NEPA assessments to independently estimate the nature and magnitude of those effects. Although agencies are not allowed under NEPA to review permits issued by the U.S. Environmental Protection Agency (EPA) or to require additional effluent permits, that restriction does not exempt them from determining the relative effects of emissions from alternate technologies, of the same emission at alternate sites, or of generating an emission relative to taking no action (Mandelker 1984). Unfortunately, experts in environmental toxicology are relatively seldom employed in NEPA assessments, partly because ecotoxicology and ecotoxicological assessment methods have been developed by the EPA and by the regulated industries in a process that is exempt from NEPA. There has been a *de facto* partitioning of the environmental assessment community into an ecotoxicological camp that is narrowly concerned with the regulation of chemicals and a NEPA camp that is concerned with everything else.

In 1969, there were no standard toxicity tests in the United States for effects on nonhuman organisms (see APHA 1965 for the best approximation of a standard method available at the time). Sophisticated toxicity tests were developed during the 1960s, but the vast majority of toxicity tests employed acute lethality as the endpoint. Now, numerous test protocols are available from the American Public Health Association (APHA 1992), American Society for Testing and Materials (ASTM 1992), and the EPA. These protocols involve lethal and sublethal effects on fish, aquatic invertebrates, algae, macrophytes, birds, terrestrial plants, terrestrial invertebrates, and even microcosms and mesocosms. Similarly, exposure models were largely based on dilution with little consideration of the environmental chemistry or intermedia transfers that dominate the rather complex models currently available. The development of these tests and models has been driven by the demands of ecotoxicological assessment, but the availability of test methods and models and of a body of toxicological and chemical data has made possible the development of improved assessment techniques.

HAZARD ASSESSMENT PARADIGM

In 1969, there was no paradigm for ecotoxicological assessments. Assessments were performed ad hoc with no consistent logic applied to selecting toxicity tests or to interpreting the relationship between the exposure estimates and test results. The principal impetus for the development of a standard paradigm was the passage of the Toxic Substances Control Act (TSCA) in 1976. The task of regulating all chemicals in commerce was so sufficiently large that most ecotoxicologists realized that some efficient means of performing assessments of toxic effects was needed. The result was the hazard assessment paradigm developed at the 1977 Pellston Workshop (Cairns et al. 1978).

The hazard assessment paradigm is based on tiered testing and assessment. In the first tier, quick and inexpensive tests were performed, and exposure models that required little or no environmental chemistry were run (e.g., an acute lethality test for a single species and a dilution factor). The results were assumed to constitute highly uncertain estimates of the maximum acceptable toxic concentration (MATC) and the expected environmental concentration (EEC), respectively (Figure 1). If the MATC was much higher than the EEC, the chemical was assumed to be safe. If the MATC was much lower than the EEC, the chemical was assumed to be unsafe. If the two values were similar, the decision was deferred; a new tier of testing, measurement, and modeling was conducted, and the assessment was repeated. The criterion for determining whether the MATC was sufficiently large relative to the EEC was the quotient of the ratio EEC/MATC, termed the "safety factor." The adequacy of the safety factor derived at each tier was a matter of professional judgment. A key assumption of the paradigm is that after a reasonable number of tiers the relative magnitudes of the true MATC and EEC would become apparent.

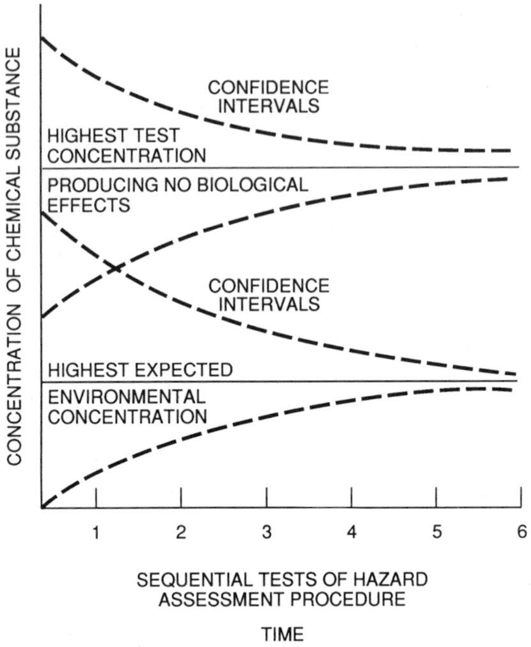

Figure 1. Diagram developed to represent the hazard assessment paradigm of tiered testing, resulting in a clear distinction between the expected concentration and a toxic concentration. (Redrawn from Cairns et al. 1979.)

RISK ASSESSMENT PARADIGM

Although the hazard assessment paradigm still dominates ecotoxicological assessment practices, the risk assessment paradigm is increasingly used. Risk assessment estimates the probability and magnitude of undesired events. It is older than hazard assessment, having originated in actuarial statistics and developed in engineering safety and human health assessments. However, it was not applied to ecological assessments until the early 1980s. It differs from hazard assessment in that it is probabilistic, requires well-defined endpoints, presents results in terms of magnitudes rather than a safe/unsafe dichotomy, does not require tiered assessment to reach a conclusion, uses formal decision criteria rather than expert judgment, and can balance risks against benefits and costs (Suter 1990a). As a result, the role of the scientist performing an assessment changes from that of a judge of acceptability to that of an analyst presenting estimates of effects and associated uncertainties to a risk manager. The paradigm for toxicological risk assessment begins with descriptive activities that define the nature and scope of the hazard, uses transport and fate models to estimate the magnitude of exposure, estimates effects in terms of exposure-response relationships, estimates risks resulting from the exposure and effects, and provides the risk estimates to a risk manager (Figure 2).

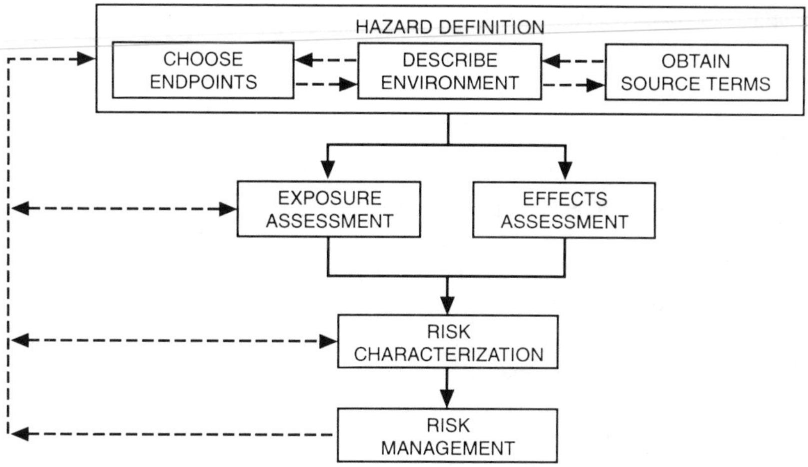

Figure 2. Diagram representing the paradigm for predictive toxicological risk assessments.

EXAMPLE

The programmatic environmental impact statement for the U.S. Army's Chemical Stockpile Disposal Program (CSDP) provides an example of the use of risk assessment techniques to assess ecotoxicological effects in a NEPA document. First, conventional engineering risk assessment techniques were used to estimate the risks of accidental releases of nerve agents during different phases of alternate disposal options (Fraize et al. 1989). Second, models and modeling approaches developed for risk assessments of toxic chemicals were applied to the resulting estimated exposures because simply comparing the estimated concentrations to test endpoints as in hazard assessments was clearly inadequate. Because standard toxicity tests use constant exposure concentrations and standard durations, they are not directly applicable to the pulsed exposure associated with a spill of nerve agent into a stream. Therefore, Tolbert and Breck (1989) used a model that assumes death to be a sum of partial contributions to an effective dose resulting from the fraction of a median lethal time to death experienced during passage of the spill. From this model, the assessors could estimate the distance downstream to the point at which a median-sensitivity fish would not be killed. Another problem that required the application of ecological risk models was the absence of avian dietary toxicity data for the nerve agents. The solution employed was to estimate avian toxicity from rat toxicity data with the use of regression models and the variance estimates from ecological risk assessment methods developed for the EPA (Barnthouse and Suter 1986). Oral LD_{50}s for mallards (*Anas platyrynchos*) and pheasants (*Phasianus colchicus*) were each regressed against rat oral LD_{50}s for organophosphate pesticides which have the same mode of action and similar structures to the nerve agents (Sigal and Suter 1989). The results were used, along with deposition rates and estimated food consumption rates, to estimate the areas within which birds entering a spill site after a spill would be killed by a dietary dose.

CONCLUSIONS

The analyses conducted for the CSDP are not ideals of ecological risk assessment because the necessary data and assessment tools are still being developed. However, they illustrate some of the advantages of the risk assessment paradigm for NEPA assessments. First, NEPA documents must compare alternatives. Hazard assessment renders a judgment of safe or unsafe, but that allows a comparison only if one option is safe and the others are not. In contrast, risk assessment attempts to estimate the magnitude of effects so that alternatives can be placed on a consistent quantitative scale (e.g., number of stream miles within which more than one half of fish are estimated to be killed). Second, NEPA documents are required to present uncertainties. The uncertainty analyses conducted in probabilistic risk assessment provide an ideal means of quantifying and presenting the consequences of uncertainty (Suter et al. 1987b; Suter 1990b). For example, the fact that the 95% prediction interval on the estimated pheasant LD_{50} covered three orders of magnitude illustrated the need for avian toxicity data, which were subsequently obtained. Third, NEPA documents are intended to be decision-support documents. Therefore, the risk assessment paradigm's assumption that there is a risk manager who is distinct from risk assessors is more appropriate to NEPA assessments than hazard assessment's assumption that the assessor determines acceptability. Finally, hazard assessment requires tiered testing to arrive at a conclusion, but NEPA documents must be able to support a decision without bringing the process to a halt to obtain more data. Risk assessment is intended to provide best estimates of effects at any level of data availability.

NEPA assessments can benefit from the use of risk assessment concepts and models. However, ecological risk assessment is still in the early stages of development. Assessors engaged in the NEPA process could make a substantial contribution to the development of ecological risk assessment.

Risk assessment references that address ecological issues include Suter (1993), Paustenbach (1990), and Cohrssen and Covello (1989).

REFERENCES

APHA (American Public Health Association). 1965. *Standard Methods for the Examination of Water and Wastewater.* 10th ed. APHA, Washington, D.C.

APHA (American Public Health Association). 1992. *Standard Methods for the Examination of Water and Wastewater.* 18th ed. APHA, Washington, D.C.

ASTM (American Society for Testing and Materials). 1992. *Annual Book of ASTM Standards, Sec. 11, Water and Environmental Technology.* ASTM, Philadelphia, PA.

Barnthouse, L. W., and G. W. Suter, II. 1986. User's Manual for Ecological Risk Assessment. ORNL-6251. Oak Ridge National Laboratory, Oak Ridge, TN.

Barnthouse, L. W., G. W. Suter, II, A. E. Rosen, and J. J. Beauchamp. 1987. Estimating responses of fish populations to toxic contaminants. *Environ. Toxicol. Chem.* 6:811–824.

Brungs, W. A. 1969. Chronic toxicity of zinc to the fathead minnow, Pimephales promelas Rafinesque. *Trans. Am. Fish. Soc.* 1969:272–279.

Cairns, J., Jr., K. L. Dickson, and A. W. Maki (eds.). 1978. Estimating the Hazard of Chemical Substances to Aquatic Life. STP 657. American Society for Testing and Materials, Philadelphia, PA.

Cohrssen, J. J., and V. T. Covello. 1989. Risk Analysis: A Guide to Principles and Methods for Analyzing Health and Environmental Risks. Council on Environmental Quality, Washington, D.C.

Fraize, W. E., R. M. Cutler, and G. F. Flanagan. 1989. The probabilistic treatment of potential accidents: What are the relative risks of lethal chemical agent releases to the atmosphere? *Environ. Prof.* 11:297–314.

Heck, W. W., O. C. Taylor, and D. T. Tingey. 1988. *Assessment of Crop Loss from Air Pollutants.* Elsevier, London.

Mandelker, D. R. 1984. *NEPA Law and Litigation.* Callaghan and Company, Wilmette, IL.

Paustenbach, D. J. (ed.). 1990. *The Risk Assessment of Environmental and Human Health Hazards: A Textbook of Case Studies.* Wiley, New York.

Sigal, L. L., and G. W. Suter, II. 1989. Potential effects of chemical agents on terrestrial resources. *Environ. Prof.* 11:376–384.

Stephan, C. E., D. I. Mount, D. J. Hanson, J. H. Gentile, G. A. Chapman, and W. A. Brungs. 1985. Guidelines for Deriving Numeric National Water Quality Criteria for the Protection of Aquatic Organisms and Their Uses. PB85-227049. U.S. Environmental Protection Agency, Duluth, MN.

Suter, G. W., II, A. E. Rosen, E. Linder, and D. F. Parkhurst. 1987a. Endpoints for responses of fish to chronic toxic exposures. *Environ. Toxicol. Chem.* 6:793–809.

Suter, G. W., II, L. W. Barnthouse, and R. V. O'Neill. 1987b. Treatment of risk in environmental impact assessment. *Environ. Manage.* 11:295–303.

Suter, G. W., II. 1990a. Environmental risk assessment/environmental hazard assessment: Similarities and differences. Vol. 13. pp. 5–15. In W. G. Landis, and W. H. van der Schalie (eds.), *Aquatic Toxicology and Risk Assessment.* American Society for Testing and Materials, Philadelphia, PA.

Suter, G. W., II. 1990b. Uncertainty in environmental risk assessment. Chap. 9. pp. 203–230. In G. M. von Furstenberg (ed.), *Acting Under Uncertainty: Multidisciplinary Conceptions.* Klewer Academic Publishers, Boston, MA.

Suter, G. W., II. 1993. *Ecological Risk Assessment,* Lewis Publishers, Chelsea, MI.

Tolbert, V. R., and J. E. Breck. 1989. Effects of chemical agent destruction on aquatic resources. *Environ. Prof.* 11:367–375.

U.S. Environmental Protection Agency (U.S. EPA). 1982. Air Quality Criteria for Particulate Matter and Sulfur Oxides, Vol. 3. EPA-600/8-82-029C,. U.S. EPA, Research Triangle Park, NC.

U.S. Environmental Protection Agency (U.S. EPA). 1986. Quality Criteria for Water 1986. EPA 440/5-86-001. U.S. EPA, Washington, D.C.

Winner, W. E., H. A. Mooney, and R. A. Goldstein. 1985. *Sulfur Dioxide and Vegetation.* Stanford University Press, Stanford, CA.

Evaluating Effects of Environmental Stress on Fish Communities Using Multiresponse Indicators

S. M. Adams, G. F. Cada, M. S. Greeley, Jr., and L. R. Shugart, Oak Ridge National Laboratory, Oak Ridge, TN

ABSTRACT

Most traditional approaches for assessing the effects of environmental stress on fish involve the generation of species lists, the estimation of densities of organisms, the use of acute or chronic laboratory tests, or the measurement of single or a few stress responses. Most of these approaches provide limited information needed to address fundamental National Environmental Policy Act (NEPA) issues such as the consideration of alternative actions and cumulative impacts. The use of multiresponse indicators of stress at several levels of biological organization permits (1) identification of biologically and ecologically relevant effects, (2) possible early detection of environmental problems (early warning indicators), (3) evaluation of the effectiveness of environmental restoration actions, and (4) possible insights into causal mechanisms between stress and effects that ultimately may be manifested at the population and community level. Responses at the biochemical, physiological, histopathological, bioenergetic, and population levels were used to assess the effects of chronic contamination loading on fish communities in some East Tennessee streams. Various biochemical/biomolecular responses provided direct evidence of toxicant exposure, while bioenergetic and histopathological indicators reflected impaired population growth and reproductive potential. Within the NEPA process, the use of multiresponse indicators can be an effective approach for addressing remedial actions and the cumulative impacts of multiple stressors on fish communities.

INTRODUCTION

NEPA requires federal agencies to consider the impacts of their actions on the quality of the environment. For those major actions that necessitate preparation of an environmental impact statement or environmental assessment, the Council on Environmental Quality's (CEQ) regulations for implementing NEPA (40 CFR 1500-1508) specify that the impacts of both the proposed action and alternatives to the proposed action be considered. The analysis of potential impacts should include not only direct effects, but also indirect and cumulative effects of proposed and alternative actions.

Traditional approaches to predicting the impacts of a proposed action on fish populations usually involve the generation of species lists and the estimation of the numerical densities and/or biomass of fish in the area affected by the project. The results of such preoperational monitoring are coupled with what is known about the possible impacts of the proposed action (e.g., from similar projects on nearby water bodies) to predict consequent changes in the abundance or biomass of the fish populations. If the action includes discharge of aqueous effluents, acute or chronic laboratory bioassays may also be performed to assess the toxicity of the discharges. These conventional studies may yield important information about the status of the aquatic resources at risk and in some of the more simple situations may be adequate to predict impacts (e.g., where the effluents are so dilute as to clearly pose no threat or so toxic as to cause substantial mortality among the species of concern).

The significance of impacts from many projects, however, is not so clearly discerned, and traditional monitoring and impact assessment approaches may not be adequate to fully describe them. For example, in many instances, fish are exposed to environmental stressors (such as a toxicant) at levels that are not immediately toxic or where effects are not obvious. Chronic stress situations such as this usually produce long-term effects and typically involve the entire reproductive lifecycle of an organism. Acute stressors, however, may affect only a short period relative to the lifecycle of an organism. Depending on its severity, chronic stress may load or limit the physiological systems of organisms, reduce growth, impair reproduction, predispose organisms to diseases, and reduce the capacity of fish to tolerate subsequent stress. These types of chronic stress effects cannot be evaluated by simple laboratory bioassays, nor would such indirect effects be obvious (at least in the short term) from postoperational monitoring of fish abundance. Even if fish abundance did measurably decline, the cause may be difficult if not impossible to determine, especially if other sources of stress also have the potential to impact the biological communities in the vicinity of the proposed action (i.e., as a result of cumulative impacts). For those NEPA projects involving remedial actions, knowledge of the cause of population-level effects is essential to designing appropriate mitigation measures.

MULTIRESPONSE INDICATORS FOR BIOMONITORING

Two of the major limitations of most biological monitoring approaches are that they do not provide early warning signals of environmental effects, and in most situations, causal relationships between a stressor (such as a pollutant) and responses at the various levels of biological organization cannot be established. Even though environmental managers and regulators would like to have a single method or measurement that satisfies all the above criteria for an effective biomonitoring program (the "Holy Grail" referred to by Cairns and van de Schalie 1980), no single variable or index is adequate for predicting population or community changes (Capuzzo 1985; Cairns and van de Schalie 1980; Heath 1987).

We have developed and applied a multiresponse or biological indicator approach for detecting and predicting stress in fish before undesirable and irreversible effects occur at the population and community level. The underlying concept of the bioindicator approach is that the effects of stress occur at the lower levels of biological organization (i.e., at the genetic, cell, and tissue level) before more severe disturbances are manifested at the population or ecosystem level (Figure 1). Effects of stress are usually expressed first at the molecular/biochemical level (Figure 1) (where pollutants, for example, affect the normal functioning of chemical processes in the body). These effects can be detected as changes in enzyme levels, cell membranes, or genetic material (DNA). Changes at these subcellular levels induce a series of structural and functional responses at the next higher level of biological organization which can impair, for example, complex processes such as hormonal regulation, metabolism, salt balance, and proper functioning of the immune system. These effects, in turn, may eventually alter the organism's ability to grow, reproduce, or survive. Ultimately, irreversible and detrimental effects may be observed at the population, community, or ecosystem levels of biological organization.

The importance of identifying and establishing causal relationships from one level of biological organization to another in environmental stress studies has been emphasized by many investigators (Sastry and Miller 1981; Bayne 1985; Capuzzo 1985; Larsson et al. 1985; Sindermann 1985). Understanding the causal relationships that link biological responses to stress with physiological attributes of animal fitness is essential before it is possible to rigorously extrapolate from effects on individuals to consequent effects on populations and communities (Bayne 1985). By measuring stress responses at various levels of biological organization, it is possible to monitor a spectrum of sensitivities to stress, specificities of effects, and other points of ecological relevance simultaneously. Understanding the relationship of these stress responses to each other should improve our capabilities for predicting population- and community-level changes before irreversible damage occurs (Adams 1990).

The principal response groups or bioindicator levels that should be investigated in a comprehensive biomonitoring study are shown in Figure 1. These response groups range from the biomolecular/biochemical level to the community and ecosystem levels and segregate along gradients of ecological relevance and response

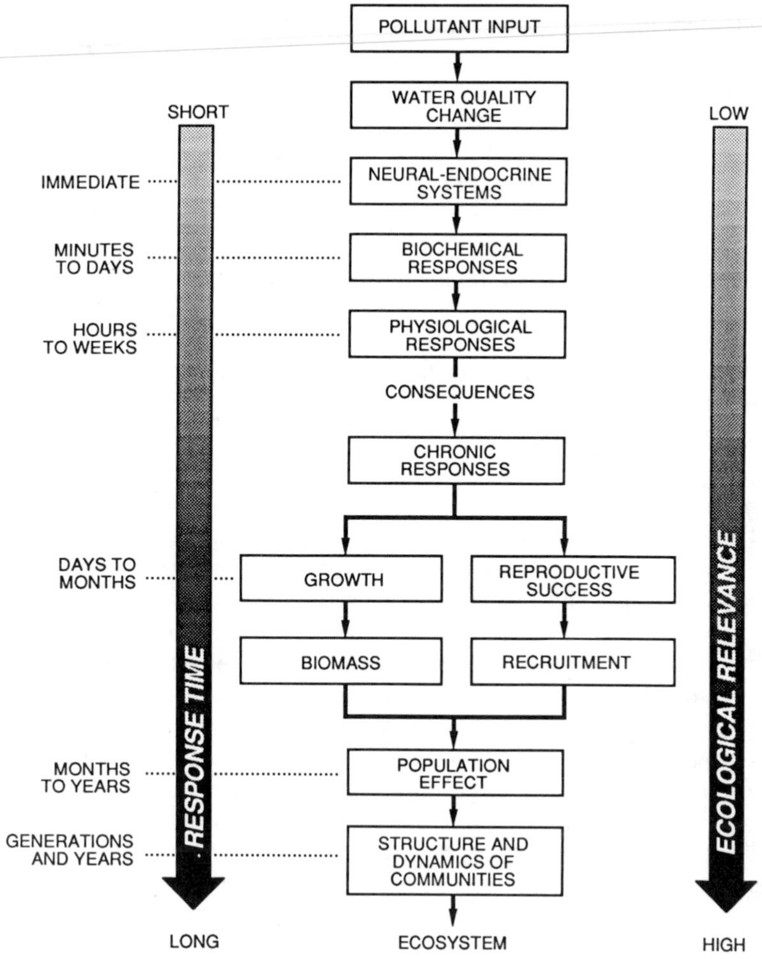

Figure 1. Levels of biological response in fish to environmental stress illustrating the continuum of these responses along gradients of response time and ecological relevance.

time. Bioindicators that reflect changes at lower organizational levels respond relatively rapidly to stress but have low ecological relevance; those that reflect conditions at higher organizational levels respond more slowly to stress but have higher ecological relevance. For example, biochemical parameters such as serum cortisol or glucose can change rapidly (within minutes) in response to an acute stressor, whereas an observable change in populations or communities may not be evident for years or for several biological generations following the imposition of a persistent environmental perturbation. Bioindicators such as histopathological or immune system parameters are intermediate in both their response times (days to months) and their level of ecological relevance.

Indicators that reflect immediate environmental changes can serve as early warning signals of potential or impending environmental effects if their causal links to long-term processes are understood (Adams 1990). Conversely, indicators of greater ecological relevance respond slowly to change; therefore, by the time these responses are observed in an aquatic system, undesirable effects at the population or community levels may have already occurred. Additionally, because many environmental factors can contribute to and influence long-term responses, the major causes of these changes may not be as readily identified or quantified if only the long-term response variables are measured.

USE AND APPLICATION OF MULTIRESPONSE INDICATORS

In terms of the practical use and application of biological indicators, no single indicator or group (response level) of indicators in itself can provide all the critical information necessary for assessing the effects of long-term stress on fish. The ideal design or application of bioindicators for assessing stress on fish would be to measure selected indicators at each of these major response levels in order to determine the causal linkages between the various levels. Once such relationships are established, predictions of long-term effects from measurements of the more short-term indicators could be made with increased reliability and certainty.

This multiresponse monitoring approach has been applied and tested in streams contaminated by various types of industrial discharges. Our most extensive use of this approach has been in a third-order East Tennessee stream that receives point-source industrial discharges of mixed contaminants. At each of four sites along the length of this stream and from a reference stream, 10 to 15 each of male and female redbreast sunfish (*Lepomis auritus*) of approximately the same size were collected by electroshocking over a 2-week period during the summer. Fish from each site were analyzed for biomolecular and biochemical responses (DNA damage and hepatic enzymes), bioenergetic indicators (growth and lipid metabolism), histopathological condition (liver condition including parasites, necrosis, melanophage aggregates, and functional liver parenchyma), condition indices (liver and visceral-somatic indices, lipid levels), and population-level responses (reproductive competence).

Elevated levels of hepatic enzymes provided direct evidence of exposure to xenobiotics, particularly in the upper two sites of the stream nearest the discharge source (Figure 2). DNA damage was four to seven times higher in the contaminated stream than at the reference site. At the intermediate levels of biological organization, effects of exposure were manifested primarily as impaired lipid metabolism, histopathological damage, and poorer body condition (Figure 2). Figure 2 demonstrates that at sites 1 and 2, where toxicant exposure is the highest (elevated hepatic enzymes), the liver-somatic index and liver parasites were also elevated, but lipid metabolism was reduced or impaired. High parasitic infestation may indicate immune system dysfunction (Anderson, 1990), while liver enlargement (increased liver-somatic index) due to hyperplasia and hypertrophy has been

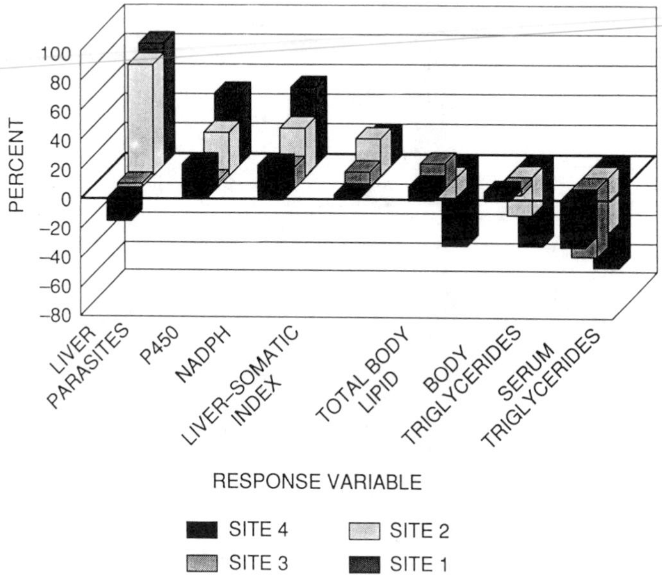

Figure 2. Relative differences in the response of liver parasite load (LPARS), hepatic enzymes (cytochrome P450 and NADPH cytochrome c reductase), the liver-somatic index (LSI), total body lipids (TLIP), body triglycerides (TRI), and blood triglycerides (BTRI) in fish from each of the four contaminated stream sites compared to the reference fish. Values above or below the zero line indicate that the response of fish at each contaminated stream site was higher or lower, respectively, than the same response for fish from the reference site. Sample sites 1–4 in the contaminated stream represent a downstream gradient from the industrial outfall (site 1) to near the stream mouth (site 4).

reported in a variety of species exposed to toxic compounds (Addison 1984; Fletcher et al. 1982).

The correlations observed between toxicant exposure and these various types of stress response in fish from the contaminated stream suggest that toxicants influence lipid metabolism either directly through metabolic processes or indirectly through the food chain. Altered lipid metabolism not only reduces the amount of physiologically available energy for growth, gonad maturation, and repair of damaged tissue, but also compromises the integrity of the immune system. Higher organizational level effects, possibly resulting from impaired metabolic function such as lipid metabolism, can also be seen as alterations in growth and reproductive competence (Figure 3). Growth potential, as reflected by the RNA/DNA ratio, is lower at all the contaminated stream sites when compared with the reference stream; and egg clutch size at the beginning of the breeding season is severely reduced at the site immediately below the industrial discharge.

The results of this study demonstrate that measuring selected stress responses at several levels of biological organization (multiresponse indicators) is necessary

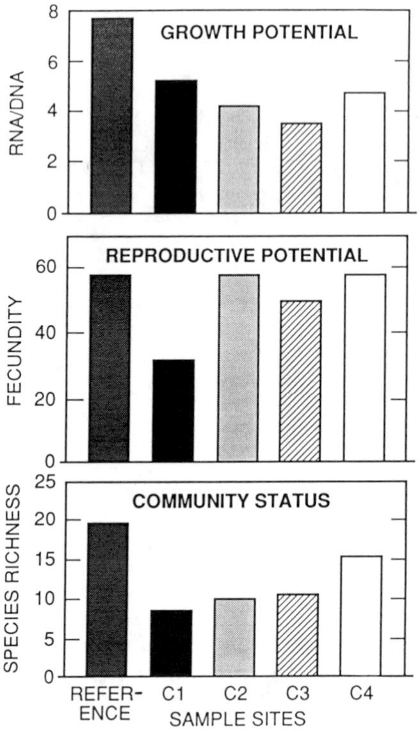

Figure 3. Growth and reproductive potential at each of the four contaminated stream sites (1–4) and the reference site.

in order to relate particular effects to specific causes (i.e., in the case of this study, contaminant exposure to impaired growth and reproductive function) and to serve as early warning indicators of potential environmental effects. The use of the more traditional biomonitoring methods in this situation (i.e., species lists, abundance data, toxicity tests) might have identified or documented a change in growth or population effects (as mediated through reproduction), but the cause of that effect could not be ascertained. In addition, by the time a change would be detected at these higher levels of biological organization (i.e., population effects), ecological damage would have already occurred, and it would essentially be too late to invoke any effective remedial actions.

CEQ regulations note that the "NEPA process is intended to help public officials make decisions that are based on understanding of environmental consequences, and take actions that protect, restore, and enhance the environment" (40 CFR 1500.1). The monitoring approach we have described here can lead to a greater understanding of the causes of impacts from proposed or ongoing actions that is difficult with most conventional monitoring techniques. The addition of a link in the chain between cause (agency action) and consequent population/ ecosystem effects can also suggest corrective measures needed to alleviate the problem, thus serving the goals that were espoused over two decades ago in the promulgation of NEPA.

REFERENCES

Adams, S. M. 1990. Status and use of biological indicators for evaluating the effects of stress on fish. pp. 1–8. In S. M. Adams (ed.), *Biological Indicators of Stress in Fish.* Symposium 8, American Fisheries Society, Bethesda, MD.

Addison, R. F. 1984. Hepatic mixed function oxidase (MFO) induction in fish as a possible biological monitoring system. pp. 51–60. In V. W. Cairns, P. V. Hodson, and J. O. Nriagu (eds.), *Contaminant Effects on Fisheries.* Wiley, New York.

Anderson, D. P. (1990). Immunological indicators: Effects of environmental stress on immune protection and disease outbreaks. pp. 38–50. In S. M. Adams (ed.), *Biological Indicators of Stress in Fish.* Symposium 8. American Fisheries Society, Bethesda, MD.

Bayne, B. L. 1985. Responses to environmental stress: Tolerance, resistance, and adaptation. pp. 331–349. In J. S. Gray, and M. E. Christiansen (eds.), *Marine Biology of Polar Regions and Effects of Stress on Marine Organisms.* Wiley, New York.

Cairns, J., and W. H. van der Schalie. 1980. Biological monitoring Part I — Early warning systems. *Water Res.* 14:1179–1196.

Capuzzo, J. M. 1985. Biological effects of petroleum hydrocarbons on marine organisms: Integration of experimental results and predictions of impacts. *Mar. Environ. Res.* 17:272–276.

Fletcher, G. L., M. J. King, J. W. Kiceniuk, and R. F. Addison. 1982. Liver hypertrophy in winter flounder following exposure to experimentally oiled sediments. *Comp. Biochem. Physiol.* 73C:457–462.

Heath, A. G. 1987. Use of physiological and biochemical measures in pollution biology. pp. 221–234. In A. G. Heath (ed.), *Water Pollution and Fish Physiology.* CRC Press, Boca Raton, FL.

Larsson, A., C. Haux, and M. Sjobeck. 1985. Fish physiology and metal pollution: Results and experiences from laboratory and field studies. *Ecotoxicol. Environ. Saf.* 9:250–281.

Sastry, A. N., and D. C. Miller. 1981. Application of biochemical and physiological responses to water quality monitoring. pp. 265–294. In F. J. Vernberg, A. Calabrease, F. Thurberg, and W. Vernberg (eds.), *Biological Monitoring of Marine Pollutants.* Academic Press, New York.

Sindermann, C. J. 1985. Keynote address: Notes of a pollution watcher. pp. 11–30. In F. J. Vernberg, F. Thurberg, A. Calabrease, and W. Vernberg (eds.), *Biological Monitoring of Marine Pollutants.* Academic Press, New York.

Geographic Information Systems: Effective Tools for Siting and Environmental Impact Assessment

M. S. Siegel, Dames & Moore, Phoenix, AZ; and D. D. Moreno, Dames & Moore, Atlanta, GA

ABSTRACT

Computer-based geographic information systems (GIS) are effective tools for conducting route selection and impact assessment studies for proposed lineal facilities. GIS offers several advantages over manual mapping and analysis methods, including

1. effective management of environmental and geophysical information
2. use of digital data available in the public domain
3. powerful data analysis operations which are difficult or costly to perform manually
4. easy and quick revisions to databases
5. high-quality cartographic products
6. scenario testing for alternative comparison
7. quantification of impacts through area measurements and statistical analyses

For a proposed highway project in Maricopa County, Arizona, a GIS was used to characterize route suitability and measure environmental impacts as part of an environmental impact statement (EIS) prepared by the U.S. Forest Service.

The project was visible and controversial; environmental issues of concern included potential disturbance of desert ecosystems, bald eagle habitat, and sensitive riparian vegetation. The proposed highway crossed the Tonto National Forest, which is heavily used for recreation by residents of the Phoenix metropolitan area.

Route suitability was characterized through a GIS-based analysis of the following factors:

1. sensitive habitats
2. cultural resources
3. visual resources
4. community/social impacts
5. construction/engineering feasibility

Candidate corridors were identified and analyzed in compliance with the National Environmental Policy Act (NEPA) and Forest Service regulations for the EIS.

INTRODUCTION

Selecting corridors for lineal facilities such as highways, power transmission lines, and pipelines presents particular challenges to environmental planners. Such studies must simultaneously address several environmental, socioeconomic, and engineering design criteria which are geographically spread across large study regions. Trade-offs between the travel use benefits of a proposed road and its projected costs — both economic and environmental — must be weighed. Furthermore, the geographic variability of physical and social conditions found along routes, as contrasted with the comparatively simple set of conditions found at a single-point site, serve to complicate the location and evaluation process.

These methodological issues can be effectively addressed through application of GISs. The purpose of this paper is to summarize, through a current case study, a GIS-based method by which corridors can be identified, evaluated, and compared.

PROJECT DESCRIPTION

The project under consideration is a highway proposed to be located in Maricopa County, AZ, crossing the Tonto National Forest. The primary purpose of the highway is to provide a shorter and faster route for motorists from the Phoenix metropolitan area headed for recreational areas north and east (Figure 1).

The project was visible and controversial from the start. Major issues of concern included the potential disturbance of desert ecosystems, bald eagle habitat, and sensitive riparian areas found along the Salt and Verde rivers. The region is heavily used for horseback riding, hiking, water-based sports, hunting, and all-terrain vehicle recreation. Environmental groups questioned whether constructing a new highway through largely undeveloped national forest lands would be justified by public need. In addition, steep and varied terrain posed major engineering constraints throughout most of the project region.

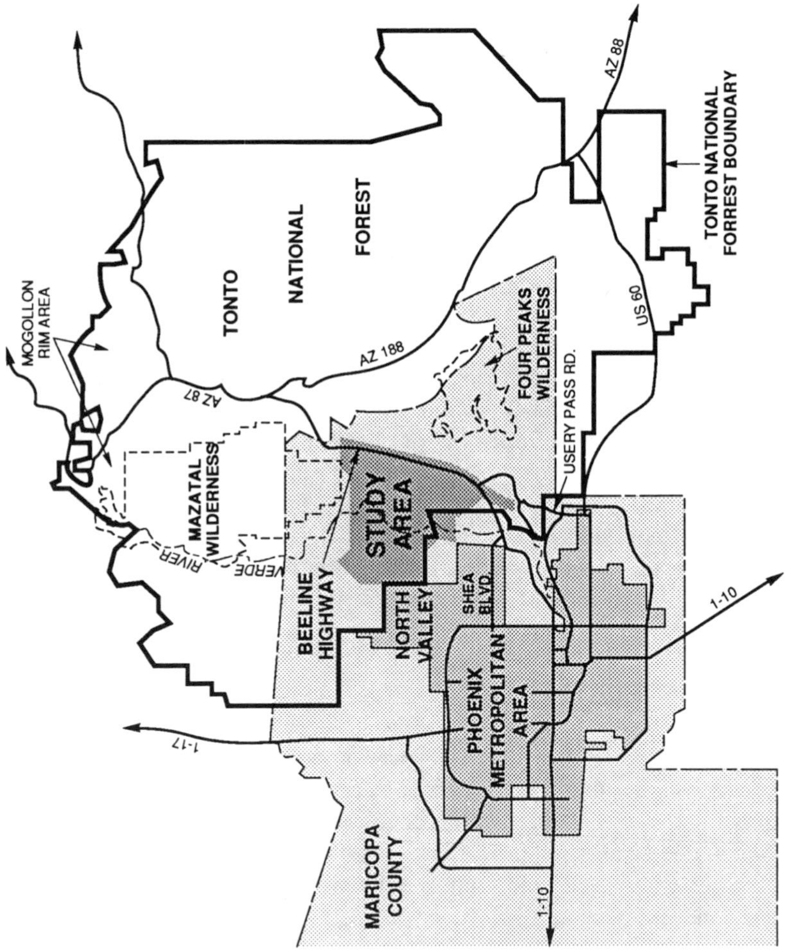

Figure 1. Project vicinity map. (From Dames & Moore.)

METHOD

The study was performed in two phases. Phase 1 sought to characterize the entire study region by relative degree of suitability for highway location based on several environmental and construction issues. Phase 2 entailed the selection of specific corridors and a more detailed comparison and evaluation of those corridors. Dames & Moore's GIS was used in both phases.

The first step was to carefully define a regional project study area, based on the location of existing roads, land jurisdictions, and a conceptual transportation plan. Within this study region, a virtually unlimited number of route solutions was initially possible. Thus, the Phase 1 methodology was designed to reduce the number of route solutions to a manageable number of candidate corridors. Regional characterization was accomplished through the development of resource suitability maps and a composite weighted overlay.

During Phase 2, candidate corridors were refined and consolidated as alternative highway routing alignments. The alternatives were then evaluated in greater detail to allow comparison between alternatives as required under NEPA/Council on Environmental Quality regulations for the EIS process. (The U.S. Forest Service was the lead federal agency.) This process entailed field investigations, detailed mapping, and specific highway location definitions. The GIS allowed environmental impacts associated with each candidate route to be quantified and compared.

PHASE 1 ROUTE SUITABILITY ANALYSIS

The Phase 1 suitability analysis was performed to (1) identify the environmental and highway engineering issues relevant to routing a highway through the Tonto National Forest and (2) design and implement a site evaluation method based on those issues. Dames & Moore resource investigators and the U.S. Forest Service interdisciplinary team took part in a 1-day "modified Delphi" session to discuss the issues, concerns, and opportunities pertinent to the project. In this session, five principal issues were addressed:

1. sensitive habitats: biological resources of concern to agencies and the public, especially riparian areas and bald eagle habitats
2. cultural resources: reflects legal protections given to archaeological and historic resources
3. visual resources: includes aesthetic impacts and the Forest Service's Visual Quality Objectives
4. community impacts: reflects the public's concern for highway traffic, land use, recreation values, noise, and air quality
5. construction/engineering: represents concern for geotechnical and design constraints

For each of the five issues, or siting factors, the appropriate resource investigators designed a specific methodology for transforming mapped input data into a composite map depicting highway location suitability for that factor (Figure 2). Each factor was also assigned a numerical weight which reflected its relative importance in the siting decision. Factor suitability maps were prepared by instructing the GIS to assign impact ratings to particular resource categories (e.g., vegetation type) and to combine impact maps into composites. The five-factor suitability maps were then combined through a weighted overlay to yield a single environmental suitability composite.

An engineering suitability map was also prepared, indicating the difficulty and costs associated with constructing a highway across floodplains, through rugged terrain, and through unfavorable geotechnical and soil conditions. Finally, the environmental and engineering suitability maps were combined and displayed in a manner which revealed relatively suitable corridors. These corridors became the project alternatives carried forward for Phase 2 EIS analyses.

PHASE 2 CORRIDOR ANALYSIS

Each of the five siting factors was addressed through more detailed corridor impact analyses in a method shown on Figure 3. For example, the cultural resources factor was evaluated through a consideration of known archaeological sites and an analysis of other observable criteria (terrain, water features, etc.) which would indicate the possible existence of undiscovered sites (Figure 4).

To measure visual impacts, a GIS-based model was designed in order to be rigorously consistent with the U.S. Forest Service Visual Resource Management System. Key components of this model were Visual Quality Objectives and Visual Absorption Capability. The GIS was used to characterize landscape types according to aspect, diversity, slope, landform complexity, and soil color contrast. The GIS also computed viewsheds from sensitive observer locations in the study region. Maps depicting visual magnitude — a measurement which considers the slope, bearing, and distance of a land plane from the viewer — were automatically calculated and displayed.

As a parallel activity, specific highway alignments for each alternative corridor were designed conceptually by engineers, and construction costs were estimated. Through the use of a personal computer-based highway design program and topographic data, the areal extent of land disturbance associated with each alternative alignment was mapped. By overlaying zones of disturbance with the resource impact maps, analysts were able to locate and quantify impacts (e.g., the amount of riparian vegetation or wildlife habitat replaced by each alignment; Figure 5). The GIS allowed such impacts to be graphically portrayed and statistically summarized, facilitating intercorridor comparison and EIS preparation.

CONCLUSION

The interdisciplinary team of environmental professionals and engineers from Dames & Moore and the U.S. Forest Service met during planning sessions to

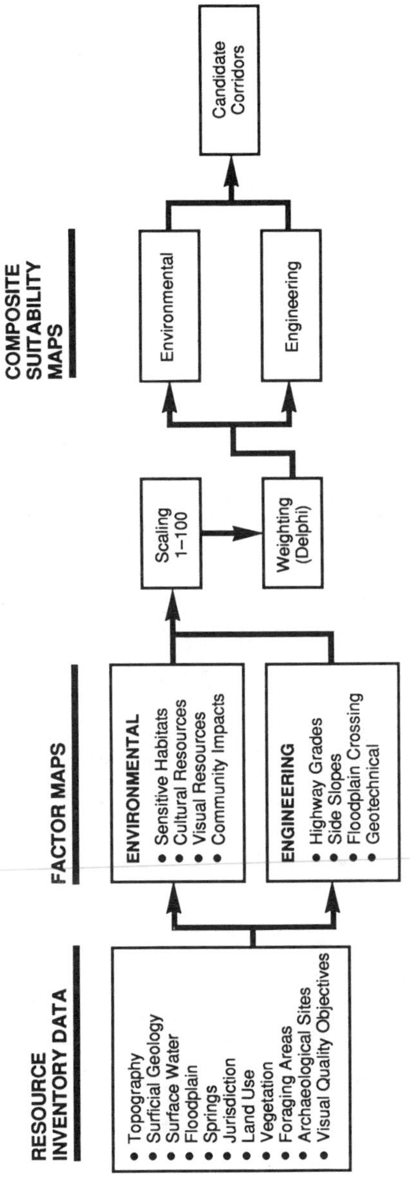

Figure 2. GIS data flow diagram. (From Dames & Moore.)

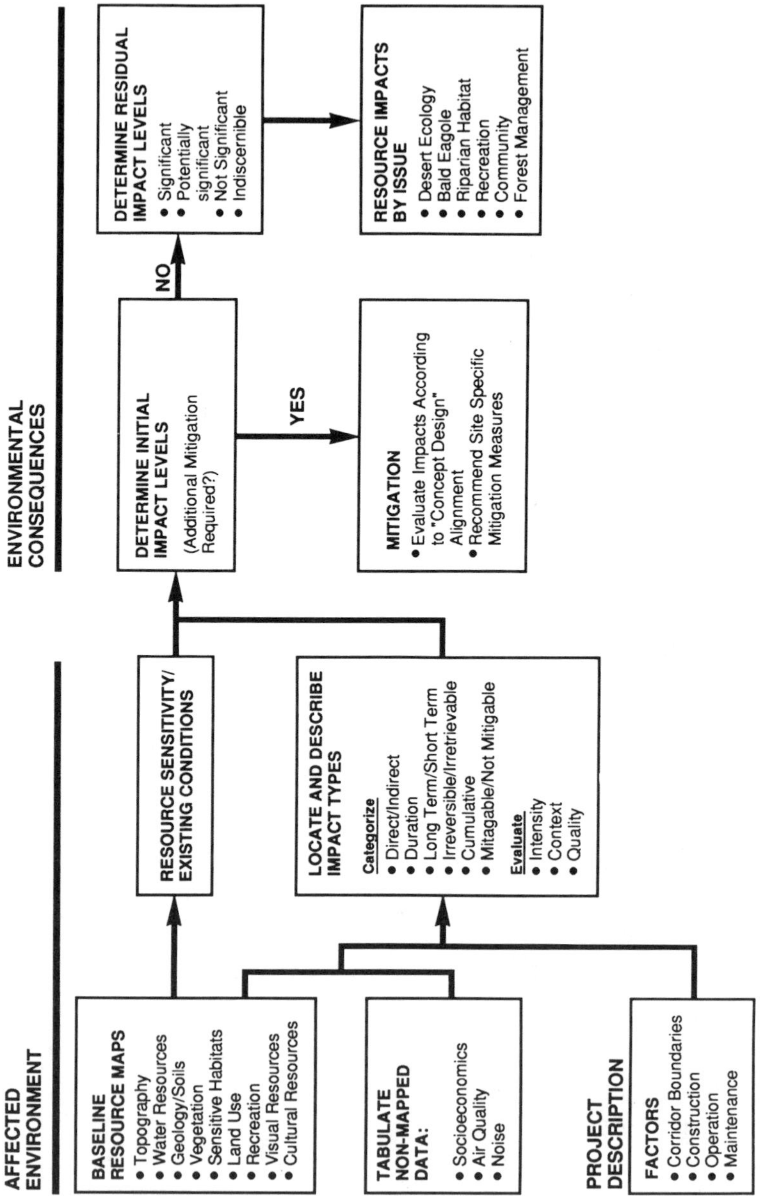

Figure 3. Impact analysis methodology. (From Dames & Moore.)

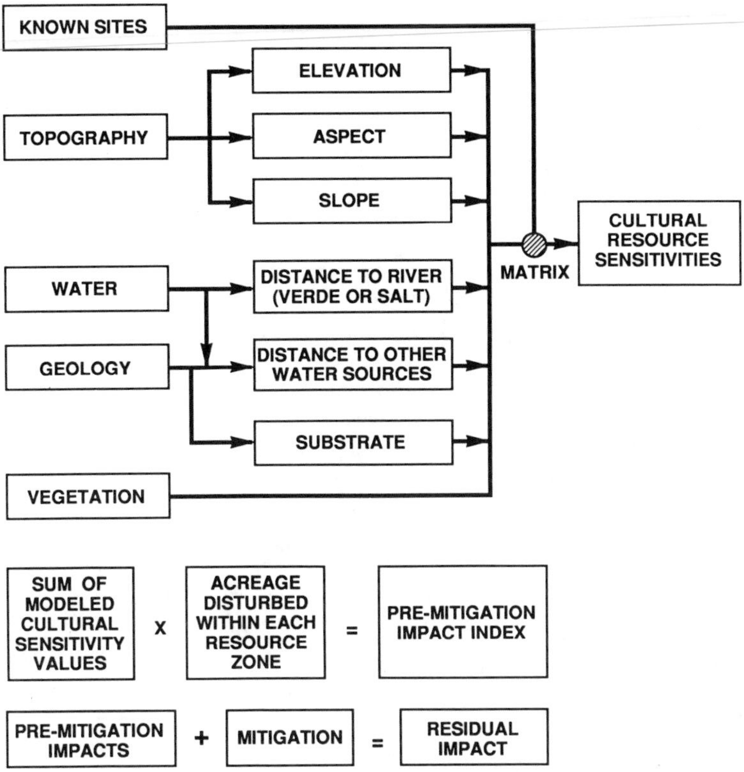

Figure 4. GIS cultural resources impact analysis method. (From Dames & Moore.)

evaluate the results of the impact analyses. In consultation with Maricopa County highway engineers, the planning team developed specific mitigation measures to reduce potential impacts. Based on the results of the GIS-based impact assessment, the nature and extent of significant impacts were determined.

Before the scheduled release of the draft EIS, Maricopa County withdrew the highway proposal. County officials stated that the degree of potential environmental impact and high cost of construction could not be justified relative to the public need for the project.

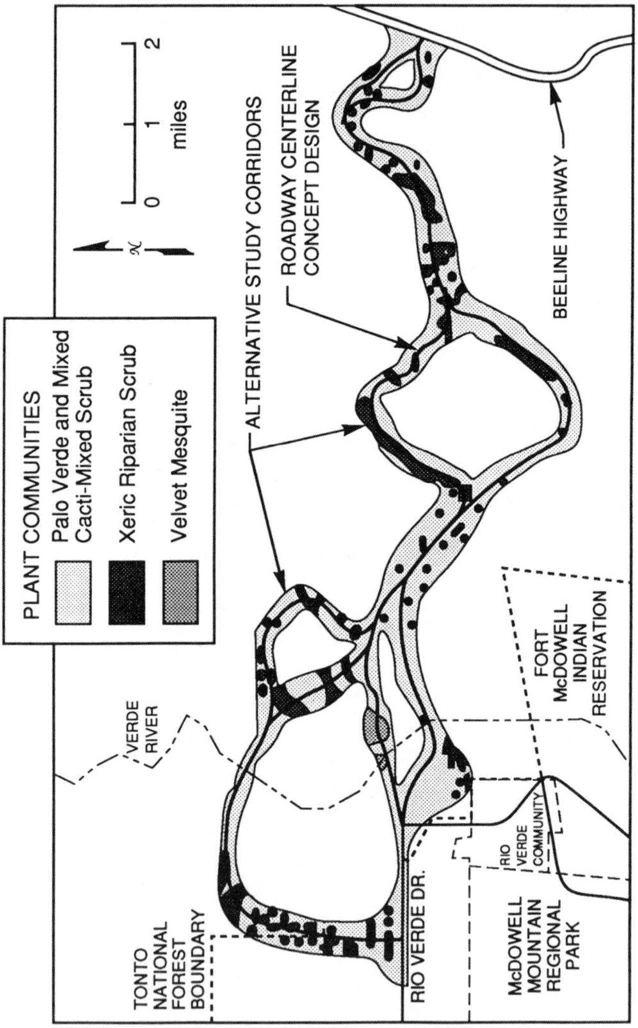

Figure 5. Plant communities in Rio Verde central corridor — impact overlay example. (From Dames & Moore.)

Applications of a Geographic Information System and Remote Sensing in Assessing Surficial Environmental Impacts of Siting and Site Characterization for a Proposed High-Level Nuclear Waste Repository at Yucca Mountain, Nevada

M. F. Winsor and D. Rousseau, Environmental Science Associates, Inc., San Francisco, CA

ABSTRACT

This paper presents the progress in one of the state of Nevada Nuclear Waste Project Office environmental investigations in process on the environmental effects of past and proposed activities of the U.S. Department of Energy (DOE) at the Yucca Mountain repository site. Both interpretation of remotely sensed imagery and development of a computerized database (GI) have assisted in assessing the extent and nature of disturbance from past siting activities. The disturbance information is being used to establish the nature of the existing environmental baseline, to develop assessments of future disturbance from DOE's extensive program of site characterization, to assist in developing mitigation, and to define strategies for identifying reclamation needs and approaches.

BACKGROUND TO THE PROJECT

Since 1976, the U.S. Department of Energy (DOE) has conducted preliminary technical suitability studies related to siting and possible development of the nation's high-level nuclear waste (HLW) repository at Yucca Mountain, NV. Those studies have disturbed the surficial environment; the significance of the impact is not fully assessed. Yucca Mountain is located in the Mojave-Great Basin Transition Desert, about 110 mi northwest of Las Vegas. The mountain itself is located partially on the Nevada Test Site (NTS) and partially on public lands administered by the U.S. Bureau of Land Management (BLM) and the U.S. Air Force (Nellis Air Force Base).

Site Selection under the Nuclear Waste Policy Act

The DOE's investigations are being carried out as part of its mission to find and develop a site to receive the nation's commercial HLW and some DOE weapons-production HLW. The Nuclear Waste Policy Act (NWPA) of 1982 formally defined this mission. The NWPA identified an objective of developing a mined geologic repository for the HLW; it established a process of site selection to include comprehensive investigations by the DOE and its contractors of technical suitability, protection of public health and safety, environmental protection, and other concerns.

The DOE carried out technical investigations at nine sites throughout the nation with a variety of different geologic materials. The well-publicized history of the DOE's progress in this program does not need to be repeated here. Briefly stated, after preliminary investigations and amid a storm of controversy, the DOE first narrowed the candidates to three sites in the western United States (Nevada, Washington, and Texas). In 1987, bending to pressure to keep momentum in the program, Congress passed the NWPA Amendments Act, which established that only one site would be characterized. The chosen site is Yucca Mountain, NV.

The selection of Yucca Mt. has generated intense opposition in Nevada from the state government, some local governmental entities, the public, and special interest groups. The state's opposition is based on the manner by which Yucca Mt. was selected, as well as technical and socioeconomic issues. The state argues that the Yucca Mt. site will prove unsuitable for a repository because of hazards posed by earthquakes, volcanism, and groundwater intrusion that would significantly compromise the integrity of the repository during the 10,000 years for which the DOE must demonstrate that the waste will be isolated from the environment. The geotechnical issues continue to be the primary focus of the Nevada Nuclear Waste Project Office's (NWPO) opposition to the HLW repository. Additional concerns raised by the state include unnecessary hazards created by the long transport routes from the primary HLW source areas in the eastern United States, socioeconomic impacts in southern Nevada, and other issues.

An important area of contention regards the need to fulfill substantive and procedural requirements of the National Environmental Policy Act (NEPA). These have been analyzed in detail by Lemons et al. (1989), Malone (1989), and Ulland and Winsor (1989). The NWPA [Sections 112(e) and 113(d)] partially exempted the HLW program from some NEPA requirements. Specifically, preliminary decisionmaking activities, defined to include activities conducted in support of recommending sites for characterization and site characterization, were exempted from an environmental impact statement (EIS) requirement of NEPA Section 102(2)(c) and from environmental review at international, state, and local levels under NEPA Sections 102(2)(E) and (F).

Despite the exemptions, other NEPA requirements remain applicable to the HLW program activities, including compliance with NEPA for repository site development [NWPA Sections 114(a) and (f)]. The NWPA and NWPA Amendments Act of 1987 limited NEPA requirements for evaluating alternatives to geologic disposal and alternative sites. The NWPA continues to require NEPA compliance, including preparing an EIS for the repository Section 114(a)(1)(D). Other NWPA provisions require environmental protection and mitigation during site characterization, site reclamation, and environmental impact mitigation if the site is found unsuitable.

The DOE's response to NWPA and NEPA requirements has not satisfied the state of Nevada. The DOE's program has been characterized as fragmented, failing to meet the intent of NEPA and the requirements of the Council on Environmental Quality (CEQ) (Lemons et al. 1989; Malone 1989). The DOE issued an environmental assessment (EA) in 1986 on the three sites then under consideration (U.S. DOE 1986). Because the EA fulfills a NWPA rather than a NEPA requirement, it is termed a statutory EA. The EA presents the DOE's conclusion that site characterization would have no significant environmental impacts. The state has contested both the finding of no significant impact and the adequacy of the information in the document on which the finding was based (State of Nevada 1985). Nevada initiated suit on the matter in federal court.

Ulland and Winsor (1989) concluded that from a regulatory compliance viewpoint, (1) the DOE environmental program and statutory EA are multidisciplinary rather than interdisciplinary, as would be required by NEPA; (2) they do not address cumulative impacts; and (3) they do not address the program in its totality of components, and, in fact, many components are yet undefined.

Beyond the issue of environmental impact documentation remains the issue of whether environmental objectives are being adequately served by the DOE's site characterization program. This question addresses issues at the heart of NEPA as the nation's comprehensive policy on the environment (Lemons et al. 1989). Part of the problem is that the DOE has not established environmental objectives. In the Site Characterization Plan (SCP) (U.S. DOE 1988b), the response to the DOE's Mission Plan Key Issue 3, the environmental issue, was put off until EIS scoping. EIS scoping is unscheduled at present. The DOE has no comprehensive environmental management program; there are neither defined goals nor implementation plans for managing the environment and resources of the NTS or for the repository program, which extends disturbances into resource areas and jurisdictions far beyond the NTS. The DOE's environmental program appears to be oriented primarily toward fulfilling an EIS preparation requirement. In the absence of defined environmental objectives and a comprehensive environmental baseline, it is difficult to explain how the DOE can assess impacts, identify significance criteria, and develop approaches to impact minimization and comprehensive mitigation.

NEVADA'S INDEPENDENT ENVIRONMENTAL REVIEW PROGRAM

Because of the preceding considerations, the state of Nevada is pursuing an independent program of evaluating environmental issues, developing environmental objectives, assessing regulatory compliance, and developing impact minimization guidelines and mitigation, including reclamation strategies (State of Nevada 1987, 1988). Despite the statutory EA's findings, the question of environmental impact significance determination for siting and site characterization remains open to further inspection.

The NWPO environmental program was established under provisions of the NWPA (Section 116), which guaranteed the state oversight of the DOE's program in Nevada for the HLW repository. Environmental Science Associates, Inc. (ESA) of San Francisco was selected in 1988 as the NWPO's prime contractor for the environmental investigations.

The Nevada environmental program is identifying disturbances in the broader study area and linking them to realistic assessments of environmental impact. This is occurring in the context of environmental management objectives rather than repository development objectives.

One of the first tasks is assessing the amount of landscape disturbance from the HLW repository program that has occurred in the past and that may be anticipated in the future. The state has sought to make these determinations for two reasons. First, the statutory EA does not address the issue of disturbance or its environmental impacts in a comprehensive manner. The state has questioned the finding of no significant impact and the adequacy of the information upon which the finding was based (State of Nevada 1985). Second, the DOE has made no firm commitment to mitigate or reclaim the impacts of their past or future disturbances. NWPA Section 113(c)(4) requires reclamation only if the site is found unsuitable. As the DOE claims that no significant impacts have occurred from past activities or are anticipated for site characterization activities, no basis exists for a mitigation requirement as covered by NWPA Sections 113(a) and 113(b)(1)(A)(iii).

Because activities disturbing the landscape were initiated as early as 1976, continuing unmitigated disturbances have already altered natural conditions and may worsen as additional disturbances are generated by site characterization. Some disturbances were created 13 years ago, and it may be assumed conservatively that another 10 years will elapse before the DOE decides to develop the repository or abandon it. For some disturbances, reclamation practices might not be put into effect for roughly 25 years.

DISTURBANCE ASSESSMENT

Given the preceding considerations, the NWPO initiated a program to identify the extent of disturbances and to develop a mitigation strategy, including reclamation needs. The fundamental question is: What disturbance has been created by past DOE activities related to repository siting and those proposed for site characterization at Yucca Mt.? To address the primary question, it is necessary to break it into its component parts.

- What elements of the project create environmental disturbance?
- Where are the past and proposed activities that create surficial environmental disturbance located?
- How can disturbances be identified?
- How can environmental change be assessed?

The continuing investigations were initiated in the summer of 1988. This paper presents results of the studies in progress and discusses some of the approaches to the assessment, focusing on the use of remote sensing and applications of a geographic information system.

What Elements of the Project Create Environmental Disturbance?

For this paper, the term "project" is operationally restricted to all siting and site characterization activities carried out by the DOE and its contractors. Thus, the benchmark is placed at about 1976 when the DOE initiated the first disturbance activities related to the HLW repository program in the Yucca Mt. area. It is bracketed on the future end at about 1999; this date is uncertain because the DOE is redesigning major elements of its site characterization program and negotiations with the state for permits have been stalled.

The activities of greatest pertinence to environmental issues are the investigations needed to answer geotechnical and hydrological questions about the suitability of the repository, as well as issues of public health and safety and radionuclide migration. Other studies relate to meteorological monitoring, cultural resources, and environmental studies themselves. Each of those activities has its own components that create environmental disturbance. The activities are too numerous and varied to describe here; see the DOE SCP for descriptions (U.S. DOE 1988b). Those past and proposed activities related to the geotechnical investigations have the potential for the most surface disruption and environmental impact. From an environmental viewpoint, the project is similar to a large exploration mining project.

ESA categorized activities described in the SCP using a classification scheme that is functional for disturbance assessment. The combined siting and site characterization programs create disturbances of the following categorized types: drill holes, trenches and test pits, seismic surveys, monitoring stations, bladed use facilities, roads and corridors, various erosional features that developed secondary to the original disturbance feature, and structures.

The SCP includes extensive subsurface activities associated with the drilling of the massive proposed exploratory shaft facility (ESF) and numerous drill holes. The subsurface disturbances may have other environmental effects, but they are excluded from the scope of this paper. However, the disturbances at the surface created by the ESF as well as other drill holes are included here.

Where Are the Past and Proposed Activities that Create Surficial Disturbances Located?

The statutory EA addressed a study area of variable size with detailed studies only in a 27.5-mi^2 area centered on the repository block and nearby surface facilities complex (SFC). The DOE's EA study area appears to have been too narrowly defined in the statutory EA. Activity descriptions and maps produced by the DOE contractors include activities reaching far beyond Yucca Mt., including, for example, seismic studies with associated disturbance features over 70 mi away in California (U.S. DOE 1988a, 1988b).

The primary activity and disturbance area has been identified by ESA in three nested geographic areas. The first area is the Core Study Area, encompassing about 86 mi^2, centered on Yucca Mt. in which the most intensive activities have occurred and are planned. A portion of the Core Study Area is identified as an area of focused baseline studies; it incorporates the first stage of detailed studies in the Core Study Area in the immediate vicinity of the ESF and SFC. In subsequent years, the detailed studies conducted in that area will be extended into the entire Core Study Area. The second area is the Cumulative Assessment Study Area, encompassing about 400 mi^2 in the vicinity of Yucca Mt. where many dispersed activities are mapped, including a broad area where environmental impacts may be expected. The third area is the Regional Study Area encompassing the outlying areas in Nevada and California in which isolated disturbance activities or regional impacts may be expected.

The nested study areas contain other land uses [many of which create environmental disturbances in their own right (e.g., mining, grazing, recreation, weapons testing)] that must be considered in cumulative impact assessment. A comprehensive view is needed to encompass cumulative disturbance so that a measure of the DOE repository program's contribution to environmental impact can be ascertained. This aspect of the ESA's disturbance assessment is in progress; it will not be reported in this paper, however, so as to focus on those disturbances directly related to the HLW repository program.

How Can Disturbances Be Identified?

To obtain complete inventories of disturbances, three routes of investigation are being pursued: DOE documents, remote sensing, and field data collection.

Document Review

First, information on activities producing disturbance were ferreted from the DOE's documents and other information sources. The DOE's SCP and related atlases are the primary information sources (U.S. DOE 1988a, 1988b). While those sources are extensive and contain much information on activities, they are not complete and have been subject to continual change. The DOE documents contain some conflicts of information. Data accuracy is not known: while some information appears to be very accurate (refined description and precise mapping), other information is vague and unmapped. The DOE's SCP is not organized by functional classes related to disturbance types that are useful for impact evaluation. Thus, the activities described in the SCP require interpretation of their potential related disturbance.

The DOE's environmental resource reports have been limited primarily to the approximately 27.5-mi^2 area of the EA and provide limited information for the Cumulative and Regional Study Areas.

Interpretation of Remote-Sensing Imagery

The tasks of classification and inventory of environmental conditions and disturbance in a large area make necessary the application of remote-sensing methods. Additionally, the state's contractors were denied access to the Nevada Test Site by the DOE for over a year, and access remains unavailable from Nellis Air Force Base lands; this situation created a dependence on remotely sensed imagery for much of the investigation.

Remote-sensing techniques are useful tools to obtain considerable information about past disturbances and existing conditions. The ESA obtained historic aerial photography (1959, 1976, and 1978) to identify landscape features prior to repository siting investigations. Large-scale (1:12,000), color, stereographic, aerial photography of the Core Study Area was flown in 1988 and provides a primary source of data for accurate measurement of indicator characteristics. Spectral imagery from high-altitude satellites is a third source of information. In each case, the images are essentially raw data that require interpretation and subsequent manipulation and analysis.

Standard air-photo interpretation was employed to identify landforms, general soil types, vegetation cover, and disturbance features. Interpreters made use of all seven standard image characteristics: shape, size, texture, pattern, shadow, tone or color, and location. Size, spacing, texture, and location were more useful indicators of vegetation and disturbance conditions than color because the desert vegetation cover is sparse, the vegetation growth forms do not identify well by color, and reflectance of the low-growing annual cover is not distinguishable from soil reflectance. The brushy plants of the Transition Desert are structurally similar and, thus, difficult to distinguish on the basis of the first six image characteristics. The interpreter used location or, more specifically, general landform classes such as colluvial slopes and alluvial fans to develop a key to visual signatures of vegetation cover. The key allows consistent guidance for further interpretive work; it is accompanied by example patterns for each signature (Cuneo 1989). Disturbance features generally stand out because of the abrupt changes in pattern from surrounding signatures. However, color sometimes was used to distinguish a disturbance feature from natural soil and bedrock exposures.

The large-scale aerial photography was supported by use of remote-sensing spectral imagery from the American Landsat 5 Thematic Mapper (TM) satellite and panchromatic imagery from the French SPOT satellite. The Environmental Research Institute of Michigan (ERIM) acquired and processed the imagery. The Landsat scene was from March 28, 1987; the SPOT scene was from March 27, 1987. ERIM processed the Landsat scene to create a false-color composite using bands 2, 3, and 4 (which correspond to the visible green, visible red, and shortwave infrared wavelengths, respectively); ERIM also used bands 3, 4, and 7 (which correspond to the visible red and shortwave infrared wavelengths) to create the shortwave infrared image. Both images cover an area about 100 mi north/south by 50 mi east/west.

The panchromatic SPOT data were processed into an orthorectified image to correct for horizontal and vertical errors created by the position of the satellite when the data were recorded. The image was sharpened through filtering (Laplacian) to improve edge contrast. The SPOT data have a higher resolution than the TM data, about 10 m compared to 30 m, respectively. The higher resolution and contrast of the planimetric SPOT image were especially helpful in locating roadways through variable terrain and in identifying other disturbances such as pads. The SPOT image also provided a synoptic view of the entire Cumulative Study Area in the plan. The orthorectified image also assisted in accurate feature placement.

The next step involved the merging of the TM and SPOT data sets to combine the advantages of higher spatial resolution of the SPOT data and greater spectral coverage of the TM data. The data were clipped to cover the Core Study Area. Data sets were merged to permit good identification of disturbance features in combination with information about soil and vegetation coverage. The result was an infrared image at much higher resolution than the original TM image. Through enhancement methods (Tasseled Cap transformation and chromatization), ERIM manipulated the overall color characteristics (spectral diversity) of the image. These enhancements were prepared in an attempt to improve digital classification of vegetation classes.

The imagery filled the need for image map graphics at several scales to display the environmental conditions in the study area. The imagery was useful as a navigational aid in the field and for identifying preliminary landform/soil and vegetation class distributions, although the exact nature of those conditions could be verified only through ground truth. The imagery clearly revealed most of the disturbance features. As an additional tool, a Nevada state plane coordinate grid (1000-ft unit) was projected onto the scene. This is used for locational reference, including data collection in the field and geographic information systems (GIS) matrix.

A preliminary attempt was made to use computer-aided techniques for classifying the vegetation cover and disturbed areas. This was accomplished by identifying 40 training sets of specific vegetative cover types for which species composition was anticipated; also selected were disturbance types and rock outcrops. The statistical separability of the training sets was insufficient to distinguish the categories through computer-aided classification of the TM scene. This was not surprising given the difficulty of separating soil and vegetation reflectance in the desert. If additional scenes were similarly created from other dates, better separation might be possible; however, funding limitations will put this study off to a later date. Some cover types were separable: the pinyon/juniper, ephemeral vegetation, and rock outcroppings.

The conversion of the TM data via the Tasseled Cap transformation and chromatic enhancement provides an additional product for manual analysis. It is possible to stratify regions of similar texture, tone, aspect, and elevation from the image. The scene provides an especially sharp image for picking out disturbance features such as roads, drill pads, and cuts. While computer-aided identification of vegetation signatures from the spectral imagery was not particularly successful, the imagery has proved useful in identifying the distributions of the vegetation series through landform, aspect, and elevational attributes revealed by the imagery and field data collection. Eight vegetation series are of particular interest: sagebrush, Great Basin mixed scrub, blackbrush, blackbrush/creosote, Mixed Transition scrub, creosote, Joshua Tree, low wash, and high wash. These form the broader framework for more detailed plant community classification.

Aspect and elevation appear to be important controlling variables of the distribution of the plant cover, as topography strongly influences localized insolation. A digital elevation model (DEM), or surface topographic model, of the study area is an important data layer for understanding the geographic distribution of many environmental variables under study. A DEM can be created through image-processing techniques at ERIM. To do this, a second SPOT image will be acquired that was taken at the same time of year as the current data set. The second image is also an off-nadir view with a look angle at least 25° different than the first image. A set of mapping polynomials is then generated for processing each image. After spatial filtering to enhance edge features, the disparity between edge features is measured and related to the actual elevation of the feature. This process is repeated iteratively at higher and higher resolutions, resulting in a DEM with horizontal errors less than 10 m and vertical errors of approximately 20 m.

Lastly, ground-level conventional photography was used to identify conditions that existed prior to a disturbance, in regards to the nature of the disturbance as well as the date of its occurrence. The photographic archive is very incomplete. No measurements were made from ground-level historic photography.

Field Data Collection

Field data collection to date has provided limited data on disturbance. ESA requested access to the Nevada Test Site in the summer of 1988, but permission was not granted until late summer 1989. This occurred at a time after Congress slashed the NWPO's budget, thereby forcing a greatly reduced field program. ESA conducted site reconnaissance studies and collected some information from the surrounding area on lands managed by the BLM to the southwest of Yucca Mt. Field investigations on the NTS were conducted in September 1989; results of that field program are still being developed. Access to the site of the exploratory shaft facility has not yet been approved by the DOE or BLM. Thus, DOE documents and interpretation of remote-sensing imagery compose the primary information sources for the bulk of information summarized in this paper.

To date, disturbance assessment has focused on past disturbance features created between 1976 and mid-1988. In summary, ESA's disturbance analysis is proceeding through a structured program to distinguish landscape features that have changed since 1976 as they appear on old and recent aerial photography or as they appear in spectral digital remotely sensed imagery. After feature identification is made, DOE documents are searched for correlation of the features with activities described by the DOE. Each feature is tagged, and the information is entered into the GIS. Not all features can be keyed to describe SCP activities. Each disturbance is identified by activity type, and associated disturbance characteristics (e.g., cut or fill area, waste handling, materials) are based on generic data derived from DOE documents. Features are mapped as accurately as possible, taking into account photographic distortion, displacement, and map scale transfer. The high-altitude orthorectified satellite imagery provides a satisfactory planimetric base map for locating disturbances, including roads that do not appear on published maps. Areal measurements are made with the use of a summarizing digital planimeter of mapped disturbances. All disturbances are first transferred to planimetric maps from photointerpreted images through the use of a KAIL projector.

Results of the Disturbance Assessment

Table 1 provides a summary of disturbance-type coverage for the Core Study Area and Cumulative Assessment Study Area. The primary sources of past disturbance were bladed roads, trails, drill holes with pads and mud pits, and bladed clearing/borrow areas. Within the study area, a total of 638 ac (about 1 mi^2) may be identified as existing "disturbed" land.

The data from past disturbances can be applied to proposed activities within the study area as reliable generic indicators of future disturbance. Given the assumed applicability of average acreage *per* disturbance type, as well as projected disturbance identified in the SCP, it is possible to estimate additional disturbance related to site characterization of approximately 520 ac. The value is very rough because the SCP lacks good definition of many activities and their location. The combined past disturbance and proposed site characterization disturbance are indicated in Table 1; the total is approximately 1160 ac of landscape change.

How Can Environmental Change Be Assessed?

The type of disturbance assessment described in this paper provides only one useful indicator of the magnitude of disturbance: total area disturbed. It falls short

Table 1. Summary of Existing and Projected Disturbance Features (Number and Acreage) from Siting and Site Characterization at Yucca Mt., NV, and Vicinity.

Disturbance Type	Numbers Currently Identified	Average Acreage	Acreage Currently Disturbed	Additional Number Expected	Additional Acreage Expected	Estimated Total Disturbance
Drill hole with pad and mud pit acres	29	2.25	65.2	47	106	171
Drill hole with pad, mud pit, and spill	2.5	2	5	5		
Drill hole (no pad)	8	0.13	1.1	150	19.5	20.6
Drill hole with water storage	1	2.25	2.25			2.25
Drill hole with pad (no pit)	24	0.98	23.5	20	19.6	43.1
Deep trench	47	0.3	13.8	42	12.6	26.4
Test pit	3	0.02	0.05	15	0.30	0.35
Infiltration pit	2	0.61	1.2	6	3.6	4.8
Erosion chain		0.1		3	0.3	0.3
"Pavement"	61	0.05	0.3	10	0.5	0.8
Reflective survey line[a]	10	1.08	10.8	?	19	30.0
Refractive survey line[a]	3	1.02	3.05	5	40	43
Refractive shock points	3	0.25	0.75	32	8	8.75
Met. station with pad	4	0.5	1.92			1.92
Met. station (no pad)	2	0.1	0.18	4	0.4	0.58
Stream gage		0.1		15	1.5	1.5
Paved parking		5		2	10	10
Bladed parking	2	5	10	4	20	30
Bladed building	1	5.26	5.26	2	10	15
Flood control structures	1	0.5	0.47			0.5
Bladed clearing/ borrow area	37	2.3	85.04	50	150	235
Paved road[a]	4		62.92			62.92
Bladed road[a]	106		233.0	30	60	293
4 trail[a]	207		117.1	150	35	152
Total			638		522	1158

[a] Linear feature.

of providing a comprehensive analysis of disturbance. Equally important, but more difficult to analyze, are volumes of cut and fill, relative location of disturbance defined by functional parameters related to environmental issues, and materials involved (import/export, wastes, introduced hazardous materials, spills of toxic materials). Subsurface impacts, which have potentially profound impact on groundwater, soils, and archaeological resources, also must be addressed. Additionally, the analysis was limited to landscape feature identification and does not address event-related disturbances such as traffic, noise, air quality emissions, and animal kills.

Relating Disturbance to Environmental Impact Assessment

An important application of the information is the development of models of disturbance and their associated impacts that can be applied to the activities described in the SCP. These are being developed for purposes of significant impact avoidance, identification of mitigation requirements, and reclamation planning. The critical question yet to be answered is: Do the identified disturbances create the potential for significant adverse environmental impacts?

Environmental impact assessment follows disturbance evaluation. The landscape disturbance assessment has yielded much good information that can form the basis of environmental impact assessment of past siting and proposed site characterization activities and repository development plans. This is an ongoing part of the state's program; results will not be reported here. However, a brief discussion of approaches is appropriate.

Impact assessment is derived by relating disturbances, singly and in combination, to environmental conditions and trends. This requires the development of good baseline information and the application of meaningful impact models to the project at hand. Baseline development is just beginning and requires field investigations in areas to which ESA has been denied access. ESA's goal is to obtain sufficient data on baseline conditions and measured effects to build impact models for each proposed activity. The impact models relate disturbance types to elements of the environment to significance criteria.

Impact significance identification must be related to site- and time-specific conditions and cumulative considerations of the relevant issues. For example, significance criteria for a particular specific species of animal cannot be identified until it is defined in relation to the broader distribution of the species, conditions of the habitat in the proposed activity area as well as in the broader range of the species, roles and niches in the ecosystem, sensitivities to disturbance, and other considerations.

A good example of this is the desert tortoise, *Gopherus agassizii*, which has been temporarily listed by the U.S. Fish and Wildlife Service in California and Nevada (U.S. DOI 1989). For California's proposed low-level radioactive waste disposal facility, U.S. Ecology, Inc. conducted evaluations of tortoise populations, habitat condition, and cumulative development impacts in an area of about 110 mi^2 to evaluate the significance of the impacts to the tortoise population at the approximately 70-ac disturbance area at the proposed site near Needles, CA.

The desert tortoise is present in the Yucca Mt. study area. The populations probably are not large, and they are located at the edge of the overall range of the tortoise. The recent listing of the tortoise places added significance to the environmental impacts of site characterization on the local tortoise populations. A reopening of the statutory EA might be in order because of this issue alone.

The DOE's proposed approach to impacts on the tortoise relies heavily on relocating individuals where disturbance activities would occur (U.S. DOE 1988c).

The approach is open to question because (1) no evidence shows that removed tortoises would survive in a new area; (2) relocated tortoises add pressure on the habitat and other tortoises in the relocation site, especially if the site is near Yucca Mt., which is at the margin of the tortoise range; (3) no easy way exists to locate tortoises that inhabit burrows for much of the year; and (4) small tortoises are not easily observed, especially by operators of large equipment. Moreover, this approach ignores considerations of habitat sensitivity and value and the potential fragmentation of the populations (with consequent break in genetic pools and transfer lines). The DOE also makes no commitment to restoring habitat suitable for the tortoise.

An approach is needed that (1) evaluates the vitality of the tortoise populations and the status of their habitat and (2) identifies sensitivities of the tortoise to disturbance and assesses the potential impacts at a given site of proposed disturbance. This would be followed by an approach that seeks avoidance of impact as a primary objective. Mitigation, such as relocation, would occur only when impact is absolutely unavoidable by other measures and would be defined in the context of larger species population and habitat management objectives. It is for such reasons that detailed analyses of environmental conditions, open evaluation of disturbances, and reevaluation of their potential impacts are needed. The DOE has characterized the SCP as a "living document," which is a useful approach for a project as large, complex, and long term as this. By the same token, the state is justified in expecting the EA to also be a living document, so that changes in the DOE program can be tied to projected disturbances and their related environmental impacts and that evolving changes in environmental parameters can be responded to with sensitivity.

These types of inventories, evaluations, and analyses are based on large volumes of evolving information in many forms (tabular data, maps, etc.). Management and manipulation of such information demand sophisticated methods that modern computer systems can efficiently provide.

GIS APPLICATIONS IN ASSESSING ENVIRONMENTAL STATUS

In studying the landscape ecology of Yucca Mt., ESA is focusing on the spatial nature of environmental features. We are interested not only in location and distribution of independently mapped features, but in how well the various feature patterns coincide or correlate. The idea is to separate landscape variables such as topography, drainage, landform, soils, and vegetation into data layers that can be combined variously. Eventually, such data can be used to develop environmental resource management plans based on an analysis of site suitability, sensitivity, or importance derived from weighting the various independent data layers and then combining them. A GIS system is designed to aid in this process.

Computerized databases are essential for handling large volumes of data. Database management systems store data in unique form (no duplications of data) and access them with speed. Relational database management systems allow users to select data or cross-tabulate data by combinations of search criteria; this capability promotes an ad hoc investigation of data relationships. Computers also benefit cartographic work; automated mapping is an efficient means of producing maps of various scales from the same data set. The GIS is a combination of database management and automated mapping. It links digitized map images (graphic data) to tabular data resident in the database. The GIS also stores the graphic data nonredundantly by defining the topology of digitized map data. Topological analysis explicitly defines the spatial relationships among the points, lines, and polygon features that make up the digitized map.

A GIS serves as a computerized system for collecting, storing, retrieving, transforming, analyzing, and displaying spatial data. With spatial data, (1) position is known with respect to a coordinate system, (2) associated attribute data (tabular data in the database) are independent of position, and (3) topological relationships can be defined.

For the Yucca Mt. project, the GIS is being used to digitize manuscript maps developed either from photointerpretation or from actual field work. These data layers include roads and disturbed areas, transect and sample locations, sampling grid, soil categories, landform categories, hydrography, vegetation signature patterns, vegetation series, plant communities, sensitive plant species, desert tortoise sign, and desert tortoise habitat. Others are to be added later as information is developed.

The base map consists of contours, political boundaries, and stream channels prepared in CAD format and then converted to the GIS system. The base map serves simply as a reference image. A raster scanned image converted to GIS format is also available as a reference image on which data can be plotted.

As discussed, a topographic model is essential for modeling the environment at the study site. ESA has developed a topographic model by hand digitizing a sample grid of elevation points from a large-scale U.S. Geological Survey topographic quadrangle. These points were interpolated and contoured using a surface modeling software system. The resultant map was imported into the GIS, and the accompanying elevation data were entered into the tabular database. A tabular database is being built parallel to the digitized maps. In addition to elevation data, the tabular database contains the following data obtained during field sampling:

- transect characteristics such as landform, elevation, aspect, percent slope, percent rock cover, soil type, and vegetation type
- percent cover of plant species at each transect
- number of wildlife captures by species at each transect

To interpret the landscape of Yucca Mt., ESA is beginning by stratifying the area based on broad regional variables like photointerpretation of large landform units. The stratified areas are then used to select sampling sites for field investigation; for example, soil samples are taken within each landform unit identified in the stratification step. The results of fieldwork are then used to develop a classification system of the feature under study, for example, a refined soils classification. After a preliminary classification system is derived, it is used to map out location of classes; for example, a detailed soils map is drawn. Once classified, the features can be inventoried and then correlated in space to determine sensitivity or relative importance (e.g., the acreage of unstable soils can be derived and located in relation to disturbed areas). Finally, management recommendations can be made that minimize impacts to sensitive areas or provide for reclamation of disturbed areas based on previous analysis of independent data layers. For example, reclamation techniques can be grouped for various combinations of soil, slope, and elevation.

For the Yucca Mt. project, the GIS aids in automating these steps of stratification, classification, inventory, analysis, and planning. Once the above data layers are digitized and the tabular field data are entered, we can combine them through overlay analysis to investigate spatial relationships in the landscape. For example, we are overlaying desert tortoise range on vegetation type, soils, and landform to derive tortoise habitat. With this understanding of habitat variables, we can recommend measures to restrict or minimize impacts to habitat or to provide guidelines for restoring damaged areas back to original habitat value.

The GIS also documents changes in environmental status over time. For example, disturbance mapping will be updated as site characterization activities proceed; updates can be compared to previous maps to calculate changed areas and to cross-reference them to vegetation type or habitat. If specific areas are undergoing reclamation, the particular restoration regime can be stored in the database along with data on success rates.

The GIS serves as a central repository for the NWPO environmental investigations of Yucca Mt. It is helping us to locate and understand environmental resources, to plan for their best management, and to track progress in maintaining or restoring them.

ESA operates pcARC/INFO software on a powerful microcomputer hardware platform. The pcARC/INFO is a version of the most widely used vector-based GIS software. ESA's DOS-operated microcomputer is an Intel 80386 running at 25 MHz with a math coprocessor. The size of the microcomputer is well-adapted to the scope of NWPO's environmental studies program. It offers the advantages of the wide range of GIS capabilities needed for baseline studies and environmental analyses without the high costs of large mainframe computers and expensive peripherals. The system also provides network capabilities with other GIS uses for the HLW repository studies of the state.

CONCLUSIONS

The use of both remote-sensing methods and database management with a GIS can be effective tools in environmental assessments and environmental management. For the HLW repository project, both tools are being applied to independently assess environmental baseline conditions, identify disturbance, and evaluate impacts. The HLW program is necessarily complex, encompasses activities in a large geographic area with a variety of biophysical environments, and is extended over a long period. Programs with these qualities are well-suited to remote sensing and GIS applications for information development and management.

Methods employing remote-sensing imagery and the GIS are useful for assessments requiring a synoptic overview and for addressing problems of classification, inventory, stratification, and environmental modeling at varying scales. Both have been instrumental in establishing the basis of an environmental baseline in advance of further disturbances that will be created by site characterization. The information is being used to develop environmental objectives that can be applied to the study area's resources in their existing state. The methods are proving additionally useful for identifying sensitivities of the resources and for modeling impacts. They are being applied for purposes of reclamation planning for the disturbances at Yucca Mt. In the future, they will be applied to monitoring and recording change in the environment. In summary, information is being developed that is assisting the state of Nevada, through its NWPA oversight role, to ensure that the DOE's HLW repository program responds to the state's concerns about environmental effects of the program. Additionally, we hope these investigations will lead to more sensitive environmental management of the HLW program at Yucca Mt. by the DOE. If that results, these investigations will have served well the substance and intent of the NEPA.

REFERENCES

Cuneo, K. 1989. Methods of Pattern Analysis of Large Scale Aerial Photography of the Yucca Mountain Core Study Area. Environmental Science Associates, Inc., San Francisco, CA (in preparation).

Lemons, J., C. Malone, and B. Piasecki. 1989. America's high-level nuclear waste repository: A case study of environmental science and public policy. Intern. J. Environ. Studies 34:25–42.

Malone, C. R. 1989. Environmental review and regulation for siting a nuclear waste repository at Yucca Mountain, Nevada. Environ. Impact Assess. Rev. 9:77–95.

Nuclear Waste Policy Act of 1982. PL 97-425, 42 USC 10101-10226.

Site Characterization Plan, Yucca Mountain Site. Nevada Research and Development Area, NV. Vols. 1–7.

State of Nevada. 1985. State of Nevada Comments on the U.S. Department of Energy Draft Environmental Assessment for the Proposed High Level Nuclear Waste Site at Yucca Mountain. Nuclear Waste Project Office, Carson City, NV.

State of Nevada. 1987. Environmental Program Planning for the Proposed High-Level Nuclear Waste Repository at Yucca Mountain, Nevada. NWPO-TR-001-87. Nuclear Waste Project Office, Carson City, NV.

State of Nevada. 1988. A Role in Environmental Compliance for the State of Nevada During Site Characterization of the Proposed High-Level Nuclear Waste Repository Site at Yucca Mountain, Nevada. NWPO-TR-008-88. Nuclear Waste Project Office, Carson City, NV.

Ulland, L. M., and M. F. Winsor. 1989. The Role of the State and Environmental Compliance in NWPA Implementation. Waste Management 1989 (proceedings) 1:167–171.

U.S. Department of Energy, Office of Civilian Radioactive Waste Management. 1986. Environmental Assessment, Yucca Mountain Site, Nevada Research and Development Area, NV. Vols. 1–3.

U.S. Department of Energy, Nevada Operations Office. 1988a. Draft NNWSI Project Site Atlas. Vol. 1.

U.S. Department of Energy, Office of Civilian Radioactive Waste Management. 1988b. Site Characterization Plan, Yucca Mountain Site. Nevada Research and Development Area, NV. Vols. 1–7.

U.S. Department of Energy. 1988c. Draft NNWSI Project Environmental Field Activity Plan for Terrestrial Ecosystems. Nevada Nuclear Waste Storage Investigations Project, Waste Management Project Office, Nevada Operations Office, Las Vegas, NV.

U.S. Department of Energy, Office of Civilian Radioactive Waste Management. 1989. Draft Reclamation Plan, Yucca Mountain Site. Nevada Research and Development Area, NV.

U.S. Department of the Interior, Fish, and Wildlife Service. 1989. 50 CFR Part 17, Endangered and Threatened Wildlife and Plants: Emergency Determination of Endangered Status for the Mojave Population of the Desert Tortoise. Emergency Rule. Fed. Regist. 54(149):32326–32331.

Big Cypress National Preserve Important Resource Areas and the NEPA Process: Integration of Geographic Information System Resource Data into National Park Service General Management Planning, Impact Assessment, and Public Involvement

N. Kuykendall and M. S. Bilecki, National Park Service, Denver, CO; and K. Klubnikin, formerly with the National Park Service, Washington, D.C.

ABSTRACT

In March 1985, the National Park Service (NPS) announced its intention to prepare a general management plan (GMP)/environmental impact statement (EIS) for Big Cypress National Preserve. The GMP process sets forth the basic management philosophy for NPS units and provides strategies for addressing issues and achieving identified management objectives over a 10– to 15-year period. Integrated into the GMP process are the National Environmental Policy Act (NEPA) requirements to provide for public disclosure of the planning and decisionmaking process and the potential environmental consequences of the proposal and alternatives.

Big Cypress National Preserve, located in southern Florida, encompasses 574,440 ac and contains one of the least developed watersheds in South Florida. The legislation that established the preserve in 1974 directed the NPS to protect the ecological integrity of the area and identified several land uses to be controlled by reasonable regulations.

The central issue is how to provide for visitor use and access without impairing limited natural and cultural resources or the beneficial experiences they provide.

Before development of the proposal and alternatives, 11 important resource areas were identified, mapped, and entered into a computer-based geographical information system (GIS). Important resource areas included eight vegetation types, habitat for four endangered and threatened species, and cultural resources.

The GIS integrated complex natural and cultural resource information (important resource areas included) needed to determine management zones and planning units, to identify planning objectives, and to assess impacts. The use of important resource areas was imperative as a tool for scoping issues, defining objectives, identifying use limits, and producing maps for newsletters and the alternatives pamphlet used in the public involvement process. This database enabled the NPS to develop the proposal and alternatives, as required by NEPA, based on the complex resources and public use of the preserve.

INTRODUCTION

In 1985, the National Park Service (NPS) began the process of preparing a general management plan (GMP)/environmental impact statement (EIS) for Big Cypress National Preserve. A GMP sets forth the basic management philosophy for NPS units and provides strategies for addressing issues and achieving identified management objectives over a 10- to 15-year period. Under NPS guidelines (NPS-2 *Planning Process Guideline* and NPS-12 *NEPA Compliance Guideline*), the GMP process is intertwined with the NEPA requirements to provide for public disclosure of the planning and decisionmaking process and the potential environmental consequences of the proposal and alternatives. A GMP identifies, within legislative parameters, reasonable use and development alternatives to ensure adequate protection of the resource values that caused Big Cypress to be included in the national park system while also allowing compatible public use and enjoyment.

Planning for the preserve was unusually complex because of the diversity of natural and cultural resources in the Big Cypress, the legislative directive to allow but control certain uses of preserve resources, the multiplicity of interest groups concerned about management of the preserve, and the intensity of their interests and emotions. In response to this highly charged political atmosphere, the NPS sought to integrate planning with NEPA impact assessment and public involvement on the basis of mutually recognized and legally defensible resource values. This was accomplished through development of designated "important resource areas" and a computerized GIS for documentation, analysis, and display.

BACKGROUND

Big Cypress National Preserve is in southern Florida, immediately north of Everglades National Park and centrally located between Miami and Naples (Figure 1). The original boundaries of the preserve, established in 1974, contain

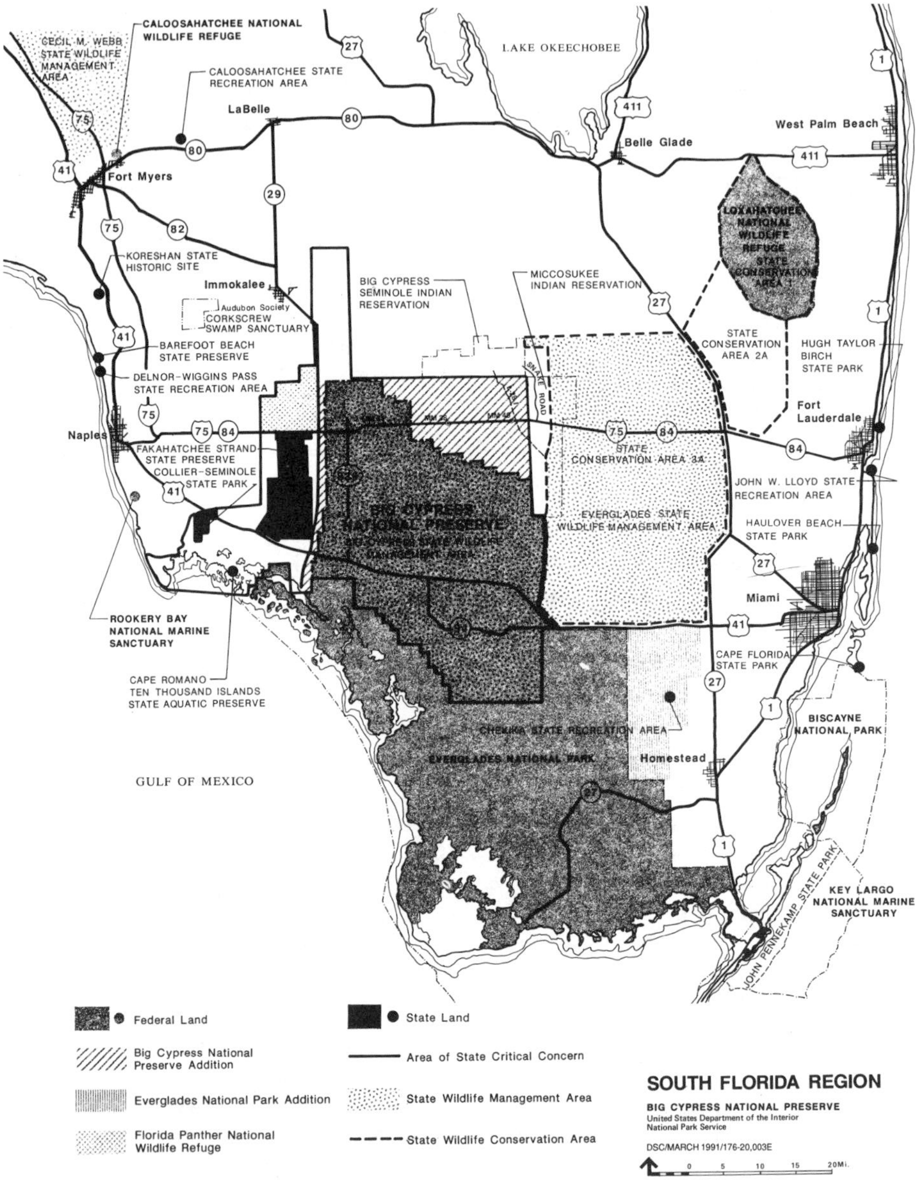

Figure 1. South Florida region.

574,440 ac and encompass one of the least developed watersheds in South Florida. Legislation in 1988 added 146,000 ac to the preserve, mostly in the adjacent Shark River watershed.

Water is a principal natural resource of the South Florida region, and 90% of Big Cypress is flooded during the wet season because of high annual rainfall (mean annual precipitation is 54 in., with approximately three fourths falling during the summer) and almost flat topography (a seaward slope of about 2 in./mi). Throughout the wet season, the water flows in a southwesterly direction through the preserve to the estuaries of western Everglades National Park.

Extensive prairies and marshes, forested swamps, and shallow sloughs characterize the preserve. Hydroperiod is the major determinant of vegetative communities, and a difference of only a few inches in elevation changes the hydroperiod and leads to the establishment of totally different plant communities. At one time, Big Cypress contained pristine cypress strands and old-growth pinelands; but by 1950, virtually all the cypress strands of commercial value and much of the pinelands within the preserve had been logged. The young cypress strands, mixed-hardwood swamps, and pinelands in the preserve today are still recovering. Big Cypress is also noted for its widespread cypress prairies — natural grasslands dotted with stunted cypress trees.

Most wildlife species common to South Florida occur within the Big Cypress watershed. In addition, ten species are listed by the U.S. Fish and Wildlife Service as threatened or endangered, and an additional 26 species are listed by the state of Florida as rare, threatened, or endangered. One of Florida's most endangered animals is the Florida panther (*Felis concolor coryi*), and it is the subject of an intensive recovery effort in the preserve and the surrounding region.

Archaeological surveys within the preserve have located 395 sites, some dating to as early as 500 B.C. (the Glades I period). Six sites have been placed on the *National Register of Historic Places*, and 12 are potentially eligible for listing.

Today, many Seminole and Miccosukee Indians depend on the preserve as a source of natural materials for housing, crafts, and other cultural and religious uses. Eleven Miccosukee villages, two ceremonial sites, and several burial sites are within the preserve's boundaries.

Oil exploration has been going on in the Big Cypress since the 1940s. Today there are two producing oil fields in the preserve. Most mineral rights and subsurface estates are privately owned.

Recreational activities in the preserve include hunting, off-road vehicle (ORV) use, fishing, camping, and hiking. White-tailed deer and feral hogs are the most popular large game animals, and hunters are the primary users of ORVs. Fishing is popular in borrow canals along major roads, and the canals are also prime locations for wildlife viewing. Campgrounds and undeveloped campsites are used mainly by hunters and winter visitors. The principal hiking trail in the preserve is the Florida National Scenic Trail, which traverses the length of the state and terminates in the Big Cypress at the Oasis ranger station. Visitor facilities include an information center at Oasis, an environmental education camp for school children at Pinecrest, two wayside picnic areas, and five primitive campgrounds.

The major tracts of nonfederal public land in the preserve are owned by the Florida State School Board and the Dade County Port Authority, which operates the Dade-Collier Training Airport (a 24,000-ac tract known as the Jetport). Some 200 small, privately owned parcels (2 ac or less) scattered throughout the preserve are classified as improved properties and are exempt from acquisition unless owners are willing to sell or the land is threatened with uses that could be detrimental to the purposes of the preserve.

LEGISLATION

When Congress passed Public Law 93-440 in 1974 to establish Big Cypress National Preserve, the concept of a "national preserve" was a new one. The intent is to protect the ecological integrity and recreational opportunities of the area while also allowing certain private land uses to the extent those uses do not interfere with the basic purpose of the preserve.

The House and Senate reports identified the natural flow of freshwater as a fundamental resource in the preserve. Freshwater flow is the key to the survival of Everglades National Park and the integrity of the entire South Florida ecosystem. Further, the report cited the natural, scenic, floral, and faunal values of Big Cypress as being worthy of national recognition and protection on their own merit. Recreation is discussed along with the natural values because the natural resources provide opportunities for recreational pursuits.

The act states that the preserve, as a unit of the national park system, is to be administered in a manner that will ensure its "natural and ecological integrity in perpetuity." The act further directs that rules and regulations necessary and appropriate to limit or control the following uses be developed:

- motorized vehicles
- exploration for and extraction of oil, gas, and other minerals
- grazing
- the draining or constructing of works or structures that alter natural watercourses
- agriculture
- hunting, fishing, and trapping
- new construction
- other uses that may need to be limited or controlled

Furthermore, the act permits the Miccosukee Tribe of Indians of Florida and members of the Seminole Tribe of Florida,

> subject to reasonable regulations ... to continue their usual and customary use and occupancy ... including hunting, fishing, and trapping on a subsistence basis and traditional tribal ceremonies.

In addition to Public Law 93-440, other federal legislation affects planning and management of the national preserve. For instance, the NPS's 1916 organic act and NEPA both direct management to conserve environmental quality for future generations. National resource protection policies contained in the Endangered Species Act, the National Historic Preservation Act, executive orders on floodplain management and wetland protection, and other regulations must also be addressed.

GENERAL MANAGEMENT PLANNING

In March 1985, the NPS announced the preparation of a GMP/EIS for Big Cypress National Preserve. An interdisciplinary team was assembled from four NPS offices: the Denver Service Center Eastern Planning Team, the Mining and Minerals Division, the Southeastern Regional Office, and the preserve.

Four public scoping meetings were held in South Florida at the beginning of the planning effort. From the start, it was apparent that the project would be highly controversial. The Big Cypress had long been viewed as Florida's last frontier, a place insulated by its rugged terrain from mainstream pressures. Traditionally, it has been a backcountry refuge where individuals can escape regulation. The reality of the modern demands on the Big Cypress is quite different from the traditional view. Although backcountry travel in the swamp is not easy, the Big Cypress is no longer isolated. In fact, the preserve is cut by two major highways connecting Miami with Naples and Fort Myers, two of the fastest growing urban centers in the nation.

Further, each public and private use mentioned in the legislation (e.g., resource protection, hunting, ORV use, oil and gas development, Native American subsistence, private landownership) has an interest group, and each interest group has its own perception of what constitutes "reasonable regulation" of each activity.

The development of the proposed action and the alternatives was guided by the identification of important resource areas, the development of a management zoning scheme based on those resources and on the NPS *Management Policies* (NPS 1989), the establishment of planning units, and the development of planning objectives. Each of these was a discrete step in the planning process, and the results are described in the following sections. The interrelationships between these steps are shown in Figure 2.

Identification of Important Resource Areas

Although the entire preserve is federally protected, the congressional directive to allow, but control, certain consumptive uses implies that some degree of disturbance to resources is acceptable so long as a core of essential resources is not compromised. One of the first tasks was to clearly define those core resources and their distribution.

Earlier, in 1983, the Energy, Mining, and Minerals Division of the NPS (now Mining and Minerals Branch) first attempted to define special resources in an internal report and map entitled *Proposed Sensitive Resource Areas* (NPS 1983). The report was intended as a guideline for the petroleum industry so as to avoid disturbing outstanding resources in the preserve and to provide a more defined framework for NPS review of oil and gas plans of operations.

The general management planning team adapted the sensitive resource areas approach for developing the plan and for guiding the environmental impact assessment. The term "sensitive resources" was changed to "important resources" because there was considerable variance in sensitivity of the resources, but all were considered to be important. Adapting the previous work involved reexam-

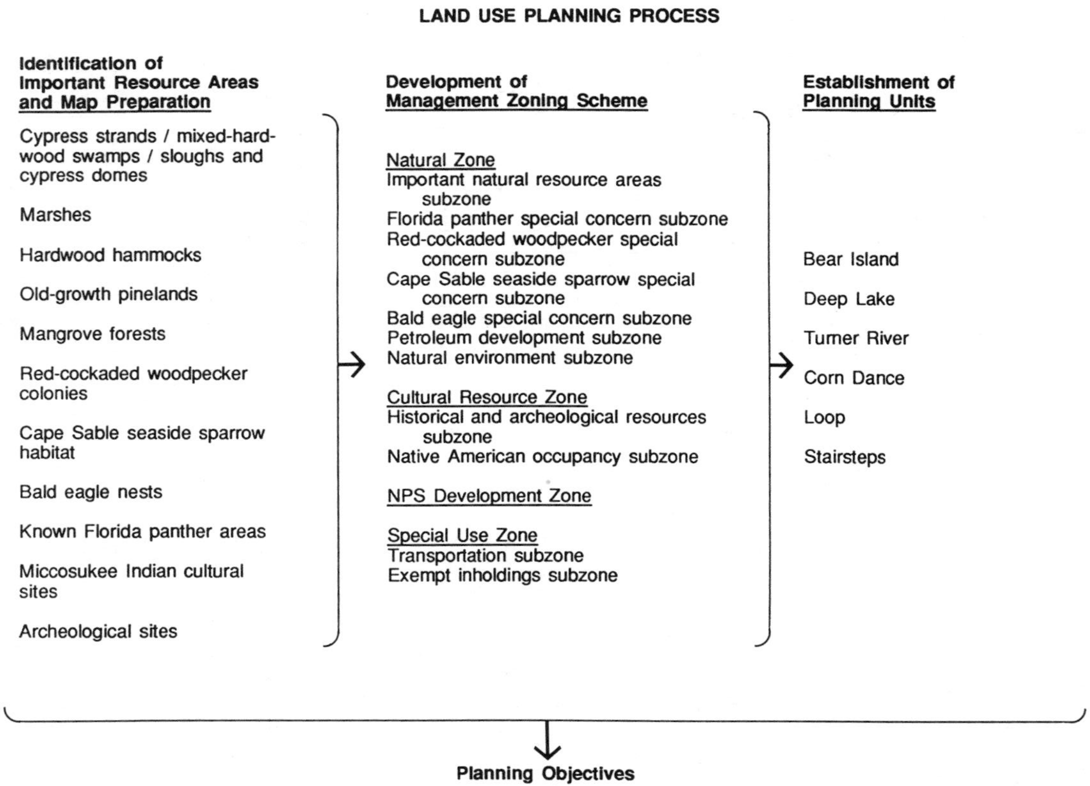

Figure 2. Land use planning process.

ining the selection criteria, selecting resources based on the criteria, and assembling and enhancing mappable data to show distribution of selected resources.

Four criteria for natural and cultural resource values were used to determine important resource areas. The criteria were directly based on the establishing legislation for the preserve and national environmental policies contained in the Endangered Species Act, the National Historic Preservation Act, the Archeological Resources Protection Act, the American Indian Religious Freedom Act, and others. The selection criteria were (1) superior examples of the natural, scenic, hydrologic, floral, faunal, and recreational values for which the preserve was established; (2) areas essential for maintaining water and quality to protect the ecological integrity of the preserve and Everglades National Park; (3) other habitat necessary for the continued survival of federally recognized threatened or endangered species of plants and wildlife; and (4) Native American cultural sites or important historic or archeological resources.

Based on these criteria, the resources listed below were selected (Figure 3). Information on the resources was assembled and mapped from NPS research, other federal and state agencies, and scientific literature. For the most part, resource information was collected from existing sources because of funding limitations. However, a detailed vegetation map of the preserve was prepared by the NPS Geographic Information Systems Division from a set of 1:63,000 color infrared aerial photographs flown in 1984 as part of the U.S. Geological Survey's National High Altitude Program. Vegetation types, including those identified as important resource values, were initially hand plotted on a series of mylar overlays. Later, vegetation was digitized as part of the preserve's computerized GIS.

Cypress Strands/Mixed-Hardwood Swamps/Sloughs and Cypress Domes

Strands, swamps, and sloughs are the major corridors of water flow in Big Cypress and are equally important to the Everglades ecosystems. Any interruption of this flow could have widespread adverse effects in both Big Cypress and Everglades. Strands, swamps, sloughs, and cypress domes contain numerous rare and protected plants. During the winter dry season, these wetlands provide water sources and habitat for wildlife.

Marshes

Marshes also have long hydroperiods, and any major alteration of water flow through this community could adversely affect both Big Cypress and Everglades. Marshes provide essential habitat for wading birds, including feeding habitat for the endangered wood stork and nesting habitat for the endangered Cape Sable seaside sparrow.

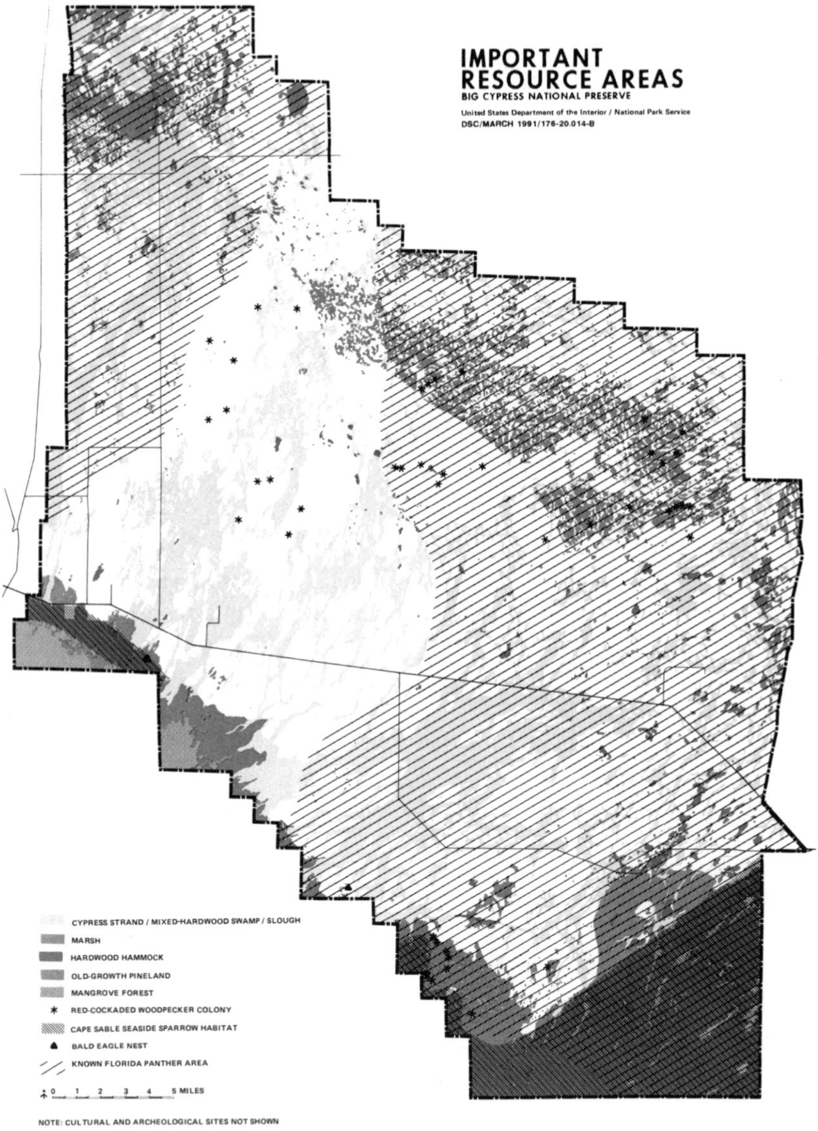

Figure 3. Important resource areas.

Hardwood Hammocks

Hammocks are small tree islands scattered throughout the preserve. Unlike other vegetation types in Big Cypress, hammocks are rarely flooded and tend to resist burning. They contain the majority of archeological sites known in the preserve. Many plant species in hammocks are listed by the state as rare, threatened, or endangered, ranging from small epiphytes to large trees. Wildlife species using hammocks include the state-protected *Liguus* tree snails and the federally listed endangered Florida panther.

Mangrove Forests

Mangrove forests help mitigate the destructive power of storm waves and contribute to estuarine food chains. They are important nurseries and spawning grounds for many estuarine and marine species. Mangroves also provide essential habitat for the manatee, American crocodile, and other protected species.

Old-Growth Pinelands

Old-growth pinelands have never been logged. This forest type was once predominant throughout the southern coastal plain, but old-growth stands have been severely reduced by logging, development, and other land uses. In Big Cypress, the old-growth pinelands are collectively among the largest such vegetative associations remaining in South Florida. Old-growth stands are essential habitat for the endangered red-cockaded woodpecker.

Old-growth pinelands were not discernible from younger-aged pine stands during development of the vegetation map. Consequently, the distribution of old-growth pinelands was estimated based on an estimate of the extent of historical logging. As indicated by red-cockaded woodpecker colony distribution, there are small, remnant old-growth stands within the range of historical logging that were not individually mapped.

Red-Cockaded Woodpecker Colonies

The red-cockaded woodpecker is a federal endangered species. There are an estimated 40 colonies in the preserve, the largest concentration known south of Lake Okeechobee. Universal transverse mercator (UTM) points of known colonies (Patterson and Robertson 1981; unpublished field data) were digitized for the preserve's GIS database.

Cape Sable Seaside Sparrow Habitat

Big Cypress is one of three remaining population centers for the Cape Sable seaside sparrow, a federally endangered species. The Cape Sable sparrow occupies only seasonally flooded, freshwater marshes. The estimated 3000 birds in Big Cypress represent about half of the known population. NPS survey and census data collected from 1978 to 1981 (Bass and Kushlan 1982) were combined and the point data generalized to indicate habitat.

Bald Eagle Nests

Three active nesting sites of the endangered bald eagle are within the preserve. The nests were mapped from preserve staff reports and digitized.

Known Florida Panther Areas

The Florida panther is under extreme risk of extinction and is federally protected as an endangered species. Survey work within the last few years has documented that panthers may use virtually the entire preserve, but they tend to consistently occupy only portions of the area. Point location data were mapped from NPS and Florida Game and Fresh Water Fish Commission surveys between 1978 and 1990 (McBride 1985; Robertson et al. 1985). Points indicated where panther sign had been found or radio-tracked panther contacts had been made. Based on the point data distribution and consultation with panther researchers, portions of the preserve where consistent panther activity was indicated were mapped as "known Florida panther areas."

Miccosukee Indian Cultural Sites

Eleven Indian villages are located along U.S. 41 and the Loop Road within the preserve. Most of the 150 individuals who live in Big Cypress are independent Miccosukee Indians who do not belong to the federally recognized tribe and who do not choose to live on reservation land. As provided in the establishing legislation, Miccosukees and Seminoles may collect traditional subsistence materials and continue their religious practices in the preserve, subject to reasonable regulations. Miccosukee sites are regarded as important resource areas, but they are not shown on the map to ensure privacy and prevent vandalism.

Archaeological Sites

NPS archaeologists have found a total of 395 sites within the preserve. These sites may contain valuable information on past occupations of the area by prehistoric people. Sites are protected under the National Historic Preservation Act of 1966, as amended, and the Archeological Resources Protection Act of 1979. Archeological site UTM coordinates provided by the NPS Southeastern Archeology Center were digitized. To prevent vandalism, the sites are not shown on the Important Resource Areas map.

While the important resource area information was being assembled, land use data were also mapped and digitized (e.g., ORV trails, grazing allotments, landownership boundaries, exempt and trespass properties, oil and gas developments). Initially, both manually generated and digitized information were produced on 1:63,000 scale mylar overlays (most data have now been digitized).

Development of Management Zones

Management zones, an NPS system of prescriptive land use designations, were developed based on the important resource areas (Figure 4). The majority of the preserve was placed within the natural zone in which lands and waters will be

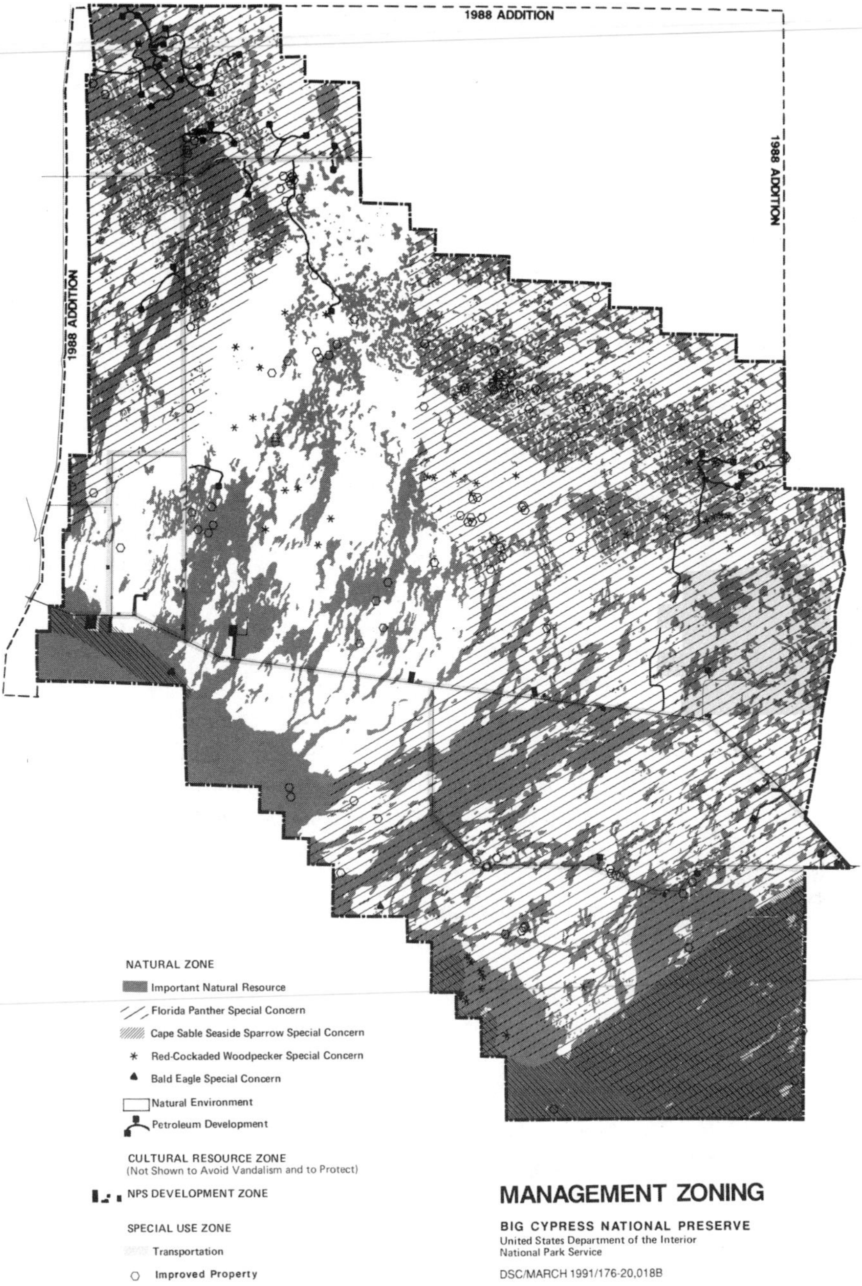

Figure 4. Management zoning.

managed primarily to conserve natural resources and processes. The natural zone was further divided into five subzones to ensure the protection of vegetation types identified as important resources, and Florida panthers, red-cockaded woodpeckers, Cape Sable seaside sparrows, and bald eagles. A petroleum development subzone was also established for areas currently occupied by oil and gas development. The petroleum development subzone was included in the natural zone because these areas are to be reclaimed and returned to a natural condition when petroleum operations cease. To protect cultural resources, a cultural zone was established with subzones for historical and archeological resources and for Native American use and occupancy. Other zones were established for NPS development and for special uses.

Delineation of Planning Units

To facilitate planning and to ensure that all planning issues and concerns were systematically addressed, the preserve was subdivided into six units (Figure 5). The planning units reflect the distribution of important resource areas and established use patterns. Roads and trails were selected as the boundaries so that the units would be clearly discernible in the field.

Establishment of Planning Objectives

Based on the legislative intent for the preserve, important resource areas, management zones, and the planning units, 29 general planning objectives were developed to guide the planning effort. Many of these objectives incorporated the need to protect important resource areas. For example, the planning objectives dealing with minerals management are to

1. Permit access for geophysical exploration, exploratory drilling, and production of private oil and gas resources, while at the same time ensuring the following:

   - the protection of important resource areas
   - the protection of air and water quality in Big Cypress National Preserve and Everglades National Park
   - the mitigation of surface disturbance
   - the restoration of abandoned sites

2. Avoid conflicts with visitor use and enjoyment, and provide for visitor safety.

Generation of Alternatives

In the fall of 1986, six preliminary alternatives were generated for managing the preserve. These ranged from continuing the status quo (which the NPS considers to be unacceptable because of insufficient regulation to protect important resource areas and other resources) to maximizing regulation of uses similarly to traditional management of national parks (managing the preserve as a national

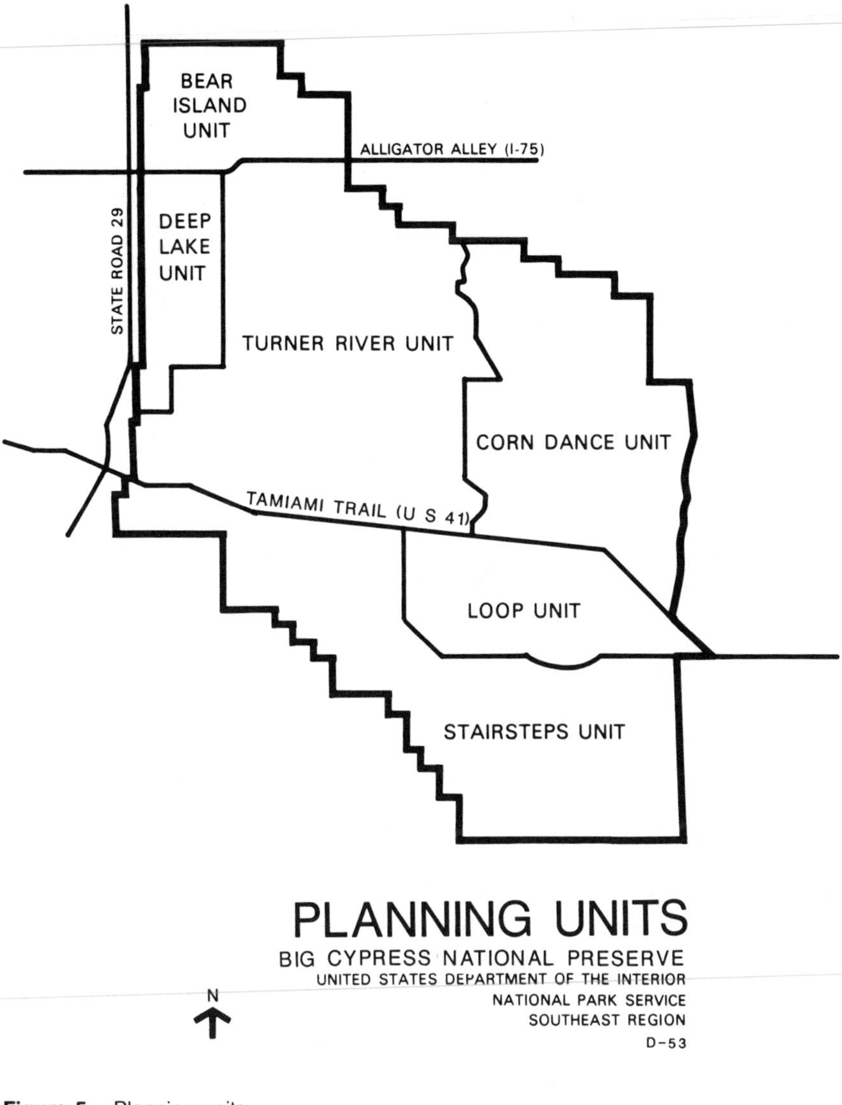

Figure 5. Planning units.

park was considered to be beyond the intent of the legislation). After public review, the alternatives were revised. To provide better clarity and separation of alternative actions, the field of alternatives was reduced from six to four by dropping two of the intermediate alternatives, and a proposed action was identified. The range of alternatives, in order of the relative amount of federal regulation of uses in the preserve, were

1. Status Quo — the continuation of existing programs (the no action alternative) with little effective regulation

2. Alternative A — somewhat increased regulation with emphasis on resources use
3. Proposed Action — intermediate increase in regulation with emphasis on resource protection
4. Alternative B — greatly increased regulation with emphasis on resource protection

Important resource areas played a significant role in formulating the proposed action and alternatives. Management alternatives for hunting, ORV access, overnight backcountry camping, and oil and gas development were based largely on the distribution of important resource areas in each planning unit. For instance, the high concentration of important resource area vegetation types and the presence of a Florida panther population in the Bear Island unit dictated a conservative management approach. Consequently, the area directly affected by future oil and gas operations under the proposal could not exceed the current acreage of roads, pads, pipelines and geographical survey lines; use of hunting dogs would be limited to only bird dogs; ORV use would be restricted to designated trails; and backcountry camping would be restricted to designated sites. Important resource areas also helped define mitigating actions for resource protection throughout the preserve. For instance, under the proposed action, future surface occupancy for exploratory drilling and production would be excluded from important resource areas, and the number of ORV trails crossing important resource areas would be reduced in all planning units.

It is difficult to adequately assess environmental impacts from the broad directions outlined in a general management plan. To provide more specific actions for environmental analysis, the proposal and the alternatives were extrapolated as scenarios to indicate how the land base at Big Cypress could be affected over the life of the plan and to identify characteristic activities and other resulting environmental effects. The scenarios were not part of the proposed or alternative actions, but rather an example of what could happen if they were implemented. Scenarios included estimates of the area occupied by developments and ORV trails; the area affected by hydrological restoration, exotic plant control, and fire management; and the area occupied or influenced by oil and gas operations. The GIS database was extremely helpful in estimating these quantities.

IMPACT ASSESSMENT

For analysis in the environmental consequences section of the draft EIS, specific impact topics were selected to focus discussion; 16 resource topics and 4 socioeconomic topics were selected. Of these, the 11 important resource areas were the focus for assessing impacts on biological and cultural resources.

The analytical capabilities of the GIS were critical for projecting, quantifying, and analyzing impacts. For example, the assessment of the economic impact of the proposed exclusion of petroleum developments within important resource areas illustrates the effectiveness of using GIS for analysis. Under the proposed action, lands not classified as important resource areas but surrounded by important resource areas and without existing road access would also be effectively off-limits to petroleum development. These hole-in-the-donut sites were quickly identified by the GIS. Combining the important resource areas with these additional inaccessible lands produced an estimate of the total surface area of the preserve effectively excluded from petroleum development. However, state-of-the-art drilling techniques would allow much of the subsurface oil and gas to be extracted below the excluded area without disturbing important resource areas. To estimate the subsurface resources potentially inaccessible because of limits on surface occupancy, it was assumed that directional drilling could effectively drain a radius of 0.9 mi under the areas off limits to surface occupancy. The GIS was programmed to estimate the relative area of inaccessible subsurface rights based on this radius. Manually, this process would have taken weeks for analysis, whereas the GIS accomplished the task in a few hours.

PUBLIC INVOLVEMENT

In September 1986, during the scoping phase of planning, a newsletter containing the important resource area map was published. About 1500 copies were distributed to the public. The newsletter summarized the legislative background for establishing the preserve and presented the selection criteria for important resource values and a brief description of each important resource area. Several persons who were normally critical of NPS management expressed appreciation for the level of information contained on the map, and the newsletter did much to establish credibility for the planning effort.

Another newsletter was published in December 1986 describing the new planning units, the planning objectives, and the six preliminary alternatives for hunting, ORV use, visitor services, and minerals management. The public was specifically asked to comment about the alternatives, and a response form was included with the newsletter. Approximately 5000 copies were distributed, and over 7300 responses were received. The responses were almost evenly divided between the two extreme alternatives. Those persons interested in resource use, such as hunters, ORV users, landowners, and petroleum interests, tended to support the status quo, while those persons interested in resource preservation, environmentalist, conservation organizations, and government agencies, tended to support much more protective management. Although most respondents seemed to accept the important resource area concept, there was little agreement on how uses should be managed to protect important resource areas. As previously mentioned, this public response was useful in narrowing the field of alternatives and in selecting the proposed action.

The draft GMP/EIS was published in August 1989 after extensive review in the NPS and Department of the Interior. Because of the length of the draft GMP/EIS and budget limitations for printing it, a newsletter was published which summarized the proposed action and the alternatives and informed the public where copies of the full document were available for review. Approximately 7000 copies of the newsletter were distributed, and about 650 copies of the full document were sent to government agencies and private organizations. The public review period closed March 1, 1990. Comments were received from 2 Indian tribes, 10 government agencies, 31 organizations, and over 4000 individuals. After consideration of the public comments and revisions to the draft, the Final Plan and EIS were published in two volumes in October 1991. The record of decision was signed on January 27, 1992.

CONCLUSIONS

The important resource area approach to planning and environmental analysis for Big Cypress National Preserve proved successful in addressing highly controversial land management issues. Designated important resource areas coupled with GIS technology gives managers and decision makers a concise method for

- scoping resource issues
- determining primary protection areas
- identifying resource use limits
- assessing the significance of potential impacts
- involving and informing the public

In keeping with NEPA, this targeted approach integrates environmental values, general management planning, impact analysis, and public involvement for decisions effecting the long-term quality of the national preserve.

The NPS is now extending the important resource area/GIS database to cover the 146,000 ac that were added to the preserve by Public Law 100-301. The same procedures that were used for the existing preserve will also be used for data collection, resource analysis, development of management and use alternatives, and future impact analysis. This procedure will be documented in an addendum to the GMP/EIS.

REFERENCES

Bass, O. L., and J. A. Kushlan. 1982. Status of the Cape Sable Seaside Sparrow. Report T-672. National Park Service South Florida Research Center, Homestead, FL.

McBride, R. 1985. Population Status of the Florida Panther in Everglades National Park and Big Cypress National Preserve. Report prepared for the National Park Service, on file Big Cypress National Preserve, Ochopee, FL.

National Park Service (NPS). 1983. *Proposed Sensitive Resource Areas.* NPS Mining and Minerals Division, Denver, CO.

National Park Service (NPS). 1989. *Management Policies.* NPS, Washington, D.C.

Patterson, G. A., and W. B. Robertson. 1981. Distribution and Habitat of the Red-Cockaded Woodpecker in Big Cypress National Preserve. Report T-613. National Park Service South Florida Research Center, Homestead, FL.

Robertson, W. A., O. L. Bass, and R. McBride. 1985. Review of Existing Information on the Population of Florida Panther in Everglades, Big Cypress, and Environs, with Suggestions for Needed Research. National Park Service South Florida Research Center, Homestead, FL.

CHAPTER 4

Social Impact Assessment and Public Involvement

Introduction and Summary

S. A. Carnes, Oak Ridge National Laboratory, Oak Ridge, TN

When Congress passed the National Environmental Policy Act (NEPA) in 1969, there was little guidance on the preparation of environmental impact statements (EIS) and the role of the public in the NEPA process. Excepting the statutory language of NEPA, which referred to impacts on the human environment, nowhere was this more evident than with respect to people. Questions such as what impacts on people should be assessed; how impacts on people should be assessed; and how people, including but not limited to those persons potentially impacted, should be involved in the assessment itself as well as NEPA's associated administrative processes were simply not addressed.

The early days of implementing NEPA did not bode well with respect to these questions. In many cases, responsible agencies did not have the requisite expertise. Agency personnel charged with preparing EISs grappled with these and other problems on an iterative basis, learning by doing and, as determined through litigation in some cases, not doing. They learned, as did we all, including contractor consultants, academicians, and private sector interests whose proposed actions triggered NEPA, in what might be called an evolutionary manner. As elaborated in Freudenburg's paper, this evolution manifested itself over the years in a number of ways. Principal among these has been a gradual movement toward increasingly sophisticated analyses of "indirect" impacts. We all learned, fairly early if not always well, how to count obvious things — people, dollars, students, and road capacities — and how to project how a proposed action might increase or decrease these numbers; we have even attempted to assess the impacts of changes in these numbers on their putative social, economic, cultural, and political phenomena, such as how the quality of education may decline with rising student–teacher ratios.

Later, and in response to both political demands and our own professional interests and commitments, we began to learn how to address more basic if less easily quantifiable impacts — impacts on community cohesion, quality of life, mental health, cultural identity, aesthetic values, and social well-being, to name just a few. We began to encounter particularly thorny problems, such as projecting the social impacts of accidents associated with "risky" activities (but were reminded by the U.S. Supreme Court in *NRC* v. *PANE* not to assess risk perception, per se), at the same time that our fellow scientists were having an equally hard time estimating the probabilities of accidents and their associated source terms, dispersion characteristics, and health and ecological effects. Moreover, we began to address and, in some cases, estimate the effects of uncertainty on our projections and assess the significance of uncertainty on the decisions to be made about proposed actions.

The other people problem, public participation in the NEPA process, is equally complex. That public participation has been instrumental in determining the evolution of NEPA assessments is the conventional wisdom. Stories abound regarding the central role of the public in reviewing and litigating NEPA assessments. Many of these stories address the public in two manifestations: the national and local public interest groups that conscientiously review and occasionally litigate, the Sierra Club, the Natural Resources Defense Council, the Environmental Defense Fund, and the Save Our 40 Acres; and the unorganized yet concerned and potentially affected members of the public. Often missing from this popular interpretation of public participation and NEPA, as pointed out in the paper by Friesema, is the critical and increasingly important role played by legally constituted representatives of the public — the interagency, interjurisdictional NEPA review offered by other federal and state agencies.

Part of the reason for this complexity regarding public participation and NEPA, then, rests with how one defines "public;" does "public" refer to the visibility of one's participation (i.e., in the full view of the public vs a more private mode) or to the representativeness of participation (i.e., is the individual participating as a citizen demonstrating his/her individual insights and perspectives or as a representative of a larger collective)? To casual observers of the NEPA scene, the public consists of those who publicly participate — those persons and groups who appear at the meetings and hearings and perhaps those who review and comment on draft environmental impact statements (EISs). These participants may be divided into two groups. First, there are those persons who feel they will be adversely or beneficially impacted by a proposed action and want to state their opposition or support, sometimes with compelling reasoned arguments and sometimes with more emotion than reason. Second, but often comprising only a small proportion of the total public participants, are "professional" public participants, those interest groups and state and federal agencies that have participated before and will participate again in the NEPA process; as a result of their past participation and obligations, they know the process, they know the roles of EISs in decisionmaking, and they know what will and will not make a difference. There is also an unseen public, unseen perhaps because the media do not consider their activities interesting or newsworthy, comprised of individuals who work within the NEPA process but outside the public's view. They are members of planning and assessment committees or representatives of local, state, and federal agencies responsible for reviewing NEPA documents and processes. They attend and participate in agency scoping and/or review meetings, where they sit across the table from lead agency representatives for hours at a time, assessing the proposed action and/or how it ought to be assessed.

Only a few people address the purposes of public participation. What does the public gain by participating, what do the EIS preparers (and the assessment) gain, and what does the decision maker gain by engaging the public in the NEPA process? Related to these questions, it seems to me that there are at least three principal purposes of public participation: improving the accountability of decision makers to the governed, improving the assessment itself by incorporating multiple perspectives on the proposed action and a more comprehensive sense of what constitutes potentially significant impacts and how to assess them, and improving the decision itself by assuring consideration of these multiple perspectives. The extent to which these three functions are fulfilled by alternative public involvement mechanisms is critical to the success of NEPA.

The papers comprising this chapter address many of these issues. Bill Freudenburg provides a sense of perspective regarding social impact assessment. He identifies the principal questions addressed by social scientists in NEPA's first 20 years and offers some speculation regarding future directions. In particular, he notes both intellectual and institutional challenges for social impact assessment. The former includes developing a more comprehensive and systematic capability to understand and assess the social impacts of new, emerging, or potential technological systems. The latter includes both a perennial social science problem — inadequate resources to conceptualize problems and collect and analyze "real, live, empirical data" — and a more fundamental problem of identifying powerful institutions in whose interest securing support for improved social impact assessment (SIA) makes sense.

Carl Petrich provides a comprehensive and lively discussion of the evolution of aesthetic assessment. His treatment, however, goes beyond aesthetic assessment by offering a compelling discussion of information processing approaches that should invigorate our understanding and appreciation of all "social" impacts. He forces us to question our often cavalier conceptualizations of social impacts by requiring a deliberate consideration of the meaning and value of *place* and how place affects other human perceptions and behavior. His paper concludes with a discussion of the challenges facing aesthetic assessment, particularly our assessment of working landscapes, in a period characterized by rapid metropolitan growth.

Benita Howell notes that our existing reliance on historic preservation legislation, executive orders, and regulations to legitimate consideration of cultural impacts results in assessments that are weak and incomplete, and she advocates expansion of cultural impact analysis through field surveys and ethnographic research. Based on case studies of SIA and public response to planning for two national recreation areas, she finds that reliance on archaeological and historic resources as the principal, and often only, indicators of cultural impacts ignores cultural conservation or the protection of living cultures. She also notes that public participation, per se, is not likely to ameliorate this condition, since many cultural minorities have economic, political, and educational disadvantages that limit their ability and will to participate. Instead, she suggests that SIA use some existing elements that now apply only in certain specific situations (e.g., the special status of cultural conservation for Native American SIA and studies of risk perception) and apply them uniformly in impact assessment.

Mark Schoepfle, Ellen Szarleta, and Susan Schexnayder present an analysis of the worth of wilderness areas that is informed by both ethnographic and survey research. In attempting to assess the social impacts of low-altitude flights over wilderness areas, the researchers explored the universe of people who cared about such impacts for one reason or another; identified their concerns through cognitive ethnography; developed alternative mitigative mechanisms; and using contingent valuation as an approach to measure revealed preferences, measured the worth of wilderness to wilderness users who had experienced overflights, users who had not experienced overflights, and nonusers. Their finding that preservation of wilderness as a public resource is important to nonusers as well as users, as measured by a willingness to pay equivalent amounts of money to mitigate adverse impacts among all three groups, belies much conventional thinking and also encourages us to attempt multimethod approaches to impact assessment.

Max Landes and Dennis Pescitelli describe the use of the analytical hierarchy process (AHP) in an assessment of the impacts of alternative routings of highway projects in the East St. Louis area. In addition to providing a systematic approach to assessing the impacts and ranking of alternative routings, this application of AHP is noteworthy for its extensive involvement of potentially affected publics in the assessment. Their approach certainly went beyond the public participation typical of most assessments, requiring significant amounts of proactive planning and structuring by professionals and substantial investments by public participants in attending meetings, assessing and ranking impacts across the alternative routings, and developing consensus. The approach is not without its problems, but the logic and comprehensiveness of the approach offer substantial benefits that may be useful to others.

Finally, Paul Friesema offers some cogent and evocative comments on the globalization of the EIS in terms of its attendant participation processes. After admitting that he is as guilty as anybody in perhaps overstating the importance of public participation in the EIS process, he presents a compelling argument that the interagency and interjurisdictional review process at work in the American EIS process may well be more transferable to other nations than our own institutional opportunities for organized and unorganized citizens to get involved in the process (e.g., attending scoping meetings, submitting written comments on draft or final EISs, and/or suing an agency over some perceived failure in the EIS). The latter may be characteristic of American political tradition and culture (or even, perhaps, of Western industrialized democracies), but may be unsuitable or unrealistic in other political and cultural traditions. Friesema also maintains that an interagency/interjurisdictional review process model may also be more effective in ensuring attention to issues of the global commons, since persons potentially affected by a local project are understandably more parochial in their perspectives and do not usually attend to regional or national, much less global, concerns.

Two Decades Later: Progress and Paradox in Socioeconomic Impact Assessment

W. R. Freudenburg, University of Wisconsin, Madison, WI

ABSTRACT AND OVERVIEW

The field of social or socioeconomic impact assessment (SIA) has accomplished a great deal in the 20 years since the passage of the National Environmental Policy Act (NEPA), but much remains to be done. For the first several years, the primary emphasis was on developing standards, guidebooks, and methods. As might be expected, this effort led to progress and a reasonable level of consensus in those subfields of SIA best lending themselves to routinization, although not in the subfields characterized by greater complexity and lower levels of funding. Arguably, more important than the development of methods has been the development of empirical findings about certain subsets of impact-creating facilities, or at least those facilities for which the anticipatory SIAs have been complemented by empirical, during-and-after investigations of the actual impacts. Notable in this regard have been displacement-inducing projects (principally rural dams and urban highways), large-scale construction projects in rural areas (particularly energy-generating facilities), and in recent years, certain risky and/or controversial facilities. Important contributions have also been made in the related areas of attitude assessment and public involvement programs, although agencies often fail to realize that measuring *attitudes toward* a project will provide only some of the information necessary to understand the *socioeconomic impacts* of the same project.

SIA faces an interrelated set of intellectual and institutional challenges. Intellectual challenges include attempting to foresee the social implications of technological changes more pervasive and subtle than those brought about by well-delimited facilities — from changes in resource management plans to the development of whole new technological systems. For resource development efforts characterized by "predictable unpredictability," the field may need to emphasize the development of improved, more systematic approaches for dealing with *uncertainty itself*, as opposed to devoting increased attention to predicting the outcomes we expect to be "most likely."

Still more daunting may be the institutional challenges not considered explicitly in this paper, but that are likely to be the focus of considerable discussion at the conference itself. While the social, cultural, psychological, and economic impacts are often the most important of the impacts of a given development project, project managers rarely recognize the importance of social science impacts before the fact; they often fail to recognize that social science research, like any scientific research, requires such resources as time; conceptual development; and real, live, empirical data. Social scientists are no better at producing "findings" out of thin air than are biologists, yet the realities of SIA often reveal social scientists being asked to "make things up" to a degree that would never be expected from biological scientists. This situation, moreover, can be self-reinforcing: the lack of necessary resources can lead to SIA research that is of poor quality, appearing to justify low levels of investment in future research.

The most important single institutional problem, however, may have to do with institutional interests. The most entrenched of the parties involved in NEPA-related debates include several who would be expected to find it in their interest for social science considerations *not* to receive full consideration. Reasons range from the parties' political orientations, to agency interests in avoiding "additional work," to the general tendency for persons trained in one set of disciplines to have difficulty appreciating the contributions available from others. Nor are environmental groups likely to champion the increased use of SIA, given that SIA may prove useful for their interests in some cases but work against their interests in others. "Integrated use" of the social sciences in the NEPA context may prove to be dependent on an appreciation for the relatively apolitical value making of decisions based on a fuller range of the relevant information.

INTRODUCTION

A conference that marks the 20th anniversary of a law or, for that matter, of a scientific field is scarcely an everyday event, so it would scarcely be appropriate for the papers prepared for such a conference to treat it as such. Instead, the occasion calls for a relatively broad summing up of what has happened to date, combined with an even broader set of speculations about the future. So it will be with this paper on social or socioeconomic impact assessment (SIA).

We are fortunate in that a number of excellent and quite detailed reviews have been done on the field during the past several years, freeing this paper, at least, from the obligation to provide such a review, doing so in broad terms, and simultaneously freeing readers from the obligation to read one. Instead, this paper will devote only the first portion of its space to a review of past work, in the broadest possible terms, discussing overarching issues and implications rather than specific findings and conclusions. The remainder of the paper will have an even broader reach, offering personal and speculative observations on the field as a whole, on the issues with which it grapples, and on the challenges it faces for the future. The more specific portions of the paper draw heavily from two other papers on which I have recently been working: one with Kurt Finsterbusch and the other with Llynn Llewellyn (Finsterbusch and Freudenburg, in press; Llewellyn and Freudenburg, 1990); the broader and more speculative portions are more fully my own. As those of you who know them might well imagine, both Dr. Finsterbusch and Dr. Llewellyn have done their best to correct errors and annoyances and have contributed a great deal that is of value, but they bear no responsibility for any ill-founded fulminations that have made it into this paper over their protests.

The first section of this paper provides a brief and general summary of progress in SIA to date; it includes a highly generalized overview of two subfields of SIA that provide insights into but should in no way be seen as representing the field as a whole. The paper turns next to a broader set of issues, challenges, and future directions, discussing three interrelated sets of issues particularly deserving of attention. The paper ends with some relatively unorthodox observations about potential future directions for the field.

A LIMITED REVIEW OF PAST WORK

SIA work can be divided into an almost infinite number of subfields; given the number and quality of other recent reviews, however, this section focuses simply on the two subfields the author knows the best (for other recent reviews, see Cortese 1982; Finsterbusch 1980, 1985; Finsterbusch and Freudenburg, forthcoming; Freudenburg 1982, 1986; Halstead et al. 1984; Murdock and Leistritz 1979; Murdock et al. 1984, 1985; Weber and Howell 1982). These two subfields provide insights into the field of SIA more broadly, partly because they mirror the changes seen within the field as a whole, although they should in no way be seen as "representing" the entire field of SIA. The first subfield comprises work that focuses on large-scale construction projects, particularly in sparsely populated areas; the second focuses on risk-inducing projects — a topic having a certain degree of overlap with the first.

Construction Projects

One significant area of SIA work has focused on social problems created by large construction projects in rural areas, particularly those projects leading to rapid community growth. Given the detailed reviews already available for this area of work (see especially Finsterbusch 1980; Freudenburg 1982a; Cortese 1982), this review will attempt only to sketch out broader trends. For this purpose, the literature on large-scale construction projects can be seen as having evolved through a series of three stages, and they will be sketched out briefly here (this discussion draws heavily from Finsterbusch and Freudenburg, forthcoming; see also Freudenburg 1986a, 1986b, 1992b; Freudenburg and Jones, in press; Seyfrit 1986).

Economic Opportunities

With relatively few exceptions (e.g., Smith et al. 1971), the literature on community growth written before 1975 referred almost entirely to the *positive* implications of growth. In the most extensive review of this early work, Summers et al. (1976) noted that the community growth resulting from rural industrialization was generally seen as beneficial, providing "an important tool for solving the twin problems of rural poverty and urban crisis." Similarly, in reviewing federal impact statements produced up to that time, Friesema and Culhane (1976, p. 343) noted, "The statements generally consider only one social consequence — the economic impact of the project" (see also the critical reviews provided by Little 1977; Freudenburg 1976). The net result is that consistent with the broader cultural disinclination to question the advantage of growth before the early 1970s (Dunlap 1982), the "early" or rural industrialization studies tended to focus on expected benefits of development while devoting relatively little attention to potential drawbacks (see also Schnaiberg 1980).

Boomtown Disruptions

Beginning in the mid-1970s, particularly after the 1973–1974 oil embargo and the subsequent development of massive projects in sparsely populated regions of the western United States and Canada, researchers paid increasing attention to social problems associated with rapid community growth. If earlier studies had drawn directly on classic economic logic, the literature in this second and shorter-lived tradition tended to draw more directly from classic sociological writings, particularly those of Durkheim and Toennies. According to the major reviews of the research performed during this second era (Cortese 1982; Freudenburg 1982a; Wilkinson et al. 1982), emphasis centered on the disruptive consequences of rapid social change in previously stable systems, particularly in the so-called "energy boomtowns." In retrospect, some of this literature represented, in part, an overreaction against the excessively favorable perspective taken by earlier work, and some of the literature (see the compilation by Davenport and Davenport 1981) was produced by human service providers whose primary focus was on helping communities and individuals cope with, rather than carefully documenting, any social problems that may have been created.

Doubting the Disruptions

If the second era represented a reaction against the first, the third, extending roughly from 1982 to the present, can be seen as a reaction against the second.

The "critical review" by Wilkinson et al. (1982) marked the eclipse of the second and the beginning of the third. While this review has been the focus of considerable criticism itself (Albrecht 1982; Finsterbusch 1982; Freudenburg 1982b; Gale 1982; Gold 1982; Murdock and Leistritz 1982), there was merit in the review's contention that much of the literature on the "boomtown disruption hypothesis" was too quick to accept assertions about the presumably negative consequences of rapid community growth. The Wilkinson et al. review has been followed, moreover, by additional empirical analyses suggesting an absence of other evidence of social problems, particularly with respect to crime (Wilkinson et al. 1984; Krannich and Greider 1985). On the other hand, it is possible that just as papers of the second era may have overreacted against the methodological problems and weaknesses of the first era, some of the work during this era could reflect an overreaction against the problems of the second.

Balanced Assessments?

It may reflect excessive hope or optimism, but the field may be moving into a fourth era: a period of empirical research that seeks to provide a balanced and comprehensive assessment of social impact phenomena, recognizing that the phenomena are considerably more complex than would be suggested by relatively simplistic emphasis on either the presumed benefits or presumed drawbacks of rapid growth.

One of the greatest difficulties with the earlier research traditions has been a tendency to treat growth as being either "all good" or "all bad." Recent work, by contrast, tends to note the importance of differential impacts (Elkind-Savatsky 1986; Flynn et al. 1983; Freudenburg 1986a, 1986c). Not only do different social groups experience different social impacts, but some evidence suggests that different social *functions* may be affected differently as well (Freudenburg 1986a; Freudenburg and Jones, in press). By taking such subtleties explicitly into account, more recent work has shown considerable progress toward scientific neutrality (Finsterbusch 1985; Murdock et al. 1985).

Risk-Inducing Projects

In a sense, of course, virtually all of the projects of traditional interest to the SIA literature can be seen as involving risks — the risks of community disruption in the energy boomtowns or the risks to physical health or property values in projects such as dams or highway developments that threaten to displace long-term residents of an area. In addition to these kinds of risks, however, many facilities are seen by concerned local residents as involving high levels of technological risks, while often appearing to be "acceptable" to technical experts who do not live nearby (e.g., Rothman and Lichter 1987); such facilities may create important socioeconomic impacts through the risks themselves. For example, the contamination at Love Canal (Levine 1982) led not only to health impacts, but also to important changes in the social structure of the affected community, including the emergence of new groups and increases in community conflict, as well as serious impacts on the credibility of social institutions (Finsterbusch 1988, 1992a; Short 1984; Dietz et al., forthcoming).

Given that research in this area is still evolving, this discussion will be an abbreviated one, being limited to those considerations having direct relevance to the measurable social impacts of risk-inducing projects. It is possible to discuss the social impacts of such projects in terms of three categories (Freudenburg 1988; Freudenburg and Pastor 1989).

The easiest category to comprehend is the first, which involves *impacts created by serious accidents*. These include accidents such as those at Bhopal or Chernobyl, which can ultimately result in literally thousands of deaths; they can also include accidents that pose little risk of direct human deaths, such as the *Exxon Valdez* incident, which created noticeable disruptions in U.S. oil prices and created a situation of jeopardy for some of the richest fishing grounds in the world. Accidents can also be "serious" even in cases where neither the human health toll nor the direct ecological damage is as extensive as in these cases. A convenient illustration is provided by the Goiânia incident in Brazil (Petterson 1988). Two scavengers had entered an abandoned medical clinic in search of scrap metal; they found a small capsule and later pried it open, releasing a mere 100 g of cesium-137. The release led eventually to 121 known cases of skin contact with the material and 4 deaths, with another 3–5 deaths expected within the next 5 years. As Petterson notes, this death toll could scarcely be considered out of line with "any other industrial accident," but just the labor costs of decontamination had exceeded $20 million (United States) within a short time, and the broader economic and social costs were far greater. Within just 2 weeks of the event being announced in the media, the wholesale value of agricultural products from the entire Brazilian state of Goias fell by 50%, and even the demand for manufactured goods (including textiles, clothing, and other finished products) was affected. This was in spite of the fact that Petterson was unable to find "even a published suggestion" that the agricultural products or manufactured goods could have been contaminated. Severe impacts were also felt through treatment and research costs, declining property values, canceled conventions, and a broader decline in the tourist trade. More than 100,000 residents lined up at monitoring stations to be checked for radioactive contamination; more than 8000 residents requested (and received) certificates that they were not contaminated, and even well-trained doctors and dentists refused to treat patients without certificates of noncontamination.

Second, *uncertainty costs* may be created for affected communities even when "nothing goes wrong." Real costs are incurred when communities invest in emergency-preparedness training or preparing of evacuation plans, when societal strains are created by inequitable distributions of technological risks, or even when individuals "invest" in the psychic costs of worrying about potential disasters (Short 1984; Baum 1987). As Freudenburg (1988, p. 44) notes, such uncertainty costs are comparable to insurance that proves in retrospect not to have been "needed": "Insurance companies keep the premiums even if the house does not burn down, the automobile is not involved in an accident, the insured person is not hospitalized, and so on." The costs are created by the *possibility* of harm, not just by the experience of harmful outcomes.

The third and perhaps most subtle category of risk-induced effects involves *"signal" incidents* (Slovic 1987) in which otherwise "minor" incidents provide the public and/or relevant officials a set of "signals" that the situation may not be fully under control. An example is provided by the accident at Three Mile Island (TMI), which was found by official investigations to have released very little radioactivity, although it did lead to significant mental health consequences for nearby populations (Dohrenwend et al. 1981; Flynn 1984). Among its other consequences, the TMI accident sent a "signal" to the policy community and to the broader public that nuclear power plants were less safe than the public had formerly been led to believe. The consequences included a significant decline in public support for nuclear power (Freudenburg and Baxter 1984, 1985), and the accident appears to have been little short of disastrous for the nuclear power industry more broadly (Lovins 1986).

ISSUES, CHALLENGES, AND FUTURE DIRECTIONS

Given that some 20 years have passed since NEPA was enacted, it is now appropriate to look to the future as well as the past. At the risk of going well beyond the data, accordingly, the remainder of this paper will discuss three sets of issues on which resolution is not yet in sight and then turn to some broader thoughts about what may still be ahead.

In terms of unresolved issues, three sets of interrelated challenges deserve special mention: Quantification, Participation, and Politics; Values and the Management of Technology; and the Challenges of Dealing with Risk and Uncertainty. For each of the three, the possibility exists that the field has been unable to resolve them to date because the issues *inherently* evade the identification of "best" solutions in a "scientific" or value-free sense.

Quantification, Participation, and Politics

Perhaps the issue identified most frequently as contentious in major reviews of SIA (Carley and Bustelo 1984; Boothroyd 1982; Bowles 1982; Freudenburg 1986b; Freudenburg and Olsen 1983) is the potential for tension between "scientific" and "political" orientations toward the field. To simplify only slightly, proponents of the "scientific" approach call for documenting impacts dispassionately and preferably quantitatively, rather than becoming "politically" involved in efforts to produce change; those proponents are found predominantly in the United States. Proponents of the "political" approach see true objectivity as unobtainable and view quantification as working to the advantage of relatively powerful groups in society and to the disadvantage of the small and unpowerful communities likely to be affected by developments. They tend instead to emphasize a participatory function for SIA, working with affected communities to help them develop their own capacities to deal with development and any attendant disruptions.

Several of the most articulate proponents of the "political" approach to SIA (Tester and Mykes 1981) are from Canada, where the roots of SIA emerge not from U.S. environmental law, but from the MacKenzie Valley Pipeline Inquiry (Berger 1977). Justice Thomas R. Berger of the British Columbia Supreme Court was appointed to examine the social, economic, and environmental impacts of the proposed pipeline, which would have brought natural gas from the Canadian Arctic to midcontinent. In the process, the pipeline would have crossed a region as large as Western Europe inhabited by only 30,000 people, half of them natives. Berger held hearings in a series of northern communities, complete with native translations, helping to educate his countrymen about their neighbors to the north, as well as to identify the implications of the pipeline itself. His final recommendation — to delay pipeline permits for 10 years to allow the settlement of native claims — did not carry the force of law, but it had major impacts on policy outcomes. At least since the time of the Berger inquiry, many SIA practitioners have argued that SIAs in general ought to place greater emphasis on changing outcomes and less on simply studying them.

Partly given that I have been identified with the "scientific" pole of this debate, readers are referred to the more detailed treatments of this topic available elsewhere (Carley and Bustelo 1984; Freudenburg 1986b). Rather than attempting to resolve it here, the following discussion offers four brief observations that have received relatively little emphasis in work to date. First, while this issue has obvious parallels to the long-standing debate in sociology between "research" and "action" (or "praxis") orientations, as well as to some of the more recent literature about the possibility of doing "value-free" research in the sociology of science more generally (Knorr-Cetina 1981; Latour and Woolgar 1986), the debates in SIA are also affected by differences in disciplinary backgrounds. In particular, SIA practitioners often come from fields such as community development or planning, where the professional's role is explicitly envisioned as that of a change agent rather than as a person who will study a process scientifically, attempting to avoid "getting involved" in it personally.

Second, the issue of client relations is a particularly thorny one for SIA, even (or perhaps especially) for proponents of the "scientific" orientation, given that the potential always exists for the SIA to become the focus of contention in court. Under U.S. environmental law, legal vulnerability for a proposal is imposed not by the creation of negative impacts, but by the failure to disclose those negative impacts in an environmental impact statement (EIS) (Meidinger and Freudenburg 1983; Llewellyn and Freudenburg 1989). The organizations contracting for SIAs, however, are often either the project proponents or else agencies that share a desire to see a project described in the most favorable light possible (Freudenburg and Keating 1982; Payne and Cluett, in press).

Third, just as difficulties can be created by questions over a proponent's influence on the content of SIA, unresolved questions remain about public participation, although these are far less frequently recognized in the literature. Even proponents of the "scientific" approach have argued for encouraging public involvement in SIA, both because local residents are such an important source of expertise on their own way of life and because of the importance, in a democracy, of having significant citizen input into decisions that may affect the broader public. Additionally, citizens often offer sources of insight that technical experts might have overlooked (Freudenburg 1988). The field has not yet fully come to grips, however, with the potential for conflict between encouraging local residents to "project" impacts for themselves, based on what they know about their own communities, and providing those same residents with "expert projections," based on empirical findings from comparable developments that have already taken place in other communities.

Fourth and finally, even when public participation is widely seen not only as appropriate but necessary, an unqualified endorsement can be tantamount to advocating that objectionable facilities be placed in the communities that raise the weakest complaints. Given the realities of public participation patterns, these are likely to be the poorest communities; this implies, in turn, that these could be the localities having the fewest resources for dealing with any problems that may result (Freudenburg and Olsen 1983).

Values and the Management of Technology

Many technological decisions require the consideration of both facts and values; even under optimal conditions, for example, risk disputes are likely to involve at least two types of questions: "How safe is the technology?" and "Is that safe enough?". The first of these two questions is, in principle, "answerable" in a scientific sense, but the second simply cannot be answered except with reference to personal values.

The "sides" involved in many technological controversies, moreover, tend to have unequal access to technical expertise. As noted by Dietz et al. (in press), industries and government agencies often have the resources necessary to "afford" the often-expensive scientific expertise that can be decisive in technological disputes, while local communities often do not, needing instead to resort to fundraising on a scale as modest as car washes and bake sales.

More broadly, local citizens and project proponents often differ in what they perceive to be the "proper" weightings of values. In a question about what is "safe enough," for example, local residents might wish for a weighting that errs on the side of safety, while the project proponent might argue for a weighting that errs on the side of reducing costs for industry. Other value differences are more subtle; for example, Johnson and Petcovic (1988) found that radiation health professionals tend to have Myers-Briggs personality profiles that favor abstract, empirical, and logical orientations and that emphasize making judgments rather than merely perceiving events. Such a personality type has important assets, but it can also lead to impatience about considerations that other types of people consider to be important, such as equity of outcomes. Also, this personality type is found in only about 1% of the U.S. population. This point throws new light on many scientists' complaints about the supposedly "illogical, emotional" reactions of affected citizens. The scientists' own personality variables, while rarely recognized as sources of "bias," may in fact be leading those professionals to make value-based and nonscientific decisions that differ dramatically from the decisions preferred by the public at large.

The parties involved are also likely to disagree about what questions should be on the agenda and how they should be discussed. In addition to genuine differences of opinion about what the "real issues" are, differences about the agenda can directly reflect differences in tactical interests. As a congressional staffer once told me, "It's something I learned from one of my law school professors: If you let me define what the case is `about,' then nine times out of ten I'll be able to win the case."

Project proponents tend to have a greater advantage through "agenda control" than is commonly recognized. As critical authors have pointed out, monetarily powerful interests often can exert a good deal of influence through "nondecision-making" — the ability to prevent an issue from getting on a policy agenda in the first place, rather than the more overt exercise of power over issues that have already appeared on the agenda. In addition, as Kunreuther and his colleagues have noted (1982), bureaucratic procedures usually give project proponents considerable control over agendas. Discussion is likely to focus on whether the proponent has complied with the agency's regulations, for example, and not on "philosophical questions" about whether a different lifestyle or technology could lead to socially superior outcomes. Finally, even the terms used to discuss the issues on the agenda are likely to work to the advantage of project proponents. As Schnaiberg (1980) has pointed out, the "production sciences," which focus on the efficiency and profitability of industrial production, tend to receive far more funding than the "impact sciences," which focus on the negative externalities and/or unforeseen implications of production systems. As a result, the data and analyses advanced by project opponents often appear to be (and may well be) "less scientific" than the data and analyses supporting project proponents. In addition, our usual terminology reflects any number of implicit value judgments. We will tend to weigh factors such as "the national need for energy," for example, against the "desire" of local communities to maintain a given way of life, rather than discussing the "need" to protect local residents against disruptions that result from the "desire" for bigger air conditioners or the "failure" to invest more in energy efficiency.

Risk and Uncertainty

Perhaps the most troublesome problem for SIAs is the need to face the irreducible difficulty of predicting the future. In some cases, SIAs are asked not only to look into the future, but into futures of almost breathtaking complexity and uncertainty. This can be seen in technology assessments for which specialists have been asked to anticipate the wide-ranging implications of everything from home computers to thermonuclear war; skeptics judge these efforts to be indistinguishable from sheer speculation. Yet the same problem is evident in "traditional" SIA; current efforts to assess the likely impacts of a high-level nuclear waste repository are being asked to project impacts 10,000 years into the future. For a comparative perspective, 10,000 years is roughly twice the age of the oldest known human civilization and is approximately comparable to the time that has elapsed since humans first started putting seeds into the ground. This act led eventually to the emergence of horticulture, then agriculture, and then — after about 5000 years — to the beginning of civilization as we know it today. Even if we focus on less daunting challenges and on the relatively developed subfields of SIA — as in projecting population changes likely to be created by the construction of well-understood facilities, in isolated rural areas where no other developments of similar magnitude are taking place — the record to date has scarcely been one to inspire confidence. In a study that examined the accuracy of demographic projections in 225 EISs prepared during the 1970s, Murdock et al. (1984) found that the average absolute error had been over 50%.

Intriguingly, the largest source of error in the Murdock et al. study came not from the social scientists, but from the engineers and physical scientists involved who routinely underestimated the number of construction workers required to build the industrial facilities in question. At the same time, however, Henshel (1982) found that demographers' projections of national population counts, while reasonably accurate under "steady-state" conditions, consistently failed to foresee socially significant changes, such as the "baby boom" of the 1940s and 1950s and the "birth dearth" of the 1960s and 1970s. The commonly prescribed antidote of increased sophistication proved not to be a helpful one; Henshel actually found the more complex and sophisticated assessments to do slightly *worse* than the more simple and straightforward ones (see also Moyer 1984).

As might be expected, specialists in the field have long grappled with the problems associated with crystal ball gazing, and three principal responses have emerged. First, numerous authors (including the present one) have emphasized the need for relevant empirical data. Theory is a way of making sense of data; it does not substitute for data (Freudenburg and Keating 1982; Freudenburg 1986b). Physical scientists often do just as poorly as social scientists at foreseeing "surprises" or predicting the future in the absence of relevant data. A widely respected geologist was literally standing on Mount St. Helens measuring the mountain's behavior when the top quarter mile of the mountain blew off in 1980; he appears to have anticipated the eruption only by the length of time required to transmit his last radio message: "Vancouver! Vancouver! This is it!". In part because of the analysis of data obtained from this first eruption, however, it became possible to predict the timing and magnitude of subsequent eruptions with what many observers found to be an impressive level of accuracy. In SIA as well, perhaps the best prediction is that the first "eruption" of any type is likely to come as a surprise, while doing the appropriate research on the first such incident can make it possible to better anticipate the incidents that follow.

The second response is that monitoring — the process or tracking social and economic changes over time — can help compensate for the things that are not known at the outset. If the children of construction workers turn out to be twice as numerous as expected, for example, then payments to the local school districts can be increased proportionately — provided that the potential for the problem was foreseen, the appropriate institutional safeguards were set up to provide additional funding, the developer stays in business, and the mechanisms work as planned.

The third response is that the SIA practitioner makes "projections" not "predictions." Rather than saying what *will* happen, the SIA practitioner projects what will happen *if* a given set of assumptions is borne out. Unfortunately, clients and affected communities often pay far more attention to the "best estimate" projections than to the caveats. As an SIA practitioner said in an informal conversation, "The locals really like those nice, solid numbers. What they usually don't realize is that I have about as much say over the size of the *actual* work force as [the local priest] has over the selection of the pope."

It helps to remember that the purpose of SIAs is to reduce the uncertainties, not to eliminate them. In fact, many SIAs are done for reasonably routine projects and are fairly adequate. Often, however, we have too little information to produce professionally credible assessments. What will be the educational implications, for example, of having personal computers in 60% of American homes, and what will the implications be for the distribution of wealth and population between rural and urban areas of the United States? How might these trends, in turn, affect the likely viability of a proposed mine or power plant? If the state of Wisconsin were to ban the use of bovine growth hormone ("bovine somatotropin") in milk production, would the action penalize the state's dairy farmers by preventing them from adopting a more efficient technology or aid the same farmers by preserving consumer confidence in the purity of the state's milk? Neither outcome, it appears, can prudently be ruled out.

At times, in short, the three typical responses of the field to the problem of uncertainty are unsatisfactory, even collectively. Providing communities and clients with caveats is a useful first step, but it is scarcely sufficient. Local officials are often forced to expand services and facilities enough to meet one population size or another, secure only in the knowledge that they will be blamed whether the facilities ultimately prove to be either "too big" or "too small." In addition, evidence from a variety of fields shows that experts routinely suffer from overconfidence, failing to provide sufficiently careful caveats, in part, because they fail to foresee many of the ways in which their "projections" might go awry (Freudenburg 1988, 1992a; Clarke 1988a, 1988b). Finally, a reliance on monitoring will make sense only in a subset of real-world situations — those in which the relevant causal mechanisms are reasonably well-understood and the likely outcomes are clustered around a single estimate. Monitoring provides far less reassurance when the probability distribution is multimodal (widely differing outcomes are about equally likely), binary (outcome depends heavily on X, where X has roughly a 50:50 chance of occurring), or completely unknown. Still other problems are presented by cases involving "unknown unknowns," when the analyst is unaware of his or her ignorance. As one practitioner put it, "It's one thing to come across a field of land mines if you know they're out there but just don't know where they're buried. It's another thing entirely when you don't even know the field is mined."

The situation calls for new approaches to the problem of uncertainty. In coping with systems when understanding is relatively high and remaining uncertainty is not likely to present major problems, it may prove appropriate to continue following the standard practices. In coping with systems that are less well-understood, however, or those for which even a maximum-likelihood estimate has little likelihood of occurring as envisioned, it may be that a completely different approach is needed. The prudent approach, in fact, may be to invest relatively little effort in producing the estimated outcome — making it clear that one is offering simply a "rough guess" — and instead devoting the majority of analytic effort to *anticipating "how much difference it makes" if the estimate is wrong* (Freudenburg 1986c, 1992b).

Notably, this would differ markedly from the kinds of sensitivity analyses done in the past. In assessing the likely impacts of a nuclear waste repository on the city of Las Vegas, for example, the "best guess" might well be that a radiological accident involving spent nuclear fuel would probably not occur within the city limits. Still, as most readers and most residents of affected communities are all too well aware, "unexpected" incidents have occurred many times in the past; Bhopal, Chernobyl, and the *Exxon Valdez* are just three recent examples of incidents not expected to occur. If an "unlikely" incident were to occur in or near Las Vegas, moreover, the consequences could be devastating for the city's tourism-based economy, even if few or no radiation-related deaths would occur. An emphasis on potential consequences of unpredictable outcomes, in short, may lead to very different planning responses than would the routine addition of "uncertainty bands" around whatever outcome the analyst judges to be "most likely."

CONCLUSIONS: A LOOK TO THE FUTURE

The field of SIA has managed to accomplish a great deal, yet much remains to be done. The record of accomplishments tends to be clearest in areas where the challenges being faced, while sometimes complex, have permitted the relatively straightforward application of technical expertise. Not surprisingly, progress has been slower in dealing with problems that inherently permit no single "right" answers, as well as in dealing with issues for which it has been impossible to develop the relevant empirical database. This may be because each individual impact situation is so unique that researchers have been unable or unwilling to ask the proper questions, or because agencies and project developers have been reluctant to devote the same kind of resources to the answering of social science questions as to the answering of biological and physical science ones. Particularly important challenges remain in documenting, quantifying, and predicting noneconomic or "*social* social impacts," especially for facilities not involving rapid or "boom" growth, for better specifying the ways in which social and bioenvironmental factors are interrelated, and in both dealing with and conveying to nonpractitioners the importance of considering the broad range of unintended consequences likely to follow from potential policy actions.

Overall, the accumulated experience of almost two decades of SIA work suggests that the field has reached a crossroads. In some areas of work, it is now possible to look forward to a period of "normal science," with improved quantification and explanation of phenomena that are now relatively well-understood. This would apply to many of the issues related to rapid community growth, both in the social and in the economic subsets of SIA. In many other areas, however, the major conclusion from work to date may be that we are only now beginning to have a grasp of what questions need to be asked; moreover, we need to come to terms with the fact that some of the questions are inherently unanswerable.

While we may be able to predict with impressive accuracy the number of new residents likely to be brought to an area by a given level of demand for work force at a new mine or power plant, the experience of many boom-then-bust communities suggests that the engineering projections of likely work force needs are subject to massive levels of uncertainty. Compounding the problem, the projects themselves, generally described in our assessments as "destined" to "go ahead as planned," have often proved to be speculative endeavors that were ultimately abandoned for reasons that range from poor planning to unforeseeable swings in the world commodity price markets (Culhane et al. 1987). While many of our SIAs have contained vague, routine assertions that the level of local impacts will be "likely to depend on the degree of local preparedness and planning," the experience of the last two decades leaves it unclear just what kinds of "local planning" might effectively prevent the disruptions created by the virtual shutdown of a region's only major industry. How should the local planning and zoning commission "prepare" for an unexpected shutdown that suddenly leaves literally 20–30% of the community's work force unemployed? As a quantitative indicator of the extent of potential dependence on world commodity trends, recent analysis of one area was able to explain well over 90% of the variance in county-level employment over a period of two decades without any reference to the kinds of factors under the control of the affected counties or, arguably, even the multinational companies whose facilities create the impacts (Gramling and Freudenburg 1990). As a broader matter, finally, even a solid foundation of empirical data from past experiences might provide little basis for making scientifically credible projections if accumulated evidence leads us to expect that the experiences of the future will have little direct comparability to those of the past.

In short, it may be that SIA is rapidly closing in on the day when the field will have done as much as is responsibly possible in foreseeing and lessening uncertainties. If so, or even if progress remains to be made, the greater challenge may be to deal more squarely with anticipating and managing the risks and uncertainties that, by their very nature, defy resolution.

In a particularly ironic twist, while one of SIA's major functions is to call attention to and assist in planning for unanticipated impacts, the continued pursuit of the approaches to SIA that have been employed in the past might have undesirable and unanticipated impacts of its own. To date, impact assessments have often relied on the assumption that a proposed development would proceed more or less as planned, and that the affected communities would not be subjected to additional changes of unknown nature, origin, or magnitude. While such assumptions were made in good faith and often were stated explicitly, they appear in retrospect to have been anything but reasonable for many communities.

In some cases, such as the Exxon oil shale development in Colorado, assessments have called for large numbers of highly paid jobs to be brought to the local economy for a period of decades, and companies have followed up by making capital investments so large by local standards as to convince even the most conservative of local skeptics that development would follow the stated trajectory; the facilities, however, were shut down literally overnight before the construction was even completed (Gulliford 1989). In other cases, such as the construction of several large-scale, coal-fired power plants, the construction phase was completed more or less on schedule, but the drop in employment at the completion of the construction phase proved to be almost as devastating to local business and housing markets as in the case of the more notorious Exxon shutdown. In still other cases, facilities such as smelters and railroads have appeared to be almost impervious to a series of economic swings over a period of several decades, thus increasing local confidence that a "mainstay of the economy" would be "here to stay," yet the facilities have fallen victim to the next swing in the economy, creating impacts all the more devastating for having arisen unexpectedly (Cottrell 1951; Gramling and Freudenburg 1990; Freudenburg 1992c).

In each of these cases and others, at least two commonalities can be noted. First, the cases demonstrate anew the time-honored prudence of the principle that it is unwise to have most of one's eggs in one basket — or much of a community's future in one economic sector. Second, they remind us that a presumably unanticipated consequence of past practice has been to make formerly isolated rural communities more dependent on the price swings of international markets, while simultaneously transferring some of the risks and uncertainties of development from the project proponents to the affected communities (Freudenburg and Gramling, 1992). Several of the town governments affected by the Exxon shutdown were the beneficiaries of agreements that required Exxon, and not the towns, to pay for capital expenditures such as water and sewage systems not "needed" either before or after the brief burst of Exxon's activities in the region; yet the local residents who placed enough faith in impact projections to make business investments in that region — and even the innocent, tax-paying citizens of less fortunate communities affected by other busts — have enjoyed no such indemnification. As citizens or community leaders (Molotch 1976) of isolated regions become increasingly desperate to attract sources of well-paid local remuneration, moreover, they may be increasingly disinclined to adopt the kinds of tough negotiating stances that could provide them with at least a degree of protection against unforeseen economic swings.

In short, the "normal science" contributions of SIA have tended in the past to simplify the management of uncertainty. The field has documented a number of consistent patterns in certain types of development, developed a reasonably respectable set of methods, and produced the kinds of research necessary for relatively dramatic improvements in the estimation of the coefficients that those methods require. Increasingly, the challenges that remain for the future may have to do not so much with efforts to make sure our projections are "right," but with understanding the most prudent steps to take when even the best of projections are virtually guaranteed to be wrong.

REFERENCES

Albrecht, S. L. 1982. Commentary. *Pac. Sociol. Rev.* 25(3):297–306.

Atherton, C. C. 1977. Legal requirements for environmental impact reporting. pp. 9–64. In J. McEvoy, III, and T. Dietz (eds.), *Handbook for Environmental Change.* Wiley, New York.

Bailey, C. 1985. The blue revolution: The impact of technological innovation on third-world fisheries. *Rural Sociol.* 5(4):259–266.

Baum, A. 1987. Toxins, technology, and natural disasters. pp. 5–53. In G. R. VandenBos, and B. K. Bryant (eds.), *Cataclysms, Crises, and Catastrophes: Psychology in Action.* American Psychological Association, Washington, D.C.

Berger, T. R. 1977. *Northern Frontier, Northern Homeland: The Report of the MacKenzie Valley Pipeline Inquiry.* Vols. 1 and 2. Supplies and Services Canada, Ottawa.

Boothroyd, P. 1982. Overview of the Issues Raised at the International Conference on Social Impact Assessment. Presented at the International Conference on Social Impact Assessment, Vancouver, B.C., Canada.

Bowles, R. T. 1982. A Quick and Dirty Profile of the Social Impact Assessment Community. Presented at the International Conference on Social Impact Assessment, Vancouver, B.C., Canada.

Bunker, S. G. 1984. Modes of extraction, unequal exchange, and the progressive underdevelopment of an extreme periphery: The Brazilian Amazon 1600–1980. *Am. J. Sociol.* 89(5):1017–1064.

Carley, M. J., and E. S. Bustelo. 1984. *Social Impact Assessment and Monitoring: A Guide to the Literature.* Westview, Boulder, CO.

Clarke, L. 1988a. Politics and bias in risk assessment. *Soc. Sci. J.* 25(2):155–165.

Clarke, L. 1988b. Explaining choices among technological risks. *Soc. Probl.* 35(1):22–35.

Cortese, C. F. 1982. The impacts of rapid growth on local organizations and community services. pp. 115–135. In B. A. Weber, and R. E. Howell (eds.), *Coping with Rapid Growth in Rural Communities.* Westview, Boulder, CO.

Cortese, C. F., and B. Jones. 1977. The sociological analysis of boom towns. *West. Sociol. Rev.* 8:76–90.

Cottrell, W. F. 1951. Death by dieselization. A case study in the reaction to technological change. *Am. Sociol. Rev.* 16:358–365.

Covello, V. T. 1983. The perceptions of technological risks: A literature review. *Technol. Forecast. Soc. Change* 23:285–297.

Creighton, J. L. 1980. Public Involvement Manual: Involving the Public in Water and Power Resources Decisions. U.S. Government Printing Office, Washington, D.C.

Culhane, P. J., H. P. Friesema, and J. A. Beecher. 1987. *Forecasts and Environmental Decisionmaking: The Content and Predictive Accuracy of Environmental Impact Statements.* Westview, Boulder, CO.

Davenport, J. A., and J. Davenport, Jr. 1979. *Boom Towns and Human Services.* University of Wyoming, Laramie.

Davenport, J., Jr., and J. A. Davenport. 1981. *The Boom Town: Problems and Promises in the Energy Vortex.* Laramie: Department of Social Work, University of Wyoming, Laramie.

Dietz, T. 1984. Social impact assessment as a tool for rangeland management. pp. 1613–1634. In *Developing Strategies for Rangeland Management.* National Resource Council/ National Academy of Sciences, Westview, Boulder, CO.

Dietz, T., P. C. Stern, and R. W. Rycroft. 1989. Definitions of conflict and the legitimation of resources: The case of environmental risk. *Sociological Forum* 4(1):47–71.

Dietz, T., R. S. Frey, and E. A. Rosa. in press. Risk, technology, and society. In R. E. Dunlap, and W. Michelson (eds.), *Handbook of Environmental Sociology.* Greenwood, Westport, CT.

Dohrenwend, B. P., B. S. Dohrenwend, G. J. Warheit, G. S. Bartlett, R. L. Goldsteen, K. Goldsteen, and J. L. Martin. 1981. Stress in the community: A report to the President's Commission on the Accident at Three Mile Island. pp. 159–174. In T. H. Moss, and D. L. Sills (eds.), *The Three Mile Island Nuclear Accident: Lessons and Implications.* New York Academy of Sciences, New York.

Dornbusch, D., et al. 1987. Impacts of Outer Continental Shelf (OCS) Development on Recreation and Tourism. U.S. Minerals Management Service, Los Angeles, CA.

Dornsife, W., P. Serie, and J. Kauffman. 1989. Including Local Leaders in Public Participation. Presented to Waste Management in March 1989. Tucson, AZ.

Dunlap, R. E. 1982. Ecological limits: Societal and sociological implications. *Contemp. Sociol.* 11(3):153–160.

Elkind-Savatsky, P. D. 1986. *Differential Social Impacts of Rural Resource Development.* Westview, Boulder, CO.

Finsterbusch, K. 1980. *Understanding Social Impacts: Assessing the Effects of Public Projects.* Sage, Beverly Hills, CA.

Finsterbusch, K. 1982. Boomtown disruption thesis: Assessment of current status. *Pac. Sociol. Rev.* 25(3):307–322.

Finsterbusch, K. 1985. State of the art in social impact assessment. *Environ. Behav.* 17(2):193–221.

Finsterbusch, K. 1988. Citizens' Encounters with Unresponsive Authorities in Obtaining Protection from Hazardous Wastes. Presented at Annual Meeting of Society for the Study of Social Problems, August, Atlanta, GA.

Finsterbusch, K. F., and W. R. Freudenburg. 1992. Social impact assessment and technology assessment. In R. E. Dunlap, and William Michelson (eds.), *Handbook of Environmental Sociology.* Greenwood, Westport, CT (in press).

Flynn, C. B. 1984. The local impacts of the accident at Three Mile Island. pp. 205–232. In W. R. Freudenburg, and E. A. Rosa (eds.), *Public Reactions to Nuclear Power: Are There Critical Masses?* Westview, Boulder, CO.

Flynn, C. B., J. H. Flynn, J. A. Chalmers, D. Pijawka, and K. Branch. 1983. An integrated methodology for large-scale development projects. pp. 55–72. In K. Finsterbusch, L. G. Llewellyn, and C. P. Wolf (eds.), *Social Impact Assessment Methods.* Sage, Beverly Hills, CA.

Freudenburg, W. R. 1976. The Social Impact of Energy Boom Development on Rural Communities: A Review of Literatures and Some Predictions. Presented at Annual Meeting of American Sociological Association, August, New York.

Freudenburg, W. R. 1979. An ounce of prevention: Another approach to mitigating the human problems of boomtowns. pp. 55-62. In *Energy Resource Development: Implications for Women and Minorities in the Intermountain West.* U.S. Commission on Civil Rights, U.S. Government Printing Office, Washington, D.C.

Freudenburg, W. R. 1981. Women and men in an energy boomtown: Adjustment, alienation, and adaptation. *Rural Sociol.* 46(2):220–244.

Freudenburg, W. R. 1982a. The impacts of rapid growth on the social and personal well-being of local community residents. pp. 137–170. In B. A. Weber, and R. E. Howell (eds.), *Coping with Rapid Growth in Rural Communities.* Westview, Boulder, CO.

Freudenburg, W. R. 1982b. Balance and bias in boomtown research. *Pac. Sociol. Rev.* 25(3):323–328.

Freudenburg, W. R. 1985. Applying sociology to policy: Social science and the environmental impact statement. *Rural Sociol.* 50(4):578–605.

Freudenburg, W. R. 1986a. Assessing the social impacts of rural resource developments: An overview. pp. 89–116. In P. D. Elkind-Savatsky (ed.), *Differential Social Impacts of Rural Resource Development.* Westview, Boulder, CO.

Freudenburg, W. R. 1986b. Social impact assessment. *Annu. Rev. Sociol.* 12:451–478.

Freudenburg, W. R. 1986c. The density of acquaintanceship: An overlooked variable in community research? *Am. J. Sociol.* 92(1):27–63.

Freudenburg, W. R. 1988. Perceived risk, real risk: Social science and the art of probabilistic risk assessment. *Science* 242(7):44–49.

Freudenburg, W. R. 1992a. Nothing recedes like success? Risk analysis and the organizational amplification of risks. *Risk: Issues Health Saf.* 3(1):1–35.

Freudenburg, W. R. 1992b. Social scientists' contributions to environmental management. *J. Soc. Issues* 45(1):133–152.

Freudenburg, W. R. 1992c. Addictive Economies: Extractive Industries and Vulnerable Localities in a Changing World Economy. *Rural Sociology* 57(3):305–332.

Freudenburg, W. R., and K. M. Keating. 1982. Increasing the impact of sociology on social impact assessment: Toward ending the inattention. *Am. Sociol.* 17(2):71–80.

Freudenburg, W. R., and D. Olsen. 1983. Public interest and political abuse: Public participation in social impact assessment. *J. Community Dev. Soc.* 14(2):67–82.

Freudenburg, W. R., and R. K. Baxter. 1984. Host community attitudes toward nuclear power plants: A reassessment. *Soc. Sci. Q.* 65(4):1129–1136.

Freudenburg, W. R., and R. K. Baxter. 1985. Nuclear reactions: Public attitudes and public policies toward nuclear power plants. *Policy Stud. Rev.* 5:96–110.

Freudenburg, W. R., and S. K. Pastor. 1989. Public Responses to Technological Risks: The Need for a Closer Look. Presented at the Annual Meeting of the Rural Sociological Society, August, Seattle, WA.

Freudenburg, W. R., and R. Gramling. 1992. Community impacts of technological change: Toward a longitudinal perspective. *Soc. Forces* 70:(4):937–955.

Freudenburg, W. R., and R. E. Jones. 1992. Deviant behavior and rapid community growth: Examining the evidence. *Rural Sociol.* 56(4):619–645.

Friesema, H. P., and P. J. Culhane. 1976. Social impacts, politics, and the environmental impact statement process. *Nat. Resour. J.* 16:339–356.

Gale, R. 1982. Commentary. *Pac. Sociol. Rev.* 25(3):339–348.

Gold, R. L. 1974. A Comparative Study of the Impact of Coal Development on the Way of Life of People in the Coal Areas in Eastern Montana and Northeastern Wyoming. Northern Great Plains Resources Program, Denver.

Gold, R. L. 1982. Commentary. *Pac. Sociol. Rev.* 25(3):349–356.

Gramling, R., and W. R. Freudenburg. 1990. A closer look at "Local Control": Communities, commodities, and the collapse of the coast. *Rural Sociol.* 55(4):541–558.

Gulliford, A. 1989. *Boomtown Blues: Colorado Oil Shale, 1885–1985.* University Press of Colorado, Niwot, CO.

Halstead, J. M., R. A. Chase, S. H. Murdock, and F. L. Leistritz. 1984. *Socioeconomic Impact Management: Design and Implementation.* Westview, Boulder, CO.

Halstead, J. M., F. L. Leistritz, D. G. Rice, D. M. Saxowsky, and R. A. Chase. 1982. Mitigating Socioeconomic Impacts of Nuclear Waste Repository Siting. North Dakota Agricultural Experiment Station, Fargo.

Henshel, R. L. 1982. Sociology and social forecasting. *Annu. Rev. Sociol.* 8:57–79.

Howell, R. E., and D. Olsen. 1981. Who Will Decide? The Role of Citizen Participation in Controversial Natural Resource and Energy Decisions. Vance Bibliographies. No. 765, Monticello, IL.

Howell, R. E., D. Olsen, M. E. Olsen, and R. E. Dunlap. 1981. Citizen Participation in Nuclear Waste Repository Siting. Western Rural Development Center, Corvallis, OR.

Johnson, R., and W. L. Petcovic. 1988. What Are Your Chances of Communicating Effectively with Technical or Non-Technical Audiences? Presented at the Annual Meeting, November, Society for Risk Analysis, Washington, D.C.

Jordan, W. S., III. 1984. Psychological harm after PANE: NEPA's requirements to consider psychological damage. *Harv. Environ. Law Rev.* 8:55–87.

Kasperson, R. E., J. Emil, C. Hohenenser, J. Kasperson, and O. Renn. 1987. Radioactive Wastes and the Social Amplification of Risks. Presented at the Waste Management '87, Tucson, AZ.

Knorr-Cetina, K. D. 1981. *The Manufacture of Knowledge: An Essay on the Constructivist and Contextual Nature of Science.* Pergamon Press, Oxford.

Krannich, R. S., and T. Greider. 1985. Rapid growth and fear of crime: A four-community comparison. *Rural Sociol.* 59(2):193–209.

Krannich, R. S., T. Greider, and R. L. Little. 1984. Personal well-being in rapid growth and stable communities: Multiple indicators and contrasting results. *Rural Sociol.* 49(4):541–552.

Kunreuther, H. 1985. Hazard Compensation and Incentive Systems: An Economic Perspective. Presented to the National Academy of Engineering Symposium on Hazards: Equity, Incentives and Compensation, June, Washington, D.C.

Kunreuther, H., J. Lathrop, and J. Linnerooth. 1982. A descriptive model of choice for siting facilities. *Behav. Sci.* 27:282–297.

Latour, B., and S. Woolgar. 1986. *Laboratory Life: The Construction of Scientific Facts.* Princeton University Press, Princeton, NJ.

Levine, A. G. 1982. Love Canal: Science, Politics, and People, Lexington, MA.

Little, R. L. 1977. Some social consequences of boom towns. *N.D. Law Rev.* 52:401–425.

Llewellyn, L. G., and W. R. Freudenburg, 1989. Legal requirements for social impact assessments: Assessing the social science fallout from Three Mile Island. *Soc. Nat. Resour.* 2(3):193–208.

Lovins, A. B. 1986. The origins of the nuclear power fiasco. pp. 7–34. In J. Byrne, and D. Rich (eds.), *The Politics of Energy Research and Development: Energy Policy Studies, Transaction.* New Brunswick, NJ.

Meidinger, E. E., and W. R. Freudenburg. 1983. The legal status of social impact assessments: Recent developments. *Environ. Sociol.* 34:30–33.

Moyer, R. 1984. The futility of forecasting. *Long Range Plann.* 17(1):65–72.

Murdock, S. H., and F. L. Leistritz. 1979. *Energy Developments in the Western United States: Impact on Rural Areas.* Praeger, New York.

Murdock, S. H., and F. L. Leistritz. 1982. Commentary. *Pac. Sociol. Rev.* 25(3):357–366.

Murdock, S. H., F. L. Leistritz, R. R. Hamm, and S.-S. Hwang. 1984. An assessment of the accuracy and utility of socio-economic impact assessments. pp. 265–296. In C. M. McKell, D. G. Browne, E. C. Cruze, W. R. Freudenburg, R. L. Perrine, and F. Roach (eds.), *Paradoxes of Western Energy Development.* Westview, Boulder, CO.

Murdock, S. H., F. L. Leistritz, and R. Hamm. 1985. The State of Socioeconomic Analysis: Limitations and Opportunities for Alternative Futures. Presented at annual meeting of Southern Association of Agricultural Scientists, February, Biloxi, MI.

National Academy of Sciences, National Research Council. 1984. *Social and Economic Aspects of Radioactive Waste Disposal: Considerations for Institutional Management.* National Academy Press, Washington, D.C.

O'Hare, M. 1977. Not on my block, you don't — Facility siting and the strategic importance of compensation. *Public Policy* 25:407–458.

Payne, B., and C. Cluett. Environmental sociology in nonacademic settings. In R. E. Dunlap, and W. Michelson (eds.), *Handbook of Environmental Sociology* Greenwood, Westport, Ct. (in press).

Pelle, E. 1987. The MRS Task Force: Economic and Noneconomic Incentives for Local Public Acceptance of a Proposed Nuclear Waste Repository. Presented at Waste Management '87, Tucson, AZ.

Petterson, J. S. 1988. The reality of perception: Demonstrable effects of perceived risk in Goiânia, Brazil. *Pract. Anthropol.* 10(3–4):8–9, 12.

Rothman, S., and S. R. Lichter. 1987. Elite ideology and risk perception in nuclear energy policy. *Am. Polit. Sci. Rev.* 81(2):383–404.

Savatsky, P. D. 1974. A legal rationale for the sociologist's role in researching social impacts. pp. 45–47. In C. P. Wolf (ed.), *Social Impact Assessment.* Dowden, Hutchinson & Ross, Stroudsburg, PA.

Schnaiberg, A. 1980. *The Environment: From Surplus to Scarcity.* Oxford University Press, New York.

Seyfrit, C. S. 1986. Migration intentions of rural youth: Testing an assumed benefit of rapid growth. *Rural Sociol.* 51(2):199–211.

Short, J. F. 1984. The social fabric at risk: Toward the social transformation of risk analysis. *Am. Sociol. Rev.* 49:711–725.

Slovic, P. 1987. Perception of risk. *Science* 236(17):280–285.

Smith, V. K., and W. H. Desvousges. 1986. The value of avoiding a LULU: Hazardous waste disposal sites. *Rev. Econ. Stat.* 48(2):293–299.

Smith, C. L., T. C. Hogg, and M. G. Reagan. 1971. Economic development: Panacea or perplexity for rural areas? *Rural Sociol.* 36(2):173–186.

Summers, G. F., S. D. Evans, F. Clemente, E. M. Beck, and J. Minkoff. 1976. *Industrial Invasion of Nonmetropolitan America: A Quarter Century of Experience.* Praeger, New York.

Tester, F. J., and W. Mykes. 1981. *Social Impact Assessment: Theory, Method and Practice.* Detselig Enterprises, Calgary, Canada.

U.S. Council on Environment Quality. 1978. Regulation for Implementing the Procedural Provisions of the National Environmental Policy Act (40 CFR 1500-1508). Washington, D.C.

Weber, B. A., and R. E. Howell. 1982. *Coping with Rapid Growth in Rural Communities.* Westview, Boulder, CO.

Wilkinson, K. P., R. R. Reynolds, Jr., J. G. Thompson, and L. M. Ostresh. 1984. Violent crimes in the western energy-development region. *Sociol. Perspect.* 27(2):241–256.

Wilkinson, K. P., J. G. Thompson, R. R. Reynolds, Jr., and L. M. Ostresh. 1982. Local social disruption and western energy development: A critical review. *Pac. Sociol. Rev.* 25(3):275–296.

Wolf, C. P. 1977. Social impact assessment: The state of the art updated. *Soc. Impact Assess.* 20:3–22.

Science and the Inherently Subjective: The Evolution of Aesthetic Assessment Since NEPA

C. H. Petrich, Oak Ridge National Laboratory, Oak Ridge, TN

ABSTRACT

Aesthetic assessment and the National Environmental Policy Act (NEPA) share common origins but have developed over the past 20 years at different rates and with different degrees of effectiveness, theoretical grounding, and public acceptability. The major currents in aesthetic assessment are outlined, with the major stumbling blocks to broader and more rigorous application highlighted. The importance and challenge of ascertaining the local valuation of aesthetic resources and the different messages that developments can convey to locals and nonlocals are described. This paper outlines the rich opportunities afforded by using an information processing approach to understand and gauge aesthetic preference and offers a brief explanation of these preferences from an evolutionary perspective. Some of the problems with garnering more effective public participation and with greater public acceptability of aesthetic assessment frameworks are explored. The paper concludes with a discussion of the challenges aesthetic assessment faces in this current period of unprecedented metropolitan growth.

NEPA AND AESTHETIC ASSESSMENT: MUTUAL ORIGINS

The roots of aesthetic assessment parallel those of the National Environmental Policy Act (NEPA); their relationship some 20 years ago was quite incestuous. For example, visual impacts from a proposed pumped storage facility for generating power for New York City raised the hackles and opened the coffers of wealthy residents along the Hudson River in Westchester County in the 1960s. The ensuing battles about the aesthetic impacts of this proposed facility and then of the transmission lines from a modification of the facility precipitated the beginnings of the Natural Resources Defense Council, one of the first activist environmentalist groups whose roots did not go back to the early fights for wilderness in this country. Many environmental historians trace the origins of NEPA to these Hudson Valley clashes. Another battle that heightened the country's sensitivity toward the environment also had roots in aesthetic issues. Walt Disney Enterprises, Inc. requested a permit from the Forest Service, an agency of the Department of Agriculture, to develop Mineral King Valley as a $35-million complex of motels, restaurants, and recreational facilities in a wilderness area in the California Sierra. The lasting legacy of the ensuing legal battle is a now-classic law review article (Stone 1972) wherein it is argued that if corporations can have standing (in the legal sense), why could not other nonhuman entities (such as trees, streams, and mountains) also have legal standing (and therefore sue to protect their rights)?

Important literature of the time also dealt with the visual resource extensively: biology Nobel Laureat René Dubos' *So Human an Animal* (1968) addressed the human genetic yearnings for certain environments; the first paperback release of Aldo Leopold's 1949 classic *A Sand County Almanac* (1970) introduced the late 1960s college audience to a special reverence for the land; Ian McHarg, a little-known landscape architecture professor from the University of Pennsylvania, burst on the environmental scene with a discussion of applied human ecology that centered heavily on visual concerns in *Design with Nature* (1969a); Luna Leopold (Aldo's son) published some of the first methodological papers on quantifying the visual resource (Leopold and Marchand 1968; Leopold 1969, 1970); and landscape architect Lawrence Halprin combined choreography with environmental planning with an emphasis on West Coast development issues in *The RSVP Cycles* (1969).

McHarg and the University of Pennsylvania's cadre of those concerned with the nation's environmental future led the country by organizing the first Earth Day in 1970. Again the connection with visual concerns about the environment were strong. As the environment became a *cause célèbre* in the late 1960s, the focus shifted to a broader front than the concerns of Rachel Carson's *Silent Spring* of earlier in the decade. Environmentalists caused legislators to ponder Aldo Leopold's dictum of 20 years earlier: "A thing is right when it tends to preserve the integrity, stability, and beauty of the biotic community. It is wrong when it tends otherwise" (1949). They gave such ideas new form, such as "Fitness and fitting are indications of health and the process of fitting is health giving" (McHarg 1981) or

> ... Beauty is not simply subjective. ... Beauty is an aspect of that which holds things together. ... It is a necessary quality of the world. If we diminish the world's beauty, we diminish the world's capacity to exist and ability to survive (Austin 1985).

In this paper, I will trace the relationship of aesthetic assessment to NEPA, where they have common roots and common goals and where aesthetic assessment has evolved away from the direct concerns of NEPA. After exploring their shared foundation and the legislative mandate to examine aesthetic issues, I discuss current conceptual issues in aesthetic assessment and how they have come to be. Included here are tools such as advanced simulation techniques, generative ideas influencing the discipline, and an examination of the scope of inputs to the discipline. Among these are anthropological sensitivity in understanding local environments, environmental psychological foundations of human transactions with landscapes, an understanding of potential evolutionary roots of human behavior as expressed through aesthetic preference, and cognitive scientific backgrounds in information processing and human decisionmaking. The problems of linking value-laden research with more presumably objective and easily measured research results in NEPA assessments are discussed as are other challenges on the horizon. Dominant among these is the evaluation of the commonplace in NEPA documents. People value, often highly so, common landscapes. Yet it is difficult to document values unless they are based on scarcity, an economist's approach. Without a convenient and broadly acceptable metric, what is the aesthetic analyst to do? I conclude the paper with a survey of other challenges comparable to that of valuing commonplace landscapes, such as the valuation and protection of working, or rural, landscapes, which are rapidly being converted to suburban uses.

NEPA AND AESTHETIC ASSESSMENT

The incestuous nature of NEPA and aesthetic concerns developed very quickly. Whereas aesthetic issues helped define and generate support for the passage of NEPA, it was NEPA that in large measure gave the necessary governmental directives that enabled the discipline of aesthetic assessment to develop a research and theoretical foundation. NEPA requires the

> Federal Government to use all practicable means... [to] ... assure for all Americans safe, healthful, productive, and aesthetically and culturally pleasing surroundings ... [and to] ... preserve important historic, cultural, and natural aspects of our national heritage, and maintain, wherever possible, an environment that supports diversity and variety of individual choice [NEPA Section 101(b)(2,4)].

To this end, federal agencies are directed to

> utilize a systematic, interdisciplinary approach that will insure the integrated use of the natural and social sciences and the environmental design arts in planning and decision making that may have an impact on man's environment [NEPA Section 101(2)(A)].

EARLY RESEARCH FOUNDATIONS OF AESTHETIC ASSESSMENT

There is a long tradition within the profession of landscape architecture of developing landscape inventories and analyses, generally with an eye to potential landscape changes (Priestley 1983). In the early part of this century, individual landscape architects developed procedures for scenic highway routing and park landscape evaluation and management. Priestley identifies the immediate forerunner of contemporary visual analysis activity as being the work at the University of Wisconsin in the early 1960s by Phil Lewis and others who prepared regional-scale visual and recreational inventories. The field as we know it today first began to take shape in about 1968 with a core literature that included Litton (1968), Fines (1968), Jacobs and Way (1968), and Linton (1968).

Both NEPA and the Forest Management Act of 1976 required the USDA Forest Service to consider visual factors in the management of its vast acreages. The early work by Litton (1968) and Shafer (1967, 1969; Shafer et al. 1969; Shafer and Mietz 1970; Shafer and Tooby 1973; Shafer and Richards 1974) highlight the Forest Service's prominent role in the discipline's early years. The field held its first major conference in only 1973 and its second not until 1979, with the Forest Service playing a major role in sponsoring and disseminating important research.

MAJOR CURRENTS IN AESTHETIC ASSESSMENT

At the signing of NEPA, there were no research results or proposals that combined environmental design with those examining the psychological and mental health responses of man to natural landscapes (Iltis et al. 1970). In the short 20-year period since this observation, this research need has been largely satisfied, although its findings have been scarcely digested, let alone implemented. As research findings, however, they have strongly influenced the theoretical underpinnings of current aesthetic impact assessment. The field of visual resource analysis and management still is, in Kuhn's (1962) terminology, at the preparadigm state, lacking any central, unifying theory or world view capable of providing a commonly agreed upon starting point for practice or a common context for research (Priestley 1983). Part of the reason for this condition is the near-intractable nature of the problems practitioners face:

> What we perceive as the environment is, in fact, a synthesis in which our current perceptions of what is actually out there are combined with a complex tapestry of associations based on our experience both of the physical world and of other people. This applies to even the most detached and rational inventory of landscape features. When it comes to analyzing subjective responses to an existing or a proposed landscape design, the problem of logical analysis becomes almost unsurmountable. It has eluded philosophers for millennia. And yet such analysis is precisely what landscape architects and other environmental designers are requiring their profession to achieve, and they have no choice but to continue in the attempt (Greenbie 1988).

Thus, aesthetics is more than the study or theory of beauty and of the psychological responses to it. It is, argues Thayer, but one component of landscape *affect*; other positive dimensions might include "feelings of community belonging, pride, health, safety, security, sense of self, oneness with nature, and emotional stability over time" (Thayer 1989).

There have been extensive summaries of the use of aesthetic analysis in landscape assessment and, to a lesser degree, in the formal aesthetic impact assessment process as well (Smardon et al. 1986; Zube et al. 1982; Wohlwill 1976; Kaplan and Kaplan 1982; Ulrich 1983; Porteous 1982; Fenton and Reser 1988; Hyman and Stiftel 1988). Most methods for performing aesthetic impact assessments involve three approaches (usually in combination): professional evaluations, predictive models, and public participation. The literature describing applications of these methodologies can be divided into two categories. The first involves assessments of large or dominating manmade alterations of the natural landscape (Jones et al. 1975; Petrich 1982, 1984). The other involves less dominating activities in more urbanized landscapes (Smardon and Hunter 1983).

Largely overlooked is how little empirical attention has been given to the experiential aspects of landscape beauty (Chenoweth and Gobster 1990). Landscape assessment research and much local, state, and federal legislation have been conceived under the fundamental assumption that beautiful landscapes provide unique opportunities for people to achieve special kinds of experiences, often called "aesthetic," that are highly valued and less likely to occur in less-beautiful places. Only recently has systematic testing of this assumption begun. Chenoweth and Gobster (1990) found that many aesthetic experiences occur in connection with ephemeral events and, not surprisingly, included senses other than vision. Because most aesthetic experiences tend to be experienced unexpectedly rather than being sought out, they tend to occur in familiar places. Counterintuitively, the unexpected is found most frequently among the familiar. This indicates the importance of managing "everyday" environments for aesthetic quality (Kaplan 1985).

Most assessment frameworks rely on descriptions of landscape qualities, not predictions of future landscape attributes. Zube (1984) noted that most researchers using numeric evaluation approaches rely on nominal or, at best, ordinal scales. For more interdisciplinary approaches, as resource management activities increasingly are, interval or ratio scale data are needed (Hobbs et al. 1989; Flug and Montgomery 1988). Unless aesthetic resource assessments are made to be commensurate with other resource evaluation schemes, they will likely be undervalued or ignored. Too heavy a reliance on numerative techniques, however, runs the risk of not capturing what should be most important about the aesthetic analysis: the experiential, affective response of the individual. For example, noted landscape observer J. B. Jackson (1984) has stated that any "landscape is beautiful when it has been or can be the scene of a significant experience in self-awareness and eventual self-knowledge." Chenoweth and Gobster (1990) have tried an approach that evaluates particular aesthetic experiences in relation to other significant life events and the changes in the overall mood of the affected individual as a result of the experience. While this may target the core of potential aesthetic effects, it does not lend itself well to measurement, especially when compared to measurements for effects on timber, rangelands, or wildlife habitat. There is hope, however, as I shall discuss below, in moving beyond the proverbial but idiosyncratic "beauty is in the eye of the beholder."

AESTHETIC RESOURCE SIMULATION TECHNIQUES

Aesthetic impact assessment relies, in part, on simulations — attempts to selectively manipulate certain characteristics of landscape scenes, while holding all other features constant — to gauge user and/or expert reactions to proposed changes in landscapes (Sheppard 1982; Killeen and Buhyoff 1983). Photomontage and professional artists' air brush techniques for incorporating changes have given way to automated digitization of two-dimensional still and video images. Three-dimensional models of varying sophistication have also proved valuable in communicating development proposals and evoking potential users. Technical innovations for increasing the effectiveness, reliability, and validity of portrayals of potential changes in the landscape are currently outpacing our ability to simulate landscape *experiences* as well as perceptions of landscape (Berleant 1988).

Since the early 1980s, computer-based simulations of proposed changes in landscapes have been significantly advanced by the introduction of high-resolution computer graphics and techniques for "frame grabbing" video signals and converting them into digital images (Bishop and Leahy 1989; Zube et al. 1987). This near-approximation of photographic quality has been further facilitated by advances in computational and display speeds, allowing the user to rapidly try different views, colors, shadow and blending routines, etc. Other advances include the depiction of "soft objects" — objects created totally within the computer — through use of geometric data in conjunction with data amplification techniques such as fractal geometry and texture mapping (Bishop and Leahy 1989). As computer hardware and software become increasingly sophisticated, greater frequency of use and reliability will only expand, augmenting the ability to assess the aesthetic impact of proposed developments. Public participation can be encouraged and enhanced through effective interactive display technologies already becoming available (Petrich 1986).

SENSE OF PLACE, LOCAL KNOWLEDGE, AND SEMIOTICS

Because NEPA documents, by their interdisciplinary nature, seek to incorporate common metrics, it is essential for aesthetic analysts first to uncover what it is "out there" that people seem to value and, if one is an astute observer, why is it valued. Only then can a metric commensurate with those of other disciplines be found so that important public perceptions and values are fairly considered and weighed in the balanced incorporation of all impacts. In this section, I explore the challenges of understanding the values people associate with their aesthetic appreciation of landscapes.

Sense of Place

The process of creating aesthetically pleasing environments encompasses nature and culture, both the making of places and things and the seeing, using, and contemplating of them (Spirn 1989). It is the interaction of nature and culture that makes a place unique. Good designers exploit this intricate interaction through understanding the processes that underlie the themes, rhythms, patterns, and order, as well as the complexities of specific cultural responses to a unique locale and time. Landscape architect Lawrence Halprin (1969) pithily captured the understanding needed to create dynamic, coherent wholes, ones that can meet changing needs and desires and connect the present with the past: "*The place is because.*"

Dick Reynolds, the chief geologist for the Sea Ranch development in northern California, illustrates this concept.

> If you can learn the history of the land over a hundred years, the knowledge can tell you what processes are at work in the region—both constructive and destructive. From these you can learn how to use the land, how to plan a development: where to locate buildings, roads and plantings. You do not necessarily have to conform to the processes at work, but at least, if you choose to go against them, you are in a better position to estimate the consequences and the costs (Halprin 1969).

The aesthetic analyst needs to be part designer to detect opportunities missed, poorly executed, or achieved. One's ability to understand how a place changed over time, to see why "the place is because," will affect how one designs a place, appreciates a place, and evaluates the aesthetic success of its design. Form and process are indivisible aspects of a single-phenomenon, three-dimensional form being an explicit point in the fourth dimension, evolutionary time (McHarg 1969b). According to architect Christopher Alexander (1964), a good designer's created form "penetrate[s] the problem so deeply that it not only solves it but illuminates it. A well-designed home not only fits its context well but also illuminates the problem of just what the context is, and thereby clarifies the life which it accommodates."[1]

Halprin (1989) refers to environmental design as being, in one sense, the bearer of a community's cultural value system. Because of this, it embodies much more than the visual or even the sensed. Such holistic pervasiveness implies that it

---

[1] Not all designers believe that process should be made visible. Begging off issues of semantics and the relative motivations of "designers" vs "artists," consider Pablo Picasso (Rusch 1970): "I want to develop the ability to do a picture in such a way that no one can ever see how it is done. To what end? What I want is that my picture should evoke nothing but emotion."

necessarily encompasses issues of context, lifestyle, and economy, all with potentially profound ecological ramifications. Hester (1989) goes beyond this to state that designers can give form to "unspeakable values, those related to sex, particularly the sensual landscape, and social change — radical, in however modest a form. ... And finally, the designer gives form to unthought values."

Goethe said, "People see what they know." Appreciated and preferred environments appear to be more strongly coded into cognitive structure in terms of familiarity and memory (Hammitt 1983). Tuan (1977) also underscores this memory-appreciation relationship, "We tend to think of remembrance as warmed-over experience, forgetting that it can itself be an exquisite pleasure."

Environments that both represent and reveal the overlapping of natural and cultural processes are rich in layered meanings, both functional and symbolic, and as a result, possess amplified potential for aesthetic experiences. Aesthetic analysts must function as part anthropologist to uncover these meanings and potentials. The shapes of this local knowledge are, in the words of anthropologist Clifford Geertz (1983), "always ineluctably local, indivisible from their instruments and their encasements." In dealing with aesthetics, the shape of local knowledge is highly charged, but not always in easily perceived ways. People's sense of security or freedom or attachment and longing for place, or "topophilia," (Tuan 1974, 1977) can make sacred the seemingly commonplace (Hester 1989).

Local Knowledge

The successful aesthetic analyst will capture people's perceptions of time and change, of who they think they are, of where they have come from and where they think they are going, as individuals, as a society, and as a species (Lynch 1972). The analyst will recognize that designs that foster and intensify the experience of temporal and spatial scales will facilitate both a reflection upon personal change and identity and a sense of unity with a larger whole.

> Design that juxtaposes and contrasts nature's order and human order prompts contemplation of what it means to be human. Design that resonates with a place's natural and cultural rhythms, that echoes, amplifies, clarifies, or extends them, contributes to a sense of rootedness in space and time (Spirn 1989).

Successfully discovering what attachments locals may have to a place is a challenging process involving what Geertz (1983) calls steering "between overinterpretation and underinterpretation, reading more into things than reason permits and less into them than it demands." Geertz cautions that learning what something means to local citizens is not best accomplished by trying to get as close as you can to them, as that is unlikely to work.

> The ethnographer does not, and in my opinion, largely cannot perceive what his informants perceive. ... In the country of the blind, who are not as unobservant as

> they look, the one-eyed is not king, he is spectator. ... The most intimate notions [of another group of people are achieved] not by imagining myself someone else, ... and then seeing what I thought, but by searching out and analyzing the symbolic forms — words, images, institutions, behaviors — in terms of which, in each place, people actually represented themselves to themselves and to one another. ... Understanding the form and pressure of, to use the dangerous word one more time, natives' inner lives is more like grasping a proverb, catching an illusion, seeing a joke — or, as I have suggested, reading a poem — than it is like achieving communion.

Semiotics

In seeking these subtle meanings, one must be concerned with how local signs signify (i.e., the local semiotics). Semiotics is necessarily a social science, a process calling on an understanding of anthropology, history, and psychology. To be of effective use, semiotics must be seen as more than a consideration of signs as a means of communications. As Geertz says, semiotics is more than deciphering codes; it is a consideration of signs as modes of thought, of idioms to be interpreted, a determination of the meanings of things for the life that surrounds them. For example, landscapes should be considered neither natural nor real anymore (Cosgrove and Daniels 1988). The state of being natural in a developed world means that through some ideology or artifice of the political system a given portion of the earth has been "allowed" to be undisturbed. It cannot literally be termed wild when it is thoroughly mapped, perhaps with trails laid out, and its "wildlife" undoubtedly managed or protected in some fashion.[2] In interpreting landscape aesthetics, the assessor must employ the notion of iconology: the application of an art history method to ascertain the underlying principles in a work of art "which reveal the basic attitude of a nation, a period, a class, [or] a religious or philosophical persuasion" (Cosgrove and Daniels 1988). The aesthetic analyst must remember that, for many, the primary function of architecture is not to create form, but to create *meaning*. Aesthetic analysis must grapple with how well such meanings are expressed and how appropriate intended meanings are to the site and the ecological and socioeconomic context of the landscape development.

Aesthetic analysts, thus, have to be acutely alert, broadly educated, and highly conversant with and sensitive to local users to grasp the ties to and meanings of these natural and cultural processes before they can evaluate the quality and appropriateness of execution. It all boils down to the ability to employ all the senses, including an appreciation of history, time, ecology, and people, in performing

---

[2] This distinction making is not just the splitting of hairs. Such issues were central in a debate among federal agencies over permission to use helicopters to facilitate the monitoring of lake chemistry in officially designated wilderness areas as part of a national acid deposition survey for which I evaluated the aesthetic dimensions of the controversy (*National Surface Water Survey: Western Wilderness Area Lakes, Environmental Assessment*. EPA 910/9-85-126. United States Environmental Protection Agency, Washington, D.C., April 1985).

site analysis in its broadest construction. I was fortunate to participate in just such a wide-ranging challenge in a precedent-setting[3] assessment of the aesthetic impact of a proposed nuclear power plant amidst the remnants of historic 19th-century Hudson River Valley landscapes (Petrich 1982, 1984).

INFORMATION AND IMAGE PROCESSING AS GUIDES TO AESTHETIC PREFERENCE

Image Processing

In this section, I trace the rapid development of aesthetic assessment. The discipline moved quickly from one of static, systematic evaluations of two-dimensional scenes to the dynamic evaluations of human preference in participatory landscapes. An extensive discussion of what may be the genetic roots of human landscape preference follows, along with elaboration of the implications of that potential behavioral orientation for current physical design and aesthetic assessment.

Simple, objective measures of environmental preference based on the physical attributes of a scene have been sought since the discipline of aesthetic assessment of landscapes was formalized in the late 1960s. The work by Elwood Shafer and others with the USDA Forest Service (see Introduction) represents an important aspect of this tradition. Luna Leopold's work (1969; Leopold and Marchand 1968), with measures of distinctiveness as proxies for scarce aesthetic resources, was largely based on relatively objective physical characteristics of streams and streamscapes. We have successfully applied this method of determining distinctiveness, and used distinctiveness itself in conjunction with other approaches, in assessing the aesthetic impact of proposed microhydropower development projects on small streams in the Owens and Upper San Joaquin River Basins of California (Petrich et al. 1989; Swihart and Petrich 1988).

In the most recent work on the efficacy of these approaches, Bishop and Leahy (1989) report the relationship of the size of a digitized image's file with the scenic attribute "complexity of the scene." They find that some procedure for edge detection that can be coupled with area and perimeter measures of the type used by Shafer and Brush (1977) may hold more promise than purely objective measures. Brush and Shafer (1975) used a regression model to predict scores for preferences for landscape scenes and correlated these with actual observed preference scores. They measured specific, objectively determined landscape attributes such as the perimeter of immediate vegetation, perimeter of intermediate vegetation, area of

---

[3] The research concluded that building the power plant in the proposed location (Greene County, NY) would entail an unacceptable negative aesthetic impact. This finding was the primary basis for the issuance of a final environmental statement by the U.S. Nuclear Regulatory Commission staff recommending that the utility's application for a construction permit to build this $1.8 billion (1977 dollars) plant be denied, an action unprecedented in the history of the nuclear power industry. No license application had ever been recommended for denial (nor have any since this action) for any environmental or safety reason, let alone aesthetics.

water, area of distant nonvegetation, and other variables on enlarged photographs. While the results showed solid correlations with the observed preferences, this "image processing" approach, as it has come to be known, has no theoretical foundation or rigor and lacks any roots in social or cultural values.

Information Processing

The information processing approach, in contrast, is based on the belief that evolving humans found the processing of spatial information to be crucial to survival. Central to its application is the concept of affordances (i.e., features or land cover configurations that largely bound what one can do in an environment). Locomotion, for example, is largely constrained by the smoothness of the terrain. Appleton (1975, 1984) developed the idea of two major affordances: prospect and refuge. Prospects provide places from which one can see, and refuges provide places where one cannot be seen. Locations providing both affordances would likely be construed as safe and useful for seeking food and shelter. Other affordances are those landscapes that provide a sense of spaciousness, conveying a feeling that the environment has ample room to roam.

Rachel and Stephen Kaplan, environmental psychologists at the University of Michigan, have examined aesthetic reactions over a period of 20 years and conclude (1989) that such reactions

> reflect neither a casual nor a trivial aspect of the human makeup. Rather, they appear to constitute a guide to human behavior that is both ancient and far-reaching. Underlying such reactions is an assessment of the environment in terms of its compatibility with human needs and purposes. Thus aesthetic reaction is an indication of an environment where effective human functioning is more likely to occur.

The Kaplans looked at the information problems confronting early humans and generalized two major cognitive processes to be involved: making sense and involvement (1982, 1989). The Kaplans see "making sense" as referring to the ability to structure an environment so that comprehension and way-finding are enhanced. By "involvement," they are referring to the process of both engaging and sustaining one's interest in an environment. Because there is survival value in both processes, they conclude that environments that foster such mental activities are more likely to be preferred.

The Kaplans also see a time dimension in human preference. In the immediate environment, landscape variables that promote making sense of the environment include those that provide order, or *coherence*. The extent to which the larger environment provides clues to way-finding and facilitates fabricating a mental map of the environment to use at a later time is known as *legibility*. To be involved in the immediate environment — to find it stimulating and interesting — requires an environment with enough but not too much information or environmental *complexity*. To accommodate the desire to engage further in an environment, the extent to which the environment seems to promise new information if one were to travel deeper into it, is a measure of its *mystery*. Rapoport and Kantor (1967) offer an example of what mystery, the most important dimension the Kaplans identified, involves.

> We may visualize a range of perceptual input from sensory deprivation (monotony) to sensory saturation (chaos). In the case of the former, there is not enough to observe, to select, to organize; there is an excess of order. In the latter, there is too much to observe, there is no relation between the elements, so that one is overwhelmed by multiplicity. In between, there is an optimal perceptual rate (an "ideal") which enables one to explore, to unfold gradually, to see, to give meaning to the environment. One needs to roam back and forth — either physically or with one's eye and mind — not taking it all in at a glance. If there is no ambiguity, the eye is attracted only once and interest is lost. If all is designed and settled, there is no opportunity to bring one's own values to the forms.

Despite some interesting differences in numerous studies with diverse economic and educational background groups, in multiple cultures, and with people of all ages, the Kaplans, their students, and others have found these four predictors of preference — coherence, legibility, complexity, and mystery — to underscore the finding that "humans interpret their environment in terms of their needs and prefer settings in which they are likely to function more effectively" (Kaplan and Kaplan 1989). Although there is consistency in the pattern of preference for natural settings, this is not to suggest a universality in what people prefer. Identifying and understanding the common patterns, the regularity of human affective response to aesthetic stimuli, are crucial to appropriate policymaking and decisionmaking (Kaplan 1988).

Miller (1984) found that the perceptual variables developed by the Kaplans provide a more easily understood causal relationship to landscape preference than do the image processing approaches. S. Kaplan (1979) points out that a potential weakness of the visual assessment technologies is that they attempt to achieve objectivity through the rating of landscape features rather than scenic preferences of potential users. It is his experience that the decisions made through use of the landscape-feature techniques often seem forced or unnatural and therefore difficult to explain to others, while decisions concerning preference are made "quickly, easily, and almost intuitively." The Kaplans (1989) maintain that "preference feels direct, immediate, and holistic. One experiences no hint that one is going through a complex, analytic, process en route to one's judgement." Kaplan (1988) argues that because such a substantial response to what humans respond to in the landscape is based on what sort of spaces are involved and the way individuals imagine moving in those spaces, approaches that emphasize only the two-dimensional picture plane are bound to miss much of what matters most to people about landscapes.

Another dimension investigated by the Kaplans (Kaplan and Talbott 1983; Kaplan and Kaplan 1989) is the capacity of different environments to support a "getting-away-from-it-all" experience. The struggle to pay attention in a cluttered and confusing environment is central to what we experience as mental fatigue.

In restorative environments, people are able to rest their minds in a way that allows them to recover their effectiveness. Restorative environments do not depend on the directed, voluntary attention of the user; rather, viewers find themselves involuntarily enjoying themselves, as in looking at a litter of puppies or visiting the zoo.[4] Restorative environments rely on a natural beauty, one that is experienced by what psychoanalytic theory calls *immediate intuition*, an indirect perception by way of the unconscious mind, drawing on memories, desires, sensory data, and other associations all working to enrich the perception almost instantaneously (Austin 1985).

This restorative potential is based on the recognition that while remote, natural areas offer the necessary privacy for restoring peace of mind, other more common areas may also offer comparable support if they afford fascinating features or activities while minimizing or eliminating auditory or visual distractions. Fascination can be assessed by the extent to which the attention paid to the environment is effortless; fascinating subjects, those that can attract our attention without our even being conscious of the process, might include rushing water, animals, waterfalls, wild flowers, caves, sunsets, or even bugs in home gardens. Such elements hold the viewer's attention but not in a riveted manner, allowing the mind room to wander. Restorative environments combine their various components into a whole that is easily perceived and that makes sense compared to user expectations.

Biologist Edward O. Wilson (1984) acknowledges that people will walk into nature to explore, hunt, and garden if given the chance: "They prefer entities that are complicated, growing, and sufficiently unpredictable to be interesting." In assessing the aesthetic impacts of hydropower development in the California Sierra, we applied these ideas by evaluating the "amount" of fascinating features or opportunities available, estimating their relative allure for the target users, assessing the coherence of the whole environs, assessing the comparison of the reality to anticipated expectations of target users, and assessing the absence of distractions (Petrich et al. 1989; Swihart and Petrich 1988).

Evolutionary Foundations of Aesthetic Preference

Pivotal to the notion of humans as information-processing animals is that we build mental models to represent significant aspects of our physical and social world. When we try to function effectively in our world, we manipulate elements of such models when we think, plan, and try to explain events. Bower and Morrow (1990) point out that it is this ability to construct and manipulate valid models of reality that provides us with our distinctive advantage over other animals, calling it "one of the crowning achievements of the human intellect." Biologist René Dubos (1968) stated

---

[4] Designers, then, need to understand that one does not necessarily create reduced-stress environments by lowering the number and complexity of environmental stimulations. The content of the environmental stimulations is critical to whether or not stress will likely be reduced; the role of natural vs manmade characteristics of the environmental elements certainly should be a prominent consideration.

> Man's responses to most environmental stimuli are profoundly affected by anticipations of the future, whether these anticipations are based on fear, factual knowledge, desire for achievement, or merely wishful thinking. Indeed, man's propensity to imagine that which does not exist, or would never come to pass without his willful and deliberate action, is the aspect of his nature that differentiates him most clearly from animals ... Many aspects of human behavior which appear incomprehensible or even irrational become meaningful when interpreted as survival of attributes that were useful when they first appeared during evolutionary development and that have persisted because the biological evolution of man was almost completed about 100,000 years ago.

Dubos reinforced his argument by pointing out that the salinity of human blood still reflects the composition of seawater from which terrestrial life originally emerged and that fight-or-flight responses are biochemical, hormonal, behavioral, or other physiological vestiges of primitive man's battle for survival with little more utility in the modern world than causing stress.

Jerome Feldman, a computer scientist working in artificial intelligence, illustrates this genetic legacy regarding the way we perceive the world (Campbell 1989):

> We do not start life with a brain that is like a book with blank pages, waiting for experience to write on them the messages that enable us to think and act intelligently. Instead, we come into the world in possession of specialized cognitive structures, little autonomous computers already organized to perform mental tasks that are much too difficult and technical to learn from scratch, by trial and error; tasks such as vision, recognizing speech sounds or a human face, perhaps even acquiring a theory of the syntax of our native language. These modular systems in the head are like the implicit knowledge of cause and effect that infants display almost at birth, systems that drive learning instead of letting learning drive them. They are processes which go on mostly below the level of awareness and are so important to survival that nature could not afford to let us simply pick them up haphazardly from experience. They are part of our inherent worldliness.

Our responses to the natural world and our aesthetic preferences are a part of this modular, systemic inheritance. "Worldliness," as used above, means understanding that it is the irrational biases of the mind, its organization of knowledge, even its irrational thinking, rather than its rules of logic, that are critical to understanding natural intelligence. Biases are systematic and deeply embedded. Thought is now seen not as a digital and local Cartesian process, but rather stream-like, highly fluid and adaptable, with thought triggering other thoughts and associations — what the artificial intelligence researchers term to be *connectionist.* While seemingly at times functioning anathema to, or even outside of logic, especially when trying to make sense of ambiguous or contradictory information, the mind is tapping into "an indescribable richness of activity that must be understood as the product of a history of living in the world that reaches back into the most remote mists of man's beginnings, and as the larger part of an intelligence that is worldly to its foundations, ... a result of the way the brain evolved" (Campbell 1989). Accommodating aesthetic preferences involves understanding the origins of the "worldliness" of our thought processes and the sometimes irrational nature of our deeply held biases and beliefs; these underlie many of our aesthetic preferences, and do so in ways that are difficult to discern. For example, biologist Wilson (1984) remarks,

> Is it unreasonable to suppose that the human mind is primed to respond most strongly to some narrowly defined qualities that had the greatest impact on survival in the past? ... For if animals choose habitats by orientation devices and prepared learning built in during generations of natural selection, it is possible that people do the same. ... [C]ertain key features of the ancient habitat match the choices made by modern human beings when they have a say in the matter.

Savanna Gestalt

Homo sapiens emerged as a distinct species in the highlands of East Africa where the moderate climate provided a physical environment very similar to that which most human beings generally consider desirable for health, comfort, and activity. The Kaplans, Dubos, Wilson, Tuan, and others have noted that this environment may not be just a comfortable environment: they each have hypothesized that it is the *preferred* human environment. This is what is meant by Wilson in the above quotation where he suggests that whenever humans can have a say in the matter, this is the environment they try to recreate.[5] Wilson (1984) calls this preferred landscape the "savanna gestalt," after the African savanna where the species *Homo* evolved for 2 million years after descending from the trees. It is the rolling grasslands with enough topographic relief to provide "prospects" and "refuges," broken with clumps or groves of trees, and punctuated with lakes or streams. It is the landscape model for such diverse areas as parks in the arid western United States, for suburban shopping malls, for cemeteries, and for the English urban park. It is, as Wilson points out, the landscape that the rich and powerful, who could move to or create any landscape anywhere, choose to develop around their prominently sited mansions, villas, temples, palaces, and corporate retreats. It is also what the most humble create around their own cabins and what suburban America has so unmistakenly imitated in its front yards. Our infatuation, over the millennia, with this landscape is, Wilson posits, a response "to a deep genetic memory of mankind's optimal environment. That given a completely free choice, people gravitate statistically toward a savanna-like environment." Orians (1980) also identifies the savanna-

[5] Dubos (1968) noted that modern man has a desire to hunt that is paradoxically compatible with love of wildlife. He notes hunting is highly satisfying for many people because it calls into play a multiplicity of physical and mental attributes that appear to be woven into the human fabric. He notes that the word paradise has been claimed to have its origin in a Persian expression which signifies a hunting ground or at least a park enclosing animals.

like environments as being environments of higher quality for humans and more productive for human existence than either wetter or drier habitats. He asserts that they have evolved as landscapes that incorporate components essential to our sense of beauty.

Zube (1984) observes that this savanna landscape is in fact a prospect-refuge landscape, "a landscape that is coherent, legible, complex, and mysterious [and evokes] a strikingly similar image — an image of a landscape of interspersed openings and wooded areas. This is the type of landscape that is frequently described as park-like."

It was the foundation for the City Beautiful movement at the turn of the century, the "quintessential emblem of a civilized, humanized natural world" (Howett 1987). It is the Jeffersonian country gentleman's environment, the landscape intermediate between the cities and the hinterlands, an agrarian ideal, and an Arcadian middle ground. There is, arguably then, an innate preference for controlled vistas or tamed nature. For raw inspiration, we prefer the wilderness, but for our art, front yards, local parks, and the daily refreshment of our spirits, we prefer this Romantic landscape archetype. A recent observer quipped that it is no coincidence that the most desirable address in New York City is Park Avenue (Fox 1989). While the existence of the savanna gestalt can likely never be proved, the three-dimensional reality that it keeps fostering by the efforts of different cultures in different millennia suggests something quite persistent. The pervasive expression of the savanna gestalt on the landscape suggests the wisdom of the old Sufi saying that thirst is the clearest proof of the existence of water. Human needs were not likely to be created by evolution except as a means to assure their fulfillment (Schmookler 1984).

CLASSICAL AESTHETICS VS ECOLOGICAL AESTHETICS

All those concerned with aesthetic preferences and the quality of landscape design do not, of course, subscribe to the design formula implicit in the savanna gestalt. Opponents do not deny that as a design framework it has produced "manifestly beautiful, delightful, useful, environmentally sensitive, and life-enhancing" landscapes (Howett 1987). Its foremost interpretation in the American landscape is by the founder of modern landscape architecture, Frederick Law Olmsted, the codesigner of Central Park and many other grand urban parks, estates, and residential developments across the country. It is so connected with Olmsted's designs that among art historians, this idyllic pastoral park is known as the Olmstedian aesthetic. It is grounded in 19th-century Romanticism, with an emphasis on seeing nature as a series of "views" worthy of being made into a picture. The framework of the picturesque dates back to 18th-century England as a middle ground position between the "sublime" and the "beautiful." While its critics acknowledge the sympathetic vibrations with perhaps primordial human longings, they see its dependence on picturesque notions as being too disengaging for the viewer. They note that the very word *scenic* derives from theatrical terms and implies "a separation, a distancing between the spectator and an environment that has been composed in some way" (Howett 1987).

The English Romantic poets virtually worshipped nature, as attested to in the period's landscape paintings, and guidebooks of the times describe the areas with the highest natural beauty as if they should be viewed as landscape paintings.[6] The legacy of this two-dimensional foundation for appreciating landscape beauty has persisted into contemporary aesthetic assessment. The Bureau of Land Management's and Forest Service's approaches to assessing scenic beauty have been criticized as being too dependent on evaluations of abstract formal landscape qualities such as form, line, and texture and on aspects of the viewer position. These are seen as "scientific" guises for what their detractors term "a quest for the picturesque" (Evernden 1981).

Many ecologists have trouble with the Olmstedian aesthetic because it relies so heavily on the visual response to nature — a contrived nature, at that — and it forces an identification of the natural world with high-art traditions that are perceived as "having been ruthlessly insensitive to the effects of certain kinds of development upon vulnerable natural systems" (Howett 1987). In its place, they advocate a more natural style of design, one accommodating competition and indigenous plant communities that reflect natural processes of plant growth, desiccation, death, and decay. They see the savanna gestalt, the Olmstedian aesthetic, as being frozen in time, static and idealized perfection[7] and as having an artificial appearance of closure (Fairbrother 1970). The advocates of a more ecologically based design paradigm do not wish to supplant the savanna gestalt, but rather to expand the range of stimulation and pleasure that landscapes can provide.

Designers who wish to create more ecologically sensitive landscapes to house their developments have adopted the term *sustainable landscapes*; such designs affect users not only by functioning to protect resources and ecosystems, but also by simultaneously symbolizing that aspect through visual, spatial, and sensory means to evoke a positive affective response (Thayer 1989).

---

[6] In fact, guidebook authors advised visitors to England's areas of natural beauty to use a landscape mirror, termed a Claude glass (for the tint in the glass that was to approximate the haze in a Claude painting), to "properly" take in the landscapes from a sufficient distance. Properly using the Claude glass required the user to *turn his/her back to the landscapes* — the ultimate in isolation from the landscape! The observer, thus, took in the environment as a series of static takes (each framed in the mirror, reversed, and reduced in size), with no sense of the sweep of the landscape or its continuity [I. D. Little. 1982. Landscape and the Claude glass. *Am. Land Forum* 3(2):4–5].

[7] Not all would agree. A further interpretation of the picturesque is given by Sidney K. Robinson's "The Picturesque in an Ancient Japanese Novel" in *Landscape Journal* 9(1):9–15, 1990. Robinson posits that the picturesque is a "complex proposition about composing the relation between power and its experience, ... a compositional move to underplay the extent of control, ... compositions that announce themselves [vs] those that slip in unnoticed. ... [The Picturesque] is ... a complex and fascinating way of composing the relation between beings and nature, between power and its absence, between things as they seem and things as they are."

COMING CHALLENGES FACING AESTHETIC ASSESSMENT

Working Landscapes

Among the strongest challenges facing aesthetic assessment in the coming years is the protection of the intermediate landscape. This is the land variously known as the working countryside, working landscape, managed landscape, humanized landscape, historic landscape, ancient landscape, cultural landscape, or heritage landscape, all terms attached by botanists, landscape historians, sociologists, historical ecologists, and other specialists who study the use and history of the land (Hiss 1989b). They distinguish "natural landscapes," those parts of the countryside which humans have scarcely affected, from "working landscapes," those lands whose function, look, character, and feel have been gently molded (not dominated) over time by successive generations of human activities in a fragile equilibrium with natural processes. Even the largest contemporary European cities have deliberately preserved their connections to their nearby countryside. Of primary importance in looking at the well-being of rural landscapes is that regions can function as enabling (even ennobling) environments only when all of the elemental environments that serve to define a region (the urban, the rural, and the wild) can offer a well-rounded range of public values (Hiss 1989b).

Unprecedented Urban Growth

This somewhat legendary American landscape, this savanna gestalt of the Jeffersonian country gentleman, is now under its most severe threat. In what Hiss (1989a) terms the first postinterstate boom, there is an ongoing acceleration in the spread into rural areas of high-rise office buildings, sprawling industrial parks, and seamless tracks of houses (Hampson 1990). In contrast to William S. Whyte's 1950s world of *The Organization Man* (suburbia with 35 million inhabitants), today's suburban population numbers over 110 million and comprises nearly one half of the country's population (urbanites account for only a third) and 60% of its metropolitan residents. The suburbs claim 60% of all metropolitan area office space and, in some areas, three of every four jobs. More than twice as many workers now commute from suburb to suburb as from suburb to city. Perimeter City, north of Atlanta, for example, employs more office workers than downtown Atlanta. According to Georgia State University geographer Truman Hartshorn (Hampson 1990), "We are now in the midst of the most rapid transformation of the city in its history. For the past 2,000 years the city had just one center. Now it has many." Suburbs have become self-sufficient, with their own stadiums, concert halls, theaters, parks, art galleries, and museums. "Suburban" Boston can now be seen to include Cape Cod; suburban, "commutable," metropolitan Washington, D.C. can include parts of West Virginia.

Minimal Skills for Coping with Explosive Growth

This postinterstate, suburban explosion is indicative of problems wherein there is a mismatch between our ability to cope and the world we inhabit. As Schmookler

(1984) said, "*Homo* has proved more *faber* than *sapiens*." To continue with the evolutionary framework raised with the savanna gestalt, we need to recognize that our ancestors had little reason, or "genetic incentive," to evolve a capacity to perceive significant long-term, large-scale changes in the environment (Wilson 1984; Ornstein and Ehrlich 1989). In a period of enormous encroachment of our metropolitan areas into the working landscape, there is no longer sufficient time to rely on the normal pace of cultural accommodation and evolution to address the attendant difficulties. As Ornstein and Ehrlich (1989) put it, "We are facing problems of a scale and speed of change for which our biology and history have left us poorly prepared." In a similar vein, in speaking of all environmental challenges of the 1990s, including global climate change, Stephen H. Schneider of the National Center for Atmospheric Research says, "The bottom line is that we insult the environment at a faster rate than we can predict the consequences. Under these conditions, surprises are certain."

The sense of connectedness to rural areas is tenuous. People rapidly discount a landscape as soon as the first scar occurs, rather like a stain "ruining" a favorite garment. Hiss (1989b) recounts a rough mathematical formula for measuring landscape degradation developed by the chairman of New York City's Audubon Society's conservation committee:

> The first 5% of development in a countryside region generally does 50% of the damage, in terms of altering people's mental geography of an area. And the second 5% of development enlarges this damage by another 50%. The environmental damage caused by the first development in a region varies tremendously from place to place, depending on the nature of the terrain and the kind of development involved, but the disproportionate initial-impact formula seems to apply across the board.

Miller (1984) empirically developed a similar observation in that he found that the type of intrusion, its meaning to the viewer, and the spatial organization of the scene were as important or, he says, even more important to landscape preference than the extent or amount of development. In addition, Simcox and Zube (1989) found that the encroachment of development into natural settings significantly weakens preferences for maintaining open spaces and, not surprisingly, fosters the desirability of more development.

Role of Aesthetic Assessment in Protecting Rural Landscapes

The role of aesthetic assessment in protecting rural landscapes hinges, then, not just on the scheme used to assess impacts of proposed developments. Groundwork must first be laid by getting communities to look at their working landscape resources systematically. Under appropriate direction, this becomes an assessment of the community's own landscape character, an inventory of its connectedness or lack thereof. It is the aggregate of these linkages that is the true regional landscape design (Hiss 1989a), not a ranking of outstanding scenic features. Schauman (1988) has facilitated a community in such an assessment, combining in a common sense way what we know about scenic characteristics with local judgments; Melnick (1983) has also explored the valuing and protection of cultural landscapes.

Because of human limitations in grasping the consequences of incremental changes and in planning decades into the future, aesthetic assessment needs to facilitate communities in exploiting the latest tools for augmenting human vision. Conventional techniques inadequately portray time and change, encouraging a continued focus on only concrete, static form. Recent computer technologies can provide the means to display slow-breaking patterns generated by natural and anthropogenic processes to facilitate the perception, manipulation, and evaluation of patterns and forms as they might emerge and change under a variety of future growth scenarios (Spirn 1989). Phil Lewis, a landscape architect at the University of Wisconsin in Madison, believes every region should have at least one place within its borders fitted out with such technologies (Hiss 1989a). These Regional Discovery Centers, as he calls them, are necessary to help people overcome their genetic bounds to the compression of time, to help people see where they have been, where they are, and — unless something changes — where they are headed.

CONCLUSIONS

After starting from common origins in the middle-to-late 1960s, aesthetic assessment has had a slower technical evolution and public acceptance than other dimensions of NEPA. The USDA Forest Service, the country's largest employer of landscape architects, gave it its first strong development and still provides support and intellectual direction. Environmental psychologists have given it its greatest academic and theoretical footing, although as a discipline it is far from having a commonly accepted paradigm. The penetration of human information processing concepts and a deeper understanding of semiotics and the attachments people hold for places signify newly staked intellectual territory for practitioners. An understanding and appreciation of the evolutionary heritage of aesthetic preference stands in the way of a more purely ecologic understanding of preference. These schools of thought are in flux, with well-known adherents on both sides, but with a wide common ground available for resolution. The discipline faces suburban growth, the *béte noire* of architects, planners, and other designers, with new tools for enabling affected publics to see where their disjointed incrementalism is leading them. With room for much intellectual and technical maturation, the challenges look manageable. Remaining most vexing is the pursuit of effective, understandable, and publicly acceptable techniques for linking values information — aesthetic impact assessment obtained through social science research methods — with landscape ecological data obtained from physical and biological science research methods.

The conceptual relationship of aesthetic assessment to NEPA is one of several possible human-landscape interactions (Zube 1987). Under NEPA, aesthetic assessment has focused on human-induced changes in the physical and biological characteristics of the landscape. In this role as an agent of change, humans are rarely considered as thinking and feeling organisms. Emphasis within aesthetic assessments under NEPA has traditionally been on the identification and mitigation of negative impacts. The positive aspects of human interactions with landscapes has received short shrift. Incorporating concepts of humans as receivers and processors of information from the landscape has broadened aesthetic assessment considerably, documenting and explaining the considerable agreement among diverse peoples about scenic landscapes. A third, underexploited, conceptualization of human interactions with landscapes is one centering on the human as a participant, transactionally, in the landscape. This view recognizes that there are reciprocal human-landscape impacts, often cumulative, often imperceptible, or unconsciously created. The interrelationship can be seen as dynamic, occurring in social, cultural, spatial, and temporal contexts that invite exploration and that provide omnidirectional sources of excessive information through multisensory modalities. From this interactive process of human-landscape give and take, landscapes take on meaning and value, the ultimate targets of aesthetic impact assessment.

REFERENCES

Alexander, C. 1964. *Notes on the Synthesis of Form.* Harvard University Press, Cambridge, MA.

Appleton, J. 1984. Prospect and refuge revisited. Landscape J. 3:91–103.

Appleton, J. 1975. *The Experience of Landscape.* Wiley, London.

Austin, R. C. 1985. Beauty: A foundation for environmental ethics. *Environ. Ethics* 7:197–208.

Berleant, A. 1988. Aesthetic perception in environmental design. pp. 84–97. In J. L. Nasar (ed.), *Environmental Aesthetics: Theory, Research, and Applications.* Cambridge University Press, Cambridge, MA.

Bishop, I. D., and P. N. A. Leahy. 1989. Assessing the visual impact of development proposals: The validity of computer simulations. *Landscape J.* 8(2):92–100.

Bower, G. H., and D. G. Morrow. 1990. Mental models in narrative comprehension. *Science* 247:44–48.

Brush, R. O., and E. L. Shafer, Jr. 1975. Application of a landscape-preference model to land management. pp. 168–182. In E. H. Zube, R. O. Brush, and J. G. Fabos (eds.), *Landscape Assessment: Values, Perceptions, and Resources.* Dowden, Hutchinson & Ross, Inc., Stroudsburg, PA.

Campbell, J. 1989. *The Improbable Machine: What the Upheavals in Artificial Intelligence Research Reveal About How the Mind Really Works.* Simon and Schuster, New York.

Chenoweth, R. E., and P. H. Gobster. 1990. The nature and ecology of aesthetic experience in the landscape. *Landscape J.* 9(1):1–8.

Cosgrove, D., and S. Daniels (eds.). 1988. *The Iconography of Landscape.* Cambridge University Press, Cambridge, MA.

Dubos, R. 1968. *So Human an Animal.* Charles Scribner's Sons, New York.

Evernden, N. 1981. The ambiguous landscape. *Geogr. Rev.* 71(2):151.

Fairbrother, N. 1970. *New Lives, New Landscapes: Planning for the 21st Century.* Alfred A. Knopf, New York.

Fenton, D. M., and J. P. Reser. 1988. The assessment of landscape quality: An integrative approach. pp. 108–119. In J. L. Nasar (ed.), *Environmental Aesthetics: Theory, Research, and Applications.* Cambridge University Press, Cambridge, MA.

Fines, K. D. 1968. Landscape evaluation: A research project in East Sussex. *Reg. Stud.* 2:41–55.

Flug, M., and R. H. Montgomery. 1988. Modeling instream recreational benefits. Paper No. 88030, Am. Water Resour. Assoc., *Water Res. Bull.* 24(5):1073–1081.

Fox, T. 1989. Using vacant land to reshape American cities. *Places* 6(1):78–81.

Geertz, C. 1983. *Local Knowledge: Further Essays in Interpretative Anthropology.* Basic Books, Inc., New York.

Greenbie, B. B. 1988. The landscape of social symbols. pp. 64–73. In J. L. Nasar (ed.), *Environmental Aesthetics: Theory, Research, and Applications.* Cambridge University Press, Cambridge, MA.

Halprin, L. 1989. Design as a value system. *Places* 6(1):60–67.

Halprin, L. 1969. *The RSVP Cycles: Creative Processes in the Human Environment.* George Braziller, Inc., New York.

Hammitt, W. E. 1983. Assessing visual preference and familiarity for a bog environment. pp. 81–96. In R. C. Smardon (ed.), *The Future of Wetlands: Assessing Visual-Cultural Values.* Allanheld, Osmun, Totowa, NJ.

Hampson, R. 1990. "Nation's Suburbs Have Changed Over the Past Four Decades." Associated Press, *The Oak Ridger*, Oak Ridge, TN, January 8.

Hester, R. T., Jr. 1989. Social values in open space design. *Places* 6(1):68–77.

Hiss, T. 1989a. "Encountering the Countryside — II." *The New Yorker*, August 28, pp. 37–63.

Hiss, T. 1989b. "Encountering the Countryside — I." *The New Yorker*, August 21, pp. 40–69.

Hobbs, B. F., E. Z. Stakhiv, and W. M. Grayman. 1989. Impact evaluation and procedures: Theory, practice and needs. *J. Water Res. Plann. Manage.* 115(1):2–21.

Howett, C. 1987. Systems, signs, sensibilities: Sources for a new landscape aesthetic. *Landscape J.* 6(1):1–12.

Hyman, E. L., and B. Stiftel. 1988. Perception and evaluation of scenic environments. pp. 111–136. In E. L. Hyman, and B. Stiftel (eds.), *Combining Facts and Values in Environmental Impact Assessment.* Westview, Boulder, CO.

Iltis, H. H., O. L. Loucks, and P. Andrews. 1970. Criteria for an optimal human environment. *Bull. At. Sci.* January, pp. 2–6.

Jackson, J. B. 1984. *Discovering the Vernacular Landscape.* Yale University Press, New Haven, CN.

Jacobs, P., and D. Way. 1968. *Visual Analysis of Landscape Development.* Department of Landscape Architecture, Harvard University, Cambridge, MA.

Jones, G. R., I. Jones, B. A. Gray, B. Parker, J. Coe, J. Burnham, and N. M. Geitner. 1975. A method for the quantification of aesthetic values for environmental decision making. *Nucl. Technol.* 23:682–713.

Kaplan, R. 1985. Nature at the doorstep: Residential satisfaction and the nearby environment. *J. Archit. Plann. Res.* 2:115–127.

Kaplan, S. 1988. Where cognition and affect meet: A theoretical analysis of preference. pp. 56–63. In J. L. Nasar (ed.), *Environmental Aesthetics: Theory, Research, and Applications.* Cambridge University Press, Cambridge, MA.

Kaplan, S. 1979. Perception and landscape: Conceptions and misconceptions. pp. 241–248. In G. Elsner, and R. Smardon (eds.), Proceedings of Our National Landscape. Pacific Southwest Forest and Range Experiment Station, Berkeley, CA.

Kaplan, S., and R. Kaplan. 1982. *Cognition and Environment: Functioning in an Uncertain World.* Praeger, New York.

Kaplan, R., and S. Kaplan. 1989. *The Experience of Nature: A Psychological Perspective.* Cambridge University Press, Cambridge, MA.

Kaplan, S., and J. F. Talbott. 1983. Psychological benefits of a wilderness experience. pp. 163–203. In I. Altman, and J. F. Wohlwill (eds.), *Human Behavior and Environment: Vol. 6. Behavior and the Natural Environment.* Plenum Press, New York.

Killeen, K., and G. Buhyoff. 1983. The relation of landscape preference to abstract topography. *J. Environ. Manage.* 17:381–392.

Kuhn, T. S. 1962. *The Structure of Scientific Revolutions.* University of Chicago Press, Chicago.

Leopold, A. 1949, 1966, 1970. *A Sand County Almanac.* Oxford University Press, Oxford (1949 and 1966) and Sierra Club/Ballantine Book, San Francisco (1970).

Leopold, L. B. 1970. Landscape esthetics. *Ekistics* 29(173):271–277.

Leopold, L. B. 1969. Quantitative Comparison of Some Aesthetic Factors Among Rivers. U.S. Geological Survey, Circular 620, Washington, D.C.

Leopold, L. B., and M. O. Marchand. 1968. On the quantitative inventory of the riverscape. *Water Res. J.* 4(4):709–717.

Linton, D. L. 1968. The assessment of scenery as a natural resource. *Scottish Geogr. Mag.* 84:213–238.

Litton, R. B. 1968. Forest Landscape Description and Inventories: A Basis for Land Planning and Design. U.S. Department of Agriculture, Forest Service, Research paper PSW–49. Pacific Southwest Forest and Range Experiment Station, Berkeley, CA.

Lynch, K. 1972. *What Time Is This Place?* The MIT Press, Cambridge, MA.

McHarg, I. L. 1969a. *Design with Nature.* Natural History Press, Garden City, NY.

McHarg, I. L. 1969b. An ecological method for landscape architecture. pp. 328–332. In P. Shepherd, and D. McKinley (eds.), *The Subversive Science: Essays Toward an Ecology of Man.* Houghton Mifflin Co., Boston.

McHarg, I. L. 1981. Human ecological planning at Pennsylvania. *Landscape Plann.* 8(2):109–120.

Melnick, R. Z. 1983. Protecting rural cultural landscapes: Finding value in the countryside. *Landscape J.* 2(2):85–97.

Miller, P. A. 1984. A comparative study of the BLM scenic quality rating procedure and landscape preference dimensions. *Landscape J.* 3(2):123–135.

Orians, G. H. 1980. Habitat selection: General theory and applications to human behavior. pp. 49–66. In J. S. Lockard (ed.), *The Evolution of Human Social Behavior.* Elsevier/North-Holland, New York.

Ornstein, R., and P. Ehrlich. 1989. *New World, New Mind: Moving Toward Conscious Evolution.* Doubleday & Co., New York.

Petrich, C. H. 1986. Expert systems: Forecasting powerful support for the designer. *Landscape Archit.* 76(3):70–74.

Petrich, C. H. 1984. EIA scoping for aesthetics: Hindsight from the Greene County Nuclear Power Plant EIS. pp. 57–92. In S. L. Hart, G. A. Enk, and W. F. Hornick (eds.), *Improving Impact Assessment: Increasing the Relevance and Utilization of Scientific and Technical Information.* Westview, Boulder, CO.

Petrich, C. H. 1982. Assessing aesthetic impacts in siting a nuclear power plant: The Case of Greene County, New York. *EIA Rev.* 3(4):311–332.

Petrich, C. H., S. F. Railsback, and M. M. Swihart. 1989. Instream flows for recreational and aesthetic resources. pp. 100–107. In *Legal, Institutional, Financial, and Environmental Aspects of Water Issues.* American Society of Civil Engineers, New York.

Porteous, D. 1982. Approaches to environmental aesthetics. *J. Environ. Psychol.* 2: 53–60.

Priestley, T. 1983. The field of visual analysis and resource management: A bibliographic analysis and perspective. *Landscape J.* 2(1):52–59.

Rapoport, A., and R. E. Kantor. 1967. Complexity and ambiguity in environmental design. *J. Am. Inst. Plann.* 33(4):210.

Rusch, C. 1970. The role of graphic activity in the design process. In G. T. Moore (ed.), *Emerging Methods in Environmental Design and Planning*. The MIT Press, Cambridge, MA.

Schauman, S. 1988. Scenic values of countryside landscapes to local residents: A Whatcom County, Washington case study. *Landscape J.* 7(1):40–46.

Schmookler, A. B. 1984. *The Parable of the Tribes: The Problem of Power in Social Evolution.* University of California Press, Berkeley.

Shafer, E. L., Jr. 1969. Perception of natural environments. *Environ. Behav.* 1:71–82.

Shafer, E. L., Jr. 1967. Forest Aesthetics — A Focal Point in Multiple-Use Management and Research. Proceedings 14 IUFRO Congress, Paper 7, Sec. 26, Munich.

Shafer, E. L., and R. O. Brush. 1977. How to measure preferences for photographs of natural landscapes. *Landscape Plann.* 4:237–256.

Shafer, E. L., Jr., and J. Mietz. 1970. It Seems Possible to Quantify Scenic Beauty in Photographs. U.S. Department of Agriculture Forest Service Research Paper NE-162. Northeastern Forest Experiment Station, Forest Service, U.S. Department of Agriculture, Upper Darby, PA.

Shafer, E. L., Jr., and T. A. Richards. 1974. A Comparison of Viewer Reactions to Outdoor Scenes and Photographs of Those Scenes. U.S. Department of Agriculture Forest Service Research Paper NE–162. Northeastern Forest Experiment Station, Forest Service, U.S. Department of Agriculture, Upper Darby, PA.

Shafer, E. L., Jr., and M. Tooby. 1973. Landscape preferences: An international replication. *J. Leisure Res.* 5(3):60–65.

Shafer, E. L., Jr., J. F. Hamilton, and E. A. Schmidt. 1969. Natural landscape preferences: A predictive model. *J. Leisure Res.* 1:1–19.

Sheppard, S. R. J. 1982. Predictive landscape portrayals: A selective research review. *Landscape J.* 1(1):9–14.

Simcox, D. E., and E. H. Zube. 1989. Public value orientations toward urban riparian landscapes. *Soc. Nat. Resour.* 2:229–239.

Smardon, R. C., and M. Hunter. 1983. Procedures and methods for wetland and coastal area visual impact assessment (VIA). pp. 171–204. In R. C. Smardon (ed.), *The Future of Wetlands: Assessing Visual-Cultural Values.* Allanheld, Osmun, Totowa, NJ.

Smardon, R. C., J. F. Palmer, and J. P. Felleman. 1986. *Foundation for Visual Project Analysis.* Wiley, New York.

Spirn, A. W. 1989. The poetics of city and nature: Toward a new aesthetic for urban design. *Places* 6(1):82–93.

Stone, C. D. 1972. Should trees have standing? — Toward legal rights for natural objects. *South. Calif. Law Rev.* 45(2):3–54.

Swihart, M. M., and C. H. Petrich. 1988. Assessing the aesthetic impacts of small hydropower development. *Environ. Prof.* 10(3):198–210.

Thayer, R. L. 1989. The experience of sustainable landscapes. *Landscape J.* 8(2):101–110.

Tuan, Y. 1977. *Space and Place.* University of Minnesota Press, Minneapolis.

Tuan, Y. 1974. *Topophilia.* Prentice-Hall, Englewood Cliffs, NJ.

Ulrich, R. S. 1983. Aesthetics and affective response to natural environment. pp. 88–125. In I. Altman, and J. F. Wohlwill (eds.), *Human Behavior and Environment: Vol. 6. Behavior and the Natural Environment,* Plenum Press, New York.

Wilson, E. O. 1984. *Biophilia.* Harvard University Press, Cambridge, MA.

Wohlwill, J. F. 1976. Environmental aesthetics: The environment as a source of affect. pp. 37–86. In I. Altman, and J. F. Wohlwill (eds.), H*uman Behavior and Environment, Vol. 1.* Plenum Press, New York.

Zube, E. H. 1987. Perceived land use patterns and landscape values. *Landscape Ecol.* 1(1):37–45.

Zube, E. H. 1984. Themes in landscape assessment theory. *Landscape J.* 3(2):104–109.

Zube, E. H., J. L. Sell, and G. Taylor. 1982. Landscape perception: Research, application and theory. *Landscape Plann.* 9:1–33.

Zube, E. H., D. E. Simcox, and C. S. Law. 1987. Perceptual landscape simulations: History and prospect. *Landscape J.* 6:62–80.

Social Impact Assessment and Cultural Conservation: Implications for Local Public Involvement in Planning

B. J. Howell, The University of Tennessee, Knoxville, TN

ABSTRACT

The National Environmental Policy Act (NEPA) generally provides a tenuous basis for assessment and mitigation of intangible sociocultural impacts of concern to local communities. The 1980 Amendments to the Historic Preservation Act authorized study of how these cultural intangibles might be better addressed. Certain recommendations arising from a report mandated by the 1980 Amendments to the Historic Preservation Act propose a reinterpretation of NEPA that might justify more extensive field surveys and ethnographic research in conjunction with social impact assessment (SIA). Case studies of SIA and public response to planning for two national recreation areas, Big South Fork and Mount Rogers, demonstrate the inconsistent treatment of these concerns under NEPA during the 1970s. Agency planning and public review procedures partly account for such inconsistency; however, in the Big South Fork project, neither SIA nor involvement of the affected public was adequate to raise cultural conservation issues unrelated to historic preservation, while better SIA and effective use of public review opportunities focused attention on cultural conservation issues at Mount Rogers. Revision of NEPA regulations and expansion of SIA to better address cultural conservation might build on three existing elements: precedents for emphasis on minority cultural concerns in Native American SIA, situational definition of critical issues through scoping, and attention shown to public perceptions and values in risk assessment.

CULTURE: A MISSING ELEMENT IN ENVIRONMENTAL POLICY

My earliest acquaintance with the process of environmental assessment and public review mandated by the National Environmental Policy Act (NEPA) came during the 1970s while I was a graduate student at the University of Kentucky. As a Sierra Club member, I attended some stormy public hearings on the U.S. Army Corps of Engineers proposal to dam the Red River Gorge in Daniel Boone National Forest. Ultimately, the project was abandoned in favor of less drastic flood control measures. The no-build decision came as a relief to residents located near the proposed project, as well as to environmentalists; however, protection of unique natural resources was the critical issue. Decision makers were not faced with weighing the relative value of local residents' livelihood and lifeways vs further urban development in the floodplain downstream. While this controversy was unfolding in Kentucky, I was also following news of the Tellico project in Tennessee, where a host of sociocultural and economic issues raised by opponents and proponents paled to insignificance in comparison with the snail darter.

NEPA alone, as it evolved in the 1970s through agency guidelines, its amendment in 1975, and the Council on Environmental Quality's (CEQ's) adoption of binding regulations in 1978, provided only a tenuous foundation for social impact assessment (SIA). As Sibley (1984) noted, the 1978 regulations define effects or impacts broadly yet interpret the "Human Environment" narrowly as the relationship of humans with the natural and physical environment. Therefore, sociocultural and economic impacts in and of themselves are not sufficient to trigger the environmental assessment process. Given the ambiguous legal standing of sociocultural issues, it is not surprising that SIA often has been treated as a minor component in the environmental impact statement (EIS), and that citizens opposed to project proposals chiefly on sociocultural grounds have sought and made use of natural environment issues as a basis for legal challenges. The snail darter provides a classic example (Wheeler and McDonald 1986).

Legislation subsequent to NEPA (e.g., the Housing and Community Development Act of 1974, the Archaeological and Historic Preservation Act of 1974) provided stronger directives for planners to account for certain of the sociocultural effects enumerated in the CEQ regulations (Section 1508.8), but substantial gaps remain. In the case of cultural impacts, the historic preservation legislation, executive orders, and regulations developed since the Antiquities Act of 1906 (Schroedl 1988) provide an unequivocal basis for addressing potential impacts of development on (1) archaeological sites and artifacts and (2) structures and places of national historic significance. Living manifestations of cultural traditions, however, fall within the scope of the Historic Preservation mandate only insofar as they can be related to these specific sites and material remains.

In a few projects planned in the 1970s, historic preservation was interpreted more broadly. One of these was the Big South Fork National River and Recreation Area. In this case, specific wording in the enabling legislation (PL93-251, Section 108) provided the rationale for expanding the scope of historic preservation; resource conservation, specifically including historic, cultural, and archaeological resources, was a primary purpose of the project. Thus, the U.S. Army Corps of Engineers was easily persuaded to fund the folklife survey that I directed in 1979–1980. Assuming that recreation development and tourism would accelerate the pace of culture change, documentation of folkways and history would mitigate the loss of local traditions. In historic preservation terms, interpreting local folklife and history for recreation area visitors could be viewed as a further step in mitigation. But a different, more comprehensive approach to local culture was needed. SIA might have played a role here, but unfortunately did not.

When I began fieldwork for the folklife survey, I was surprised to learn how inadequate, from my perspective, SIA for the Big South Fork project had been. Virtually no contact had been made with residents of the project area, in spite of the finding that adverse social impacts would be most significant for them. Three years after the final EIS (U.S. Army Corps of Engineers 1976) was published, the folklife survey began generating data on land ownership and land use patterns, household composition, sources of income, community organization, and the like — information pertinent to assessing impacts of relocation and mitigating them through a well-planned and well-executed relocation program. In a paper drafted while I was still in the field but not published until some years later (Howell 1984), I discussed how the folklife project, with the opportunities it provided for contact with the local population, might profitably have been coordinated with SIA. But the introduction to the folklife survey report (Howell 1981, p. 8) clearly expressed my frustration with the "culture gap" I perceived in environmental planning research and provided a rationale for mitigation recommendations that went beyond documentation. I wrote,

> Discussion of current economic and social conditions (in the EIS) was quite limited and was based on county-wide statistical data and a few interviews with large-scale land owners. The report did not address the immediate situation of the 34 households (some 125 individuals) believed to be living within the proposed BSFNRRA boundaries in 1975, nor did it assess the possible social and cultural impacts of their impending relocation. It has been necessary for the folklife study to explore these issues in connection with developing an understanding of historic and present cultural adaptations ..., assessing the nature and viability of surviving traditional culture elements, and making useful recommendations directed toward minimizing adverse impacts and enhancing preservation and perpetuation of the traditional culture.
>
> Museum collection and documentation is one approach to heritage preservation....The broader folklife approach ... looks beyond cultural artifacts to examine the behavior associated with the changing traditional culture. It traces the thread of continuity between past and present and finds viable traditions that can be perpetuated *in vivo*, not merely preserved *in vitro*. Such a study is the first step in designing development policies that build on the past to support diversity and variety of individual choice and thereby help to maintain a living cultural heritage in rapidly changing areas like the Big South Fork.

CULTURAL CONSERVATION: REDEFINING THE SCOPE OF NEPA

Cultural Conservation: The Protection of Cultural Heritage in the United States (Loomis 1983), also known as the Cultural Conservation report, resulted from a Congressional directive included in the 1980 Amendments to the Historic Preservation Act (Title III, Section 502). Recognizing that existing legislation and regulations neglected nonmaterial aspects of cultural heritage, Congress commissioned a study to provide recommendations for legislative and administrative action to support efforts "to preserve, conserve, and encourage the continuation of the diverse prehistoric, historic, ethnic, and folk cultural traditions that underlie and are a living expression of our American heritage." For the most part, the Cultural Conservation report proposes natural extensions of historic preservation activities to incorporate a broader array of cultural phenomena; however, certain recommendations in the report propose a reinterpretation of NEPA that might justify more extensive field surveys and ethnographic research in SIA (Howell 1983).

In particular, the Cultural Conservation report states that a national program of cultural conservation should "increase the application of knowledge about community life and values to decision-making in environmental planning and design of impact mitigation projects" (Loomis 1983, p. 71). It specifically recommends that administrative action be taken to "strengthen the role of the CEQ by including consideration of folklife and related traditional lifeways among the factors discussed in environmental impact assessments and statements" (Loomis 1983, p. 76).

NEPA certainly offers ample scope for such revision. If one reads the objectives set forth by Congress in Title I (Section 101b) rather than the regulations CEQ developed to implement the act, cultural conservation seems to fall squarely within NEPA's purview. Two of the six stated objectives (second and fourth objectives) speak directly to cultural conservation issues:

1. Ensure for all Americans safe, healthful, productive, and esthetically and culturally pleasing surroundings.
2. Preserve important historic, cultural, and natural aspects of our national heritage, and maintain, wherever possible, an environment which supports diversity, and variety of individual choice.

I was, of course, deliberately echoing this last language in the introduction to my Big South Fork report.

SIA practitioners, like the historic preservation community, perceived a "culture gap" in the CEQ regulations interpreting NEPA. Critical appraisals of SIA published while the Cultural Conservation report was in preparation pointed out the need for more qualitative data. Soderstrom (1981, p. ix) suggested ethnographic "side studies" coordinated with SIA, while Peterson and Gemmell (1981, p. 387) called for full integration of ethnographic techniques into SIA. Wolf (1978, p. 185) noted that SIA often fails to address the impacts that local residents perceive as most critical, a problem several practitioners felt could be addressed through participant observation research (Finsterbusch 1981; Soderstrom 1981).

Even during NEPA's first decade, it was becoming apparent that an environmental assessment and public review process played out through legal challenges in the courts was a costly and inefficient means of identifying and resolving conflicting public interests. But efforts to avoid lengthy legal battles by improving the level and quality of public involvement in planning only underscored the necessity to identify all segments of the public affected by a proposed project and to develop an understanding of each group's lifeways, values, and attitudes toward the project as a basis for achieving compromise outside the courtroom. Practitioners of SIA concurred fully with cultural conservationists on this point (e.g., Peterson and Gemmell 1981, p. 393; Soderstrom 1981, pp. 72–73; Willeke 1981).

If the cultural conservation mandate contained in NEPA is to be honored, cultural issues must be addressed adequately in impact studies and culturally distinctive communities must be heard during the planning process. The following cases illustrate the variable outcomes of public attempts to raise cultural conservation issues in the absence of clear regulations that speak to the cultural conservation intent of NEPA.

CULTURAL CONSERVATION ISSUES IN TWO FEDERAL LAND MANAGEMENT PROJECTS

Because cultural conservation concerns fall into a gap between SIA and historic preservation research, these issues in project planning usually emerge only indirectly during public review and response to the draft EIS and project plan. As a result, chance factors determine whether the public's cultural conservation concerns are acknowledged and whether specific issues are resolved. Outcomes depend greatly on organizational factors (e.g., the regulations and procedures particular federal bureaucracies have developed in response to NEPA and how effectively citizens organize to use available opportunities for review and comment). Treatment of cultural conservation issues related to two national recreation areas, Big South Fork and Mount Rogers, provide apt illustration because the projects had similar goals and scope and affected similar populations in rural Appalachia.

MISSED OPPORTUNITIES FOR CULTURAL CONSERVATION AT BIG SOUTH FORK

The Big South Fork National River and Recreation Area, located about 80 miles northwest of Knoxville astride the Tennessee–Kentucky state line, was authorized in 1974. Its planning prompted response from two different contingents of citizens: environmentalists affiliated with state and national organizations and local proponents of tourism development. Citizens' efforts to have a voice in planning and management decisions were complicated to some extent because two federal agencies were involved: the U.S. Army Corps of Engineers, Nashville District, as lead agency in planning, and the National Park Service, which ultimately would manage the area.

While environmental assessment and planning involved only limited contact with local citizens other than public officials, the planners' site visits revealed considerable evidence of environmental abuse — slag heaps and polluted streams left behind by abandoned coal mines, erosion caused by clear-cutting and off-road vehicles, garbage dumps, and evidence of poaching and pothunting. These observations could only validate the planners' belief that federal ownership and regulation were essential for the area; they certainly were not conducive to a sympathetic view of local culture. Yet environmentalists from Knoxville and Oak Ridge who were concerned about the environmental impacts of too much development fared little better than local residents in their efforts to be heard.

Their organization, Tennessee Citizens for Wilderness Planning, took the lead in organizing a Big South Fork Preservation Coalition with other environmental and historic preservation organizations. They directed a position paper to the U.S. Army Corps of Engineers expressing the group's concerns as early as 1975 (U.S. Army Corps of Engineers 1976, pp. D28–36). The coalition sought to minimize construction of high-impact recreational facilities on federal land, but they also recommended excluding five areas of plateau farmland and residences, judging (accurately) that their condemnation would only serve to arouse local hostility toward the project. These upland areas were not critical to watershed restoration, but would provide ideal sites for campgrounds, visitor centers, and the like. It would have been legitimate to raise the cultural conservation issue in response to the enabling legislation's explicit call for cultural as well as natural resource conservation. Critics might have pointed out the absurdity of dismantling traditional farms and displacing farm families as a strategy for "conserving and interpreting the area's unique cultural and historic values." However, the environmentalists did not make common cause with affected landowners by bringing them into the Big South Fork Preservation Coalition, nor did they broaden the scope of their conservation concerns to include the cultural landscape, not surprising in light of the many traces of human blight I enumerated earlier.

Local residents shared the coalition's concern about inessential condemnation of private land, but they generally were at odds with environmentalists' desire for low-impact forms of recreation with only modest construction. Resident landowners and tenants facing condemnation or eviction could only hope that long-term economic benefits of tourism might compensate for their immediate losses; local political and business leaders who supported the recreation area were unequivocal proponents of development. They wanted the U.S. Army Corps of Engineers to build facilities that would appeal to car-bound visitors as well as outdoor enthusiasts. Although this segment of the local population did not form the Big South Fork Development Association until early 1979, after the environmental assessment process had been completed and the final version of the project plan was in preparation, many of its members were among the local leaders who had been consulted during the perfunctory SIA. Theirs was the only local perspective on sociocultural costs and benefits that had been heard during project planning.

Neither the Development Association nor the Preservation Coalition effectively raised the issues of cultural conservation that should have been addressed in the Big South Fork project, issues that inadequate SIA left untouched and the folklife survey addressed only belatedly and tangentially. The environmentalists wanted the recreation area boundaries redrawn to save farms, but they did not actively work with farmers to pursue this goal. Thus, the plight of the families most directly affected by the recreation area was largely neglected by planners and organized public groups alike, while the families themselves lacked the skills, economic resources, and political power required to organize effective participation in the NEPA process on their own. A few landowners attempted to contest condemnation of their property; however, questions about the adequacy of the EIS that were raised informally never crystallized into legal action. Most important, the planning process never came to grips with the issue of inequitable distribution of the project's costs and benefits.

THE SPECIAL CASE OF HISTORIC RUGBY

In the historic English colony of Rugby, TN, adjacent to the Big South Fork recreation area boundary, historic preservation was obviously an issue; in that situation, the U.S. Army Corps of Engineers addressed associated cultural conservation issues more than adequately. A private organization, Historic Rugby, Inc., already was well along in a program to restore or reconstruct many of the community's Victorian homes and public buildings. The recreation area plan called for the U.S. Army Corps of Engineers to reconstruct the 1880s Tabard Inn in Rugby to be managed by Park Service concessioners as a lodge for Tennessee visitors.

Rugby had been placed on the National Register of Historic Places in 1972 and was named a State Historic District the same year. Thus, it was clear that under existing historic preservation legislation, the U.S. Army Corps of Engineers had an obligation to assess the impacts of additional tourism generated by the recreation area and to assist Historic Rugby in developing a plan to mitigate those impacts.

This project involved more than historic preservation, however, because the corporate enterprise, Historic Rugby, is embedded in a community that includes private homes, not all of which date to the original colony. Some community members have made substantial personal and financial commitments to Historic Rugby, Inc. over many years. But the community also encompasses newcomers who have built permanent retirement homes or summer retreats, as well as members of native Tennessee families whose ancestors lived in the area before the English colonists arrived. Together these people make Rugby a living community, something like the one its founder Thomas Hughes envisioned, rather than a museum of Victorian buildings.

The U.S. Army Corps of Engineers underwrote a needs assessment and community plan for Rugby. This entailed working closely with community residents, first to supplement the usual resource inventories with interviews and later to develop ideas for the plan through a series of open public meetings and work sessions with a steering committee of 20 residents. The steering committee represented the various groups within the community, not only the preservationists. Committee meetings brought into the open conflicts and long-standing resentments between old-timers and newcomers, between supporters of Historic Rugby and residents who resented tourists invading their privacy. Their debates continued in public meetings, which were well attended and sometimes stormy, but during 18 months of talking (between July 1980 and December 1981), the factions negotiated a plan for orderly community development and zoning (Building Conservation Technology).

RECOGNITION OF CULTURAL CONSERVATION ISSUES AT MOUNT ROGERS

When the Mount Rogers National Recreation Area was authorized by Congress in 1966, over half of its proposed 154,000 ac already were part of the Jefferson National Forest. Although the Forest Service planned to increase its landholdings, private inholdings would continue. Congressman Pat Jennings, who pushed the Mount Rogers enabling legislation through Congress, had long been an enthusiastic proponent of tourism as an economic development strategy for what he called the "Whitetop Wonderland" of southwestern Virginia.

In the winter of 1967, local citizens organized a seminar to discuss the project and its potential economic benefits with the Forest Service planner assigned to oversee the project. They later formed the Mount Rogers Citizens Development Corporation (MRCDC) to promote private economic development in response to new opportunities the recreation area would create.

In this pre-NEPA era, the Forest Service established close contact with MRCDC as a means of keeping the public informed on the planning process, gathering public reaction and input on the plan, and encouraging local interest in land use planning and formation of a regional planning commission. While there was some concern that haphazard, unregulated development around the recreation area could spoil the assets that attracted visitors, the local government and business interests represented in MRCDC unanimously favored development.

The Forest Service produced an ambitious plan in response to their congressional mandate and the aspirations of the pro-development local constituency. It included a 63-mi scenic highway along the Iron Mountain range, linking Interstates 77 and 81; a ski slope, winter sports complex, and year-round resort at Whitetop Mountain; eight recreation areas for camping and picnicking; and seven impoundments for water supply and recreation purposes. Visitation was expected to reach 1 million by 1976 and 5 million by the year 2000. An information brochure (U.S. Forest Service 1969) summarized the plan, beginning with the reassurance that there would be "public recreation areas on National Forest lands with complementing developments on the intermingled private lands." The brochure continued, "The adoption of a rural motif for the private land will preserve the idea of early American rural life and use the natural advantages of the area to attract millions of persons each year."

The Forest Service and local citizens enjoyed a solid consensus on the plan until 1974, when condemnation proceedings were begun to acquire additional acreage, small rural properties and tracts that local businessmen had envisioned developing privately. Meanwhile, the demography and social climate of southwest Virginia had been changing. By the mid-1970s, significant numbers of well-educated, environmentally conscious exurbanites had moved into the area, attracted by the rural peace and quiet. These newcomers had the will and the organizational skills to launch a campaign against what they saw as overdevelopment, particularly the scenic highway and the ski resort proposals. By acquiring additional acreage, the Forest Service hoped to forestall undesirable private development of inholdings that would threaten the "rural America" character of the area, one of its most valuable assets for current residents and prospective visitors alike, but this action did not reassure those who feared overdevelopment. Rather, they led the opposition to increased land acquisition and used this issue to draw long-time local residents, some of whom had originally supported the recreation area, into the opposition movement.

Meanwhile, the Forest Service had revised its planning procedures consistent with NEPA and the National Forest Management Act (P.L. 94-588). In 1978, a formal EIS and draft plan presenting five alternatives was transmitted to the CEQ and released for public review (U.S.F.S. 1977). Potentially adverse sociocultural impacts were discussed at some length in the draft EIS. These included land acquisition and displacement of residents, some of whom would suffer psychologically and socially; organizational as well as economic problems associated with increased demand for governmental, social, and economic services; pressure from outside developers to acquire sites in neighborhoods and communities adjacent to the recreation area; rising land costs and property taxes; loss of farms and income from part-time farming; and cultural changes affecting quality of life, including restrictions on land use which would affect traditional leisure activities. This discussion was careful to point out the considerable social diversity of the area and its implications. While, in general, the recreation area would only intensify already established trends toward a more urban lifestyle, communities and citizens located nearest to proposed developments, and thus most directly affected by potential adverse impacts, were the most traditional, the most tied to place and neighbors, the most likely to find sociocultural change traumatic, and the least prepared to take advantage of economic opportunities presented by the project. The researcher summed up the situation in a follow-up report to the Forest Service: "The greatest impact problems will be in the more isolated and less mainstream locales where elements of earlier cultural themes thrive" (Rhyne 1979, Appendix A).

While the SIA implied that the cultural conservation factor should be considered in decisionmaking, the ensuing public review process gave opponents of intensive development the opportunity to demonstrate how much importance they attached to these issues. Local activists quickly established a grassroots organization, Citizens for Southwest Virginia, to respond to the draft EIS and plan. They secured the aid of an attorney with personal ties to the area and experience in Washington; prepared a detailed 54–page response (Blanton 1978); and orchestrated an effective letter writing and petition campaign which involved a broad socioeconomic cross-section of local citizens, from traditional farmers to newcomers with experience in environmental organizations, community development work, and Washington bureaucracies. With the help of their congressman and support from environmental organizations like the Sierra Club and Trout Unlimited, this group sought and obtained a 3-month extension of the deadline for response to the draft EIS and plan. During the response period, Forest Service representatives met with civic organizations, political bodies, special interest groups, and interested citizens; they participated in open discussions on radio and TV interview and call-in shows; they held bus tours of the area, including two weekend "workshop tours" attended by representatives of national environmental and conservation organizations (Blankenship). These events were efforts to regain support for the preferred action alternative, which in essence retained all of the development features of the 1969 preliminary plan, but the Forest Service also intended to honor the spirit as well as the letter of the NEPA process.

Based on content analysis of the massive response received during the public comment period and additional interviewing done by the SIA researcher in the summer of 1979, the Forest Service identified ten areas of controversy which were thoroughly discussed in the final EIS and master plan (U.S.F.S. 1980). Public concerns included several issues related to cultural conservation: the overall scope and intensity of development, anticipated social and cultural change, economic growth, and land acquisition. Residents of the area other than business people expressed strong concern that too much development, particularly the ski slope and scenic highway, would lead to tawdry commercialism and undermine the welfare of residents (Rhyne 1979). Five million visitors per year seemed too many to absorb without adverse impacts of the sort that already were occurring in connection with Western North Carolina tourism development (Beaver 1982; Bingham 1978).

Public response to the EIS was not just collected and analyzed, but taken seriously. The 1980 plan discarded the full development alternative and substituted the low-density recreation alternative, a plan which would better conserve both natural and cultural resources. The scenic highway and ski area were eliminated, stream impoundments and fully equipped camping spaces were reduced in number, and scenic easements were proposed as an alternative to fee simple acquisition of many inholdings.

Continuous planning procedures adopted by the Forest Service have led to reappraisal of plans and timetables for development of the Mount Rogers recreation area during the past decade. Pushed by both environmental concerns and federal budgetary constraints, modifications have further scaled back both the intensity and pace of development. While this turn of events has caused many of the recreation area's early proponents to lose interest in it, a recent survey (Howell 1988) showed critics of the 1978 draft plan to be generally satisfied with the scale and pace of development, with Forest Service management of the area, and with the agency's efforts to involve the public in decisionmaking since that time.

EXPANDING SIA TO INCLUDE CULTURAL CONSERVATION CONCERNS

The discussion of sociocultural impacts included in the Mount Rogers EIS pointed to a number of issues relevant to cultural conservation that were thoroughly aired through public response to the 1978 draft plan. In contrast, SIA and planning procedures for the Big South Fork project provided too little exposure to the local situation for inequitable social impacts and other cultural conservation issues to become apparent, nor was public response to the plan effective in raising issues that planners and contract researchers had overlooked. It is simplistic to think that the standard procedures to obtain public review of plans (i.e., public hearings and comment periods) reach all segments of the public whose interests are at stake. To be sure, Citizens for Southwest Virginia organized quickly, mobilized a broad base of local support that cross-cut social and economic differences, enlisted the aid of national environmentalist organizations, and had the legal and bureaucratic expertise needed to understand and take advantage of the NEPA process. But more often, as at Big South Fork, public response simply excludes cultural minorities with economic, political, and educational disadvantages that limit their will and ability to participate. SIA, not the vagaries of public participation, must be responsible for identifying all segments of the public that may be affected by a proposed development and clarifying how they may be differentially impacted.

If, as the Cultural Conservation report recommends, CEQ regulations ultimately are revised to define the human environment more broadly and give sociocultural factors more weight in environmental planning, what might the SIA of the future look like? Perhaps the change required is not so much a matter of inventing new procedures and research strategies as combining existing elements, some of which now apply only in certain specific situations, and applying them across the board. This discussion will conclude by identifying three such building blocks that already exist in SIA.

NATIVE AMERICAN SIA: PRECEDENTS FOR SPECIAL EMPHASIS ON MINORITY CULTURAL CONCERNS

Because the federal government recognizes a special relationship between itself and tribal governments that does not apply to other ethnic minorities, Native American SIA has led the way in interjecting cultural conservation concerns into SIA. The 1978 CEQ regulations (Section 55989) speak directly to the consultative role Indian peoples should have in SIAs whenever projects will affect people living on reservations. Further, the American Indian Religious Freedom Act (P.L.

92-469) provides for access to sacred sites and use of sacred resources even when these are located beyond reservation land. On this basis, Stoffle and his associates (1982) were able to help Southern Paiute obtain recognition of intangible cultural resources that were threatened by electric power development (Stoffle et al. 1982), and more recently, Reimensnyder (1990) worked with Eastern Cherokee and Forest Service personnel to have sacred resources given appropriate consideration in planning for the Nantahala National Forest.

Tribes are unique among minority groups in another respect as well, however. As controllers of significant natural resources, they encompass both sides of the equation — affected community and decision makers who must weigh the merits of resource development. Like other project managers, tribal leaders have need of SIA, but not simply as a *pro forma* exercise to fulfill federal requirements. Their efforts to collaborate with researchers in defining the scope and content of SIA and collecting data relevant to their needs are described in the monograph Indian SIA (Geisler et al. 1982). These case studies suggest how other minority communities might work with social scientists to develop credible quantitative and qualitative ethnographic and ethnohistoric data to substantiate their cultural conservation concerns. Active participation in the research process rather than passively waiting for an opportunity to respond to a finished EIS is a seemingly radical departure from current practice, but, in fact, scoping recognizes the need for just such collaboration with affected communities.

SCOPING: SITUATIONAL DEFINITION OF CRITICAL ISSUES

Federal initiatives to enhance cultural conservation call for official recognition of something natives have recognized all along — that maintenance of cultural identity is an important factor in quality of life. Recent guides for the conduct of SIA stress the need for scoping to identify the sociocultural issues that are salient for decisionmaking in a particular case. Scoping provides a logical entrance point for cultural conservation to be incorporated into SIA where appropriate, because good scoping entails defining the affected area, surveying its communities to identify all potentially affected stakeholder groups, and obtaining input from group representatives to help determine what topical coverage is most critical before the SIA team develops its research strategy. Only if cultural conservation issues become apparent early in the planning process can their policy implications be examined; otherwise, cultural conservation activity will remain simply a modestly expanded form of documentation to supplement historic preservation efforts.

RISK ASSESSMENT: FACTORING IN PUBLIC PERCEPTIONS AND VALUES

Anthropologists working in the field of risk assessment have begun to demonstrate the relevance of sociocultural context to planners and managers and to other members of the risk assessment team (Wolfe 1988). Experience has forced the scientific and technical community concerned with risk to acknowledge that public perceptions of risk are seldom congruent with the quantitative, probabilistic assessments of environmental hazards worked out by natural scientists and that better public education is not the key to eliminating incongruities (Fischhoff et al. 1981). The public's assessment of risk draws upon basic cultural values attached to intangible aspects of that elusive variable, quality of life. Public concerns thus cannot be understood without ethnographic insight into culture, nor can environmental policy conflicts pitting government and industry against a distrustful public be resolved except by finding acceptable compromises between competing value sets.

Because cultural conservation is similarly a matter of values, ethnographic techniques applied in risk studies set a useful precedent of dual attention to external and internal assessment criteria, what anthropologists refer to as etic and emic analysis. As in risk assessment, cultural impact assessment requires ethnographic research to develop basic data — in this case, to identify culturally distinctive groups and survey their extant culture patterns to determine the likelihood of cultural impacts and the potential for mitigation through cultural conservation efforts. But more importantly, ethnographic research is necessary to provide insight into group values (e.g., how much value group members themselves attach to maintenance of cultural tradition in comparison with other, potentially competing values, such as innovation and economic development).

Cultural conservation may or may not be at issue in a specific case; to uphold the principle of individual choice espoused by NEPA, that judgment ultimately must be made by members of the affected communities themselves, not by outside experts. But where cultural conservation is a significant local concern, SIA must become flexible enough to address that concern and effectively introduce it into planning considerations, whatever the proposed project and whatever community's lifeways and values are at stake.

ACKNOWLEDGMENTS

Discussion of the Big South Fork situation is based on documentary research and personal contact between 1978 and 1985 with local people, Corps of Engineers, National Park Service personnel, and other contract researchers. I conducted the folk culture survey mentioned in this paper between 1978 and 1980, sponsored by the Southeast Archeological Center, National Park Service (contract CX500090902), and continued to maintain a residence in the area until 1985.

Discussion of the Mount Rogers situation is based on analysis of published documents, Forest Service working papers, and responses to the draft plan and EIS on file at the recreation area headquarters in Marion, VA. I am grateful to Charles Blankenship, Jefferson National Forest Planner, for supporting my interest in using Mount Rogers as a comparative case and for arranging access to unpublished documents and correspondence in 1987. I am also grateful to Area Ranger Larry Grimes, who encouraged me to return to the Mount Rogers area for survey research and interviewing May–July 1988 and provided housing through the recreation area's volunteer program. Research in the spring of 1988 was made possible by released time from The University of Tennessee, Knoxville.

REFERENCES

Beaver, P. D. 1982. Appalachian families, landownership, and public policy. pp. 146–154. In R. L. Hall, and C. B. Stack (eds.), *Holding on to the Land and the Lord: Kinship, Ritual, Land Tenure, and Social Policy in the Rural South.* University of Georgia, Athens.

Bingham, E. 1978. The impact of recreational development on pioneer lifestyles in southern Appalachia. pp. 57–70. In H. Lewis et al. (eds.), *Colonialism in Modern America: The Appalachian Case.* Appalachian Consortium Press, Boone, NC.

Blankenship, C. A. The Mount Rogers National Recreation Area: A History of Land Use Planning and Public Involvement. Typescript.

Blanton, W. (ed.). 1978. Response of Citizens for Southwest Virginia to the Draft Environmental Impact Statement for Mount Rogers National Recreation Area and Mount Rogers Scenic Highway. Typescript.

Building Conservation Technology. Master Plan for the Development, Management and Protection of the Rugby Colony Historic Area. U.S. Army Corps of Engineers, Nashville, TN.

Finsterbusch, K. 1981. The potential role of social impact assessment in instituting public policies. pp. 2–12. In K. Finsterbusch, and C. P. Wolf (eds.), *Methodology of Social Impact Assessment.* 2d ed. Hutchinson-Ross, Stroudsburg, PA.

Fischhoff, B., et al. 1981. *Acceptable Risk.* Cambridge University Press, New York.

Geisler, C. C., et al. (eds.). 1982. Indian SIA: The Social Impact Assessment of Rapid Resource Development on Native Peoples. Natural Resource Sociology Research Lab Monograph No. 3. University of Michigan, Ann Arbor.

Howell, B. J. 1981. *A Survey of Folklife Along the Big South Fork of the Cumberland River. Report of Investigations No. 30.* The University of Tennessee Anthropology Department, Knoxville.

Howell, B. J. 1983. Implications of the cultural conservation report for social impact assessment. *Hum. Organ.* 42(4):346–350.

Howell, B. J. 1984. Folklife research in environmental planning. pp. 127–140. In W. Millsap (ed.), *Applied Social Science for Environmental Planning.* Westview, Boulder, CO.

Howell, B. J. 1988. A Ten Year Report Card for Mount Rogers National Recreation Area: Public Assessment of Forest Service Management. Prepared October 1988 for the Area Ranger, MRNRA (manuscript available from author).

Loomis, O. H. (coordinator). 1983. Cultural Conservation: The Protection of Cultural Heritage in the United States. Library of Congress, Washington, D.C.

Peterson, G. L., and R. S. Gemmell. 1981. Social impact assessment: Comments on the state-of-the-art. pp. 386–399. In K. Finsterbusch, and C. P. Wolf (eds.), *Methodology of Social Impact Assessment.* 2d ed. Hutchinson-Ross, Stroudsburg, PA.

Reimensnyder, B. L. 1990. Cherokee Sacred Sites in the Appalachians. pp. 107–117. In B. J. Howell (ed.), Cultural Heritage Conservation in the American South. Southern Anthropological Society Proceedings No. 23.

Rhyne, E. H. 1979. Social Impact Analysis for the Proposed Mt. Rogers National Recreation Area. Typescript.

Schroedl, G. F. (compiler). 1988. *United States Government Documents Pertaining to Archaeological Resource Management: A Handbook.* The University of Tennessee Anthropology Department, Knoxville.

Sibley, B. J. 1984. A decade of NEPA: Federal agency responses to the National Environmental Policy Act. pp. 221–239. In W. Millsap (ed.), *Applied Social Science for Environmental Planning.* Westview, Boulder, CO.

Soderstrom, E. J. 1981. *Social Impact Assessment: Experimental Methods and Approaches.* Praeger, New York.

Stoffle, R. W., et al. 1982. Southern Paiute Peoples' Responses to Energy Proposals. pp. 107–134. In C. C. Geisler, et al. (eds.), Indian SIA: The Social Impact Assessment of Rapid Resource Development on Native Peoples. Natural Resource Sociology Research Lab Monograph No. 3. University of Michigan, Ann Arbor.

U.S. Army Corps of Engineers, Nashville District. 1976. Big South Fork National River and Recreation Area Final Environmental Impact Statement. U.S. Army Corps of Engineers, Nashville, TN.

U.S. Forest Service, Jefferson National Forest. 1969. Mount Rogers National Recreation Area: A Review. U.S. Forest Service, Jefferson National Forest, Roanoke, VA.

U.S. Forest Service, Jefferson National Forest. 1977. Mount Rogers National Recreation Area Draft Environmental Statement. U.S. Forest Service, Jefferson National Forest, Roanoke, VA.

U.S. Forest Service, Jefferson National Forest. 1980. Mount Rogers National Recreation Area Final Management Plan and Environmental Impact Statement. U.S. Forest Service, Jefferson National Forest, Roanoke, VA.

Wheeler, W. B., and M. J. McDonald. 1986. *TVA and the Tellico Dam, 1936–1979. A Bureaucratic Crisis in Post-Industrial America.* The University of Tennessee, Knoxville.

Willeke, G. F. 1981. Identifying publics in social impact assessment. pp. 305–311. In K. Finsterbusch, and C. P. Wolf (eds.), *Methodology of Social Impact Assessment.* 2d ed. Hutchinson-Ross, Stroudsburg, PA.

Wolf, C. P. 1978. The cultural impact statement. pp. 178–193. In R. S. Dickens, Jr., and C. E. Hill (eds.), *Cultural Resources: Planning and Management.* Westview, Boulder, CO.

Wolfe, A. K. (ed.). 1988. Anthropology in environmental risk studies. *Pract. Anthropol.* (special issue) 10(3–4):4.

How Severe Is Severe: Public Involvement and Systematic Understanding of Wilderness as a Resource

G. M. Schoepfle, General Accounting Office, Washington, D.C.; E. J. Szarleta, Canisius College, Buffalo, NY; and S. Schexnayder, Oak Ridge National Laboratory, Oak Ridge, TN

ABSTRACT

Wilderness is a culturally defined resource whose preservation is threatened by an increasing number of demands for the use of public lands. The military is one of these competitors, and the potential impacts of low altitude training flights on wilderness character and experience have aroused public concern. Because these impacts are difficult to measure and define directly, social scientists have devoted increasing attention to measuring them indirectly through people's revealed preferences. Methods for defining and measuring these preferences, however, are beset with bias. This article reports on how researchers can use cognitive ethnography and contingent valuation to address bias. Ethnography can describe accurately the knowledge base and social context from which various segments of the public determine their preferences. Contingent valuation can establish the validity of the ethnographic results over a wider population.

This article also describes briefly the implications of this combination of approaches toward public involvement and proposes how public involvement can be a resource in determining the costs and benefits of environmental changes. Ability to determine these changes will be important in the context of implementing assessments under NEPA.

INTRODUCTION

Wilderness lands in the United States have been set aside by law as resources available for the enjoyment of the public. They are to be protected from logging, mining, cattle grazing, recreational development, and other consumptive uses. Increasing public concern is being raised about the effects on wilderness character and the wilderness experience of low altitude flights (LAFs) conducted for training purposes by the military. As a result, the protection of wilderness resources is receiving increasing attention through the National Environmental Policy Act (NEPA).

Although the potential impacts of LAFs are of increasing public concern, it is difficult to define or measure them. That is, wilderness is a culturally defined resource, and an understanding of the public's perceptions of these impacts is vital. However, these perceptions often depend on a host of individual, social, and environmental contextual features. These features, in turn, may be beyond the scope of the environmental impact assessment process. As a result, it may be desirable to describe or measure peoples' preferences for protecting wilderness resources rather than the impacts of LAFs themselves. Preferences describe how much people like or dislike a resource's quality or availability, or potential changes in this quality or availability. By studying peoples' preferences for maintaining the quality of clean air, for example, it is possible for planners and policy makers to determine indirectly the perceived impacts of air pollution on the enjoyment of national parks or wilderness.

Studying such preferences may be a potentially useful tool for inferring the impacts of LAFs on wilderness and thus, a valuable contribution to the environmental impact assessment process outlined in NEPA. To make full use of such studies, it is still necessary to connect them systematically to the impacts of LAFs on wilderness perceived by the public. After summarizing the historical and cultural issues behind wilderness preservation, and explaining how researchers can apply the study of preferences to explain wilderness as a land use issue, this paper presents the approach, methods, and results obtained from a study of the impacts of military LAFs on the wilderness. It presents results of ethnographic research to show that an understanding of both perceived impacts and preferences is incomplete without an understanding of the social and political context that includes the military, federal agencies entrusted with the care of wilderness lands, wilderness preservation advocates, and coalitions of rural citizens from areas overflown by the military. Finally, the article presents results of contingent valuation that links ethnographic findings on preferences to a wider population.

WHAT IS WILDERNESS AND WHY IS IT AN ISSUE?

There is a total of 29,000 mi^2 of land designated as wilderness in the continental United States, excluding Alaska. According to the Wilderness Act of 1964, wilderness is not only an area "untrammeled by man and his works," but an area where a governmental regulatory and administrative apparatus exists to ensure that it remains untrammeled.

> A wilderness, in contrast with those areas where man and his own works dominate the landscape, is hereby recognized as an area where the earth and its community of life are untrammeled by man, where man himself is a visitor who does not remain. An area of wilderness is further defined to mean ... an area of undeveloped federal land retaining its primeval character and influence, without permanent improvements or human habitation, which is protected and managed so as to preserve its natural conditions [Pub. L. 88-577, Section 2(c)].

The movement to preserve these lands arose as part of the Romantic movement in the 19th century. As American settlement and development spread westward, artists, philosophers, writers, naturalists, scientists and other intellectual leaders articulated a need to preserve areas from consumptive land uses such as mining, farming, logging, cattle grazing, and settlement. These leaders wanted these pristine areas set aside to afford an appreciation of nature unaffected by human activities (Watkins 1989).

By the mid-19th century, these intellectual leaders pressed for legislation that would protect these public lands from further development. This legislation culminated in the establishment of the National Forest, National Park, and Wilderness systems, and marked the beginning of a public lands use planning program. At the turn of the century, the movement split into two overlapping branches: conservationist and preservationist. The conservationists espouse the protection of public lands for conserving water, forest, grassland and mineral resources, and to prevent pollution and erosion. Adherents to this movement have been willing to trade off the pristine nature of some areas for the development of dams for flood control and long-term consumptive uses (Watkins 1989). To protect the pristine nature of these lands, preservationists advocate the prevention of even relatively unobtrusive uses and recreational development in national parks and other areas.

Bitter disputes broke out between these two branches over wilderness preservation. As early as 1906, the dispute over the construction of the Hetch-Hetchy Dam on the Yosemite River divided preservationists such as John Muir against conservationists such as Gifford Pinchot (Watkins 1989). Pinchot and his allies advocated building the dam to supply water for San Francisco, which had just been stricken by the great earthquake. Muir and his allies opposed the dam, advocating the preservation of Yosemite Valley and its environs. Similar conflicts exist today over controlled grazing and logging of public lands, and a quick resolution is not in sight.

The federal government is a major actor in this conflict over the use of public lands. It is represented by the Department of Agriculture and Department of the Interior. Under the Department of Agriculture is the U.S. Forest Service. Under the Department of the Interior are the National Park Service, U.S. Fish and Wildlife Service, and the Bureau of Land Management. All these services administer designated wilderness lands in addition to other lands within their respective programs.

These branches of the government have often found themselves in the role of mediator between conservationists and preservationists. At the same time, they have had to mediate between these two and advocates for even more consumptive uses of public lands as well. Three factors further complicate this mediation role. First, as steward of public lands and resources, the federal government must protect these lands for the greatest good to the greatest number of people (Mitchell 1990). Because consumptive and nonconsumptive uses are incompatible, however, federal attempts to reach compromise often result in the federal government being seen as "taking sides."

Second, agencies such as the U.S. Forest Service support themselves to some degree through user fees. Some federal officials and members of the public thus see the federal agencies as favoring the use of public lands by the interests able to pay the highest fees. The U.S. Forest Service, for example, is portrayed by some as favoring logging and other consumptive interests over nonconsumptive interests such as backpacking because the former can pay higher fees (O'Toole 1988).

Third, the federal government is often seen as a facilitator of private sector development more than as a steward of public resources (Edelstein 1988). This tendency became clearest to the public during the Reagan Administration when the Department of the Interior, under Secretary James Watt, tried to transfer large sections of public lands into private corporate hands (Radford 1986).

These debates have become increasingly important to public lands use policy. Their importance increases further as economies organized around consumptive uses such as mining, grazing, and logging continue their decline (Marston 1988). With its own agenda for public lands use, the military now looms increasingly important as changes in their tactics and technology require the use of more land and airspace (New York Times 1990). These land uses affect the environmental impact assessment process and arouse concerns often beyond the scope of the assessment process.

WILDERNESS AND PREFERENCE

Natural resource economists have become increasingly involved in public land use issues and thus, issues pertaining to wilderness use. Resource economists have studied the value of wilderness to compare the costs and benefits of its preservation with other public land use. Wilderness is a public good, but it is hard to measure monetarily for three reasons: (1) no individual or group owns it as private property, (2) it has multiple uses and values, and (3) wilderness recreational uses are nonconsumptive (i.e., use of these resources does not deplete the amount remaining). Thus, wilderness has a potential existence value (i.e., it could be valuable to people, regardless of whether or not they use the lands themselves, because it is part of the U.S. national heritage).

Contingent valuation has been used widely by resource economists and pubic lands use planners. It elicits the preferences people have for certain economic goods that cannot be bought or sold in the market. The elicitation centers around a set of hypothetical questions, or scenarios, which first describe the good itself and then a change either in the good's quality or availability. The respondents are then asked how much they are willing to pay for a positive change or for the prevention or mitigation of a negative change. Alternatively, they may be asked how much they are willing to accept as compensation for a negative change (Mitchell and Carson 1989). Finally, respondents are presented with a payment vehicle that identifies how the respondent will pay. Payment vehicles may include taxes, entry fees, insurance premiums, and other mechanisms that are acceptable and familiar to the public.

Researchers then aggregate the willingness to pay or accept responses. This aggregation is a monetary amount that represents a public's value for either enhancing or mitigating a degradation of a good's quality or availability. This amount can then be compared to the costs of enhancement or mitigation in a cost-benefit analysis. Known as the Pareto approach (Sugden and Williams 1978), balancing the costs and benefits requires assuming that everybody is content with the current or existing distribution (Freeman 1979; Sugden and Williams 1978). Although this assumption is questionable under many research circumstances, it is nevertheless often maintained.

Contingent valuation has achieved popularity recently because it translates preferences into an easily measured monetary amount. Methodological and theoretical problems in the form of hypothetical bias, however, beset this ease of measurement. Bias can result when the respondents' understanding of a question differs from that of the researcher in ways that the researcher does not expect. Answers can be distorted in a particular direction as a result (Cantor et al., forthcoming). To deal with this bias, the questions must be worded in ways that are meaningful to the respondents' knowledge, experience, and concerns, and are intelligible to the respondent and not ambiguous in their meaning. Similarly, the payment vehicle must not be objectionable to the respondent.

If the questions fail the first requirement, the respondents may answer the question accurately enough, but the answer may have nothing to do with what he or she does in real life (Mitchell and Carson 1989). If the second requirement is not met, the respondents may be forced either to guess wildly, supply the answer which they assume that the researcher wants, "read between the lines" to see what answer they should make, or in other ways guess at the intent of the information in the scenario. Objectionable payment vehicles may provoke the respondent into refusing to cooperate with the survey altogether.

Theoretical issues contribute further to hypothetical bias in three ways. First, a theoretical assumption underlying contingent valuation is that people can state their willingness to pay as individuals. Individuals are not supposed to answer a question by considering their role within a wider social context. This may be an unrealistic assumption under many circumstances. Second, contingent valuation also assumes that the good can be understood as an abstract entity not embedded in social context. Respondents are thus supposed to value wilderness resources without regard to who may be using them or to the kinds of people living in their vicinity.

Third, as a monetary measure, contingent valuation elicits willingness to pay for changes in nonmarket goods as if they were part of a market transaction system. The respondent is not asked whether or not he or she thinks the distribution of a good's quality or availability within this hypothetical market is fair. Some respondents may perceive the system as unfair, however, when a change in a good's quality or availability forces them to bear certain costs, such as environmental degradation, while others who may support or benefit from the change do not. Their response to a willingness to pay or accept question thus may not represent market behavior; it may be a form of demand for reparation or similar nonmarket payment (Mitchell and Carson 1989).

THE USE OF ETHNOGRAPHY IN ANALYZING THE IMPACTS OF THE MILITARY ON WILDERNESS

Cognitive ethnography addresses problems of bias arising from social context and fairness. Like any ethnography it describes a community's way of life holistically. It differs from other ethnography in concentrating primarily on studying the knowledge people have for adapting to their society (Goodenough 1956). This knowledge is either consciously verbalized or preconscious. Knowledge is preconscious when an individual does not have to think about what he or she is doing. Referred to as "know how" (Chomsky 1959, 1969), examples of such knowledge include using a typewriter or playing a piano. It differs from "knowing what" (i.e., knowledge an individual can think about consciously and discuss). In some instances, know how is difficult for an individual to discuss; indeed, discussion may inhibit know how in the same way that thinking about using a typewriter while using one can inhibit typing. Nevertheless, individuals can think or talk about preconscious knowledge under certain contexts (Khilstrom 1987), and the knowledge is thus accessible to interviewers if the questions are framed properly (Werner and Schoepfle 1987).

Researchers thus rely primarily on interviews to elicit the total range of knowledge, or domain, of a given subject. The result is a body of terms and discourse about this domain. Relying on discourse requires a logically elegant form of recorded interview analysis. The first step in the analysis is to identify a set of key terms. The next step is to identify a set of attributes that define those terms. The third step is to identify the logical relationships among terms and their definitions. These relationships include taxonomy (i.e., *X* is a kind of *Y*), sequence (i.e., first *X* and then *Y*), and modifications of these relationships, such as implication (if *X* then *Y*), cause-effect (*X* causes *Y*, or *Y* only if *X*), part-whole (*X* is a part of *Y*), and other more complex logical relationships among terms and their attributes (Werner and Schoepfle 1987).

One characteristic of this ethnographic approach is a sampling procedure that relies on the selection of a small number of individuals. Researchers interview these individuals intensively and in great detail. By analyzing these interviews, researchers can define individuals by the attributes which describe social relationships existing between and among them. The researchers can then link these individuals within a network (Mitchell 1969; Werner and Schoepfle 1987).

The network allows researchers to describe a social context systematically. Individuals can be defined by attributes which describe them as members of social groups. The attributes can thus describe the similarities among individuals which may highlight their membership to a single group, or differences which describe their membership in contrasting groups. The same similarities and contrasts can define the groups themselves. Together, attributes of similarity and contrast describe taxonomies, or classifications, of groups and individuals within a network (Werner and Schoepfle 1987). Finally, these individuals can then define classifications and definitions of adverse and beneficial impacts. Impacts are positive or negative effects of a change (i.e., changes for the better or the worse) and can be defined and classified in the words people themselves use.

Thus, cognitive ethnography can address at least some of the theoretical and methodological issues associated with hypothetical bias in four ways. First, it provides a systematic sampling procedure that allows the researchers to identify networks of individuals who are knowledgeable about the beneficial and adverse impacts of a change in the environment. Second, it provides a means of eliciting definitions of the social context within which people recognize and formulate their knowledge of beneficial and adverse impacts. Third, it provides the means of eliciting definitions of the hidden costs themselves. Fourth, it provides the means of eliciting the mitigative measures to these impacts. These measures could make the change in the environment fairer to all affected members of the public.

In the following sections, taxonomies resulting from application of the ethnographic approach to LAFs are presented as tree diagrams. Also, attributes — quoted phrases from different individuals — will be presented to provide further information about a term. In some cases, attributes from several speakers are presented together. In other cases, paragraph quotes are used to support the researchers' interpretations.

Sampling

As part of the research, initial sampling involved the selection of 12 case studies of specific Air Force flight paths throughout the United States. Of these 12 flight paths, 5 were located above areas designated as wilderness. In the following discussion, data from underneath one of these routes, SR-300, an Air Force low-flying plane route, are described. SR-300 is located over west central Nevada and eastern California. The sampling involved a network approach and proceeded through the process illustrated in Figure 1.

Researchers identified local federal officials from the Bureau of Land Management, U.S. Forest Service, National Park Service, and U.S. Fish and Wildlife Service and then interviewed these officials.[1] They asked them to list local individuals active in wilderness preservation. Researchers also contacted officials of national-level and regional chapters of wilderness preservation advo-

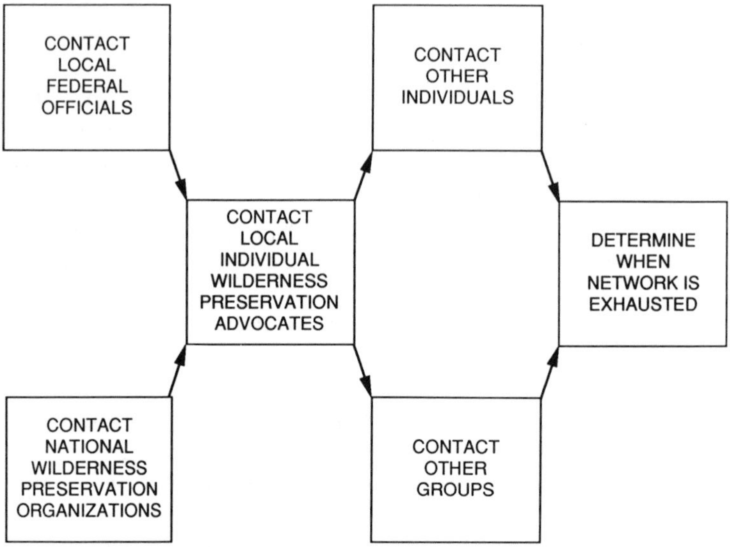

Figure 1. Network approach.

cacy groups, such as the Sierra Club, Wilderness Society, Audubon Society, and Nature Conservancy. They interviewed these officials the same way. Researchers then interviewed the local wilderness preservation advocates and others who were associated with public lands use and requested still more references. They continued this sequence until the list of key informants was exhausted.[2]

The research team asked these officials and private citizens to list individuals whom the research team could interview and the different categories to which these individuals belonged. The goal of these questions was to determine, with minimal bias, the kinds of organizations and groups with whom these individuals identified or were identified, the kinds of public and private positions these groups and individuals assumed with respect to wilderness use, and how these positions

[1] According to the 1964 Wilderness Act, these officials are responsible for administering federal lands under their jurisdiction which are designated as wilderness. This also conforms with the Council on Environmental Quality guidelines for consultation with federal officials. We simply make this consultation more systematic here.

[2] It is rarely possible to reach the entire boundary of a network. In this particular case, for example, only a few state legislators and representatives of mining, logging, and cattle interests were contacted. After a few preliminary interviews, it became clear that their concerns with military flights over wilderness did not differ from those that would emerge in response to flights occurring over nonwilderness areas.

compared and contrasted. By using this approach, the research team intended to select speakers who were the most informed and articulate about the impacts to wilderness character and wilderness experience resulting from LAFs.

CATEGORIES OF IMPACTS

For all those interviewed,[3] the major categories of impacts included those to wilderness character, wilderness experience, and caretaker activities. Definitions of impacts to wilderness character depend on defining wilderness character itself. Perceived impacts to the wilderness experience can be divided into categories denoting solitude, enjoyment of wildlife, and safety. Impacts to caretaker activities involve disrupting interagency relations between federal caretakers in the Department of the Interior or U.S. Forest Service and the military.

For some speakers, violation of wilderness character is a violation of principle. The following attributes help to define what various speakers meant by the impacts of LAFs on wilderness character.

1. The aim of going to the wilderness is to escape to a place not impacted by man.
2. Low altitude military aircraft voids the possibility of doing that.
3. The person with the least amount of tolerance is being wedged out. There are fewer places where the hard core type can go for the wilderness experience.
4. It is like the Mona Lisa with a mustache drawn on it.

Attributes 1 and 2 show a setting not affected by man. They imply that an intrusion ruins this setting completely in principle. Attributes 3 and 4 explain the principle in more detail. Many speakers compared their plight as wilderness users to an endangered part of the biosphere. They maintained that the environmental impact assessment process should consider a rare or vulnerable species or parts of a community as integral parts of a system rather than considering only the effects on the welfare of the majority. Just as all parts of a biological community should be considered, so should minorities within human communities. Solitude is the most important element because it is a state one normally assumes when enjoying any aspect of wilderness. Solitude is the removal of the individual from the day-to-day stress of industrialized society. He or she can then achieve a balance between intellectual endeavors constrained by everyday job requirements and unguided reflection associated with wilderness use.

The following are attributes several speakers used in defining wilderness experience.

---

[3] The term "speaker" will be used to refer to individuals interviewed through ethnographic approaches. Similar terms include "informant" and "consultant" and denote people who have been requested to discuss a wide domain of their knowledge. The term "respondent" will be used to refer to those who responded to the contingent valuation questionnaire in particular and survey questionnaires in general (see below).

1. I get out and put things in perspective.
2. Paperwork and writing and reading in good combination [with wilderness use] is the optimum.
3. I feel more balanced.
4. I get above the smog and noise and intrusions of trivial things.
5. I feel recharged and alert.
6. Out there [in the wilderness] there is complete relaxation, losing tension.

The effects of military LAFs on the enjoyment of wildlife involved concerns about the disruption of wildlife staging, nesting, breeding, and migration. Some officials expressed concern about the potential dangers to some users through the possibility of avalanches, startling climbers, or startling horses. The isolation of wilderness areas would also make pilot rescue difficult if there were a crash or other accident. Although speakers maintained that these impacts could occur, it was difficult to document when and where they actually happened.

Impacts to federal caretaker activities centered around problems of interagency coordination with the military. Forest Service and Department of the Interior officials were concerned, for example, about the possibility of patrols or fire-fighting planes colliding with military planes. They occasionally complained about the military failing to honor commitments to limit the number of flights over certain areas, to agree to a floor altitude, and to keep flights away from areas in which users were vulnerable to a disruption of their activities.

Overall, although speakers expressed considerable concern about environmental degradation, wilderness users were reluctant to claim that these flights ruined the wilderness experience. As one speaker observed:

> You have a four day trip out in Central Nevada with a jet that blasts by you once for a period of five seconds. That five seconds sticks in your mind probably as much as the rest of the four day trip; and that's not going to kill anybody; it's not going to cause anybody to lose a month's salary or anything like that; but it's a stress. It's something I don't like and it irritates me ...
>
> It [the wilderness trip] isn't ruined. You don't come back and say "Bummer! I had a great time and all of a sudden the jet flew over and ruined my trip." I think people are more resilient than that.

The complaints about the military's LAFs over wilderness areas involved not only direct impacts to wilderness character and experience, but also the failure of the military to respond to the concerns of the wilderness users in a straightforward manner. As one user summed it up:

> People just don't like it. It's more or less a question of the military saying "screw you guys; you're not very important. We're going to try to do what we have to do to make you happier but it's not going to affect our decision. The decision has already been made to do this."

Not surprisingly, then, understanding people's perceptions of LAF impacts on wilderness requires more than a description of direct impacts to wilderness character, experience, or caretaker activities. It is necessary to shift attention to the social context.

Social Context

Wilderness users maintained that a proper understanding of their position required that the researchers interview the coalitions of which the wilderness users are a part and that have also objected to LAFs. Referred to here as "a very unusual coalition of people," these people were often members of umbrella organizations, such as Citizen Alert, in Nevada. Citizen Alert already existed before the issue of military use of airspace arose. Begun in 1979, the group originated to protest the MX Missile Racetrack and also opposes the designation of lands for a nuclear waste repository. They have joined other organizations throughout the United States in advocating the rights of rural citizens such as farmers and ranchers against resource development that would endanger the local environment and these peoples' customary uses of it.

For purposes of explaining the issue of military LAFs, speakers divided the coalition into wilderness users, who were referred to as "environmentalists," and others, who were referred to as "normally in opposition to environmentalists." The tree diagram in Figure 2 shows how the environmentalists and their erstwhile opponents constitute the two major classifications of the coalition. It also diagrams how environmentalists, in turn, can be classified into two major groups: the main line and radical environmentalists.

The discussion immediately following will describe what both the environmentalists and their former opponents have in common. Discussion is then directed toward how the environmentalists and their opponents differ and how the environmentalists, in turn, differ among themselves. What the groups have in common is important to understanding the true impacts to the public from the LAFs. The groups' differences highlight how they have achieved compromise and consensus despite important differences in outlook toward how public lands ought to be used. In the ensuing discussion, those who may sometimes disagree with the environmentalists will be referred to as rural rights advocates.

Similarities

The attributes shown in Figure 3 comprise a definition of the overall coalition. The first two attributes in Figure 3 indicate that the environmentalists and others, whatever their differences, are united in their position toward the military. The remaining attributes summarize the position itself in more detail.

Impacts on quality of life and helplessness (Attributes 3 and 4 in Figure 3) pertain little to direct effects such as health and personal safety. Instead, the concern is about the disruption of peaceful rural lives.

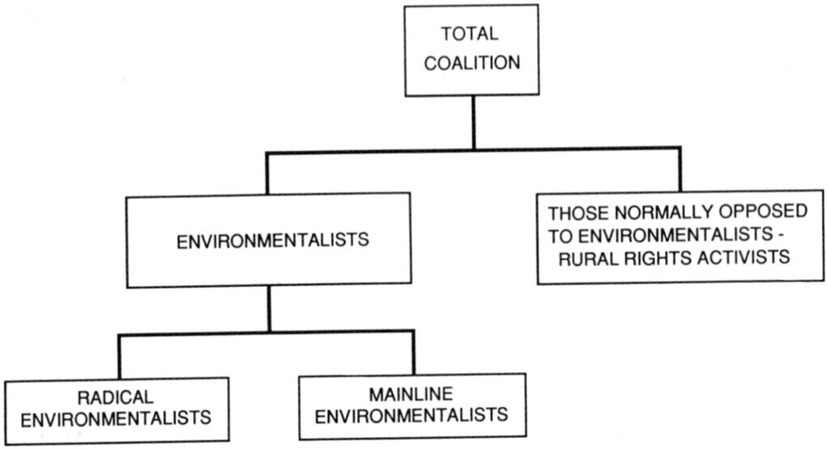

Figure 2. A very unusual coalition of people.

> What affects people the most when they are either overflown or they're under an [officially designated flight route], is that they have a very strong sense that they've lost control over their lives and property. There's something going on that they're really helpless to deal with, that is making their lives really uncomfortable if not intolerable.

In addition, they have neither involvement nor recourse in the decisions which affect them.

Inverse condemnation (Attribute 5 in Figure 3) is a legal concept in which use of land or airspace degrades property even though the property itself is not directly affected (Lorenz 1982). As a result of this inverse condemnation, the rural residents bear the adverse impacts of military training without compensation for damages.

> In other words, the kind of idea is that you [i.e., the military or anyone else] can gain public land simply by making it unusable for other purposes. ... It's so simple in a very basic sense. An individual living in a rural area has every right that any other citizen has, and what is happening now disabuses him of his ... rights.

Failure to compensate implies that those who benefit, presumably from enhanced national defense capabilities, gain at the expense of those who suffer adverse impacts.

Members of the coalition question the legitimate need for LAFs (Attribute 6 in Figure 3). They maintain that the military's need for land and airspace is based less on genuine program objectives than on interservice rivalry and poor planning. One speaker summarized the process as "go find your own sand box" and explained:

1. They were all speaking with the same voice on this particular issue.

2. You don't get the kind of classic division of forces in this sense .... the idea is that the environmentalists in the classic sense are involved, almost to the same degree as the other groups who would be considered opponents of the environmentalists, at least in other issues, wilderness and that kind of stuff.

3. They believe that low altitude flights can't be done over people, parks or wilderness without degrading quality of life.

4. [They believe that] there's something going on that they're powerless to deal with that is making their lives really uncomfortable if not intolerable.

5. [They believe that] this kind of stuff already amounts to a form of inverse condemnation.

6. [They believe that] they're [the military] not satisfied with the millions of acres they already have under withdrawal, they just simply move in and take over.

7. [They are sure that] there's been extensive illegal activities.

8. [Public meetings were] a waste of time. The FAA had no intention of even keeping a record of what we said.

Figure 3. Attributes of rural rights advocates and environmentalists — "A Very Unusual Coalition of People."

> [For example], where ... the Navy and the Air Force are working together [the Air Force will] suddenly tell Navy, "go find your own sandbox." ... The [Navy] are simply ordered out and told to find their own place. In other words, they have the epitome of inter-service rivalry going on even today. The idea is just plain empire building. They're not satisfied with the millions of acres they already have under withdrawal, they just simply move in and take over.

In effect, the military has failed to document to these members of the public a legitimate need for the land and airspace it wishes to use.

Coalition members maintain that the inverse condemnation proceeds with such secrecy that the military violates policies under which other federal agencies discharge their public trust (see Attribute 7 in Figure 3). The federal agencies are seen as cooperating with the military because they do not protest these violations. Advocates cited how they caught the military installing facilities which would denigrate the status of a potential wilderness area (i.e., a wilderness study area).

Coalition members reported having seen their attempts either to confront activities whose legality they question or to gather adequate information blocked or ignored by the military and other federal agencies (see Attribute 8 in Figure 3). First, coalition members perceive that the military and the Federal Aviation Administration restrict scoping processes and other public meetings to occasions in which concerns and requests for information are noted, but no effort is made to respond to either. Advocates cited an example in which federal officials failed to provide a court reporter; the rural rights advocates had to do so instead. Second, coalition members believe that the military conducts publicity campaigns that do not address concerns and queries, but appeal to a sentiment of patriotism. These are seen as attempts to gain legitimacy at the expense of those who protest.

> There are studies ... that show that when you've, what the Air Force calls 'orientated' a population, which just means that you've gone in and educated them and told them why you're doing these overflights and what they can expect—your complaint rate is less than half, or even lower than it would be without doing that. The euphemisms like the "Sound of Freedom," the "Sound of Security" are ways of conditioning people to accept a certain amount of turmoil in their lives, and feel it's for a good reason.

Third, coalition members feel that the military portrays the concerns of the protesters as trivial by branding the leaders as troublemakers. The protesters maintain that they, for their part, have consistently documented their concerns, have resorted to confrontation only to highlight the illegality of the military, and have highlighted these illegalities successfully in many cases.

Differences

Although the rural rights advocates and environmentalists have their similarities, differences are also important. The major differences are between the environmentalists and the rural rights advocates and among the environmentalists themselves. Both levels of differences are important because they highlight how groups of people with differing views have worked together to achieve consensus toward the military and compromise among themselves.

The principal differences between the environmentalists and the rural rights advocates appear to be based on the fact that private property issues and consumptive use of public lands are important to rural advocates. The following illustrates the position regarding compensation for wilderness advocates.

> If the military wants to continue [expand their use of airspace], they have to develop some type of program whereby, either they have to deal with every property owner on every route — this still doesn't deal with the wilderness issues and that sort of thing — but as far as private property, they have to deal with them, and [work out with them] whether they decide to sell their area, or lease it, or [allow the property owners to] say, yeah, we'll give it to you for 20 years, or whatever...

Legislative goals of wilderness preservation directed toward nonconsumptive uses of the land are important to the environmentalists. Thus, in the past, environmentalists have clashed with other rural residents over other land use issues. For example, they have opposed various grazing practices by ranchers, logging by timber interests, mining by mining interests, and various development projects proposed by Native Americans for their lands.

The environmentalists differ among themselves according to the degree to which they are willing to compromise with those who advocate consumptive uses of public lands. Radical environmentalists oppose consumptive use on the basis of principle. They are seen as unwilling to cooperate with other local interests and as engaging in disruptive activities rather than working through established political and legal channels. Main line environmentalists described themselves as willing to work with local and national wilderness preservation groups and rural rights advocates. They are perceived as more flexible in negotiating with the military and other groups. The tree diagram in Figure 4 defines environmentalists in general and contrasts main line and radical environmentalists. Interestingly, the term "radical environmentalist" was generally used as a disclaimer by the main line environmentalists. The researchers did not encounter anyone who fit the definition of radical environmentalist during their field work.[4]

Both levels of contrast show that neither the main line environmentalists nor rural rights advocates have reached complete agreement on many issues. They both recognize their differences, but have identified the positions toward the military on which they do agree. On the issues upon which they do not agree, they have decided not to confront each other and to operate through established channels of political representation. Both groups appear to have considered the practicalities of political representation as a very important determinant in the position they have taken. Thus, they are concerned about not only insuring that their position will be part of regional policy, but also remaining in the decision process itself.

Mitigative Measures

The rural rights advocates have responded to the problems with the military by recommending mitigative measures and lobbying for legislation to promote them. For them, mitigation must ensure that the military involves the public in any plans that might result in inverse condemnation of private and public lands or airspace. The military must similarly make explicit to the public the reasons it needs to have this additional land and airspace. The following attributes illustrate their position.

1. [The rural resident is] having this problem because of the military's failure to internalize the costs of these operations.
2. The full cost of military training is not reflected, because the military doesn't pay for all that they use and take.

---

[4] Opponents of environmentalists have accused some so-called radical environmentalists of "monkeywrenching" (i.e., vandalism). However, the numbers of such vandals and purposefulness and pervasiveness are unclear. Indeed, some environmentalists maintain accusations of "monkeywrenching" are attempts by advocates of mining, ranching, and logging to discredit environmental concerns.

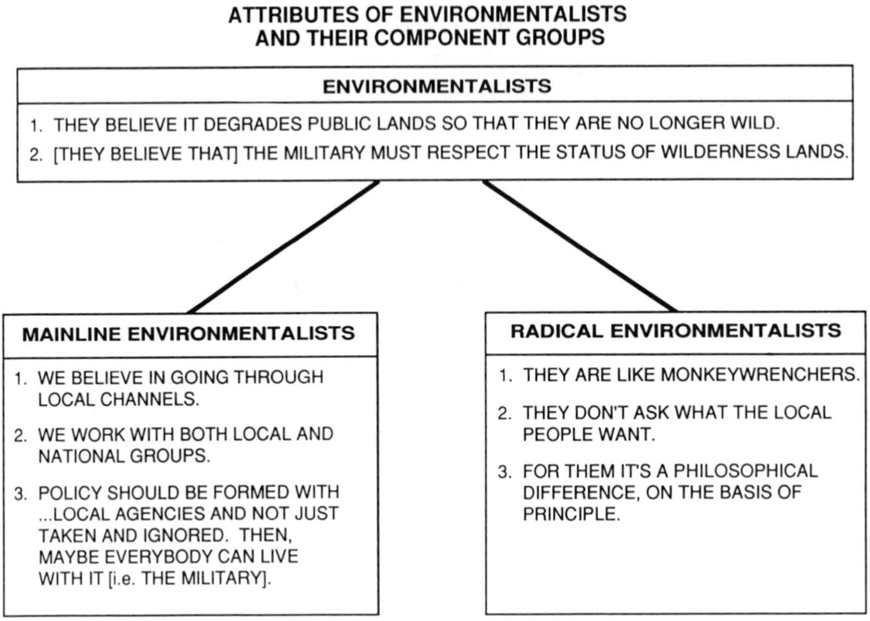

Figure 4. Attributes of environmentalists and their component groups.

3. If [internalizing the full cost] was done, the problem, to a large extent, would vanish, in my opinion.

Mitigative measures themselves include purchasing the airspace from the people affected, developing "sacrifice areas," and using more flight simulation. Speakers evaluated them according to how realistic they were, as well as how desirable.

Purchasing the airspace was considered appealing from the standpoint of principle. As one speaker maintained, "If they're going to [use airspace] and people don't have a choice, then they have to be paid for their airspace."[5] According to some speakers, not everybody might require such compensation, and many might not object at all to the use of airspace by the military. Nevertheless, they should have an opportunity to ask for compensation. Both the rural rights advocates and military officials, however, acknowledged the possibility that people would demand compensation regardless of whether or not they were adversely affected. Thus, some maintained that the approach, while morally acceptable, might be difficult to implement.

The second option, national sacrifice areas, involves purchasing the land upon which people reside, moving the people out if they wish, and using the airspace above solely for military purposes. The measure was considered good because

[5] Actually, there is no way for the military to purchase airspace. In all likelihood, the military would be able to offer some sort of compensation indirectly through another federal agency.

it would help make explicit the hidden costs to the people, and the people could be adequately compensated. Training could then proceed smoothly. The measure was considered disadvantageous because it would be a wide-sweeping measure that would still engender considerable opposition from those directly affected. There would still be winners and losers.

The third option, the use of flight simulation, involved no flying at all. Parallel to this approach were recommendations for the Air Force to fly over the Pacific Ocean where nobody was living. Neither the Air Force nor the rural rights advocates considered total reliance on either option as realistic. Although they maintained that members of the public often recommended them, these measures would not satisfy the needs of the military. As one rural rights advocate observed about flying over the Pacific Ocean, "They could also use the Pacific Ocean to fly over, but there are no mountains and canyons over the Pacific."

Many environmentalists were sympathetic with the goals and mitigative measures outlined by the rural rights advocates. The military should assure the public, through congressional review, that they are making efficient use of the airspace that they already have. If they need more land and airspace, they should document for public review why they need it. They also recommended the following specific mitigative measures. First, if more airspace is needed, wilderness character should be protected. The military should replace into the public trust land and airspace which it no longer needs. Through public involvement, wilderness character could be protected by limiting the areas over which flights occurred and by scheduling the flights to interfere minimally with periods of peak visitor use, wildlife staging and migration, severe fire danger, and other critical periods.

CONTINGENT VALUATION

This ethnography highlights elements needed to deal with bias in the study of preferences. It identifies the people to whom issues of wilderness impacts are meaningful and who are most proficient at articulating the impacts. Thus, researchers can elicit the language and meaning of the impacts. Second, by describing the social context within which the impacts are articulated, ethnography is more likely to describe the full range of impacts. These impacts refer to more than wilderness experience or wilderness character. They include the fairness with which the federal government involves the public and the way in which the military proposes to use public lands. Third, by eliciting the mitigative measures, ethnography provides part of the indirect means of studying these impacts to fairness in public involvement and to public land use issues.

The contingent valuation attempted to measure desirability, fairness, and practicality for a national sample. Studying a national sample was necessary for two reasons. First, as the contrasts between the environmentalists and rural rights advocates show, people's preferences for mitigative actions are highly dependent on their assessment of social and political realities within which they make their decisions. A better understanding of this social and political context will show whether people's preferences toward wilderness preservation are based on what they consider desirable or on what they consider realistic. Second, the ethnography also suggests that the coalition interviewed under SR-300 in Nevada had over 10 years of experience dealing with the military and the federal government on other matters. It is therefore important for social scientists to determine the degree to which other people in the United States, with different experiences, would express the same preferences as those indicated by the coalition living under SR-300.

Contingent Valuation Design

The contingent valuation questionnaire consisted of four parts: the scenarios, the willingness to pay question, the payment vehicle, and background questions. The first three are part of any contingent valuation questionnaire; the background questions reflect additions prescribed by the ethnography.

Scenarios

The scenarios were preceded by an introduction that included a description of a change in the quality of the resource, affirmation that these flights are considered essential to the maintenance of military preparedness, and affirmation that policy alternatives will require trade-offs between what the military would like to do and what people are willing to tolerate. For each scenario, the change in quality of wilderness experience is represented by changes in the altitudes, speeds, and the effects of military LAFs over wilderness areas.

Following this general description, the respondent is presented with the three scenarios themselves. These scenarios describe ways of mitigating adverse effects to wilderness enjoyment from military training LAFs. Although these scenarios were derived from the ethnographic research, they were worded in a way that would remove respondents from the immediate context of the case study areas yet still make the questions meaningful. The following are summaries of the scenarios:

1. *Restrict flying to Air Force-owned lands.* The military appropriates all lands (e.g., federal, private, or corporate owned) under certain airspaces, compensates the users at fair market value, and eliminates public access to them. They become, in effect, military reservations or "sacrifice areas."
2. *Restrict flying to nonwilderness area.* Public access to land under all airspace is not restricted, but the military is restricted from use of wilderness areas, thus requiring the establishment of new airspace.
3. *Restrict flying over wilderness areas.* LAF activities over national parks and wilderness lands are restricted to a minimum altitude of 2000 ft above ground level, no more frequently than one flight a week, and to an area amounting to no more than 50% of the designated land area of the national park or wilderness. Because of these restrictions, the Air Force may legitimately require more airspace.

Each of these three scenarios provided the respondents with contextual information about the adverse effects to wilderness enjoyment resulting from military use. The respondents were reminded that each mitigative approach involves additional costs (e.g., in the form of designating additional airspace).

Willingness to Pay and Payment Vehicle

The survey then ascertained the respondents' willingness to pay for each kind of mitigative action by asking "would you approve of $X from your federal tax payments being used annually to support and manage a program to [action] over National Parks and Federal Wilderness Areas?" The dollar amount was a randomly generated monetary value between 20 and $200. The action corresponded to each of the three hypothetical scenarios.

Phrasing this question incorporated both theoretical issues in economics and statistical considerations regarding starting point bias. The theoretical issues emerged during a focus group discussion intended to evaluate how meaningful a draft form of this survey was to the public. Speakers warned the researchers that they would refuse to cooperate if the payment vehicle involved raising taxes. They were opposed to any form of tax increase, even a hypothetical one. The researchers responded by proposing a transfer among uses of taxes already paid. Although this response makes the willingness to pay question more acceptable to the respondents, it does so at the expense of theoretical rigor. Mitchell and Carson (1989) observed that if people do not have to consider an immediate impact to their pocketbooks, they will tend to express higher willingness to pay. We acknowledge this possibility, but considered the trade-off necessary in order to obtain respondent cooperation.

Starting point bias (i.e., determining the appropriate level of payment) results from suggesting to a respondent a payment amount which is higher or lower than what they would have considered without a starting point. If the starting point bias exists, the answers may thus not be true representations of what the respondent would prefer to pay. Starting point bias occurs particularly when iterative bidding procedures are used. In iterative bidding. the interviewer and respondent take turns proposing payment and indicating acceptance of payment (Cummings et al. 1986).

The researchers decided on a dichotomous choice bid or a so-called "take it or leave it" approach rather than iterative bidding. Here, the respondent is offered a monetary sum. The researchers assume that a "yes" answer means that the respondents are willing to pay at least this sum. A "no" answer is assumed to mean that the respondent is willing to pay less than this sum or is unwilling to pay any sum at all. This response has been accepted as a viable alternative to starting point bias (Bishop and Heberlein 1979; Boyle et al. 1985) because the respondent can simply say yes or no. It is also more suitable for the mailed questionnaire used in this research.

The use of a dichotomous choice bidding process necessitated the use of the logit function to estimate the dollar amount people say they are willing to pay.

Background Questions

Background questions elicit information about factors that could affect the respondent's willingness to pay. These questions include respondent's exposure to flights, attitudes toward flights during their everyday lives and during wilderness experience, willingness to avoid military flights over wilderness altogether, how familiar they are with the effects of these flights, and whether or not they express concern about the environment. Concern for the environment is a global term that suggests that the respondent is concerned with impacts other than direct effects on wilderness use. In addition to these questions, socioeconomic and demographic questions were also included to obtain information about age, sex, household income, military service, and geographic location.

Results of the Contingent Valuation

Researchers drew a national sample of 12,000 from a demographically balanced population of 250,000 panel members. These members were accessible to Market Facts, Inc., a marketing research firm. The research firm then mailed the questionnaire to these panelists.

The analysts received 8900 responses. From this sample, they drew subsamples of wilderness users who experienced LAFs, wilderness users who had not experienced them, and nonusers. Because the logit function was used, a minimum of 300 individuals from each of these categories was needed to guarantee the reliability of the statistical analysis. Of the 8900 responses, 323 users had experienced LAFs; a stratified random sample of the remainder of the total sample was drawn, composed of 334 users who had not experienced flights and 340 nonusers. The sample of 997 is thus used for statistical analysis.

The researchers hypothesized that five factors would influence the willingness to pay for the hypothetical policy changes described in the three scenarios: (1) wilderness users who experienced overflights would be willing to pay more for mitigative measures than wilderness users who had not, and both groups would be willing to pay more than nonusers; (2) individuals indicating concern for the environment would be willing to pay more than those who did not; (3) past military service is negatively related to willingness to pay; (4) the higher price tag is negatively related to an individual's willingness to pay; and (5) household income level is positively related to the willingness to pay.

Hypotheses 1 and 2 would support the contention that actions taken to mitigate wilderness degradation are valued primarily by wilderness users. They were formulated from the ethnographic study. Hypothesis 3 (derived from the focus group discussion) supports the contention that those with military service backgrounds would tend to be more supportive of military initiatives (and less willing to pay). Hypotheses 4 and 5 are economic considerations.

A preliminary test determined no significant difference in the willingness to pay across the three strata, i.e., users who had experienced LAFs, users who had not, and nonusers. Thus, the willingness to pay to mitigate adverse impacts arising from military LAFs does not appear to differ between any of the wilderness users and the nonusers. Hence, the logit equation estimated is of the following form:

$$\ln(\mathrm{WTP}) = \ln(\mathrm{Constant}) + b_1 \ln(\mathrm{Price\ Tag}) + b_2 \ln(\mathrm{Concern\ for\ the\ Environment}) + b_3 \ln(\mathrm{Military\ Service}) + b_4 \ln(\mathrm{Income})$$

In this equation, WTP, or willingness to pay, is an indicator of an individual's response to the price tag offered in the payment vehicle. It is not interpreted as a valuation, but rather the *probability* that the individual would pay the maximum dollar amount rather than forgo the opportunity to implement the alternatives described in the scenarios. The variables "concern for the environment" and "military service" have been discussed above. The variable "income" represents household income in 1988.

The result of the estimation revealed that for Scenario 1 (the "restrict flying to Air Force owned lands" or "sacrifice area" scenario), only environmental concern and income were significant (at the 0.5 level); price and past military service were not. For Scenario 2 (the "restrict flying to nonwilderness areas" scenario), environmental concern and past military service were significant, but income was not. For Scenario 3 (the "restrict flying over wilderness lands" scenario), all variables were significant.

Furthermore, the results of the contingent valuation survey questions confirmed the ethnography for a larger population of wilderness users in two major ways. First, responses to open-ended questions about annoyance showed that users who experienced these flights tended to cite as bothersome the violation of solitude, followed by wilderness character, noise, and wildlife disturbances. Nonusers and users who had not experienced flights tended to cite noise. Thus, the respondents defined direct impacts of LAFs over wilderness as an issue pertaining to wilderness, not as part of a wider annoyance of LAFs over places of residence, work, or nonwilderness recreation.

Second, the evaluation of the trade-offs suggested in the ethnography is similar to the ranking of the scenarios. Average willingness to pay for Scenario 1, the "sacrifice areas," was $121.72. Scenario 2, exclusion from wilderness, was favored slightly more (WTP = $125.19). Scenario 3, the limiting of flights, was favored the most (WTP = $132.07).

Respondents were also asked how acceptable was each scenario as a means of reducing conflict between the needs of the military and adverse environmental impacts. Over half (51.1%) considered Scenario 1 unacceptable, and 42.5% considered Scenario 2 and 30.8% considered Scenario 3 as unacceptable. An average of only 15.2% found any of them very acceptable, and 24.9% found them somewhat acceptable. In short, they appear to be making the best of what they see as a bad situation and are judging the scenarios on the basis of practicality as well as desirability.

It is the way that the survey results did not support the ethnography that we also find important. First, since the willingness to pay did not differ significantly among the three different strata, the results suggest that preservation of wilderness as a public resource is important to the public at large, not to a small group of users. These results seem to refute what some in the ethnography claimed and what might be assumed by some decision makers — that only users attach a value to wilderness. The public may indeed attach an existence value to wilderness whether or not they use it.

Finally, it should be noted that the total willingness to pay is a substantial amount of money. If all persons 18 years and older in the United States were to pay the sum indicated for average willingness to pay in each of the three scenarios, the total amounts would be approximately 21.9, 22.5, and $23.8 billion, respectively. Although these averages are only about 10% of the $233.3 billion spent by American households for all kinds of recreation, it is almost 14 times the $1.6 billion budgeted by the federal government (U.S. Statistical Abstract 1987) for the upkeep of national parks and wilderness areas. Thus, although both the aggregate and average willingness to pay must be interpreted with caution, particularly because of the payment vehicle design (i.e., the reallocation of existing tax revenues rather than increased taxes), the amounts reported are substantial enough to warrant further investigation.

CONCLUSION

We acknowledged that defining and measuring direct impacts to wilderness character and wilderness experience are difficult, and we proposed that measuring preferences for these resources may provide one reasonable approach to a solution. We further proposed that the measurement of preferences is most easily achieved by eliciting from people how much they value mitigating adverse impacts to the resources. Such a value could be stated as if these mitigative measures could be bought and sold on the market. However, respondents were asked, in effect, to make their best judgement of what to do about a change in the environment with which they may have had no direct experience or which may not happen at all. Moreover, they were being asked to judge this change in market terms, regardless of moral issues, social responsibility, or the feasibility of the particular transfer payment (i.e., the reallocation of existing revenues rather than additional taxes). The possibilities of bias, distortion, and outright refusal to cooperate are thus myriad.

We then proposed to address the problem of bias by the use of cognitive ethnography, a research approach used by anthropologists and linguists to describe the knowledge base of people through the language they use. Through this approach, we conducted a case study to identify the people who had an interest in wilderness preservation.

In addition to describing how they perceived the impacts, speakers in the ethnography also described other individuals and organizations who were important in influencing how their perceptions were formed and articulated. By considering the perceptions of these others, we were able to identify impacts to wilderness character and experience, and to the public involvement processes and wider public land use issues. By considering this expanded set of impacts, it was possible to identify mitigative measures.

The contingent valuation showed that information reflecting public involvement and public lands use could be incorporated into studies of preference. This combination of the ethnography and the survey showed that wilderness preservation advocates are advocates for a public good. The ethnography had already shown that at least some wilderness protection organizations had, in a sense, earned the right to such public advocacy through consensus and compromise with rural rights advocates with whom they have been in opposition on other matters regarding public lands use. The contingent valuation established this validity for a wider population. It demonstrates that the public finds the results of such consensus and compromise acceptable when wilderness is confronted with such initiatives.

Historically, wilderness preservation advocates have clashed with the conservationists, and the debates continue as new projects are proposed for America's dwindling public lands. As far as the military is concerned, however, wilderness preservation advocates find common cause with rural rights advocates, many of whom may be conservationists and consumptive use advocates rather than preservationists.

To involve these advocates for a public good, the environmental impact assessment process will have to become concerned with more systematic and meaningful public involvement. Social science methods are available to meet this need. Eventually, such involvement will make more explicit many hidden adverse (and maybe even beneficial) impacts and thus costs (and benefits). Although this explicitness may upset or delay many projects, or may result in potentially fundamental changes in the proposed action, failure to acknowledge public involvement may be far more harmful in the long run, even if an admittedly small proportion of the public is directly affected.

REFERENCES

Associated Press Release. 1990. "Environmentalists and Ranchers Fight Pentagon Bid for More Land." *New York Times,* January 4.

Bishop, R. C., and T. A. Heberlein. 1979. Measuring values of extra-market goods: Are indirect measures biased? *Am. J. Agric. Econ.* 61(5):926–930.

Boyle, K. J., R. C. Bishop, and M. P. Welsh. 1985. Starting point bias in contingent valuation surveys. *Land Econ.* 61:188–194.

Cantor, R. A., G. M. Schoepfle, and E. J. Szarleta. 1992. Sources and consequences of hypothetical bias in economic analyses of risk behavior. In B. J. Garrick, and W. Gekler (eds.), *The Analysis, Communication, and Perception of Risk.* Plenum Press, New York (forthcoming).

Chomsky, N. 1959. Review of verbal behavior by B. F. Skinner. *Language* 36:26–58.

Chomsky, N. 1969. *Language and Mind.* Harcourt, Brace and World, New York.

Cummings, R. G., D. S. Brookshire, and W. D. Schulze. 1986. *Valuing Environmental Goods: An Assessment of the Contingent Valuation Method.* Rowman and Allanheld, Totowa, NJ.

Edelstein, M. R. 1988. *Contaminated Communities: The Social and Psychological Impacts of Residential Toxic Exposure.* Westview, Boulder, CO.

Freeman, M. A. 1979. *The Benefits of Environmental Improvement: Theory and Practice.* Johns Hopkins University Press, Baltimore, MD.

Goodenough, W. H. 1956. Componential analysis and the study of meaning. *Language* 38:195–216.

Hanemann, W. M. 1984. Welfare evaluation and contingent valuation experiment with discrete responses. *J. Agric. Econ.* 66:332–341.

Kealy, M. J., J. F. Dovidio, and M. L. Rockel. 1988. Accuracy in valuation as a matter of degree. *Land Econ.* 64:2:158–171.

Khilstrom, J. F. 1987. The cognitive unconscious. *Science* 237:1445–1452.

Lorenz, F. M. 1982. Determination of just compensation in cases of inverse condemnation based on aircraft overflight. M.S., U.S. Marine Corps.

Marston, E. 1988. Cutting the apron strings. *High Country News* 20:12–13.

Mitchell, J. C. 1969. Theoretical orientations in African urban studies. *The Social Anthropology of Complex Societies.* Tavistock, London.

Mitchell, J. G. 1990. War in the woods II: West side story. *Audubon* 92(1):82–121.

Mitchell, R. C., and R. T. Carson. 1989. Using Surveys to Value Public Goods: The Contingent Valuation Method, Resources for the Future. Washington, D.C.

O'Toole, R. 1988. Reforming the Forest Service. Washington, D.C., Island Press. Reviewed in Bruce Farling, Can the Forest Service Turn Entrepreneurial? *High Country News* 20(13):14–15.

Radford, J. 1986. *The Chaco Coal Scandal: The People's Victory Over James Watt.* Rhombus Publishing Co., Corrales, NM.

Sugden, R., and A. Williams. 1978. *The Principles of Practical Cost-Benefit Analysis.* Oxford University Press, Oxford.

U.S. Statistical Abstract. 1987. Table 490, pp. 304–305.

Watkins, T. H. 1989. Untrammeled by man: The making of the wilderness act of 1964. *Audubon* 91(6):74–91.

Werner, O., and G. M. Schoepfle. 1987. *Systematic Fieldwork, Vol. 1: Foundations of Ethnography and Interviewing.* Sage Publications, Beverly Hills, CA.

Using the Analytical Hierarchy Process in NEPA-Based Public Involvement: A Profile of Success

M. M. Landes and D. R. Pescitelli, Illinois Department of Transportation, Springfield, IL

ABSTRACT

Illinois highway planners successfully used the Analytical Hierarchy Process (AHP) and EXPERT CHOICE™ microcomputer software to evaluate five alternate highway locations and determine the optimal location for a new 22-mi freeway in the Metro-East area of St. Louis. These techniques were used in conjunction with a structured public involvement process that included four citizen groups, each with a different focus, and an advisory committee made up of representatives of these groups. The results gave the rank and relative preference of each location for each of the citizen groups, as well as an overall benefit-cost ratio for each alternate. This information aided planners, engineers, and managers designing the location of the highway and helped citizens better articulate their own concerns and understand the interests of others. This case study demonstrates AHP's adaptation to the National Environmental Policy Act (NEPA) process in an exemplary program of successful public involvement for a major public works project.

HIGHWAY PROJECT OVERVIEW

The project involves construction of approximately 22 mi of four-lane, fully access-controlled divided highway in Madison County, IL. Madison County lies in the southwestern part of the state and is part of the St. Louis metropolitan area.

Its 1980 population was 248,000; the population of communities in the study corridor is approximately 98,000.

The proposed project was developed in response to existing and projected traffic volumes for the study area. Consequently, the two major beneficial impacts of the action would be (1) to relieve traffic congestion on existing routes through a redistribution of traffic patterns and (2) to promote fast, safe, and efficient travel within and through the study area by providing a continuous north-south route. Furthermore, the completion of the proposed highway and the subsequent change in traffic patterns would encourage economic growth and development in adjacent communities. Future commercial, light industrial, and residential development would be influenced by the location of the new highway.

Although alternatives other than building a new highway were considered, this paper focuses on five geographic alternates for building such a highway. The five alternates considered share a common alignment in the southern and northern parts of the project corridor. In the middle portion of the corridor, Alternates 1–4 are in a western locale, whereas Alternate 5 has an eastern location. The western alternates cross an existing urbanized area, while the eastern alternate crosses primarily agricultural lands. The eastern and western alternates differ significantly in their effects on the natural and manmade environment, as well as in the traffic and transportation service they will provide to the area. Although the western alternates would provide better traffic service to the most congested portion of the study area, these alternates would also result in a much higher number of residential and commercial displacements. The eastern alternate would have a much greater effect on agricultural operations.

The social impact of the proposed action would vary depending on the alternate selected. In general, the western alternates, which cross the more urbanized parts of the study area, would result in more residential and commercial displacements, create more neighborhood disruption, and cause proximity impacts such as increased noise levels for residences left near the new highway. The major impacts of the alternate alignments are shown in Table 1.

OVERVIEW OF THE ANALYTICAL HIERARCHY PROCESS

The Analytical Hierarchy Process (AHP) is a decisionmaking technique developed at the Wharton School of Business by Dr. Thomas L. Saaty, an applied mathematician. The AHP is based principally on utility theory, which allows a decision maker to express preferences for alternate courses of action with respect to defined evaluation criteria. Evaluation criteria are first ranked in order of importance according to a numerical scale, and then the same scale is used to measure the preference of one alternative over another for each of the evaluation criteria. These judgments are developed using pairwise comparisons among all the criteria and alternatives being evaluated. Using complex matrix algebra, normalized numerical scores that capture all these intensity judgments are com-

Table 1. Important Highway Project Impacts Used in Environmental Impact Statement and Analytical Hierarchy Process.

Growth and Development Characteristics
Retention of existing business and industry
Expansion of existing business and industry
Recruitment of new business and industry
Expansion of existing residential developments
Development of new residential subdivisions
Community Aspects
Increased public revenues
Increased ability to control land use in accordance with plans
Increased multicommunity cohesion and regional identity
Enhanced access to public facilities and services
Environmental Impacts
Streams and wetlands
Surface waters
Pollution (including air and noise)
Visual impacts (including views of and from the road)
Wildlife
Woodlands
Agriculture Impacts
Prime farmland converted to other uses
Severance management zones created (acreage)
Conversion of designated agricultural preservation areas to other uses
Adverse travel by farmers
Farm residential displacements
Farm building displacements
Socioeconomic Impacts
Urban displacements (residential, commercial, public facilities)
Schools (including adverse impacts on school district populations, adverse travel for buses, and visual and audible intrusions after mitigation)
Neighborhoods (including severance, proximity impacts, and adverse travel)

Note: Impacts not included above would either not occur or would be insignificant for any of the alternate highway locations under consideration.

puted for each alternative under consideration. These final scores reflect the relative preference of each alternative under consideration.

The AHP consists of several parts. The first step is construction of a functional hierarchy as a method to organize and display the complexity of factors and issues related to an overall goal. The hierarchy is constructed using a series of levels including the goal, evaluation criteria, subcriteria, and alternates (see Figure 1).

Once the hierarchy is defined and relationships of each element in each level are defined, pairwise comparisons of all elements in a level are made with respect to the factors in the hierarchy directly above them. Saaty (1986) has developed a nine-point scale that is used in making these comparisons (see Table 2).

Finally, to obtain the overall relative rank of alternatives for a goal or decision, the results of the paired comparisons or judgments are mathematically combined (i.e., synthesized). Besides reporting the relative priorities of the alternatives, the process yields an inconsistency index that points out important discrepancies in judgments so that they can be reconsidered.

For projects such as this one, two separate hierarchies can be constructed: one for benefits and another for costs or adverse effects. Then, a ratio of benefit-to-

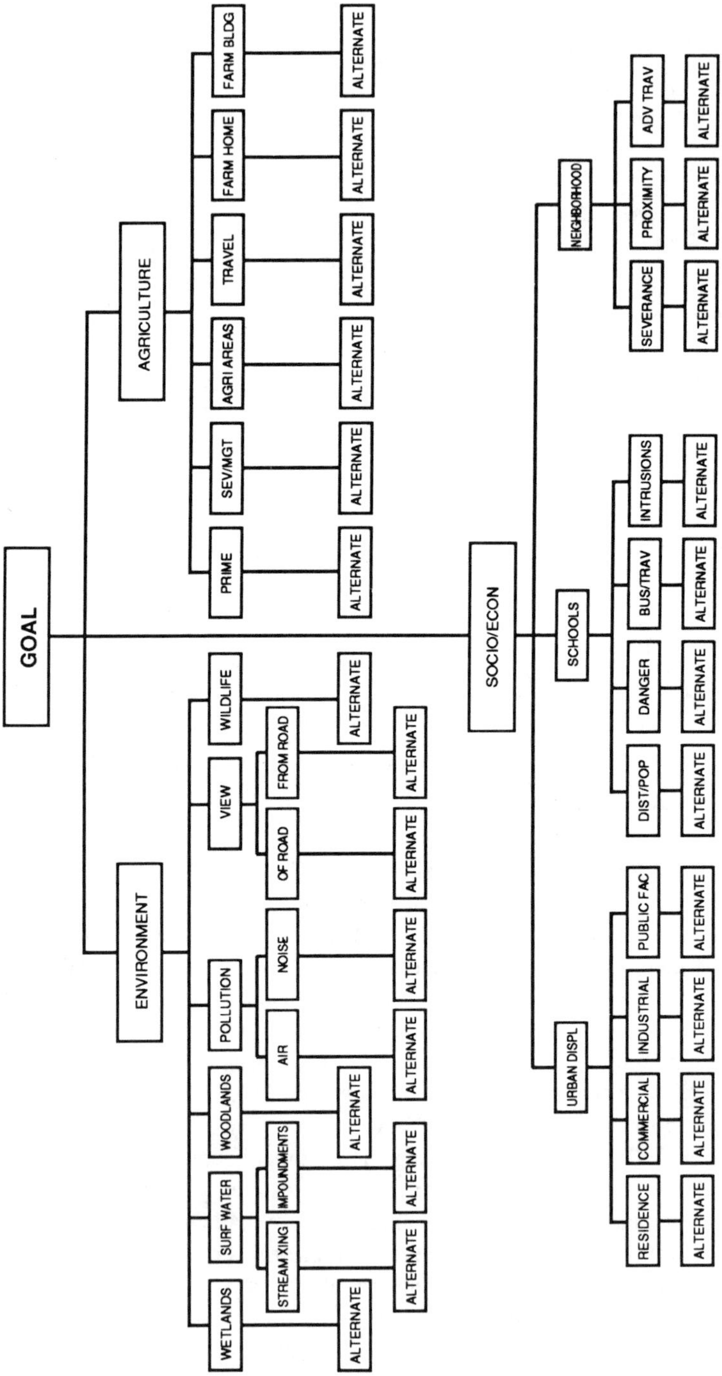

Figure 1. Generalized functional hierarchy.

Table 2. Intensity Scale Used in Assigning Importance or Preference Ratings in AHP Pairwise Comparisons.

Intensity of Importance	Definition	Explanation
1	Equal importance	Two activities contribute equally to the objective
3	Weak importance of one over another	Experience and judgment slightly favor one activity over another
5	Essential or strong importance	Experience and judgment strongly favor one activity over another
7	Demonstrated importance	An activity is strongly favored, and its dominance is demonstrated in practice
9	Absolute importance	The evidence favoring one activity over another is of the highest possible order of affirmation
2, 4, 6, 8	Intermediate values between the two adjacent judgments	

Source: Saaty, T. L. 1986. *Decision Making for Leaders.* University of Pittsburgh. With permission.

cost scores for each alternative can be computed. These scores help to identify the relative trade-offs involved in selecting one alternative over another.

The AHP theory and calculation procedures are made more accessible by user-friendly software for the personal computer developed by Dr. Ernest Forman. The program, called EXPERT CHOICE™, is available from Decision Support Software, Inc. of McLean, VA.

INFORMATION EXCHANGE REVIEW

The NEPA procedures are intended to ensure that environmental information is available to public officials and citizens before decisions are made. The policy of the Federal Highway Administration (FHWA) is that decisions should be made in the best overall public interest and that other agencies and the public should be involved early enough in the decisionmaking process to influence studies and final actions.

To be effective, public involvement needs to be an integral part of the highway planning process, beginning at the earliest stages and ensuring adequate opportunity for citizen input and exchange of viewpoints throughout project development. The FHWA's position is that public involvement is an integral part of the environmental impact assessment process of gathering data concerning social, economic, and environmental impacts of proposed actions. At times, data on social impacts in particular are available only from the public (U.S. Department of Transportation 1988).

To enhance public participation for the FAP 413 highway project, the Illinois Department of Transportation (IDOT) began organizing an information exchange in July 1986. The information exchange consisted of two components: four citizen working groups and an advisory committee (see Figure 2).

The goals of the information exchange were

- to allow groups with divergent viewpoints equal access to project information and individuals planning the project
- to provide a means for the IDOT to obtain public concerns and possible solutions
- to attempt to reach a consensus for the optimal highway location

With these goals in mind, project planners at IDOT identified organized groups and other interests potentially affected by the location of the proposed highway. These groups and interests were combined according to their common concerns into groups small enough to be effective as a forum for discussion and work. Accordingly, the Public Interests, Agriculture, Public Officials, and Growth and Development working groups were established.

The Public Interests Working Group was formed to represent environmental, community, and neighborhood interests. The Agriculture Working Group was organized to represent interests concerned with the conservation and preservation of agricultural resources. The Growth and Development Working Group was created to represent local and regional organizations and interests concerned with land use and transportation planning; economic development; residential, commercial, and industrial development; and local and regional commerce. Finally, the Public Officials Working Group was organized to represent local elected public officials.

The overall objectives of the working groups were to (1) evaluate the alternate highway locations developed by the IDOT and discuss and comment on impacts, methods of analysis, and evaluation procedures and (2) assess the relative merits of the alternates, identify advantages and disadvantages of them, and suggest ways to make the alternates more acceptable.

To simplify the process and to shorten the time required to accomplish the objectives of the working groups, the groups limited their assessment of the alternate highway locations to their areas of interest and expertise. A brief constitution and set of bylaws governed the activities of each working group and the advisory committee.

The advisory committee was formed to integrate the interests and the values of the working groups into the planning process, to facilitate the information exchange, and to guide the overall process. Each of the working groups selected two individuals to represent its interests on the advisory committee.

USING AHP

The six basic steps in using the AHP follow:

- identify the goal or objective
- identify the alternatives and the evaluation criteria

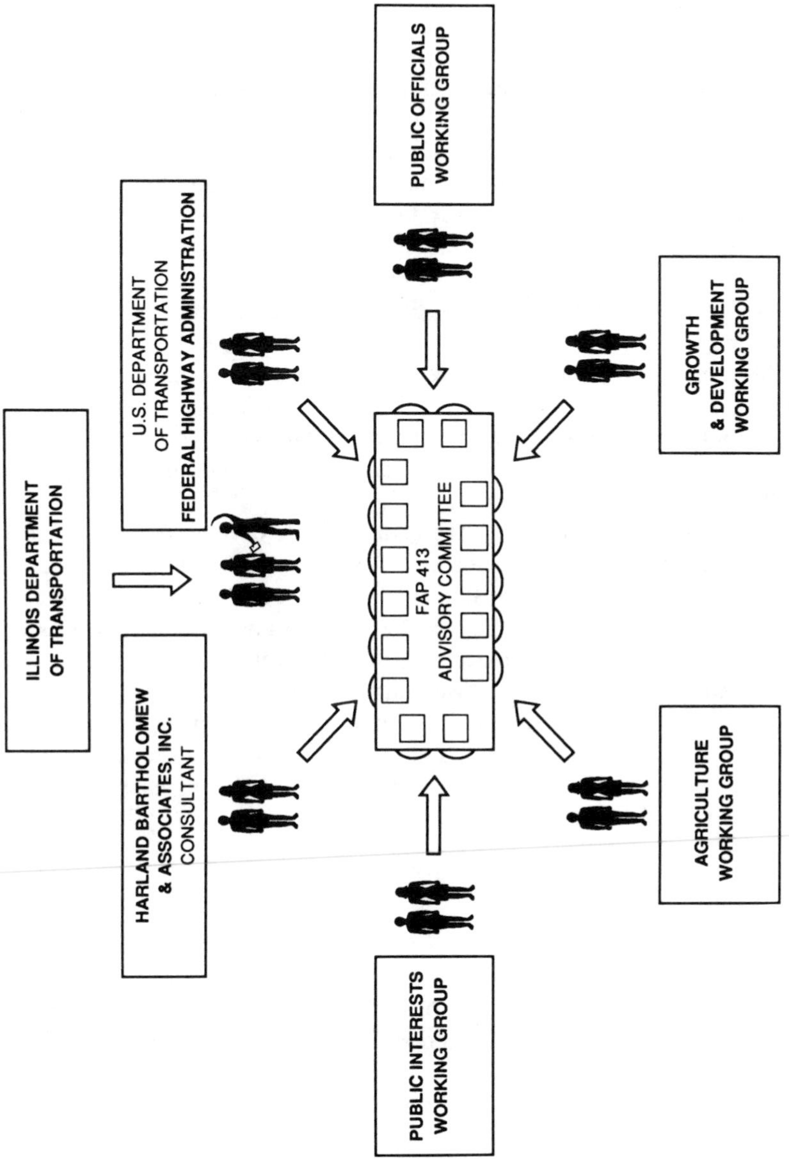

Figure 2. Information exchange.

- research the alternatives
- construct the hierarchy
- make judgments
- synthesize the judgments

Identify the Goal or Objective

The goal was stated as "finding the best location for the highway." However, because any of the alternative locations would have costs as well as benefits, the problem was restated with two dimensions instead of one to accommodate the analytical hierarchy process and to facilitate a benefit-cost trade-off analysis. The goal thus became "finding the highway location with the greatest benefits" and "finding the highway location with the greatest costs."

Identify the Alternatives and the Evaluation Criteria

In this step, project planners analyzed the objectives of each of the four working groups and listed the social, economic, and environmental effects to be addressed in the environmental impact statement. This analysis yielded a list of factors used in building the hierarchies to evaluate the alternate highway locations (see Table 1).

Numerous possible locations for the highway were developed and evaluated as part of the project planning process. These alternates consisted of unique segments that when combined into all possible combinations between the termini, constituted 80 different possible highway locations. An interdisciplinary team of engineers, planners, and environmental specialists evaluated and analyzed the potential alternates. Several alternates were eliminated for obvious social, economic, or environmental conflicts; similarity to other more desirable segments; deficiencies in geometric characteristics; or the ability to provide a desired level of traffic service. Following several lengthy comparative analysis sessions and three public informational meetings, five alternates remained for evaluation by the working groups.

Research the Alternatives

Research on the alternates consisted of various quantitative measures consistent with the specified evaluation criteria. These measures included cost, engineering data, and a broad spectrum of social, economic, and environmental impacts. The information became more specific and accurate as the number of alignments under consideration decreased and possible locations became more precise. The most important comparative data for the five alternate locations are listed in Table 3.

Construct the Hierarchy

The next step was to construct the hierarchies from the previously identified evaluation criteria and alternate locations. The goal statement for the first hierarchy

Table 3. Comparative Data for Alternate Highway Locations.

Item Length of Highway	Alt.1 20.6	Alt.2 20.6	Alt.3 20.8	Alt.4 20.8	Alt.5 24.1
Right-of-Way Needed (acres)					
Prime and important farmland	650	657	652	657	872
Other farmland	166	162	183	181	172
Residential land	57	53	59	55	36
Commercial land	13	15	14	15	12
Public/quasi-public land[a]	14	10	14	10	7
Recreational land	13	13	13	13	14
Other land	32	32	30	30	11
Existing rights-of-way[b]	68	70	71	73	72
Total	1013	1012	1036	1034	1196
Displacements					
Nonfarm residential	98	110	102	114	29
Farm residential	8	8	5	5	13
Commercial buildings	11	13	11	13	4
Businesses	12	21	12	21	4
Public/quasi-public	1	1	1	1	1
Agriculture					
Affected[c] farms	82	82	78	78	99
Severed farms	50	53	48	51	60
Ecological Resources					
Woodlands taken (acres)	189	187	181	179	193
Wetlands taken (acres)					
Emergent	0.25	0.25	0.25	0.25	0.25
Forested	7.0	7.0	7.0	7.0	4.8
Wetlands affected[c]					
Emergent	1	1	1	1	1
Forested	4	4	4	4	4
Major stream crossings	3	3	3	3	3
Other Properties Affected[c]					
Churches	5	5	5	5	2
Schools	3	3	3	3	1
Private recreation areas	3	3	3	3	2
Estimated Cost ($million)					
Right-of-way	14.0	16.2	14.1	16.3	7.6
Construction	178.8	176.7	178.2	176.1	172.0
Total	192.8	192.9	192.3	192.4	179.6

[a] Refers to churches, schools, hospitals, and other similar land uses.
[b] Rights-of-way on existing public roads and railroads.
[c] "Affected" connotes properties from which right-of-way will be required.

was to find the highway location with the greatest benefits. These benefits consisted of growth and development characteristics and desirable community enhancement factors (see Figure 3). The other hierarchy's goal was to find the highway location with the greatest cost (see Figure 4). These costs consisted of environmental impacts, adverse socioeconomic impacts, and agricultural impacts.

At first, the hierarchy goal of finding the alternate location with the greatest costs appears counterintuitive. Specifically, a more logical goal might seem to be finding the alternate location with the least cost. It is possible and proper to construct a hierarchy to yield this information. However, when constructed in this way, the numerical relative preference scores of AHP cannot be used to develop

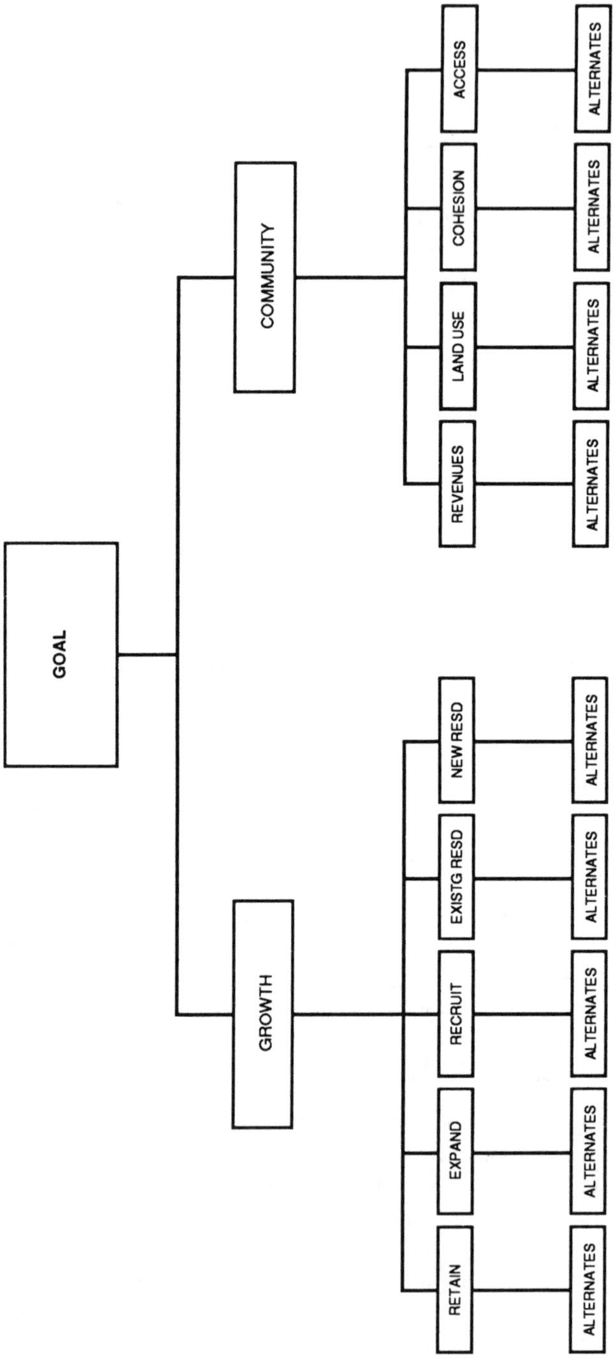

Figure 3. Find the alternate with the greatest benefits.

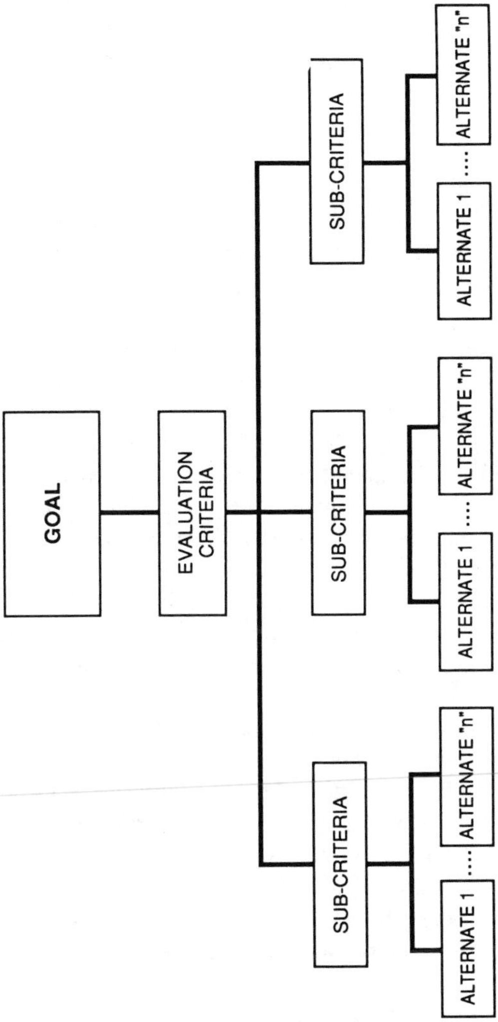

Figure 4. Find the alternate with the greatest costs.

benefit-to-cost ratios or other marginal benefit-cost measures. The analysis undertaken in this project is consistent with the ultimate goal of finding not the most beneficial or the least costly location, but rather identifying the one that yields the most benefits for the costs involved.

Absent from either the benefits or costs hierarchies were quantitative measures of road user benefits and construction costs. These quantitative factors were addressed in a separate analysis that involved calculating the net present benefits after allowing for construction and maintenance costs, consistent with the time value of money (Illinois Department of Transportation 1989).

Make Judgements

The working groups then examined the portions of the hierarchies for which they were responsible and made pairwise comparisons of alternatives with respect to the applicable criteria and subcriteria. This involved examining pairs of evaluation criteria and assessing the relative importance of one over another using Saaty's numerical intensity scale.

For example, refer to the Environmental Impacts evaluation criteria evaluated by the Public Interests Working Group in Table 4. The first phase of the judgments involved establishing the relative importance of the impact subcriteria. Each participating member of the working group made pairwise comparisons of these factors and rated the relative importance (or equality) of one vs another, exhausting all the possible paired combinations. These judgments were recorded on forms and returned to IDOT staff, who then entered the numerical equivalents of these judgments into EXPERT CHOICE™ using Saaty's intensity scale. A built-in geometric mean calculator in the program yielded a single average judgment for each comparison pair. These average judgments were then synthesized mathematically to produce the relative importance rankings for each set of evaluation criteria shown in Table 4.

The next phase of judgments consisted of making pairwise comparisons of alternative highway locations with respect to each subcriterion listed in Table 4. Similarly, the relative preference of one alternate location vs another was made for all possible combinations of alternates with respect to each subcriterion listed.

Where quantitative data were available, they were used in defining the preference ratios for alternates. For those factors not readily measured quantitatively, the judgments consisted of subjective responses based on past knowledge or experience of the evaluators. Similar to the exercise in the first phase of making judgments, forms were given to group members who participated in the exercise. EXPERT CHOICE™ was then used to calculate the geometric mean of the subjective responses. These mean responses became the values entered into the matrix of preference and importance scores, which were then mathematically synthesized.

Synthesize the Judgements

With EXPERT CHOICE™, a synthesis of judgments can be made at any level in the hierarchy. Mathematically, synthesis involves calculating the eigenvectors

Table 4. Intermediate Synthesis of Highway Evaluation Criteria by Working Groups.

Benefits	Relative Importance
Growth and Development Working Group	
Growth and Development Characteristics	
Recruitment of new business and industry	27
Development of new residential development	24
Expansion of existing businesses and industry	20
Retention of existing businesses and industry	15
Expansion of existing residential development	14
Total	100
Public Officials Working Group	
Community Management Concerns	
Increased public revenues	31
Enhanced access to public facilities and services	30
Increased sense of regional identity	21
Increased ability to control land use in accordance with plan	18
Total	100
Costs	**Relative Importance**
Public Interests Working Group	
Environmental Impacts	
Increased air and noise pollution for affected neighborhoods	51
Adverse visual impacts on affected neighborhoods	14
Impacts on wildlife	11
Impacts on streams and wetlands	09
Impacts on surface waters	08
Impacts on woodlands	07
Total	100
Socioeconomic Impacts	
Adverse impacts on neighborhoods	63
Adverse impacts on educational facilities	20
Displacements in urban areas	17
Total	100
Agriculture Working Group	
Agricultural impacts	
Farm buildings displaced	22
Farmer adverse travel	20
Farm homes displaced	18
Amount of prime farmland taken	15
Impacts on agricultural preservation zones	14
Amount of severance management zones created	11
Total	100

for all the preference and importance matrices and then normalizing these to add up to 100. Table 4 presents the results of the intermediate synthesis that showed the relative importance of the evaluation subcriteria on a scale of 1 to 100. The results were then synthesized to show the relative preference for highway locations of each of the four working groups. For this analysis, each of the major evaluation criteria (i.e., environmental impacts, social impacts, growth and development characteristics, and others) was assumed to have equal importance. The results are shown in Table 5.

Table 5. Synthesis of Judgments Showing Relative Rank of Alternate Highway Locations for Each Evaluation Criterion.

Alternates	1	2	3	4	5
Benefits: Growth & Development Working Group					
Growth and Development Characteristics					
Recruitment of new business and industry	37	22	21	15	05
Opportunities for new residential development	17	17	31	25	10
Expansion of existing business and industry	42	24	17	13	04
Retention of existing business and industry	37	30	14	15	04
Expansion of existing residential development	25	20	28	22	05
Overall ranking for growth and development	32	22	22	18	06
Benefits: Public Officials Working Group					
Community Management Concerns					
Increased public revenues	42	28	15	11	04
Enhanced access to public facilities and services	53	25	13	07	03
Increased sense of regional identity	49	25	15	08	04
Ability to control land use in accordance with plans	46	23	10	10	10
Overall ranking for community management	47	25	14	09	05
Costs: Public Interests Working Group					
Environmental Impacts					
Increased air and noise pollution	40	12	34	11	03
Adverse visual impacts	37	15	32	13	03
Impacts on wildlife	28	11	29	13	19
Impacts on streams and wetlands	19	18	19	18	26
Impacts on surface waters	20	20	17	16	27
Impacts on woodlands	20	20	20	20	21
Overall adverse environmental impact ranking	33	14	29	13	10
Socioeconomic Impacts					
Adverse impacts on neighborhoods	36	13	35	13	03
Adverse impacts on educational facilities	35	14	34	14	03
Displacements in urban areas	21	25	22	26	06
Overall adverse socioeconomic impact ranking	33	15	33	15	04
Costs: Agriculture Working Group					
Agricultural Impacts					
Farm buildings displaced	17	17	15	15	36
Farmer adverse travel	16	16	20	20	27
Farm homes displaced	21	15	15	15	33
Amount of prime farmland taken	19	19	19	19	24
Impacts on agricultural preservation zones	15	15	15	15	39
Severance management acreage created	17	18	18	20	27
Overall adverse agricultural impact ranking	18	17	17	17	31

Note: Some rows may not add to 100 due to rounding.

BENEFIT-COST ANALYSIS

Because the costs and benefits were contained in separate hierarchies, AHP and EXPERT CHOICE™ were then used to compute single summary benefits and costs scores for each of the five alternates. As noted above, these scores assume that adverse effects (costs) on the natural environment, social, and agricultural concerns are weighted equally. Similarly, positive effects on growth and development and community concerns are rated equally. The scores for each alternate are listed in Table 6. A higher benefits score denotes greater benefits; a higher costs score denotes greater adverse effects.

Table 6. Benefit-Cost Analysis for Alternate Highway Locations Using Ranking Scores Derived from Analytical Hierarchy Process.

Alternate Location	Benefits Score	Costs Score	Benefit to Cost Ratio
No. 1 (western location)	0.393	0.281	1.4
No. 2 (western location)	0.239	0.153	1.6
No. 3 (western location)	0.180	0.264	0.7
No. 4 (western location)	0.134	0.152	0.4
No. 5 (eastern location)	0.054	0.150	0.4

The results show that the sole eastern location, Alternate 5, has the lowest level of benefits and costs associated with any of the alternates. Alternates 1 and 3 are considered to have greater adverse costs than Alternates 2 and 4 because they each contain a segment that traverses a residential neighborhood, causing greater displacement of residents and leaving a substantial number of dwellings adjacent to the new highway. Alternate 5 has a costs score close to those of Alternates 2 and 4, primarily because it has significantly greater adverse impacts on prime farmland and agricultural operations than do any of the four western alternates.

By calculating a relative benefits-to-costs (B:C) ratio, the scores indicate that Alternates 1 and 2 have greater associated benefits than costs. Alternate 2 is not the most beneficial location, but it has a better B:C ratio because it has greater benefits for the costs involved. On the basis of these figures, Alternate 2 would be slightly preferable to Alternate 1, whereas Alternates 3 and 4 would have questionable value. Alternate 5 is clearly deficient because its perceived benefit level is substantially lower than those associated with any of the western locations (i.e., Alternates 1–4).

BENEFITS OF USING AHP AND EXPERT CHOICE™

AHP enables decision makers to systematically examine the simultaneous interaction of salient variables in complex, unstructured situations. It helps decision makers identify and set priorities on the basis of objectives and goals by forcing methodical evaluations that articulate the value systems of all those involved in making the decision.

It allows complex problems to be broken down into more manageable components. This is of significant benefit when trying to communicate the complexities of highway location planning to public officials and interested citizens. After completing this exercise, most participants had become more aware of the balancing of competing concerns that is a hallmark of the highway planning process.

AHP accepts subjective judgments based on knowledge and experience, as well as objective judgments based on quantitative indicators. This feature makes it accessible for matters that may not be highly quantifiable but are of serious concern to affected parties and decision makers. This is particularly relevant for some of the environmental and social effects.

In a public involvement program, the AHP offers a means to represent various interests in a balanced participation. If groups cannot be convened to develop consensus judgments, each group member can be given a questionnaire to complete. The geometric mean of the judgments can then be used as a proxy measure of the group's consensus.

The EXPERT CHOICE™ software provides a facile means of documenting all judgments and decisions through printed reports.

CONCLUSION

Applying AHP in the planning process for a major new highway in a heavily urbanized area was successful. More specifically, it demonstrated how citizens and government officials can use this process to evaluate, assess, and combine judgments in an environment of participatory decisionmaking.

The process accepts numerical values as well as subjective judgments, demonstrating its flexibility while ensuring that important nonquantitative variables are not lost or minimized in the decisionmaking process. The paired comparisons force a systematic examination of all issues, and an inconsistency ratio algorithm draws attention to inconsistent judgments.

In most public involvement programs, the process is as important as the final result. Similarly, the process involved in AHP is as important as the resulting numerical scores. This process enforces discipline on complex decisions by drawing out the value systems of all participants. This fosters greater understanding of others' concerns and aptly demonstrates that weighty compromises are often necessary if progress is to occur in our modern urban society.

REFERENCES

Illinois Department of Transportation, Division of Highways. 1989. Draft Environmental Impact Statement for FAP Route 413, Madison County, Illinois. Illinois Department of Transportation, Springfield, IL.

Saaty, T. L. 1986. *Decision Making for Leaders.* University of Pittsburgh.

U.S. Department of Transportation, Federal Highway Administration, Office of Environmental Policy, Community Values Branch (HEV 12). 1988. Public Involvement. Washington, D.C.

The Future of NEPA: The Review Process in a Global Era

H. Paul Friesema, Northwestern University, Evanston, IL

This conference is not just a review of 20 years of experience under the National Environmental Policy Act (NEPA). It is also a forum to look ahead. So we may begin by anticipating future directions with the environmental impact statement (EIS) process. I foresee the globalization of the EIS process in two ways:

1. Adoption of EIS-like processes will occur around the globe by the European Community, World Bank, Eastern Bloc, and others.
2. Issues of the global commons will become an increasing focus in domestic EISs. EISs on plans for power plants, outer continental shelf (OCS) drilling, timber harvesting, and the like will be evaluated in terms of the world problems of global warming, tropical rain forests, endangered wildlife, and destruction of oceans and fisheries, among others.

As those of you who write EISs will know, these predictions are of a type which are the safest to make. To a significant degree, these innovations have already occurred. Both of these tendencies to globalize the EIS process are affected by, and in turn affect, the participation requirements and processes associated with environmental assessment. This paper examines the globalization of the EIS in terms of the participation processes.

ADOPTION OF EIS-LIKE PROCESSES AROUND THE GLOBE

I am not alone in noting a tendency for American advocates of NEPA to assert that participation is the key to the effectiveness of the process in the United States. Despite such assertions, other entities adopting the analytics of the EIS process often do not include recognizable public participation procedures, especially the judicial oversight and easy access to the courts by project opponents — often seen as the hallmark of the U.S. process. Many American students of the EIS process have lamented the underdevelopment of the participation processes in foreign adoptions of environmental assessment procedures and have asserted that without effective public participation the whole process will fail. But is that likely to be true? In what ways is public participation such a key to the success of NEPA? Are there alternative ways to achieve the oversight and adversarial review of prediction and analyses which we hold so dear?

The truth is that despite lots of discussion of public participation under NEPA, the consequences of participation have not been systematically studied, even acknowledging the good work by people like Jeanne Neinabor and Paul Sabatier, the puttering around by Paul Culhane and me, and others. We have produced some good stories about the consequences of litigation and other participation in individual cases and with some specific agencies. But the systemic consequences of public participation are still a little vague. Nonetheless, there is a conventional wisdom (one I contribute to) that public participation is the key to the success of the EIS process. Success is often measured in terms of decisions or outcomes which are approved by environmental advocates. Of course, I do not think there can be much dispute that the participation and review process has been very significant. But I suspect that the consequences of the direct involvement of citizen and interest groups in reviewing and challenging EISs is a little overstated. In the first place, I think it is useful to disaggregate the participation aspects created by NEPA and the CEQ regulations implementing the act.

The most visible part of the public participation–public review process is the institutional opportunities for organized or unorganized citizens to get involved in the EIS issue — from attending a scoping meeting to submitting written comments on draft or final EISs to suing an agency over some perceived failure in the EIS. But there is a parallel review process. The parallel process is the interagency, interjurisdictional review process. Other agencies and governments within our federal system are routinely incorporated into the NEPA process. Because of the diversity of agency and jurisdictional interests, much of the advocacy role we attribute to interests groups and concerned citizens is also performed by agencies and jurisdictions which have the opportunity and interest to review a lead agency's proposal. Sometimes agencies and jurisdictions work informally in conjunction with interest groups, and sometimes they do not. It seems to me that interest group and interjurisdictional oversight are somewhat interchangeable.

Although it is rarely discussed in the NEPA literature, interjurisdictional review can be very consequential. I would guess that far more projects have been fundamentally altered because of the interagency and interjurisdictional oversight than have been because of direct interest group involvement. The role of the Environmental Protection Agency (EPA) in nuclear power plant licensing, the Fish and Wildlife Service in reviewing U.S. Army Corps of Engineers projects, the National Oceanic and Atmospheric Administration (NOAA) on economic development projects, the state of California on OCS leasing, or the state of Utah on deployment of MX missiles all come to mind. Although reviewing agencies and jurisdictions often do not have quite the freedom to litigate EISs that well-healed interest groups might have (but how many of them do you know?), they have other avenues to enforce their interests, including, among others, withholding permits and invoking interagency negotiations and appeals.

Any survey of a random sample of EISs would reveal far more interjurisdictional involvement than interest group involvement in the NEPA process, even excluding the *pro forma* acknowledgments of receipt of EISs. Many final EISs appear with virtually no citizen or interest group participation reported. But very few final EISs appear without some extensive and consequential interjurisdictional critiques. In some important ways, the agencies are more effective reviewers than are most private advocates. One reason that agencies often offer the most potent critiques of projects is because of the peculiarities of the EIS process. One of those peculiarities is that much of the participation under NEPA comes in the form of submission of written scientifically and technically based comments. The other peculiarity is in the tricky relationship between critiquing the project itself and critiquing the document that analyzes the project. Becoming an effective advocate, using the NEPA process, is something of an art form. Most citizens and interest groups are only activated around a small number of issues or a single issue. By contrast, many jurisdictional actors become involved in so many EIS reviews that they develop some expertise with the process.

As NEPA is exported, I think that there are a lot of things worth considering, and perhaps studying, in comparing interjurisdictional advocacy and interest group/citizen advocacy under NEPA. But I also think enough is clear to raise some questions about the necessity of private and interest group involvement in the EIS process. It may be easier for world bodies, and some other countries, to incorporate the interjurisdictional and intergovernmental review and oversight aspect of NEPA than it is for them to incorporate American-style public participation. The American experience suggests that interagency and interjurisdictional review can be quite potent.

ISSUES OF THE GLOBAL COMMONS IN DOMESTIC EISs

As we become increasingly aware of global interdependencies, it is becoming apparent that EISs need to consider cumulative and global impacts more and more. But some of the peculiarities of public participation under NEPA work against considering global interdependence. Although national environmental interest groups with global concerns frequently participate in local EISs, most of the active commenters on EISs are people or groups based in the immediate area where a project will occur. Despite bumper stickers proclaiming "Think globally, act locally," most locally based actors are primarily interested in issues close to home. People living in Knoxville or New York might see a central connection between timber harvesting in Oregon and the rain forest of the Amazon. People in Oregon will see that as well, but their primary concern will be the fate of the forests of the northwest.

A few years ago, Tom Armentano, Paul Culhane, and I were reviewing EISs on fossil-fueled power plants to see how the issue of acid rain was being considered. We noted a peculiarity. The most substantial citizen and interest group concern over acid rain issues was reflected in EISs for power plants located in the northeast, where acid rain concern was the greatest. Those same types of advocates (based in the northeast) were not very involved in overseeing EISs for fossil-fueled power plants to be located in the midwest, where the acid precipitation was generated. There were commenters in the midwest, of course, but their agenda mostly concerned local impacts.

The use of scoping processes works to focus EIS attention upon local issues as well. Once again, interjurisdictional and interagency review may be extremely important in raising the global concerns for particular projects coming under NEPA review.

Although the measure of success of the NEPA process should be the quality of the decision and not the quality of the documents supporting a decision, the EIS documents and their review are of great importance beyond their influence over particular outcomes. EISs and their review are important avenues for raising issues and getting concerns on public agendas. The NEPA process is very important in both raising and maintaining public understanding about environmental issues and problems — consciousness raising. The NEPA process is also an important institutional means of incorporating scientific concerns into public decisions, and the peer review aspect of the public/interagency comments seems to be central to improving the science in EISs. So the participation requirements of NEPA are of the greatest importance. They will be critical in the years ahead as environmental assessment becomes globalized. These remarks are designed to emphasize that there are really two parallel and somewhat interchangeable review processes at work under NEPA and that the interjurisdictional review process may be particularly important in the next decades.

CHAPTER 5

CUMULATIVE IMPACT ASSESSMENT

Introduction

R. B. McLean, Oak Ridge National Laboratory, Oak Ridge, TN; and T. N. Russo, Federal Energy Regulatory Commission, Washington, D.C.

There seems to be no difference of opinion that cumulative impact assessment is difficult and often suffers from lack of information to make it work well. The authors in this chapter emphasize that cumulative impact assessment needs the big picture approach, one that demands an understanding of ecosystem function and human environments on which potential effects might occur. The baseline, against which impacts are evaluated, changes and thus is in need of systematic monitoring as well as the centralization of baseline data. Multiple agency management of the resources in a region is usually not coordinated, which often results in an inaccurate picture of reasonably foreseeable actions as well as conflicts in resource planning. Despite these drawbacks, there are tools to be used in cumulative assessment that help sidestep these problems and allow the assessment team to be responsible to the mandates of the National Environmental Policy Act (NEPA). Examples are given that illustrate the techniques of breaking complex actions into manageable units and large geographical regions into representative areas. Lack of data and understanding of the physical characteristics of systems are compensated for by using reasonable bounding analyses. Mitigation can then be suggested to minimize impacts for these subunits, and the subunits can be reassembled to complete the big picture. When the resources of concern are interdependent in complex ways, there is added challenge in capturing these interactions in an evaluation. Two papers illustrate matrix methodologies that aid in accomplishing this and reemphasize the need to know the ecosystems of concern as well as the importance of agency involvement and coordination. The final message in this chapter is that cumulative assessment that steps back to see the big picture, evaluates alternatives within this framework, and incorporates the input from all interested parties can help to sort out complex and emotional issues so the decision makers can make informed decisions.

Toward Defining and Assessing Cumulative Impacts: Practical and Theoretical Considerations

C. K. Contant, Graduate Program in Urban and Regional Planning, The University of Iowa, Iowa City, IA; and L. L. Wiggins,[1] Department of Urban Studies and Planning, Massachusetts Institute of Technology, Cambridge, MA

ABSTRACT

Analysts, decision makers, and practitioners have struggled with the assessment of cumulative impacts of projects and development proposals. Precise definitions have been difficult to operationalize and predictive assessment methods have been slow to develop. Nonetheless, the cumulative nature of many environmental impacts has become an increasingly important, relevant, and visible topic to the public. In this paper, we review the progress made in analyzing cumulative environmental impacts, highlight some of the major challenges remaining to be addressed, and propose particular areas where improvements may be possible.

We first make a careful study of the definitions of cumulative impact. Following the establishment of this conceptual base, we turn to an evaluation of methodologies useful for conducting cumulative impact analyses. Finally, we examine potential improvements to current cumulative impact assessment by focusing on (1) improving the monitoring and prediction of actions and impacts over space and time; (2) increasing the knowledge of the responses of environmental systems to development perturbations, including synergistic and indirect effects; and (3) developing

[1]L. L. Wiggins is currently with the Center for Urban Policy Research, Rutgers University, New Brunswick, NJ.

management systems that provide the appropriate responses to actions that produce significant cumulative effects.

INTRODUCTION

Traditional environmental impact analyses have focused primarily on examining the consequences of a single activity. Consequences are viewed as the impacts of that action on the components in the environment that are thought to be most worthy of attention or protection in a given context. By expanding both the scope and scale of the analysis, cumulative impact assessment examines the positive and negative consequences of multiple activities and sources of impact on a larger set of environmental components (Clark 1986).

Several current environmental problems provide examples of how many small, repeated actions can produce cumulatively significant consequences. Consider the recent attention given to the worldwide importance of such concerns as global climate change and biodiversity. International attention has also been focused on the causes and consequences of ozone depletion and acid rain (Wooley and Wappett 1982). On the local level, many communities have become aware of the cumulative impacts of development on transportation congestion, urban infrastructure (such as water, sewers, and landfills), and the overall quality of urban life.

The importance and currency of the problem of cumulative impacts seems to be undeniable; yet several scientific and administrative limitations have severely constrained the incorporation of cumulative impact considerations into project-level decisionmaking. Therefore, the ability to document, predict, and control the cumulative consequences of development activities remains the focus of continuing discussion and investigation.

In this paper, we discuss both the conceptual foundations and the practical requirements of cumulative impact assessment. We begin our investigation of cumulative impacts by considering the role of regulatory language, interpretations by the courts, and studies by academics and practicing environmental analysts in defining the nature and complexity of cumulative impacts. Next, we evaluate methodologies for conducting cumulative impact assessments by comparing them with an *ideal* technique. Finally, we suggest three directions for improving cumulative impact assessment: monitoring actions and impacts over space and time, modeling impacts in complex systems, and managing impacts through case-by-case decisions.

DEFINING CUMULATIVE IMPACTS AND THEIR ASSESSMENT

Expanding an environmental analysis to include cumulative impacts adds several new concepts for consideration. In this section, statements from regulatory language, interpretations provided by the courts, and insights provided by academics and practitioners are reviewed. These new dimensions form the foundation for analyzing cumulative impacts.

Regulatory Definitions

Cumulative impacts have been recognized for decades (Vlachos 1985; Dickert and Tuttle 1985). Procedurally, the need for their assessment first appeared in the 1973 guidelines for implementing the National Environmental Policy Act (NEPA) and was formally required in the 1978 regulations for NEPA (CEQ 1978). As part of the total environmental impact of new development actions, cumulative impacts are viewed as

> the impact on the environment which results from the incremental impact of an action when added to other past, present, and reasonably foreseeable future actions regardless of what agency (Federal or non-Federal) or person undertakes such other actions (CEQ 1978, p. 56004).

Important cumulative consequences are noted particularly for individually minor but collectively significant actions taking place over time.

Regulatory references to cumulative impacts highlight three fundamental issues about development actions, their cumulative consequences, and their assessment. First, developmental pressure on environmental systems is expected to produce impacts that accumulate over time and space. These impacts accumulate when the incremental or added effects from one action combine with other effects of past or future projects, regardless of how small or insignificant the increment. Second, cumulative impacts can be produced by a combination of actions that may produce intense developmental pressure on the environment in an area. These significant cumulative actions could be small or large, similar or different, part of a larger project or unconnected. The third issue addressed in these regulations establishes the manner in which cumulative impacts are viewed. For each individual action, a cumulative impact analysis forces a review of the project's impacts in the context of its location, other present or anticipated developmental activities, and present and anticipated environmental conditions.

Although these regulatory issues are easy to understand, numerous procedural problems have arisen when using them to assess project impacts. Consequently, several lawsuits have resulted from conflicting views of the proper application of these concepts.

Role of the Courts

Since the introduction of environmental impact analysis requirements in the early 1970s, the U.S. courts have been actively involved in defining, clarifying, and shaping environmental policy. The courts provided early guidance in defining cumulative actions and their impacts, but the emphasis of the courts later shifted to detailing the proper level of analysis for cumulative considerations. Both aspects clarify the discussion of cumulative impacts.

In the early 1970s, several cases brought the concept of cumulative impacts to the courts' attention. Interestingly, little guidance was given by these cases in determining what constitutes cumulative impacts; rather, the courts concentrated on the procedural considerations of what combination of actions should be regarded as *cumulative actions* or those actions that produce significant cumulative effects. In *Scientists' Institute for Public Information, Inc.* v. *Atomic Energy Commission* (ELR 1973a), the notion of *irretrievable commitment of resources* became the critical test to determine whether a given action was connected to previous or subsequent actions. If a decision on a pending action represented a commitment to a whole program or project, the entire project should be viewed together, not incrementally. This *whole project* approach was designed to prevent the segmentation of large projects to the level where separated parts of the total project had impacts that could be considered individually insignificant (Merson and Eastman 1980).

Another rule used to determine the adequate scope of analysis for a project was developed in *Indian Lookout Alliance* v. *Volpe* (ELR 1973b). The court concluded that if a proposed segment of a project had independent utility or significance, then an analysis could be prepared that was limited to that segment alone. This rule allowed large projects to be segmented into *independently significant* parts to avoid a close analysis of the overall or cumulative impact of the entire project. Many cases followed, with the courts trying to rectify the seeming contradiction between the *irretrievable commitment* and *independent significance* rules (e.g., *Minnesota Public Interest Research Group* v. *Butz* and *Trout Unlimited* v. *Morton*).

Attempting to resolve the contradiction, the court in *Sierra Club* v. *Morton* (ELR 1975) provided the agency responsible for preparing a project analysis with the discretion to determine what constituted a cumulative action. A balancing test was proposed to identify when a *whole project* analysis would be necessary. This case became a turning point for interpreting cumulative impact requirements because it allowed agencies to exercise their discretion in determining appropriate levels of analysis. Upon appeal, however, the case was further clarified by the U.S. Supreme Court in *Kleppe* v. *Sierra Club* (ELR 1976). The *Kleppe* case found that only fully proposed projects need to be included in cumulative analyses and that contemplated or anticipated projects need not be included. Strong reaction to this strict interpretation of the 1973 NEPA guidelines led to the precisely defined language in the 1978 NEPA regulations regarding the inclusion of "reasonable foreseeable future actions." Equally important in the court's findings on *Kleppe*, however, is the bold recognition that for some natural resource situations basin boundaries may be a more appropriate level of cumulative impact analysis than the smaller project area itself.

A more recent case, *Thomas* v. *Peterson* (ELR 1985), further refined what constitutes cumulative impacts, cumulative actions, and the need for larger-scale project or program analyses. The *Thomas* case established *close dependency* as the criterion to determine if projects can be *tiered* and then analyzed in the context of programmatic or regional assessments (Barney 1981; James et al. 1983). Further, it upheld the definitions promulgated in the 1978 NEPA regulations and highlighted the importance of cumulative impacts as the controlling factor in determining impact analysis scope and timing (Hapke 1985). In fact, this case established the use of cumulative impact considerations for several specific purposes: to determine the need for programmatic or regional impact analyses, to prevent piecemeal applications of projects with significant impacts, and to choose ecological rather than jurisdictional boundaries for project analysis.

States with *little NEPAs* have also experienced litigation on cumulative impact issues. In *San Franciscans for Reasonable Growth* v. *City and County of San Francisco*, the court evaluated the environmental analyses for four pending downtown highrise buildings and found that by ignoring the existence of the other buildings currently in the permit process, each analysis had provided inadequate discussions of cumulative impacts. In fact, when the court performed its own analysis, significant cumulative impacts were found on traffic and many other municipal services (Selmi 1984).

Enriching the Definition

Other key aspects of cumulative impacts have been documented by researchers and practitioners in the past decade. The first of these additional considerations concerns the impact of an activity on an entire environmental or human system. Although the specific and direct impact of an action may be insignificant or undetectable to one aspect of the system, it may become significant when felt throughout the rest of the system components (Vlachos 1985; Stakhiv 1988). Systemic processes of accumulation and delayed response (Baskerville 1986) reflect the types of characteristics that can result in significant impacts from individually minor but repeated effects.

Two other conditions incorporate the responses of environmental systems to cumulative perturbations. It has been noted in many studies of wetland systems, for example, that *threshold* levels exist (Preston and Bedford 1988). These thresholds correspond to the point at which further system perturbations, no matter how small, will result in major deterioration or collapse. Additionally, systems may not always respond linearly when impacts accumulate. The presence of these nonlinear functional attributes (Bedford and Preston 1988) forms an added dimension for investigation in cumulative impact studies. Both conditions provide new pathways by which cumulative impacts occur and must be addressed.

A final aspect of cumulative impact considerations is the synergistic nature of some effects in the environment. Synergism reflects the observation that effects on the whole system are often more than the sum of the parts in terms of magnitude, intensity, severity, or complexity (Vlachos 1985). Although synergism may account for some of the previously noted nonlinear functional relationships, it also accounts for those impacts resulting from the interaction, combination, and new patterns of effects.

Applying this natural system perspective to defining cumulative impacts, Baskerville (1986) characterizes cumulative impacts into three groups. His first form of cumulative impact results from repeated *incremental insults to the system.* Each increment adds to the previous increments over time, in either a linear or nonlinear fashion.

Baskerville's second form of cumulative impact is illustrated by the situation in which a single action or some limited set of actions results in a system change in structure or dynamic. An example of this type of impact is the delayed response caused by system changes that take place after exposure to a cancer-causing agent. The initial exposure incident might have appeared benign, but its introduction into the system created changes in the system structure, and these changes produced significant effects much later in time.

The third form of cumulative impact noted by Baskerville corresponds to the accumulation of impact by cycling over space and time. He cites the clear-cutting of forest as an example. Although the impacts of one clear-cut area may be partially mitigated by natural processes of recovery, these new species of vegetation and animals may not be the proper ones. As the natural balance is shifted by clear-cutting and recovery activity over time throughout a forest, the impact moves through various cycles, migrates around the forest, and changes the overall nature of the forest in a cumulative manner.

Beanlands and others (1986) added to the clarity of the definitions by suggesting that projects produced significant cumulative effects when impacts become time crowded or space crowded. The concept of crowding refers to the effect resulting from the inability of a natural system to recover from an earlier or close perturbation before a new one is present. Further, they introduced the term *nibbling* to denote the impact resulting from the incremental insult of repeated actions on an area over time.

In our research (Contant 1984), we attempted to draw the link between developmental activities and the cumulative impacts resulting from them. We suggested that cumulative impacts were the result of additive and aggregative actions producing impacts that accumulate incrementally or synergistically over time and space. Additive actions are repeated similar activities, while aggregative actions are groupings of dissimilar activities. Impacts of both sets of actions accumulate in an incremental (nibbling) fashion and/or combine synergistically (interactively) to produce effects other than those occurring directly or indirectly from actions. Implicit in this view of synergistic impacts is the recognition that some effects may be delayed over time as systemic changes occur.

Subsequently, we have added the concept of growth induction to reflect the fact that the introduction of certain activities can accelerate or decelerate the rate of development of new activities or can result in a stronger interaction among environmental parameters. This characterization of growth-induced impact recognizes the precedent-setting effect of some activities in stimulating even greater development than previously anticipated. In addition, this concept captures the fact that certain activities may produce impacts that act as catalysts to produce greater system-wide effects than would be expected otherwise.

When viewed collectively, these attempts at understanding cumulative impacts suggest that there are five aspects inherent in such a definition. First, similar and dissimilar actions occur to produce cumulative consequences. Second, the impacts of these actions accumulate over time and space. Third, this accumulation takes place in both incremental and synergistic ways. Synergistic or system-altering effects may produce impacts that are much more than the simple addition of incremental effects. Fourth, some activities may produce major changes in system behavior that either occur immediately or are delayed. Finally, some actions are growth inducing and, as a result, represent significant changes in the impetus for future activity or the form of environmental system response. Each of these five aspects describes ways by which actions can produce cumulatively significant negative or positive impacts.

ANALYSIS OF CUMULATIVE IMPACTS

Conceptual definitions of cumulative environmental effects provide the foundation for identifying, predicting, and evaluating cumulative impacts. In this section, we identify the operational difficulties associated with using the conceptual definitions described above, then we propose an *ideal* impact assessment methodology that reflects both these conceptual requirements and operational concerns. Finally, we describe and critically evaluate three methodologies that have been used or are proposed for use in cumulative impact assessment.

Operational Problems with Definitions of Cumulative Impacts

Conceptually and intuitively, these five aspects of cumulative impacts are readily understandable and seemingly complete. When these aspects are made operational in an environmental assessment, however, numerous problems are encountered. These problems result from difficulties associated with (1) monitoring present and past activities and predicting future actions; (2) scientific shortcomings in understanding and predicting natural systems' behaviors within and across media and over time; and (3) organizational, legal, and jurisdictional conflicts and limitations.

One factor identified in all definitions of cumulative impacts is the accumulation of impacts over time and space for a variety of projects and activities. To identify this incremental impact for one project, it must be possible to predict with some certainty what reasonably foreseeable future actions will be undertaken. It is also necessary to provide a baseline by monitoring past and current actions. Further, actions must be classified as similar, dissimilar, connected, unconnected, or growth inducing. This requirement to monitor past and current actions and to predict future actions for a large affected area often exceeds the capabilities of most governmental entities (Contant 1984; Stakhiv 1988).

The second category of operational problems reflects a series of scientific challenges that underlie cumulative impact identification and prediction. For many media (atmospheric, terrestrial, aquatic, and others), scientific knowledge of impact processes within that media is still relatively limited. In addition, for those impacts that cross media or natural systems, the mechanisms of transfer and impact are not well-understood. Under these conditions of limited knowledge, predicting the impacts of even one activity on one natural system taxes the capabilities of most scientific models. Identifying the cumulative impacts of several actions on one natural system further complicates the problem. Finally, assessing the interactive or synergistic impacts of many actions across a variety of systems or environmental media is virtually impossible.

Some hope is held out for identifying the responses of physical, chemical, and biological systems to perturbations (Roots 1986). Studies of wetland systems (Bedford and Preston 1988; Stakhiv 1988), atmospheric systems (Clark 1986), and aquatic ecosystems (Proett 1977) have provided meaningful input into understanding the impact pathways and accumulative nature of effects. At present, however, the need for information of this kind far exceeds the capabilities of providing it.

A final category of operational difficulties includes the conflicts often present among organizations, jurisdictions, and their boundaries. This set of difficulties relates directly to the differences between the scale of management activities and the scale of environmental consequences. Impacts may occur across a variety of natural environments, each under the jurisdiction of separate entities. Alternatively, many organizations may have jurisdiction over parts of a geographic region within which an impact occurs. A further complication occurs when numerous entities control an environmental media but have responsibilities for different activities that affect that media (Hirsch 1988). Each of these *institutional albatrosses* (Dickert and Tuttle 1985) presents profound problems for operationalizing the conceptual definitions of cumulative impacts.

Requirements for *Ideal* Cumulative Impact Assessment Methods

Despite the operational difficulties in assessing the existence and severity of cumulative impacts, methodologies must be developed to assess these impacts and to provide appropriate responses to them. We propose that an idealized impact assessment methodology must accomplish several specific tasks. For clarity, we have grouped these tasks into three sets: monitoring, modeling, and management. These sets also reflect many of the same methodological issues noted in our discussion of operational difficulties. First, an ideal methodology must *monitor* development activity over time and space and allow for the mechanisms by which past and present development may lead to changing growth patterns over time. In addition, the methodology must monitor changes in environmental parameters in the appropriate area over time. These data serve as baseline values against which new impacts can be compared.

Second, an impact assessment methodology must provide clear and accurate *models* of the responses of natural systems affected by expected development activities. These models should describe the impact of a perturbation, the ability of the system to recover from the impact, the interactive or synergistic impacts within and across systems, as well as the threshold or delayed effects that may occur. The methodology must also provide a clear and justifiable forecast of probable future actions, preferably with some range of uncertainty (perhaps with associated probabilities) attached to the forecast.

Third, effective *management* systems must be able to evaluate these cumulative developmental effects and translate this evaluation into action on a particular proposed project. Specifically, this management stage takes the impacts predicted earlier and views them in the context of established policies or goals. Based on the severity of the impacts, management strategies and actions can be implemented in the case-by-case review of a project's cumulative impacts. This combination of setting goals, identifying severity of impacts, and designing and implementing appropriate responses forms the core of this management element (Contant 1984).

These three components of an ideal cumulative impact assessment method provide the foundation for evaluating the three assessment procedures that have been used or proposed for use. In the subsequent methodological review, each methodology is described along with one or more illustrative examples of using the technique to assess and control cumulative impacts. Following these descriptions, we critique the capability of each methodology to monitor actions and impacts over space and time, to model the wide range of natural system effects, and to manage activities that produce a significant cumulative effect.

Programmatic Assessments

The earliest and most common attempts at cumulative impact assessment involve grouping numerous probable development activities together and analyzing their *collective* impact on the environment. These studies typically employ either an analysis of several projects combined into a *scenario* of expected development in a geographic area (regional assessments) or an assessment of an entire program of related or similar activities (programmatic assessments). Regional assessments examine all types of development expected in a particular geographical area, while programmatic assessments study the impacts of related or similar projects expected to occur as part of a larger program of activities (such as coal mining, gas leasing, etc.). In both cases, some future time horizon is determined, anticipated development is identified, and expected impacts of the development are forecast.

To address the cumulative consequences of these grouped projects, the severity of the aggregated effects of all development is evaluated. If severe, the group of projects is modified or parts of projects are removed to reduce the total impact. Decisions on future development projects, therefore, are based on the projects' consistency with anticipated development scenarios for the region or the program. Therefore, by explicitly identifying expected future development, programmatic assessments put particular project activities and their impacts into a broader geographical, environmental, and developmental context.

This approach to cumulative impact assessment relies on the notion of *tiering* that was suggested in NEPA regulations (CEQ 1978). Assessments are prepared at critical points in the planning process, and subsequent project proposals are evaluated against the earlier assessment. Merson and Eastman (1980, p. 564) suggest the following:

> The program statement has a number of advantages. It provides an occasion for a more exhaustive consideration of effects and alternatives than would be practicable in a statement of an individual action. It ensures consideration of cumulative impacts that might be slighted in a case-by-case analysis.

There are many examples of regional and programmatic assessments. A typical early example is the environmental assessment performed by the Baltimore District of the U.S. Army Corps of Engineers on proposed projects on two tributaries to the Chesapeake Bay (COE 1975); a large number of pending permit applications were grouped together and their impacts analyzed. A variety of other regional studies have been performed to determine the cumulative impacts of potential land development activities. Cumulative impact studies of the San Francisco downtown area, Lake Tahoe shore zone, Louisiana coastal zone, and Chesapeake Bay development plan identified probable growth scenarios and their likely environmental effects. In several of these studies (Phillips, Brandt, Reddick-McDonald and Grete, Inc. 1978; Maryland Department of State Planning 1972), however, no mechanisms were proposed to determine or manage excessive growth. In another one of these studies (Environmental Science Associates, Inc. 1983), mitigation measures were created to ensure conformance with existing policy requirements. In the final example (Center for Wetland Resources 1977), excessive cumulative impacts were used to argue for the development of policies to protect against these effects.

Several other recent examples of programmatic assessments have addressed the cumulative impacts of anticipated developments in mining (Hawkes et al. 1989), hydropower (Proett 1977; Emery 1989), and natural gas exploration (Smith et al. 1989). These programmatic assessments often do not occur before development but are demanded once it is perceived that a critical impact threshold is reached by the addition of one more project to existing development levels. In these situations, programmatic assessments are used to determine if the cumulative impacts of many past similar activities exceed acceptable levels. If so, no further development can occur.

Several theoretical and operational problems are associated with programmatic assessments. Conceptually, programmatic assessments depend heavily on the accurate prediction of future actions and impacts; but if these predictions are inaccurate, a project's impacts can be viewed within the wrong context. This severely limits the use of programmatic assessments for cumulative impact analysis. Further, programmatic assessments produce snapshot views of development activities and impacts for some point in the future. There is no clear mechanism to provide ongoing monitoring over time to document actual development actions, impacts, or growth patterns and to reassess the direct or cumulative impacts.

Finally, programmatic assessments cannot be used effectively as tools to control or manage cumulative impacts. While management approaches can be used to control the total impact of a package or group of expected activities, programmatic assessments do not provide the information necessary to manage or control an individual project's contribution to increasingly severe cumulative consequences or to suggest strategies to improve individual incremental decisionmaking.

Operationally, there are a variety of problems with regional or programmatic assessments. Hapke (1985) points out some of the problems with *tiering*. He suggests (p. 10,296) that agencies have frequently been seen to jump "from the general EIS on the trunk of the planning tree directly to site-specific actions in the upper leafy branches, without any supporting branches in between." He suggests the use of intermediate assessments between programmatic and project-specific ones. Operationally, the preparation of assessments for the trunk, branches, subbranches, and leaves rapidly becomes infeasible.

A final operational criticism of using this approach for cumulative impact analysis rests on the recognition that programmatic assessments often are performed only when a critical level of development or impact is expected to be exceeded. Although this criticism does not fault the process, it does suggest that as a tool for anticipating and managing cumulative impacts, the approach may be somewhat less than adequate or timely.

Suitability Studies

A second category of methods used to perform cumulative impact assessment relies on the use of suitability studies to determine appropriate areas for development. Rather than forecasting and analyzing the impacts of individual projects on the environment, suitability analyses examine the characteristics of a region and identify areas that are appropriate for or sensitive to different types of development. As a result, cumulative impacts are addressed in the context of the natural system's ability to withstand developmental pressure.

To determine suitability ratings, overlay maps are created with *layers* or *coverages* for various relevant factors. The resulting composite map identifies whether an area is suitable or unsuitable for a particular type of land use (e.g., residential or industrial) based on the interaction of characteristics such as slope stability, water quality, or vegetation (McHarg 1969). These suitability ratings can be used to express the additive and synergistic responses of the environment even in the absence of more sophisticated quantitative models. Therefore, as an operational method for cumulative impact assessment, suitability studies attempt to identify the fundamental natural system responses to development activity. Further, suitability analyses focus concern on the locational attributes of cumulative impacts and identify potential cumulative impact problems *before* they occur. A typical example of this methodology is the set of suitability wetland studies conducted by the Portland District of the U.S. Army Corps of Engineers (COE 1976). *Profiles* were created from existing data for the environmental, visual, social, and economic resources of the wetland areas. Thirteen activity types were identified, and matrices were developed to describe the direct (not cumulative) environmental impacts of these activity types on the critical parameters. Each parcel within the wetland was evaluated in terms of how well-suited it was for each activity type. Activities were then rated as either acceptable, unacceptable, acceptable with conditions, or requiring the preparation of an environmental impact assessment.

There are, however, a variety of problems identified with this approach. The selection of factors and scales used is often extremely subjective. Stronger criticisms have been leveled at the actual overlay process, since careless analysts will often add quantities with incommensurate units [see Hopkins (1977) for further criticisms]. In response to this criticism, a variety of different scaling and weighting techniques have been suggested. Although the data requirements for these analyses are intensive, recent improvements in computer technologies may alleviate some of the effort and expense.

In addition, several other limitations are associated with suitability analysis as a technique to analyze cumulative impacts. Other than its attempt to identify suitable areas for certain generic types of development, this approach makes no prediction of specific future actions or impacts. Further, management of impacts is limited to *go/no go* determinations on pending projects; the definition of policies, strategies, or specific managerial actions is inconsistent with performing or implementing suitability analyses. Finally, in practice, little or no monitoring occurs in conjunction with suitability studies. Therefore, no accounting for development rates, impacts, or changing physical conditions can be incorporated into a cumulative impact analysis.

Carrying Capacity Studies

A final category of past attempts at cumulative impact assessment is referred to here as *carrying capacity studies*. These studies attempt to recognize that inherent limits exist for many environmental and socioeconomic factors, and these thresholds impose constraints on development. At the heart of the analysis is the definition of carrying capacity: "the ability of a natural or man-made system to absorb population growth or physical development without significant degradation or breakdown" (Schneider et al. 1978, p. 1). Carrying capacity studies identify these natural constraints on development and provide mechanisms to monitor the incremental use of any unused capacity.

The general approach to cumulative impact assessment using carrying capacity studies begins with an identification of potentially limiting factors to development. Equations are formulated to relate development to impacts for each of these limiting factors. These *linkage* equations are used to identify numerical limits to development (*threshold levels*) imposed by each limiting factor. The carrying capacity of an area is defined by the most restrictive level of development associated with the limiting factors. Proposals for new projects are evaluated in terms of how they affect the remaining capacity of the limiting factors. In this way, the cumulative effects of individual decisions are accounted for in a systematic way.

Many examples of carrying capacity studies for cumulative impact assessment are found in air and water quality control programs (Raffle 1978; Zener 1981). Forecasts of air and water quality impacts from proposed projects are performed with sophisticated physical or mathematical models. Impacts are analyzed by comparing them to *standards* which describe numerical health and safety limits. If impacts are below standards, capacities are not exceeded, and the proposal is authorized. If impacts are more severe, mitigation measures (e.g., emission offsets) become necessary for project approval. Cumulative impacts, in these instances, consist of the sum of impacts from individual projects. Accumulated impacts must remain below established limits with mechanisms established to ensure that fact.

Some carrying capacity studies cannot rely on quantitative, physical limits to restrict development; instead, limits are based on subjective, goal-oriented statements of public opinion. Strategies appropriate to consider cumulative impacts in these instances rely on intense information gathering efforts, including consultation and coordination among outside groups, the applicant, and the regulatory agency. An example of this form of capacity study is the use of growth-control measures in California cities based on transportation level-of-service indicators.

Compared with the other two general methodologies described above, the carrying capacity technique fares better in addressing the major issues associated with cumulative impact analysis. It focuses on monitoring activity and impacts over time, it documents and models impacts within a particular environmental system, and it provides explicit means for managing direct and cumulative impacts. The methodology, however, does suffer from some operational and conceptual weaknesses. Because of its focus on modeling system thresholds and impacts, the technique is severely limited by the level of scientific understanding of a natural system. Further, when implemented, the methodology tends to concentrate on only one media; therefore, it typically does not deal well with multimedia, interactive, or synergistic impacts. Again, however, this criticism is primarily a weakness of science, not of the methodology per se.

One critical limitation of the approach relates to its ability to manage development. While a carrying capacity approach provides a readily adaptable impact management framework through the identification of unused capacity, it does not provide a systematic way to define an orderly, preference-based or goal-directed use of that remaining capacity. Actions are limited to decisions on and modifications to first-come, first-served project requests.

In sum, three generic methodologies have been used to incorporate cumulative impact considerations into project-level decisions. Each approach provides particular insights into the problem, yet none responds to all of the requirements of an ideal approach.

IMPROVING CUMULATIVE IMPACT ASSESSMENT

Because existing methodologies often fall short when judged against an ideal methodology for cumulative impact analysis, we propose several improvements to cumulative impact assessment. These potential gains include improvements to

(1) monitoring actions and impacts over time and space, (2) scientific modeling of complex natural systems, and (3) the management of cumulative impacts through decisions about specific actions.

Potential Improvements to Monitoring Efforts

One rapidly changing technology stands out in its potential to improve monitoring over space and time. Well-designed and carefully implemented geographic information systems (GIS) could improve monitoring capabilities for many types of development activities and some environmental parameters. For urban development in particular, vector-based systems could tie attribute data about current land use, proposed development, zoning variance requests, and permit information to specific parcels. This ability to summarize information about particular land use changes over time by type of action, and to display this data spatially could improve the typical urban environmental impact assessment.[1]

Clearly, GIS technology allows easier spatial aggregation and disaggregation of mapped information over a variety of noncoincident boundaries. This capability has the potential of improving the analysis of cumulative impacts on natural regions, such as watersheds, by agencies whose jurisdictions are defined by political boundaries that do not correspond to natural features. Further, with statewide GIS coverage of natural features such as soils, vegetation, and land use, there is also an enhanced potential for sharing of important environmental data among different levels of government. Two unanswered questions remain: whether the level of accuracy of statewide mapping efforts will be sufficient for local planning needs and whether the cost and effort involved in developing accurate and detailed maps for a variety of attributes over a large geographic area will exceed the resources of most jurisdictions. At this point, scaling up small prototype systems to large regions is still costly and difficult.

For monitoring development information over time, GIS technology has already proven useful with several research groups reporting the use of raster-based systems to analyze changes in watersheds. One example of this use of GIS examined a series of digitized aerial photos over a 50-year period (Dickert and Tuttle 1985) to evaluate changes in land coverage and the related impacts on sedimentation. Another study (Johnston et al. 1988) compared aerial photos from two time periods and combined the raster-based data for vegetation with vector-based information on streams, soils, and topography. Other capabilities of the GIS were used to define buffer zones around linear and polygonal features and to measure various distances between water bodies and land use zones.

[1] For example, in the litigation of the environmental impact statements for San Francisco high-rises discussed earlier, a properly designed GIS system could have answered the court's question about the total square footage in high-rises under construction, permitted and not yet started, and under review. Admittedly, a parcel database without graphic capability could also have answered this question. The added advantage of a GIS would be the ability to see quickly the spatial pattern of such similar land uses.

The main problem with current GIS technology in the analysis of impacts over time is the tendency of many systems either to be snapshots at one point in time or to not carry forward the chronological information in the database. For example, many parcel-based GIS systems are designed to carry only the present land use information, ownership, and other similar current data. In the interest of currency and timely updating, changes to the attribute database fields often wipe out the very history that would be useful to the analysis of cumulative impacts over time (e.g., old parcel boundaries are updated when a new subdivision is built). Having more accurate and complete historical information with spatial detail could improve our ability to forecast future actions and their likely environmental consequences. Keeping track of historical data, however, is not without its costs in both system design and implementation. Data requirements will be magnified by many orders if this suggestion is implemented.

Potential Improvements to Modeling Efforts

A number of challenges remain in improving the modeling efforts necessary for the appropriate consideration of cumulative impacts. Although the understanding of some natural systems exceeds that of others, there is very little doubt that the level of knowledge is still quite low for determining the response of environmental systems to developmental perturbations. First, we know very little about the ways a development action is transferred into stimuli to the various environmental systems affected. Second, we have some information, but in most cases quite limited information, on the total response of the system to the stimuli. Third, we know little about most system characteristics of interest to cumulative impact analysis such as thresholds, nonlinear responses, recovery processes, delayed responses, or structural and systemic alterations. Finally, we lack information to determine the synergistic effects of a variety of stimuli on one system or the interaction of effects between systems.

In the absence of these and related aspects of scientific information, it becomes difficult to produce relevant and accurate models to predict the impacts of a proposed project. Nonetheless, several modeling efforts have provided simple, and in some cases elegant, estimates of the impacts of new activity on a particular environmental system (see, for example, Dickert and Tuttle 1985; Bedford and Preston 1988). Other attempts at understanding interactive impacts have used matrix techniques to identify the increased impact expected by interactions between particular, highly interrelated environmental systems (see, for example, Emery 1986).

Major advancements in understanding and modeling environmental system behaviors may be possible with continued studies into natural ecosystems (Hunsaker 1989); these studies attempt to identify system responses, recovery time, and synergistic or interactive effects through intensive data collection and modeling. Results of these investigations could add the level of detail and sophistication needed for improved modeling efforts.

GIS technology may also be of use in the improvement of existing models because it makes the traditional overlay analysis — one of the core methodologies for combining the effects of a variety of physical attributes measured over space — extremely easy to conduct. A stronger argument for the ability of GIS to improve modeling may lie in its capabilities for providing buffer zones, for measuring distances between development actions, or for conducting spatial queries interactively.

To date, GIS technology has been notably lacking in sophisticated models of development actions and their impacts (Harris 1989), including air or water dispersion models. Current research at the U.S. Environmental Protection Agency is aimed, in part, at evaluating the potential of GIS technology in these models. Another area for potential research is whether existing complex models could be incorporated inside a GIS, exist in their current form with a strong link to a GIS for display, or perhaps should not be linked at all because of the potential complexity of such an integrated system.

Potential Improvements to Management Efforts

A brief review of the difficulties inherent in analyzing cumulative impacts indicates that the majority of the problems arise in the management or control of cumulative consequences of individual projects. These problems often result from inadequate statements of policy goals or directives, mismatch of jurisdictional and impact boundaries, or a general unwillingness to control incremental effects until some major impact occurs. Some of these problems have potential solutions that require adaptations of existing planning or assessment techniques; others call for major changes in the way society manages impacts on environmental systems.

Mitigation has been used extensively as a strategy to control the adverse effects of proposed development. Identifiable impacts are ameliorated either through in-kind restoration or through monetary contributions. In the context of cumulative impacts, mitigation measures are used often to reduce project incremental impacts to the lowest level possible by altering the project design or to provide adjustments for negative impacts. Conceptually, the idea is that if mitigation lessens project impacts, then the sum of very small impacts should be cumulatively very small. However, some systems do not behave in such a simple manner. Therefore, mitigation has been criticized as a tool to manage continued development that may cumulatively stress a system beyond its capacity, particularly if inadequate monitoring or enforcement exists (Dickert and Tuttle 1985). Without complete reduction of impacts to zero, mitigation simply reduces the additive effects of cumulative impacts, yielding a delayed but not eliminated negative effect.

To provide a management system that is more responsive to cumulative concerns, we proposed a graduated scale of involvement for case-by-case decision making (Contant 1984). Our approach recognizes that more information, discussion, coordination, and consideration of cumulative impacts is necessary in cases where thresholds are likely to be reached or where knowledge is lacking or when policies cannot be precisely defined. Alternatively, when thresholds are distant, impacts are well-understood, and when policies clear and precise, simple administrative strategies could be used to ensure cumulative impact management.

This graduated level of management is explicitly intended to incorporate cumulative impact concerns into case-by-case analyses required of most jurisdictions. Conceptually, this approach reflects an increasing need for control and involvement as project incremental impacts become cumulatively significant when viewed over time and space. When put into practice, this approach performed quite well over the limited time of its usage. It may hold some promise for further application and investigation.

Unfortunately, no single management approach seems to respond adequately to all cumulative impact considerations. Organizations performing the impact analyses are seldom responsible for the entire area in which the impact occurs, nor are they solely responsible for the control of those impacts. This mismatch between impact area boundary and management/jurisdictional boundary, at a minimum, calls for improvements in the cooperation and coordination between all the entities involved. At best, new entities could be defined based on the boundaries of environmental systems.

Further, the jurisdiction over which managerial control must be exercised is often far smaller than the impact area or affected area. Therefore, even the most well-developed efforts to control cumulative impacts within a jurisdiction can be thwarted by inaction by a nearby entity within the impact area. Consider, for example, the case involving efforts to alleviate traffic congestion. Controls imposed by one city will be ineffective if the neighboring cities which generate travel impose no similar restrictions. Therefore, adequate consideration of cumulative impacts requires area-wide coordination and anticipation of new development. Cumulative impacts that are felt at a basin-wide or regional scale (Stakhiv 1988) can only be addressed through planning processes directing development at that same scale.

This call for regional management of cumulative impacts is not as simple as one might think, however. Cowart (1986), in his discussion of Vermont's Act 250, suggests that a state land use program is not enough. It must also include the proper planning processes that define the relevant policy goals, determine appropriate management strategies, and adopt the proper control actions. Only under these conditions can an enlightened and proper control of cumulative consequences occur.

In sum, there have been few successes at managing cumulative impacts. Some ideas hold promise, such as mitigation along with proper monitoring or development of a graduated system of analysis of cumulative impacts. More fundamental changes may be necessary to reflect the proper match of management to the actual level and scale of impact. This call for regional planning, and its related demand for control of private action seems to be the unpopular conclusion drawn by those interested in monitoring, predicting, and managing cumulative environmental effects.

CONCLUSIONS

In this paper, we have attempted to provide a summary of the state of the art for cumulative impact analysis. We have summarized the various components that form cumulative impacts. We have also identified many of the operational difficulties associated with putting these conceptual definitions into action. Further, we examined existing techniques used for cumulative impact analysis and assessed their strengths and weaknesses.

This review of the present state of cumulative impact analysis generated two major sets of findings. First, certain elements are essential in the performance of cumulative impact assessment and control. Second, several new technologies or techniques may respond to current limitations and provide some solutions to analysis problems. Specifically, we have suggested that three elements are necessary to address and control cumulative environmental impacts: detailed monitoring, accurate modeling, and effective management.

Monitoring efforts provide the critical baseline information to describe past and present conditions. Further, monitoring is essential to understanding growth patterns and anticipating future development actions. The second element, modeling, provides the link between development actions and environmental system responses. Modeling efforts in the context of cumulative impact considerations must explore, in particular, system thresholds, synergistic effects, and interactions among media, as well as the more classical stimuli-response behaviors of environmental systems. Finally, management of cumulative impacts is essential due to the mismatch between the *cumulative* level of analysis and the *individual* case-by-case decisionmaking environment. Therefore, management efforts must reconcile the societal demands for control of cumulative adverse impacts with the similar need for increased development as reflected in individual project planning and construction.

New technologies and techniques may hold some promise for improving cumulative impact analysis. Geographic information systems may aid in monitoring over time and space and in providing an interactive mechanism for modeling. Ecosystem analyses may dramatically improve our understanding of critical environmental systems, their characteristics, and their behaviors under developmental stress. Management approaches, including mitigation and graduated levels of analysis, should also aid in translating spatial and temporal constraints into individual project-level decision criteria.

Suggestions such as these may improve the monitoring, modeling, and management of development actions that produce adverse cumulative impacts. Many challenges remain, however. Further research is necessary to provide answers to many questions. How useful and effective are the tools suggested in this paper to improve impact analyses? How will new scientific knowledge be translated for input into organizations and decisionmaking processes? Can cumulative impacts be controlled through existing legal and institutional arrangements? Continued investigations must be conducted into these theoretical and practical questions concerning the analysis of cumulative impacts.

REFERENCES

Barney, P. E. 1981. The programmatic environmental impact statement and the National Environmental Policy Act regulations. *Land Water Law Rev.* 16(1):1–31.

Baskerville, G. 1986. Some scientific issues in cumulative environmental impact assessment. pp. 9–14. In G. E. Beanlands, et al. (eds.), *Cumulative Environmental Effects: A Binational Perspective.* Minister of Supply and Services, Canada.

Beanlands, G. E., W. J. Erckmann, G. H. Orians, J. O'Riordan, D. Policansky, M. H. Sadar, and B. Sadler (eds.). 1986. *Cumulative Environmental Effects: A Binational Perspective.* The Canadian Environmental Assessment Research Council and the United States National Research Council, Minister of Supply and Services, Canada.

Bedford, B. L., and E. M. Preston. 1988. Developing the scientific basis for assessing cumulative effects of wetland loss and degradation on landscape functions: Status, perspectives, and prospects. *Environ. Manage.* 12(5):751–771.

Center for Wetland Resources. 1977. Cumulative Impact Studies in the Louisiana Coastal Zone: Eutrophication, Land Loss. Louisiana State University, Baton Rouge.

Clark, W. C. 1986. The cumulative impacts of human activities on the atmosphere. pp. 113–123. In G. E. Beanlands, et al. (eds.), *Cumulative Environmental Effects: A Binational Perspective.* Minister of Supply and Services, Canada.

Contant, C. K. 1984. Cumulative impact assessment: Design and evaluation of an approach for the Corps Permit Program at the San Francisco District. Ph.D. dissertation. Stanford University, Stanford, CA.

Cowart, R. H. 1986. Vermont's Act 250 after 15 years: Can the permit system address cumulative impacts? *Environ. Impact Assess. Rev.* 6:135–144.

CEQ (Council on Environmental Quality). 1978. National Environmental Policy Act: Final regulations. *Fed. Regist.* 43(230):55978–56007.

COE (U.S. Army Corps of Engineers). 1975. Environmental Assessment: Proposed Projects on Spa and Back Creeks. COE, Baltimore, MD.

COE (U.S. Army Corps of Engineers). 1976. Siletz Wetlands Review. Portland District, Portland, OR.

Dickert, T. G., and A. E. Tuttle. 1985. Cumulative impact assessment in environmental planning: A Coastal Wetland Watershed example. *Environ. Impact Assess. Rev.* 5:37–64.

Emery, L. 1989. FERC's Experience in Analyzing Cumulative Impacts. Poster presented at The Scientific Challenges of NEPA: Future Directions Based on 20 Years of Experience Conference, October 24–27. Knoxville, TN.

Emery, R. M. 1986. Impact interaction potential: A basin-wide algorithm for assessing cumulative impacts for hydropower projects. *J. Environ. Manage.* 23:341–360.

Scientists' Institute for Public Information, Inc. v *Atomic Energy Commission. Environ. Law Rep.* 3:20525 (1973a).

Indian Lookout Alliance v. *Volpe. Environ. Law Rep.* 3:20739 (1973b).

Sierra Club v. *Morton. Environ. Law Rep.* 5:20463 (1975).

Kleppe v. *Sierra Club. Environ. Law Rep.* 6:20532 (1976).

Thomas v. *Peterson. Environ. Law Rep.* 15:20225-20230 (1985).

Environmental Science Associates, Inc. 1983. Growth Management Alternatives for Downtown San Francisco: Downtown EIR Consultant's Report. Department of City Planning, San Francisco, CA.

Hapke, P. 1985. *Thomas* v. *Peterson*: The Ninth Circuit Breathes New Life into CEQ's Cumulative and Connected Actions Regulations. *Environ. Law Rep.* 15:10289–10297.

Harris, B. 1989. Beyond geographic information systems: Computers and the planning professional. *J. Am. Plann. Assoc.* 54:85–90.

Hawkes, C. L., J. B. Beattie, S. L. Cain, S. R. Culver, P. L. Jenkins, T. L. Johnson, M. T. Reynolds, D. L. Vance-Miller, R. L. West, and S. L. Wynn. 1989. Quantitative Cumulative Impact Analysis, Resource Protection Goals, and a Geographic Information System Used to Evaluate Impacts of Places and Lode Mining in Three Alaska National Parks. Paper presented at the Scientific Challenges of NEPA: Future Directions Based on 20 Years of Experience Conference, October 24–27. Knoxville, TN.

Hirsch, A. 1988. Regulatory context for cumulative impact research. *Environ. Manage.* 12(5):715–723.

Hopkins, L. D. 1977. Methods for generating land suitability maps: A comparative evaluation. *J. Am. Inst. Planners* 43:386–400.

Hunsaker, C. T. 1989. Ecosystem Assessment Methods for Cumulative Effects at Regional and Global Scales. Paper presented at the Scientific Challenges of NEPA: Future Directions Based on 20 Years of Experience Conference, October 24–27. Knoxville, TN.

James, T. E., S. C. Ballard, and M. D. Devine. 1983. Regional environmental assessments for policymaking and research and development planning. *Environ. Impact Assess. Rev.* 4(1):9–24.

Johnston, C. A., N. E. Detenbeck, J. P. Bonde, and G. J. Niemi. 1988. Geographic information systems for cumulative impact assessment. *Photogramm. Eng. Remote Sensing* 54:1609–1615.

Maryland Department of State Planning. 1972. Maryland Chesapeake Bay Study. State of Maryland, Annapolis, MD.

McHarg, I. L. 1969. *Design with Nature.* Natural History Press, New York.

Merson, A., and K. Eastman. 1980. Cumulative impact assessment of western energy development: Will it happen? *Univ. Colo. Law Rev.* 51:551–586.

Phillips, Brandt, Reddick-McDonald and Grete, Inc. 1978. The Cumulative Impacts of Shorezone Development at Lake Tahoe. California State Lands Commission, Sacramento, CA.

Preston, E. M., and B. L. Bedford. 1988. Evaluating cumulative effects on wetlands functions: A conceptual overview and generic framework. *Environ. Manage.* 12(5):565–583.

Proett, M. A. 1977. Cumulative impacts of hydroelectric development: Beyond the cluster impact assessment procedure. *Harv. Environ. Law Rev.* 11:77–146.

Raffle, B. I. 1978. The New Clean Air Act — Getting Clean and Staying Clean. *Environ. Rep.,* Monograph No. 26, Vol. 8.

Roots, E. F. 1986. Closing remarks: A current assessment of cumulative assessment. pp. 149–160. In G. E. Beanlands, et al. (eds.), *Cumulative Environmental Effects: A Binational Perspective.* Minister of Supply and Services, Canada.

Schneider, D. M., D. R. Godschalk, and N. Axler. 1978. The Carrying Capacity Concept as a Planning Tool. Report No. 338. Planning Advisory Service, American Planning Association, Chicago, IL.

Selmi, D. P. 1984. The judicial development of the California Environmental Quality Act. *Univ. Calif., Davis, Law Rev.* 18:199–285.

Smith, W. H. B., L. J. Boberschmidt, G. A. Malone, T. S. McDowell, J. Hildreth, and P. Bradley. 1989. Cumulative Impact Analysis of Natural Gas Resource Development in the Coastal Waters of Alabama and Mississippi. Paper presented at the Scientific Challenges of NEPA: Future Directions Based on 20 Years of Experience Conference, October 24–27. Knoxville, TN.

Stakhiv, E. Z. 1988. An evaluation paradigm for cumulative impact analysis. *Environ. Manage.* 12(5):725–748.

Vlachos, E. 1985. Assessing long-range cumulative impacts. pp. 49–80. In V. T. Covello, et al. (eds.), *Environmental Impact Assessment, Technology Assessment, and Risk Analysis.* Springer-Verlag, Berlin.

Wooley, D. R., and J. Wappett. 1982. Cumulative impacts and the Clean Air Act: An acid rain strategy. *Albany Law Rev.* 47:37–61.

Zener, R. V. 1981. *Guide to Federal Environmental Law.* Practicing Law Institute, New York.

Assessing Cumulative Impact on Fish and Wildlife in the Salmon River Basin, Idaho

J. S. Irving,[1] U.S. Department of Energy, Argonne National Laboratory, Argonne, IL; and M. B. Bain,[2] U.S. Fish and Wildlife Service, Auburn University, Auburn, AL

ABSTRACT

The National Environmental Policy Act of 1969 (NEPA) alluded to cumulative impacts, although no formal definition was recognized until 1978 when the Council on Environmental Quality (CEQ) addressed the issue. Subsequently, several legislative acts, federal and state regulations, and court rulings required that cumulative impacts studies be included in environmental impact assessments. Attempts to include cumulative impacts in environmental impact assessments, however, did not begin until the early 1980s. One such effort began when the Federal Energy Regulatory Commission (FERC) received over 1200 applications for hydroelectric projects in the Pacific Northwest. Federal and state agencies, Indian tribes, and environmental groups became concerned that numerous small developments could have potentially significant cumulative impacts on fish and wildlife resources. In response to this concern, FERC developed the Cluster Impact Assessment Procedure (CIAP) which consists of (1) public scoping meetings; (2) interactive workshops designed to identify projects with potential for cumulative effects, resources of concern, and available data; and (3) preparation of a NEPA document (EA or EIS). The procedure was modified to assess the cumulative impacts of 15 hydroelectric projects in the Salmon River Basin, ID. The methodology achieved its primary objective of evaluating the impact of multiple hydroelectric developments on fish and wildlife resources.

[1]J. S. Irving is currently with EG&G Idago, Inc., Idaho Falls, ID.
[2]M. B. Bain is currently with the U.S. Fish and Wildlife Service at Cornell University, Ithaca, NY.

However, the paucity and low quality of data limited the analysis. In addition, the use of evaluative techniques to express and analyze impacts and interactions among proposed projects hindered acceptance of the conclusions. Notwithstanding these problems, the cumulative impact study provided a basis for decision makers to incorporate the potential impact of multiple projects into the hydropower licensing process.

INTRODUCTION

The traditional approach to environmental impact assessment has been to identify the effect of a single development project on individual resources of public interest. Little effort has been made to evaluate the impact of multiple projects on multiple resources. The term *cumulative impact assessment* is often used to refer to a holistic approach to environmental analysis and planning. The National Environmental Policy Act (NEPA) indirectly addressed cumulative impact by referring to interrelations of all components of the natural environment. The Council on Enviornmental Quality (CEQ) defined cumulative impact as the incremental impact of multiple current and future actions with individually minor but collectively significant effects (40 CFR Pts. 1508.7 and 1508.8). Cumulative impact can be concisely defined as the total effect of multiple land uses and developments, including their interrelationships, on the environment. This definition, and current usage of the term, implies that the total effect of several separate projects may be different from the simple sum of single-project impacts.

Cumulative impacts have been recognized in several federal legislative acts (e.g., NEPA Northwest Power Act, Endangered Species Act, and the Federal Water Pollution Control Act), federal regulations (such as those by CEQ referred to above), and court rulings (Horak et al. 1983). The scientific community has widely accepted the influence of an interacting set of factors on the well-being of a species ever since Hutchinson (1957) introduced the multidimensional niche concept. The biological basis for considering multiple factors and their interactions has been recently reviewed (Vernberg 1978; Livingston 1979; Coats and Miller 1981; Sheehan 1984; Reed et al. 1984); and reviews of impact assessment practices (Rosenburg et al. 1981; Beanlands and Duinker 1984; Orians 1986) have specifically identified the lack of cumulative effect considerations as a significant shortcoming. Despite legal and scientific recognition of the need for cumulative impact analysis, there is little indication that progress had occurred before 1985 (Vlachos 1985), one reason being the absence of suitable assessment methods (Contant and Ortolano 1985; Paquet and Witmer 1985). Reviews of existing methods that could be used indicate that none effectively addresses multiple projects, multiple resources, and impact interactions (Horak et al. 1983; Vlachos 1985). A few studies (e.g., Cada and McLean 1985; Leathe et al. 1985) addressed the impacts of multiple projects, but only aggregated or summed these impacts without explicitly considering the environmental impact associated with interactions among the projects.

Consequently, this area of environmental analysis is only beginning to develop conceptually and in practice.

In the early 1980s, increasing electricity rates and demand, as well as incentives in the Public Utilities Regulatory Policy Act, resulted in well over a thousand applications for small-scale hydroelectric developments being filed with FERC (FERC 1984). The public and several federal and state agencies voiced concern that the combined effect of numerous small-scale hydroelectric developments could severely impact valuable fish and wildlife resources. The concern was not so much for the impact from many single projects as it was for the potential combined effects (i.e., cumulative impacts) of several projects potentially affecting an important fish or wildlife population. In response to these concerns, the Federal Energy Regulatory Commission (FERC) proposed the Cluster Impact Assessment Procedure (CIAP) (FERC 1985a). This procedure included many aspects of an earlier FERC cumulative impact study conducted in the San Joaquin River Valley (FERC 1985b). The CIAP was primarily a schedule of interactive workshops intended to determine the number of proposed projects, to identify target fish and wildlife resources for analysis, to define important components of the target resources, and to determine sources and availability of data.

Initial applications of the CIAP were conducted in three western river basins: the Owens (California), the Snohomish (Washington), and the Salmon (Idaho). The Owens CIAP application was conducted essentially as proposed by FERC (FERC 1985b). The Salmon and Snohomish CIAP applications varied considerably from the FERC-defined procedure, although a series of workshops were retained. Based on the responses of the public and government agencies to the FERC request for comments (FERC 1985a) on the CIAP methodology and on input at scoping meetings and workshops, the original CIAP methodology was supplemented (Witmer et al. 1987) with a structured multiple analysis method (Bain et al. 1986, 1989).

This paper describes an application of a structured multiproject assessment method in the context of the CIAP to evaluate the cumulative impact of 15 small-scale hydroelectric projects in the Salmon River Basin of Idaho (FERC 1987). The Salmon River is part of the Columbia River system, the major river basin in the Pacific Northwest. The headwaters of the Salmon River provide important spawning and rearing habitat for salmon and steelhead trout. The areas surrounding these headwaters provides habitat for large mammals such as elk, mule deer, and the gray wolf, a threatened and endangered species. Although this assessment study involved several aquatic and terrestrial target resources analyses, details for chinook salmon will be used as an example.

PROCEDURE

The CIAP process, as proposed by FERC (FERC 1985a), includes four steps: (1) geographic scoping, (2) resource scoping, (3) multiple-project assessment, and

(4) documentation. The purpose of geographic scoping is to identify target resources (e.g., fish and wildlife species, special habitats) that could be affected in a cumulative manner and the proposed projects that could have a cumulative impact on target resources. The resource scoping step finalizes the list of target resources and identifies components of the target resources for analysis. Target resource components are distinct attributes considered to be directly related to the well-being or quality of the target resource (e.g., spawning habitat for chinook salmon, calving areas for elk and mule deer, or the impact to prey animals for the gray wolf). The multiple-project assessment step is the part of the CIAP added by Argonne National Laboratory (ANL) staff[1] and includes (1) assigning impact values to each resource component, (2) assessing impact interaction among projects, (3) integrating impacts for configurations of proposed projects using matrix calculations, and (4) determining criteria for selecting configurations for detailed evaluation.

Geographic Scoping

The geographic scoping meeting lasted 1 week and involved approximately 50 scientists from the U.S Bureau of Land Management, U.S. Fish and Wildlife Service, U.S. Environmental Protection Agency, U.S. Geological Survey, U.S. Forest Service, National Marine Fisheries Service, Idaho Department of Fish and Game, Idaho State Historical Society, Columbia River Inter-Tribal Fish Commission, Northwest Power Planning Commission, several tribal representatives, businesses, organizations, and individuals. Discussions focused on what resources could be cumulatively impacted by two or more proposed projects. The initial target resources considered were chinook salmon (*Oncorhynchus tshawytscha*), sockeye salmon (*O. nerka*), steelhead trout (*O. mykiss*), westslope cutthroat trout (*O. clarki*), elk (*Cervus elaphus*), white-tailed deer (*Odocoilues virginanus*), mule deer (*O. hemionus*), and soils (stability). Fish and wildlife agencies (e.g., Idaho State Department of Fish and Game, U.S. Fish and Wildlife Service) argued for inclusion of rainbow trout (*Oncorhynchus mykiss*), brook trout (*Salvelinus fontinalis*), and bull trout (*S. confluentus*). Other target resources discussed were water quality, visual quality, recreation, land use, and cultural resources. Only white-tailed deer were considered unaffected by the proposed projects in a cumulative manner.

Eighteen proposed projects in the Salmon River Basin were conceivably appropriate for the analysis. However, the meeting group determined that three of the projects did not pose any potential for cumulative adverse impacts to target resources and could be studied independently. The remaining 15 projects, located in the Lower Salmon River, Little Salmon River, and South Fork Salmon River subbasins, were included in the cumulative impact study because of their potential to cause cumulative impacts on the target resources (Figure 1).

[1] Argonne National Laboratory, acting as an extension of the FERC staff, considered an environmental assessment on the proposed hydroelectric projects in the Salmon River Basin, ID.

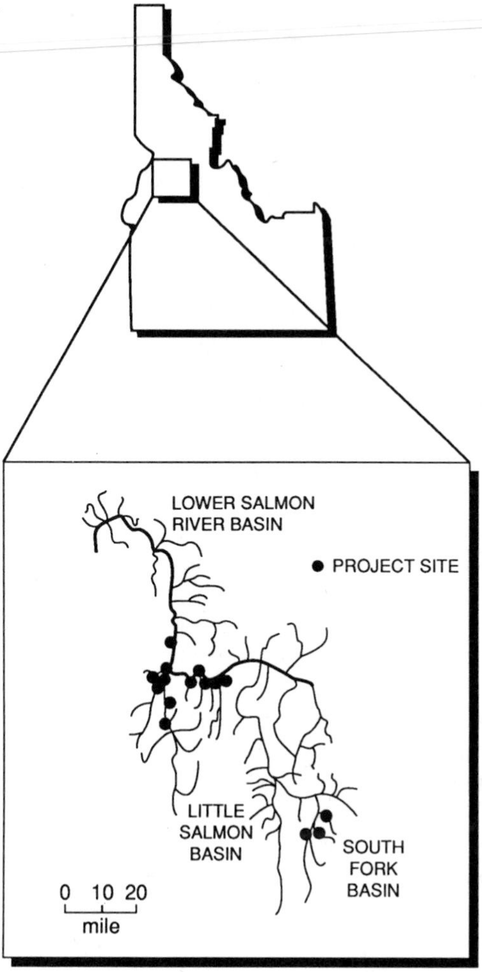

Figure 1. Location of proposed hydroelectric projects in the Salmon River Basin, ID.

Resource Scoping

The resource scoping meeting lasted 3 days and involved approximately 40 scientists representing federal and state agencies, developers, utilities, and conservation groups. Assessment of impacts to important resources requires the identification of resource components that describe the species lifecycle or habitat. Target resources and resource components do not need to include all items to model the environment; however, target resources should be significant elements of the environment that may be affected by project developments. An evaluation of each resource component requires an integrated analysis of several factors that actually describe the physical characteristics of that lifecycle or habitat (e.g., spawning, incubation, rearing). During the resource scoping workshop, a final list of target resources was developed; it included chinook salmon, steelhead trout, resident trout,[2] elk and mule deer, gray wolf (*Canus lupus*),[3] and riparian habitat. Resource components were identified for each target resource, although we will emphasize chinook salmon to illustrate the remaining study steps.

The major activities associated with hydroelectric developments include the placement of facilities in or near streams and the alteration of streamflow characteristics. Impacts to fish (salmon) and aquatic environments were categorized into three major groups: (1) impacts from alteration of streamflow, (2) impacts from interference with migration and movement of salmon within the stream, and (3) impacts from alternation of sedimentation and bedload movement (salmon spawning habitat).

Changes in sedimentation and streamflow were probably the most important impacts associated with the construction and operation of the proposed hydroelectric developments. Many of the proposed projects occur in areas of unstable soil types or where past disturbance has increased the baseline sedimentation in the streams and rivers. Changes in streamflow may affect all life stages of the chinook salmon (spawning, incubation, rearing, and migration). Many other impacts were associated with hydroelectric development (e.g., loss of cover, changes in water temperature, decrease in dissolved oxygen), but it was determined that concentration on the effects of sedimentation and streamflow during the matrix analysis was sufficient to assess significant impacts.

The chinook salmon resource components analyzed further were (1) spawning/incubation habitat, (2) juvenile rearing habitat, (3) adult holding habitat, (4) migration/movement disruption, and (5) sediment/transport. Similar rationale was used to determine resource components for the resident trout, elk and mule deer, gray wolf, and riparian habitat.

Component impact values represent impact magnitudes on a standardized scale. Use of a common scale simplifies the combining of component values into one impact value for a target resource. Without some form of standardization, component values would vary in magnitude, depending on the units involved. Any standardized scale can be used, and in this study, a 0–4 range (no impact to very high impact) was selected. Impact-level criteria for chinook salmon components of spawning/incubation, juvenile rearing, and adult holding habitat are shown in Table 1. Similar criteria were developed for other target resources and components.

Multiple-Project Assessment

The multiple-project assessment step was conducted by ANL staff using resources, components, and impact criteria identified in scoping meetings. For

[2] Rainbow,cutthroat, and bull trout were grouped together as resident trout for this analysis.

[3] The gray wolf was added as a target resource following the matrix-technical workshop because of new information and its status as an endangered species.

Table 1. Description of Impact Level Criteria for Chinook Salmon Resource Components: Spawning/Incubation, Juvenile Rearing, and Adult Holding Habitat.[a]

Impact Levels[b]	Description of Impact Levels[c]
4 (High)	>25% decrease in weighted usable area (WUA) or if WUA not available, then <30% of the mean annual flow
3 (Moderate)	>15–25% decrease in WUA or if WUA not available, then 30 to <60% (April–September) or 30–<40% (October–March) of the mean annual flow
2 (Low)	>5–15% decrease in WUA or if WUA not available, then 60 to <80% (April–September) and 40–<80% (October–March) of the mean annual flow
1 (Negligible)	>0–5% decrease in WUA or if WUA not available, then 80–100% of the mean annual flow
0 (None)	0% or an increase in WUA or if WUA not available, then 100 or >100% of the mean annual flow

[a] Weighted usable area and mean annual flows generated from the applicants' information were used to assign impact values (0, 1, 2, 3, or 4) for increasing levels of impact.
[b] When using the percentage of the mean annual flow, impacts levels were adjusted downward by 1 unit of impact (e.g., 3 to 2) if only a limited amount of anadromous fish habitat was available and by 2 if there was no or very little anadromous fish habitat present.
[c] Where possible, impact levels were assigned using information from approved instream flow modeling study results. If the study results were not available or not approved, then the percentage of the mean annual flow was used to assign impact levels.

each proposed project and target resource, impact levels were determined. For chinook salmon, and the Riordan Creek Project,[4] impact values were developed as follows assuming full implementation of recommended mitigation. Based on surveys conducted by the applicant, it was concluded that chinook salmon do not use Riordan Creek for spawning or adult holding habitat. Therefore, the spawning/incubation and adult holding components were assigned an impact level of 0 (Table 2). Based on the same surveys, juvenile chinook salmon were observed and captured in the lower part of Riordan Creek. The recommended instream flow would decrease the weighted usable area between 0 and 5% from that of a normal water year. Therefore, an impact level of 1 was assigned to the juvenile rearing habitat component based on analysis of instream flow information. The risk of upstream obstruction, impediment, or loss of juvenile chinook salmon from the powerhouse discharge would be low, assuming full implementation of mitigation. An impassable barrier prevents adult salmon from reaching the diversion dam. Therefore, the migration/movement component was assigned an impact level of 1. Based on analysis of the development plans, soils and geology, and streamflow and water quality, we concluded that the Riordan Creek drainage is relatively undisturbed with a low sedimentation potential and a low to moderate mass-wasting potential. Therefore, an impact level of 2 was assigned to the sediment/transport component. In summary, impact levels were assigned based on existing conditions and expected impacts, with recommended mitigation, then compared

[4] The proposed Riordan Creek Hydroelectric Project (FERC Preliminary License Number 6433) is located on Riordan Creek, a tributary of Johnson Creek in the South Fork of the Salmon River Basin, ID.

Table 2. Summary of Impact Levels for the Chinook Salmon Resource Components for the 15 Proposed Projects with Staff-Recommended Mitigation.

	Impact Levels by Resource Component[a]				
Project[b]	Spawning/ Incubation	Juvenile Rearing	Adult Holding	Migration/ Movement	Sediment/ Transport
Riordan Creek	0	1	0	1	2
Ditch Creek	0	0	0	0	1
Trapper Creek	0	1	0	1	2
Fall Creek	0	0	0	1	3
Lower Squaw Creek	0	0	0	0	1
China Creek	0	0	0	0	0
Upper Squaw Creek	0	0	0	0	3
Grave Creek	1	1	1	2	4
Lower Hat Creek	1	1	1	0	2
Elkhorn Creek	1	1	1	2	3
Partridge Creek	1	1	1	2	3
Lake Creek	1	1	1	2	2
Allison Creek	1	1	1	2	3
Shingle Creek	1	1	1	1	2
French Creek	1	1	1	2	4

[a] Impact levels are 0 = none, 1 = negligible, 2 = low, 3 = moderate, and 4 = high.
[b] Proposed hydroelectric projects in the Salmon River Basin, ID.

with the criteria for each target resource component.

To assess cumulative impacts of multiple projects, we used a model-based methodology that accounts for interactions among project impacts. The model format is matrix oriented and accepts information from any discipline. Matrix algebra is used to compute values representing cumulative impact on each target resource for every project configuration (combination) of the proposed projects. Essentially, a relative cumulative impact score for each configuration is computed. The general formula can be simply stated as

$$\text{Total impact} = \text{sum of project impacts} \pm \text{interaction impacts}$$

This general computation was applied to all possible configurations of the developments under consideration. The interaction impact could cause the total impact to be either greater or less than the project-specific impacts. At this point in the analysis, all configurations are screened separately for each target resource to reduce the number of potential project configuration.

To identify when multiple-project impacts interact, the following question was considered for all project-by-project pairs: Can the level of impact of one project affect the level of impact of another project? If the answer is no, the impact of the projects on the component is strictly additive. If the answer is yes, then there

[5] Interactions that result in impacts beyond the sum of project-specific impacts (i.e., in addition to strictly additive impacts).

Table 3. Interaction Coefficients and Criteria for the Migration/Movement Component of the Chinook Salmon Interaction Matrix.

Interaction Coefficient	Criteria
0.0	No project interaction on migration/movement of target resource
0.1	Project interaction on migration/movement possible but not likely to occur with negligible impact to target resource
0.5	Project interaction on migration/movement likely to occur with low to moderate potential impact to target resource
1.0	Project interaction on migration/movement likely to occur with high or severe potential impact to target resource

is an interaction among project impacts, which can be either *supraadditive*[5] or *infraadditive*.[6] Two interaction coefficients for the model were selected to represent impact interactions between one project and any other in the configuration. For one pair of projects, A and B, one coefficient would represent the effect of A on B and another coefficient the effect of B on A. The interaction is supraadditive if the coefficient is positive and infraadditive if it is negative. An example of interaction coefficient criteria for the chinook salmon is shown in Table 3.

In assessment studies involving many projects, the model calculations become cumbersome. Therefore, project-by-project matrices were completed, and a computer program was developed to execute calculations (details in Bain et al. 1986). The basic matrix computations are illustrated with an example of one target resource, three resource components, and two projects in Figure 2. No weights are used to place emphasis on any resource component so the component matrix and adjusted component matrix are the same (Figure 2). The adjusted component matrix is summed across resources components to derive the weighted sums for each project. An interaction matrix is used to derive an interaction effects matrix, which is then summed across resources components to derive the interaction effects sum. Additive, supraadditive, and infraadditive effects for each project are accounted for by adding the weighted and interaction effects sums. A total cumulative impact score is derived by adding across projects, and this score is used as a relative index of cumulative impact for the two-project configuration. Also, although impact values are summed across target resource components, they are not summed across target resources. This prevents problems with averaging impact scores across target resources.

The utility of the model depends on whether the potentially numerous project configurations can be ranked so that some manageable subset can be considered further. Ranking would be simple if only one target resource was involved. With multiple target resources, no single ranking of configurations can be obtained. A total of 35,767 project configurations was possible with the 15 proposed hydroelectric projects in the Salmon River Basin. The ANL staff developed

[6] Interactions that result in impacts that are less than the sum of project-specific impacts (i.e., causing the total impact to be less than strictly additive).

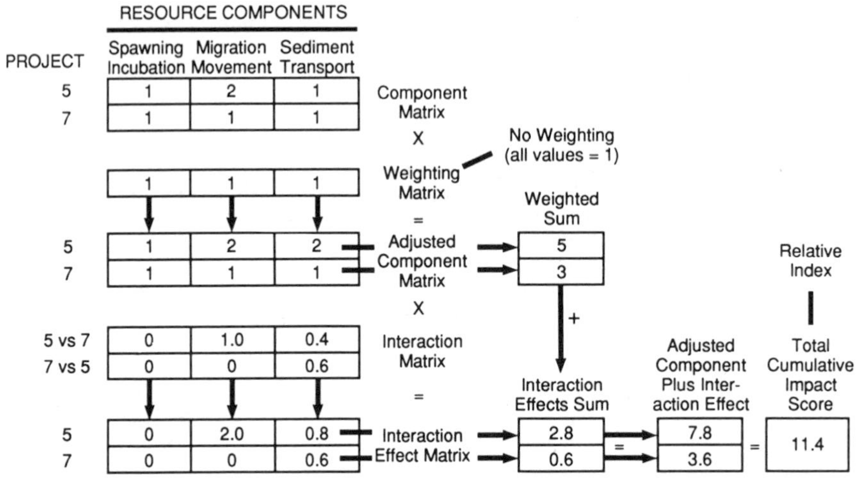

Figure 2. Example of cumulative impact computations for a target resource with three resource components and two projects.

cumulative impact scores for all these configurations for all target resources using the computerized model. Screening this large set of configurations by multiple criteria reduced it to a manageable subset.

Screening criteria were developed for all target resources and are illustrated here for chinook salmon. Each project configuration was compared with screening criteria and placed into one of three management scenarios. The configuration with the largest number of projects meeting all the criteria for a particular scenario was selected for discussion of impacts.

Scenario A, the resource agency management scenario, is a development strategy resulting in no or negligible impact to chinook salmon in the basin. Management policies of the Idaho Department of Fish and Game, National Marine Fisheries Services, and U.S. Fish and Wildlife Service are directed at permitting no or negligible impacts to salmon populations and habitats in the basin. The strictest application of this strategy would allow no projects to be developed if they affected anadromous fish populations or habitats. In practice, consultation and negotiations on project design and location, minimum flows, and other mitigative measures between project proponents and resource agencies have lead to the development of projects that would have negligible impacts to chinook salmon. That is, the projects would not have a significant impact on chinook populations or habitats.

For Scenario A, the following criteria were used to screen all project configurations: (1) a single resource component could not exceed an impact level of 1 (negligible level of impact), and (2) the cumulative impact score could not exceed five times the number of projects in the configuration. Therefore, project configurations with individual impact levels exceeding 1 or with cumulative impact scores greater than five times the number of projects in the configuration [e.g., greater than 75 (15 projects × 5 components = 75 relative impact score)] would be excluded from Scenario A.

As proposed, none of the projects met the above criteria. Implementation of recommended mitigative measures would reduce impacts for all proposed projects and result in three projects meeting the none or negligible impact criterion. The other 12 projects exceeded the first criteria (impact level greater than 1) in one or more resource components and were eliminated from further discussion under Scenario A.

Thus, for the chinook salmon target resource, Scenario A includes seven project configurations: three configurations involving one different project each (three one-way configurations), three configurations involving different combinations of two projects each (three two-way configurations), and one configuration involving three projects (one three-way configuration). The three-way configuration was the largest combination of projects meeting the criteria for Scenario A and, therefore, was used to represent the scenario.

Scenario B, the biological threshold management scenario, represents a strategy resulting in an insignificant biological impact. For Scenario B, the following criteria were used to screen all project configurations: (1) a single resource component could not exceed an impact level of 2 (low level of impact), and (2) the cumulative impact score could not exceed ten times that number of projects in the configuration. Seven projects exceeded the first criteria (impact level greater than 2) in one or more of the resource components and were eliminated from further discussion under Scenario B. The remaining eight projects met the low-impact criterion. Scenario B includes 248 configurations (8 one-way, 28 two-way, 56 three-way, 70 four-way, 56 five-way, 28 six-way, 8 seven-way, and 1 eight-way) and is represented by the eight-way project configuration.

Scenario C, the unrestricted development scenario, is a development strategy that would not restrict the amount of cumulative impact that occurs to chinook salmon. This strategy includes all 15 proposed projects.

Each target resource was evaluated similarly using these same development scenarios. Screening by microcomputer allowed trials with all target resource criteria set at different levels. Iterations of the screening process with successively more selective criteria helped to fine-tune the subset of configurations.

Documentation Phase

A detailed description and concise summary of the impacts associated with any recommended configuration are important, since relative ratings and indexes are used in much of the analysis and evaluation work. Those reviewing or using the study recommendations need a clear disclosure of the anticipated environmental impacts in terms familiar to them. Although text descriptions can be used to elaborate details, a summary (preferable a one-page table of major points) is needed to convey the magnitude of the impacts and their probability of occurrence for each target resource.

Table 4. Largest Project Configuration Under Each Development Management Scenario.[a]

	Scenario		
	A	B	C
Target resource	Resource agency management	Biological threshold management	Unrestricted management
Chinook salmon	B,E,F[b]	A,B,C,E,F,I,L,N	All projects
Steelhead trout	B,E	A,B,C,E,I,L,N	All projects
Resident trout	B,E	A,B,C,E,I,L,N	All projects
Elk/mule deer	All projects	All projects	All projects
Gray wolf	All projects	All projects	All projects
Riparian habitat	All projects	All projects	All projects

[a] The entries in this table represent the largest combination of proposed projects that could be developed (with implementation of staff-recommended mitigation) under each of the three developments scenarios (A, B, C) for each target resource.
[b] Project codes: A = Riordan Creek, B = Ditch Creek, C = Trapper Creek, D = Fall Creek, E = Lower Squaw Creek, F = China Creek, G = Upper Squaw Creek, H = Grave Creek, I = Lower Hat Creek, J = Elkhorn Creek, K = Partridge Creek, L = Lake Creek, M = Allison Creek, N = Shingle Creek, and O = French Creek.

In the Salmon River Basin analysis, none of the projects proposed by the applicants would result in insignificant impacts across all target resources (Table 4). Consequently, each project as proposed had a potential to contribute to the cumulative impacts occurring to one or more of the target resources. With appropriate mitigation, however, seven of the proposed projects would not individually cause significant impacts to any of the target resources. Additionally, if the recommended mitigation (e.g., staggering construction of certain sets of configurations, erosion control plans) would be implemented, there would be no significant interactions between projects. The remaining eight projects would cause significant cumulative impacts to several target resources even with recommended mitigation.

Using the three management scenarios, we evaluated the combined impact of various combinations of the 15 proposed projects to determine what levels of cumulative impacts would occur. A two-project configuration met the Scenario A criteria across all target resources (Table 4). These two projects, with recommended mitigation, would have a negligible level of impact to any target resource. This conclusion was reached because for each target resource, the number of individual animals and the amount of habitat affected (both in absolute terms and relative to the total present in the basin) would be negligible. No critical areas would be affected, and impacts to any specific resource component would be negligible. Because the scenario included two widely separated projects, no or negligible interaction would occur.

A seven-project configuration met the Scenario B criteria across all target resources (Table 4). These seven projects, with recommended mitigation, would have a low level of impact to any target resource. The number of animals and habitat affected would be low, not exceeding any biological threshold. No critical areas would be affected, and impacts to any specific resource component would be low. However, Scenario B does include groups of projects that would have a high potential for interaction (mainly from sedimentation). Staggering the construction of projects with moderate or high interaction potential would reduce cumulative impact from interaction to low levels.

We concluded that under Scenario C, the cumulative impacts to all target resources except elk, mule deer, and riparian habitat would be significant. We reached this conclusion because moderate or moderate to high levels of impact would occur to several resource components for each target resource and because of the high potential for interaction (mainly from sedimentation) among many of the proposed projects. Populations and habitats of chinook salmon would be significantly reduced beyond reductions that were already occurring in the basin.

We concluded, that with proper mitigation, including that of staggering project construction, impacts would be low for those projects meeting the requirements of Scenario B (Table 4). However, based on economic analyses that were not part of the environmental study, four projects did not have a positive net economic benefit. Therefore, we recommended the configuration of three projects meeting biological threshold management Scenario B criteria and having a positive net economic benefit (Table 4).

DISCUSSION

The success of any impact assessment methodology is largely determined by the extent to which assumptions and simplifications can be defended. From its onset, the CIAP met with broad-based resistance from developers, environmental groups, and federal and state agencies. Many practical and procedural questions were raised concerning the CIAP and the multiple-project analysis model we used. How would preliminary permits be incorporated into the process? How could the analysis be conducted with inadequate information? Would the averaging of impact values mask the true impacts across projects or target resources? What does the cumulative impact score mean? Despite these kinds of questions, FERC directed its "staff to proceed with the CIAP as it has been specified ... making such modifications to the CIAP as are appropriate ... ."[7] Modification of the CIAP, while addressing many of the above issues, never did fully allay the concerns of the developers, environmental groups, or agencies. Although a great deal of time and effort was spent attempting to eliminate perceived problems, little debate actually focused on basic assumptions and appropriate simplifications.

One of the major strengths of the CIAP was the workshops and meetings scheduled early in the process. Designed to be interactive, the workshops and meetings solicited comments and suggestions from developers, environmental groups, and state and federal agencies. These meetings were designed to collect

[7] Memorandum to the commission from the Office of General Counsel, Federal Energy Regulatory Commission, April 18, 1985.

baseline information, to scrutinize method assumptions, and to determine the structure and scope of the analysis. The identification of project clusters, target resources, resource components, and impact criteria was used to scope the analysis. While the workshops and meetings succeeded for some of these purposes, it failed at others. The workshops and meetings provided an opportunity for political statements and agency posturing. An exchange of biological information was replaced by position statements by developers, environmental groups, and federal and state agencies. A *court hearing* type of atmosphere contributed to this largely unproductive exchange. In the end, the workshops and meetings were used to state positions and criticisms. They were not conducive to a genuine debate and defense of the CIAP method.

Although not a weakness of the methodology, inadequate information hindered the acceptance of the CIAP's conclusions. The methodology was capable of handling both qualitative and quantitative information. With the use of evaluative techniques, impact values (ranging from 0 to 4) were assigned to a wide range of information. This dimensionless scale circumvents the problem of limited quantitative data and poorly understood resource-impact relationships. When an evaluative approach is used, the development of appropriate impact criteria is key to the success of the methodology. The impact assessment team was challenged, and sometimes divided, by the task of criteria development, but the involved agencies and organizations devoted relatively little attention to this difficult technical step.

An important component of our analysis that was never widely appreciated was the multiproject assessment model. The model was just another step in the analysis, a tool, used to derive the cumulative impact score. In this case, the cumulative impact score was used to develop a *relative* ranking of the many different project configurations. Workshop participants never fully scrutinized the role of the impact criteria, matrix analysis (impact scores), and project screening (cumulative impact scores). Instead they developed suspicions about the assessment method solely from the model inputs and final results.

Another difficult task was assigning of nonadditive interactions (i.e., supraadditive and infraadditive). Little is known of the actual interactions among environmental factors and resources. Without good information, a qualitative system of incorporating interaction was used. Although this simplification still accounted for project interactions, it may have been unacceptably simplified and therefore may not parallel actual resource responses. Basic research is needed on how biological resource populations are affected by numerous, spatially dispersed changes in their environment. Until biological responses are even superficially known on a landscape level, analysis methods like the one we used will not have a firm biological basis.

The documentation phase was the final and most visible part of the process. This phase evaluated the environmental impacts of select configurations of projects identified using impact levels and project screening. This was accomplished using actual biological information, not impact levels. This phase was the key to linking the impact levels to the actual potential project impacts. The impact levels, criteria, and interaction coefficients merely provided the assessment team with a small, manageable subset of potential project configurations to describe in full detail. Probably the greatest achievement of this study was our ability to shift discussion and analyses from individual projects to specific configurations of projects (i.e., numerous single impacts to complex scenarios with a cumulative impact). By considering configurations of proposed projects, we were able to make one of the first comprehensive attempts at addressing the cumulative impact issue.

REFERENCES

Bain, M. B., J. S. Irving, R. D. Olsen, E. A. Stull, and G. W. Witmer. 1986. Cumulative Impact Assessment: Evaluating the Environmental Effects of Multiple Human Developments. ANL/EES-TM-309. Argonne National Laboratory Report, Argonne, IL.

Bain, M. B., J. S. Irving, and G. W. Witmer. 1989. An approach to assessing cumulative impacts on fish and wildlife. In *Freshwater Wetlands and Wildlife: Perspectives on Natural, Managed, and Degraded Ecosystems.* U.S. Department of Energy, Office of Scientific and Technical Information.

Beanlands, G. E., and P. N. Duinker. 1984. An ecological framework for environmental impact assessment. *J. Environ. Manage.* 18:267–277.

Cada, G. F., and R. B. McLean. 1985. An approach for assessing the impacts on fisheries of basin-wide hydroelectric development. pp. 367–372. In F. W. Olsen, R. G. White, and R. H. Hamre (eds.), *Proceedings, Symposium on Small Hydroelectric and Fisheries.* American Fisheries Society, Bethesda, MD.

Coats, R. N., and T. O. Miller. 1981. Cumulative silvicultural impacts on watersheds: A hydrologic and regulatory dilemma. *Environ. Manage.* 5:147–160.

Contant, C. K., and L. Ortolano. 1985. Evaluating a cumulative impact assessment approach. *Water Resour. Res.* 21:1313–1318.

Federal Energy Regulatory Commission. 1984. Commission Directive to Staff. Issued 20 December 1984. U.S. Federal Energy Regulatory Commission, Washington, D.C.

Federal Energy Regulatory Commission. 1985a. Procedures for assessing hydroelectric projects clustered in river basins: Request for comments. *Fed. Regist.* 50:3385–3403.

Federal Energy Regulatory Commission. 1985b. Final Environmental Impact Analysis of Small-Scale Hydroelectric Development in Selected Watersheds in the Upper San Joaquin River Basin, California. FERC/EIA-001. Federal Energy Regulatory Commission, Office of Hydroelectric Licensing, Washington, D.C.

Federal Energy Regulatory Commission. 1987. Final Environmental Impact Statement, Salmon River Basin: Fifteen Hydroelectric Projects, Idaho. FERC/FEIS-0044. Federal Energy Regulatory Commission, Washington, D.C.

Horak, G. C., E. C. Vlachos, and E. W. Cline. 1983. Methodological guidance for assessing cumulative impacts on fish and wildlife. Prepared by Dynamac Corporation, Fort Collins, Colorado, for the Eastern Energy and Land Use Team. Contract 14-16-009-81-058. U.S. Fish and Wildlife Service, Kearneysville, WV.

Hutchinson, G. E. 1957. Concluding remarks. *Cold Springs Harbor Symp. Quant. Biol.* 22:415–427.

Leathe, S. A., M. D. Enk, and P. J. Graham. 1985. An evaluation of the potential cumulative bioeconomic impacts of proposed small-scale hydro development on fisheries of the Swan River drainage, Montana. pp. 377–387. In F. W. Olson, R. G. White, and R. H. Hamre (eds.), *Proceedings, Symposium on Small Hydroelectric and Fisheries*. American Fisheries Society, Bethesda, MD.

Livingston, R. J. 1979. Multiple factor interactions and stress in coastal systems: A review of experimental approaches and field implications. pp. 389–411. In W. B. Vernberg, et al. (eds.), *Marine Pollution: Functional Responses*. Academic Press, New York.

Orians, G. H. 1986. *Ecological Knowledge and Environment Problem-Solving*. Committee on the Application of Ecological Theory to Environmental Problems, National Research Council, National Academy Press, Washington, D.C.

Paquet, P. J., and G. W. Witmer. 1985. Cumulative impact of small hydroelectric developments: An overview of the issues. pp. 343–345. In F. W. Olson, R. G. White, and R. H. Hamre (eds.), *Proceedings, Symposium on Small Hydroelectric and Fisheries*. American Fisheries Society, Bethesda, MD.

Reed, R. M., J. W. Webb, and G. F. Cada. 1984. Siting energy projects: The need to consider cumulative impacts. In *Proceedings, Facility Siting and Routing Conference, 1984: Energy and Environment, Banff, Alberta, Canada, April 15–18, 1984*. Environment Canada, Ottawa, Ontario.

Rosenburg, D. M., et al. 1981. Recent trends in environmental impact assessment. *Can. J. Fish. Aquatic Sci.* 38:591–624.

Sheehan, P. J. 1984. Effects on individuals and populations. pp. 23–50. In P. J. Sheehan, et al. (eds.), *Effects of Pollutants at the Ecosystem Level*. Wiley, New York.

Vernberg, F. J. 1978. Multiple-Factor and Synergistic Stresses in Aquatic Systems in Energy and Environmental Stress in Aquatic Systems. J. H. Thorp, and J. W. Gibbons, eds. U.S. Department of Energy Report CONF-771114. U.S. Department of Energy, Washington, D.C.

Vlachos, E. C. 1985. Assessing long-range cumulative impacts. In V. T. Covello, et al. (eds.), *Environmental Impact Assessment, Technology Assessment, and Risk Analysis*. Springer-Verlag, New York.

Witmer, G. W., M. B. Bain, J. S. Irving, R. L. Kruger, T. A. O'Neil, R. D. Olsen, and E. A. Stull. 1987. Cumulative impact assessment: Application of a methodology. Waterpower 1987. In Proceedings, International Conference on Hydropower/American Society of Civil Engineers, Portland, Oregon, August 19–21, 1987.

Use of Programmatic EISs in Support of Cumulative Impact Assessment

A. Myslicki, Science Applications International Corporation, McLean, VA

ABSTRACT

Programmatic environmental impact statements (EISs) can be an extremely useful resource in efforts to comply with the cumulative impact provisions of the Council on Environmental Quality (CEQ) regulations (40 CFR 1508.7), especially when many similar yet minor actions are expected over time. The overall value of programmatic EISs primarily results from the ready availability of data and analytical resources and inherent cost economies associated with access to relevant and reduced information. Using three programmatic EISs as examples, this paper examines how the analyses in the EISs were used to consider impacts for similar individual actions that collectively resulted in potentially significant cumulative impacts. The discussion includes a step-by-step approach to cumulative analysis and development of corresponding mitigation measures. This approach uses statistically representative models to scale up individual actions to regional or national levels. The paper also discusses representative models that are available to assess impacts at the regional level, including transport and fate models.

INTRODUCTION

Need for Cumulative Impact Analysis

The Council on Environmental Quality (CEQ) regulations require that all federal agencies consider cumulative impacts in their environmental analyses. CEQ regulation 40 CFR 1508.7 defines cumulative impact as "the impact on the environment which results from the incremental impact of the action when added to other past, present, and reasonably foreseeable actions regardless of what agency (federal or nonfederal) or person undertakes such other actions." "Action" is further defined in 40 CFR 1508.25 to include connected actions and similar actions.

Over the last 5 years, plaintiffs have increasingly focused on the issue of cumulative effects. The U.S. Forest Service (USFS) and the U.S. Army Corps of Engineers, for example, have lost several recent cases when an environmental assessment (EA) failed to broadly define cumulative effects and accept the relatively low threshold for concluding that an effect is reasonably foreseeable. In *National Wildlife Federation* v. *United States Forest Service*, 592 F. Supp 931 (D. Roe. 1984), *modified in part*, 801 F.2d 360 (9th Cir. 1986), the court held that the National Environmental Policy Act (NEPA) required preparation of an EIS analyzing the cumulative impacts of timber sales in a USFS 7-year plan and could not rely on individual EAs and Finding of No Significant Impacts for each timber sale.

The question is not whether cumulative impacts should be considered but how. This paper discusses the usefulness of using programmatic EISs to assess cumulative impacts of a specific subset of federal actions. Programmatic EISs are very useful for analyzing impacts when a number of similar yet often minor actions are expected over time. This could range from dredge and fill projects to any kind of small development projects expected within a watershed.

Advantages of Using Programmatic EISs to Look at Cumulative Impacts

Using programmatic EISs to look at cumulative impacts allows the agency time for comprehensive planning. A programmatic EIS may be the only place that impacts across diverse geographic areas have the opportunity to be considered. When these issues are evaluated from the programmatic perspective, mitigation measures can be developed for both site-specific and cumulative impacts, often negating the need to conduct EISs on future site-specific action. Therefore, although such EISs initially take additional time, they can speed program implementation overall.

Disadvantages of Using Programmatic EISs to Look at Cumulative Impacts

A major disadvantage of conducting a programmatic EIS is that it provides opponents with a vehicle for stopping the entire program. Successful challenges to the U.S. Department of the Interior's coal leasing program and outer continental shelf leasing program and the most recent USFS Spotted Owl protection program illustrate this possibility. The other disadvantage is that in the short term it does cost more time and money. Cumulative impact assessment can be expensive, and it may reveal problems that are difficult to deal with. There are also important scientific limitations to our ability to predict cumulative impacts. Further, it is sometimes difficult to know how to analyze a number of minor yet diverse actions in a single document and how to find, collect, and use data at the cumulative level to input into predictive models.

Major Points Illustrated by the Case Studies

Three major issues seem to be of most concern when producing programmatic EISs that also take a hard look at cumulative impacts.

1. Development of scenarios for characterization of potential reasonably foreseeable actions. The case studies illustrate the use of model scenarios to overcome this problem.
2. The availability of models for analyzing cumulative impacts for different resource elements. The case studies identify a number of models and discuss how to deal with incomplete and unavailable information and how to use representative rather than actual data.
3. Delays caused by the use of programmatic EISs. The case studies show that programmatic EISs can actually speed implementation of the program by allowing the early identification and mitigation of significant impacts (both cumulative and site specific).

The three case studies that illustrate these points are

- Final EIS on the Comprehensive Impacts of Permit Decisions Under Tennessee Federal Program (OSMRE 1985)
- Final EIS on Managing Competing and Unwanted Vegetation (USFS 1988)
- Draft EIS on the National Boll Weevil Cooperative Control Program (APHIS 1989)

THREE CASE STUDIES

Office of Surface Mining Reclamation and Enforcement (OSMRE) Tennessee Coal Mining EIS

This EIS points out how the development of model scenarios can be effectively used in the analysis of cumulative effects. It also shows how a programmatic analysis of cumulative impacts and the subsequent development of mitigation measures can negate the need for preparation of future EISs.

Scope of the EIS

The purpose of this EIS was to analyze the cumulative impacts of decisions by the OSMRE on permit applications for coal mining (both surface and underground) under the federal program for Tennessee. The scope of analysis included deficient existing permits, new applications for revisions/modifications of existing permits, permit applications for new mines that had not been processed, and the 40–50 permit applications for new mines expected to be filed each year. The EIS analyzed the approval or denial of federal permits for all these mining operations.

Developing Representative Actions

A major step in this complex cumulative analysis was to break the actions down into representative model actions. First, however, we wanted to divide the area for analysis into something more manageable than the entire Tennessee coal field. We have found that this bottom-up approach — starting with small representative actions and scaling up to larger regional actions — works well in defining cumulative impacts. In the end, we based the analysis on 19 representative model mines in five mining regions (Figure 1) that are typical of mining operations in Tennessee. Each model mine was used to evaluate the potential impacts of a particular type of mining operation on the surrounding environment. The cumulative impacts are then analyzed by aggregating the impacts from several model mine types over the regional areas.

We have used this process of model actions or representative scenarios and regions of analysis in many EISs and believe that it serves to focus what can otherwise turn into very generalized statements about probable impacts at the programmatic level. The model actions try to define the potential range of differences in possible actions based on public issues and concerns. For example, in the Bureau of Land Management (BLM) (BLM 1989) and USFS Pacific Northwest Competing Vegetation Management Program (USFS 1988) EISs, which are discussed later, we broke down the potential actions into six different scenarios based on chemical application method, distance to people, distance to water, and distance to wildlife. In the Boll Weevil EIS (APHIS 1989), we found that the analysis was best differentiated along regional lines because the issues of concern (exposure to people and aquatic organisms) were different along regional lines and a wide range of different types of actions was not contemplated.

Selecting Regions for Analysis. The process used to select regions of analysis and model mines in the Tennessee EIS illustrates the analytical process that can be used to break down many different types of complex actions. The regions of analysis were selected on the basis that mining operations within an individual region have similar physical and environmental characteristics and thus would have similar impacts. In defining the regions, the following methods of dividing the Tennessee bituminous coal field were evaluated.

1. considering the coal field as one region
2. dividing the coal field into a northern and a southern region divided at the Cumberland Plateau overthrust
3. using watershed drainage basins to delineate the coal field in accordance with U.S. Geological Survey water resource units
4. using regions based on Bailey's description of ecosystems within the coal field (We have found Bailey's classification of ecosystems to be very useful on a number of projects.)

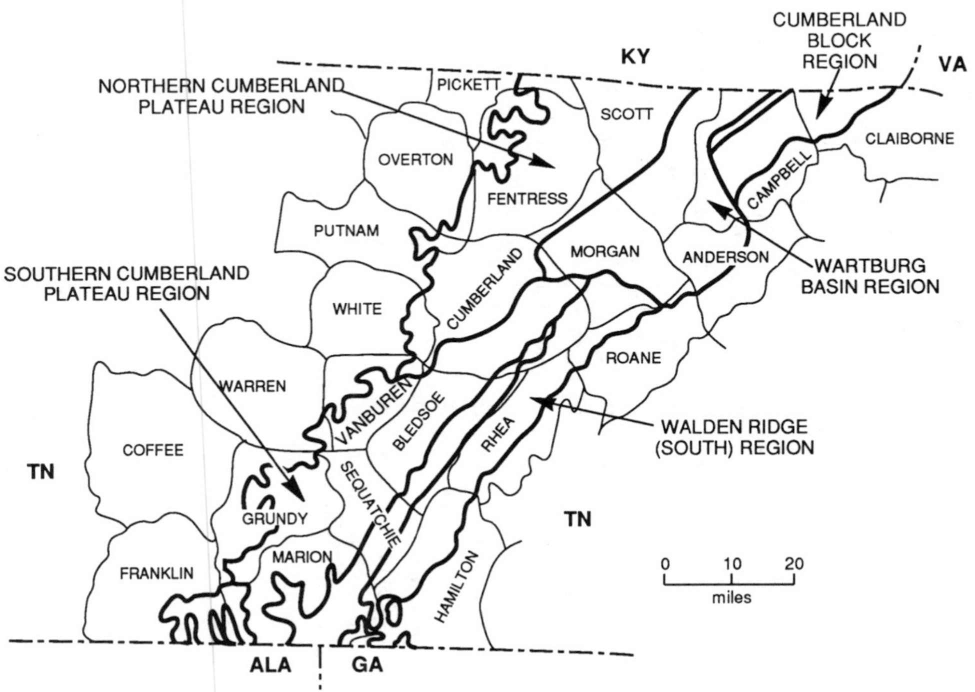

Figure 1. Regional map of the Tennessee bituminous coal field. (From U.S. Department of the Interior. 1985. pp. 3–12. In Final Environmental Impact Statement. OSM-EIS-18. Comprehensive Impacts of Permit Decisions Under Tennessee Federal Program, Washington, D.C.)

5. dividing the coal field into geographic regions based on similarity of physical conditions which, in turn, would affect the environmental impacts of mining

Each method had both advantages and disadvantages relative to the environmental analysis. However, careful evaluation by the interdisciplinary team lead to the choice of geographic regions based on physical conditions and environmental impacts as the most appropriate method for delineating regions. The regional boundaries were based on structural features and coal seam characteristics in the coal field (Figure 1). Each of the five regions could be described differently in terms of mining types, production rates, current and historic mining patterns, total surface disturbance from mining, current land use, geography, topography, known coal reserves, overburden characteristics, terrestrial and aquatic species, geohydrology, and surface water drainage patterns.

Characteristics of the Model Mines. For each region, three to five model mines were developed to represent the range of mining activities in that region and thus to analyze the range of impacts of individual permitted mines. Model mines were developed by using a modified Delphi technique. A team of experts with extensive experience in the Tennessee coal fields (including both OSMRE and industry consultants) was assembled; this team developed and described the different characteristics of the model mines, including underground mines, mountaintop removal mines, contour mines, auger mines, and area mines; in all, the team settled on 19 model mines. Some examples of the different characteristics of the model mines are shown in Table 1.

The first part of the analysis investigated the impacts associated with the approval or denial of individual model mines, and it identified three major impacts: hydrology (especially impacts to wells), noise and dust from haul roads, and impacts on wildlife within hardwood forest communities. The next section analyzed the cumulative impacts of the collective issuance of permits by OSMRE over a 5-year period within each region.

Cumulative Impacts

It is interesting to note how concerns at the individual permit level translated into cumulative effects. Cumulative impacts to hydrology were the first concern. One of the most valuable results of this analysis was to identify a coal seam with very high potential for acid-forming materials. The EIS interdisciplinary team labeled the Sewanee coal seam "hot as a firecracker." Special attention was then give to all permit applicants proposing to mine this seam, and, in fact, in later years it was part of the reason OSMRE made a partial unsuitability determination in the Rock Creek Watershed. Impacts to surface water could have been analyzed using the OSMRE Cumulative Hydrologic Impact Assessment model; however, after examining the type of mining predicted over the next 5-year period, the use of models was not considered necessary.

In most regions, 80–90% of the operations permitted over the next 5 years would be on previously mined lands, the majority of which have experienced inadequate surface and groundwater protection. Cumulative improvements to

Table 1. Model Mines in Tennessee Coal Fields.

Characteristics of Region in which Model Mine Is Located	Cumberland Block Region	Wartburg Basin Region	N. Cumberland Plateau Region	S. Cumberland Plateau Region	Walden Ridge South Region
Model Area Mines					
Coal seams commonly mined[a]	7, 9	8	4, 6	4, 5	3, 4, 5
Size of permit area[b] (acres)	60	35	80	200	30
Annual disturbance[b] (acres)	18	7	16	28	6
Existing land use[c]	R, M, T	R, P, M	R, P, T	M, T	M, T
Potential for toxic overburden[d] (low-medium-high)	M–H	L–M	L–M	M-H	M–H
Coal production capacity[b] (tons/year)	100,000	50,000	50,000	150,000	50,000
Model Auger Mines					
Coal seams commonly mined[a]	9, 10, 11	11, 14, 17, 18	—	—	—
Annual disturbance[b] (acres)	26	20	—	—	—
Proximity to dwellings[b] (miles)	2	4	—	—	—
Potential for toxic overburden[d] (low-medium-high)	L–M	L–M	—	—	—

[a] Coal seams: (1) Bon Air, (2) Wilder, (3) Richland, (4) Sewanee, (5) Lantana, (6) Nemo, (7) Rex, (8) Poplar Creek, (9) Coal Creek, (10) Blue Gem, (11) Jellico, (12) Jordan, (13) Pioneer, (14) Windrock, (15) Big Mary, (16) Red Ash, (17) Walnut Mountain, (18) Pewee, (19) Split, (20) Grassy Spring, and (21) Cold Gap.
[b] Number represents average.
[c] Land use — (M) Mining, (T) Timber, (I) Industrial, (P) Pasture, and (R) Residential.
[d] Toxicity level based on overburden pH characteristics: low potential pH >6; medium potential pH 5–6; high potential pH <5.

Source: U.S. Department of the Interior. 1985. pp. 4–10 and 4–12. In Final Environmental Impact Statement. DSM-EIS-18. Comprehensive Impacts of Permit Decision Under Tennessee Federal Program, Washington, D.C.

water quality and quantity would result when these areas are remined and reclaimed in accordance with the Tennessee Federal Program. In addition, to access the cumulative offsite impacts, the U.S. Geological Survey monitoring points on the major tributaries were selected as close as possible to their exit points from the regions. Existing data at only one of the five regional exit points — Walden Ridge South — indicated significant cumulative offsite adverse impacts on a regional basis with regard to current water quality standards. Because the Walden Ridge South Region has a naturally acidic pH regime and low buffering capacity, the addition of comparatively small volumes of acid drainage could cause a measurable depression of pH. This finding indicated that during permit review, these problems would require special attention. In fact, North Chickamauaga Creek was later determined unsuitable for mining.

The concern for the removal of hardwood forests and commercial forest land was also examined. Because much of the forest land is being logged, the disturbance caused by coal mining was not found to be significant, especially when lands would be reclaimed to timber or pasture which would eventually return to the original forested ecosystem.

Finally, there was the concern for cumulative impacts from dust and noise from coal haul trucks. When modeling showed that particulate standards would be violated within a short distance of unpaved roads, OSMRE developed mitigation measures for noise and dust on unpaved roads within the permit area.

Results

The process of examining the cumulative impacts of issuing coal mining permits in Tennessee proved useful to the OSMRE Tennessee regulatory authority in several ways. First, it identified several sensitive areas and coal seams that otherwise would not have been brought to the attention of the permit writers. Second, it allowed a broad look at the type of permitting anticipated over the next 5 years (mostly remining) and the formulation of strategies for dealing with this type of permitting. Finally, because of the attention given to individual actions in the EIS, OSMRE has not had to write any EISs for actions covered by the EIS in the last 4 years.

BLM and Forest Service Pacific Northwest Vegetation Management EISs

These EISs illustrate how model actions can be grouped in order to analyze cumulative impacts over time and large geographic areas. Both EISs are also good examples of how looking at cumulative impacts can lead to the development of mitigation measures that ensure individual actions will not have a significant impact on the human environment.

Scope of the EISs

The BLM and USFS began preparation of a worst-case analysis to examine the cumulative effects of their vegetation management activities, including the use of herbicides, in the Pacific Northwest Region. At first, both agencies were using similar methods to examine the effects of their vegetation management practices on the environment, but early in 1986, they decided to take different paths to complete that analysis and the NEPA process. This discussion focuses on the quantitative risk analysis of human health effects that was similar for both EISs.

Discussion of Impacts from Representative Actions

Two major public concerns were voiced repeatedly throughout the EIS process. First, there was the concern regarding the potential human health effects from the use of herbicides — specifically, their potential to cause cancer. Second, there was great concern about the effect of not using herbicides on the ability to produce high quality commercial timber and thus the effect on timber-dependent communities.

The bottom-up approach was used in this programmatic EIS. First, the potential actions were separated into representative actions. In this case, one of the driving forces in setting up these model actions was to ensure that the potential range of human exposures would be analyzed. A number of sophisticated transport and fate models were used to calculate the estimated doses to humans from drift, drinking water, eating fish and berries, and contact with vegetation for each scenario. Potential human health effects were quantified for each chemical for each scenario based on the margin-of-safety approach for systemic effects such as liver or kidney damage and the probability of getting cancer in a lifetime for oncogenic effects. Based on these calculations, more than 100 mitigation measures were developed to ensure an acceptable level of risk to the public.

Cumulative Impacts

The two major concerns, human health and economics, were both analyzed for cumulative impacts.

Impacts on Human Health. Estimates were made of the number of acres potentially treated with each herbicide and the number of times a person could be exposed in a lifetime based on the particular type of model action. For example, for forest reforestation projects, it was estimated that a person could be exposed 3 times during the 70-year rotation and that under worst-case assumptions they could be exposed from ten different operations or 30 times during their lifetime.

Interestingly, both the USFS and BLM, although acknowledging that people could be exposed to the same herbicides from agricultural and other nonfederal activities, chose to use the concept of additive risk for looking at cumulative effects. Therefore, the probability of getting cancer in a person's lifetime is expressed as the additional risk of cancer from the implementation of the USFS or BLM vegetation management program. For example, the cumulative effects on a fisherman from the use of the herbicide 2,4-D over the entire USFS program would be approximately 1 in 100 million.

Table 2. Changes in Number of Jobs in Oregon and Washington.[a]

Alternative A	–900	–0.03% of total employment
Alternative B	None	Total employment = 2,827,000[b]
Alternative C	–17,600	–0.62% of total employment
Alternative D	–1,800	–0.06% of total employment
Alternative E	–800	–0.03% of total employment
Alternative F	–2,500	–0.09% of total employment
Alternative G	+600	+0.02% of total employment
Alternative H[b]	–500	–0.02% of total employment

[a] Information was generated using the IMPLAN local economic impact model. "Jobs" include permanent and temporary positions, part-time and full-time positions, without discrimination. Figures shown represent average annual conditions over the coming decade. Includes direct, indirect, and induced employment resulting from changes in vegetation management practices, changes in the allowable sale quantity, and changes in livestock grazing.
[b] Preferred alternative.

The risk of systemic effects, such as organ damage or reproductive effects, is expressed by comparing an estimated daily dose with a daily dose that was considered safe in animals over their treatment period. (The daily exposure is divided by the animal no-observed-effect level to yield a margin of safety.) Therefore, because it is not possible to quantify the effects of repeated exposure, the cumulative effects must be stated in a qualitative manner. In most cases, this was done by comparing the number of doses that a person would potentially receive over a lifetime.

Economic Impacts. The USFS examined the cumulative effects of the programmatic alternatives for management of unwanted and competing vegetation on the local economies of Oregon and Washington. Changes in the management of unwanted vegetation change the work that is done on the national forests and change the amount of commodities the forests can be expected to produce. The USFS used the economic input-output model IMPLAN, discussed in detail in the next section, to quantify the cumulative impacts from implementation of the eight alternatives on average annual conditions over the next 10 years. The principal measures of annual regional economic effects were changes in jobs, changes in personal income, and changes in payments to local governments. The predicted changes in numbers of jobs in Oregon and Washington is shown in Table 2. There was some concern about the use of IMPLAN because some researchers believe that its technical coefficients may no longer be accurate because of changes in the labor component of lumber and other outputs. Instead of discarding the model, USFS acknowledged this potential problem, explained it in detail, and completed the analysis.

Results

The USFS EIS, prepared pursuant to a court injunction, has had the original injunction lifted. The use of the programmatic-type EIS allowed BLM and USFS to develop extensive mitigation measures based on both site-specific and cumulative effects that should ensure that the vast majority of all site-specific operations will not have a significant impact on the human environment.

The USFS and BLM used the programmatic EIS process very differently. Because USFS viewed the process as an opportunity to build consensus among interested parties, their process took over a year longer, cost a good deal more, and in the end placed more restriction on the use of herbicides and prescribed burning. The Forest Service published a list of over 50 mitigation measures in the Record of Decision; and although the program is still held up in the administrative appeals process, these mitigation measures should also ensure that site-specific operations will not have a significant impact on the human environment.

National Boll Weevil Cooperative Control Program EIS

This EIS illustrates the techniques that can be used to analyze a program that has cumulative effects over both time and a large geographic area. It also shows how large-scale models can be used without the need for prohibitively expensive data collection.

Scope of the EIS

This programmatic EIS describes alternative methods of boll weevil control to be used in a national program in cooperation with the U.S. Department of Agriculture (USDA) Animal and Plant Health Inspection Service (APHIS). Under the preferred alternative, program cooperators would strive for eradication of boll weevil populations across the Cotton Belt of the United States.

Before undertaking preparation of this programmatic EIS, APHIS had been conducting their analyses and their policy analysis and planning on a state-by-state level. The preparation of the programmatic EIS allowed APHIS to take a step back and look at the cumulative effects and problems of the program over a 20-year period. Indeed, one of the many policy decisions that came up during preparation of the EIS was how long did APHIS estimate that it would take to complete the entire Cotton Belt: 15 years? 20 years? Also, the process allowed the sharing of information across state programs, and some regional differences that had no basis in the biology of the boll weevil were eliminated.

Cumulative Impacts

The cumulative impacts of concern in the boll weevil control program are the impacts to human health from use of insecticides and the impacts to aquatic organisms, particularly fish.

Cumulative Human Health Impacts. The calculation of human health impacts used the same methodology previously described for the BLM and USFS EISs and produced some interesting results.

Because the control program was estimated to move across the cotton belt over a 20-year period and because cotton growers would continue to use large amount of insecticides until the program was implemented in their state and was successful, cumulative human health risks varied significantly from region to region. For example, northern Alabama may initiate the National Boll Weevil Cooperative

Control Program in 1991. Based on existing practices, residents and workers will be exposed relatively infrequently, 1.3 applications per year for the next 2 years. Assuming that 3.5 more years of treatment is needed to eradicate the boll weevil, exposures in the northern Alabama region could occur for approximately 5.5 years. In contrast, the boll weevil program in the southern Texas region may not begin until the year 2008. Assuming that existing practices continue until the program begins, residents and workers in this region may be exposed to an average of six treatments per year for 19 years prior to program implementation. This exposure, combined with the additional program exposure of 34 years will result in 2123 years of exposure. Therefore, the cumulative analysis indicated to decision makers that the faster the eradication program moved across the Cotton Belt the less the cumulative risk to human health.

Cumulative Impacts to Aquatic Organisms. The cumulative effects of soil buildup and insecticide fate for six representative locations and ten different soil types were predicted using the Groundwater Loading Effects of Agricultural Management Systems (GLEAMS) model developed in the SUDA Agricultural Research Service (Leonard et al. 1987, 1988). The GLEAMS modeling estimated the amount of insecticide lost to runoff of both sediment and water, percolation, and the amount that remains in the soil. Figure 2 shows an example of the output GLEAMS generated showing the buildup of the insecticides in the second and third year of the program in two different soil types. Table 3 shows an example of the predicted insecticide losses in runoff water and adsorbed to eroded sediments for the 2-year storms. During preliminary analyses, both 10-year and 100-year storms were also modeled to determine what scenarios would best represent an extreme situation. Although a higher mass of insecticide runs off with higher volumes of rain during the larger storm events, 2-year storms caused the highest insecticide concentration in almost all cases. Leaching of the insecticide into the groundwater was not found to be a problem.

The Exposure Analysis Modeling System (EXAMS) (Burns et al. 1982) was used to determine the effects of agricultural runoff containing insecticides considered for use in the EIS. EXAMS II is a set of mathematical models that simulates the most important factors that contribute to the degradation and transport of a chemical in an aquatic environment. EXAMS II requires a three-dimensional definition of the environment to be modeled. The environments selected for modeling included the Red River Basin in Texas and Oklahoma, the Tennessee River Basin in Alabama and Tennessee, the Sunflower River Basin in Mississippi, the Gila River Basin in Arizona, and the Flint River Basin in Georgia.

The amount of data needed to use all of the input parameters available for EXAMS to these five river basins would have been prohibitively expensive, and because of the uncertainties in variation in streamflows and river channels, it would have been of questionable value. Instead, generalized river characteristics were used for each of the five river basins, and the differences within each river basin were mainly represented by the different physical and chemical parameters of the insecticides.

Once the EXAMS environments had been created for the individual river basins and the chemistry parameters have been defined, the amount of insecticides

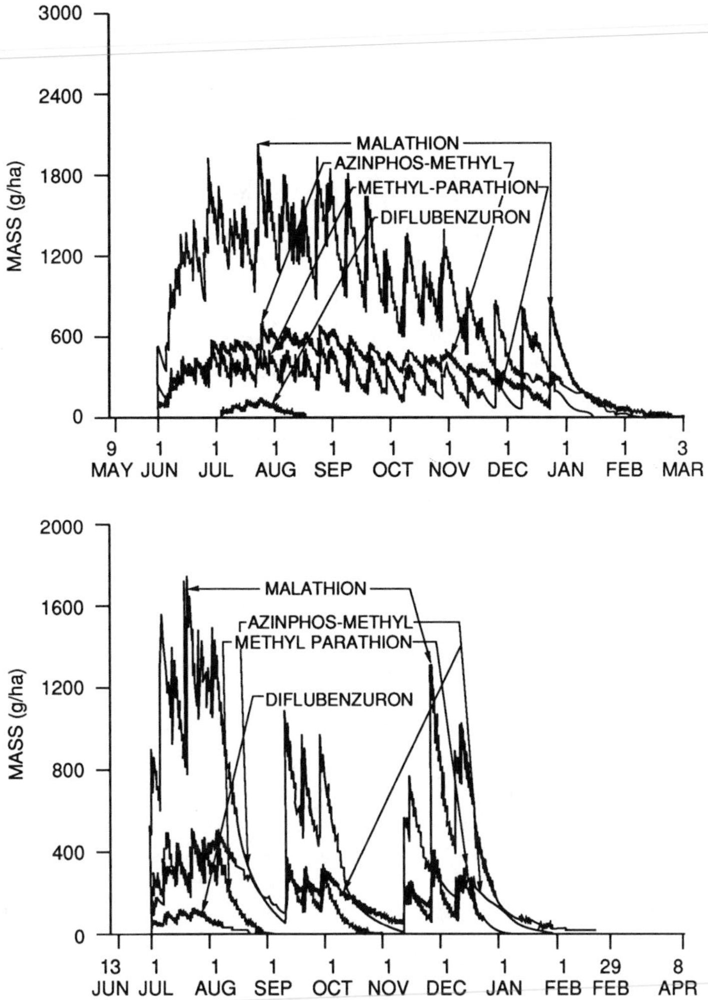

Figure 2. GLEAMS graphical output for the third year of an eradication program on Alabama Decatur silt loam.

introduced to the system must be determined. The loadings were estimated by using the runoff values calculated by GLEAMS and determining the acreage of treated cotton fields in counties within the drainage basin. A number of conservative assumptions were made to estimate the insecticide loadings. For example, it was assumed that the entire basin was sprayed at the same time. In addition, this analysis conservatively assumed that the insecticides applied in the furthest reaches of a river basin would reach the main truck without being degraded; realistically, some degradation would occur.

The insecticide chemistry variables, river dimensions, compartmental storm flows, biological variables, and predicted insecticide loading, were used to run EXAMS II. Table 4 shows an example of the maximum predicted river

Table 3. Predicted Pesticide Losses for a 2-Year Storm: Runoff Water, Eroded Soil, and Leachate under a Beltwide Eradication Program.

State/soil[a]	Azinphos-methyl		Diflubenzuron		Malathion		Methyl parathion	
	Water (mg/L)	Soil (μg/g)	Water (mg/L)	Soil (μg/g)	Water (mg/L)	Soil (μg/g)	Water (mg/L)	Soil (μg/g)
AL/loam	0.25	4.5	0.02	1.1	0.05	0.3	0.04	5.8
AZ/loam	No runoff or leaching — no pesticide losses							
AZ/clay	0.36	7.7	0.02	1.4	0.10	0.9	0.03	5.8
GA/sand	0.25	13.5	0.02	3.9	0.06	0.0	0.04	18.3
MS/loam	0.31	3.1	0.02	0.8	0.08	0.3	0.05	3.5
MS/clay	0.28	7.5	0.02	2.0	0.07	0.7	0.04	8.6
TX/c/loam	0.21	2.8	0.02	3.1	0.05	0.3	0.07	7.0
TX/c/clay	0.28	6.9	0.01	1.1	0.07	0.7	0.02	4.5
TX/p/sand	0.25	7.6	0.02	2.2	0.06	0.7	0.04	10.3
TX/p/clay	0.27	9.4	0.01	1.6	0.05	0.7	0.02	6.5

Note: No leaching was predicted under any 2-year storm scenario.

[a] AL/loam — Alabama, Decatur silt loam; AZ/loam — Arizona Shontik sandy loam; AZ/clay — Arizona, Trix clay; GA/sand — Georgia, Cowerts loamy sand; MS/loam — Mississippi, Dubbs very fine sandy loam; MS/clay — Mississippi, Sharkey and Alligator clays; TX/c/loam — Texas, Coastal, Orelia sandy clay loam; TX/c/clay — Texas, Coastal, Victoria clay; TX/p/sand — Texas, Rolling Plains, Miles fine sandy loam; TX/p/clay — Texas, Rolling Plains, Abilene clay loam.

concentration in any compartment 7 days after a 2-year storm, which was found to be the highest concentration overall. The criteria that Environmental Protection Agency (EPA) uses in ecological risk assessment (EPA 1986) were used as the basis for judging the risk to different representative species in the five river basins. Under these extremely conservative assumptions, a number of aquatic species, especially invertebrates, would be at risk (Table 4).

RESULTS

The decision to prepare a programmatic EIS on the boll weevil program allowed APHIS the time to examine the potential cumulative impacts of the program and to develop detailed mitigation measures. Analysis was completed of the effects on broad river basins through the use of representative data. Gathering of actual data would have taken exorbitant amounts of both time and money and would not have added substantially to the understanding of cumulative effects. The preparation of this EIS has not held up APHIS's existing eradication program.

TWO MODELS WITH WIDESPREAD APPLICATION FOR CUMULATIVE IMPACT ANALYSIS

IMPLAN

The USFS developed and uses IMPLAN to estimate the effects of their various actions on employment, income, population, and other parameters at the country level and any higher aggregate affected area. Because IMPLAN automatically considers all sectors of the economy, it is possible to calculate the multiplier or ripple-down effects from any federal action. Many other federal agencies are now using this model of its flexibility to look at a range of economic effects across large geographic areas and over time. It was recently updated with 1987 economic data and, therefore, is capable of very accurate predictions. There are a number of books and instructional manuals written about IMPLAN which are available from the USFS's Fort Collins office.

EXAMS II

EXAMS was developed at EPA's Athens Environmental Research Laboratory. It can currently be accessed through EPA's Graphical Exposure Modeling System. EXAMS II, the most recent enhanced version, can be used to model the transport and fate of a wide variety of chemicals (it was originally designed for organics) in aquatic environments. A range of biological, chemical, and physical factors are considered. The program requires input of (1) ecosystem variables, (2) chemical

Table 4. Summary of Highest Risks to Aquatic Species in Rivers Modeled.

Insecticide	Tennessee River			Flint River			Sunflower River		
	Fish	Invertebrate	Amphibian	Fish	Invertebrate	Amphibian	Fish	Invertebrate	Amphibian
Eradication									
Malathion	C	A	C	C	A	C	A	A	A
Azinphos-methyl	C	A	C	B	A	C	A	A	C
Diflubenzuron	C	C	ND	C	C	ND	C	A	ND
Methyl parathion	C	A	C	C	A	C	C	A	C
Suppression									
Malathion	C	A	C	C	A	C	A	A	B
Azinphos-methyl	C	A	C	B	A	C	A	A	C
Diflubenzuron									
Methyl parathion	C	A	C	C	A	C	C	A	C

variables, (3) chemical application loading variables, and (4) model operation variables. The aquatic environments used by EXAMS II can accept up to 50 compartments per system. Therefore, small reaches of a stream to a large river system can be analyzed for cumulative effects.

CONCLUSIONS

As illustrated by the three case studies, programmatic EISs can be useful tools for looking at cumulative impacts. (Actually, programmatic EISs are sometimes the only mechanism for addressing cumulative impacts at the national or regional level.) The three EISs cited in this paper also show that complex situations can be simplified by using model scenarios to represent the range of reasonably foreseeable actions. Finally, and perhaps most importantly, the example EISs have shown that the use of programmatic EISs to examine cumulative impacts can actually speed implementation of the program by allowing the early identification and mitigation of significant impacts, both cumulative and site specific.

REFERENCES

Burns, L. A., D. M. Cline, and R. R. Lassiter. 1982. *Exposure Analysis Modeling System (EXAMS): User Manual and System Documentation.* U.S. Environmental Protection Agency, Environmental Research Laboratory, Athens, GA.

Burns, L. A., and D. M. Cline. 1985. *Exposure Analysis Modeling System: Reference Manual for EXAMS II.* U.S. Environmental Protection Agency, Environmental Research Laboratory, Athens, GA.

Leonard, R. A., W. G. Knisel, and D. A. Still. 1987. GLEAMS: Groundwater loading effects of agricultural management systems. *Trans. Am. Soc. Agric. Eng.* 30(5):1403–1418.

Leonard, R. A., et al. 1988. Modeling pesticide metabolite transport with GLEAMS. pp. 255–262. In *Proceedings on Planning Now for Irrigation and Drainage.* IR Div/ASCE, Lincoln, NE.

U.S. Department of Agriculture, Animal, and Plant Health Inspection Service (APHIS). 1989. Draft Environmental Impact Statement: National Boll Weevil Cooperative Control Program. U.S. Department of Agriculture, Washington, D.C.

U.S. Department of the Interior, Bureau of Land Management (BLM). 1989. Draft Environmental Impact Statement: Western Oregon Program — Management of Competing Vegetation. U.S. Department of the Interior, Bureau of Land Management, Oregon State Office.

U.S. Environmental Protection Agency (EPA). 1986. Standard Evaluation Procedures: Ecological Risk Assessment. Hazard Evaluation Division, Washington, D.C.

U.S. Office of Surface Mining Reclamation and Enforcement (OSMRE). 1985. Final Environmental Impact Statement: Comprehensive Impacts of Permit Decisions under Tennessee Federal Program. OSM-EIS-18. U.S. Department of the Interior, Wahington, D.C.

U.S. Forest Service (USFS). 1988. Final Environmental Impact Statement: Managing Competing and Unwanted Vegetation. U.S. Department of Agriculture, Forest Service, Pacific Northwest Region.

Cumulative Impacts Assessment and Management Planning: Lessons Learned to Date

S. C. Williamson, U.S. Fish and Wildlife Service, National Ecology Research Center, Fort Collins, CO

ABSTRACT

Cumulative impacts assessment should be bound closely with management planning for an ecosystem of concern and should consist of scoping and analysis of impacts from the past to the present. The management planning should consist of interpretation and direction of impacts of ongoing and near-future actions. When dealing with many problems in a complex situation, the recommended course of action for cumulative impacts assessment is to emphasize scientific, cause-effect understanding and communication; stress measurable overall action toward progressive goals; use a generation-long, ecosystem-level, problem-solving, and solution-achieving process; and ratify an interagency collaborative drive toward cumulative improvement of the situation.

Selection of a strategy for dealing with each priority cumulative impacts problem should be based on the mitigation options of restoration, impact avoidance, or impact minimization. The major objectives of cumulative impacts assessment and management planning should be to generate logical, scientific, and timely problem analyses; bring agencies together collaboratively to develop an overall management plan and proactive, measurable resource goals; and meld those results into comprehensive species and habitat maintenance and enhancement blueprints for the ecosystem of concern. Natural resource agencies can soon anticipate a shift from scrutinizing individual permits, licenses, and assessments within an ecosystem of concern to a new capability of providing ecosystem-level guidance. The public can expect an active increase in positive ecological impacts and reduction in negative impacts as a result of cumulative impacts assessment and management.

INTRODUCTION

The growing awareness of cumulative impacts is accompanied by some puzzlement as to how they should be addressed. There is general agreement, however, that cumulative impacts (also known as cumulative effects) are a serious ecological challenge as typified by (1) significant deterioration of major ecosystems (estuaries, lakes, and rivers), (2) fragmentation and loss of critical habitats (wetland complexes), and (3) long-term population declines (anadromous fish and migratory waterfowl).

The foremost cumulative impacts concern of natural resource agencies has been the negative effects of multiple human actions (interacting with each other and with natural events) within a major and highly valued ecosystem. For example, San Francisco Bay is considered the major estuary in the United States that has been modified the most by human activity (Nichols et al. 1986). Of the original 140,000 ha of freshwater marsh and 80,000 ha of saltwater marsh, only 12,500 ha (6%) remain. Sediment attributed to hydraulic mining debris has been deposited in central San Francisco Bay to a depth of 25 cm. Of the historic flow of the river system, 40% has been removed annually for local consumption upstream and within the delta, while another 24% has been exported annually for municipal and agricultural consumption. Maximum annual concentrations of sulfates and nitrates in the San Joaquin River have increased threefold and fivefold since 1950. Approximately 100 invertebrate species have been introduced; nearly all macroinvertebrates on the inner shallows of the bay are introduced species. Commercial fisheries for only herring and anchovy still exist; the former commercial fisheries for salmon, sturgeon, introduced striped bass (*Morone saxatilis*), and Dungeness crab (*Cancer magister*) have halted (Nichols et al. 1986).

Forman and Godron (1986) described a progression of landscape ecology degradation (beginning with the most sensitive): (1) relative species abundance changes, (2) sensitive species disappear and native species diversity decreases, (3) nonnative species colonize, (4) biomass and cover decrease, (5) production decreases, and (6) erosion increases. In San Francisco Bay, all of these changes have been observed. Although progress has been made on water quality attributes such as dissolved oxygen and enteric bacteria concentrations, the major changes in the estuary (sediment deposition, loss of wetlands habitat, population declines of many fishes, and introduction of exotic species) occurred decades ago and former high quality conditions have been forgotten. Nichols et al. (1986) concluded that further improvement in water quality alone is not likely to have a significant positive effect on these major changes.

For the past decade, the term "cumulative impact" has been used merely in conjunction with assessment (i.e., scoping and analysis). The process described in this document is intended to commit assessment to management planning needs (i.e., interpretation and direction of total cumulative impacts) in an affected ecosystem. Solitary cumulative impacts assessment may be a decreed, one-time assignment; cumulative impacts assessment in combination with management planning should be a proactive, long-term process. Assessing one cumulative impact (the incremental impact only without the rest of the cumulative impacts to date in the affected ecosystem) has been relatively unsuccessful. Assessing cumulative actions (again without the rest of the cumulative impacts to date in the affected ecosystem) similarly has been relatively unsuccessful. Assessing cumulative impacts (the total impacts to date of past actions and natural events) can be and has been accomplished with various levels of success. We have learned to recommend cumulative impacts assessment and management planning because of its greater potential for achieving long-term goals.

Definition

When determining the scope of an environmental impact study, federal agencies are required by the Council on Environmental Quality (CEQ) regulations to consider three types of actions and three types of impacts (40 CFR 1508). The three types of actions to be considered are cumulative actions (actions that when viewed with other proposed actions have cumulatively significant impacts), connected actions (closely related actions that may be triggering or interdependent), and similar actions (actions that have similarities such as common timing or geography). The three impacts that should be discussed in the same impact statement are direct, indirect, and cumulative impacts.

This paper is specifically concerned with ecological cumulative impacts assessment (concentrating on impacts up to the present). The CEQ regulations, first published in 1978, provide definitions that can be summarized as follows:

- Cumulative impact is the impact on the environment of the incremental impact of the action.
- Cumulative impacts are the total of the incremental impacts of past actions and present actions on the environment.
- *Environment* equals the effects on natural resources and on the components, structures, and functioning of affected ecosystems.
- *Effects* are synonymous with impacts, and the total effect of an action may, on balance, be either beneficial or detrimental.

Translation

Most of the terminology of cumulative impacts assessment is relatively new and subject to various interpretations. In numerous workshops, the specific wording of definitions and distinctions has proven to be necessary, but not universally acceptable. In this paper, the following distinctions and definitions are used.

- Cumulative impacts are the combined effects of all human actions and natural events on the ecological environment (Salwasser and Samson 1985).

- Cumulative actions (plural) assessment is scoping and analysis of the total impacts of multiple proposed actions on the affected ecosystem and is not the subject of this paper.
- Cumulative impact assessment is scoping and analysis of the incremental impact of one past action on the affected ecosystem and also is not the subject of this paper.
- Cumulative impacts (plural) assessment is scoping and analysis of the total impacts of past actions and natural events on the affected ecosystem.
- Cumulative impacts management planning is interpretation and direction of the total impacts of current actions and multiple proposed actions on the affected ecosystem.

Typology

We have found that typologies of cumulative impacts create a good deal of research interest, but, like definitions, the pursuit of a definitive typology may turn into a tangent from cumulative impacts assessment. The typology of cumulative impacts presented by the National Research Council's Committee on the Applications of Ecological Theory to Environmental Problems (1986) is adopted for this paper. That typology recognized the following cumulative impacts: time-crowded perturbations, space-crowded perturbations, synergisms, indirect effects, nibbling, threshold developments, and lag effects.

An ecosystem of concern is usually characterized by substantial reductions in populations, lower or discontinued harvest of several important fish and wildlife species, and substantial declines in the quality and quantity of several critical habitats. Several human actions cause the declines, and the declines are probably irreparable in the near future without corrective actions. Multiple causes of these declines is one of the major difficulties with cumulative impacts assessment and management planning projects. Cumulative impacts assessment within an ecosystem of concern should first connect multiple ecological causes (due to human actions and interrelated natural events) to historic and current state of the affected ecosystem (represented by habitat components, structures, and functioning) and then to numerous observed effects on natural resources (particularly fish and wildlife).

Conventional Environmental Impact Assessment

A classical environmental impact assessment is motivated by a proposed project; it focuses on and describes many site-specific environmental effects attributable to the project (an individual development action or one interrelated set of development actions) (Truett et al. 1993). A cumulative impacts assessment is generally driven by resource declines or concern over possible declines; it should focus on an ecosystem of concern and provide an overview of major species and habitat problems and the causes of the problems. The CEQ directed that environmental impact assessment consider cumulative impacts; however, according to Granholm et al. (1987), numerous institutional difficulties have been found with the practice of including cumulative impacts assessment as part of the environmental impact assessment process, including (1) determining appropriate timing, costs and level of effort; (2) apportioning the cost and responsibility for the assessment and mitigation among participants; (3) coordinating assessment of different types of projects that cross agency jurisdictions; (4) selecting appropriate methods and development scenarios for a particular assessment; (5) limited history of application of most of the appropriate methods in a regulatory context; and (6) identifying specific roles for project proponents and other interested parties. Making ecological cumulative impacts assessment part of an environmental impact assessment has been difficult and ineffective; the best use for a cumulative impacts assessment has been in management planning for an ecosystem of concern.

Environmental impact assessment focuses on inventorying and analyzing individual project effects; cumulative impacts assessment for management planning should focus on understanding the ecosystem involved and formulating management programs to solve ecological problems. Recognizing that no agency has the overall authority to regulate, design, or plan for all the aspects of cumulative impacts, a cumulative impacts assessment should look at a much larger geographic area than typically used for evaluating an individual development action. Cumulative impacts are a pervasive problem that requires a different way of doing business from just the review of individual federal projects, permits, or licenses (Muir et al. 1990).

In environmental impact assessment, decreasing the negative effects of individual development actions (minimizing impacts to no net loss when possible) is a desirable near-term strategy; in cumulative impacts assessment, more can be accomplished through striving to increase the positive effects of total development actions (improving the ecosystem when the opportunity presents itself). The individual elements of cumulative impacts cannot be regulated well on a project basis, but overall impacts can be assessed and managed (Burns 1991). Cumulative impacts assessment, as described here, can lead to comprehensive ecosystem guidance with information feedback from implementation, monitoring, and evaluation.

Why Do Cumulative Impacts Assessment?

Cumulative impacts assessment is most appropriate when dealing with many ecological causes and effects. Cumulative impacts assessment and management planning should be used in the most difficult ecological situations that encompass cumulative causality (started by multiple human actions and natural events), cumulative system effects (followed by decline of multiple habitats), cumulative fish and wildlife population effects (resulting in declines of multiple species), and cumulative restoration (rejuvenated by multiple human actions).

To be truly effective, cumulative impacts assessment and management planning should not only investigate and decrease the ongoing negative effects of human actions, but should also concentrate on exploring and obtaining a more positive overall impact. The ecological challenge of cumulative impacts assessment and management planning in the future is to identify what should be done in terms of ecological changes, rather than merely what should not be done.

DESIGN OF SUCCESSFUL ASSESSMENT

Advocates of traditional methods in educational and governmental institutions have created unfocused, time-consuming, misguided, and narrowly defined assessments (National Research Council of the United States 1986; Canadian Environmental Assessment Research Council and United States National Research Council 1986). The evolution of cumulative impacts assessment methods has been constrained by the reluctance to accept responsibility for cumulative impacts assessment and management planning. Under these conditions, jurisdictional problems have overwhelmed the process, and the broad spatial and temporal bounds necessary for managing cumulative impacts are not incorporated into the assessments. It is important for effective cumulative impacts assessment and management planning to emphasize not numerous small assessments or a single, final plan, but an ongoing, regional, long-term strategy and planning process (National Research Council of the United States 1986; Canadian Environmental Assessment Research Council and United States National Research Council 1986).

Numerous theoretical, analytical, and institutional impediments hinder cumulative impacts assessment (Dickert and Tuttle 1985; Meehan and Webber 1985). Gosselink et al. (1990) pointed out that regulatory agencies have difficulty in dealing with cumulative impacts because the environment in which impacts interact is complex; changes may not be measurable individually; site-specific reviews do not represent a large enough geographical area (i.e., entire watershed or river basin) and ignore the time line; and regulatory agencies find effective, concerted action difficult. The recommended ways to deal with these four difficulties are based on a dozen case studies, the results of which are summarized in the sections that follow.

- Emphasize scientific, cause-effect understanding and communication of the overall situation, each problem, and problem interactions.
- Stress measurable overall action toward progressive goals for each problem.
- Use a generation-long, ecosystem-level, problem-solving, and solution-generating process.
- Ratify an interagency collaborative drive toward cumulative improvement of the overall situation.

Collaborative

One essential point in conducting a cumulative impacts assessment and management planning project is gaining early consensus among the concerned natural resource agencies and institutions, particularly on whether to conduct such an assessment and on a strategy for addressing the ecosystem of concern. Collaboration with other regulatory agencies is essential to a successful cumulative impacts assessment and management planning project because the responsibility for natural resources typically rests with many local, state, and federal agencies.

At least one subject matter expert from each of the concerned natural resource agencies should be involved in the scoping and analysis phases of a cumulative impacts assessment and management planning project. Agency differences can be minimized, and support can be gained from sharing information and understanding of technical issues. Management users from the concerned natural resource management agencies should be involved in the early design of the assessment and again later in the interpretation and direction phases. This cooperation creates a sense of ownership, commitment, and responsibility in the participants and their agency and promotes greater coordination, cooperation, and consensus among the natural resource agencies.

Goal Oriented

Goal orientation forces a cumulative impacts assessment and management planning project to be purposeful and focused. Setting quantitative, measurable, time-dependent goals implies that society has deemed particular ecological resources or conditions as desirable and that management agencies are committed to conserving, protecting, or enhancing those resources and conditions. For a cumulative impacts assessment and management planning project to be purposeful, it should be directed toward increasing some of the resources above current status, not just maintaining status quo or avoiding deterioration thresholds. In particular, deterioration threshold evasion (impact minimization) is not a desirable way to deal with cumulative impacts in an ecosystem of concern given the opportunity of stabilizing (impact avoidance) or managing upward (restoration).

The Canadian Environmental Assessment Research Council and United States National Research Council (1986) questioned whether cumulative effects could be managed without a comprehensive set of societal goals. To improve cumulative impacts assessments and make them useful in regulatory decisionmaking, explicit societal goals should be defined and made part of a comprehensive, future-oriented planning process (Horak et al. 1983; Stakhiv 1986; Canadian Environmental Assessment Research Council and United States National Research Council 1986). A key point of cumulative impacts management planning is that strategic policy decisions should be made and goals for resources of concern set in the assessment before management planning takes place.

Problem Solving Process

Because each situation differs in complexity, amount of available and usable data, and degree of understanding of the problems and ecological processes, an extensive education with ecological problem solving is needed for agencies to gain the skill, knowledge, and technology required to successfully assess and manage cumulative impacts. It is important to start cumulative impacts assessment from the effects (species and habitat problems) side instead of the causes (development actions and natural events) side and take a problem-solving and solution-generating total ecosystem view. The advantages of a problem-solving approach are that it encourages concentration of effort; a thorough search for an unbiased statement of the situation and specific problems; an incremental and sequential analysis; and identification and selection of realistic, feasible, and economical solutions. Ecological problem solving is a key element in a successful cumulative impacts assessment (Salwasser and Samson 1985; National Research Council of the United States 1986).

The complexity of many cumulative impacts problems corroborates the assertion that cumulative impacts assessment cannot be accomplished by a method or technique developed to apply in all cases. According to the Canadian Environmental Assessment Research Council (1988), cumulative impacts assessment underlines the need for a long-term, well-organized approach to resolve the problem's scientific and institutional aspects. A successful cumulative impacts assessment should employ a problem-solving process that can be applied intensively to a wide range of situations and that utilizes adaptively the most appropriate methods and techniques.

Scientific Cause and Effect

Cumulative impacts assessment requires a high order of analysis and interpretation of cause-effect linkages; new concepts and alternative thinking processes to restructure the problem; new techniques to aggregate diverse impacts; and a holistic, integrative perspective (Horak et al. 1983). Granholm et al. (1987) concluded that new methods are needed to deal with the complexities of multidisciplinary systems and that available techniques such as group problem solving, area assessment, and simulation modeling are used either ineffectively or not at all. Because cumulative impacts assessment, unlike traditional environmental impact assessment, is a form of pattern analysis and must detect and analyze trends, cumulative impacts assessment needs scientific understanding of cause and effect (Canadian Environmental Assessment Research Council and the United States National Research Council 1986).

Many cumulative impacts assessment efforts, no matter how potentially sound analytically, degenerate before they begin because of the lack of four prerequisites for successful management. To be effective in cumulative impacts assessment, use both a problem-solving process and scientific cause and effect; to be effective in cumulative impacts management planning, use both goal setting and collaboration. The major ecosystem-level success stories (e.g., Lake Washington, Lake Erie, Lake Michigan, Potomac River) have had those necessary ingredients for success.

RECOMMENDED PROCESS

The recommended cumulative impacts assessment and management planning process should use the following steps: (1) in the scoping phase, define the ecological situation in specific terms of individual problem statements and select one strategy for each problem; (2) in the analysis phase, investigate and document the problems and their causes in detail using the best available data and analytical tools and then set several goals; (3) in the interpretation phase, develop and document options, estimate changes using mathematical models, and develop a plan; and (4) in the direction phase, implement and incrementally improve the management plan and systematically evaluate, improve, and update the problem statements, data, analytical tools, and mathematical models.

It has been useful to distinguish cumulative impacts assessment (Steps 1 and 2) as the portion of the time horizon from the past to the present and cumulative impacts management planning (Steps 3 and 4) as the portion of the time horizon from the present to the future. Step 1 focuses on qualitative problem descriptions and is intended to accomplish problem identification, clarification, and expression. Establishing appropriate temporal, spatial, and political boundaries is difficult, but it is critical to the success of a cumulative impacts assessment (Lee and Gosselink 1988). Concern about cumulative impacts by federal natural resource regulatory agencies has been pronounced in areas that are moderately large and complex (entire ecosystems with a focus on aquatic and wetland habitat). Generally, a multiagency group of natural resource management experts should be gathered to work collaboratively in a workshop setting. The group identifies important ecological problems contributing to the overall situation, agrees on problem statements, and documents those problems using the relevant scientific literature. Careful statement of each problem goes a long way toward stimulating action on its solution.

Step 2 provides quantitative problem analyses and goal statements that are technically and scientifically credible. The status and historic trends of the priority resources are documented, graphed, and mapped. Based on an evaluation of the best data, literature, and scientific judgment available, early problem statements are accepted, modified, or rejected. The importance of causal factors is evaluated. Data gaps, research needs, and preferred predictive mathematical models are identified. Specific management goals are generated and supported, both scientifically and institutionally. For example, in an early restoration planning workshop for Commencement Bay, WA, the natural resource trust agencies developed the following problem and goal statement: "Virtually none (less than 1%) of the original 10 km^2 (2,470 acres) of subaerial wetlands in the Commencement Bay-lower Puyallup River ecosystem remain. By 2005, restore at least x-y acres (some numbers between 10% and 50%) of these wetlands in that ecosystem."

In Step 3, the focus is on defining management opportunities. The quantitative analyses from Step 2 should be used to identify the most important causal factors in each problem. Effective alternative actions that may achieve the goals are identified and evaluated. Determine which of the actions identified above are ecologically, politically, institutionally, economically, and legally feasible, and identify the mechanisms through which effective actions can be implemented. Each agency's responsibilities should be identified, and the ability of agencies to have a significant positive effect should be evaluated. Several alternative management plans should be evaluated with the mathematical model for achieving the resource goals. The recommended plan should contain the set of effective actions that optimally achieve the multiple goals for the priority resources.

At several points during the assessment process, subjective value judgments must be made with reference to some framework of social values. It has proven essential to deliberate collaboratively on the ramifications of each possible strategy and gain interagency consensus early in the scoping of the problem. Strategy selection should be based on the CEQ's five options for mitigation and depends on society's *acceptable standards* for ecological resources: (1) where the current ecological condition is below acceptable standards, a restoration strategy is appropriate; (2) where the current condition is about equal to acceptable standards, a strategy of impact avoidance (no net loss of habitat) is usually chosen; and (3) where the current condition is above acceptable standards, a strategy of allowing some decline from current conditions by impact minimization will work. Impact minimization is generally the current strategy of the natural resource agencies concerned about cumulative impacts assessment. Just an agreement on the most desirable strategy for each problem is frequently a major advancement for the agencies involved.

RECENT HISTORY

Considering the number of articles being published, interest in cumulative impacts is increasing. The first two articles with the term were published in 1975. From 1975 to 1980, the publication rate was between one and four papers per year; from 1981 to 1984, between 6 and 13 papers per year; and from 1985 to 1988, between 11 and 37 papers per year (Williamson and Hamilton 1989).

Project Background

A cumulative impacts assessment project was initiated by the U.S. Fish and Wildlife Service in 1984 at the Western Energy and Land Use Team (now the National Ecology Research Center). The project's systems analysis approach involved (1) understanding Ecological Services' user needs (Williamson et al. 1986); (2) conducting real world analyses of cumulative impacts problems using prototype trials (e.g., Williamson et al. 1987); and (3) developing and refining an assessment process. Collaboration with other cumulative impacts assessment researchers was emphasized in interagency conferences to advance cumulative impacts assessment (see Hunsaker and Williamson 1992).

To develop and improve a Fish and Wildlife Service approach, we undertook several cumulative impacts assessment case studies (by observing the work of other agencies) and prototype trials (by conducting them jointly with U.S. Fish and Wildlife Service's Ecological Services field offices across the country). Some of these are described below. The process developed through these stages has been characterized as a classical planning process for the purpose of ecological problem solving.

Understanding Barge Navigation Effects on Rivers

The field office in Cookeville, TN used a scientific, cause-effect network diagram to prepare comments on barge traffic permit applications in the Ohio, Tennessee, and Cumberland rivers. The diagram met the needs of the U.S. Army Corps of Engineers and the field office in determining principal resources of concern, problems, and causal pathways. The major contribution of this project was the successful use of cause-effect network analysis. The diagram did not provide a quantitative analysis, but it did provide a mechanism for understanding and communication between agencies about important factors and a framework for tracking potential effects of barge traffic. The Corps' office asked that future efforts also provide such a cause-effect network diagram. The cause-effect network diagram was later used by the Annapolis field office and the National Fisheries Center–Great Lakes to specifically describe several major omissions in a barge traffic simulation model prepared under contract for the U.S. Army Corps of Engineers.

Designing Better Oil Field Development in Alaska

The Fish and Wildlife Service's regional and field offices developed planning aid documents to minimize the cumulative impacts of oil and gas development on the wet tundra of the North Slope coastal plain of Alaska (Meehan and Webber 1985; Walker et al. 1987). The regional office applied the cumulative impacts assessment and management planning process to the Colville Delta oil field with the intention of extending the lessons learned at the Prudhoe Bay oil field. For example, in the wettest parts of the oil field, resultant flooding and thermokarst were found over more than twice the area covered by roads and other construction (Walker et al. 1987). Alaskan oil development provided an opportunity to study cumulative impacts on a well-defined terrestrial scale and in a relatively pristine habitat resource.

Planning Restoration of Chesapeake Bay

The Annapolis field office conducted a cumulative impacts assessment and management planning project in accord with the Chesapeake Bay Restoration Plan. The Environmental Protection Agency has the lead role in the bay restoration program and has emphasized restoration of water quality (e.g., nitrogen loading, dissolved oxygen concentration) in the bay. One of the documents (Flemer et al. 1983) prepared as part of the bay restoration program is so good that it can serve as a template for the report of a cumulative impacts assessment. In conducting a prototype cumulative impacts assessment, workshop participants defined problems, identified important cause-effect relationships, and developed preliminary remedial action plans (Williamson et al. 1987). As problems were examined through cause-effect network analysis, there was a clear movement away from problem statements focusing on development actions (near the start of causal chains) and fish and wildlife species (near the end of effect chains); when problem identification was based on habitats (the hub of causes and effects), the assessment focused clearly on ecological goal attainment and remedial action management planning. As a consequence of the assessment, a 70% decline in distribution of native submerged aquatic vegetation across the bay proper and increased amounts of suspended particulates in historically productive watersheds were identified as keystone problems to be dealt with. The Fish and Wildlife Service has chosen to pursue an emphasis on living resources (e.g., indigenous species of submerged aquatic vegetation) as opposed to an emphasis on water quality.

Guiding Growth in an Urbanized Estuary

The Daphne, AL field office conducted a Mobile Bay cumulative impacts assessment and management planning project. The assessment and planning project contains four major elements: (1) a cause-effect network analysis; (2) a status and trends analysis; (3) goal setting for bay resources by the natural resource management agencies; and (4) development of a coordinated action agenda. In goal-setting work for eight problems, each element had some combination of the following: current action goals (things that can be done immediately or that should continue), management-related information goals (things that need to be done to improve and maintain our understanding of the bay system), and 20-year attainment goals (what do we want the bay to look like in 20 years?). The state of Alabama was concurrently conducting a cumulative impacts assessment for Mobile Bay. The field office's work (with the other natural resource agencies) was adopted and carried forward as the state's recommended approach and results for Mobile Bay. The advisory group for the state's project has come to the conclusion that project-by-project cumulative impact assessments (see earlier definition) are ineffectual and that an ecosystem-level cumulative impacts assessment (leading to comprehensive ecosystem-level guidance) is advantageous.

Analyzing a Great Lakes Connecting Waterways Ecosystem

The ecosystem of concern is the connecting waterways between Lake Huron and Lake Erie consisting of Lake St. Clair and the St. Clair and Detroit rivers. In a multiagency (federal, Michigan, Ontario, and Great Lakes Fishery Commission) workshop sponsored by the Fish and Wildlife Service's National Fisheries Center — Great Lakes and the East Lansing, MI field office, cause-effect network diagrams were constructed for four ecological problems in each of the waterways. Using a mathematical matrix method, relative importance values for cause-effect relationships were assigned by subject matter experts at the workshop. With institutional mechanisms for making decisions and implementing actions already in place (National Research Council of the United States and The Royal Society of Canada 1985), this cumulative impacts assessment and management planning project emphasized technical enhancement of assessment methods, particularly matrices and simulation modeling.

Cluster Impact Assessment Procedure

In 1985, the Federal Energy Regulatory Commission contracted with Argonne National Laboratory to develop the Cluster Impact Assessment Procedure to evaluate the potential cumulative impacts of multiple small-scale hydroelectric projects (Bain et al. 1986). The Cluster Impact Assessment Procedure was used to identify geographic areas of concern in the Snohomish River (Washington) and Salmon River (Idaho) watersheds, determine projects that could have adverse effects on target resources, and conduct a multiple-project (cluster) impact analysis. The cumulative actions assessment (see earlier definition) was intractable for cumulative impacts assessment (GAO 1988). The natural resource agencies found the mathematical matrix approach unduly complex and biologically unacceptable. In both river basins, the commission issued preliminary permits without requiring site-specific information for assessment of cumulative impacts or a comprehensive plan (Feldman 1988). The GAO (1988) found that preparation of a comprehensive river basin plan could have been a major help in resolving disagreements between the Federal Energy Regulatory Commission and the natural resource agencies over the way to carry out a cumulative impacts assessment.

Avoiding Impacts on Salmon Habitat

The Snohomish Guidelines for the Evaluation of Hydropower Projects were designed to avoid further loss of salmonid habitats and populations due to 600 proposed (mostly small scale) hydroelectric development projects in Washington's Snohomish River basin. This cumulative impacts management planning project (see earlier definition) addressed each project through individual project siting, design, operating, and timing specifications for the project applications. The Snohomish guidelines (Stout 1988) were used in a situation where no further deterioration or loss of habitat has been accepted (and legislated) as society's intention. The guidelines were developed by a coalition of concerned Native Americans, Washington state agencies, and federal natural resource agencies.

Bottomland Hardwoods

The Fish and Wildlife Service conducted three workshops for the Environmental Protection Agency on the ecological impacts of bottomland hardwood deforestation. Cumulative impacts analysis at large watershed or river basin levels became an important part of the project. The participants came to the conclusions that ecological goals are essential, that goals frequently are not available at scales that match cumulative impacts problems, and that a means of nonincremental analysis is needed for cumulative impacts assessment (Gosselink et al. 1990). Setting goals other than no change or no further loss was difficult because it involved society's preferences as well as practical aspects of regulation. Gosselink and Lee (1989) described a landscape ecology approach and its use in a cumulative impacts assessment that involves habitat fragmentation and loss. They hypothesized that individual features are not as important as the pattern, and it is the key features of the pattern which must be identified to conserve biotic diversity and the broad functional values associated with these ecosystems.

FUTURE NEEDS

The general problem facing the U.S. Fish and Wildlife Service's field offices in trying to assess cumulative impacts in 1985 was that cumulative impacts assessments were not happening (Williamson et al. 1986). That problem could best be redressed by convincing the responsible entities, both inside and outside the Fish and Wildlife Service, to conduct those assessments. In addition, a process and methods were needed to provide the technical capability to conduct a cumulative impacts assessment based on the resources, not just regulations. Two causes of the identified problems were technical shortcomings of individual assessment tools and institutional hesitancy to try something new.

From a tentative schedule for the ten highest priority actions for addressing cumulative impacts (Williamson et al. 1986), the first six actions have been effected with various levels of effort and success. The other four actions have the potential of moving federal agencies from attempted cumulative impact assessment to the more productive area of cumulative impacts assessment and management planning. In their order of importance, the suggested actions are (1) review and enlarge agency goals and policies for habitats and species; (2) conduct cumulative impacts assessments for fish, wildlife, and habitat resources of national and regional concern; (3) develop a federal interagency council to foster better cumulative impacts assessments; and (4) make ecological monitoring and project follow-up functions a natural resource agency responsibility.

Institutional reluctance to conduct cumulative impacts assessment and management planning projects has decreased considerably but needs to be reduced further (Muir et al. 1990). An interagency analysis of the differences between an incremental impact viewpoint (i.e., environmental impact assessment) and a total impacts viewpoint (i.e., cumulative impacts assessment) would help. Undertaking sound environmental impact assessment without the regional context and cumulative changes is difficult (Canadian Environmental Assessment Research Council 1988; GAO 1988). Federal agencies could improve individual analyses of permits and licenses by providing a framework within which to evaluate them; this can be done by highlighting ecosystem-based, collaborative cumulative impacts assessment and management planning (Stakhiv 1988; GAO 1988; Gosselink et al. 1990).

An example of such an approach is the Environmental Protection Agency's National Estuary Program. The program employs collaborative problem solving to balance conflicting uses while restoring or maintaining the estuary's environmental quality. Because of their early entry into cumulative impacts work, the Chesapeake Bay and Great Lakes programs are frequently used as models for cumulative impacts assessment and management planning; the participating natural resource agencies have learned how to get the desired results in less time and with less money. The National Estuary Program stresses focusing on the most significant problems, using existing data, emphasizing applied research, funding specifically targeted basic research, and employing demonstrated management strategies. Nichols et al. (1986) observed that the future well-being of the urbanized estuaries depends on achieving an increased understanding of each one's physical, chemical, and biological processes and how specific human activities affect those processes; meanwhile, economically important actions are considered without sufficient quantitative understanding of an action's effects on the estuary.

Through coalition and commitment of the responsible natural resource agencies, those agencies can jointly do a more effective job of comprehensive ecological planning and management. So far in cumulative impacts assessment and management planning projects, the responsible natural resource agencies have included state departments of fish and wildlife, state and local departments of natural resources, ecological research institutes, Native American tribal councils, the U.S. Army Corps of Engineers, Environmental Protection Agency, Fish and Wildlife Service, National Oceanic and Atmospheric Administration, National Marine Fisheries Service, and Canadian federal and provincial natural resource agencies. Cumulative impacts assessment and management planning can shed light on these agencies' unified activities that will attract funding for ecological programs.

The real contribution of mitigation and reclamation actions to achieving society's ecological goals could be improved by using cumulative impacts assessment and management. When the natural resource conservation, regulatory, and land management agencies ratify an interagency collaborative drive toward cumulative improvement of the overall situation, they should be able to move toward several management goals simultaneously. With cumulative impacts assessment and management planning, the Fish and Wildlife Service has been able to promote such positive aspects of management, mitigation, and reclamation and avoid negative, adversarial, and confrontational situations.

We have progressed considerably from the baseline study approach that emerged shortly after passage of NEPA as the primary response of ecologists to multiple-species concerns and as the major supplier of information for the environmental impact assessment process (Truett et al. 1993). Twenty years later, we have learned that you cannot effectively regulate individual minor contributions to cumulative impacts, but that you can plan for them in the aggregate. It is no longer a question whether we should conduct cumulative impacts assessment and management planning projects. We should! The question is also not how can we best conduct such a project. That depends on the situation, as described here. The question now is: "How can we acquire the support for conducting cumulative impacts assessment and management planning projects?" Instead of continuing to rely on the ability of American citizens and institutions to respond to individual ecological crises as they are recognized and popularized, a technological capability for coordinated, effective action through cumulative impacts assessment and management planning should be developed before ecological problems reach crisis proportions.

ACKNOWLEDGMENTS

Special thanks are extended to all of the U.S. Fish and Wildlife Service personnel who participated in developing the information presented in this document.

REFERENCES

Bain, M. B., J. S. Irving, R. D. Olsen, E. A. Stull, and G. W. Witmer. 1986. Cumulative Impact Assessment: Evaluating the Environmental Effects of Multiple Human Developments. ANL/EES-TM-309. Argonne National Laboratory, Argonne, IL.

Burns, D. C. 1991. Cumulative effects of small modifications to habitat. *Fisheries* 16:12–17.

Canadian Environmental Assessment Research Council. 1988. The Assessment of Cumulative Effects: A Research Prospectus. Cat. No. En 106-10/88. Minister of Supply and Services, Canada.

Canadian Environmental Assessment Research Council and United States National Research Council. 1986. Cumulative Environmental Effects: A Binational Perspective. Cat. No. En 106-2/85. Minister of Supply and Services, Canada.

Dickert, T. G., and A. E. Tuttle. 1985. Cumulative impact assessment in environmental planning: A coastal wetland watershed example. *Environ. Impact Assess. Rev.* 5:37–64.

Feldman, M. D. 1988. *National Wildlife Federation* v. *FERC* and *Washington State Department of Fisheries* v. *FERC*: Federal Energy Regulatory Commission ignores Ninth Circuit rebuke on hydropower permitting. *Ecol. Law Q.* 15:3–360.

Flemer, D. A., G. B. Mackiernan, W. Nehlsen, V. K. Tippie, R. B. Biggs, D. Blaylock, N. H. Burger, L. C. Davidson, D. Haberman, K. S. Price, and J. L. Taft. 1983. *Chesapeake Bay: A Profile of Environmental Change.* U.S. Environmental Protection Agency, Philadelphia, PA.

Forman, R. T. T., and M. Godron. 1986. *Landscape Ecology.* Wiley, New York.

Gosselink, J. G., and L. C. Lee. 1989. Cumulative impact assessment in bottomland hardwood forests. *Wetlands* 9:89–174.

Gosselink, J. G., L. C. Lee, and T. A. Muir (eds.). 1990. *Ecological Processes and Cumulative Impacts: Illustrated by Bottomland Hardwood Wetland Ecosystems.* Lewis Publishers, Inc., Chelsea, MI.

Granholm, S. L., E. Gerstler, R. R. Everitt, D. P. Bernard, and E. C. Vlachos. 1987. Issues, Methods, and Institutional Processes for Assessing Cumulative Biological Impacts. Prepared for Pacific Gas and Electric Company, San Ramon, CA.

Horak, G. C., E. C. Vlachos, and E. W. Cline. 1983. Methodological Guidance for Assessing Cumulative Impacts on Fish and Wildlife. Contract No. 14-16-0009-81-058. Prepared for U.S. Fish and Wildlife Service, Eastern Energy and Land Use Team, Kearneysville, WV.

Hunsaker, C., and S. Williamson. 1992. Techniques for assessing cumulative impacts. pp. 49–51. In *Making Decisions on Cumulative Environmental Impacts: A Conceptual Framework.* World Wildlife Fund, Washington, D.C.

Lee, L. C., and J. G. Gosselink. 1988. Cumulative impacts on wetlands: Linking scientific assessments and regulatory alternatives. *Environ. Manage.* 12:591–602.

Meehan, R., and P. J. Webber. 1985. Towards an Understanding and Assessment of the Cumulative Impacts of Alaskan North Slope Oil and Gas Development. U.S. Environmental Protection Agency under U.S. Department of Energy Interagency Agreement No. DE-A106-84RL10584. U.S. Environmental Protection Agency, Washington, D.C.

Muir, T. A., C. Rhodes, and J. G. Gosselink. 1990. Federal statutes and programs relating to cumulative impacts in wetlands. pp. 223–236. In J. G. Gosselink, L. C. Lee, and T. A. Muir (eds.), *Ecological Processes and Cumulative Impacts.* Lewis Publishers, Inc., Chelsea, MI.

National Research Council of the United States, Committee on the Applications of Ecological Theory to Environmental Problems. 1986. *Ecological Knowledge and Environmental Problem-Solving: Concepts and Case Studies.* National Academy Press, Washington, D.C.

National Research Council of the United States and The Royal Society of Canada. 1985. *The Great Lakes Water Quality Agreement: An Evolving Instrument for Ecosystem Management.* National Academy Press, Washington, D.C.

Nichols, F. H., J. E. Cloern, S. N. Luoma, and D. H. Peterson. 1986. The modification of an estuary. *Science* 231:567–573.

Salwasser, H., and F. B. Samson. 1985. Cumulative effects analysis: An advance in wildlife planning and management. *Trans. North Am. Wildl. Nat. Resour. Conf.* 50:313–321.

Stakhiv, E. Z. 1986. Cumulative impact analysis for regulating decisionmaking. pp. 213–222. In J. A. Kusler, and P. Riexinger (eds.), *Proceedings, National Wetland Assessment Symposium.* ASWM Tech. Rep. 1. Association of State Wetland Managers, Chester, VT.

Stakhiv, E. Z. 1988. An evaluation paradigm for cumulative impact analysis. *Environ. Manage.* 12:725–748.

Stout, D. J. 1988. Preventing cumulative impacts: The Washington experience. pp. 204–206. In J. A. Kusler, M. L. Quammen, and G. Brooks (eds.), *Proceedings, National Wetland Symposium: Mitigation of Impacts and Losses.* ASWM Tech. Rep. 3. Association of State Wetland Managers, Chester, VT.

Truett, J. C., H. L. Short, and S. C. Williamson. 1993. Ecological impact assessment. In T. Brookout (ed.), *Wildlife Management Techniques.* 5th ed. The Wildlife Society, Inc., Washington, D.C.

U.S. General Accounting Office. 1988. Energy Regulation: Opportunities for Strengthening Hydropower Cumulative Impact Assessments. GAO-RCED-88-82. Report to the Chairman, Committee on Energy and Commerce, House of Representatives, Washington, D.C.

Walker, D. A., P. J. Webber, E. F. Binnian, K. R. Everitt, N. D. Lederer, E. A. Nordstrand, and M. D. Walker. 1987. Cumulative impacts of oil fields on northern Alaskan landscapes. *Science* 238:757–761.

Williamson, S. C., and K. Hamilton. 1989. Annotated bibliography of ecological cumulative impacts assessment. *U.S. Fish and Wildlife Serv. Biol. Rep.* 89(11).

Williamson, S. C., C. L. Armour, and R. L. Johnson. 1986. Preparing a FWS cumulative impacts program: January 1985 workshop proceedings. *U.S. Fish and Wildlife Serv. Biol. Rep.* 85(11.2).

Williamson, S. C., C. L. Armour, G. W. Kinser, S. L. Funderburk, and T. N. Hall. 1987. Cumulative impacts assessment: An application to Chesapeake Bay. *Trans. North Am. Wildl. Nat. Resour. Conf.* 52:377–388.

Proposed Methodology to Assess Cumulative Impacts of Hydroelectric Development in the Columbia River Basin

K. E. LaGory, E. A. Stull, and W. S. Vinikour, Argonne National Laboratory, Argonne, IL

ABSTRACT

A methodology to assess the cumulative impacts of hydroelectric development on multiple resources was developed and proposed for use in the Columbia River Basin. The methodology is a matrix-based procedure that uses models of the response of populations or resources to project-induced environmental change within project impact zones. Cumulative impacts are calculated based on estimations of single-project effects, first-order interactions among projects, shared project features, and an estimation of the impacts of existing projects. The methodology is most effective when quantitative models are used; however, it is possible to assess cumulative impact using this methodology when less detailed or only qualitative information is available. To demonstrate the use of this methodology, a hypothetical example was developed for the effects of several small developments in a watershed that contained salmon spawning areas.

INTRODUCTION

Hydroelectric development has occurred at a rapid pace in the Pacific Northwest over the last several decades. Concern over the impacts of development resulted in the enactment of the Pacific Northwest Electric Power Planning and Conservation Act of 1980. This legislation led to the development of the Columbia River Basin Fish and Wildlife Program, under which the Bonneville Power

Administration (BPA) was directed to develop criteria and methods for assessing the cumulative environmental effects of multiple hydroelectric developments within the Columbia River Basin. The BPA contracted Argonne National Laboratory to study these cumulative effects issues and to develop a methodology to assess cumulative effects.

Most recent applications for development have been for small-scale projects (10 MW or less) on high-gradient headwater streams. The methodology described in this paper enables the consideration of numerous projects and a comparison of a variety of development scenarios. The details of the methodology have been published in a two-volume report (Stull et al. 1987a; Stull et al. 1987b) which also discusses key species and habitats in the region, the types of effects caused by hydroelectric development, and available methods of cumulative impact assessment. A similar methodology was developed by Argonne National Laboratory to assess cumulative effects (Bain et al. 1986) and was applied to hydroelectric development in the Snohomish River Basin, WA (FERC 1986) and Salmon River Basin, ID (FERC 1987; Irving and Bain, this volume, Chap. 5). The two methodologies are similar, but the methodology described in this paper provides an explicit procedure for determining project interaction.

CONCEPTUAL FRAMEWORK OF THE METHODOLOGY

We define cumulative effect as an environmental change resulting from the accumulation and interaction of the effects of one action with effects of one or more other actions occurring on a common resource. Thus, cumulative effects result from incremental changes as well as interactive effects among projects. A cumulative effect can be conceptualized as being the sum of single-project effects and the effects of interactions among projects.

Project Impact Zones

The methodology we proposed is based on the premise that nonadditive cumulative effects are derived from the modification of single-project effects when other projects are present (i.e., whenever interactions between projects occur). These interactions can occur whenever the impact zones of two or more projects overlap and impacts occur contemporaneously. The project impact zone is the geographical area affected by a project that contains the species or resource under consideration. For example, the project impact zone for sedimentation effects on salmon is the length of stream over which sediment deposition patterns are altered. Figure 1 illustrates the impact zones of five hydroelectric projects that produce elevated sediment loads within a drainage. In this example, the impact zones of Projects 3, 4, and 5 do not overlap, and thus these projects do not interact. On the other hand, the impact zones of Projects 1 and 2 do overlap, and interaction of these projects is possible depending on the nature of biological response to environmental change within this area of overlap.

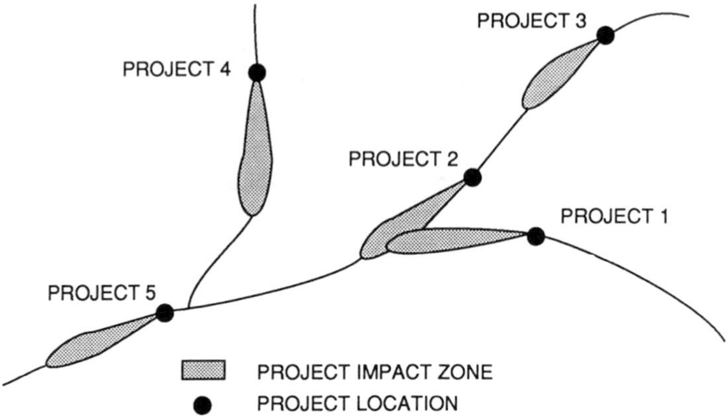

Figure 1. Project impact zones for five hypothetical projects within a river basin.

Biological Response and Types of Cumulative Effects

Cumulative effects can be either additive, supraadditive, or infraadditive, depending on the nature of the biological response to environmental change (Figure 2). Additive cumulative effects occur if the response is linear over the range of change being considered. Thus, an additive cumulative effect is the same as the sum of the single-project effects. If the cumulative effect is additive, the projects essentially do not interact. A cumulative effect is supraadditive if the response to the total environmental change is greater than the sum of the responses to single projects; it is infraadditive if the response to total environmental change is less than the sum of single-project effects. Complex environmental responses (e.g., sigmoidal) can result in supraadditive or infraadditive cumulative effects depending on the range over which change occurs.

Interaction Coefficients

Interaction coefficients measure the departure from the additive condition for each pair of projects; they are equal to 0 when two projects do not interact, greater than 0 when the interaction is supraadditive, and less than 0 when the interaction is infraadditive. The magnitude of interaction coefficients is based on the degree of overlap between project impact zones and the response of the species or resource to single-project effects and the effects of the two projects together. The formula for calculating interaction coefficients is

$$C_{1,2} = \frac{O_{1,2}}{Z_1} \times \frac{R_{1,2o} - \left(R_{1o} + R_{2o}\right)}{2R_{1o}}$$

where $C_{1,2}$ = an interaction coefficient, in this case, the effect of Project 2 on Project 1

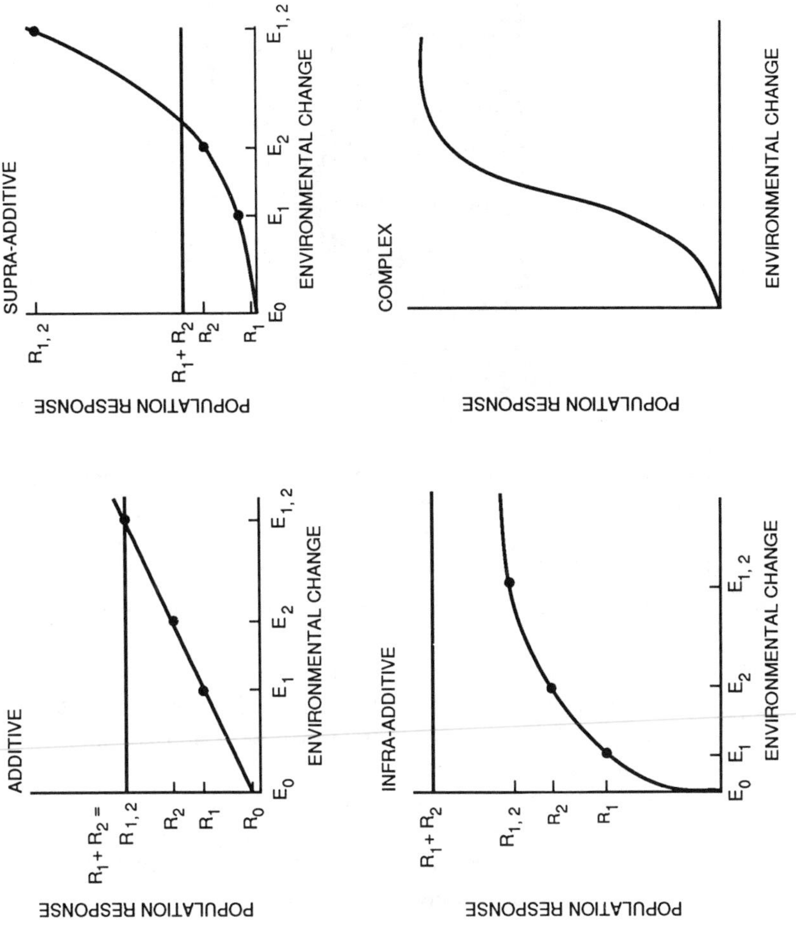

Figure 2. Examples of biological response to environmental change.

$$C_{1,2} = \frac{O_{1,2}}{Z_1} \times \frac{R_{1,2_0} - (R_{1_0} + R_{2_0})}{2R_{1_0}}$$

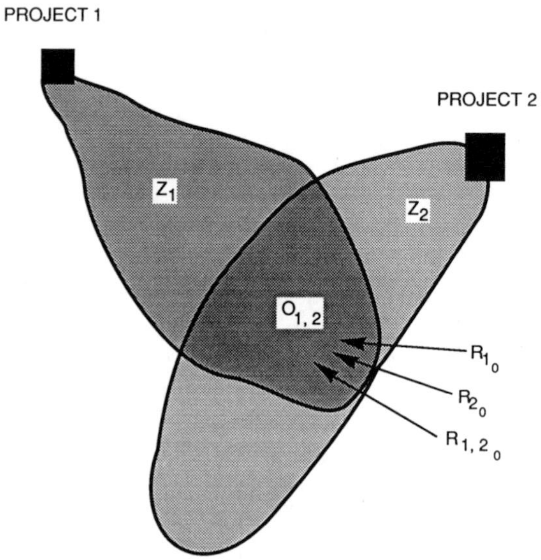

Figure 3. Graphical depiction of overlapping project impact zones and terms used in calculating interaction coefficients.

$O_{1,2}$ = the area of overlap between the project impact zones of Projects 1 and 2

Z_1 = the area of the impact zone of Project 1

$R_{1,2o}$ = the predicted response of the species or resource to both projects in the area of overlap of project impact zones if both projects are built

R_{1o} = The predicted response of the species or resource to Project 1 in the area of overlap if only Project 1 is built

R_{2o} = the predicted response of the species or resource to Project 2 in the area of overlap if only Project 2 is built

Figure 3 provides a graphical representation of the sources of terms used in calculating interaction coefficients. The term $O_{1,2}/Z_1$ is the proportion of the impact zone of Project 1 that is overlapped by the impact zone of Project 2. The second term in the equation ($[R_{1,2o} - (R_{1o} + R_{2o})]/2R_{1o}$) is the proportional relationship between the response of the species to both projects relative to the response of the species to Project 1 alone. These terms weight the interactions according to the size of the area affected as well as the magnitude of cumulative and single-project effects.

Matrices Used in Cumulative Impact Assessment

Our methodology uses matrices for accumulating incremental single-project effects and interactions between project pairs into a total that represents the cumulative effect of all projects acting together. Three matrices are used to calculate cumulative effect: the impact matrix, interaction matrix, and product matrix.

The impact matrix is a $1 \times N$ matrix that contains the estimates of single-project effects (I_i) for each of the N projects being considered:

$$\text{Impact Matrix} \quad \begin{bmatrix} I_1 & I_2 & \cdots & I_n \end{bmatrix}$$

The interaction matrix is an $N \times N$ matrix that contains the interaction coefficients ($C_{i,j}$) for each pair of projects:

$$\text{Interaction Matrix} \quad \begin{bmatrix} 1 & C_{1,2} & \cdots & C_{1,n} \\ C_{2,1} & 1 & \cdots & C_{2,n} \\ \cdot & \cdot & \cdots & \cdot \\ \cdot & \cdot & \cdots & \cdot \\ \cdot & \cdot & \cdots & \cdot \\ C_{n,1} & C_{n,2} & \cdots & 1 \end{bmatrix}$$

As described above, the interaction coefficients represent the ability of each project to modify the effects of another.

The product of these two matrices is used to calculate total cumulative effect of the N projects:

$$\text{Product Matrix} \quad \begin{bmatrix} I_1 & I_1C_{1,2} & \cdots & I_1C_{1,n} \\ + & + & \cdots & + \\ I_2C_{2,1} & I_2 & \cdots & I_2C_{2,n} \\ + & + & \cdots & + \\ \cdot & \cdot & \cdots & \cdot \\ \cdot & \cdot & \cdots & \cdot \\ \cdot & \cdot & \cdots & \cdot \\ + & + & \cdots & + \\ I_nC_{n,1} & I_nC_{n,2} & \cdots & I_n \end{bmatrix}$$

$$= \left[\sum_{i=1}^{n} \left(I_n C_{n,1} \right) \quad \sum_{i=1}^{n} \left(I_n C_{n,2} \right) \cdots \sum_{i=1}^{n} I_n C_{n,n} \right]$$

Cumulative effect equals the sum of the elements of this $1 \times N$ matrix.

The following example illustrates the use of these matrices for the effects of five projects on one resource:

Impact Matrix

Effect of project:	A	B	C	D	E
	[5	10	12	50	20]

Interaction Matrix

Effect of project:		A	B	C	D	E
On project:	A	1	0.2	0	0	0
	B	0.5	1	0	0	0
	C	0	0	1	0	0
	D	0	0	0	1	0
	E	0	0	0	0	1
Product Matrix =		[10	11	12	50	20]

Cumulative effect = sum of elements of product matrix = 103

In this example, the sum of single-project effects (i.e., the sum of the elements within the impact matrix) equals 97, and therefore, the cumulative effects of these five projects are supraadditive and the project interactions account for approximately 6% of the cumulative effect.

DESCRIPTION OF THE PROPOSED METHODOLOGY

Using the methodology described here, a cumulative effects assessment is accomplished by performing a series of six steps, resulting in an estimate of the impact of multiple projects acting together on a single species or resource. This procedure must be repeated for each species or resource under consideration. The six steps are

1. Determine the relationships among projects.
2. Perform single-project assessments.
3. Calculate interaction coefficients.
4. Calculate unadjusted cumulative effect.
5. Adjust for the effects of shared project features.
6. Incorporate the effects of existing projects.

The use of the methodology is demonstrated below using a hypothetical example of three hydroelectric projects located within the same river basin. The cumulative effects of changes in sediment load on salmon recruitment are assessed.

Determining the Relationships Among Projects

This step is necessary for delineating the project impact zones of each project and the areas of impact zone overlap and for providing a conceptualization of the nature of project interaction. In determining the relationships among projects, it is best to begin by developing a map that shows the locations of projects, all associated project features, the distribution of the species or resource being considered, and important habitats. Three aspects of project relationships should be considered: geographical, temporal, and ecological.

A determination of geographical relationship simply places projects in a physical context relative to one another. This information can be used to determine distances between projects, topographical relationships, and ultimately ecological relationships. Temporal relationships of project construction and operation activities are important because impacts may be short-lived, and unless these are contemporaneous, project interaction may be minimal. In addition, the magnitude of impacts may vary seasonally as a result of climate or the biological characteristics of species. Understanding the ecological relationships among projects, based on the overlap of project impact zones, is essential to an understanding of cumulative effect. Impact zones should be based on the ecology of the species under consideration and the nature of the impacts expected. Impact zones can be some area of habitat, a reach of a stream, or an entire population. The degree of project impact zone overlap and the ecology of the population or resource within this area of overlap determines the nature of project interaction.

Figure 4 illustrates the relationships among the three hypothetical projects in our example. These three projects are located within the same watershed; they would be built concurrently; and the project impact zones of each overlap to some extent. Projects B and C would share a segment of access road that could affect fish in the adjacent stream. Salmon are known to spawn within the impact zones of each project, and each project would contribute sediment to downstream areas.

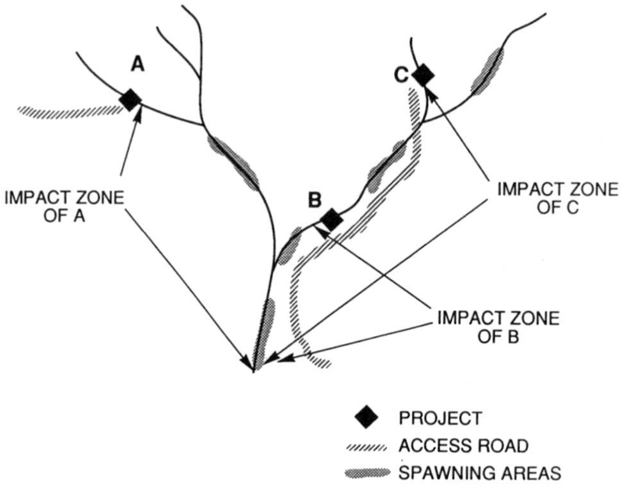

Figure 4. Example map showing the location of three hypothetical projects and their associated facilities. Each project would contribute sediment downstream and thus affect salmon recruitment. The project impact zones of each, relative to sedimentation effects, are shown.

Performing Single-Project Assessments

Once impact zones are established, a model is used to determine the response of the species or resource to the environmental change induced by each project within the project impact zone. The model must be the same as that used in the calculation of interaction coefficients.

For ease of interpretation, the effects of projects should be expressed in units that directly reflect the magnitude of the effect (e.g., the number of adult individuals lost, the acres of habitat lost) rather than evaluative criteria. This will facilitate interpretation of the assessment and enable placement of the assessment in the context of established management goals.

In this assessment, we use a model that describes the relationship between percent emergence of salmon fry and the amount of fine sediment in the substrate. Stowell et al. (1983) determined that an increase in fine sediment caused a reduction in the emergence of salmon fry and a subsequent loss of recruits to the population (Figure 5). By using this relationship to predict changes in sediment load that result from construction and operation of the proposed facilities, we can predict changes in percent emergence and then estimate the number of recruits lost to the population. Based on our calculations, Project A, B, or C, if constructed alone, would result in the loss of 160, 98, or 145 recruits, respectively.

Calculating Interaction Coefficients

Interaction coefficients are calculated for all pairwise permutations of projects. For three projects (A, B, and C), these are AB, BA, AC, CA, BC, and CB. The

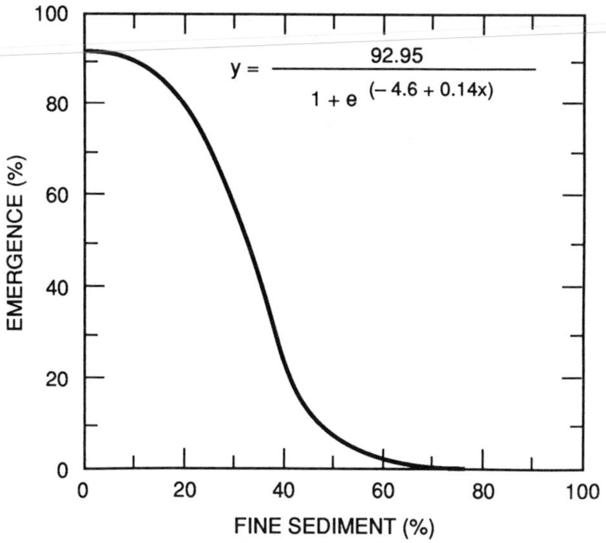

Figure 5. The relationship between the percent emergence of salmon fry relative to the percent of fine sediment in the stream substrate. (From Stowell, R., et al. 1983. Guide for Predicting Salmonid Response to Sediment Yields in Idaho Batholith Watersheds. U.S. Forest Service, Northern and Intermountain Regions.)

formula for calculating the interaction coefficient that describes the influence of one project on another is

$$C_{1,2} = \frac{O_{1,2}}{Z_1} \times \frac{R_{1,2o} - \left(R_{1o} + R_{2o}\right)}{2R_{1o}}$$

The terms in this equation were defined and discussed above.

Calculating the Degree of Project Overlap

The extent of project impact zones and overlap can be measured in several different ways, each with implicit assumptions about the distributions of the population, preimpact environmental conditions, and project-induced environmental change (Table 1). The method chosen will depend on the level of understanding of the species under consideration and the amount of site-specific information available.

Table 2 provides the estimates of overlap determined using different methods of calculating overlap for an assessment of the effects of project construction on a hypothetical salmon population. As can be seen, overlap ratios and thus interaction coefficients can vary considerably according to the method used. Although the simplest method is based on the area affected, the safest and most accurate method (but also the most difficult) is based on the impacts within project impact zones.

Table 1. Assumptions in the Calculation of Project Overlap Ratios.

	Assumed Distribution		
Measurement	**Population or Habitat**	**Pre-Impact Conditions**	**Environmental Change**
Area	Even	Even	Even
Habitat	Uneven	Even	Even
Population	Uneven	Uneven	Even
Impact	Uneven	Uneven	Uneven

Table 2. Examples of Various Calculations of Project Overlap.

		Impact Zone			Overlap Ratio	
Measurement	**Example**	**A**	**B**	**Overlap**	**A**	**B**
Area	Acres	21	15	6	0.29	0.40
Habitat	Acres of habitat	7	15	3	0.43	0.20
Population	Number of fish before impacts	402	1273	242	0.32	0.19
Impact	Number of fish lost	161	8	6,1	0.04	0.13

Determining Response Terms

Response terms are calculated or determined from some model. Such a model can be univariate or multivariate, quantitative or qualitative. Obviously, the accuracy of the assessment depends on the applicability of the model chosen. In our example of the effects of sediment on fish, we must choose or develop a model that represents this relationship. We use the same model we used for determining single-project effects above. Using the predicted increases in sediment load and this model, we calculate the response terms needed to calculate interaction coefficients.

Adjustments to the Interaction Coefficient Formula

Two situations require an adjustment to the formula for interaction coefficients: shared project features and impact segmentation.

For the case of shared project features, adjustment must be made because each single-project assessment contains the effect of the shared feature. Thus, the formula for interaction coefficient becomes

$$C_{1,2} = \frac{O_{1,2}}{Z_1} \times \frac{R_{1,2o} - \left(R_{1o} + R_{2o}\right) + R_{so}}{2R_{1o}}$$

Table 3. Example Worksheet for the Calculation of Interaction Coefficients for Four Projects that Affect Salmon Recruitment.

Project		Impact Zone		Response Term				Interaction Coefficient
1	2	$O_{1,2}$	Z_1	$R_{1,2o}$	R_{1o}	R_{2o}	R_{so}	$C_{1,2}$
A	B	5.2	12.4	26.2	11.1	12.1	0.0	0.0567
A	C	5.2	12.4	42.2	11.1	18.2	0.0	0.2437
B	A	5.2	7.5	26.2	12.1	11.1	0.0	0.0860
B	C	7.5	7.5	47.8	12.1	18.2	6.4	0.9876
C	A	5.2	15.2	42.2	18.2	11.1	0.0	0.1212
C	B	7.5	15.2	47.8	18.2	12.1	6.4	0.3240

where R_{so} is the response of the species or resource in the area of overlap to the effect of the shared project feature. In our example, Projects B and C share a segment of road that would affect the salmon population in a nearby reach of stream.

If the population, environmental conditions, or impacts are unevenly distributed in the overlap zone, segmenting overlap zones to more accurately reflect these distributions should be considered. The formula for interaction coefficients with n segments is

$$C_{1,2} = \sum_{p=1}^{n} \frac{O_{1,2p}}{Z_1} \times \frac{R_{1,2op} - \left(R_{1op} + R_{2op}\right) + R_{sop}}{2R_{1op}}$$

where p represents the segment under consideration. Essentially, an interaction coefficient is calculated for each segment of the impact zone, and these coefficients are summed to provide an overall interaction coefficient.

Calculating the Interaction Coefficients

Once the impact zones and response terms are calculated, determining the interaction coefficients is a simple process which can be accomplished, in most cases, using a spreadsheet. Table 3 is a sample worksheet for the three projects being considered in our example. In this example, impact zone values ($O_{1,2}$ and Z_1) are calculated based on the amount of spawning habitat present within the entire impact zone of each project and the amount within areas of overlap between project pairs. Response terms ($R_{1,2o}$, R_{1o}, R_{2o}, and R_{so}) represent the expected decrease in percent emergence that results from project-induced changes in fine sediment as discussed above.

Calculating Unadjusted Cumulative Effect

As discussed above, three matrices are used in the calculation of cumulative effect: the impact matrix, interaction matrix, and product matrix. For our example,

$$\underset{\text{Impact Matrix}}{\begin{bmatrix}160 & 98 & 145\end{bmatrix}} \times \underset{\text{Interaction Matrix}}{\begin{bmatrix}1 & 0.0567 & 0.2437 \\ 0.0860 & 1 & 0.9876 \\ 0.1212 & 0.3240 & 1\end{bmatrix}} = \underset{\text{Product Matrix}}{\begin{bmatrix}186 & 154.1 & 280.8\end{bmatrix}}$$

The sum of the elements of the product matrix equals the unadjusted cumulative effect. In this case, unadjusted cumulative effect = 620.9 salmon lost. This unadjusted cumulative effect is larger than the sum of the single-project impacts (160 + 98 + 145 = 403 salmon lost).

Adjusting for the Effects of Shared Project Features

Each of the single-project impacts within the impact matrix will include the impacts of features shared by other projects; thus, the unadjusted cumulative effect value will contain an overrepresentation of the effect of those features. To remove replication of impacts, the following adjustment should be made.

$$\text{Cumulative effect} = U - \sum_{i=1}^{n} I_i (S_i - 1)$$

where U = unadjusted cumulative effect, calculated as the sum of the elements in the product matrix
I_i = impact of shared project feature i
S_i = number of projects sharing feature i

In our example, only one project feature is shared (a segment of access road by Projects B and C):

$$\text{Cumulative effect} = 620.9 - 52(2 - 1) = 568.9 \text{ salmon lost}$$

Thus, for the three-project example given, the interactive effect of the three projects accounts for 29% of the predicted cumulative effect.

Incorporating the Effects of Existing Projects

The effects of existing projects should be incorporated into an assessment of cumulative effect whenever appropriate. For some impacts (e.g., construction effects), the species may have recovered from the impact and returned to predevelopment levels. For other impacts (e.g., sedimentation effects or passage mortality), the existing project may continue to adversely affect the species. In this latter case, existing projects have effectively changed the baseline from that which existed before development occurred.

Ongoing effects of existing projects normally would be incorporated into the assessment of the single-project effects of each proposed project. These single-project assessments would be based on an environmental baseline that incorporates the effects of the existing project. For instance, if existing projects in our hypothetical watershed contributed sediment to the impact zones of proposed projects, these existing projects would affect the baseline sediment load used in our calculations of percent fry emergence.

If, as suggested above, the effects of existing projects are incorporated into the new environmental baseline of proposed projects, the losses attributable to proposed projects will often be less than would have occurred had the existing projects never been built. A cumulative assessment that evaluates total losses, including those of existing projects, should be calculated using the following formula modified to incorporate these effects.

$$\text{Cumulative effect} = U - \sum_{i=1}^{n} I_i\left(S_i - 1\right) + \sum_{j=1}^{m} I_j$$

where I_j = impact of existing Project j
m = number of existing projects that affect the species or resource being considered in the assessment

DISCUSSION

Our methodology is based on the assumption that the impacts of hydroelectric development on fish and wildlife can be assessed using models of biological response to project-induced environmental change. The methodology does not place any restrictions on the type of models used. Qualitative, univariate, or multivariate models can be used. For the methodology to be effective, however, the models used should be sensitive to nonadditive relationships in order to allow an estimation of project interaction. In many cases, however, population response models may not be currently available.

The estimate of cumulative effect produced using our methodology will vary depending upon the types of models used in calculating single-project effects and interaction coefficients. In our example, we used a model whose output was the number of salmon recruits lost. Using such a quantitative model would enable a determination of the significance of cumulative effect based on preestablished management goals for a species' population. Quantitative models based on evaluative criteria or indices also can be used, but a determination of significance necessarily will be more subjective. The methodology is flexible enough, however, to utilize either type of model.

The Habitat Evaluation Procedures (HEP) (U.S. Fish and Wildlife Service 1980) and the Instream Flow Incremental Methodology (IFIM) (Bovee 1982), developed by the U.S. Fish and Wildlife Service, provide alternatives to population models. Both HEP and IFIM use habitat suitability index models that are based on the relationships between environmental conditions and habitat suitability. Habitat suitability index models are currently available for a wide variety of fish and wildlife species and could be used effectively with our methodology to assess the cumulative impact of projects.

The methodology is also more effective if relatively detailed site-specific information is available for each project. It is necessary that each project have the same type of information available to allow comparisons of impacts and to allow calculation of interaction coefficients. As mentioned above, the type of information available can affect the magnitude of the overlap ratio ($O_{1,2}/Z_1$) and therefore that of the interaction coefficient. Each type of information also has associated assumptions about the distributions of the population, habitat, preimpact environmental conditions, and project-induced environmental change. These assumptions must be taken into account, and a decision should be made as to the validity of these assumptions and the seriousness of violation.

Synergisms among different types of impacts can occur and complicate any calculation of cumulative effect. For instance, temperature and sediment may have a complex relationship in terms of their effect on the survival of eggs or alevins in redds. If several hydropower projects were to significantly affect both sediment and temperature, there would be a potential for a synergistic effect to occur. If, however, in the assessment of cumulative effects, two univariate response models were used (one for sediment effects and one for temperature effects), each effect would be accumulated separately, and this synergistic relationship would be ignored. The use of a multivariate-response model that included temperature effects would be necessary to account for this relationship.

Temporal changes in the severity and types of impacts also add complexity to cumulative effects assessment. In many instances, most impacts will be most severe during the first several years following project construction and then will level off to some stable value. Several different approaches can be taken when dealing with the effects of time. These approaches include (1) obtaining the average impact over the period of interest, (2) limiting the assessment to the period after the population or resource has stabilized, or (3) using a model that incorporates time-dependent processes.

In order to place bounds on estimates of cumulative effect, minimum and maximum values for single-project effects, response terms, and impact zones could be used in the calculations. Sensitivity analyses could also be performed to determine the range of cumulative effect estimates that could be reasonably expected. Sensitivity of the assessment to changes in input will depend to a large extent on the nature of the population's or resource's response to environmental change, as well as the range of environmental change that could be reasonably expected to occur.

As the above discussion indicates, several decisions must be made in applying our methodology, and these decisions will affect the eventual outcome of the cumulative assessment. For this reason, we recommend that any group performing such an assessment incorporate some mechanism for negotiation or decisionmaking. The Adaptive Environmental Assessment and Management methodology (Holling 1978) is ideal for providing a decisionmaking framework and for developing single-project assessment models that can be used to produce information for the cumulative assessment. Such a framework would be essential to successful application of our methodology to any major river basin or subbasin.

REFERENCES

Bain, M. B., J. S. Irving, R. D. Olsen, E. A. Stull, and G. W. Witmer. 1986. Cumulative Impact Assessment: Evaluating the Environmental Effects of Multiple Human Developments. ANL/EES-TM-309. Argonne National Laboratory, Argonne, IL.

Bovee, K. D. 1982. A Guide to Stream Habitat Analysis Using the Instream Flow Incremental Methodology. Instream Flow Information Paper 12. FWS/OBS-82/26. U.S. Fish and Wildlife Service, Western Energy and Land Use Team, Fort Collins, CO.

Federal Energy Regulatory Commission. 1986. Snohomish River Basin, Seven Hydroelectric Projects, Washington. FERC/FEIS-0042. Federal Energy Regulatory Commission, Washington, D.C.

Federal Energy Regulatory Commission. 1987. Salmon River Basin, Fifteen Hydroelectric Projects, Idaho. FERC/FEIS-0044. Federal Energy Regulatory Commission, Washington, D.C.

Holling, C. S. 1978. *Adaptive Environmental Assessment and Management.* Wiley, New York.

Stowell, R., A. Espinosa, T. C. Bjornn, W. S. Platts, D. C. Burns, and J. S. Irving. 1983. Guide for Predicting Salmonid Response to Sediment Yields in Idaho Batholith Watersheds. U.S. Forest Service, Northern and Intermountain Regions.

Stull, E. A., M. B. Bain, J. S. Irving, K. E. LaGory, and G. W. Witmer. 1987a. Methodologies for Assessing the Cumulative Environmental Effects of Hydroelectric Development on Fish and Wildlife in the Columbia River Basin — Volume 1: Recommendations. DOE/BP-19461-3. Final Report to Bonneville Power Administration, Portland, OR.

Stull, E. A., K. E. LaGory, and W. S. Vinikour. 1987b. Methodologies for Assessing the Cumulative Environmental Effects of Hydroelectric Development on Fish and Wildlife in the Columbia River Basin — Volume 2: Example and Procedural Guidelines. DOE/BP-19461-4. Final Report to Bonneville Power Administration, Portland, OR.

U.S. Fish and Wildlife Service. 1980. Habitat Evaluation Procedures. ESM 102. U.S. Fish and Wildlife Service, Division of Ecological Services, Washington, D.C.

Walking Through Walls: Using NEPA's Cumulative Impacts Concept to Reconcile Single-Issue Environmental Statutes

D. R. Pescitelli, Illinois Department of Transportation, Springfield, IL; and D. L. Merida, U.S. Department of Transportation, Little Rock, AR

ABSTRACT

The number and nature of single-issue environmental laws and Executive Orders often place project planners into situations wherein competing demands can stall even the best-planned public project. This paper describes a case study of a highly controversial highway project to demonstrate how the cumulative impacts concept associated with the National Environmental Policy Act (NEPA)-based analyses can be used to balance environmental concerns and reconcile competing legal requirements.

The project triggered involvement of several different environmental and cultural resource protection laws, and the resulting litigation frustrated completion of the project for 7 years.

A team of planners and engineers used the cumulative impacts concept as a way to reconcile the requirements of special-issue environmental laws with the stringent requirements of Section 4(f) of the Department of Transportation Act of 1966. Cumulative impacts used in the analysis included visual impacts, compatibility with existing terrain, agricultural impacts, effects on threatened and endangered species, wildlife habitat impacts, and effects on historic and archaeological resources.

This extension of the cumulative impacts approach to Section 4(f) was challenged in litigation but was eventually endorsed by both federal district and appellate courts.

This validation demonstrates how the cumulative impacts concept can eliminate artificial barriers between single environmental issues and thus facilitate a wide-ranging balancing of environmental effects consistent with the spirit of NEPA.

PROJECT AND ITS SETTING

Location

The project involved construction of a four-lane, controlled-access highway between the communities of Jacksonville and Barry in west-central Illinois. The highway is approximately 52 mi long and crosses the Illinois River midway between these two communities. The highway is part of Federal Aid Primary Route 408 (FAP 408), commonly called the Central Illinois Expressway (CIE), which extends 100 mi from Springfield to the Mississippi River. The purpose of the project is to provide better traffic service and to connect the communities of west-central Illinois with the well-developed transportation network to the east (Figure 1). The part of the project at issue was construction of a crossing over the Illinois River at a location between Valley City and Florence (Figure 2).

Setting

The prominent landform in the project area is a formation of limestone bluffs along the west side of the Illinois River. These bluffs rise sharply to a height of about 200 ft above the floodplain on the east side of the river. Near the river, the bluffs are dissected by a number of ravines. Farther west, the terrain opens into flat to gently rolling uplands that are drained by three small streams.

The area east of the river is a 3-mi-wide floodplain with high bluffs on the eastern limits. These eastern bluffs rise to approximately 140 ft above the floodplain.

In the floodplain east of the river, the predominant land use is agriculture. On the west side of the river, land cover consists primarily of bottomland and upland forest, shrubland, grassland, and a modest amount of agricultural use. In the flatter upland areas, agricultural uses predominate.

The project area is marked by numerous and significant prehistoric and historic resources. The first of these is the Burnt Hill Multiple Resource Area (MRA), a series of 68 habitation sites spanning eight prehistoric cultures from Paleo Indian (9000–7500 B.C.) to Mississippian (A.D. 900–1400). The MRA is a 23-mi-long, 1-mi-wide transect along the highway centerline which runs across the dissected uplands from Barry eastward to the bluff crest along the east bank of the Illinois River near the community of Valley City. The area affected by the project includes one of the few intact Hopewell Indian burial mound complexes remaining in the lower Illinois River region. The project area is also near some of the earliest pioneer settlements west of the Illinois River.

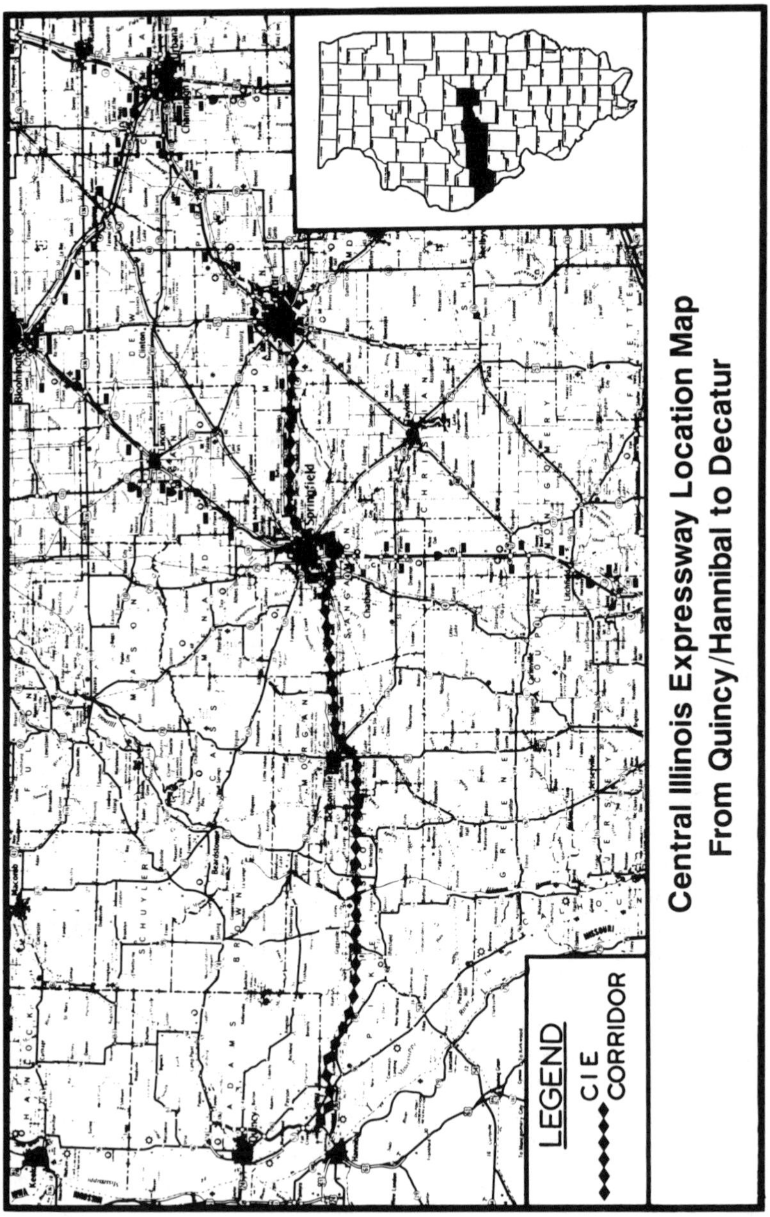

Figure 1. Central Illinois Expressway location map from Quincy/Hannibal to Decatur.

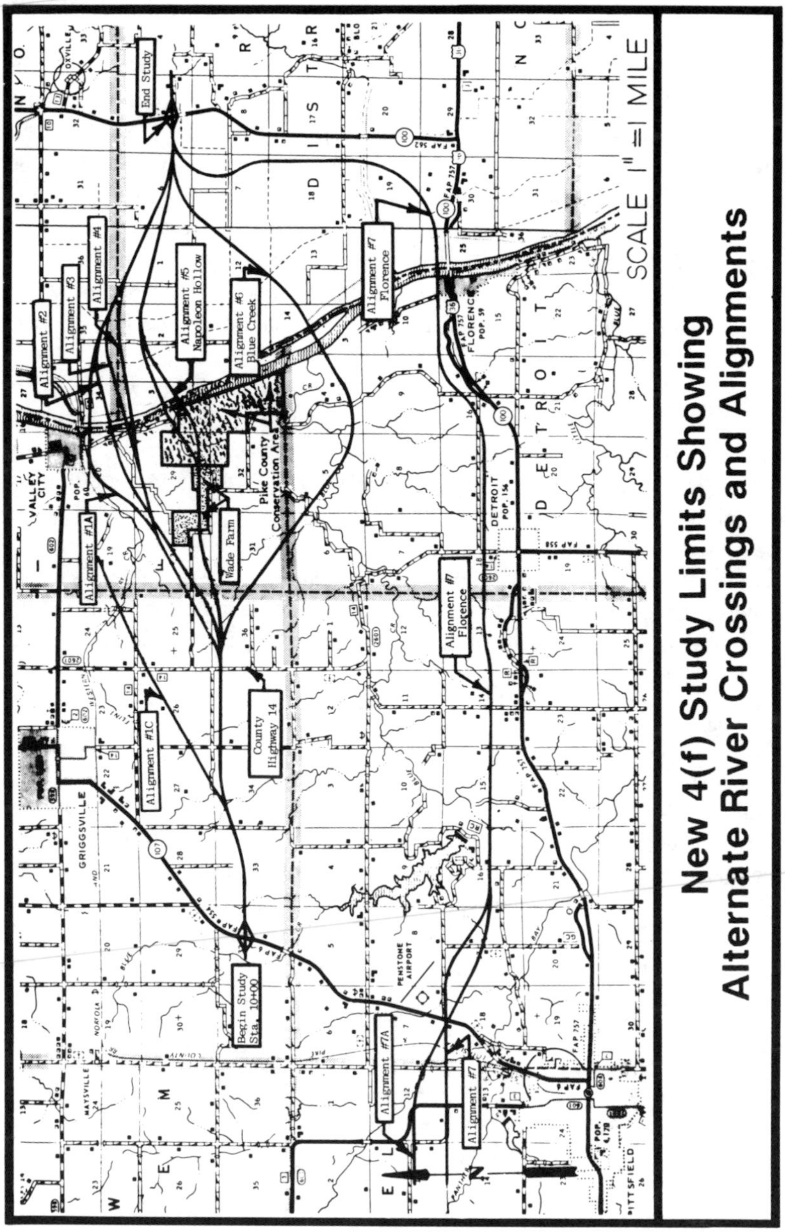

Figure 2. New 4(f) study limits showing alternate river crossings and alignments.

The proposed location of the river crossing affected two properties which are subject to the provisions of Section 4(f) of the Department of Transportation Act of 1966. The first of these, the Pike County Conservation Area (PCCA) including Big Blue Island, is 862 ac of state-owned land used for hunting and wildlife protection. Over three fourths of the PCCA consists of upland and bottomland forests. Nine major ravines occur in the PCCA. One of these, known locally as Napoleon Hollow, was selected as the site of the proposed river crossing because it is a natural, 4000-ft-long and 600-ft-wide opening in the bluffs on the west side of the river. A split alignment was proposed with the east- and west-bound lanes traversing the ravine walls on either side of the hollow. Napoleon Hollow became the focus of controversy concerning the alleged presence of the American bald eagle, a federal endangered species. The hollow also contains two archaeological sites of prehistoric Indian settlements and burial mounds which have been determined to be eligible for listing in the *National Register of Historic Places.*

Another Section 4(f) resource, the Sam Wade property, consists of 190 acres of wooded and cultivated land directly west of the PCCA. The house and barn are located on a 100-ac part of the original 160-ac farmstead. The farm is privately owned; there is no public access. Approximately 70 ac of the farm is cultivated, and the remaining acreage consists of upland woods. In 1978, the Keeper of the *National Register*, over the objections of the Illinois State Historic Preservation Officer, found the Wade property eligible for the *National Register*. The eligibility was granted because of the cut-stone farmhouse, which exemplified early 19th-century construction, and because of the association of the farm with the agrarian development of Pike County.

History

Planning for the project began as early as August 1969, when the U.S. Bureau of Public Roads [precursor of the Federal Highway Administration (FHWA)] approved the CIE corridor. Before the PCCA was purchased by the State of Illinois in 1970 for conservation purposes, the Illinois Department of Transportation (IDOT) began detailed studies of a river crossing that located the CIE through Napoleon Hollow and the PCCA and across the Wade farm. In March 1971, the U.S. Supreme Court handed down its definitive Section 4(f) decision in *Citizens to Preserve Overton Park* v. *Volpe* (401 U.S. 402).

In November 1971, a combined draft environmental impact statement (EIS) and Section 4(f) report on the CIE was approved for circulation by the FHWA. Public hearings were held in the spring of 1972, and the U.S. Secretary of Transportation approved the combined report in April 1974. In June of that year, the FHWA granted design approval to IDOT for the section of the CIE between Jacksonville and Barry. Construction commenced at Jacksonville in late 1975 and proceeded for 12 mi westward.

Shortly after their farm was determined eligible for the *National Register of Historic Places*, the Wade landowners began objecting to the location of the highway through the PCCA and their property. While discussions proceeded between IDOT and the Wades, land purchases and construction of the highway continued on other sections not affected.

In June 1980, the Wades filed suit to block any further land acquisition and construction activity for the highway. The complaint charged that the EIS and Section 4(f) report did not meet legal requirements and were thus invalid. The U.S. District Court for the Northern District of Illinois placed a preliminary injunction on IDOT barring construction of the twin bridges at the Napoleon Hollow location, followed by a permanent injunction in April 1983. In addition, the Wades and local naturalists claimed that the hollow was being used as a wintering roost site by the American bald eagle and used this claim to argue that the final EIS for the project was deficient. The latter charge drew the attention of the news media and prominent regional bald eagle and conservation groups; interviews, editorials, and feature stories appeared frequently in the local press.

During this time, a court-appointed independent researcher undertook winter studies in 1980 and 1981 aimed at proving or disproving the contention that bald eagles were using the hollow through which the highway would go. In another study, researchers under contract to IDOT captured two Indiana bats, another species on the federal threatened and endangered list, in Napoleon Hollow in June 1983. Although the capture of these males did not indicate the presence of a nursery colony in the hollow, they did heighten the controversy which had by this time completely enveloped the project. These new studies failed to confirm either the existence of bald eagle night roost trees in the hollow or the existence of Indiana bat maternity colonies in the area affected by the proposed highway location.

In its decision in *Wade* v. *Lewis* [561 F. Supp. 913 (1983)], the U.S. District Court ruled that (1) the proposed use of certain federal funding sources to build the bridges was improper and (2) the Section 4(f) statement as approved by the U.S. Secretary of Transportation did not adequately demonstrate the absence of prudent and feasible alternatives to routing the highway through the PCCA and the Wade property. Although the court found the original EIS was inadequate, it ruled that the administrative record, as supplemented by new endangered species and natural resource studies, was sustained under the National Environmental Policy Act (NEPA) as not arbitrary or capricious.

The court directed the IDOT and FHWA to undertake a new analysis of alternate river crossings with regard to the requirements of Section 4(f). The stage was thus set for new studies which would have to grapple with the stringent and often conflicting requirements of several different single-issue environmental laws and regulations. Although NEPA-based considerations appeared at this juncture to be of minor importance, the cumulative impacts doctrine so well established in NEPA was to play a decisive role in the remaining history of this project.

RELEVANT ENVIRONMENTAL LEGISLATION

NEPA

NEPA (42 U.S.C. 4321-7) established procedural requirements relative to "major Federal actions significantly affecting the quality of the human environment" and introduced the EIS as the principal means of addressing the environmental impacts of a proposed action. Since 1970, federal agencies have used the NEPA process as the framework for implementing a full range of environmental requirements. The EIS is prepared early in the project development process and is the key decision document in assessing the relative merits of reasonable alternatives that have been identified. The EIS serves to assess social, environmental, and economic impacts of a proposed action and to develop mitigation measures for adverse impacts. It also provides a means for coordination with other agencies and the public to ensure a full disclosure and consideration of a proposed action's costs and benefits.

NEPA also established the Council on Environmental Quality (CEQ) as an advisory unit within the Executive Office of the President. The CEQ, under Executive Order 11514, developed guidelines to be used by federal agencies in preparing EISs. On May 24, 1977, Executive Order 11991 amended the earlier order and directed the CEQ to issue regulations which would be binding on all federal agencies. The CEQ regulations have been revised since their initial issuance, but still prescribe the procedures for implementing NEPA.

The regulations require three types of documents, depending on the scope and magnitude of the proposed action. Major actions significantly affecting the quality of the human environment require an EIS as noted above. Actions which clearly have little or no impact may be "categorically excluded" from the process with minimal documentation. Where the impact of the proposed action is unclear, an environmental assessment is prepared to assess the impacts. If the impacts are not significant, a finding of no significant impact is prepared. If the impacts are significant, an EIS is prepared.

NEPA requires the consideration of all social, economic, and environmental impacts. It requires a balancing of all impacts to determine the environmentally preferable alternative for the proposed action. Special or single-interest environmental statutes attempt to circumvent the evaluation required by NEPA by elevating their issues of concern to special status.

Section 4(f)

Section 4(f) of the Department of Transportation Act of 1966 (49 U.S.C. 303) restricts the use of any publicly owned land — in particular, a significant park, recreation area, wildlife and waterfowl refuge, or any land from a significant historic site (i.e., one that is either on or eligible for the *National Register of Historic Places*). This act requires consultation with the secretaries of the Interior,

Housing and Urban Development, and Agriculture, as well as the official having jurisdiction over the land. The statute requires a specific finding that there are no feasible and prudent alternatives to the use of such land and that the proposed action includes all possible planning to minimize harm before the FHWA can approve the use of these lands for highway purposes. Agency discretion in making this finding is very limited because of numerous court cases, including *Citizens to Preserve Overton Park* v. *Volpe*.

Section 106

Section 106 of the National Historic Preservation Act of 1966 (16 U.S.C. 470ff) provides that prior to approval of the expenditure of any federal funds, agencies shall "take into account the effect of the undertaking on any district, site, building, structure, or object that is included in the *National Register of Historic Places*." The involved federal agency "shall afford the Advisory Council on Historic Preservation ... a reasonable opportunity to comment." On July 12, 1978, former President Jimmy Carter directed the Advisory Council to develop implementing procedures for Section 106. These complex regulations establish an intricate *comment* procedure by which federal agencies obtain information and opinions from the Advisory Council. These regulations, among other things, establish criteria of effect and adverse effect. They describe in extensive detail the responsibilities of federal agencies, the State Historic Preservation Officer, and the Advisory Council on Historic Preservation.

Agricultural Preservation Laws

The purpose of the Federal Farmland Protection Policy Act of 1981 (7 U.S.C. 4201-9) is to "minimize the extent to which Federal programs contribute to the unnecessary and irreversible conversion of farmland to nonagricultural uses, and to assure that Federal programs are administered in a manner that, to the extent practicable, will be compatible with State, unit of local government, and private programs and policies to protect farmland." The act required the U.S. Department of Agriculture to develop criteria to identify the effects of federal programs on the conversion of farmland to nonagricultural uses. Federal agencies must use these criteria and document compliance on form "AD 1006." If adverse effects are anticipated, federal agencies are required to consider alternatives to lessen the effects.

Threatened and Endangered Species Act

The Endangered Species Act of 1973, as amended (16 U.S.C. 1531-43), prohibits federal agencies from taking any action that will jeopardize the continued existence of any threatened or endangered species or will result in the destruction or adverse modification of the habitat of such species. The 1978 amendments to the Act provide for exemptions from the requirements by decision of a cabinet-level committee.

HOW CUMULATIVE IMPACTS WERE USED IN RECONCILING SINGLE-ISSUE ENVIRONMENTAL LAWS WITH SECTION 4(F)

Development of Alternatives

As part of its Section 4(f) reevaluation, IDOT planners considered 7 main alignments and 14 subalignments, including the originally proposed crossing through Napoleon Hollow and the PCCA (IDOT 1975). Some of these were dropped from further study because of engineering problems and unacceptable design features. Alternative highway locations carried into detailed analysis are shown in Figure 2. The termini of the alternate locations investigated were the interchanges of the CIE with state highways on either side of the river. The location of the east interchange was established by the fact that grading had already occurred up to the nearest intersecting state highway at the time of the injunction. On the west side, the alternative locations joined the CIE at the nearest state route (SR 107). In addition to the originally proposed location through the Wade Farm, the study evaluated five additional locations through or around the Wade Farm. These alternate locations were evaluated for the purpose of finding the one that did the least total harm to both Section 4(f) properties (i.e., the PCCA and Wade Farm) considered together.

Cumulative Impacts Concept and Section 4(f) Requirements

The environmental impacts of each of the remaining alternatives were assessed in detail. Because of the numerous legal requirements triggered by this project, the IDOT study team approached the problem of determining whether the alternative locations were prudent and feasible by focusing on the cumulative environmental impacts of these alternatives. These impacts included compatibility with existing terrain, agricultural land impacts, impacts on threatened and endangered species, wildlife habitat impacts, visual effects, and effects on historic and archaeological resources.

In essence, the approach employed looked at the prudent and feasible test of Section 4(f) only once for each alternative alignment. Rather than determining whether each individual impact rendered an alignment imprudent, the test focused on determining whether the cumulative environmental effects of each alternative alignment rendered it imprudent. This approach, when used in the context of the stringent tests of Section 4(f) and existing case law on this statute, was a relatively novel legal argument that had not been strictly tested in the courts of jurisdiction.

Cumulative Impacts Used in Analysis

Six alternate locations north of the PCCA were eliminated from consideration because of engineering deficiencies which led to possible safety problems and drainage constraints. These alternates did not meet the feasibility test of Section 4(f). Other alternate locations to the south of the PCCA were discarded because they either involved other Section 4(f) resources or were too circuitous and involved longitudinal floodplain crossings.

Four other locations near Alternate 5 which passed near Napoleon Hollow but not through Section 4(f) resources were considered to be imprudent because of their extraordinary cumulative environmental impacts. These impacts included excessive additional excavation in locations that did not use the natural opening in the bluffs afforded by Napoleon Hollow. This particular impact is interesting from the standpoint that massive excavation is not considered a usual occurrence in the development of roads in Illinois. In areas with mountainous or hilly terrain, the amount of excavation that was considered extraordinary in the locale of west-central Illinois may be encountered more frequently and thus may not be considered to have the extraordinary significance that was attached to it in this project. Furthermore, local soils encountered in this project were highly erodible, a factor which lent additional importance to the problems of excavating into the bluffs.

This same excessive excavation also was considered to have adverse visual impacts and impacts on wildlife. The undesirable visual effects would have occurred because the large opening in the bluffs would have to be stabilized with concrete retaining walls. Although similar walls were also required with the Napoleon Hollow crossing, their impact was mitigated by the split alignment which effectively fragmented the massive bulk of the highway into two separate and less visually intrusive sections. The *concrete canyons* which would have been created on a location other than the one proposed were considered to be visually objectionable and incompatible with the existing terrain. They also would have effectively cut off natural on-land movement patterns for wildlife in the area, resulting in increased wildlife mortality from collisions with vehicles. The location through Napoleon Hollow avoided this impact because its split-alignment design was able to bridge rather than fill larger natural ravines, thereby leaving them open for wildlife movement below the highway.

Also playing a prominent role in the accretion of cumulative impacts for these alternates was their impact on agricultural land and operations. Of particular concern was the effect of alignments that diagonally traversed agricultural lands. This diagonal severance of farm plots causes adverse impacts on the ability of farmers to maintain productivity on land near the highway. This diminished productivity is caused by problems regarding seeding and fertilizer and herbicide application which occur when cultivating equipment cannot be turned at the end of evenly aligned rows. Furthermore, farmers must often use circuitous routes on frontage roads and overpasses to get to their fields on the other side of the highway. Whereas these impacts may be less prominent in other states, they are a particularly strong concern to members of the farming community in prominent agricultural states such as Illinois. This importance is emphasized by the existence of federal and state statutes concerning preservation of prime agricultural lands and minimization of adverse impacts on agricultural operations.

Finally, some of the remaining alternate locations would have had adverse effects on important archaeological sites, including one Indian burial mound with religious significance. Specifically, the diaries of Joseph Smith describe this mound as the burial place of an ancient warrior king prominent in the mythological prehistory of the Mormon religion.

The consideration of each of these impacts was mandated in some fashion by single-issue environmental laws and regulations. The cumulative significance of them was based in substantial part on their particular relevancy to and prominence in the local project setting.

RESULTS

Plaintiff's Objection

This cumulative impacts methodology did not go unchallenged by the plaintiffs. In their brief in *Wade* v. *Dole* [631 F. 2d 1100 (1986)], the sequel to *Wade* v. *Lewis*, they directly disputed this approach:

> In considering these factors, there is no finding that any one factor supports a finding of imprudence. Even cumulatively, these impacts cannot be utilized to satisfy this requirement of a finding of extraordinary magnitude to support a decision of imprudency with respect to [an] alignment.

Court's Endorsement

In its decision in *Wade* v. *Dole*, the U.S. District Court for the Northern District of Illinois rejected the plaintiffs' argument and explicitly endorsed the use of the cumulative impacts concept.

> Though the plaintiff does not directly challenge the practice of cumulating impacts, it appears to consider this approach improper. Throughout its briefs, the plaintiff objects to the elimination of alternatives on the basis that no single factor, considered separately, would establish that the alternative is not feasible or prudent ...
>
> The cumulative impact approach is consistent with the plain meaning of *prudent*, the policies expressed in *Overton Park*, and the regulations implementing the National Environmental Policy Act ... Further, it recognizes the reality that in nature such factors as safety, wildlife mortality, flooding, and farm severance do not occur in isolation. We conclude, therefore, that adverse impacts which considered separately would not be *truly unusual factors (Overton Park* at *413*) may support a finding of *not prudent* when considered collectively.

Ruling on the plaintiffs' appeal of this decision, the U.S. Court of Appeals for the Seventh Circuit similarly endorsed the cumulative effects concept as applied to the *prudent* standard of Section 4(f).

> A prudent judgment by an agency is one that takes into account everything important that matters. A cumulation of small problems may add up to a sufficient reason to

use §4(f) lands. It would be imprudent to build around the park if the Secretary were convinced that the aggregate injuries caused by doing so exceeded those caused by reducing the size of the park. Even a featherweight drawback plays some role. No feather weighs very much, but a ton of feathers still weighs as much as a 2,000 pound block of lead. [*Eagle Foundation* v. *Dole*; 813 F. 2d 798 (1987)]

Similar Litigation and Results

Other cases involving similar cumulation of impacts and Section 4(f) are very limited in number. Furthermore, the approach of using cumulative impacts in these cases was not developed as explicitly and challenged as directly as in *Wade* v. *Dole*.

Two such cases which bear the most resemblance to *Wade* are *Citizens to Preserve Wilderness Park, Inc., et al.* v. *Adams* [543 F. Supp. 21 (1981)] and *Stop H-3 Association* v. *Dole* (740 F. 2d 1442). In the first of these, the U.S. District Court of Nebraska permitted the use of park land subject to Section 4(f) because an alternative to its use had problems which included a less beneficial effect on reducing congestion, more air and noise pollution impacts on nearby residents, bisection of a school district, removal of trees, proximity to three cemeteries, and deficient geometric configurations which could be unsafe for motorists. The court considered all these factors in combination as being *truly unusual* and sufficient for finding that there was no prudent alternative to the use of park land for highway purposes.

In the second case cited, *H-3*, the U.S. Court of Appeals for the Ninth Circuit issued a ruling with a result opposite to that in *Citizens* when it found that an alternative to using park land could not be considered imprudent because it had cumulative impacts which included the dislocation of a church, 4 businesses, and 31 residences; increased air, noise, and visual pollution in the vicinity; $42 million in additional costs to construct; and less desirable geometric characteristics. In this case, however, it was not the process of considering impacts cumulatively that was challenged, but rather the decision that these particular impacts were significant enough to collectively justify the use of park land subject to the requirements of Section 4(f).

Wade v. *Dole* remains a definitive case which shows how cumulation of impacts can be sufficient to overcome the stringent restrictions of Section 4(f). Accordingly, *Wade* is a noteworthy example of how impacts mandated for consideration by several single-issue environmental laws were used in a cumulative sense to meet the requirements of another single-issue law.

CONCLUSIONS

The single-issue environmental laws passed during the last few years, coupled with those already on the statute books, have dramatically complicated the job of capital development agencies striving to build and reconstruct the nation's infrastructure. As these laws are passed, each environmental issue or natural resource is singled out for special consideration, and agencies are directed to convert these resources or lands to other uses only as a last resort. However, if this trend continues, the nation may eventually exhaust its supply of *other places* to build highways, bridges, sewage treatment plants, landfills, and hazardous waste disposal sites. For instance, should prime agricultural land be converted to other uses at the expense of wetlands or vice versa? Should urban neighborhoods be disrupted in order to save wildlife habitat? The list of special protected uses continues to grow, and all are considered important enough to avoid unless absolutely necessary.

Accordingly, when faced with the task of justifying the use of a protected resource, project planners are likely to cumulate the impacts on all other protected resources that would occur with avoidance alternatives. Because they are legislatively mandated to give special consideration to a wide range of protected resources, the ability to argue persuasively to use any one of them is enhanced by the sum of potential impacts on the others. Arguably, if all resources are special, then none is special.

This balancing of impacts can be justified under the aegis of the cumulative impacts concept so clearly stated in implementing regulations for NEPA promulgated by the CEQ and so effectively used in litigation such as that described above. However, because of differences in geography and social value systems, universal guidelines on when the scales are tipped toward use or nonuse of a resource are unattainable. This situation is likely to result in even more litigation that will challenge decisions resulting in the significant use of any protected resource. In this litigation, development agencies can be expected in many cases to rely on cumulation of impacts to argue that these decisions were not arbitrary or capricious based on circumstances unique to each project. When walled in by single-issue environmental laws and regulations, decision makers are likely to find that the cumulative impacts approach will give them the means to walk through these walls and effectively discharge their responsibilities.

REFERENCE

Illinois Department of Transportation. 1975. Federal Aid Primary Route 408; Illinois Route 100 in Scott County to Illinois Route 107 in Pike County; Section 4(f) Evaluation FHWA-IL-EIS-72-07-F/4(f), Springfield, IL.

CHAPTER 6

Regional and Global Analysis

Introduction

D. S. Shriner, Oak Ridge National Laboratory, Oak Ridge, TN

This chapter was organized in recognition of the fact that the global community will increasingly address actions that have regional and, potentially, global ramifications. Federal actions that involve or relate to climate change, water resources, air pollution, species diversity, reforestation/deforestation, and stratospheric ozone depletion are examples of continuing societal challenges which can manifest themselves at regional to global spatial scales.

As the spatial scale of potential impacts of an action increases, the probability of cumulative impacts of multiple projects/actions within a geographic region increases. One issue which arises as a result of the potential for cumulative impacts is whether appropriate databases will exist based on previous or future National Environmental Policy Act (NEPA) site-specific or programmatic assessments to adequately address the data needs for cumulative impact assessment at regional and global scales. Perhaps as equally limiting as data availability to the process of impact assessment at large spatial scales is the availability of methodological approaches capable of and appropriate to extrapolation in space and time. Two papers in this chapter specifically address these methodological barriers to large-scale impact assessment and the specific actions on a national level which must be undertaken in order to adequately address questions of cumulative impact and large spatial scales.

Once information and methodological constraints to large-scale assessment have been addressed, issues such as transboundary transport of pollutants and U.S. development in foreign jurisdictions may more easily be addressed than they currently can be. Technologically, the time is relatively near when global-scale environmental data bases could exist in standardized format and of uniform quality. Institutionally, technological and methodological developments of the nature outlined in this chapter will be necessary, even essential, to implementation of anything approaching a uniform global environmental policy equivalent to NEPA.

This chapter contains a series of contributed papers that provide insight into issues (such as global climate change and maintenance of biodiversity) and regions (the Arctic) where new methods and data resources are needed to strengthen the scientific base upon which impact assessments are made. While this series of papers does not answer all of the questions raised, it does provide a thoughtful discussion of methods development approaches and needs. Finally, a U.S. national program aimed at developing a uniform environmental monitoring database for the continental United States is outlined. If successful, this program may prove to be a valuable prototype for development of a uniform global environmental monitoring database and assessment approach.

In their paper, Cushman et al. discuss the phenomenon of global climate change; how actions being considered, ranging from the licensing of a single facility to the implementation of a broad federal policy, could relate to global climate change; and how the NEPA process can accommodate global climate change implications now and in the future. The paper also summarizes guidance from the Council on Environmental Quality concerning the incorporation of the issue of climate change in the NEPA process.

Biodiversity, as an issue emerging in both scientific and public concern, is difficult to evaluate because of the complex nature of effects which occur slowly and which tend to be cumulative in nature. In their paper, Henderson et al. suggest the need for an expanded interpretation of NEPA which could recognize particular attributes or indicators of biodiversity to be addressed in the environmental assessment and review process. The authors discuss reasons for valuing biodiversity and examine some of the current legislation that addresses biodiversity in terms of potential protection of biodiversity. Finally, they present a hierarchical framework of compositional and structural biodiversity from which measurable indicators can be selected which could be used in the environmental assessment and review process.

In her paper, Johnson discusses the Arctic as a specific region where unusual aspects of the Arctic environment, serious deficiencies in data, and a lack of Arctic expertise can combine to make implementation of the NEPA process especially difficult. The 1989 findings of the Arctic Research Commission identified a need to improve the scientific and coordination aspects of the environmental impact statement (EIS) process. The most critical deficiency identified was the absence of impartial, external quality control mechanisms for the data and information used in the stages of scoping, synthesis, EIS preparation, and follow-up monitoring programs.

In her paper, Hunsaker proposes that regional ecological risk assessment provides a useful approach to assist scientists in accomplishing the task of assessing cumulative impacts. Critical issues such as spatial heterogeneity, boundary definition, and data aggregation are discussed, and an example illustrates the importance of integrated databases, modeling, and boundary definition at the regional scale.

The Environmental Monitoring and Assessment Program (EMAP) is a comprehensive interagency, interdisciplinary effort initiated in 1990 to provide a strategic approach to the long-term assessment of extent, magnitude, and location of change in environmental condition at regional and national spatial scales. In their paper, Saul et al. outline EMAP plans to monitor indicators of ecological condition in aquatic and terrestrial ecosystems as well as air and deposition. Assessments will integrate and analyze data collected within the program along with other available information to identify possible causes of observed changes in environmental condition. As such, EMAP will provide a foundation for national environmental protection which has potential for international expansion to support global monitoring and regulation.

Global Climate Change and NEPA Analyses

R. M. Cushman, D. B. Hunsaker, Jr., M. S. Salk, and R. M. Reed, Oak Ridge National Laboratory, Oak Ridge, TN

ABSTRACT

Human activities such as energy production and use, industrial activity, and land use change are expected to cause a global climate change that would have local and regional manifestations during the next century. Although the resulting impacts are not yet known with certainty, potential effects on agriculture, water, forests, ecosystems, fisheries, coastal areas (from rising sea level), and other environmental resources have been predicted.

The National Environmental Policy Act (NEPA) provides for consideration of such topics as global climate change. However, the implementation of the environmental impact statement (EIS) requirements of NEPA, which has been the major activity under the law since it was passed, has seldom been used to address the issue. Climate change and its consequent effects have two important implications for NEPA determinations: (1) the potential for an action, either individually or in concert with other actions, to alter climate must be assessed and (2) the potential for future climate change to alter the "baseline" environment (and thus to affect the action or to alter the impact of the action) must be assessed, even if the action under consideration will not in itself contribute to climate change.

In this paper we evaluate the climate change issue (including the uncertainty of the temporal and spatial distribution of impacts) in a NEPA context. We discuss the kinds of actions to which NEPA is applicable, the types of analyses that might be appropriate, and the problems they might involve. In particular, the opportunities and limitations for addressing climate change under current Council on Environmental Quality (CEQ) regulations through the environmental assessment (EA)/EIS process are addressed. We also discuss changes that have been proposed for NEPA and its implementing regulations and how they could affect the analysis of global climate change. EISs at the program or policy level, rather than at the site-specific project level, appear to be the most appropriate for addressing the contributions of actions to global climate change; and EISs for actions that will continue for several decades or longer are most appropriate for addressing the potential effects of climate change on actions.

INTRODUCTION

In the past few years, the issue of global climate change has attracted the attention of the public. Although the issue has been analyzed in the technical literature for decades (e.g., Callendar 1938; Revelle and Suess 1957), discussion now appears frequently in newspapers and popular magazines and on television. Responses to and prevention of global climate change are also now being discussed in the political arena both within the United States and internationally. It has been argued that the planning process of the U.S. federal government should reflect a sensitivity to this issue — specifically that procedures to implement the National Environmental Policy Act (NEPA) (42 USC 4321) should take into account global climate change where appropriate (e.g., Prickett and Wirth 1989).

Two principal aspects of global climate change warrant consideration in a NEPA context:

1. An action under consideration, ranging from the licensing of a single facility to the implementation of a broad federal policy, may affect global climate, either individually or in conjunction with other actions.
2. Although the action itself may not directly affect global climate, it and its impacts may in some way be affected by the regional or local manifestations of global climate change.

In this paper, we discuss the phenomenon of global climate change, how actions being considered could relate to global climate change, and how the NEPA process can accommodate consideration of actions having global climate change implications now and in the future. We also summarize current bills in Congress pertaining to NEPA and climate change and guidance from the Council on Environmental Quality (CEQ) concerning the incorporation of the issue of climate change into the NEPA process.

GLOBAL CLIMATE CHANGE

The Greenhouse Effect

It is now widely understood that human activities, specifically fossil-fuel emissions, industrial activities, and land use changes, have the potential for changing global climate. Most climatologists predict that these changes could occur over the next few decades or century; there is even some assertion that the first manifestations of global climate change have been observed already, given the recent run of warm years (Jones et al. 1988). The primary mechanism for man-induced global climate change is the increasing concentration of greenhouse gases in the atmosphere — that is, gases that trap infrared radiation at the earth's surface (instead of allowing it to escape to space), thereby raising the temperature of the earth and the lower atmosphere (i.e., the troposphere). Gases that are currently accumulating in the atmosphere, thereby contributing to a global climate change, include carbon dioxide (CO_2), methane, nitrous oxide, chlorofluorocarbons (CFCs), and (although with less certainty as to its increasing trend) tropospheric ozone.

Greenhouse Gases — Trends and Sources

At this time, CO_2 accounts for the majority of the greenhouse effect, but the combined effect of other trace gases (especially methane, nitrous oxide, CFCs, and tropospheric ozone) could soon equal or exceed the potential warming from CO_2 alone (Ramanathan et al. 1985). As of 1980, concentrations of CO_2 had increased from a preindustrial (1880) level of 270 to 339 ppm, methane from 1.15 to 1.65 ppm, nitrous oxide from 285 to 300 ppb, and CFCs from 0 to 0.18 ppb (CFC-11) and 0.28 ppb (CFC-12) (Ramanathan et al. 1985). Current increases in atmospheric concentrations have been estimated as follows: CO_2, 0.4% per year; methane, 1.1 ± 0.1% per year; and nitrous oxide, 0.2–0.3% per year (Wuebbles and Edmonds 1988). It is less certain whether or how much the atmospheric concentration of ozone has risen or will rise (Ramanathan et al. 1985; Wuebbles and Edmonds 1988). Although CFC concentrations increased by about 6% per year from 1977 to 1986, the Montreal Protocol (United Nations Environment Programme 1987) is expected to limit emissions by the 1990s (Wigley 1988).

To put into perspective the contributions to the problem from individual projects, programs, or policies, it is useful to quantify the magnitude of global emissions of greenhouse gases. The 1988 total global emission of carbon from fossil-fuel combustion was 5.7 billion metric tons, of which half is accounted for by the United States, the former U.S.S.R., and China (Table 1). The typical annual carbon emission from a 1000-MW coal-fired power plant, operating 70% of the time at 38% thermal efficiency, is about 1.4 million metric tons — only 0.02% of the global total from fossil-fuel combustion.

The net global carbon emission from deforestation has been estimated to be about 1.8 billion metric tons per year (Houghton et al. 1987). Of this total, the annual emission is about 1.7 billion metric tons from tropical land use change (Table 1) and about 0.1 billion metric tons from that in temperate and boreal regions (Houghton et al. 1987). In terms of the land area deforested, estimates for 1980 derived from various sources ranged from 5.4 to 8.5 million ha for the tropical Americas, 3.1 to 10.9 million ha for tropical Africa, and 0.9 to 9.4 million ha for tropical Asia (Houghton et al. 1987), with the world total ranging from 9.4 to 28.0 million ha. Only about half of the annual total carbon emitted to the

Table 1. Sources of Carbon Emissions, Globally, and by Country.

Fossil-Fuel Combustion[a] (10^6 metric tons C/year)			Tropical Land Use Change[b] (10^6 metric tons C/year)		
1.	United States.	1301	1.	Brazil	336
2.	U.S.S.R.	1067	2.	Indonesia	192
3.	China	582	3.	Colombia	123
4.	Japan	259	4.	Ivory Coast	101
5.	West Germany	178	5.	Thailand	95
6.	India	158	6.	Laos	85
7.	United Kingdom	151	7.	Nigeria	59
8.	Poland	123	8.	Philippines	57
9.	Canada	118	9.	Burma	51
10.	Italy	93	10.	Peru	45
	World total	5743		Tropical Americas	665
				Tropical Africa	373
				Tropical Asia	621
				All tropics	1659

[a] Estimate for 1988; data from Marland (1990).
[b] Estimate for 1980; data from Houghton et al. (1987), Table 4.

atmosphere each year from fossil-fuel combustion and deforestation appears in the annual increase in the atmospheric concentration of CO_2. Further research into carbon uptake by the oceans and terrestrial biosphere is needed to determine the sinks for the other half (i.e., to balance the global carbon budget).

The budgets for the other trace gases are not as well known as that for CO_2. Recent estimates of annual fluxes of methane to the atmosphere range from 200 to 600 million metric tons, primarily from fossil fuels (30–123 million metric tons), wetlands and tundra (21–146 million metric tons), ruminants (100 to 200 million metric tons), biomass burning (25–110 million metric tons), rice production (25–137 million metric tons), and termites (5–50 million metric tons) (Ehhalt 1985; Wahlen et al. 1989).

CFCs, on the other hand, are entirely manmade. Production of CFC-11 and CFC-12 in 1986 was about 0.38 million and 0.49 million metric tons, respectively; by the next century, production cutbacks to 20 to 65% of the 1986 level are possible, consistent with the Montreal Protocol (Wigley 1988). Even lower production and emission levels are possible if more countries adhere to the Montreal Protocol and if more stringent production limits are established. CFCs are predominantly used as refrigerants, foams, and propellants (Wuebbles and Edmonds 1988).

Direct emissions of ozone to the atmosphere from human activities are not a significant source of that gas; ozone, however, can be produced chemically in the troposphere by reactions involving nitrogen oxides, CO_2, methane, and other hydrocarbons (Wuebbles and Edmonds 1988). In addition to natural sources (e.g., oceans, estuaries, and natural soils) totalling 8.5 ± 3.5 million metric tons of nitrogen per year, nitrous oxide has several manmade sources: fossil-fuel combustion [estimated to be 4.0 ± 1.0 million metric tons of nitrogen per year, although the most recent research (e.g., Sloan and Laird 1990) suggests that sampling artifacts may have caused a considerable overestimation of nitrous-oxide emissions from this source], biomass burning (0.7 ± 0.2 million metric tons of nitrogen per year), fertilized soils (0.8 ± 0.2 million metric tons of nitrogen per year), and cultivated natural soils (1.5 ± 0.5 million metric tons of nitrogen per year) (Wuebbles and Edmonds 1988).

Expected Climate Change

Simulations by general circulation models (GCMs) of the climatic response to a doubling of atmospheric CO_2 or the equivalent combination of all the greenhouse gases — a reference point commonly studied by climate modelers and one that could be realized during the next century — suggest that the global average temperature would rise by a few degrees Celsius and the global hydrologic cycle would accelerate, resulting in more total evaporation and precipitation (Mitchell 1989). Although more difficult to predict, changes are also expected in the frequency and intensity of storms (Love 1988). Various models disagree on the regional distribution of temperature and precipitation changes (Grotch 1988), although several models indicate that many temperate zones may become drier as well as warmer. There is even greater uncertainty about the transient response of climate as CO_2 doubling is approached. The equivalent doubling of atmospheric CO_2 will likely occur in the next century, but because of the thermal lag of the oceans, the associated warming could well be delayed by several decades. Whether or not a greenhouse warming has already begun, future climate change appears likely given the changes that have already occurred in the sources and sinks of greenhouse gases.

In addition to changes in temperature and precipitation, sea level is expected to rise because glaciers and polar ice sheets will melt and ocean water will expand as it warms. Estimates of global sea level rise by the year 2100 have ranged from <1 to >3 m. A recent report by the National Research Council (1987) considered a range from 0.5 to 1.5 m to be most likely.

Impacts of Climate Change

Many studies have projected the potential impacts of climate change on environmental resources. For example, Smit et al. (1989) concluded that grain and oilseed agriculture in northern Ontario, Canada could benefit from the predicted climate change, while crops in the southern parts of that province could be negatively affected. Flaschka et al. (1987) predicted that a warmer and drier climate could markedly reduce runoff in four streams in the Great Basin region of the United States. Dobson et al. (1989), summarizing the presentations at a 1988 conference, reported a broad concern that climate change could threaten global biodiversity. Pastor and Post (1988) described potential changes in biomass, species composition, and other characteristics of northern forests. Stevenson et al. (1986) projected a major loss of coastal wetlands from an accelerating rise in sea level. Frye (1983) noted that climate change could affect marine fisheries (e.g., declining stocks leading to increasing costs per unit of fish landed) and concluded that new management approaches would have to be developed. Linder and Gibbs (1986) determined that climate change could affect demand for electricity and generating capability. These examples show that a number of environmental resources could be sensitive to climate change.

Current impact assessment of climate change has some significant methodological limitations. Major uncertainties remain concerning future trace gas emissions and atmospheric concentrations. The current GCMs have severe limitations: clouds and surface hydrology (runoff and soil moisture) are poorly represented, the circulation of the oceans has not been linked with the atmospheric models, and GCMs do not yet provide credible climate simulations on a local or regional scale (Hillel and Rosenzweig 1989). Uncertainties remain concerning the spatial and temporal patterns of future climate and sea level; the response of individual resource sectors to changing climate; the interactive effects of climate and atmospheric CO_2 on vegetation; the interactions between resource sectors within a region and the interactions between regions; and the future course of societal institutions, demographics, and technology. Therefore, the findings of climate change impact studies should be interpreted as statements of the sensitivity of various resources to climate change rather than as predictions.

CONSIDERING CLIMATE CHANGE IN THE NEPA PROCESS

NEPA and the Climate Change Issue

The most frequently cited part of NEPA is Section 102(2)(c), which requires federal agencies to prepare EISs on "every recommendation or report on proposals for legislation and other major Federal actions significantly affecting the quality of the human environment." To be included in these statements is a consideration of "[t]he relationship between local short-term uses of man's environment and the maintenance and enhancement of long-term productivity" [Section 102(2)(c)(iv)]. NEPA includes several statements that would pertain to climate change as well as to other global environmental problems. For example, among its purposes (Section 2) is to prevent or eliminate damage to the environment and biosphere. Furthermore, the *Declaration of National Environmental Policy* [Section 101(a) of NEPA] states that

> The Congress, recognizing the profound impact of man's activity on the interrelations of all components of the natural environment, ... and recognizing further the critical importance of restoring and maintaining environmental quality to the overall welfare and development of man, declares that it is the continuing policy of the Federal Government ... to use all practicable means and measures ... to create and maintain conditions under which man and nature can exist in productive harmony ...

To carry out the policy stated above, Section 101(b) says

> ... it is the continuing responsibility of the Federal Government to use all practicable means...to improve and coordinate Federal plans, functions, programs, and resources to the end that the Nation may —
>
> ... fulfill the responsibilities of each generation as trustee of the environment for succeeding generations;
>
> ... [and]
>
> ... attain the widest range of beneficial uses of the environment without degradation, risk to health or safety, or other undesirable and unintended consequences ...

Congress further authorized and directed [Section 102(2)(b)] that all federal agencies were to identify and develop methods and procedures to "insure that presently unquantified environmental amenities and values may be given appropriate consideration in decisionmaking along with economic and technical considerations." Finally, Section 102(2)(f) directs federal agencies to

> Recognize the worldwide and long-range character of environmental problems and ... lend appropriate support to initiatives, resolutions, and programs designed to maximize international cooperation in anticipating and preventing a decline in the quality of mankind's world environment ...

Thus, the breadth of NEPA is such that it covers timely concerns, such as climate change, that were not a matter of concern when it was enacted (Bear 1989; Prickett and Wirth 1989).

Limitations of NEPA with Respect to the Climate Change Issue

Although NEPA has been a major factor in requiring federal agencies to incorporate environmental considerations into their decisionmaking processes, it has several shortcomings that limit its usefulness. First, environmental impacts are just one factor to be considered by federal agencies in their decisionmaking processes. Section 102(2)(b) of NEPA says environmental values should be balanced with the economic and technical aspects of a project, but it does not require that the environmentally preferred alternative be selected. Thus, if agencies have adequately considered environmental impacts, they can determine that economic or technical factors overshadow environmental concerns and select a proposed alternative that is not the most protective of the environment.

A second shortcoming is that NEPA does not apply to past federal actions, just future ones. Specifically, in terms of sinks and sources of greenhouse gases, changes that have already occurred are often considered to be major contributors to the problem. Thus, NEPA would have no effect on those causes of global climate change. A third shortcoming of NEPA is its lack of mandated follow-up activities. A federal agency can analyze potential impacts and propose mitigation

Table 2. Bills Introduced in the 101st Congress (as of July 1989) Considering Both Global Climate Change and NEPA.

Bill	Prime Sponsor	Short Title (or Purpose)
HR 980	Jones	Global Environmental Research and Policy Act of 1989
HR 1003	Davis	To consider impacts of federal actions on the oceans and Great Lakes
HR 1113	Studds	To consider impacts of federal actions on the global environment
HR 2777	Pelosi	International Banking Environmental Protection Act of 1989
HR 3332	Jones	Global Environmental Research and Policy Act of 1989
HR 3847	Conyers	Department of Environmental Protection Act; Federal Facilities Compliance Act of 1990
HR 3977	Conte	Antarctica Protection and Conservation Act of 1990
HR 4210	Jones	Antarctica Environmental Protection, Clean-Up, and Liability Act of 1990
HR 4514	Vento	Antarctica World Park and Protection Act of 1990
S 201	Gore	World Environmental Policy Act of 1989
S 333	Leahy	Global Environmental Protection Act of 1989
S 676	Baucus	Global Environmental Protection Act of 1989
S 1045	Symms	National Environmental Policy on International Financing Act of 1989
S 1089	Lautenberg	To clarify national policy
S 2006	Glenn	Department of the Environment Act of 1990
S 2368	Moynihan	National Biological Diversity Conservation and Environmental Research Act of 1990
S 2571	Gore	Antarctic Environmental Protection Act of 1990
S 2575	Kerry	Antarctica Protection Act of 1990

measures to make the impacts acceptable. However, there is no requirement that monitoring take place to determine whether the impacts actually occur and whether the mitigation measures taken (if any) minimized or prevented adverse impacts. Finally, NEPA is written very broadly so that it can be construed to require consideration of many topics not recognized as important in the late 1960s, global climate change being just one example. However, such broadening of the law could be subject to court challenge.

Legislation Proposed in the 101st Congress Concerning Climate Change and NEPA

As of July 30, 1990, almost 120 bills had been introduced in the 101st Congress dealing in some manner with global climate change. Table 2 lists the bills that would amend NEPA or specifically require consideration of global climate change in NEPA documents. The discussion that follows will focus on how climate change and NEPA would be linked in those bills.

Some bills (HR 1113, HR 3847, and S 1089) would amend NEPA to generally clarify national environmental policy and to reauthorize and reinvigorate the CEQ. Several bills would amend NEPA to require federal agencies to specifically consider the impact of their actions, including extraterritorial actions (S 1089, HR 3332, and HR 3847), on the global environment, either in general (HR 980) or by listing specific issues to be considered (HR 1003, HR 1113, HR 3332, HR 3847,

HR 3977, HR 4210, HR 4514, S 2571, S 2575, and S 2762), including the oceans, the Great Lakes, the atmosphere, geographic areas outside the United States, Antarctica, and other aspects of the global environment. Some bills (HR 980, HR 1113, HR 3332, and HR 3847) would require the CEQ to incorporate these requirements into its regulations within 1 year. Thus, if these bills become law, CEQ's NEPA regulations would be broadened to include guidance for assessing the impacts (including the cumulative impacts) of proposed major federal actions on global climate change, sea level rise, depletion of the ozone layer, transboundary pollution, and other phenomena of international environmental concern within, as well as beyond, the jurisdiction of the United States and its territories and possessions.

Bills S 333 and S 676 (both entitled the Global Environmental Protection Act of 1989) would add a section to NEPA called Title III — Global Protection, Subtitle A — Atmospheric Protection. This new part, which would be cited as the Atmospheric Protection Act of 1989, would require federal agencies to include consideration of the environmental impact of "pollutants, substances, products or practices which may contribute to global climate change ... or trace gas modification of the atmosphere" in every recommendation, report, or proposal for legislation and other major actions significantly affecting the quality of the human environment.

Bill HR 980 would establish a Council on Global Environmental Policy (CGEP) to advise the President on global change policies and to recommend the research goals, priorities, and approaches that would most effectively provide a basis for policy decisions on global environmental change. The CEQ chairman would also be chairman of CGEP. Bill S 201 would replace the existing CEQ, which was created by Section 202 of NEPA, with a Council on World Environmental Policy to be chaired by the Administrator of the Environmental Protection Agency (EPA). The new council would continue to perform all the duties, authorities, and functions of the current CEQ in addition to assuming new responsibilities detailed in Section 103 of S 201.

Bill HR 3332 (the Global Environment Research and Policy Act of 1989) would establish a committee to develop a National Global Change Research Plan. This plan would set up a 10–year program to coordinate oceanographic, atmospheric, terrestrial, and polar research programs on processes and factors that contribute to global change. The act would also direct the CEQ to advise the President on policies relating to global change. The purposes of the act would be to (1) to provide for a national research plan which when implemented will contribute to our ability to understand, assess, predict, and respond to human-induced and natural process of global change and (2) establish a mechanism to coordinate development of national policies to abate, mitigate, and adapt to the impact of global change.

Several bills would establish a national environmental policy on the U.S. participation in multilateral development banks. Bill HR 2777 (the International Banking Environmental Protection Act of 1989) would not specifically amend NEPA, but would require that NEPA-type environmental assessments be prepared before the U.S. Executive Director of a multilateral development bank could vote in favor of any proposed bank action that would have a significant effect on the human environment. Bill S 2006 (the Department of the Environment Act of 1990) has similar provisions. Like HR 2777, it would not amend NEPA, but it would require that the environmental efforts of borrowing countries be considered when multinational development banks issue loans or other financial or technical assistance. Both Title III (International Financing) of S 676 (the Global Environmental Protection Act of 1989) and S 1045 (the National Environmental Policy on International Financing Act of 1989) would amend NEPA to require environmental impact assessments of potential international financial activities that might significantly affect the quality of the human environment before U.S. representatives could vote on such requests. Bill S 1045 does not explicitly mention global climate change. Title III of S 676 discusses global climate change but not in Section 302, which would amend NEPA.

Hearings have been held by various Senate and House committees and subcommittees of the 101st Congress on both general global environmental topics (e.g., global warming, ozone depletion, the role of oceans in global warming, and possible climatic surprises caused by global warming) and specific bills. Because numerous bills are being considered and amended during the congressional deliberations, it is uncertain what, if anything, will be passed by the 101st Congress on these topics. However, no matter what happens in the 101st Congress, the topic will undoubtedly continue to be of interest and, thus, will likely be considered by future sessions of Congress.

Guidance from the President's Council on Environmental Quality

Since 1987, the CEQ has been concerned about global climate change and its relationship to NEPA. In July 1989, the CEQ issued draft guidance on considering climate change in NEPA documents for review and comment (CEQ 1989) Comments have been received and will be considered in revising the guidance in the future. The following summarizes the draft guidance from the CEQ.

A key aspect of the draft guidance is that global warming is a *reasonably foreseeable* impact of emissions of greenhouse gases that therefore must be considered in future NEPA documents. The draft guidance directs federal agencies to determine whether or not, and the extent to which, their actions affect emissions of greenhouse gases and to consider whether the actions they take may be affected by climate-induced environmental change. To determine the potential of their actions to affect climate, the guidance directs federal agencies to immediately review the extent to which their activities contribute to the emission of greenhouse gases and, thus, to global climate change. The CEQ recommends that agencies focus their examination on programs because, in most cases, analysis of the impacts of individual projects would not provide meaningful information. In particular, the draft guidance recommends considering long-range energy, transportation, and forest management actions. The CEQ also directs federal agencies to review ongoing programs to determine the extent to which such programs may contribute to global warming. As noted in the guidance, existing NEPA documents may need to be supplemented or new programmatic documents prepared to substantiate the results of this review.

To determine the effects of climate change on federal actions, the draft guidance notes that requisite information is likely to be unavailable and directs agencies to deal with this lack of information in accordance with 40 CFR 1502.22, which deals with *incomplete or unavailable information.* This section of the regulations indicates that an EIS should state that the information is unavailable, give its relevance to evaluating the impacts, summarize the appropriate existing scientific knowledge in the area, and evaluate impacts using generally accepted theoretical approaches or research methods. The CEQ also directs agencies to pay close attention to ongoing research in the area of climate change impacts. The draft guidance directs agencies to identify the actions that are most sensitive to climate change impacts (e.g., long-range decisions concerning agriculture, forestry, and coastal zone resources, as well as decisions regarding sites for proposed facilities).

Analysis of the Climate Change Issue in the NEPA Process

Two primary issues are germane to a consideration of climate change in the NEPA process: the potential effects of a given proposed action on climate change and the potential indirect effects resulting from a proposed action that was affected by a changing climate (i.e., the *affected environment* may change, which in turn could alter the environmental effects of a proposed action). As discussed in the previous section and as stated in the CEQ regulations (40 CFR 1500.1), NEPA's goal is to help public officials make decisions based on an understanding of environmental consequences and to take actions that protect, restore, and enhance the environment. The *NEPA process* refers to all measures necessary for compliance with the requirements of Section 2 and Title I of NEPA (40 CFR 1508.21). Given the draft CEQ guidance, the language of NEPA, the CEQ regulations, and bills being considered in Congress, climate analyses will be appearing in the NEPA process with increasing frequency. This section addresses general issues associated with climate change and NEPA.

Effects of Proposed Actions on Global Climate

Principal reasons for assessing the effects of proposed actions on global climate and the effects of climate change on proposed actions, as suggested by the goals of NEPA and CEQ regulations, are to inform a decision maker of the potential climate change consequences associated with an action (i.e., disclosure) and to help the decision maker take actions to minimize potential adverse climate change impacts (i.e., mitigation). Meeting these goals for the climate change issue involves challenges not typically associated with potential impacts evaluated in NEPA documents.

A key aspect of evaluating a proposed action's effects on climate in conjunction with NEPA analyses is conducting the analyses at the appropriate level of the federal action (e.g., legislation, policy, program, or project) that would allow effective mitigative actions to be taken. It is important to keep in mind some of the important characteristics of the climate change issue as described in the previous sections: global scale, possible time frame for effects of at least 10–30 years, and many anthropogenic (and biogenic) contributors to climate change. Assessing the potential impacts of proposed actions on climate in NEPA documents needs to be done in a manner that is designed to reflect these characteristics.

Requiring the evaluation of a proposed action's effects on climate may not necessarily provide the information needed to make a decision that will help offset anthropogenically induced adverse changes in global climate. At the project level, analysis of potential impacts on climate is not likely to provide meaningful information (other than disclosure) for a decision on mitigative actions to be taken for at least two reasons: (1) one project's contributions to the sources and sinks of greenhouse gases is likely to be very small (e.g., as noted above, carbon emissions from one 1000-MW coal-fired power plant are about 0.02% of the 1988 global total from fossil-fuel combustion), and thus any mitigative actions would have very little, if any, effect and (2) the current state of the art of climate simulation models is not sufficiently refined to tell us if one project's emissions would trigger a cumulative impact (i.e., when added to emissions from other sources) that would warrant mitigation (i.e., we do not know if a significant adverse impact would be triggered by one project's emissions).

Analysis of impacts at the program or policy stage would likely provide slightly more useful information than at the project level because the potential changes in sources and sinks of greenhouse gases would be larger percentages of baseline conditions. However, even policies and programs are likely to be contributing rather than determining factors in global climate. As shown in Table 1, the United States contributes less than one fourth of total global carbon emissions from fossil-fuel combustion; thus, there is an *upper bound* to the potential effectiveness of laws, such as NEPA, of the United States alone in dealing with global climate. Furthermore, not all of the U.S. carbon emissions shown in Table 1 are subject to NEPA, such as automobiles and private sector enterprises, without federal involvement. For example, transportation (primarily automobiles and light trucks) is responsible for about one third of U.S. carbon emissions (Schneider 1989), and the manufacture and use of these vehicles is not uniformly subject to NEPA. Pedersen (1988) notes that coal-fired electric generating plants, another major contributor to U.S. carbon emissions, can often be built without any NEPA analyses. Thus, policy-level or program-level analyses may provide useful information in mitigating adverse climate changes, but they will not be a panacea.

Fortunately, policy or program EISs are not the only NEPA vehicles available for dealing with the climate change issue. Another option is a legislative EIS which is described in the CEQ regulations (40 CFR 1506.8) but which has received limited use in the almost two decades of NEPA implementation (Bear 1989). Because most federal actions stem in some way from legislation, a legislative EIS would provide a sufficiently large umbrella for anticipating and evaluating potential impacts from policies, programs, plans, and projects. Possible drawbacks to this approach are that (1) many assumptions would be required to translate broadly worded proposed statutes into more discrete actions whose impacts could then be assessed and (2) these assumed actions may not resemble those actually resulting from the legislation. It would thus be important to formulate assumptions such that the impact analyses provide an upper bound to potential effects. Legislative EISs thus offer a potential for addressing an environmental impact of large geographic scale at an early enough stage in the decision process to allow decisions to be made that can affect many policies, programs, plans, and projects, thereby allowing a decision maker to take appropriate actions for avoiding or minimizing potentially significant adverse impacts to global climate.

Analysis of the climate change issue in NEPA documents would result in the most benefit by being done at the stage in the decision process that will affect the largest scale of sources and sinks of greenhouse gases, which, as illustrated above with the example of the coal-fired power plant, would be at the program or policy level or higher (i.e., a legislative EIS). Current regulations implementing NEPA call for agencies to use *program, policy, or plan environmental impact statements* and to tier *from statements of broad scope to those of narrower scope* (40 CFR 1500.4). However, agencies are not required under current regulations to analyze large-scale impacts in policy, program, or legislative EISs before proceeding to preparation of NEPA documents covering smaller geographic scales. Requiring climate change analyses in NEPA documents may only lead to increased appearance of climate change impacts in project-specific documents. Requiring analysis of broad-scale impacts in legislative, policy, or program EISs would provide more meaningful information to the decision process and is more likely to lead to the implementation of corrective actions to mitigate climate change than is conducting climate impact analyses in project-specific EISs.

How does one *do a climate impact assessment* or *consider climate change* in a NEPA document? Since the analysis of climate change in NEPA documents is fairly young, it is premature at this time to answer the above question in detail; however, some general comments can be made. Of principal interest is a tiered methodology that provides for a more in-depth analysis of potential impacts if a proposed action has a greater potential to influence climate. The following paragraphs discuss this methodology.

Perhaps the most simplistic level of analysis would be to use changes in sources and sinks of greenhouse gases as indicators of potential for climate change. If these changes are small percentages of a given baseline, then there is little potential for measurable effect (not considering cumulative effects, however) from the proposed action. Baselines that could be used include sources/sinks from the same type of source category in the United States, sources/sinks from all categories in the United States, or sources/sinks from all categories worldwide. Comparison of a change in sources and sinks against a baseline could be used to define a threshold above which more detailed analyses would be conducted (e.g., if an action has the potential to affect less than 1% of current emissions, it would undoubtedly have a negligible effect on climate, and it would not represent a significant opportunity for a decision maker to mitigate adverse climate changes). The threshold also could be tied to uncertainties in the estimated strengths of sources and sinks of greenhouse gases. If the uncertainty in emissions was, for example, 10%, then estimated changes in emissions of 10% or less may not be significant and thus would not warrant impact analysis. A change of more than 10% should warrant an assessment, perhaps using one or more climate models.

The next tier in the methodology is more sophisticated and involves the use of simulation models such as GCMs and climate resource impact models. This approach would be used if the screening level analyses based on changes in the strengths of sources and sinks suggest a need for further analyses. Simulation models would be used to provide a general understanding of the relationships among the strengths of sources and sinks of greenhouse gases and resultant climate effects. Such analyses would suggest source/sink strengths that are important with respect to climate change and which could be used as a basis for comparison in assessing potential effects from proposed actions. The advantages offered by this approach over the first tier of analysis are (1) the geographic location (and timing) of changes in sources and sinks can be factored into the analyses and (2) the spirit of cumulative impact analysis can be more completely addressed due to the consideration of other (outside of the proposed action) sources and sinks of greenhouse gases.

The third tier of analysis would involve the use of GCMs or climate resource impact models for specific proposed actions. Because the use of these tools is resource intensive and requires special expertise, it should be reserved for proposed actions that clearly (based on analyses conducted in the first two tiers) have the potential to affect climate. The principal advantage of this tier is that it would provide an impact analysis unique to a given proposed action, rather than trying to *gage* potential impacts from generic analyses; it would also reflect the advantages of the previous tier.

Given that we can do some type of impact analysis, it is important to discuss the use of results of such an analysis by a decision maker. Expecting a decision maker to take action to protect, enhance, and restore the environment with respect to climate presents an interesting challenge because no consensus exists on the desired state of the climate. In other words, there is a potential for an impact in the absence of standards or guidelines on the condition of the environment that is desired as a result of taking mitigative actions; this situation is unlike that associated with potential impacts from changes in air quality, water quality, or radiation levels where criteria are available for evaluating the state of the environment. One possible definition of *protecting* the environment with respect to climate would be to minimize any net increase in the strengths of sources of greenhouse gases above a specified baseline. However, what is expected in terms of *restoring* or *enhancing* the environment? The greatest realistic hope here is that decision makers will minimize the types of actions that change climate.

We conducted an examination of recent EISs to identify the potential for considering global climate change in NEPA documents. Table 3 lists types of recent federal agency projects that were the subject of EISs in late 1988 and 1989.

This table is based on a listing of EIS titles maintained at the Northwestern University Transportation Library in Evanston, IL as of September 1989. The listing was screened to select types of projects that might involve potential effects on global climate change. Several generalizations can be made based on the information in Table 3.

1. Most EISs are prepared for specific projects, which by themselves have limited potential to affect climate change on a global scale. Relatively few programmatic EISs are prepared that would serve as an appropriate instrument for addressing cumulative effects of an agency's actions on climate change.
2. Most federal agencies engage in actions that have some potential for affecting climate change. The most obvious types of actions are those that involve major construction projects or resource exploitation where significant land is cleared. Such actions not only remove CO_2 sinks in the form of vegetation and associated soils, but they also involve the short-term use of heavy equipment that use fossil fuels and therefore contribute greenhouse gases to the atmosphere. For some agencies (e.g., the Federal Highway Administration and the Federal Aviation Agency), approval and implementation of major construction projects is a significant aspect of their agency's activities. The cumulative contribution of these actions to global climate changes is worthy of consideration.
3. A number of agencies' actions also indirectly contribute greenhouse gases to the atmosphere by promoting the development of transportation systems or the use of technologies that depend on fossil fuels. For example, the development of highway systems and airports encourages the growth of transportation and unavoidably the use of fossil fuels. On the other hand, development of urban mass transit systems leads to a more efficient transportation system and thus contributes to a more efficient and reduced use of fossil fuels per capita. Government programs that promote the mining and use of coal and other fossil fuels also contribute to the increase in greenhouse gases.
4. Agencies such as the Fish and Wildlife Service, the Forest Service, the Bureau of Land Management, and the National Park Service that are involved with the proposal, establishment, and management of wildlife refuges, wilderness areas, or other natural areas contribute to the preservation of natural vegetation and ecosystems that act as CO_2 sinks. Actions by these agencies, therefore, can have a net beneficial influence on the global climate situation.

The above discussion emphasizes direct impacts associated with a particular project; however, cumulative impacts also need to be considered when addressing climate change. Cumulative impact assessment is mandated in the CEQ regulations and has been addressed by the courts [e.g., *Fritiofson* v. *Alexander* 772 F. 2d 1225 (5th Cir. 1985) and *Natural Resources Defense Council* v. *Hodel* F. 2d (C.A.D.C. 1988)] and various reports and studies (e.g., CEQ 1988). The nature of the climate change issue, however, presents complications. The CEQ regulations define cumulative impact as effects resulting from the incremental impact of the action when added to other past, present, and reasonably foreseeable future actions, regardless of what agency (federal or nonfederal) undertakes such other actions. Cumulative impacts can result from individually minor but collectively significant actions taking place over a period of time. The *Fritiofson* v. *Alexander* decision additionally stated that cumulative impact assessment required the iden-

Table 3. Examples of Recent Federal Agency Environmental Impact Statements in Which the Issue of Climate Change Could Have Been Addressed.

Agency	Project	Possible Effects on Greenhouse Gas Emissions: Increase	Decrease
Bureau of Indian Affairs	Biomass-fueled power plant	a, b	c, d
Bureau of Land Management	Geothermal development	a, b	d
	Trans-Alaska gas system	e, b, f	
	Oil and gas development	e, b, f	
	New transmission line	e, b, f	
	Pipeline	e, b, f	
Department of Defense	Base expansions and modifications	e, b	
	Missile testing	g	
	Steam-generating facility	a, e, b	
	Dam and reservoir	b	c
Department of Energy	Clean coal technology development	a, e, b, f	d
	Management of hazardous and radioactive wastes	e, b	
Environmental Protection Agency	Lignite mine	e, b, f	
Federal Aviation Agency	Airport construction and expansion	e, b, f	
Federal Energy Regulatory Commission	Small-scale hydropower development	e, b	d
Federal Highway Administration	Highway construction and modification	e, b, f	
Fish and Wildlife Service	National Wildlife Refuge plans		h
	Acquisition of holdings		h
	Conservation plans		h
Forest Service	Implementing forest plans	e, b, f	c, h
	Timber sales	e, b	c
	Exploratory drilling	e, b, f	
	Fire recovery projects	e	c
	Open-pit mine development	e, b, f	c
	Vegetation management for reforestation (pest management)	e *e,b,f*	
Housing and Urban Development	Housing developments	e, b	
Mineral Management Service	Gulf of Mexico sales	e, b, f	
	OCS mining programs	e, b, f	
National Aeronautic and Space Administration	Shuttle development	g	
Office of Surface Mining	Mining permits	e, b, f	
Soil Conservation Service	Watershed plans	b	h
Urban Mass Transportation Authority	Mass transit development	e, b, f	d

Note: a — Direct contribution of a greenhouse gas to the atmosphere. b — Removal of CO_2 sink by clearing vegetation. c — Creation of a CO_2 sink by planting a crop or creating a water body. d — Energy source that replaces fossil-fuel use or increases the efficiency of such use. e — Short-term contribution of greenhouse gas to atmosphere from use of equipment during construction or clearing. f — Indirect contribution of greenhouse gas by promoting use of fossil fuels. g — Short-term use of fossil fuel. h — Long-term protection of CO_2 sink by preservation and enhancement of natural areas.

tification of the impacts from the proposed action, characterization of the region of influence for these impacts, identification of other contributors to these impacts in this region, and summation of the magnitudes of the potential effects — a straightforward concept for many assessments. In light of these recommendations, climate change and cumulative impacts introduce some interesting considerations. First, because climate change is a global problem, is the region of influence indeed the world? If so, it would be difficult, if not impossible, to identify all of the actions contributing to the impact in the region of influence (many of which are outside of NEPA's current purview, but which would still need to be factored into the analysis under current NEPA regulations). Identifying present contributors to sources and sinks of greenhouse gases would involve an extensive data collection exercise; however, the results would be applicable to and useful for virtually any cumulative impacts climate analysis and would thus not need to be repeated for each NEPA document. Identifying future contributors would also entail extensive work, including updating estimates of sources and sinks.

Assuming that the information needs can be met, what would constitute cumulative impacts analysis for climate change? The same types of ideas discussed for assessing direct effects of proposed actions would apply here also. Screening-level analyses could be conducted, based on changes in sources and sinks; results from these would be used in turn to determine the need for more sophisticated analyses using models. As climate models become more refined, it may be possible to identify situations where a direct impact of small magnitude warrants mitigation due to cumulative impacts.

Effects of Climate Change on Proposed Actions

The potential effects of global climate change on proposed actions are also within the scope of NEPA. In this case, the location of the proposed action vis-à-vis anticipated changes in the *affected environment* is important. Thus, the type of action and the extent to which it is a source or sink of greenhouse gases are not of principal concern. Rather, the location and timing of an action are of primary interest. Is a proposed action to be located in an area expected to undergo a significant change in existing conditions as a result of climate change? Will the action be around long enough to be affected by the changes? Will these changes affect potential impacts that have been identified for a given proposed action? In terms of location, coastal regions are at the forefront of public attention because climate change could result in sea level rise, which in turn could affect implementation of an action with resultant potential impacts. Areas of interest, however, need not be confined to the coasts. Meteorological aberrations triggered by a changing climate could also affect the interior of the United States through drought conditions or flooding.

None of the locational aspects mentioned is by itself cause for concern, unless the action proposed is going to extend long enough to be affected. Although there are important facilities located on the coast, such as the U.S. Strategic Petroleum

Reserve storage facilities located near the Gulf Coast (e.g., the Bryan Mound facility near Freeport, TX), the projected operational time frame for these facilities is less than 50 years, and therefore, they may not be appreciably affected by climate change. Nuclear power plants could also be affected by flooding resulting from sea level rise, but again the time frame is such that effects on currently operating facilities would be minimal. Long-term activities, such as sites for disposal of low-level radioactive and/or hazardous wastes, would be more likely to be affected. Major changes in the hydrological regime of these sites could affect the impacts resulting from waste disposal.

What, then, needs to be done in NEPA documents? If the proposed action under evaluation is projected to be operational in a time frame that could experience impacts from climate change, then potential changes in climate should be highlighted, as should their resultant impacts. The analysis of climate is also location dependent. That is, the action itself may not affect climate, but it is proposed for a location that may be subject to climate change, which in turn could affect potential impacts of the action. The climate analysis would be presented in the *Affected Environment* (40 CFR 1502.10) section of the EIS to describe the postulated future climate of the proposed site(s). Many of the same tools discussed previously (e.g., GCMs and climate resource impact models) would be applicable here. Once the future climate scenario is identified, potential impacts can be assessed in the appropriate sections of the document. The need for a climate forecast and assessment of resulting potential impacts would be most relevant for actions that could be affected by changes in climatic conditions (e.g., a change in rainfall, sea level, or groundwater).

The triggers for inclusion of climate change in NEPA documents can differ for assessing the impacts of a proposed action on climate and for assessing the impacts of climate change on a proposed action. In the former case, the trigger for conducting the analysis is whether the proposed action (legislative, program, policy, or project) significantly contributes to sources and sinks of greenhouse gases; one possible level of significance is a change greater than the uncertainty in current inventories of strengths of sources and sinks. In the latter case, the trigger is whether the action would exist long enough to be affected by climate change, whether it would be in an area subject to significant climate change, and whether such change would exacerbate or ameliorate potential impacts of the proposed action.

CONCLUSIONS

Despite substantial uncertainties, the prospects for global climate change and its resulting environmental impacts are likely enough that consideration in the NEPA process is warranted. Especially on a program or policy level (rather than on a site-specific project level), federal actions can and should be evaluated in terms of the extent to which the action may contribute to global climate change.

Furthermore, it is appropriate to consider the extent to which global climate change may alter the baseline environment in which an action is taking place that will not affect global climate per se. Given the many individual small sources of greenhouse gases, analysis of the cumulative impacts of an action, in concert with other actions, may be warranted.

The NEPA procedures for EIS preparation already accommodate evaluation of the climate change issue, especially under requirements to consider incomplete or unavailable information. Guidance drafted by the CEQ and pending federal legislation make it increasingly likely that consideration of climate change will have to be incorporated in the NEPA process. Addressing climate change in NEPA documents will be technically challenging and will be worthwhile because it will result in better (i.e., more informed) decisions.

Incorporating climate change in NEPA documents is not likely to be a panacea because of (1) the technical difficulties in making accurate predictions, (2) the lack of a requirement to implement the environmentally preferred alternative identified in NEPA documents, (3) the small fraction of global emissions of greenhouse gases contributed by actions under the purview of NEPA, and (4) the application of NEPA mainly to actions not yet taken (the changes in sources and sinks of greenhouse gases that have already occurred are likely a large part of the problem). Global climate change is a function of cumulative actions, and our capability to quantify such activities is only now developing. Furthermore, continuation of the current lack of mandated follow-up activities under NEPA could result in a situation in which potential effects are analyzed, but implementation of mitigation is not confirmed or its effectiveness not evaluated. Both of these situations would reduce NEPA's effectiveness in minimizing or preventing adverse environmental impacts. Actions in other countries are not considered, nor are current (i.e., *grandfathered*) state, local, or private-sector actions. Nevertheless, the problem of global climate change is a result of a series of individually small actions, and any solution will similarly result from a series of separate steps. For the United States — which contributes so much to total fossil fuel use, develops so much technology, and, through its international linkages and activities, affects so much of the world — consideration of its actions in the NEPA process is a very important step.

Perhaps the key to understanding NEPA's potential in dealing with the climate change issue lies not in the *NEPA process*, CEQ guidance, EIS preparation, or global climate models, but rather in a little-recognized provision in the language of NEPA. Over 20 years ago, NEPA's authors put forth the currently popular notion of *think globally, act locally* when they wrote in Section 101(c) that *each person should enjoy a healthful environment and that each person has a responsibility to contribute to the preservation and enhancement of the environment.* Implementing Section 101(c) of NEPA may be the biggest challenge in dealing with climate change in the coming years. Preparing informative and technically sound NEPA documents is an important first step in involving the public so that the individual responsibilities identified in Section 101(c) are indeed carried out.

ACKNOWLEDGMENTS

We thank Johnnie B. Cannon, J. Warren Webb, and two anonymous reviewers for their comments on this manuscript. We thank Lydia S. Corrill for her technical editing. Environmental Sciences Division Publication No. 3537.

REFERENCES

Bear, D. 1989. NEPA at 19: A primer on an "old" law with solutions to new problems. *Environ. Law Rep.* 19(2):10060–10069.

Callendar, G. S. 1938. The artificial production of carbon dioxide and its influence on temperature. *Q. J. R. Meteorol. Soc.* 64:223–240.

Council on Environmental Quality (CEQ). 1988. Workshop on Cumulative Impacts Assessment, November 30, 1988. Washington, D.C.

Council on Environmental Quality (CEQ). 1989. Draft of Guidance to Federal Agencies Regarding Consideration of Global Climate Change in Preparation of Environmental Documents. Washington, D.C.

Dobson, A., A. Jolly, and D. Rubenstein. 1989. The greenhouse effect and biological diversity. *Trends Ecol. Evol.* 4(3):64–68.

Ehhalt, D. H. 1985. Methane in the global atmosphere. *Environment* 27(10):6–12 and 30–33.

Flaschka, I., C. W. Stockton, and W. R. Boggess. 1987. Climatic variation and surface water resources in the great basin region. *Water Res. Bull.* 23(1):47–57.

Frye, R. 1983. Climatic change and fisheries management. *Nat. Resour. J.* 23:77–96.

Grotch, S. L. 1988. Regional Intercomparisons of General Circulation Model Predictions and Historical Climate Data. DOE/NBB-0084. U.S. Department of Energy, Washington, D.C.

Hillel, D., and C. Rosenzweig. 1989. The Greenhouse Effect and Its Implications Regarding Global Agriculture. Res. Bull. No. 724. Massachusetts Agricultural Experiment Station, University of Massachusetts, Amherst, MA.

Houghton, R. A., R. D. Boone, J. R. Fruci, J. E. Hobbie, J. M. Melillo, C. A. Palm, B. J. Peterson, G. R. Shaver, G. M. Woodwell, B. Moore, D. L. Skole, and N. Myers. 1987. The flux of carbon from terrestrial ecosystems to the atmosphere in 1980 due to changes in land use: Geographic Distribution of the Global Flux. *Tellus* 39B(1–2):122–139.

Jones, P. D., T. M. L. Wigley, C. K. Folland, D. E. Parker, J. K. Angell, S. Lebedeff, and J. E. Hansen. 1988. Evidence for global warming in the past decade. *Nature* 332:790.

Linder, K. P., and M. J. Gibbs. 1986. The Potential Effects of Climate Change on Electric Utilities: New York Case Study Results. ICF Incorporated, Washington, D.C.

Love, G. 1988. Cyclone storm surges: Post greenhouse. pp. 202–215. In G. I. Pearman (ed.), *Greenhouse*. CSIRO Publications, East Melbourne, Victoria, Australia.

Marland, G. 1990. CO_2 emissions. pp. 92–113. In Trends '90: A Compendium of Data on Global Change. ORNL/CDIAC-36. Oak Ridge National Laboratory, Oak Ridge, TN.

Mitchell, J. F. B. 1989. The greenhouse effect and climate change. *Rev. Geophys.* 27(1):115–139.

National Research Council. 1987. *Responding to Changes in Sea Level.* National Academy Press, Washington, D.C.

Pastor, J., and W. M. Post. 1988. Response of northern forests to CO_2-induced climate change. *Nature* 334:55–58.

Pedersen, W. F., Jr. 1988. Does NEPA provide a means of addressing climate change? pp. 297–299. In *Preparing for Climate Change.* Government Institute, Inc., Rockville, MD.

Prickett, G. T., and D. A. Wirth. 1989. Environmental impact statements and climate change. *Environment* 31(2):44–45.

Ramanathan, V., R. J. Cicerone, H. B. Singh, and J. T. Kiehl. 1985. Trace gas trends and their potential role in climate change. *J. Geophys. Res.* 90(D3):5547–5566.

Revelle, R., and H. E. Suess. 1957. Carbon dioxide exchange between atmosphere and ocean and the question of an increase of atmospheric CO_2 during the past decades. *Tellus* 9:18.

Schneider, C. 1989. Preventing climate change. *Issues Sci. Technol.* 5(4):55–62.

Sloan, S. A., and C. K. Laird. 1990. Measurements of nitrous oxide emissions from P.F. fired power stations. *Atmos. Environ.* 24A(5):1199–1206.

Smit, B., M. Brklacich, R. B. Stewart, R. McBride, M. Brown, and D. Bond. 1989. Sensitivity of crop yields and land resource potential to climatic change in Ontario, Canada. *Clim. Change* 14:153–174.

Stevenson, J. C., L. G. Ward, and M. S. Kearney. 1986. Vertical accretion in marshes with varying rates of sea level rise. pp. 241–259. In D. A. Wolfe (ed.), *Estuarine Variability.* Academic Press, San Diego, Calif.

United Nations Environment Programme. 1987. Montreal Protocol on Substances that Deplete the Ozone Layer. Nairobi, Kenya.

Wahlen, M., N. Tanaka, R. Henry, B. Deck, J. Zeglen, J. S. Vogel, J. Southon, A. Shemesh, R. Fairbanks, and W. Broecker. 1989. Carbon-14 in methane sources and in atmospheric methane: The contribution from fossil carbon. *Science* 245:286–290.

Wigley, T. M. L. 1988. Future CFC concentrations under the Montreal protocol and their greenhouse-effect implications. *Nature* 335:333–335.

Wuebbles, D. J., and J. Edmonds. 1988. A Primer on Greenhouse Gases. DOE/NBB-0083. U.S. Department of Energy, Washington, D.C.

Can NEPA Protect Biodiversity?

S. Henderson, NSI Technology Services, Inc., U.S. EPA Environmental Research Laboratory, Corvallis, OR; R. F. Noss, U.S. EPA Environmental Research Laboratory, Office of Research and Development, Corvallis, OR; P. Ross, U.S. Environmental Protection Agency, Office of Federal Activities, Washington, D.C.

ABSTRACT

Biodiversity has emerged as a prominent issue in the scientific and conservation communities, and is of increasing concern to the general public. As with other "new" environmental problems (e.g., global climate change, stratospheric ozone depletion), biodiversity is difficult to evaluate because it involves slow, cumulative, complex effects that are unquestionably serious but difficult to document. Unlike traditional pollution concerns, these new problems are not confined to one location or tied to one development or industry but are regional or global in scope. Given scientific uncertainty about cause-effect relationships, it is very difficult to formulate environmental policy or regulations that adequately address the problem. Biodiversity is also a very broad issue, involving aspects of species richness, species composition, genetic variation, habitat structure, landscape pattern, and ecological and evolutionary processes. Although components of biodiversity are addressed by various pieces of environmental legislation, a comprehensive approach is lacking. A new interpretation of the National Environmental Policy Act (NEPA) could involve the recognition of particular attributes or indicators of biodiversity to address in the environmental assessment and review process. We present a hierarchical framework of compositional and structural biodiversity from which measurable indicators can be selected.

INTRODUCTION

The birth of the environmental movement in this country reflected a public outcry over the effects of "the profound influences of population growth, high-density urbanization, industrial expansion, resource exploitation, and new and expanding technological advances" (NEPA 1970) on the natural environment. Increased environmental awareness and legislation have led to the expenditure of billions of dollars in an attempt to clean up our environment since the passage of the NEPA 20 years ago. Today, we find an environment so altered and degraded that we wonder what those billions of dollars accomplished. Although there have been local successes in fighting pollution, a close look at our planet today suggests that we are losing the war against environmental degradation.

As scientists and policy makers learn more about pollution and human impacts on the environment, they are discovering a new generation of environmental problems (e.g., loss of biodiversity, massive deforestation, global climate change, stratospheric ozone depletion) that typically involve less visible pollutants and highly diffuse sources. Unlike traditional localized pollution episodes, these new problems are not confined to one location or tied to one development or industry but are regional or global in scope. They are fundamentally a problem of too many people using too many resources.

Loss of biodiversity has emerged as one of these new environmental concerns. It has become a prominent issue in the scientific and conservation communities and is of increasing concern to the general public. As with other new environmental problems, loss of biodiversity is difficult to evaluate because it can involve slow, cumulative, complex effects that are serious but uncertain in detail and often difficult to document.

In this paper, we will first discuss biodiversity, what it is and why we value it, and then examine some of the current environmental legislation that addresses biodiversity. NEPA will be reviewed in terms of its potential to protect biodiversity. Finally, we will suggest an interpretation of NEPA involving the recognition of indicators of biodiversity that could be used in the environmental assessment and review process.

WHAT IS BIODIVERSITY?

Simply and succinctly, biodiversity refers to the "variety of life and its processes" (H. Salwasser, personal communication). If we try to understand the meaning behind these words, we quickly abandon simplicity and find ourselves mired in a topic of overwhelming complexity. Ecologists generally study diversity at several levels of biological organization, most notably genetic, species, ecosystem, and landscape. Genetic diversity refers to the diversity of genetic material within species; species diversity is the diversity of species within a defined area; ecosystem diversity is the diversity of interacting plant and animal species in natural communities and their relationships with the physical environment. The spatial arrangement of ecosystems across a landscape is an important higher-level

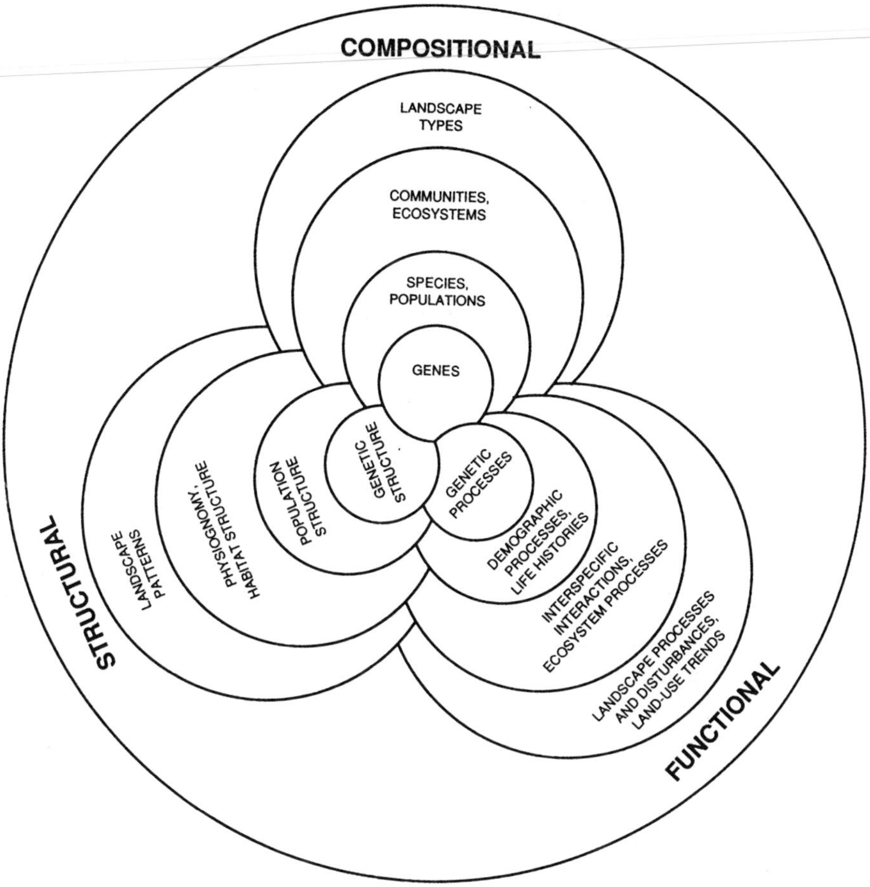

Figure 1. Interdependence and connectivity of different levels of biological organization within structural, functional, and compositional spheres. (From R. F. Noss. 1983. A regional landscape approach to maintain diversity. *Bioscience* 33:700–706. With permission.)

aspect of biodiversity.

To assess diversity at these levels is no mean feat. However, to comprehend fully the diversity of biota, we must go further and consider three primary attributes of ecosystems that determine biodiversity: composition, structure, and function (Franklin et al. 1981). Composition has to do with the identity and variety of elements in a collection and includes species lists and measures of species and genetic diversity. Structure refers to the biotic and abiotic habitat elements or patterns of a system. Function involves ecological and evolutionary processes, including gene flow, disturbances, and nutrient cycling (Noss 1990). Figure 1 shows the interdependence and connectivity of biodiversity at different levels of organization within each attribute.

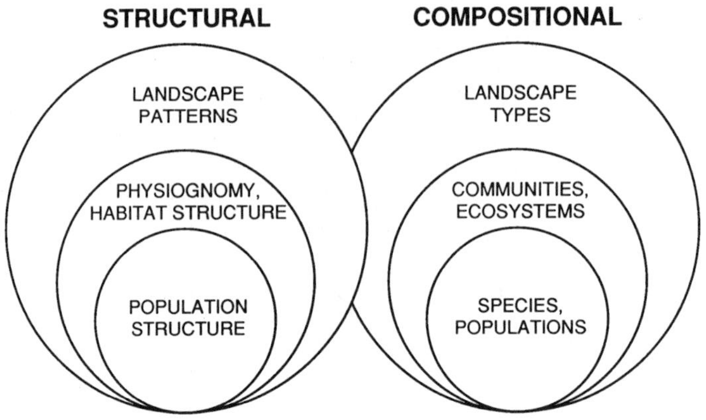

Figure 2. Hierarchical levels and attributes of biological diversity best suited for environmental assessment. (Adapted from R. F. Noss. 1983. A regional landscape approach to maintain diversity. *Bioscience* 33:700–706.)

For the purposes of this paper, we will focus on those hierarchical levels and attributes best suited for environmental assessment given our current state of knowledge — that is, species, ecosystem, and landscape diversity within the spheres of compositional and structural diversity (Figure 2). In evaluating composition and structure, it is useful to keep in mind the ecological and evolutionary processes that underlie these more tangible attributes. Our knowledge of biodiversity is limited, but we cannot wait for all the puzzle pieces to fit. We must start sharpening those tools we do have for potential use in environmental assessment.

VALUES OF AND THREATS TO BIODIVERSITY

The basic issue that drives all concerns about biodiversity is the accelerating and irreplaceable loss of genes, species, populations, and ecosystems. Associated with this loss is the loss of products obtained from nature, current or potential disruption of essential ecological processes and services, and loss of options for biological and cultural adaptation to an uncertain future. In addition, a growing number of people believe that species, ecosystems, and other elements of biodiversity (i.e., the natural world) are valuable in and of themselves.

Patterns and processes of biodiversity currently are deteriorating in most regions of the world, a problem that many biologists consider to be of crisis proportions. Although some threats to biodiversity, such as the enhanced greenhouse effect, are global in scope, biodiversity is being lost mostly due to local and regional impacts that are addressable in an environmental assessment process.

Estimates of the annual global rate of species extinction range from 1000 to 10,000 times that before human intervention (Wilson 1988). Many species not yet in immediate danger of extinction have been severely reduced in numbers and geographic range. In addition to species loss, communities are changing as sensitive species decline and are replaced by opportunistic weedy species, often exotics. Some ecosystems are disappearing altogether. Endangered ecosystems in the United States include tropical hardwoods (in Florida and Hawaii), tallgrass prairie, Palouse prairie, long-leaf wiregrass communities, Florida scrub, and old-growth forests of all types. Biodiversity faces threats primarily from the following stressors, all of which are caused or aggravated by human population growth and resource consumption:

1. habitat modification
   - physical alteration of habitat, especially activities that lead to habitat fragmentation
2. pollutants
   - abiotic (i.e., heavy metals, acid rain, organic contaminants)
   - biotic (i.e., invasive, introduced, bioengineered organisms)
3. climate change
   - anthropogenic-induced changes (short term)
   - natural fluctuations (long term)
4. exploitation (i.e., harvest, removal, harassment).

CURRENT LEGISLATION ADDRESSING BIODIVERSITY

Given the degree of scientific uncertainty surrounding biodiversity, it will be challenging to formulate environmental policy and regulations that adequately address this issue. Although most of what we now call biodiversity has been of concern to biologists, conservationists, and governments for a long time, biodiversity is relatively new as a self-contained policy issue. At present, policy makers are searching for the appropriate regulatory traction to address the biodiversity issue.

In the last few decades, there has been a plethora of environmental legislation and regulation. Many of these laws address important aspects of biodiversity. They generally fall into four categories:

1. species protection laws (e.g., Endangered Species Act, Marine Mammal Act)
2. "special" area designation laws (e.g., Wilderness Act, National Park Organic Act of 1916)
3. public lands management laws (e.g., National Forest Management Act, Federal Land Policy and Management Act)
4. pollution abatement laws (e.g., Clean Air Act, the Clean Water Act, and FIFRA)

By addressing components of biodiversity in a fragmented and often indirect way, these laws have provided only piecemeal protection. As evidenced by scientific data, we are losing species at an alarming rate, global climate change may render special designated areas useless for the species or communities they were set up to protect, and whole ecosystems are degraded by widespread pollutants far from their source.

In 1986, the Office of Technology Assessment (OTA) reviewed the biodiversity issue and recommended specific actions to Congress. Concluding that current environmental legislation addressing biodiversity was inadequate, the OTA report (OTA 1987) recommended that Congress strengthen the national commitment to maintaining biodiversity. The report further states that, although incomplete, enough information exists to define priorities for the maintenance of biodiversity. In response to the OTA report, the National Biological Diversity Conservation and Environmental Research Act (HR 1268) was introduced by Congressman James Scheuer. This is the first major piece of environmental legislation that addresses biodiversity holistically. Among other measures, HR 1268 would amend the National Environmental Policy Act (NEPA) to require assessment of biodiversity in environmental impact statements (EISs). Because the status of HR 1268 is uncertain and our biological resources cannot be put on hold until comprehensive legislation is in place, it is prudent to evaluate NEPA for its ability to encourage protection of biodiversity.

NEPA AND THE CONSERVATION OF BIODIVERSITY

The goal of NEPA is to promote the health of all biological resources; therefore, it is an appropriate legal mechanism through which to promote the conservation of biological diversity. NEPA requires proposed actions to preserve "important historic, cultural and natural aspects of our national heritage, and maintain, wherever possible, *an environment which supports diversity*" [italics added]. We believe that "diversity" can properly be interpreted as including biodiversity.

Central to NEPA is the EIS through which "environmental amenities and values...[are] given appropriate consideration in decision making along with economic and technical considerations" (Carlson 1988). All federal agencies are required to prepare detailed statements assessing the environmental impacts of and alternatives to major federal actions significantly affecting the environment. A secondary mechanism requires all federal agencies to review their authorizing statutes, regulations, and policies and procedures to ensure compliance with the "intent, purposes, and procedures set forth in this Act."

Although most EISs address only a few aspects of biological diversity (i.e., threatened and endangered species), reductions or alterations of biological diversity are clearly environmental impacts. EISs do not examine the cumulative spatial pattern of impacts in relation to the long-term maintenance of biodiversity. For example, we know that habitat fragmentation is one of the most serious threats to biodiversity, as it leads to diminished contiguous habitat area, isolation of populations, and detrimental physical and biological impacts (edge effects) on isolated habitat patches from surrounding areas. Although single actions may not, in themselves, seriously compromise the biodiversity of a given region, cumulative isolation and fragmentation of natural areas from multiple actions can ultimately lead to severe reduction in biodiversity. Within the context of the NEPA EIS review process, impacts should be interpreted to include such cumulative effects.

It is likely that the Council on Environmental Quality (CEQ) will soon be directed to prepare guidelines for the consideration of biological diversity in EISs. A report from the Senate Environment Committee (1988) directed the CEQ to determine whether biological diversity was being adequately addressed in EISs and to prepare new guidelines if necessary. HR 1268 would amend NEPA to "... require explicit assessment of effects on biological diversity in environmental impact statements." Section 5 of HR 1268 would clarify the diversity provision in Section 101(b) of NEPA by inserting "including biological diversity" after "which supports diversity."

In light of these potential amendments to NEPA and given NEPA's current broad mandate to protect the environment, guidelines for the consideration of biodiversity in EISs should be developed now. It is important to note that aspects of biodiversity are addressed in the NEPA process; bills or legislation would simply make this focus explicit and comprehensive.

HOW IS BIODIVERSITY MEASURED?

It is one thing to mandate that diversity be considered, it is quite another to determine how diversity will be considered. How will those people involved in the actual writing and reviewing of EISs be able to give more than cursory attention to biodiversity? By what yardstick will biodiversity be measured? Because the real world is complex and constantly changing, it is difficult to measure biodiversity in a meaningful way. Unlike the physical sciences, ecology has few, if any, laws. Clear and generally accepted measures of biodiversity do not exist.

Measuring biodiversity is more than a numbers game. Most ecologists will agree that "richness," the number of species, ecosystems, habitats, or any other group of things in a defined area is not, by itself, an appropriate measure of biodiversity (Noss 1990). In fact, species richness can be a misleading indicator of biotic value if many species in the sample are exotic, "weedy," or highly tolerant of human disturbance or pollution (Noss 1983; Norse et al. 1986).

Because it is impractical, if not impossible, to measure accurately the total composition and diversity of biota in a given area, shortcuts must be taken. The identification and evaluation of measurable surrogates or indicators are necessary for the assessment of biodiversity. If we crystallize the components of biodiversity into measurable elements, it is possible to identify particular indicators to use in the environmental assessment and review process.

INDICATORS

Probably the best known use of an indicator is the coal miner's canary. In this example, miners took a caged canary into the mine with them to detect toxic fumes. An unconscious canary indicated the presence of lethal fumes and alerted the miners in time to make a life-saving retreat from the mine. The miner's canary is one-dimensional in its usefulness as an indicator. The canary could not warn of a potential cave-in, a hungry bear in a coal mine waiting just ahead, or a potentially fatal equipment malfunction.

Ideally, environmental assessments should consider a range of ecological parameters at several levels of organization (Karr et al. 1986). Like the canary, no one indicator can do it all. We suggest a characterization of biodiversity that identifies the major components at several levels of organization. Table 1 is a compilation of terrestrial biodiversity indicators at the compositional and structural levels. As with most categorizations, some boxes in Table 1 overlap, and distinctions are somewhat arbitrary. The table may be a useful checklist of biodiversity attributes to consider in preparing or reviewing EISs and could be used as a framework for selecting specific indicators by which to evaluate biodiversity in any geographical area of concern. Such a framework would allow us to (1) characterize the biodiversity of an area in a way comparable to other areas and (2) identify specific aspects of biodiversity that can be used as indicators to monitor and assess the overall status or health of an ecosystem.

SELECTING INDICATORS

Different indicators are useful for different environmental assessment questions. An important point in the selection of appropriate indicators is the question, "What are we assessing and why?" In application, many indicators will be specific to ecosystems. For example, coarse woody debris is a structural element critical to biodiversity in many old-growth forests such as in the Pacific Northwest, but it may not be important in more open-structured habitats, including forest types subject to frequent fires (Noss 1990).

The following is a list of characteristics that should be considered when choosing indicators. Ideal indicators would be (Cook 1976; Sheehan 1984; Munn 1988)

1. sufficiently sensitive to provide an early warning of change (hypersensitive to stress)
2. distributed over a broad geographical area or otherwise widely applicable
3. capable of providing a continuous assessment over a wide range of stress
4. relatively independent of sample size
5. easy to measure, collect, assay, and/or calculate (in the case of an index)
6. able to differentiate between natural cycles or trends and those induced by anthropogenic stress
7. relevant to ecologically significant phenomena

No single indicator will possess all of these desirable properties; a set of complementary indicators is required. Ideally, indicators would be selected by biologists familiar with local and regional ecology but who have no vested interest in the outcome of the environmental assessment.

Once indicators have been selected and measured, the status or health of the population or ecosystem can be evaluated and the information incorporated in the section on the affected environment of the EIS. Predictions can be made about how the proposed project and its alternatives may affect the identified indicators and the underlying endpoint of biodiversity. Then, environmental consequences, including impacts on biodiversity, can be better appreciated.

The impacts of federal projects on biodiversity can be assessed adequately only if those impacts are monitored during the course of the project and beyond. Ongoing monitoring goes beyond current NEPA requirements; however, this is an important part in the assessment of impacts to biodiversity. If the indicators chosen in an EIS are monitored in the affected environment and in a nearby control site (such as a nature reserve), then impacts can be assessed statistically, particularly if impacted and control samples can be replicated.

CONCLUSION

Biodiversity has been thrust forward as a new environmental concern. Biodiversity encompasses many concerns for which assessment methodologies are not yet well-developed. This creates difficulties in setting goals or standards for factoring biodiversity into decisionmaking.

NEPA as an appropriate legal mechanism through which to ensure the conservation of biological diversity is addressed in federal decisions. NEPA mandates that all federal agencies prepare detailed EISs assessing the environmental impacts of and alternatives to major federal actions significantly affecting the environment. Measuring biodiversity by particular relevant indicators provides information not only for an assessment of ecosystem health, but also on possible direct, indirect, and cumulative effects of a proposed federal action on biological diversity as an endpoint in itself.

We do not suggest that NEPA alone can protect biodiversity. A major shortcoming of NEPA is that it applies only to federal activities and does not apply to the local/state/private sector. Also, courts have held that EIS requirements are primarily procedural and responsible for documenting impact for disclosure to the public rather than requiring protection.

Finally, we could rephrase our title question from "Can NEPA protect biodiversity?" to "Will NEPA protect biodiversity?". Certainly protecting biodiversity will be one of the major scientific challenges faced by NEPA in the coming years.

ACKNOWLEDGMENTS

We wish to thank Don Phillips, Blair Csuti, Ann Hairston, and two anonymous reviewers for their insightful reviews and comments.

REFERENCES

Carlson, C. 1988. NEPA and the conservation of biological diversity. *Environ. Law* 19(1).

Cook, S. E. K. 1976. Quest for an index of community structure sensitive to water pollution. *Environ. Pollut.* 11:269–288.

Franklin, J. F., K. Cromack, W. Denison, A. McKee, C. Maser, J. Sedell, F. Swanson, and G. Juday. 1981. Ecological Characteristics of Old-Growth Douglas-Fir Forests. USDA Forest Service General Technical Report PNW–118. Pacific Northwest Forest and Range Experiment Station, Portland, OR.

Karr, J. R., K. D. Fausch, P. L. Angermeier, P. R. Yant, and I. J. Schlosser. 1986. Assessing Biological Integrity in Running Waters: A Method and Its Rationale. Illinois Natural History Survey Special Publication No. 5. Champaign, IL.

Munn, R. E. 1988. The design of integrated monitoring systems to provide early indications of environmental/ecological changes. *Environ. Monit. Assess.* 11:203–217.

Norse, E. A., K. L. Rosenbaum, D. S. Wilcove, B. A. Wilcox, W. H. Romme, D. W. Johnston, and M. L. Stout. 1986. *Conserving Biological Diversity in Our National Forests.* The Wilderness Society, Washington, D.C.

Noss, R. F. 1983. A regional landscape approach to maintain diversity. *BioScience* 33:700–706.

Noss, R. F. 1992. Indicators for monitoring biodiversity: A hierarchical approach. *Conservation Biology* (in press).

Office of Technology Assessment (OTA). 1987. Technologies to Maintain Biological Diversity. Office of Technology Assessment, Office of Technology Assessment, Washington, D.C.

Salwasser, H. 1992. Personal communication.

Sheehan, P. J. 1984. Effects on community and ecosystem structure and dynamics. pp. 51–99. In P. J. Sheehan, D. R. Miller, G. C. Butler, and P. Boudreau (eds.), *Effects of Pollutants at the Ecosystem Level.* Wiley, New York.

Wilson, E. O. (ed.). 1988. *Biodiversity.* National Academy Press, Washington, D.C.

Improvements to the Scientific Content of Arctic Environmental Impact Statements

P. L. Johnson, U.S. Arctic Research Commission, Washington, D.C.

ABSTRACT

During 1989, the Arctic Research Commission undertook a critical review of the current state of environmental impact evaluation in the U.S. Arctic. The National Environmental Policy Act (NEPA) — the basic national charter for the protection of the environment — requires that an environmental impact statement (EIS) be prepared for legislation and for major federal actions significantly affecting the quality of the human environment. Specifically, the commission sought to determine if the scientific and technical aspects of the NEPA process as practiced in the Arctic could be improved.

The commission finds that the NEPA process can be lengthy, costly, and frustrating and can lead to litigation. These concerns are more severe in arctic regions such as Alaska because of the unusual aspects of the arctic environment as well as a serious lack of data and arctic expertise.

The commission concluded that there is no need to change the law or regulations of NEPA. However, there is a need to improve the scientific and coordination aspects of the EIS process. The most critical deficiency is the absence of impartial, external quality-control mechanisms for the data and information used in the stages of scoping, synthesis and EIS preparation, and follow-up monitoring programs.

Based on its findings, the commission recommends:

1. Impartial, external scientific and engineering review mechanisms should be established for each of the following stages: (a) the scoping plan, (b)

synthesis and preparation of the EIS, and (c) environmental monitoring programs.
2. The external reviewers should be carefully selected for their expertise in the relevant fields and independence of the parties affected.
3. The external reviewers should be protected from litigation and compensated at prevailing federal rates.
4. The full text of all reviews should become part of the public record.
5. Agencies should strongly encourage publication of all significant new information in the refereed scientific or technical literature.
6. The identification, review, and resolution of environmental issues should be channeled through a single coordination team established for that purpose.
7. The Council on Environmental Quality (CEQ) should issue a memorandum (similar to its April 10, 1981, memorandum entitled "Scoping Guidance") setting guidelines for the implementation of the external review process and for the establishment of the coordination team.

INTRODUCTION

Environmental Analysis in the Arctic

The National Environmental Policy Act (NEPA) requires that an environmental impact statement (EIS) be prepared for every recommended proposal for legislation and other major federal actions significantly affecting the quality of the human environment. NEPA is the basic national charter for the protection of the environment (Anderson 1974), and regulations to implement the NEPA process have been issued by the Council on Environmental Quality (CEQ) in the Executive Office of the President (CEQ 1981a, b, 1984, 1986). Over the years, the process has been reinforced and expanded by the courts, and generic environmental research and development has developed in response to the increasing demands for environmental information.

The EIS process — from initiation by a private organization or federal agency; through scoping, assessment, impact statement, and public review; and to decision, permit, and subsequent monitoring — is a lengthy, complex, and often expensive sequence. It can create controversy and result in litigation. In theory, if not always in practice, these problems should not arise if there are well-defined issues, relevant and reliable data, and unambiguous conclusions. These requirements are not often met. The difficulties and frustrations are widespread but are particularly severe for cases involving the high arctic regions of Alaska.

On several occasions in the recent past, the U.S. Arctic Research Commission has been approached with questions related to environmental assessment and the EIS process. Because accurate scientific and technical knowledge and adequate data bases are such a fundamental requirement, the commission decided to determine what useful role, if any, it could play in helping to improve the scientific and technical aspects of the process within the stipulations of existing laws and regulations. Although not explicitly mandated in the Arctic Research and Policy Act of 1984 (Pub. L. 98-373), such an activity of the commission would address one important element of the U.S. Arctic Policy set forth in National Security Directive Number 90 by the president, namely to support "sound and rational development in the Arctic region, while minimizing adverse effects on the environment."

To accomplish its task, the commission reviewed NEPA and its implementing regulations; interviewed a number of individuals involved in the NEPA process in the public and private sectors; received views and suggestions from its advisors; and requested testimony from selected Alaskan representatives of industry, federal agencies, and state and local governments in a public session held in Anchorage on June 27, 1989. In this review and analysis, the commission focused on three basic issues: the quality of data and analysis, public participation in the process, and the timing involved in preparing environmental assessments and impact statements.

FINDINGS

The commission finds that there is no need for any changes in the law and the regulations of NEPA. However, there is room for improvement in the scientific and technological aspects of the EIS process.

There was a general consensus in the commission that the Arctic clearly merits special consideration in the EIS process. Five principal arguments were given.

1. Many of the environmental concerns in the Arctic deal with some of its unique aspects, such as permafrost, sea ice, a large-amplitude photocycle, and short and intensive growing seasons. The effects of manmade environmental insult are aggravated by the relatively small number of species in arctic ecosystems and the slowness of environmental recovery (environmental fragility). The result is that there are few comparable precedents on which to base EIS predictions in Alaska and that the environmental consequences of erroneous predictions can be far more serious and long lasting than in temperate regions.
2. In the Arctic, including the Alaskan Arctic, there is a serious lack of data and information concerning the physical and biological (ecological) environment covering long periods of time on a decadal scale.
3. Much of the needed data are available only in proprietary reports and in the "gray" literature.
4. The cost of data acquisition in the Arctic is much higher than equivalent operations in the other 49 states, and the process takes much longer.
5. The number of qualified arctic investigators is small. In particular, the limited number of contractors and consultants with expertise in the U.S. Arctic, especially in environmental biology, restricts the possibilities of finding "untainted" contractors for the EIS process.

The scoping process is underused; the agencies may wait until all documents are ready for review before seeking public involvement, although this violates the intent of NEPA and CEQ regulations. Few people show up during scoping hearings; independent scientists are often not interested or urged to participate. The scoping process is sometimes perceived to be organized to support decisions already made. Inadequate scoping delays the resolution of controversies.

The EIS process is frustrating because it is so long and easily becomes politicized. EIS documents are too big; a lot of the information included is redundant or only peripherally related to the case but is included in anticipation of possible litigation. EISs are often viewed as supporting agency opinions rather than being the basis for such opinions.

Budgetary limitation is a significant concern to the participating agencies with the result that limited staff are overburdened and the possibilities of obtaining additional information are limited. Public disclosure of all relevant technical information may also be resisted in preparing an EIS, even though the U.S. Supreme Court has affirmed the concept of the EIS as a full-disclosure document.

In a number of agencies, the science that goes into an EIS is seldom critically reviewed, and there are few checks and balances at the agency level. Clear standards are seldom given to the participants, and the commercial relationship between consultant and client often limits credibility of contractor reports. Too many nonverifiable hypotheses and unstated assumptions are included in the EIS, and much of the documentation is based on the "gray literature." In recent years, there has been a decline in participation of university, national laboratory, and agency scientists in impact-related research. This has led to a decline in the quality control of data in EIS documents.

By and large, the most critical element in the EIS process is the quality of the scientific information and, specifically, the objectivity with which it is evaluated in the context of the specific issues. In particular, there is a need for external reviews within the assessment process to be competent scientifically and as objective as possible. Such reviews should focus on the quality and relevance of the data and information used, the methods of analysis applied, and the correctness of the scientific and technical conclusions reached.

An independent external review can ensure that there is good information based on scientifically credible data and appropriate analysis, that gaps in knowledge are identified, and that all issues of relevance are addressed. Independent review can also help by eliminating "red herrings," trivial issues, and faulty analysis, strengthening the whole EIS process.

In addition to external scientific and engineering review, formal appointment of coordination teams of representatives of the affected and knowledgeable organizations similar to the approach recently used in Alaska by the Minerals Management Service can be very beneficial. In 1988, Alaska and the Department of the Interior appointed a task force to conduct the review process of proposed offshore gold mining near Nome, AK. The federal and state members (five each) were assisted by advisors from local organizations and the private sector. The resulting EIS did not lead to litigation but resolved issues through discussion.

Channeling the identification, review, and resolution of environmental issues through a single coordination team comprised of representatives from affected federal, state, and local government agencies and advisors from interested academic and special interest groups will have the following benefits:

1. It will establish a means to systematically identify, review, and resolve environmental issues and provide a logical complementary context to implement external reviews at appropriate stages.
2. It can focus the external reviews on scientific as well as socioeconomic concerns of the state and local organizations.
3. It will promote a more cooperative and positive approach to development of technical guidelines and procedures for safe, effective, environmentally sound development of the natural resources under terms acceptable to federal, state, and local governments.

CONCLUSION

The commission concludes that there is a need for external reviews within the EIS process by competent and impartial scientists and engineers. Such reviews should focus on the quality and relevance of the data and information used, the methods of analysis applied, and the correctness of the scientific and technical conclusions reached. Furthermore, coordination of the EIS process and implementation of external review would benefit from the appointment of a coordination team with representatives from affected federal, state, and local government agencies and advisors from interested academic and special interest groups concerned with the preparation of an EIS.

RECOMMENDATIONS

1. To improve the EIS process, impartial external scientific and engineering review mechanisms should be established for each of the following three stages: the scoping plan, synthesis and preparation of the EIS, and environmental monitoring programs.

   The scoping process is intended to define the questions of concern to be addressed in the EIS. An early and independent review of the agency scoping plan for the scoping phase will lead to more appropriately framed questions and therefore to a more effective process. Consequently, this independent review process should commence as early as possible in the assessment process.

   External review of the quality of data and methods of analysis during synthesis and preparation of the EIS is especially important. This independent review must be done before the draft EIS is released for public comment. Reviewers should not focus on results, but upon the methods used to obtain them. Such disinterested review can bring credibility to the process and greatly reduce subsequent delay, litigation, and attendant costs. Some decisions stipulate that an environmental monitoring program is to become part of the project. Peer review of the design of the monitoring procedures will help ensure accurate and usable results.

2. The external reviewers should be carefully selected for their expertise in the relevant fields and independence of the parties affected.

   Comprehensive external review is estimated to require four to six reviewers and to last 45–60 days; however, this additional time could be reduced if the review is conducted concurrently with the other tasks by the agency. Different reviewers may be necessary for each of the three stages recommended, depending on the proposed project.

   Sources of expertise will vary with the subject; but referrals can be sought from organizations such as the Alaska Science and Technology Foundation, the Arctic Research Consortium of the United States, the Arctic Research Commission, or the Marine Mammal Commission. In addition, many scientific societies and professional organizations maintain computerized rosters of their members coded by areas of expertise which also could be used as a source. Of all the above possibilities, the commission suggests that a contract to a university consortium for external reviews would be a desirable mechanism to enhance general credibility of the EIS process.

3. Since the EIS decision rests with the lead agency which is not bound by the outcome of the scientific reviews, the external reviewers should be protected from litigation. To achieve timely response, reviewers should be compensated at prevailing federal daily consulting rates.

4. The full text of all review comments should be provided to the official responsible for the final decision and to cooperating agencies and become part of the public record. Negative reviews could lead to (a) revision of the scoping plan, (b) reanalysis of the available data, (c) determination that the data or information available are incomplete or unreliable, (d) redesign of the project, or (e) redesign of the monitoring requirements.

5. Agencies should stipulate that all significant new information obtained by their contractors and by their own agency personnel in the course of an EIS process be published in the refereed scientific or technical literature.

6. The identification, review, and resolution of environmental issues should be channeled through a single coordination team established for that purpose by the affected federal, state, and local government agencies and assisted by advisors from interested academic and special interest organizations. Greater use of the coordination team concept from the very beginning of a project will help foster less confrontational development of federal EISs and enhance state and local government participation, cooperation, and technical assistance in the NEPA process.

7. To implement the proposed external, third-party expert reviews in the NEPA process, agencies which carry lead agency responsibility must modify their internal guidelines and procedures. The CEQ should issue a memorandum (similar to its April 30, 1981 memorandum entitled "Scoping Guidance") setting

guidelines for the implementation of the external review processes and for the establishment of a coordination team. Each agency, subject to CEQ concurrence, should develop criteria and accompanying instructions for the reviewers and the coordination team. Ample precedence and procedural models for external reviews exist in the federal agencies.

ACKNOWLEDGMENTS

The findings and recommendations herein reported are the work of the U.S. Arctic Research Commission. Appreciation is expressed to the commissioners, advisors, and many officials from both public and private organizations who contributed ideas and comments.

REFERENCES

Anderson, F. R. 1974. The National Environmental Policy Act. pp. 238–419. In E. L. Dolgin, and T. G. P. Guilbert (eds.), *Federal Environmental Law*. West Publishing Co., St. Paul, MN.

Council on Environmental Quality (CEQ). 1981a. Questions and Answers About NEPA Regulations. Memorandum for Federal EPA Liaisons, Federal, State, and Local Officials and Other Persons Involved in the NEPA Process. March 16, 1981.

Council on Environmental Quality (CEQ). 1981b. Scoping Guidance. Memorandum for Federal General Counsels, NEPA Liaisons, Federal State and Local Officials and Other Persons Involved in the NEPA Process. April 30, 1981.

Council on Environmental Quality (CEQ). 1984. National Environmental Policy Act (NEPA) Implementation Procedures; Appendices I, II, and III; Final Rule. *Fed. Regist.* 49(247):49750–49782.

Council on Environmental Quality (CEQ). 1986. Regulations for Implementing the Procedural Provisions of the National Environmental Policy Act (40 CFR 1500-1508). July 1, 1986.

Ecosystem Assessment Methods for Cumulative Effects at the Regional Scale

C. T. Hunsaker, Oak Ridge National Laboratory, Oak Ridge, TN

ABSTRACT

Environmental issues such as nonpoint-source pollution, acid rain, reduced biodiversity, land use change, and climate change have widespread ecological impacts and require an integrated assessment approach. Since 1978, the implementing regulations for the National Environmental Policy Act (NEPA) have required assessment of potential cumulative environmental impacts. Current environmental issues have encouraged ecologists to improve their understanding of ecosystem process and function at several spatial scales. However, management activities usually occur at the local scale, and there is little consideration of the potential impacts to the environmental quality of a region.

This paper proposes that regional ecological risk assessment provides a useful approach for assisting scientists in accomplishing the task of assessing cumulative impacts. Critical issues such as spatial heterogeneity, boundary definition, and data aggregation are discussed. Examples from an assessment of acidic deposition effects on fish in Adirondack lakes illustrate the importance of integrated databases, associated modeling efforts, and boundary definition at the regional scale.

INTRODUCTION

Effective management of our natural resources requires a holistic approach to environmental assessments. Since 1978, the implementing regulations for the National Environmental Policy Act (NEPA) have required assessment of potential cumulative environmental impacts. Cumulative impact assessment, effects assessment for programmatic environmental impact statements (PEIS), and ecological risk assessment share some common goals and needs when applied to large geographic areas or regions. In the United States, a region can range in size from an area the size of several counties to several states. A region should contain a certain degree of homogeneity with respect to the characteristics used to define it (de Blij 1978). The goals of these assessments include making informed decisions and protecting or managing the environment at large geographic scales. The needs of cumulative, programmatic, and risk assessments at the regional scale include (1) regional and national integrated databases, (2) monitoring that characterizes conditions at several spatial scales, (3) quantified relationships between landscape structure and function, (4) mechanistic understanding of the controls on landscape functions at several spatial scales, and (5) models for several spatial and temporal scales.

The common goals of assessments for cumulative impacts, PEISs, and ecological risk are to make informed decisions and to protect or manage the environment for large geographic areas. A cumulative impact assessment should qualitatively or quantitatively assess "the impact on the environment which results from the incremental impact of the action when added to other past, present, and reasonably foreseeable future actions" (CEQ 1978). The PEIS is appropriate for general matters or related actions that are similar in nature or broad in scope and have cumulative impacts (Sigal and Webb 1989). Myslicki (this volume, Chap. 5) points out the advantages of using PEISs to look at cumulative impacts. A regional risk assessment should evaluate the aggregate influence of multiple disturbances on the total resource as bounded by the region of influence for the hazard of interest. Risk assessment goes beyond a cumulative or programmatic assessment in that it must quantify the probability of impact and the associated uncertainty. Thus, a regional ecological risk assessment is the extreme quantification of a cumulative or programmatic assessment and represents what assessments should be striving to achieve.

This paper proposes that a regional ecological risk assessment provides a useful approach for assisting scientists in accomplishing the task of assessing cumulative impacts. A risk assessment approach is independent of scale (i.e., the components of the assessment are developed for the appropriate space and time scales of each individual assessment). The Canadian Environmental Assessment Research Council and the U.S. National Research Council (1986) stated, "neither scientists nor institutions are working at the temporal and spatial scales needed for the assessment of cumulative effects." Thankfully, this statement is no longer true; however, often the research and analyses are being developed in different disciplines. The theory and analyses from landscape ecology (Turner 1989) and research and tools from geography, both of which focus on spatial scale, when combined with a risk assessment approach, hold great promise for future NEPA assessments. This paper discusses in detail the importance of defining the regions/subregions for assessment; this activity relates to our need to quantify relationships between landscape structure and function and to understand the mechanisms that control landscape functions at different spatial scales.

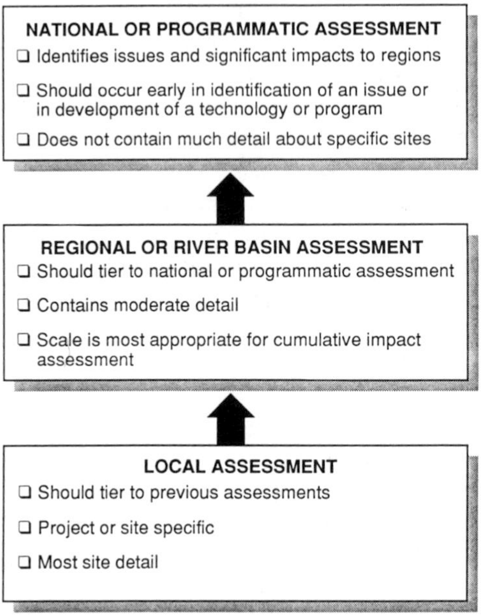

Figure 1. Relationships between national or programmatic, regional, and local assessments. Arrows show the direction for tiering between these assessments.

CUMULATIVE IMPACT ASSESSMENT APPROACH

Spatial and temporal scales are important to the understanding, analysis, and management of cumulative effects. As with regional risk assessments, adequate cumulative impact assessments require an understanding of the contributions to assessment uncertainty from boundary definition (geographic area for assessment), data resolution and aggregation, and spatial heterogeneity of a resource within the assessment area. Cumulative impacts are best addressed at the regional scale, while both national and regional scales are appropriate for programmatic issues (Figure 1). Often cumulative impacts are too complex or extensive to adequately address in most local assessments except in a qualitative manner, unless a regional assessment is available from which to tier. National or programmatic assessments, because of their breadth and often a lack of integrated databases, cannot be expected to address cumulative impacts in much detail (Cada and Hunsaker 1990; FERC 1988a). Several environmental impact statements (EIS) for hydropower development illustrate the ability to quantify cumulative effects at a regional or river basin scale (FERC 1985a, 1985b, 1986, 1988a, 1988b).

The approach outlined for regional ecological risk assessment (Hunsaker et al. 1990) can be used for cumulative impact assessment and programmatic assessments for large geographic areas. Risk assessments can be thought of as having two distinct phases: the definition phase and the solution phase

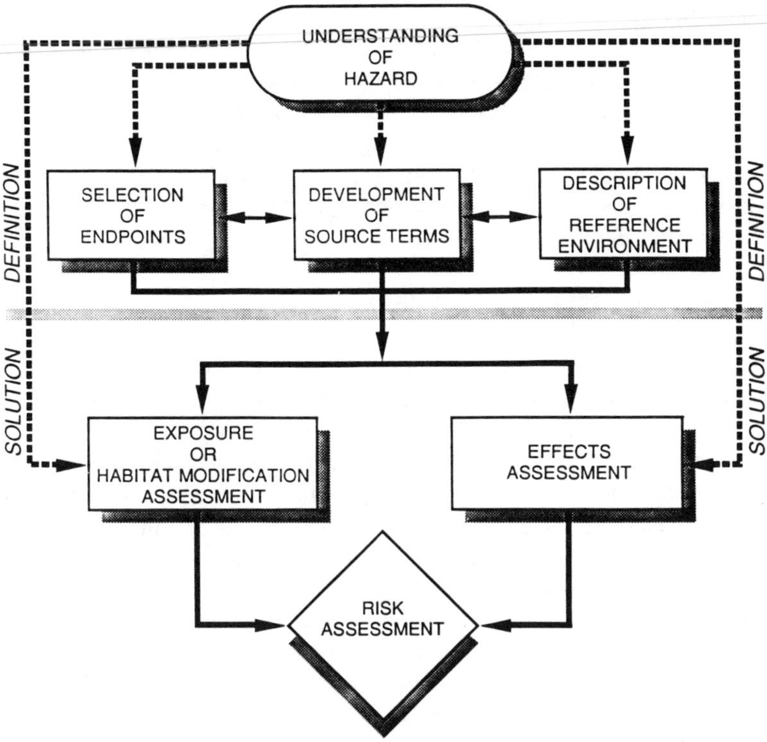

Figure 2. The two phases of regional risk assessment: the hazard definition or scoping of the problem and the problem solution. (From Hunsaker et al. 1990. Regional ecological risk assessment. *Environ. Manage.* 14(3):325–332. With permission.)

(Figure 2). In the definition phase, the endpoint (entity and its quality of concern), source terms (source and associated magnitude of hazard), and geographic area for assessment (reference environment) are defined. The definition of these three elements should be an iterative process, and the understanding of the hazard should take into account not only the ecological processes of interest, but also the social, economic, and institutional processes significant to the hazard. In the solution phase, exposure and effect are assessed and then combined to determine the risk or probability of a negative event happening.

SELECTION OF REGIONS AND SUBREGIONS FOR ASSESSMENT

For any cumulative effects assessment, the assessor should consider the contributions to assessment uncertainty from boundary definition, data resolution and aggregation, and spatial heterogeneity. For risk assessment, such uncertainty should be quantified. The importance of boundary definition, selection of regions and subregions in this paper, is discussed in detail because appropriate definition of the assessment region can reduce uncertainty; it is also related to data resolution, data aggregation, and spatial heterogeneity. Throughout the discussion, points are illustrated with examples from a demonstration assessment of atmospheric pollutant effects on aquatic ecosystems. The ability to provide such a refined example is possible because of years of research and analyses by many scientists, primarily funded by the National Acid Precipitation Assessment Program (NAPAP 1990).

Demonstration Data and Models

Recent international agreements for controlling atmospheric pollutants have focused on methods to identify and map the distribution and amounts of deposition of atmospheric pollutants that do/do not cause significant harmful effects on the environment (critical loads) (CLRTAP 1989). The United States has emphasized the need to develop critical loads for appropriate geographic areas (i.e., regions/subregions whose endpoints or resources of interest respond in a similar way to the hazard of interest). The examples used in this paper come from that effort to develop and demonstrate an assessment approach for determining and mapping critical loads in the United States. In these examples, the assessment region or reference environment is the Adirondack region in New York State. The endpoint is the proportion of lakes with brook trout, and the source of the acidic deposition hazard is sulfate deposition. The sulfate exposures are for current deposition and 50% of current deposition. The lake acidification model results shown are from an empirical, steady-state model (Henriksen 1984). The probability of fish presence was predicted from surface water pH using an empirical fish response model. These predictions assume that brook trout once existed in the lakes (Baker et al. 1988, 1990).[1]

Three data sets of lake water chemistry are used to evaluate the robustness of the assumption that the variability between the response of lakes to a hazard within a subregion is smaller than between subregions. The Eastern Lake Survey (ELS) (Linthurst et al. 1986) database provides a statistically derived population for lakes ranging from about 4 ha to 2000 ha. In ELS, 128 lakes occur in the Adirondacks. The Direct Delayed Response Project (DDRP) (Turner et al. 1990a) was developed within the statistical sampling frame of the ELS and was designed to project the long-term effects of specified levels of sulfur deposition on a sensitive subset of the ELS. Thirty-seven DDRP lakes occur in the Adirondacks. The Adirondack Lakes Survey (ALSC 1989) has 1280 lakes and is a census of

[1] Model development has been a major part of the 10-year NAPAP effort. Models have been extensively applied to assess the regional effects of acidic decomposition (NAPAP 1990; Turner et al. 1990b; Sullivan 1990; Baker et al. 1990; Thornton et al. 1990). In particular, watershed models have been used both alone and in combination with fish response models to project changes in water chemistry (Thornton et al. 1990) and in the suitability of waters for fish (Baker et al. 1990) resulting from deposition-driven changes in acid-base chemistry. These analyses have been performed for relatively large regions like the Adirondack Mountains.

Adirondack lakes 1 ha in size and larger. Using lake location, each lake in each database was assigned to its appropriate subregion, and cumulative frequency distributions for each subregion and each database were developed for the probability of fish being present.

Selection of Assessment Subregions

The influence of different databases (resource populations) and current and predicted data distributions for a subregion scheme are discussed with regard to selecting subregions for cumulative effects assessments. I use several existing subregion schemes to illustrate how one can evaluate the appropriateness or usefulness of subregions for an assessment. Regions are divided into subregions to improve the results of the assessment and provide geographic perspective for the policy maker. Three possible ways to divide the region (Adirondacks in the example) into subregions are illustrated using field data (Hunsaker et al. 1986) (Figures 3 through 5).[2]

Aquatic studies classically use the watershed as a physiographic unit of assessment. The Adirondacks can be divided into three large river basins (Figure 3). The Upper Hudson River and the Lake Ontario–St. Lawrence River basins are different with respect to fish presence for all three data sets. The proportion of lakes with a high probability of fish presence is much less for the Lake Ontario–St. Lawrence River basin. This probably results from the basin coinciding with the region of high sulfate deposition in the western Adirondacks. For the purposes of this example, differences in data distributions are determined in a qualitative manner; in actual practice, the cumulative frequency distributions would be drawn with confidence bands to determine statistical differences in distributions. High-elevation lakes tend to have lower pH values and thus are less likely to have trout present (Hunsaker et al. 1986). A useful subregion boundary occurs at the 600-m elevation contour for the DDRP and ELS databases (Figure 4); however, this relationship does not hold for the ALSC. Since the effect of acid deposition on lakes can be affected by soil processes, soil groups provide another logical group of subregions. The haplorthods-haplaquods and cryorthods-cryaquods show different patterns for the endpoint of fish presence for the ELS and the ALSC. The latter are cool soils with high elevations and less buffering capacity; thus, they have a high probability of having low pH lakes and no trout. Soils have complex patterns and the problem of having too many subregions for available data is also illustrated in Figure 5 with the DDRP database.

Data and subregion definition can both contribute to analytic uncertainty. The results of an effects model are likely to be different when different populations

---

[2] In Figures 3 through 5, pH and fish presence are presented as cumulative proportions. Cumulative frequencies have been converted to cumulative proportions to facilitate comparisons between data sets and subregions. The curves depict the proportion of lakes having a probability of pH or fish presence of x or less. To read a curve for a given subregion using a given data set, pick a value on the horizontal axis and read the proportion of lakes on the vertical axis with a probability of x or less.

CURRENT CONDITIONS

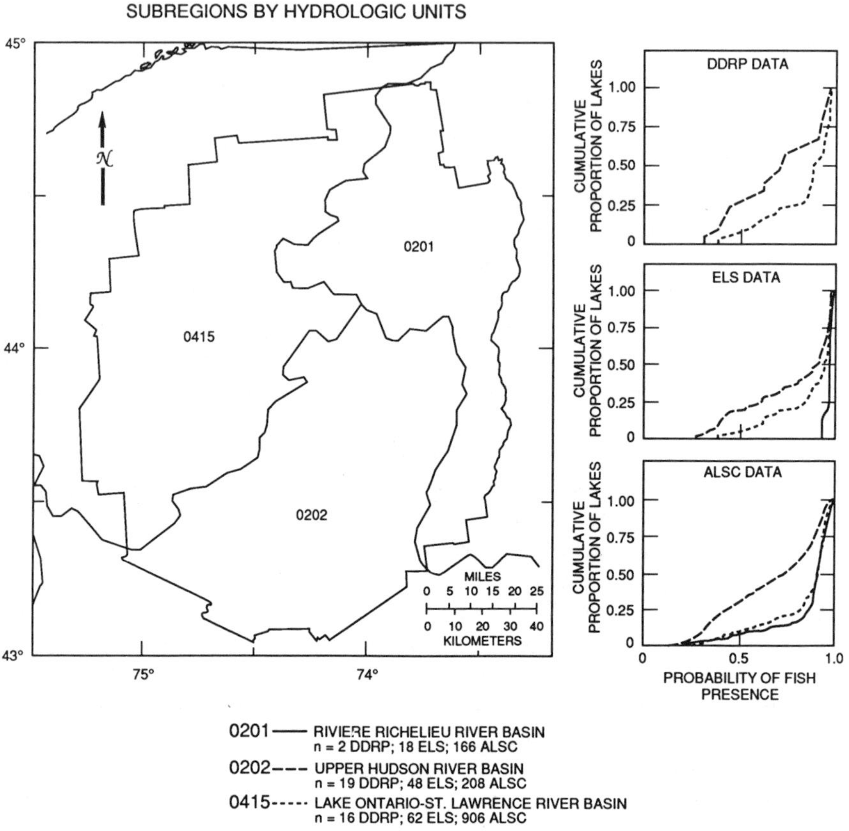

Figure 3. Current probability of brook trout being present in Adirondack lakes within watershed subregions according to major river basins.

of resources are used, such as the different data sets for lakes (ALSC, ELS, and DDRP). For example, using the river basins as subregions, the Henriksen model predicted fairly different subregional responses to deposition loads for each database (Figure 6). Model results using lake populations from ELS and ALSC predicted that lakes in the Lake Ontario basin would have a higher probability of fish being present under a 50% deposition reduction from current levels than lakes in the other two basins. When DDRP lakes were used, the model predicted the opposite. For current deposition, the model predicts the Lake Ontario basin to have the highest probability of fish presence for all the lake populations.

The subregion schemes captured spatial differences in response under different deposition scenarios with varying degrees of success. This point is illustrated using the Henriksen model and the ALSC database. This database contains the largest number of lakes, and the Henriksen model could be applied to this data

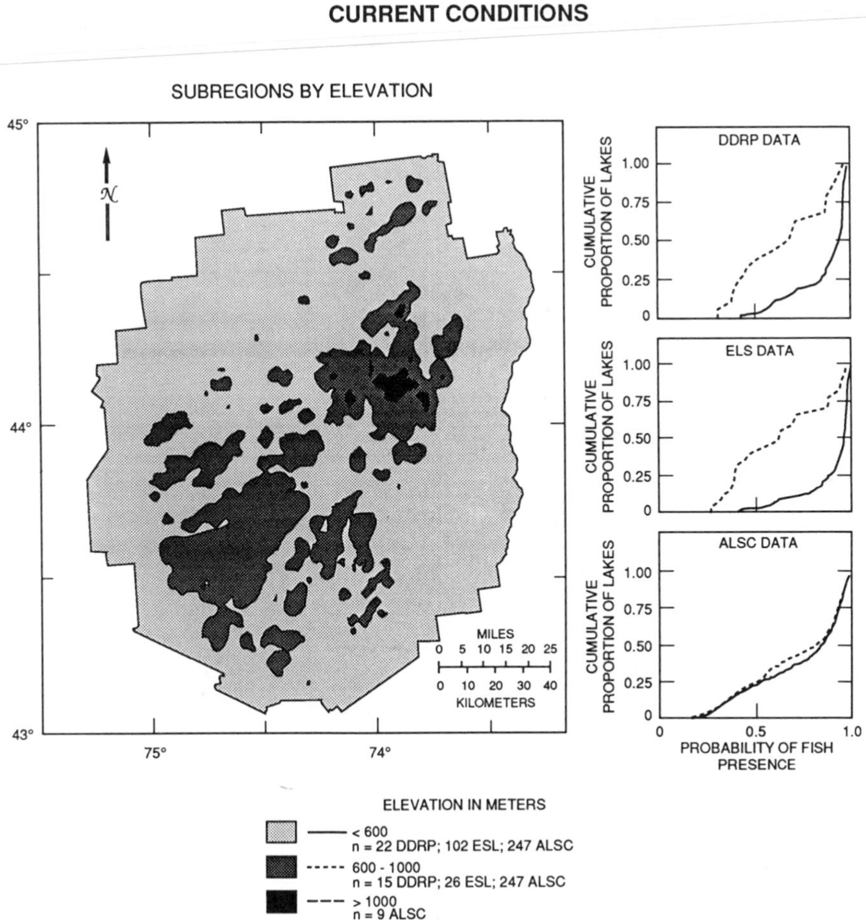

Figure 4. Current probability of brook trout being present in Adirondack lakes within elevation subregions.

set. The soil order and the river basin subregion schemes best captured spatial differences in response as shown by the vertical separation of the cumulative frequency curves for the subregions (Figure 7). In an actual assessment, confidence bounds would be calculated and graphed for the cumulative frequency distributions and would be used to determine if results were significantly different for different lake populations or subregion designations. For the elevation subregion scheme, lakes in the low- and medium-elevation subregions had very similar response patterns to deposition scenarios for the Henriksen model and the ALSC database. This lack of distinction is supported by the field data for the ALSC; however, field data for ELS and DDRP show a distinction between elevation subregions. If resources in a subregion are not responding differently to an exposure than resources in an adjacent subregion, there may be no reason

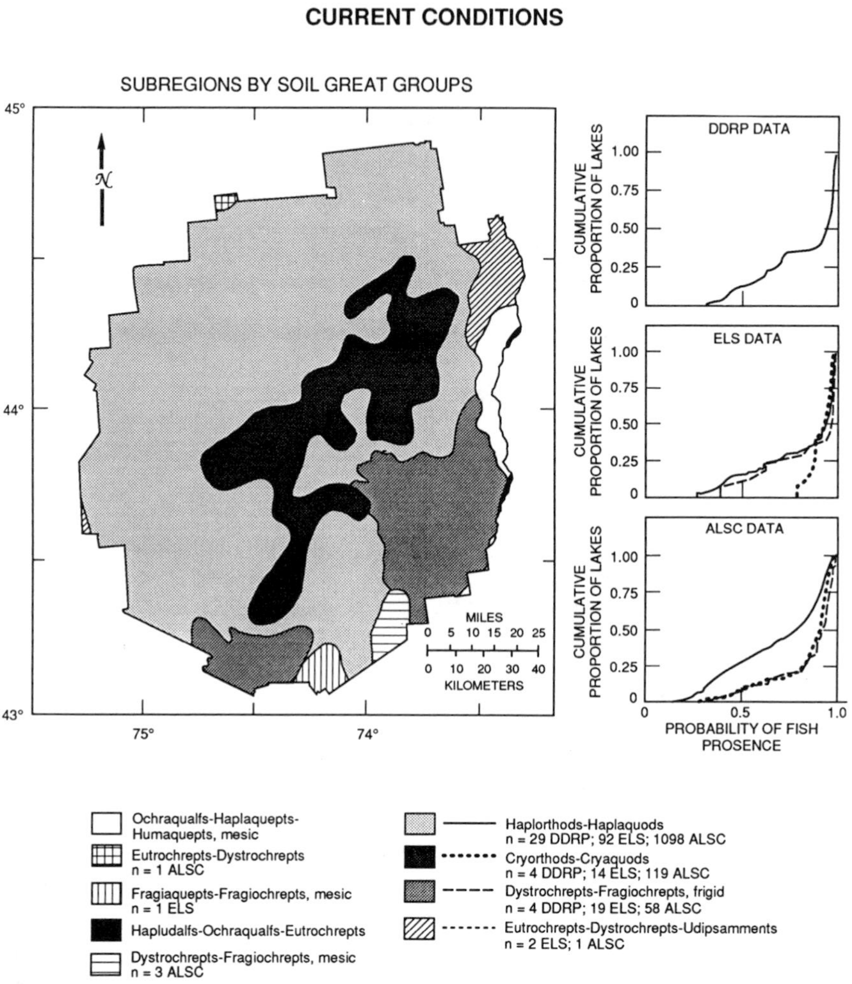

Figure 5. Current probability of brook trout being present in Adirondack lakes within soil subregions. Cumulative frequencies are not shown for subregions with less than ten lakes.

to keep separate subregions. Thus, one can conclude that selection of appropriate subregions for an assessment can differ depending on the databases and data type (field monitoring vs model predictions) used. Confidence can be increased in a subregion scheme if it is supported by both field data and model predictions.

DISCUSSION AND CONCLUSIONS

The approach as outlined by Hunsaker et al. (1990) for regional ecological risk assessment is useful for scoping and performing both cumulative and

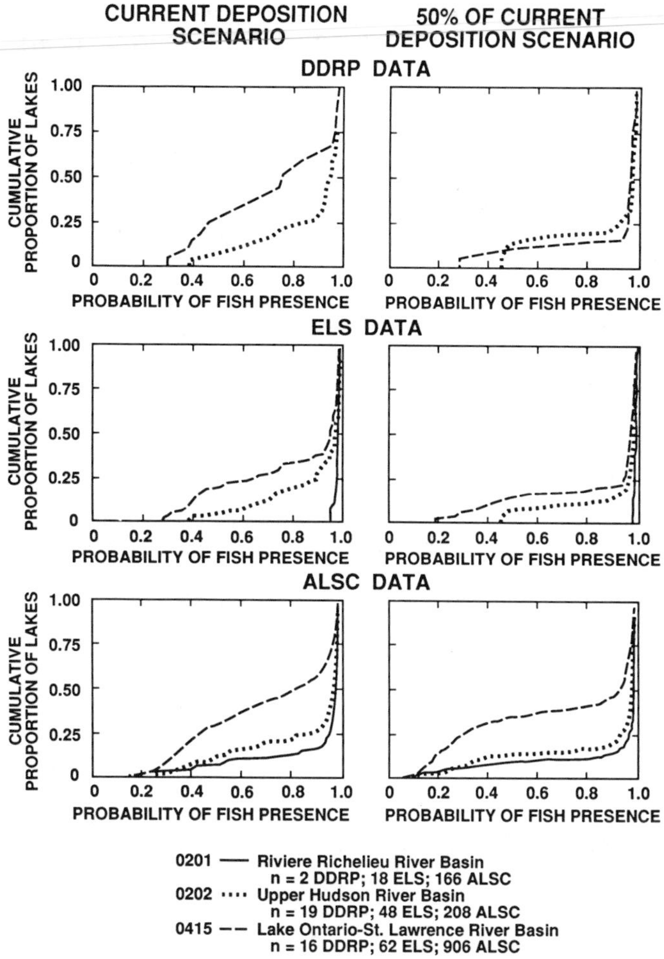

Figure 6. Henriksen model results for presence of brook trout with current deposition and reduced deposition for three databases. Lakes were assigned to river basin. Cumulative frequencies are not shown for subregions with less than ten lakes.

programmatic effects assessments at the regional scale. The definition of the assessment region and subregions is an important component of the assessment process. As shown by the Adirondack examples, different data sets may suggest somewhat different subregion schemes. Defining regions and subregions can improve the assessment by giving policy makers a geographic context and by capturing the spatial variability of endpoint responses. The use of ecologically functional subregions should improve the cost-benefit ratio for control and the accuracy of the sensitivity predictions by fine tuning effects models. Even a logical subregion scheme is not useful if it is so complex that sample size within subregions becomes too small for statistical confidence. Of course, the risk assessment approach stresses quantification of effects and uncertainty; such

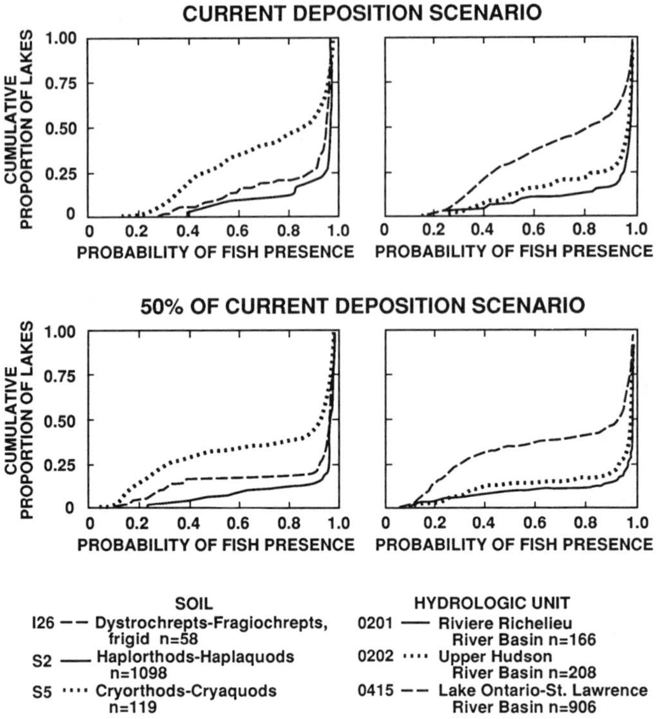

Figure 7. Henriksen model results for presence of brook trout with current deposition and reduced deposition. Lakes were assigned to soil and river basin subregions. The number of lakes used was 1280 from the Adirondack Lakes Survey Corporation (ALSC) database. Cumulative frequencies are not shown for subregions with less than ten lakes.

quantification should be our ultimate goal because it will provide policy makers and the public with an objective way to make decisions when cumulative effects are involved.

It is always a challenge to present in an understandable manner the analyses and results of complex assessments to policy makers and the general public. This task is only exacerbated for regional and cumulative assessments where large amounts of data, large geographic areas, and more quantitative methods are the norm. I believe that dose-response curves, cumulative frequency distributions, and maps are very important tools for illustrating cumulative effects analyses. Tools that will improve our ability to perform regional and cumulative assessments include geographic information systems, improved application of remote sensing data, and landscape indices that capture landscape patterns relevant to ecological processes (O'Neill et al. 1988). The availability of integrated databases is one of the factors most hindering our ability to perform these assessments. A recent emphasis on ecological monitoring at the regional and national scales (Hunsaker and Carpenter 1990) and revisions to national monitoring programs (Hirsch,

Alley, and Wilber 1988) are an encouraging sign that such databases may exist in the future. Both consistent and comprehensive long-term monitoring are needed at the correct scales for cumulative effects assessments.

As outlined in Figure 1, there is a logical spatial hierarchy that is sometimes neglected in the preparation of impact assessments. As Myslicki (this volume, Chap. 5) comments, "many times a programmatic EIS is the only place that impacts across diverse geographic areas have the opportunity to be considered." Effects assessments for programmatic and cumulative assessments addressing a large geographic area share some common goals and needs, and a regional ecological risk assessment approach is suitable for these assessments.

Control of impacts often occurs at the local scale. Thus, if cumulative impacts are best addressed at the regional scale, we must prevent the EIS process from becoming an analysis without context. This can be achieved by regional and programmatic planning as performed by associations of city and county governments, river basin commissions, state planning activities, national monitoring and assessments, and follow-up audits of NEPA documents.

ACKNOWLEDGMENTS

The author wishes to thank her many coworkers for their contributions to the projects that provided the basis and impetus for this paper. A special thanks goes to Robin Graham, Glenn Suter, Larry Barnthouse, and Glenn Cada of the Environmental Sciences Division at Oak Ridge National Laboratory. I acknowledge the work of the many scientists who contributed to the U.S. Critical Loads project and the National Acid Precipitation and Assessment project. I especially thank J. W. Elwood for runs of Henriksen's model, J. P. Baker for use of her brook trout model, and P. Ringold and G. R. Holdren for project management. The use of chemistry data is also appreciated: Adirondack Lake Survey Corporation and U.S. Environmental Protection Agency (EPA) Direct/Delayed Response Project and Eastern Lake Survey. Research was sponsored jointly by the EPA under Interagency Agreement 1824-B014-A7 and the Environmental Sciences Division, Office of Health and Environmental Research, U.S. Department of Energy, under contract DE-AC05-84OR21400 with Martin Marietta Energy Systems, Inc. Although this research was partially funded by EPA, it has not been subjected to EPA review and, therefore, does not necessarily reflect the views of EPA and no official endorsement should be inferred.

REFERENCES

Adirondack Lakes Survey Corporation (ALSC). 1989. Adirondack lakes study (1984–1987): An Evaluation of Fish Communities and Water Chemistry. ALSC, Albany, NY.

Baker, J. P., C. S. Creager, S. W. Christensen, and L. Godbout. 1988. Identification of Critical Values for Effects of Acidification on Fish Populations. Report No. I296-I22-7/31/88-01F. U.S. Environmental Protection Agency, Washington, D.C.

Baker, J. P., D. P. Bernard, S. W. Christensen, M. J. Sale, J. Freda, K. Heltcher, D. Marmorek, L. Rowe, P. Scanlon, G. Suter, W. Warren-Hicks, and P. Welbourn. 1990. Biological Effects of Changes in Surface Water Acid-Base Chemistry. State-of-Science/Technology Report 13. National Acid Precipitation Assessment Program, Washington, D.C.

Cada, G. F., and C. T. Hunsaker. 1990. Cumulative impacts of hydropower development: Reaching a watershed in impact assessment. *Environ. Prof.* 12:2–8.

Canadian Environmental Assessment Research Council and the U.S. National Research Council (eds.). 1986. Cumulative Environmental Effects: A Binational Perspective. Canada Environmental Assessment Research Council, Hull, Quebec.

Council on Environmental Quality (CEQ). 1978. Regulations for Implementing the Procedural Provisions of the National Environmental Policy Act. 40 CFR 1500-08. CEQ, Washington, D.C.

Convention on Long-Range Transboundary Air Pollution, Task Force on Mapping Critical Levels/Loads (CLRTAP). 1989. Methodologies and Criteria for Mapping Critical Levels/Loads and Geographical Areas Where They Are Exceeded, Draft Manual. Workshop on Mapping, November 6–9, 1989. Bad Harzburg, Federal Republic of Germany.

de Blij, H. J. 1978. *Geography: Regions and Concepts. Introduction.* John Wiley & Sons, New York.

Federal Energy Regulatory Commission (FERC). 1985a. Procedures for assessing hydropower projects clusters in river basins. Request for comments. Docket No. EL85-19-000. *Fed. Regist.* 50(16):3385–3403.

Federal Energy Regulatory Commission (FERC). 1985b. Final Environmental Impact Analysis of Small-Scale Hydroelectric Development in Selected Watersheds in the Upper San Joaquin River Basin, California. FERC/EIA-0001. Office of Hydropower Licensing, Washington, D.C.

Federal Energy Regulatory Commission (FERC). 1986. Owens River Basin, Seven Hydroelectric Projects — California. Final Environmental Impact Statement. FERC/EIS-0041. Office of Hydropower Licensing, Washington, D.C.

Federal Energy Regulatory Commission (FERC). 1988a. Hydroelectric Development in the Upper Ohio River Basin. Final Environmental Impact Statement. FERC/EIS-0051. Office of Hydropower Licensing, Washington, D.C.

Federal Energy Regulatory Commission (FERC). 1988b. Public Utility Regulatory Policies Act Benefits at New Dams and Diversions. Final Staff Report Evaluating Environmental and Economic Effects. Docket No. EL87-9. Office of Hydropower Licensing, Washington, D.C.

Henriksen, A. 1984. Changes in base cation concentrations due to freshwater acidification. *Verh. Internat. Verein. Limnol.* 22:692–698.

Hirsch, R. M., W. M. Alley, and W. G. Wilber. 1988. Concepts for a National Water-Quality Assessment Program. U.S. Geological Survey Circular 1021, Washington, D.C.

Hunsaker, C. T., S. W. Christensen, J. J. Beauchamp, R. J. Olson, R. S. Turner, and J. L. Malanchuk. 1986. Empirical Relationship Between Watershed Attributes and Headwater Chemistry in the Adirondack Region. ORNL/TM-9838. Oak Ridge National Laboratory, Oak Ridge, TN.

Hunsaker, C. T., and D. E. Carpenter (eds.). 1990. Ecological Indicators, Environmental Monitoring and Assessment Program. EPA 600/3-90/060. U.S. Environmental Protection Agency, Office of Research and Development, Research Triangle Park, NC.

Hunsaker, C. T., R. L. Graham, G. W. Suter, R. V. O'Neill, B. L. Jackson, and L. W. Barnthouse. 1990. Regional ecological risk assessment. *Environ. Manage.* 14(3):325–332.

Linthurst, R. A., D. H. Landers, J. M. Eilers, D. F. Brakke, W. S. Overton, E. P. Meier, and R. E. Crowe. 1986. Characteristics of Lakes in the Eastern United States. Vol. 1: Population Descriptions and Physiochemical Relationships. EPA-600/4-86-007A. U.S. Environmental Protection Agency, Washington, D.C.

Myslicki, A. 1993. The use of programmatic EISs in support of cumulative impact assessment. In S. G. Hildebrand, and J.B. Cannon (eds.), *The Scientific Challenges of NEPA: Future Directions Based on 20 Years of Experience*. Chap. 5. Lewis Publishers, Inc., Chelsea, MI.

NAPAP (National Acid Precipitation and Assessment Program). 1990. 1989 Annual Report to the President and Congress and Findings Update. NAPAP, Washington, D.C.

O'Neill, R. V., J. R. Krummel, R. H. Gardner, G. Sugihara, B. Jackson, D. L. DeAngelis, B. Milne, M. G. Turner, B. Zygmutt, S. Christensen, V. H. Dale, and R. L. Graham. 1988. Indices of landscape pattern. *Landscape Ecol.* 1:153–162.

Sigal, L. L., and J. W. Webb. 1989. The programmatic environmental impact statement: Its purpose and use. *Environ. Prof.* 11:14–24.

Sullivan, T. J. 1990. Historical Changes in Surface Water Acid-Base Chemistry in Response to Acidic Deposition. State-of-Science Technology Report 11. National Acid Precipitation Assessment Program, Washington, D.C.

Thornton, K., D. Marmorek, and P. Ryan. 1990. Methods for Projecting Future Changes in Surface Water Acid-Base Chemistry. State-of-Science Technology Report 14. National Acid Precipitation Assessment Program, Washington, D.C.

Turner, M. G. 1989. Landscape ecology: The effect of pattern on process. *Annu. Rev. Ecol. Syst.* 20:171–197.

Turner, R. S., C. C. Brandt, D. D. Schmoyer, J. C. Goyert, K. D. Van Hoesen, L. J. Allison, G. R. Holdren, P. W. Shaffer, M. G. Johnson, D. A. Lammers, J. J. Lee, M. R. Church, M. L. Papp, and L. J. Blume. 1990a. Direct/Delayed Response Project Data Base Users' Guide. ORNL/TM-10369. Oak Ridge National Laboratory, Oak Ridge, TN.

Turner, R. S., R. B. Cook, H. V. Miegroet, D. W. Johnson, J. W. Elwood, O. P. Bricker, S. E. Lindberg, and G. M. Hornberger. 1990b. Watershed and Lake Processes Affecting Surface Water Acid-Base Chemistry. State-of-Science Technology Report 10. National Acid Precipitation Assessment Program, Washington, D.C.

Environmental Monitoring and Assessment: A National Priority

G. E. Saul and K. W. Thornton, FTN Associates, Ltd., Little Rock, AR; and
R. A. Linthurst, U.S. Environmental Protection Agency, Washington, D.C.

ABSTRACT

The environmental impact statement (EIS) process, promulgated to meet the requirements of the National Environmental Policy Act (NEPA) of 1969, has forced decision makers to consider environmental consequences of proposed legislation and regulation. Although the EIS process has been effective at maintaining or enhancing the quality of many local environments, we now face more serious levels of environmental deterioration from continuing, persistent, and cumulative impacts of pollution at regional, national, and global scales. Problems such as acid rain, nonpoint-source pollution, and global climate change require a new generation of environmental monitoring and assessment strategies to provide critically needed, policy-relevant ecological information to policy makers and decision makers.

The Environmental Monitoring and Assessment Program (EMAP), a comprehensive interagency, interdisciplinary effort initiated in 1990, provides a strategic approach to identify and assess the extent, magnitude, and location of degradation or improvement of environmental condition at regional and national scales. EMAP will monitor indicators of ecological condition in aquatic ecosystems (estuaries, wetlands, rivers and streams, lakes and reservoirs, Great Lakes), terrestrial ecosystems (agroecosystems, arid lands, forests), as well as air and deposition. Assessments, which integrate and analyze data collected within EMAP with other available information, will be used to identify possible causes of improving or deteriorating environmental condition. This information can be used by decision makers who must set environmental policy, environmental managers who must set priorities in research and monitoring programs, analysts who require an objective basis for evaluating the overall effectiveness of environmental policy, and the public who want to know about the condition of the nation's environment. EMAP's focus on the biosphere from a global perspective represents the next step in environmental monitoring in the United States. Because EMAP is fully consistent with NEPA, it provides a foundation for national environmental protection with international links to global monitoring and regulation.

INTRODUCTION

Since its enactment into law in January 1970, the National Environmental Policy Act (NEPA) of 1969 (Pub. L. 91-190) has been the action forcing mechanism compelling federal decision makers to consider environmental consequences of proposed legislation or regulations and to act to preserve and enhance the nation's environment in promulgating these regulations (Kennedy 1988). The environmental impact statement (EIS) process, promulgated to meet NEPA requirements, has had a tremendous impact on project evaluation, planning, and development. The EIS process has been effective in that some proposed actions with potentially significant adverse environmental impacts were never initiated or were stopped; many impacts have been ameliorated, and some impacts have been mitigated. This process has focused attention on the quality of many local environments.

The EIS process has clearly demonstrated the need for increased understanding of the structure and function of ecological systems and their interrelationships. During the first 20 years of NEPA, assessment research and technology have advanced greatly. We have significantly improved our knowledge of ecological processes and our ability to assess and predict site-specific environmental and ecological impacts of proposed actions. Despite these improvements, however, we still face continued environmental deterioration. Continuing, persistent, and cumulative impacts from pollution sources on larger scales appear to be gradually eroding the quality of our ecosystems at landscape, regional, and global scales. Increasingly, reports raise concerns about regional-, national-, and global-scale problems such as acid rain, nonpoint-source pollution, and global climate change.

Environmental protection programs have been estimated to cost more than $70 billion annually (ES&T 1989), yet the means to assess their effects on the environment over the long term are only now being developed. Addressing this new generation of environmental problems requires a major shift in the overall approach to environmental protection and in the strategy for developing the knowledge necessary to design and implement environmental protection programs. Assessing whether the condition of the nation's ecological resources is improving or degrading requires data on large geographic scales over long time frames. Comparability of data among geographic regions and over extended time periods is critical. Meeting this need by simply aggregating data from many individual, local, and short-term networks has proven difficult, if not impossible (Suter et al. 1988). Site-specific approaches generally are inadequate for assessing the magnitude, extent, current status, and trends in ecological condition at regional or national scales.

New approaches are needed to assess the impact of proposed regulatory programs on the environment at these scales. Local perturbations differ from regional, national, and global perturbations with respect to temporal scales and dominant processes (O'Neill et al. 1986). Differences in assessment objectives, methods, measurement resolution, indicator selection, and sampling extent make the use of existing information systems for assessing large-scale problems difficult. Establishing baseline conditions against which future changes can be documented with confidence has become a national priority with the increasing complexity, scale, and social importance of existing and emerging environmental issues (Likens 1983).

Recognizing the broad base of support for better environmental surveillance, the U.S. Environmental Protection Agency's (EPA) Science Advisory Board (SAB) recommended in 1988 that a program be implemented within the EPA to monitor ecological status and trends and to develop methods for anticipating emerging problems before they reach crisis proportions (U.S. EPA 1988). The EPA was urged to become proactive in ecological monitoring because its unique regulatory role requires that quantitative scientific assessments be made of the effects of pollutants that travel through multiple media and ecosystems (U.S. EPA 1988).

ENVIRONMENTAL MONITORING AND ASSESSMENT PROGRAM

In response to the recommendations by the SAB, Congress, and the public, the EPA is designing the Environmental Monitoring and Assessment Program (EMAP). EMAP is intended to be an umbrella program under which the EPA can work as part of an interagency effort (i.e., federal, state, local, and private agencies) to monitor and assess the extent, magnitude, and location of degradation or improvement in environmental condition at national, regional, and subregional scales. Simultaneous monitoring of pollutant exposure, environmental condition, and change will permit the identification of possible causes of deteriorating or improving ecological conditions. When fully implemented, EMAP will answer the following critical questions for policy and decision makers and the public:

- What is the current status, extent, and geographic distribution of our ecological resources (e.g., estuaries, lakes, streams, wetlands, forests, grasslands, arid lands)?
- What proportions of these resources are degrading or improving, where, and at what rate?
- To what levels of environmental stress and pollutants are these ecological resources exposed?
- What are the possible causes of adverse or improving conditions?
- What ecological resources are at risk?

- Are adversely affected ecosystems responding as expected to control and mitigation programs?

EMAP will provide the EPA Administrator, the Congress, and the public with statistical data summaries and periodic interpretive reports on status and trends in indicators of ecological condition. Because sound decisionmaking must consider the uncertainty associated with quantitative information, all EMAP status and trends estimates will include statistically based uncertainty statements.

General Approach

EMAP will monitor indicators of ecological conditions in near-coastal, wetlands, inland surface waters, and terrestrial ecosystems, as well as indicators of atmospheric stress. Pollutant exposure and ecological response indicators to be monitored include those related to multiple and cumulative pollutant interactions, regional air pollution and acidic deposition, habitat loss and modification, nonpoint-source pollutant loadings to inland surface waters and estuaries, and changes in radiant energy inputs and climate.

Environmental monitoring data are collected by the EPA to meet the requirements of a variety of regulatory programs. Many federal agencies collect environmental data specifically to manage particular ecological resources. Efficient execution of the EPA's mandate to protect the nation's ecosystems requires, therefore, that EMAP complement, supplement, and integrate data and expertise from the regulatory offices within EPA and from other resource management agencies. EMAP is intended to be the umbrella program for integrating data from other ongoing programs. Interagency coordination is actively being pursued with other federal agencies such as the Department of Interior, Department of Commerce, and Department of Agriculture. This coordination avoids duplicative monitoring efforts, facilitates exchange of existing and future data, and increases the expertise available to quantify and understand observed status and trends. EMAP also will draw upon the expertise and activities of the EPA Regional Offices, states, and the international scientific and academic community.

Ecological monitoring programs of the 1990s and beyond must be able to respond and adapt to new issues and perspectives in detecting trends and patterns in environmental change (Linthurst et al. 1986). These demands will be met by EMAP through a flexible design that can accommodate future questions and objectives, as well as changing criteria of performance and scientific capability. Further, EMAP's design will encourage analysis, review, and reporting processes that foster discovery of unanticipated results and promote the widespread dissemination of scientifically sound information. Periodic evaluations of the program's direction and emphasis will be key in maintaining its viability and relevance, while retaining the continuity of the basic data sets. These evaluations will serve to preclude the "aging" that typically hinders long-term monitoring efforts.

Planning and Design

The major 1990–1991 activities around which EMAP is being developed are

- evaluation and testing of indicators of ecological condition; design and evaluation of integrated statistical monitoring networks and protocols for collecting status and trends data on indicators
- nationwide characterization of ecological resources to establish a baseline for monitoring and assessment
- implementation of regional-scale surveys to define the current status of our estuarine and forest resources

Although EMAP ultimately will monitor ecosystems in all categories, early emphasis is being placed on testing and implementing the program in estuaries, inland surface waters, wetlands, and forests. Because precipitation and air quality are two important factors influencing ecosystems, EMAP will contribute to the evaluation and maintenance of the atmospheric deposition network (i.e., the National Trends Network/National Dry Deposition Network, Sisterson et al. 1990). These ecosystems and deposition networks offer immediate opportunities to demonstrate the EMAP approach.

Indicator Evaluation and Testing

EMAP will evaluate and monitor indicators that describe the overall condition of an ecosystem. In order to provide useful information, EMAP must monitor ecological conditions in a way that has value for decision makers and the public. Indicators, therefore, must reflect ecological characteristics clearly valued by society. Measurement methods must be standardized and quality assured so that spatial patterns and temporal trends in condition within and among regions can be accurately assessed (Schaeffer et al. 1988). Three categories of indicators will be evaluated: response, exposure-habitat, and stressor (Hunsaker and Carpenter 1990) (Figure 1).

Response indicators are measurements that quantify the overall biological condition of ecosystems by measuring either organisms, populations, communities, or ecosystem processes as they relate to endpoints of concern. Examples include signs of gross pathology (e.g., the appearance of tumors in fish or visible damage to tree canopies); the status of organisms that are particularly sensitive to pollutants or populations of species important to sportsmen, commercial interests, or naturalists; and indices of community structure and biodiversity.

Exposure-habitat indicators are physical, chemical, and biological measurements that reflect pollutant exposure, habitat degradation, or other causes of poor ecosystem condition. Examples include ambient pollutant concentrations, acidic deposition rates, bioaccumulation of toxics in plant and animal tissues, ionizing or electromagnetic radiation, and measurements of habitat condition or availability (e.g., siltation of bottom habitat and vegetative canopy complexity).

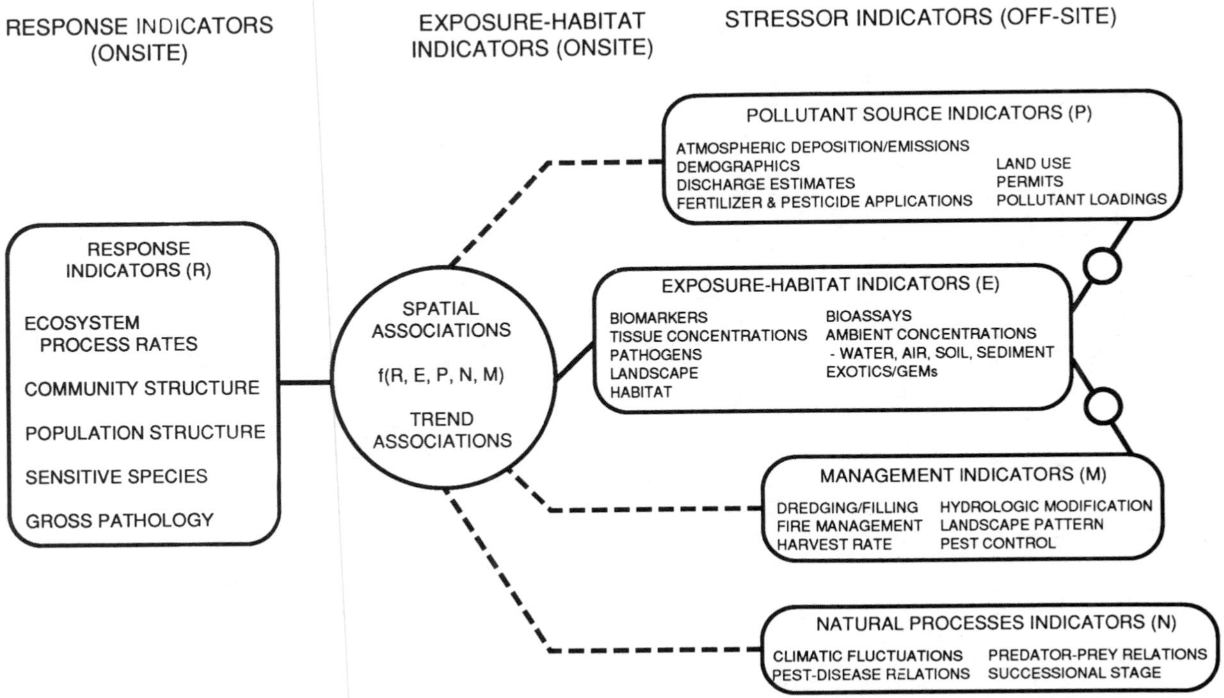

Figure 1. Conceptual diagram of EMAP indicator relationships. The circles indicate that analysis is by statistical association, rather than explicit (causal) mathematical relationships. (From Hunsaker, C. T., and D. E. Carpenter, Eds. 1990. Environmental Monitoring and Assessment Program: Ecological Indicators. EPA 600/3-90/060. U.S. Environmental Protection Agency, Washington, D.C.)

Stressor indicators are measurements of pollutant sources, management actions, and natural processes that lead to changes in pollutant exposure and habitat condition. Examples include coal production, changing population densities, pesticide applications, pollutant emission inventories, and land use.

The strategy is to measure suites of indicators in all categories for each ecosystem type. The suite of response indicators should reflect adverse effects of both anticipated and unanticipated environmental stresses (e.g., new pollutants). Criteria must be developed for each response indicator to identify when conditions change from acceptable or desirable to unacceptable or undesirable. Criteria could be based on attainable conditions under "best management practices" as observed at regional reference sites — that is, relatively undisturbed sites that are typical of an ecoregion (Hughes et al. 1986).

A suite of exposure indicators will be used to determine whether ecosystems have been exposed to environmental stress and what might be the possible causes of poor condition. For example, undesirably low diversity in stream fish communities across a region might be related to the presence of toxics in sediments, siltation of bottom habitat, insufficient flow, low pH, or bioaccumulation of toxics. In this example, stressor indicators that might be examined in diagnosing the cause could include the number and type of industrial discharges, farmed acreage or construction activity, water withdrawals, presence of mine spoils or acidic deposition, and regional pesticide applications. Because of the scale of EMAP, diagnosis and identification of stressors are not aimed at individual sources, but at populations of sources that reflect stress and exposure on a regional basis.

The focus on biotic indicators to assess condition rather than pollutant sources or ambient concentrations is unusual in environmental monitoring and reflects the unique goals of EMAP. While compliance monitoring and remediation involve identifying, with a high degree of confidence, polluting activities and pollutant concentrations that can be linked unequivocally to individual polluters, EMAP will use suites of indicators to assess the conditions of multiple ecological systems across regions, coupled with an evaluation of associated pollutant sources and other anthropogenic environmental disturbances. Expected benefits of EMAP's regional approach to environmental monitoring include improved capabilities to detect emerging problems and to identify ecological resources most in need of additional research, assessment, or remediation. Regional monitoring and assessment will play a critical role in determining whether current environmental regulations are adequately protecting the nation's ecological resources.

Network Design

EMAP has been designed to provide statistically reliable estimates of ecological status and trends at regional and national levels with known certainty. Meeting this goal has required the development of an integrated statistical monitoring framework that provided the basis for determining and reporting on ecological indicators at various geographic scales. The design is adaptable to monitoring on regional as well as continental and global scales, and it permits statistical correlation and association analyses to examine spatial and temporal patterns of response, exposure, and stressor indicators. Such a framework also enables the incorporation or substitution of data from existing monitoring sites and networks. The design is adaptable and flexible and can accommodate changes in the spatial extent of an ecological resource and address current and emerging issues (Overton et al. 1990).

A uniform, systematic grid has been constructed for identifying sampling sites (Figure 2). This grid was subdivided into planar subgrids to achieve the appropriate sampling density for the scale of resolution (e.g., national, regional, or subregional) desired for an assessment of the condition of ecological resources. Currently, a subgrid for the continental United States that includes approximately 12,600 sites is being evaluated. Ecological resources will be identified and characterized at each of these sites, and their number and areal extent will be determined (Figure 3). Field sampling of suites of indicators will be conducted on a random subset of sites. There will be a unique subset of field sites or sampling units associated with each ecosystem type during an appropriate index period to provide an annual census of ecological conditions.

The hierarchical nature of the grid frame allows several kinds of indicators in different ecosystem categories to be measured at multiple levels of resolution (Figure 4). Ecosystem distribution and extent, which can be mapped using satellite and aerial remote imagery, can be measured at a higher grid density (i.e., more sampling sites) because sampling costs are relatively low, thereby providing greater resolution for subregional-scale reporting. Field measurements, for which sampling costs are higher, will be collected at lower grid densities (i.e., fewer sampling sites); the hierarchical structure, however, provides enhanced statistical resolution for reporting on regional status and trends. Subregional issues concerning either geographically restricted categories (e.g., small estuaries or high-elevation forests) or special interest categories (e.g., lakes sensitive to acidic deposition) can be addressed by increasing the subregional grid density. The results from these high-interest, smaller-scale issues can be related to and reported within the context of the regional and national resource estimates.

Landscape Characterization

Landscape characterization is the documentation of the principal components of landscape structure (i.e., the physical environment, biological composition, and human activity patterns) in a geographic area. EMAP will characterize the national landscape by mapping and quantifying landscape features (e.g., wetlands, soils, land uses) in areas associated with EMAP sampling units. The use of remote sensing technology (e.g., satellite and aerial photography) supported by cartographic analysis and demographics will permit a quantitative estimation of various landscape features and, over time, changes in these features that might be related to human activities and pollutants.

Landscape characterization also will provide detailed information needed for the network design. For example, the different types of lakes, streams, wetlands, forests, and other resources associated with each grid point will be identified so

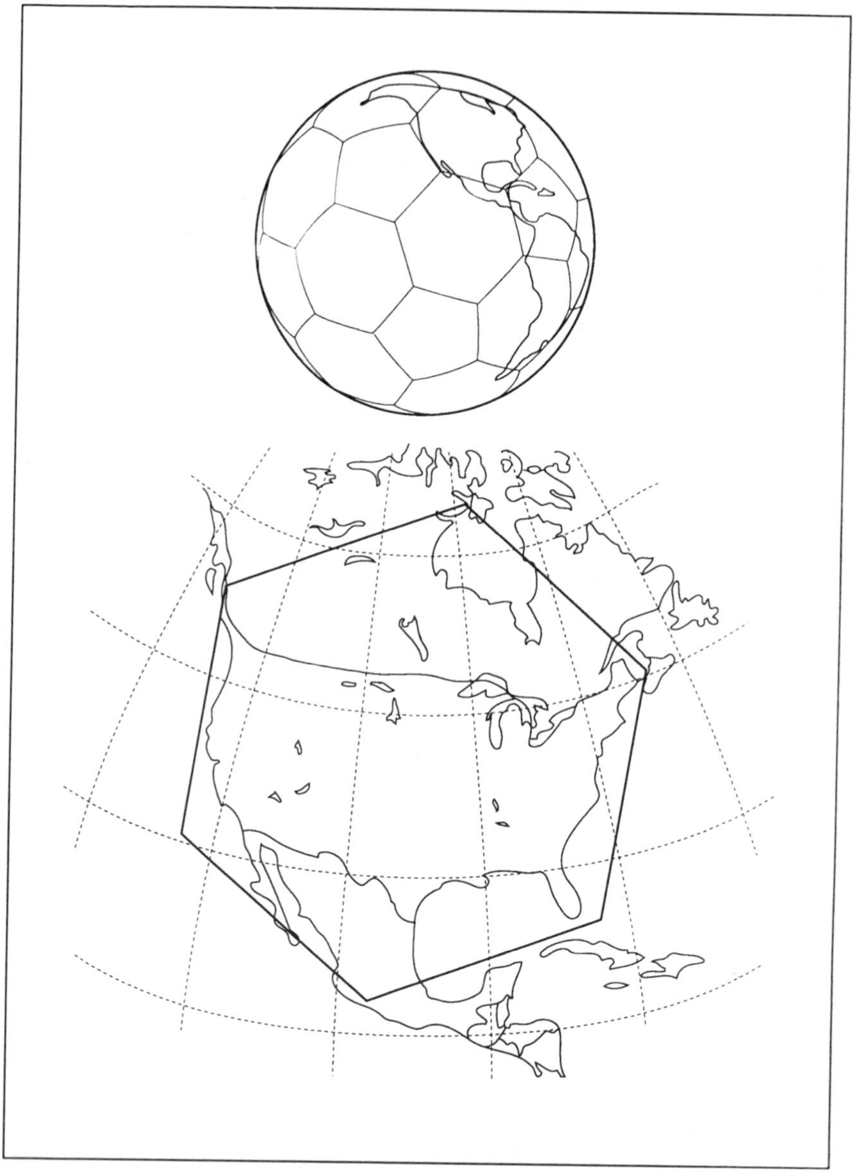

Figure 2. Projection of a truncated icosahedron model as a soccer ball pattern onto the earth with an enlargement of the projection onto the United States. (Adapted from Overton et al. 1990.)

that a subset of sampling units can be randomly selected for field sampling. Characterization activities will include the development of classification procedures and protocols so that ecological resources can be classified into various subcategories of interest (e.g., dominant forest stands, hydrologic lake types). Characterization will describe the physical and spatial aspects of the environment

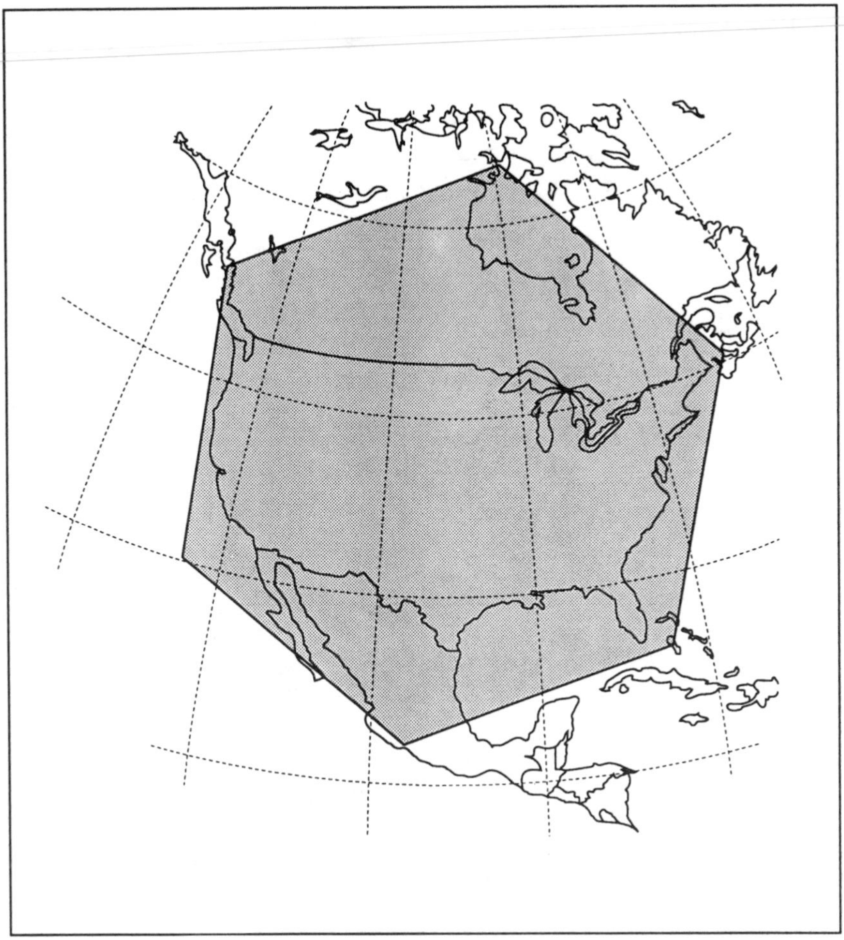

Figure 3. Approximate placement of proposed 12,600 sites in EMAP baseline sampling network for conterminous United States.

that reflect habitat modification. Examples include soil type, physiographic features, edge to interior ratios, contagion, and fractal dimension, which indicate landscape pattern (Turner 1989). For example, small patches of forest might be more susceptible to air pollutants or pesticides imported from adjacent agricultural land use than larger patches. Characterization also will compile data on stressor indicators that can be identified from remote sensing and mapped data, including land use, mining activities, population centers, transportation and power corridors, and other anthropogenic disturbance.

National assessments of status and trends of the condition of ecosystems involve knowing not only what percentage of a resource is in desirable or acceptable condition, but also how much of that resource exists. Certain aspects of landscape data are necessary to diagnose possible causes of undesirable condition in response

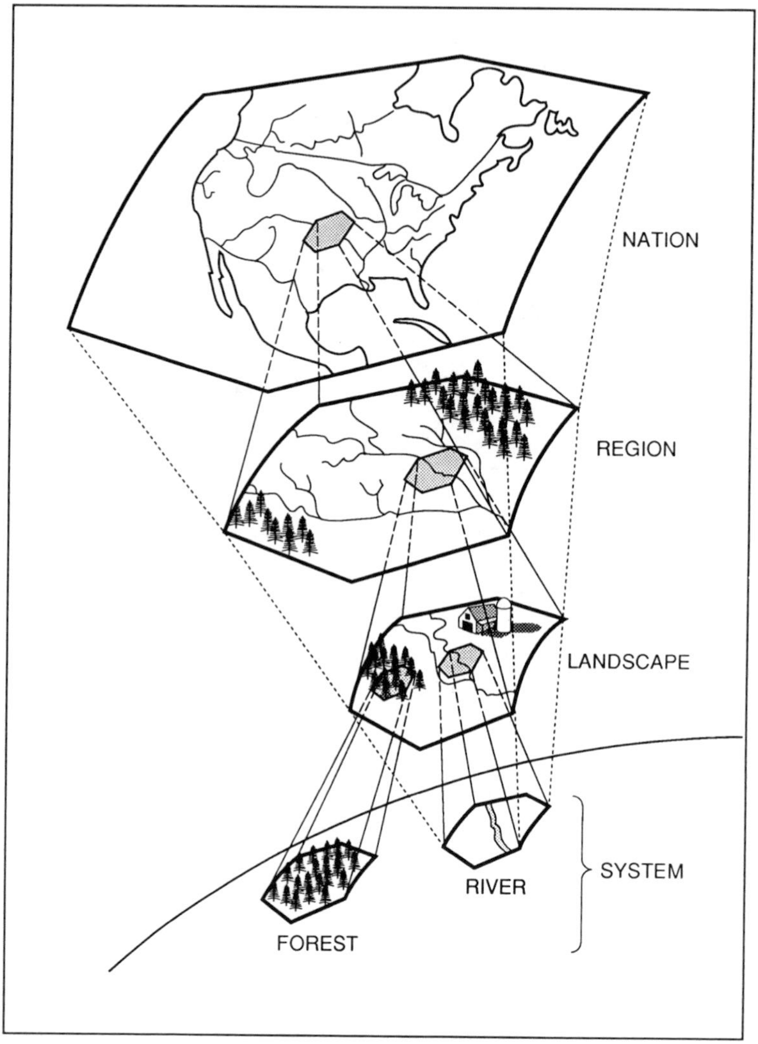

Figure 4. Hierarchial grid frame of EMAP sampling design permits measurements at multiple levels of resolution.

indicators. Some types of wetlands are being lost at an alarming rate; conversion and loss of other types of ecosystems also are occurring. Such changes may be of particular interest if correlated with pollutant exposure or other human-induced stressors. For most ecosystems, there are few, if any, comprehensive national databases from which quantitative estimates of ecosystem extent and changes in condition can be made with known confidence.

EMAP will compile, manage, and update these data in Geographic Information System format. A standardized characterization approach and a landscape information network common to all ecosystems will be used to optimize cost and

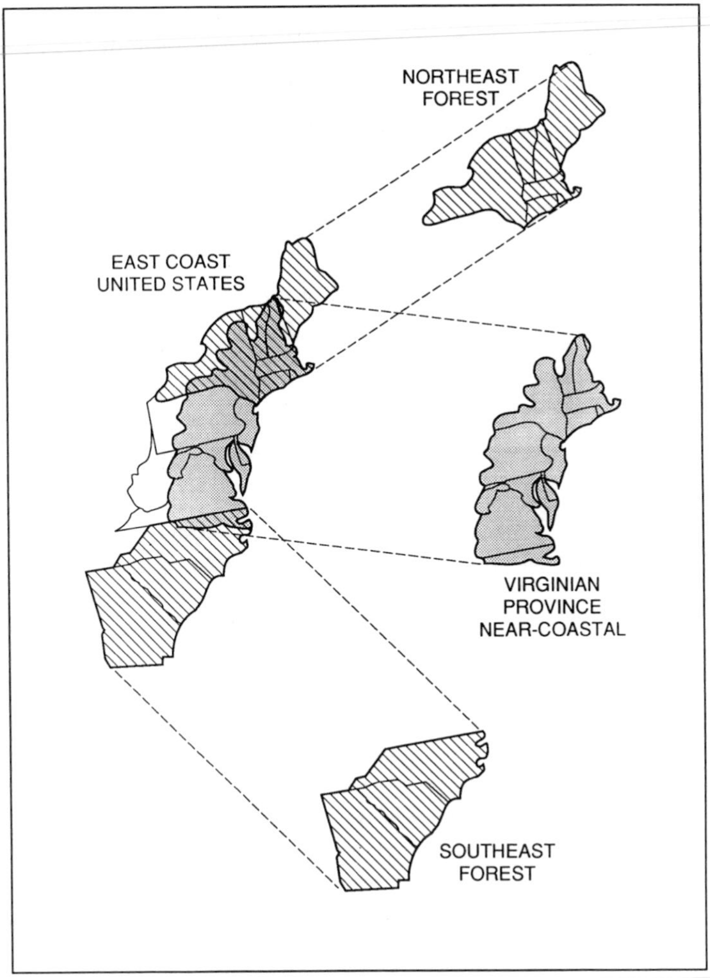

Figure 5. Northeastern and southeastern forested regions (diagonal) and Virginian Province near-coastal region (stipple) in which Fiscal Year 1990 demonstration projects will be conducted.

data sharing and to ensure common format and consistency. Through interagency collaboration, EMAP will establish design requirements for the integrated characterization database, including acceptance criteria for baseline data, consistent classification detail and accuracy, and suitable spatial and temporal resolution to distinguish landscape attributes of particular interest.

Demonstration Project

Demonstration projects will be conducted in northeastern and southeastern forests and near-coastal estuaries in the Virginian Province in Fiscal Year 1990

(Figure 5). Information obtained from these demonstration projects will be used to refine the EMAP sampling design, and the studies themselves will serve as a model for implementing EMAP in other study areas and types of ecosystems.

These demonstration projects have five goals:

- Evaluate the utility, sensitivity, and applicability of the EMAP indicators on a regional scale.
- Determine the effectiveness of the EMAP network design for quantifying the extent and magnitude of poor ecological condition in these environments.
- Demonstrate the usefulness of results for planning, priority setting, and determining the effectiveness of pollution control actions.
- Develop standardized methods for indicator measurements that can be transferred to other study areas and made available for other monitoring efforts.
- Identify and resolve logistical issues associated with implementing the network design.

The strategy for accomplishing the above goals is to field test the EMAP indicators and network design through a demonstration study in northeastern and southeastern forests and in the estuaries of the Mid-Atlantic area of the United States. The projects are being cooperatively conducted with the (USDA) Forest Service's Forest Health Monitoring Program and the National Oceanic and Atmospheric Administration's (NOAA) National Status and Trends Program, respectively. The northeastern and southeastern forests were selected because of the importance of forest resources in these two regions and the concern over the possible effects of criteria air pollutants and to coincide with the initiation of the USDA Forest Health Monitoring program in these two regions. The Mid-Atlantic estuarine study area was chosen because adverse pollutant impacts are evident; contaminants are present in the water, sediments, and biota; the vitality of many living resources are threatened; and seven of the area's larger estuaries are included in the EPA's National Estuary Program and NOAA's Status and Trends Program.

FUTURE ENVIRONMENTAL MONITORING

Ecological monitoring in the 1990s and beyond must evolve to address the new generation of cumulative and persistent environmental concerns occurring at regional, national, and global scales. EMAP represents a long-term commitment to periodically assess and document the condition of the nation's ecological resources. EMAP satisfies a national need and a national priority for environmental information and is fully consistent with the purposes, mandate, and data needs of the NEPA process, especially when actions involving large spatial scales are proposed. EMAP will provide critically needed information to decision makers for the setting of environmental policy. Program managers may use EMAP information to assign priorities in research and monitoring efforts; scientists can focus research to broaden our general understanding of ecosystem types; environmental managers and analysts can use EMAP data to objectively evaluate the effectiveness of the nation's environmental policies. The coupling of EMAP data with regional or national socioeconomic data will allow for the preparation of multidimensional, large-scale environmental impact assessments. EMAP's focus on the biosphere from a global perspective represents the future of environmental monitoring in the United States and, by being fully consistent with NEPA, provides the foundation for national environmental protection with international links to global monitoring and regulation.

REFERENCES

ES&T. 1989. Environmental index. *Environ. Sci. Technol.* 23:1067.

Hughes, R., D. Larsen, and J. Omernik. 1986. Regional reference sites: A method for assessing stream potentials. *Environ. Manage.* 10:629–635.

Hunsaker, C. T., and D. E. Carpenter, Eds. 1990. Environmental Monitoring and Assessment Program: Ecological Indicators. EPA 600/3-90/060. U.S. Environmental Protection Agency, Washington, D.C.

Kennedy, W. 1988. Environmental impact assessment in North America, Western Europe: What has worked, where, how, and why? *Int. Environ. Rep. (BNA)*, pp. 257–262.

Likens, G. E. 1983. A priority for ecological research. *Bull. Ecol. Soc. Am.* 64:234–243.

Linthurst, R. A., K. Thornton, P. Kellar, and D. Landers. 1986. Long-term monitoring of acidification trends in lakes: A regional perspective. pp. 6–16. In *Monitoring and Managing Environmental Impact: American and Soviet Perspectives.*

O'Neill, R. V., D. L. DeAngelis, J. B. Waide, and T. F. H. Allen. 1986. *A Hierarchical Concept of Ecosystems.* Princeton University Press, Princeton, NJ.

Overton, W. S., D. L. Stevens, C. B. Pereira, D. White, and T. Olsen. 1990. Design Report for EMAP, Environmental Monitoring and Assessment Program. Draft Report. Environmental Research Laboratory — Corvallis. EPA, Corvallis, OR.

Schaeffer, D. J., E. E. Herricks, and H. W. Kerster. 1988. Ecosystem health: I. Measuring ecosystem health. *Environ. Manage.* 12:445–455.

Sisterson, D. L., V. C. Bowersox, A. R. Olsen, T. P. Meyers, and R. L. Vong. 1990. Deposition Monitoring: Methods and Results State-of-the-Science/Technology Report No. 6. National Acid Precipitation Assessment Program (in press).

Suter, G. W., II, L. W. Barnthouse, and R. V. O'Neill. 1988. Treatment of risk in environmental impact assessment. *Environ. Manage.* 11:295–303.

Turner, M. G. 1989. Landscape ecology: The effect of pattern on process. *Annu. Rev. Ecol. Syst.* 20:171–197.

U.S. EPA. 1988. Future Risk: Research Strategies of the 1990s. SAB-EC-88-040. Science Advisory Board. U.S. Environmental Protection Agency, Washington, D.C.

CHAPTER 7

NEPA Follow-up

Introduction and Summary

D. B. Hunsaker, Jr., Oak Ridge National Laboratory, Oak Ridge, TN; and Allan Hirsch,[1] Midwest Research Institute, Falls Church, VA

Reviewing the National Environmental Policy Act (NEPA) experience over the last 20 years, it is clear that one of its weakest aspects has been what happens after an environmental assessment (EA) or environmental impact statement (EIS) has been completed (i.e., postproject follow-up). The focus of attention to date has been on preparing documents. There has been debate and controversy involving agencies, research scientists, environmental organizations, and the courts concerning what constitutes an adequate NEPA document. The President's Council on Environmental Quality (CEQ) and the individual agencies have developed regulations and guidance that deal with document preparation, often in great detail. However, there are relatively few specific procedures or guidelines with respect to postproject monitoring and inspection. Postproject follow-up is a very important component of the process, and for several reasons, as discussed below, it needs to be strengthened if NEPA's full environmental benefits are to be realized.

First, postproject monitoring is necessary to assure that the environmental safeguards and mitigation measures adopted in the finding of no significant impact (FONSI) or record of decision (ROD) will actually be carried out satisfactorily. A NEPA document that bases its prediction of ultimate environmental impact on the assumption that a program of mitigation will be carried out can have little validity unless there is some assurance that the specific measures identified will actually be implemented and maintained. Considerable evidence exists that typically such assurance does not exist. As one review of federal agency

[1] Deceased.

postproject activities points out, "It is largely an act of faith to categorically accept that environmental mitigation — however well intentioned or well planned in the EIS — will actually be brought to fruition during the development of the approved project. It is equally an act of faith to presume that agencies will devote as much time and talent to inspecting and enforcing environmental mitigation measures as they do in carrying out the project" (Bassin and West 1986).

There are exceptions to this general observation, as the papers in this chapter on the I-476 Environmental Monitor and on the Tennessee Valley Authority's (TVA) postproject process illustrate. However, unfortunately, such programs are not typical examples; they are the exceptions. We are not suggesting here that environmental stipulations and procedures in the FONSI, EIS, or ROD are systematically provided for by the responsible agencies. As the I-476 experience points out, it may take intervention by an outside agency, by citizens, or environmental groups to assure compliance.

There is a second and quite different reason for conducting postproject analyses; we need feedback with respect to the adequacy and accuracy of preproject predictions if we are to learn from experience. This aspect of postdocument work is probably even more neglected than is compliance monitoring. The emphasis in the first two decades of NEPA implementation has been on supporting the studies necessary to make predictions concerning the impacts of future projects, but not looking back to check the accuracy of those predictions. It is most difficult to get funding for retrospective analyses. There are few incentives for an agency to examine its predictions and assumptions and demonstrate that they were wrong.

Yet we all know that environmental predictions are often imprecise. Many predictions in EISs are, at best, qualitative in nature. Some of these shortcomings stem from the inherent difficulty of forecasting situations involving many variables within the natural environment and the current state of the science. Others relate to uncertain social and economic changes resulting from project development that may lead to second- and third-order effects with even greater consequences than the primary effects.

Sometimes, the environmental impact assessment may be relatively straightforward, involving discrete projects or technologies with well-understood effects. Often, this is not the case, and we find ourselves dealing with situations with complex or little-understood impacts, where our capability to predict the future is very limited. The controversy that led to amendment of the CEQ's *worst case* regulations several years ago (see 40 CFR 1502.22) and the growing interest in incorporating formal risk assessment into NEPA documents reflect recognition that the environmental impact assessment often involves major uncertainties.

Testing the accuracy of our predictions through monitoring of actual impacts would help implement components of NEPA beyond the requirement to prepare a *detailed statement*. For example, one of the action-forcing requirements spelled out in Section 102(2)(H) of NEPA is to *initiate and utilize ecological information in the planning and development of resource-oriented projects.* Postproject monitoring of impacts would provide information helpful to identifying and understanding the linkages present in ecological systems. Information from the monitoring of actual impacts would also be helpful in complying with the information needs of Sections (F) and (G) of Section 102(2) of NEPA (which deal with recognizing the worldwide and long-range character of environmental problems and with providing advice and information useful in restoring, maintaining, and enhancing the quality of the environment, respectively).

In the final analysis, effective postproject studies could encourage and contribute to more adaptive environmental management, which would result in improved protection of natural resources and the environment — the ultimate objective of NEPA. The work of Holling (1978), Walters (1986), and others points to numerous examples where ecosystems do not behave in ways predicted or expected, and stresses that surprise is inherent in dealing with natural systems. Management policies often go awry because of overreliance on initial predictions and failure to build in provision for subsequent adjustment.

This argues for the importance of moving toward more adaptive approaches in which initial decisions are carefully monitored and adjusted in accordance with actual results. Not all aspects of a project may lend themselves to an incremental, adaptive approach. Once a highway is constructed, for example, it may be impossible to adjust for fundamental errors in its siting. On the other hand, the mitigation plan could be based on a more flexible adaptive approach in which the realities of environmental and social responses set the direction, rather than on total reliance of an initial prediction which may be, at best, a shot in the dark.

Fortunately, there are signs that recognition of the importance of follow-up studies and monitoring is increasing. There are monitoring requirements in the NEPA amendment recently brought before the Congress. Anne Miller's paper on the activities of the Economic Commission for Europe shows that the matter is receiving international attention as well. The U.S. Department of Energy (DOE), in its final rule implementing NEPA procedures (DOE 1992), included a requirement to prepare mitigation action plans that describe the planning and implementation of (a) measures designed to mitigate adverse environmental impacts associated with actions identified in a ROD, and (b) measures identified in a FONSI as necessary to render potential impacts not significant. These plans would be prepared for EAs that culminated in FONSIs based on the DOE's commitment to take mitigative actions and for RODs that contain mitigation. Furthermore, on the state level, the California Environmental Quality Act was amended in 1988 to require that agencies adopting mitigated negative declarations (the California equivalent of the FONSI) must affirm that approved mitigation measures are actually implemented before a project is approved (Remy et al. 1989).

The papers in this chapter will examine the wide range of issues associated with postproject follow-up studies. The experience reflected in those papers can make an important contribution as agencies move toward improving this neglected aspect of the NEPA process. The principal message emphasized by the papers in this chapter is that adequate provision for follow-up studies should be made an integral part of the NEPA process. The case studies described herein show that

1. An investment in NEPA follow-up activities can yield significant environmental benefits. Dodds' paper on I-476 shows how a substantial follow-up effort

can achieve environmental results on a complex and controversial project by assuring that the commitments contained in the NEPA documents are met. Clearly, not every project will require or could justify an effort of this intensity, but the I-476 example shows what can be accomplished when post-EIS activities are taken seriously.

2. NEPA follow-up can yield an extra measure of environmental protection beyond that which would be necessary to comply with environmental regulations. It is sometimes observed that NEPA has lost some of its importance because of the subsequent enactment of a wide range of environmental laws that provide adequate protection through requirements for compliance with regulations and standards. However, Broili's paper on the TVA's well-developed system for tracking NEPA commitments makes an important point. The TVA system distinguishes NEPA commitments from those already required under existing law, such as compliance with environmental permits. Adherence to legal requirements such as air and water quality standards can be achieved with or without NEPA. Yet there are a number of other more discretionary and preventive aspects of environmental protection that only tend to get addressed under the NEPA umbrella. Monitoring for NEPA compliance can provide this additional dimension to environmental protection.
3. Two important aspects of mitigation and NEPA follow-up are determining whether or not the mitigation was actually implemented and, if it was implemented, whether or not it was as effective as planned. These two issues are thoroughly addressed in this chapter. The paper by Barbara Young assesses the extent to which mitigation measures identified in EISs for proposed U.S. Army Corps of Engineers projects were actually implemented in the completed projects. It was found that about 65% of 212 mitigation measures evaluated in the U.S. Army Corps of Engineers' EISs were completed as outlined in the EIS or in a manner that went beyond that described in the EIS. The remaining 35% were either inoperable, partially complete, or incomplete. The paper by Roelle and Manci examines the effectiveness of mitigation, once implemented, for projects involving impacts to fish and wildlife resources. They found that in 61 case studies where all of the mitigation recommendations for a single development action were evaluated, mitigation was considered effective for 54%, partially effective for 30%, and ineffective for 16% of the recommendations. The results by Young and Roelle and Manci show (at least for the case studies examined) that some mitigation measures are implemented and that they can be as effective as planned.
4. The essence of the environmental assessment process is prediction. The limitations of environmental science and the complexity and difficulty of predicting environmental outcomes, however, raise real questions about the reliability of our EISs when they deal with other than routine or well-understood impacts. Substantive information on the predictive effectiveness of EISs is difficult to obtain. Roelle and Manci's review of studies of mitigation measures indicates just how limited and how subjective that information is. To date, we have not yet undertaken the studies that would enable us to use the enormous investment in NEPA analyses as a learning experience through which we could improve our predictive and environmental management capabilities.
5. Last, we can improve future predictions of environmental impacts by retrospective education on the shortcomings of present analyses. Bernard, Hunsaker, and Marmorek discuss meeting this goal through the use of three general types of tools: structured impact hypotheses, environmental audits, and environmental

monitoring. Interweaving these tools into the framework of the environmental impact assessment will promote evolution of the accuracy of impact predictions. After examining over 2000 predicted impacts in 11 environmental impact assessment documents prepared for Canadian hydroelectric projects, they conclude that effects monitoring leads to an expanding body of information and improved understanding of mechanisms, both of which contribute to formulating more accurate impact predictions. In addition, structured impact hypotheses play an important role in this process by both clarifying the basis for impact predictions and assisting in impact auditing procedures. The Economic Commission for Europe findings, reported on by Miller, also address this issue by recommending that a preliminary plan for postproject analysis should be prepared during the environmental assessment phase. In that way, preproject analyses can be better designed, and the relationship between prediction and feedback mechanisms can be strengthened.

The papers in this chapter suggest an important direction for more effective NEPA project assessment for the future. The ideal project would involve testable predictions of environmental outcomes based upon best available information, adequate administrative provisions to assure compliance with commitments made in the environmental review, monitoring to assess environmental response, and maintenance of maximum flexibility in project design to permit subsequent adaptation to deal with unexpected results.

REFERENCES

Bassin, N. J., and D. R. West. 1986. Analysis of Selected Agencies Post-Final Environmental Impact Statement Commitments. Unpublished report prepared for Office of Federal Activities. U.S. Environmental Protection Agency, Washington, D.C.

Holling, C. S. (ed.). 1978. *Adaptive Environmental Assessment and Management.* International Institute for Applied Systems Analysis, Wiley, New York.

Remy, M. H., T. A. Thomas, S. E. Duggan, and J. G. Moose. 1989. *Guide to the California Environmental Quality Act (CEQA).* 1989 ed. Solano Press, Point Arena, CA.

U.S. Department of Energy (DOE). 1990. 10 CFR Part 1021. National Environmental Policy Act Implementing Procedures; Proposed Rule. *Fed. Regist.* 55(213):46444–46464.

Walters, C. 1986. *Adaptive Management of Renewable Resources.* MacMillan Publishing Co., New York.

IN MEMORIAM

As this book went to press, members of the NEPA community were saddened by the news of the death of Dr. Allan Hirsch. Allan had been one of the driving forces who shaped the evolution of NEPA over the last two decades. In recent years, Dr. Hirsch played a key role in raising national consciousness of the need to conduct follow-up studies as part of the NEPA process. His concern for verifying implementation of mitigation measures and for evaluating actual environmental impacts in light of predictions has been particularly noteworthy. With the death of Dr. Hirsch, the NEPA world has lost a scientist, a policymaker, a regulator, and a philosopher.

Identification, Tracking, and Closure of NEPA Commitments: The Tennessee Valley Authority Process

R. T. Broili, Environmental Quality Staff, Tennessee Valley Authority, Knoxville, TN

ABSTRACT

The Tennessee Valley Authority (TVA) is a corporate agency of the federal government with a broad array of activities, including the stewardship of the Tennessee River system for purposes of navigation, flood control, and the production of hydroelectricity. The TVA is one of the largest producers of electricity in the world, supplying electricity to approximately 3 million residential customers, 50 industries, and several federal nuclear, aerospace, and military installations through a combination of hydro, coal-fired, and nuclear plants. The TVA service area covers 80,000 mi^2 in parts of seven states. Additionally, the TVA is heavily involved in numerous resource development programs.

As a federal agency, the TVA is subject to compliance with the National Environmental Policy Act (NEPA). The TVA guidelines for NEPA compliance have been modified several times over the past 20 years. Guideline modifications adopted in the early 1980s include a requirement for identifying commitments, including mitigation measures, made in NEPA documents and ensuring that these commitments are instituted. As a result of this action, the TVA has developed a process for tracking commitments and an administrative process for determining when the commitments have been fulfilled.

This paper provides information on this process and how it has operated to date. Problems encountered and suggested improvements to such a system are also presented. Finally, the question of how such a process relates to the willingness of federal agencies to fulfill the spirit of NEPA is discussed.

WHAT IS THE TENNESSEE VALLEY AUTHORITY?

The Tennessee Valley Authority (TVA) is a corporate agency of the federal government with a broad array of responsibilities, including the stewardship of the Tennessee River system for purposes of navigation, flood control, and the production of hydroelectricity. The TVA is one of the largest producers of electricity in the world, supplying electricity to approximately 3.5 million residential customers, 50 industries, and several federal nuclear, aerospace, and military installations through a combination of hydroelectricity, coal-fired, and nuclear plants. The TVA service area covers 80,000 mi^2 in parts of seven states. Additionally, the TVA is heavily involved in numerous resource development and protection programs which include

- the National Fertilizer and Environmental Research Center in Muscle Shoals, Alabama
- economic development (industrial and community components)
- land management (11,000 mi of shoreline and 250,000 ac under TVA custody and control)
- waste management (solid and hazardous waste handling and waste minimization components)
- forestry
- fisheries
- wildlife
- air quality
- water quality
- Land Between The Lakes (64,000-ac outdoor recreation, resource conservation, and environmental education center)

TVA AND NEPA

The National Environmental Policy Act (NEPA) affects virtually every decision that the TVA makes that can result in impacts to the physical environment. Considering the broad array of TVA projects, programs, and facilities, the challenge to the agency to effectively implement such a mandate is great. Some of these programs involve promoting or making actions possible by nonfederal entities. These also fall under NEPA, since they are carried out with the TVA's consent or assistance. The TVA has internal procedures which provide guidance for compliance with NEPA. These procedures have been reviewed and approved by the President's Council on Environmental Quality (CEQ). These procedures have been revised three times (1974, 1981, and 1983) since initial procedures were developed in 1971 in response to the CEQ guidelines.

TVA procedures for NEPA compliance contain a section entitled *Mitigation Commitment Identification, Auditing and Reporting*. The requirements set forth in this section form the basis for TVA's mitigation-related activities as part of the NEPA process. These activities may best be characterized as identification, tracking, and closure. The requirements contained in the TVA procedures can be paraphrased as follows:

- All significant measures planned to minimize or mitigate expected environmental impacts shall be identified in the environmental assessment (EA) or the environmental impact statement (EIS) (or in some lower form of NEPA-related review) and compiled in a commitment list.
- Each commitment shall be assigned to a responsible organization within the TVA.
- Periodic reports on the status of the commitment shall be made.
- Management activities, including auditing, are to be conducted to ensure that commitments are met.
- For circumstances that warrant modifying or deleting previously made commitments, the responsible organization must receive approval from appropriate corporate entities.

Those familiar with the diversity of NEPA procedures among federal entities may recognize the existence of such a procedure for mitigation as a variance from the federal agency norm. The TVA has this element in its NEPA procedures because the CEQ regulations were interpreted by TVA staff in the late 1970s and early 1980s to mandate such a requirement, and when the TVA procedures were revised in 1981, this section was included and subsequently approved by CEQ. The language in the CEQ regulations which prompted this action by the TVA is contained in Section 1505.2, *Record of Decision in Cases Requiring Environmental Impact Statements*, which states that a "monitoring and enforcement program shall be adopted and summarized where applicable for any mitigation." The TVA's response to this requirement was to institutionalize an administrative mechanism for ensuring that appropriate mitigation was conducted for all NEPA reviews, not just those involving EISs. This includes mitigation resulting from EAs and many lower-level NEPA review documents. It should be noted that the TVA has adopted a process in which projects that are normally categorically excluded may require some level of project-specific review and may have environmental conditions placed on them if they have characteristics which can result in even minor impacts.

MITIGATION AND NEPA COMMITMENT

To understand the TVA's system for identifying, tracking, and closing commitments, one has to first define mitigation and subsequently understand what the TVA considers to be a NEPA commitment. A mitigation commitment is defined in the CEQ regulations, Section 1508.20, to include any specific action which is proposed in a NEPA document for the purpose of achieving one or more of the following results:

- avoiding the impact altogether by not taking a certain action or parts of an action
- minimizing impacts by limiting the degree or magnitude of the action and its implementation
- rectifying the impact by repairing, rehabilitating, or restoring the affected environment
- reducing or eliminating the impact over time by preservation and maintenance operations during the life of the action
- compensating for the impact by replacing or providing substitute resources or environments

A NEPA commitment, as it has been historically employed by the TVA, has a somewhat broader context than a mitigation commitment. A NEPA commitment, in addition to being a mitigation commitment, may also be a commitment for future NEPA action which may result from the activity under evaluation. For instance, the TVA sometimes provides long-term leases of property under its custody or control to local government entities for development of industrial parks. The NEPA review of such an action will generally contain a commitment that requires proposals for subsequent industrial development of specific industries to be subject to an additional NEPA review of a specific industry's activities. This is basically identical to the tiering process as defined in the CEQ regulations, Section 1508.28. When such a commitment is pertinent, it is contained not only in the appropriate NEPA document but also in any contractural documents developed to convey the lease.

A NEPA commitment may also take the form of a commitment for future nonregulatory-driven consultation with internal or external entities before, during, or after taking an action. For example, if a road is to be built through an environmentally sensitive area, a commitment may be contained in a NEPA document requiring the TVA or external agency staff with sensitive resource expertise to be present during parts of the construction to ensure that the sensitive resource is protected. Such a commitment is often the difference between good intentions and good results.

In addition to having some broad definition of what a NEPA commitment is, there is also a recognition in the agency of what an appropriate NEPA commitment is not. First, the TVA does not consider an action already required by law to be a NEPA commitment. An example of such an action is obtaining and/or meeting the normal standards of an environmental permit. The TVA considers this a given requirement which has to be carried out regardless of NEPA and, consequently, is represented in the baseline of environmental impact resulting from an action as opposed to being a commitment to mitigate beyond the baseline.

The mere statement that best management practices (BMPs) are to be employed to mitigate adverse impacts may not be considered to be a NEPA commitment. For instance, the TVA finds the statement that BMPs will be employed for erosion and sediment control to be inappropriate as a NEPA commitment.

There are various measures which may be employed to minimize erosion and sediment contamination from a site-specific action. Consequently, a site-specific erosion control plan is necessary for achieving desired results and thus becomes the appropriate NEPA commitment. In contrast, the TVA has specific procedures for carrying out BMPs for handling asbestos, PCBs, and general hazardous waste at its fossil and hydroelectric plants already prepared both generically and on a facility-specific basis and contained in the facility Spill Prevention Control and Countermeasure (SPCC) Plan. These are considered to contain enough specificity to be referenced as a NEPA commitment.

The TVA also tries to avoid including *vague statements of good intentions* as NEPA commitments. This issue was dealt with by the judiciary in *Preservation Coalition, Inc.* v. *Pierce* (667 F 2d 851 [9th Cir. 1982]). Our experience is that the success of implementing commitments is greatly enhanced by specificity and that vague statements result in vague actions.

IDENTIFYING NEPA COMMITMENTS

Project managers are made aware during preliminary NEPA consultation with agency central environmental staff of the nature of commitments and how statements made in the NEPA document may be considered to be commitments. During the initial development of NEPA document information, mitigation measures may be suggested or negotiated between project planning staff and in-house resource specialists or external entities (Figure 1). Scoping also presents a major opportunity to identify commitments. Although the process of developing mitigation measures and other NEPA commitments is conducted throughout project planning, no effort is made to extract these commitments into a specific commitment list until at least a rough draft NEPA document is completed.

When a rough draft document is completed, project staff, with guidance from agency central environmental staff, prepares a list of commitments extracted from the text. This list is appended to EA and EIS documents. For lesser levels of review, any commitments are an integral part of the review document. Obviously, the commitment must be consistent with the general definition of a NEPA commitment discussed above. In situations where external entities may be carrying out the action in question, NEPA commitments may be inserted as contract conditions or deed provisions.

This method of identifying commitments has evolved over the past decade. In the early stages of commitment identification, project managers would go through a NEPA document and extract all statements which contained the terms *will* or *shall* for inclusion in a commitment list. This practice resulted in very large numbers of commitments for some projects. Many of these commitments had little relation to what is now considered a NEPA commitment. This system proved largely unworkable because of the sheer mass of commitments and the lack of specificity of individual commitments. The thrust of changes in this system has been to limit commitments to understandable, specific actions which conscientious management can reasonably implement.

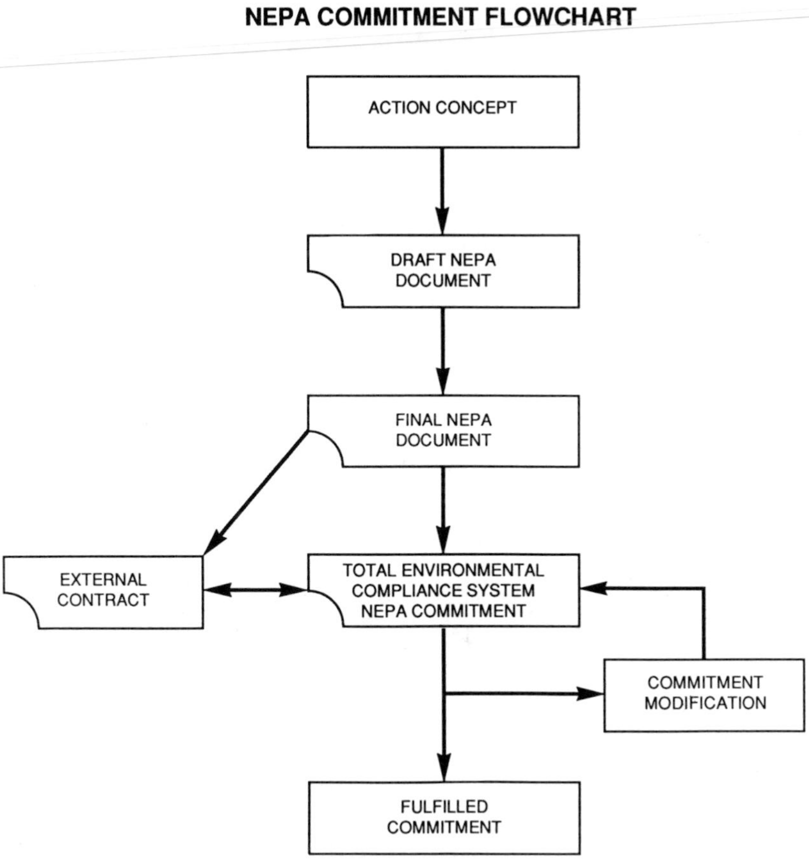

Figure 1. NEPA commitment flow chart.

TRACKING NEPA COMMITMENTS

Concurrent with finalization of a NEPA document, commitments are entered into the TVA's Total Environmental Compliance System (TECS) for the purpose of tracking them through to completion and closure. The TECS is a computer-based information system which contains (1) all environmentally related regulatory commitments (including NEPA commitments), violations, discharge requirements, and subsequent actions and (2) all in-house environmental audit findings and subsequent actions. Figure 2 is a flowchart of the database structure of the TECS. Before a NEPA commitment is entered into the TECS, an effort is made to detail the commitments to a level that presents a workable management plan for accomplishing an action. Figure 3 presents the various information fields available for entering information on a NEPA commitment. All of these fields are not necessarily used, the level of details being dependent on the nature of the commitment.

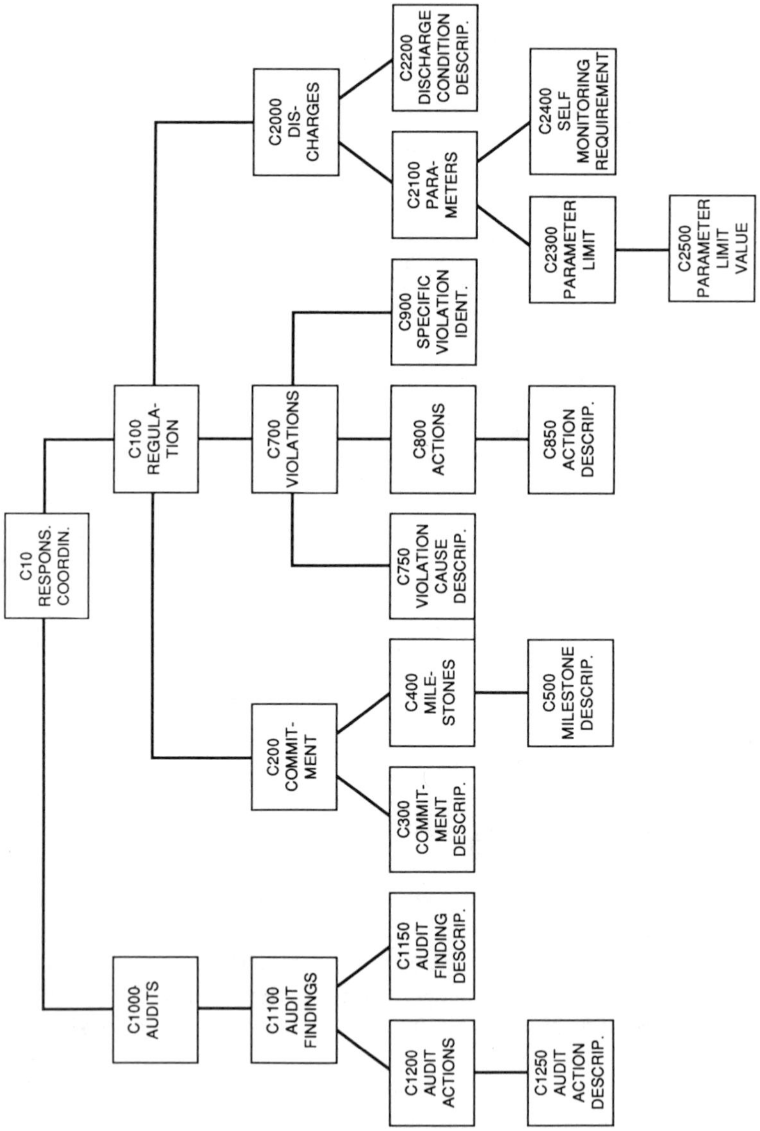

Figure 2. Total environmental compliance system (TECS) database structure.

Fields to be completed for EISs:

101 Category ID — always "3"
102 EIS identifier — list first the type of assessment:
EA — environmental assessment
FES — final environmental statement
ES — environmental statement
FEIS — final environmental impact statement

List second the facility using standard TVA abbreviation if possible. List last the date of the final document. Example: FEIS BRF 02/01/79

104 Name of EIS
106 Date document was finalized
108 Date commitment is to be completed
110 Date commitment list was finalized
112 Date commitment list was last sent out for review
122 Abbreviation for type of project
124 Type of EIS: EA = environmental assessment; DE = draft EIS; FE = final EIS; ND = negative determination
126 Number of EIS version: 0 = initial, etc.
202 Commitment number
204 Date commitment is due
206 Date commitment was met
208 Date staff was officially notified commitment was met
216 Abbreviated version of commitment
226 Organization responsible for commitment
234 Code representing commitment category
242 Page reference of commitment
302 Commitment/comment line number
304 Commitment/comment
402 Milestone number
408 Date milestone is to be completed
410 Date milestone was completed
420 Individual responsible for accomplishing milestone
422 Section responsible for accomplishing milestone
424 Branch responsible for accomplishing milestone
426 Division responsible for accomplishing milestone
428 Office responsible for accomplishing milestone
502 Line number for milestone
504 Milestone

For input instructions, see detailed explanations in the computer dictionary.

Figure 3. Fields to be completed for EISs.

Examples of project commitment lists or plans as they are contained in the TECS are shown in Figures 4 and 5. Figure 4 is a list of commitments for which no action has been taken, action dates set, or responsible organization assigned. This will be accomplished before project implementation begins. Figure 5 is a commitment plan which has already been completed with milestone data and completion dates included.

TVA ENVIRONMENTAL QUALITY NEPA RELATED COMMITMENT PLAN

EA PICKWICK CHANNEL, DATE OF DOCUMENT: 05/01/88 ENVIRONMENTAL ASSESSMENT — PICKWICK CHANNEL MODIFICATION PROJECT — TENNESSEE RIVER — HARDIN COUNTY, TENNESSEE — TVA/ONRED/EQS-88/4 — NEPA REVIEW BY BROILI

COM NO	COORD. ORG	DATE DUE/ COMPL.	COMMITMENT	COMMENT
001	WRES		THE CONTRACTOR WILL SUBMIT A SPILL PREVENTION CONTROL AND COUNTERMEASURES PLAN FOR ANY STORAGE OF PETROLEUM PRODUCTS ON SITE. A PLAN OUTLINING STORAGE AND USE OF EXPLOSIVE MATERIAL WILL ALSO BE REQUIRED.	
002	WRES		DREDGED MATERIAL WILL NOT BE BARGED ACROSS THE TWO MUSSEL BEDS LOCATED ALONG THE RIGHT BANK SHORELINE, CONSERVATIVELY CONSIDERED AS TENNESSEE RIVER MILES 202.5 TO 203.7 AND 205.0 TO 206.5. ANY DREDGED MATERIAL DEPOSITED ALONG THE SHORELINE FRONTING THE TWO MUSSEL BEDS WILL BE PLACED FROM THE BANK.	
003	WRES		ALL MUSSELS WILL BE REMOVED FROM THE AREA TO BE DREDGED AND FROM A BUFFER ZONE AROUND THE DREDGED AREA, THE WIDTH OF WHICH WILL BE DETERMINED FROM RESULTS OF A BLAST EFFECTS TEST. BOTH THE BLAST EFFECTS TEST AND MUSSEL REMOVAL WORK WILL BE COORDINATED WITH THE U.S. FISH AND WILDLIFE SERVICE (FWS) AND THE TENNESSEE WILDLIFE RESOURCES AGENCY (TWRA). MUSSELS REMOVED FROM THE PROJECT AREA WILL BE RELOCATED TO SUITABLE HABITATS IN UPPER KENTUCKY RESERVOIR SELECTED BY FWS AND TWRA.	

MILESTONES

NO.	MILESTONES	DATE DUE	DATE COMPL.	RESP. ORG.
A.	PREPARE WORK PLANS FOR BLAST EFFECTS TEST AND RELOCATION OF THE MUSSELS.			
B.	COORDINATE WORKPLANS WITH FWS AND TWRA.			
C.	PERFORM BLAST EFFECTS TEST AND DEFINE AREA FROM WHICH MUSSELS WILL BE REMOVED.			
D.	RELOCATE MUSSELS.			

Figure 4. The TVA environmental quality NEPA-related commitment plan (Pickwick Channel).

TVA ENVIRONMENTAL QUALITY NEPA RELATED COMMITMENT PLAN

CE29, DATE OF DOCUMENT: 08/09/82 COLBERT STEAM PLANT — NEW ASH DISPOSAL POND — NEPA REVIEW BY GENGOZIAN

COM NO	COORD. ORG	DATE DUE/ COMPL.	COMMITMENT	COMMENT
001	F&H	010183 050484	BEFORE CONSTRUCTION COMMENCES IT WILL BE DEMONSTRATED THAT GROUND AND SURFACE WATERS WILL NOT BE SIGNIFICANTLY IMPACTED.	MILESTONE A, B, AND C, REPORT WAS SUBMITTED TO THE STATE JUNE 1982; 8/4/82 AL ONSITE VISIT 8/18/82 REPORT FROM ALABAMA; MILESTONE D—ACCORDING TO MY DISCUSSIONS WITH TERRY MANSEILL (F&H PR) THE SOILS DATA WERE COLLECTED IN MAY 1982 AND PROVIDED TO W RES IN AUGUST 1982; MILESTONE F—11/16/82 BETSON TO MANSEILL, SUMMARY OF WATER SUPPLY AVAILABILITY; MILESTONE H—REPORT TO BE AVAILABLE AUGUST 1983; MILESTONE I, J, K—PLANS SUBMITTED 11/29/82, EL-ASHRY TO ALABAMA AND APPROVED 12/14/82, ALABAMA TO EL-ASHRY.5/4/84, GENGOZIAN TO MARSH, MILESTONE H OF COMMITMENT 001 CAN BE CLOSED. THE DIVISION OF AIR AND WATER RESOURCES HAS PREPARED A FINAL REPORT WHICH DOCUMENTS ITS CONCLUSIONS ON THE BASELINE GROUNDWATER MONITORING STUDY AS WELL AS RECOMMENDING FUTURE MONITORING. WE WILL BE DISCUSSING NEEDS FOR FUTURE MONITORING WITH THE APPROPRIATE PARTIES IN THE NEAR FUTURE.

MILESTONES

NO.	MILESTONES	DATE DUE	DATE COMPL.	RESP. ORG.
A	PREPARE REPORT ON PROPOSED IMPOUNDMENT AS DESCRIBED IN 2.A. OF THE AWIC CONSTRUCTION GUIDELINES (FOR) INDUSTRIAL AND MUNICIPAL SURFACE IMPOUNDMENTS.	06/01/82	04/05/83	F&H PR
B	SUBMIT REPORT REFERENCED IN 1.A. ABOVE.	06/01/82	04/05/83	EQS
C	ARRANGE AN ONSITE GEOLOGICAL/HYDROLOGIC REVIEW BY AWIC.	09/01/82	08/04/82	EQS
D	SOILS DATA WILL BE COLLECTED AND PROVIDED TO W RES.	08/01/82	04/05/83	OEDC
E	DETERMINE AND EVALUATE STRUCTURAL STABILITY UNDERLYING POND SITE AND PROVIDE REPORT TO F&H PR AND EQS FOR REVIEW.	08/01/82	04/05/83	OEDC
F	IDENTIFY WATER SUPPLY AVAILABILITY AND GROUNDWATER USE WITHIN ONE MILE RADIUS OF RESERVATION AND PROVIDE TO F&H PR.	10/01/82	04/05/83	W RES
G	INSTALL APPROPRIATE GROUND WATER MONITORING WELL SYSTEM.	10/01/82	11/16/82	OEDC
H	PROVIDE GROUND WATER MONITORING WELL SYSTEM DATA ALONG WITH CONCLUSIONS AND RECOMMENDATIONS TOF&H PR AND EQS.	01/01/83	05/04/84	W RES
I	PREPARE ENGINEERING REPORT.	10/01/82	04/05/83	F&H PR
J	SUBMIT ENGINEERING REPORT.	10/01/82	04/05/83	EQS
K	OBTAIN AWIC APPROVAL TO CONSTRUCT.	01/01/83	04/05/83	EQS

Figure 5. The TVA environmental quality NEPA-related commitment plan (Colbert Steam Plant).

COM NO	COORD. ORG	DATE DUE/ COMPL.	COMMITMENT	COMMENT
002	F&H PR	010186 022184	TVA WILL OPERATE THE ASH DISPOSAL POND SUCH THAT IT DOES NOT SIGNIFICANTLY IMPACT GROUND AND SURFACE WATERS.	MILESTONES A AND B — BASED ON CONVERSATION WITH STEVE WELLS ON 4/5/83, MODIFICATIONS TO THE NPDES PERMIT AND A SUBSEQUENT REVISION WERE SENT 2/20/81 AND 3/19/82, RESPECTIVELY. 2/21/84, GENGOZIAN TO MARSH, F&H PR BEGAN USING THE ASH POND ON 9/14/83 (MILESTONE F). VERBAL APPROVAL FOR USE OF THE POND WAS OBTAINED BY S. WELLS FROM SONJA MASSEY OF ADEM 9/16/83. THE APPROVAL WAS CONTINGENT UPON TVA SUBMITTING A LETTER OF CERTIFICATION BY A PROFESSIONAL ENGR. THAT THE POND WAS CONSTRUCTED ACCORDING TO PREVIOUSLY SUBMITTED PLANS & SPECS (MILESTONE E). THIS WAS DONE 11/14/83 (MILESTONES C & D). ACCORDING TO MR. WELLS, NO OTHER FORMAL CORRESPONDENCE FROM ADEM IS EXPECTED.

MILESTONES

NO.	MILESTONES	DATE DUE	DATE COMPL.	RESP. ORG.
A	PREPARE MODIFICATION TO THE NPDES PERMIT OR AMEND REAPPLICATION PACKAGE.	10/01/82	04/05/83	F&H PR
B	SUBMIT THE MODIFICATION TO NPDES PERMIT OR THE AMENDED REAPPLICATION PACKAGE TO THE STATE OF ALABAMA 180 DAYS PRIOR TO DISCHARGE.	10/01/82	04/05/83	EQS
C	PREPARE CERTIFICATION TO AWIC THAT PROJECT HAS BEEN CONSTRUCTED AS REQUIRED BY AWIC AND AS PRESENTED IN THE PLANS AND SPECIFICATIONS (INITIATOR AS-BUILT).	10/01/85	02/21/84	OEDC
D	SUBMIT CERTIFICATION TO AWIC.	10/01/85	02/21/84	EQS
E	OBTAIN AWIC APPROVAL TO USE.	12/01/85	02/21/84	EQS
F	BEGIN OPERATION.	01/01/86	02/21/84	F&H PR

Figure 5. (continued).

The TECS first began to track NEPA commitments in 1981. At that time, all old NEPA commitments which could be identified from existing TVA NEPA documents were entered into the system. Some commitments involved documents released as early as 1974. As of October 1989, the system contained 172 projects with a total of 1594 commitments. Of these, 387 commitments are still considered open and involve 109 projects. Of the 109 projects, 40 involve EISs, 31 involve EAs, the remainder being sub-EA-level reviews.

The range of commitments made for individual projects is 1–391. In recent years, the trend has been toward fewer, more precise commitments; and the average number of commitments for recently prepared EAs and EISs averages approximately ten.

MODIFYING AND DELETING

Situations occasionally arise where commitments are modified or deleted. When project components are modified or deleted, there may be either no need for the commitments or the need to modify the commitment. This can be done at the request of the project manager but first requires concurrence of the TVA General Counsel and the Director of Environmental Quality. Before such changes, appropriate resource specialists within and/or outside the agency are consulted.

CLOSING COMMITMENTS

Commitments are closed when the action to fulfill a commitment is completed and verification of this completion is provided to the Director of Environmental Quality. This verification can occur in two ways. First, it can be determined through periodic audits performed by the Environmental Auditing Department which may examine a project site and determine that the action has been completed. The audit report can then recommend a commitment be closed. A written memorandum from the project manager or responsible organization is still required to document closure, and a written concurrence from the Director of Environmental Quality finalizes the paper trail.

The other pathway is through the Environmental Compliance Department which, like the Environmental Auditing Department, is staff to the Director of Environmental Quality. An appropriate Environmental Compliance staff member can recommend that a commitment be closed based on personal investigation or documentation from the project manager or responsible organization. The same paper trail is utilized except there is no audit report involved. The Environmental Compliance Department staff members routinely review commitment status of projects assigned to them as part of their position responsibilities.

Upon closure of the commitment, closure is appropriately noted in the TECS. The TECS is not purged of the commitment, but the commitment is maintained for historical purposes.

FUTURE NEEDS

The TVA feels it has made considerable progress developing and operating a system for NEPA commitment identification, tracking, and closure. From an environmental management standpoint, there is still need for improvement. A more concise working definition of what constitutes a NEPA commitment is desirable. There are old commitments in the TECS which need to be cleaned up and closed out. The most important need is to constantly make project and program managers aware that the NEPA commitment process does not end with the finalization of a NEPA document but with implementation of a project that appropriately minimizes or avoids environmental damage.

The I-476 Environmental Monitor: An Effective Approach for Monitoring Implementation of NEPA Documents

P. J. Dodds, KCI Technologies, Inc., Harrisburg, PA

ABSTRACT

In 1981, the President's Council on Environmental Quality (CEQ) required that an independent environmental monitor be appointed to oversee design and construction of I-476, a new 16.9-mi highway in southeastern Pennsylvania. KCI Technologies, Inc. was appointed as that monitor and charged with three broad responsibilities:

1. to act as liaison between the public and the Pennsylvania Department of Transportation (PENNDOT)
2. to review design and construction activities to ensure compliance with commitments
3. to act as an independent source of environmental expertise

Since 1982, the monitor has administered a program that included meeting with the public, reviewing design plans and reports, reviewing construction activities, attending public and resource agency meetings, and publishing a monthly newsletter.

The monitoring process has required effective interactive cooperation by PENNDOT, the designers, and the contractors. It has assured that commitments in the National Environmental Policy Act (NEPA) documents have been satisfied. These environmental services cost less than 1% of the total project cost.

INTRODUCTION

Regulations promulgated by the President's Council on Environmental Quality (CEQ) for implementing NEPA briefly discuss the use of follow-up monitoring programs for National Environmental Policy Act (NEPA) documents (CEQ 1978). These regulations do not require such a program but suggest that one may be used where appropriate. Under Sections 1505-2(c) and 1505-3 of 40 CFR 1505-08, a monitoring program may be used to ensure the implementation of mitigation measures and to ensure that decisions and commitments made in the environmental documents are carried out.

This paper describes an approach currently being used to monitor the implementation of the environmental impact statement (EIS) for a highway in southeastern Pennsylvania. The primary focus of this monitoring program is to ensure the compliance of the environmental commitments made in the environmental documents.

PROJECT LOCATION

Interstate 476 (I-476), commonly known as the Blue Route, is a proposed highway to be located in the western suburbs of Philadelphia (Figure 1). This north-south expressway will travel from the Pennsylvania Turnpike in Montgomery County to I-95 in Delaware County for a length of 21.5 mi. Most of the corridor for the highway lies within or near the stream valleys of the Crum Creek, Darby Creek, and Ithan Creek watersheds. Within this corridor remains much of the wooded open space of an otherwise densely populated suburb. It contains many historical, recreational, and environmental resources which are considered to be of significant public value. Residential areas border the corridor, and in many situations the alignment *threads the needle* between homes and sensitive environmental features.

PROJECT HISTORY

The Blue Route alignment was selected by the U.S. Bureau of Public Roads in 1963 (Larson 1984). Final design for the highway began in 1964 and construction started in 1967. However, as a result of NEPA and Section 4(f) of the U.S. Department of Transportation Act of 1966, it became necessary to reevaluate the project. At that time, only 5.1 mi of the expressway had been built.

To comply with the new federal environmental protection laws, the Pennsylvania Department of Transportation (PENNDOT) prepared in 1975 a draft EIS for the corridor from I-95 to I-76. After submitting a draft of the Final EIS/4(f) Statement to the Federal Highway Administration (FHWA) in 1978, the FHWA requested

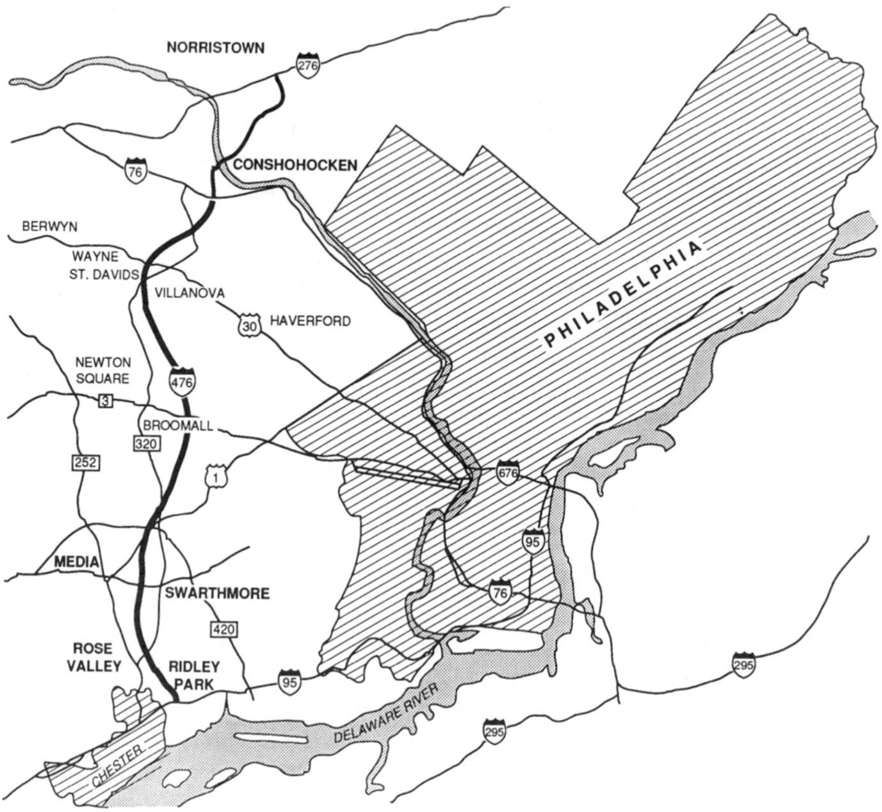

Figure 1. Location of the Blue Route (I-476) corridor and environmental monitor involvement.

PENNDOT to reevaluate the project. A new Final EIS/4(f) Statement addressing these issues received FHWA approval in 1980. The proposed action was then published in the *Federal Register* (PENNDOT 1984).

The U.S. Department of the Interior (DOI), however, expressed concern about the significant impacts such a project could have on the stream valleys and recreational areas within the corridor. These concerns were referred by the DOI to the CEQ for resolution. With the assistance of the CEQ, a compromise was reached between the U.S. Department of Transportation (DOT) and DOI. The agreement incorporated additional commitments and requirements to address and mitigate the concerns and impacts of the project. Of special significance was the commitment to appoint an environmental consultant to oversee both the design and construction phases of the project from I-95 in Chester, PA to I-76 in Conshochocken, PA (Figure 1). This consultant would act as an environmental monitor to ensure that environmental commitments were kept and public concerns addressed.

ENVIRONMENTAL MONITOR RESPONSIBILITIES

Three broad charges were established for the monitor's role:

1. to act as liaison between the public and PENNDOT during design and construction to ensure a free exchange of information and that public concerns are addressed in a timely manner
2. to review design and construction activities to ensure that the environmental commitments and requirements are incorporated
3. to act as an independent source of environmental expertise to provide advice and recommendations on ways to incorporate the commitments and reduce project impacts

Liasion Activities

The monitor has an office centrally located to the corridor which is supplied with all environmental documents, design plans and reports, and final construction drawings. Individuals, as well as citizen and environmental groups, visit the office to study the documents and to express any concerns or issues about the project. To date, there have been over 4000 such contacts. The monitor also meets with the public (individuals, civic organizations, etc.) at their homes and along the proposed project alignment to discuss design and construction related issues. The monitor's liaison activities extend to assisting the public in understanding design constraints, plan development, and construction operations.

Because the Blue Route is an interstate highway with parkway-like goals, some of the objectives for the project were at times in conflict with each other. The design had to be optimized to address all the objectives, and it was our role to ensure that the public understood this. As a result, not all of the concerns raised were resolved in favor of the individual or group registering that concern.

Public concerns and issues reported to the environmental monitor are relayed to PENNDOT, their design consultants, and contractors for their consideration (Figure 2). The monitor tracks each issue until it is addressed by PENNDOT. To assist us in tracking public issues and concerns, we maintain detailed daily contact logs, public contact records, PENNDOT and design consultants' contact records, and public meeting minutes.

The environmental monitor is available to the public through community meetings and as a speaker service to interested groups and organizations. During the design phase, we attended 190 community meetings and field walks. At each of these meetings, the monitor kept detailed minutes to record the issues as well as any resolution and commitments made by PENNDOT and its design consultants. To date the monitor has also had over 100 speaking engagements with civic organizations, community associations, and environmental groups.

The public is kept abreast of design and construction activities and of the issues and their resolution through our monthly newsletter. The newsletter, which has a readership of over 5000, summarizes the project status, major issues encountered, issues resolved, special features, recommendations for further mitigation by the

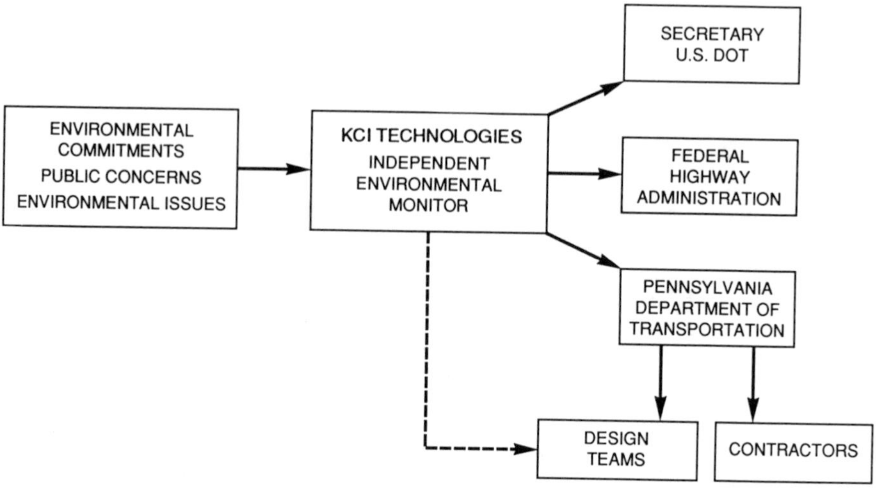

Figure 2. Relationship of environmental monitor to the Pennsylvania Department of Transportation, Design Consultants, Contractors, Federal Highway Administration, and the U.S. Department of Transportation.

monitor, and any previously unresolved issues. Issues were not removed from this list until they were either addressed on the final design plans or during construction. During the design phase, the newsletter summarized the proceedings of the individual community meetings, the issues raised at these meetings, PENNDOT's response, and the monitor's assessment of PENNDOT regarding its compliance with the commitments.

This liaison role not only ensures that the major issues raised by local governments, citizen groups, and environmental groups are addressed, but that the concerns of the individual are also given proper consideration.

Design and Construction Review

Design

In the preparation of environmental documents, PENNDOT made numerous commitments to mitigate the impacts of the highway. To ensure that these commitments are complied with in the design and construction phases of the project, we first compiled a list of the many commitments. A checklist was also developed for commitments made at meetings with the public and resource agencies. All of the commitments were regularly reviewed for compliance before completion of final design. This provided a mechanism whereby problems could be identified early in design. For example, our review found a discrepancy between the Task Force Alignment Plan (the selected alternative in the EIS) and the preliminary alignment design behind a residential area. We took this issue to PENNDOT with the recommendation that the alignment be shifted at the least to match the selected alternate. We also suggested that PENNDOT study the feasibility of a greater shift, since the proximity of the alignment was a major concern of one neighborhood. A larger shift was studied and made within the engineering constraints of that area.

During design, the commitment review process was also facilitated by monthly design review meetings with PENNDOT, informal status meetings with the design consultants, and specific topics meetings (Figure 3); there were 141 such meetings in the design period. These meetings allowed the monitor to discuss design progress and public concerns, to review design plans and see how they incorporated the environmental commitments, and to provide an opportunity to make recommendations to further reduce project impacts. Monthly reports were prepared and sent to PENNDOT which summarized design and public issues recorded during the month, our recommendations for addressing the issues, PENNDOT's response, and any follow-up action that was needed or taken.

The monitor reviewed all plans to ensure not only that state-of-the-art methods were being used to incorporate the commitments but also that appropriate mitigation measures were developed and to recommend measures to further reduce the impacts of the highway. Topics that the monitor reviewed include:

- highway design
- stormwater management
- erosion and sediment control
- wetland mitigation
- stream relocation mitigation
- wildlife habitat enhancement
- landscaping and architectural treatments
- protection of 4(f) areas
- noise abatement
- maintenance and protection of traffic
- hydrology and hydraulics
- floodplain issues
- highway lighting
- water quality

Because the highway alignment was within or adjacent to a stream valley, one of the issues echoed repeatedly by the public was the effect of highway runoff on water quality. The monitor had identified this issue to PENNDOT early in the design process. The monitor first provided PENNDOT with a set of recommended mitigation measures to address the potential impacts of highway runoff on both water quantity and quality. Meetings were then held with PENNDOT and the design consultants to discuss the incorporation of water quality mitigation measures into the stormwater management plans.

Construction

Before construction began, the monitor reviewed the final contract drawings and specifications to ensure that any changes made did not conflict with the commitments and/or create additional impacts. Early in the design phase, we

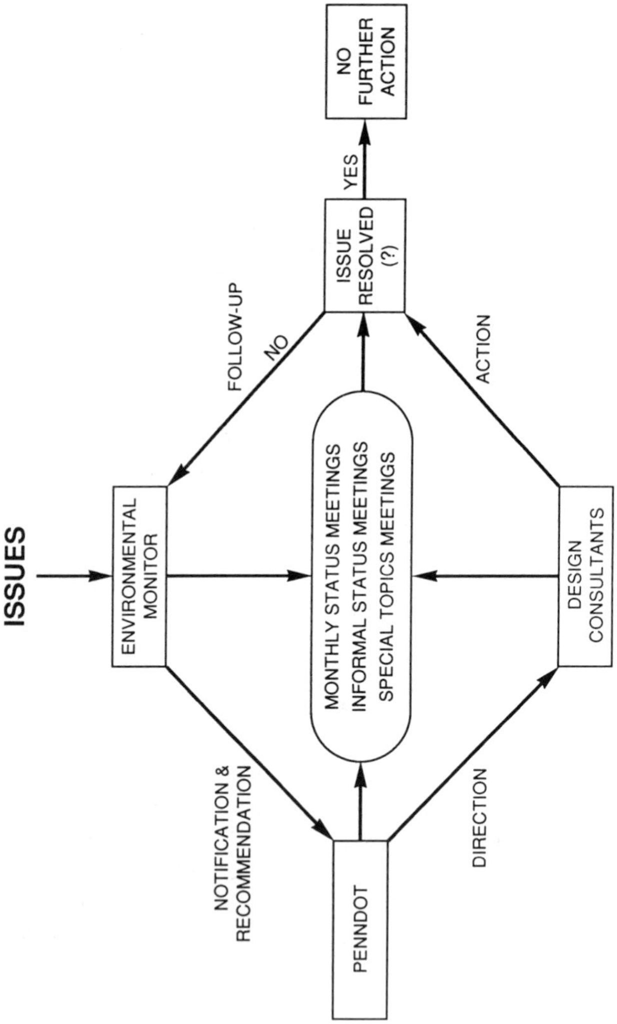

Figure 3. Design review interaction with the Pennsylvania Department of Transportation.

reviewed PENNDOT's own specifications regarding environmental issues. Our preconstruction review also assessed areas that may be particularly sensitive to construction activities. The monitor listed these critical areas, potential problems that could occur, and possible remediation. This information is utilized by our staff during construction monitoring to anticipate and prevent potential problems from occurring.

Each day, the monitor makes a number of circuits through each construction section to ensure compliance with the plans, specifications, special provisions, environmental commitments, and environmental permits. Additional inspections are made for sensitive areas or areas of potential impacts, including the critical areas mentioned above. This includes daily review of erosion and sediment control, protection of existing vegetation and habitat area (wetland and upland), stream and wetland mitigation, environmental permit issues, dust and air quality, water quality, construction noise, landscaping, maintenance and protection of traffic, and items related to public safety (e.g., pedestrian access).

Each section monitor must be ready to respond to any problem or issue that arises on the construction site. We must (1) make an assessment of the problem, (2) recommend to PENNDOT the best solution, and (3) coordinate with PENNDOT and the contractor to ensure that the problem is satisfactorily addressed.

These problems may include any unforeseen environmental and community issues that were not addressed during design or circumstances where the design has been found to be inadequate because of the dynamic nature of the construction site. Basically, the section monitors must be onsite environmental problem solvers. For example, during the construction of a three-cell box culvert, the contractor needed to remove a downstream dike used to channel the streamflow away from the construction area of the culvert. The contractor proposed removal of the dike via the stream. This action would have required a number of crossings by equipment across the stream channel. To comply with the commitments and to minimize stream impacts, the monitor recommended an alternate procedure that did not require equipment in the stream. This procedure was incorporated, thereby significantly minimizing impacts to the stream.

To facilitate review of construction issues and their resolutions, the monitor uses environmental logs, environmental memorandums to PENNDOT, and monthly reports which summarize the issues, action taken, and action needed. All environmental and community issues raised by the monitor are discussed at monthly construction meetings, weekly environmental concerns meetings, and field walks with PENNDOT and the contractor (Figure 4).

Reporting and Commitment Resolution

The monitor prepares a separate final report for each phase of design and construction. These documents describe public concerns, incorporation of the environmental commitments, and final resolution. Such reports necessitate that an accurate account of all issues and PENNDOT's response to them be recorded. As described earlier, a checklist of whether or not the environmental commitments

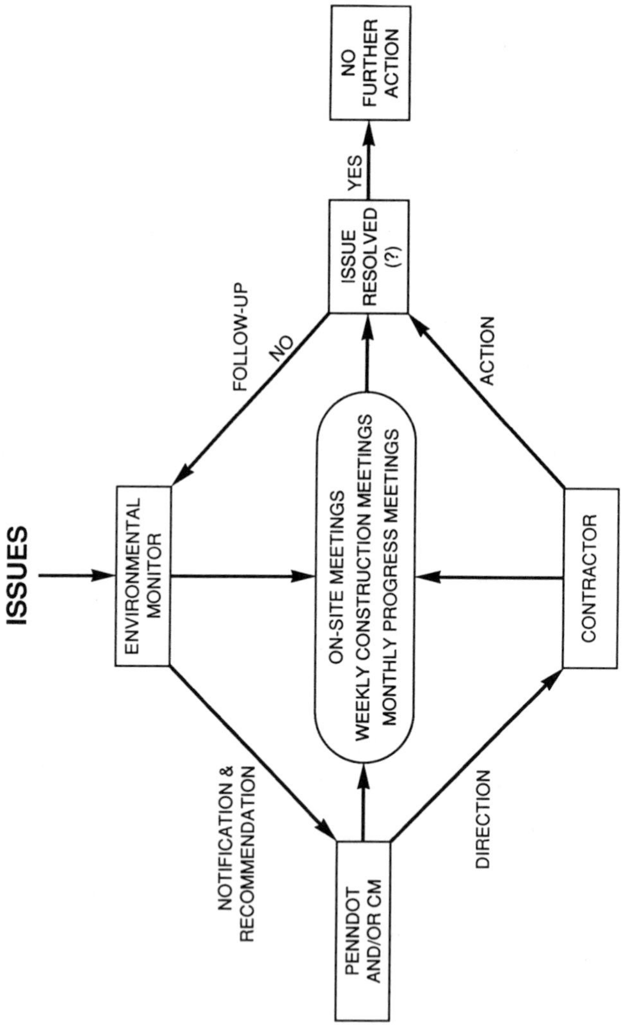

Figure 4. Construction review interaction with the Pennsylvania Department of Transportation.

have been met is also kept. If a commitment has not been met, then the reasons for such action must be valid to the satisfaction of the environmental monitor. If the environmental monitor feels that PENNDOT has not satisfactorily resolved an issue regarding the commitments, then the monitor may go directly to the FHWA for resolution (Figure 1). If the monitor is still not satisfied, then we may consult with the office of the U.S. Secretary of Transportation.

SOURCE OF INDEPENDENT ENVIRONMENTAL EXPERTISE

Because of the many environmental and community issues associated with a project of this magnitude, the environmental monitor also acts as an independent source of environmental expertise. The monitor advised the design teams and PENNDOT on ways to integrate the environmental commitments into design to ensure that they are implemented during construction. We have evaluated the design in concert with public issues and are evaluating the construction activities daily to determine whether or not additional mitigation is necessary and whether or not project impacts can be reduced even more. This function overlaps with the other responsibilities previously discussed. Other examples of our environmental application to the project include the following:

- Independent Noise Analysis — The representatives of an adjacent community requested that the monitor conduct an independent noise study to check PENNDOT's findings. We conducted the study, held meetings with the community, conducted some additional analyses in response to their concerns, and provided the community with a report for their review and comment before submitting it to PENNDOT.
- Water Quality Monitoring Program — A water quality program was developed by the environmental monitor to address the increasing concern that water quality would be significantly impacted during and after construction. This program includes sampling for turbidity and solids, heavy metals, and general parameters such as alkalinity, nutrients, pH, and dissolved oxygen. In addition, water quality sites upstream and downstream of two landfills in the corridor are monitored since the highway either passes through or near them. This program includes sampling before construction, during construction, and for 1 year after construction. The intent is to update baseline conditions documented in the EIS, provide an indication of any construction-related impacts, and document recovery after construction.
- Benthic Monitoring of Relocated Stream Reaches — In response to public concern about the impacts of stream relocation on biological communities, we are conducting an evaluation of two relocated stream reaches within the corridor. This evaluation will include benthic sampling within the relocated segment, upstream of the relocation, and upstream of any construction influence. The monitoring will focus on macroinvertebrate establishment within the relocations.
- Hazardous Waste Sampling — The monitor provides expertise in the area of hazardous waste in the event that the contractor encounters suspect materials and makes recommendations to PENNDOT on how they should proceed.

- Wetland Mitigation Monitoring — Two large wetland creation sites are being monitored to determine groundwater levels and vegetation establishment. Groundwater levels are measured to ensure adequate soil moisture conditions for hydrophytic plants. Circular plots have been established along transects to determine qualitatively the percent cover of plant species, the type of plants, the plants' wetland indicator status, and their vigor.
- Monitoring Construction Activities — Monitoring construction activities is done to ensure that designated protected areas such as existing vegetation and wetlands are not disturbed. Before construction began, the monitor walked each section with PENNDOT, the contractor, and the design consultant to designate areas to be protected. These areas were subsequently fenced off before the contractor began clearing activities.

DISCUSSION

The activities and responsibilities of the monitor have extended well beyond the original emphasis of the environmental monitor's role of monitoring the implementation of the commitments made in the EIS. Indeed, the 1981 Record of Decision stressed the commitments, and the scope for the environmental monitor's role stated that the monitor is limited to providing recommendations and information as they related to the environmental commitments; however, as the process evolved, it became much more. Our monitoring program has incorporated elements of another type of follow-up monitoring which has a different set of objectives: a monitoring and assessment program that conducts analyses to validate impacts and to provide information that could help to prevent or mitigate impacts on future similar projects. Studies such as water quality sampling, noise monitoring, and others have helped us to identify issues, address public concerns, and ensure that the commitments are being adequately met. We have found that there is a necessary exchange of information between the two types of follow-up monitoring programs.

The complex interaction of design, construction, environmental and public issues, and the number of involved parties dictate that the monitoring process cannot be done lightly. The monitoring process must have well thought-out procedures for addressing public concerns, identifying problems, and working toward a resolution of those problems. It requires that there be effective interaction with and cooperation by the owner (in this case, PENNDOT), the designers, and the contractors. This interaction is highlighted by Figures 2–4. Listed below are the advantages of the process and areas that need improvement, along with recommendations for future environmental monitor projects based on our experience to date.

Advantages

- provides a timely response to public concerns
- provides updated information to the public on design and construction issues

- ensures compliance with environmental commitments
- ensures that mitigation measures are properly implemented
- ensures compliance with state and federal environmental permits

Ensuring compliance refers to notifying PENNDOT of existing and/or potential noncompliance issues and providing recommendations to address them. The monitor is not responsible for enforcing the permits; that is the jurisdiction of PENNDOT and the regulatory agencies.

- identifying unforeseen environmental problems, design oversights, etc.
- providing advice and recommendations to reduce project impacts and to mitigate environmental problems
- provides cost-effective protection of the environment

The total cost for the I-476 project is approximately $700 million. The services of the monitor cost $3.8 million only, about 0.5% of the total project cost. The type and magnitude of the environmental services provided at such low cost are important considerations if follow-up monitoring programs are proposed for other projects.

Areas Needing Improvement

Perception of Our Role

Despite numerous reports and advertisements about our role, people have perceived us as being the EPA, officials of PENNDOT, and the Pennsylvania Department of Environmental Resources. This has led to some misunderstanding about the nature of our responsibilities and respective authority. Future monitoring projects should ensure that the role of the monitor is clear to the public as well as to the proponent's personnel.

Inspection vs Overseeing

As the monitor, our role is to oversee PENNDOT. During construction, we initially found ourselves doing routine daily inspection work (e.g., silt fence maintenance items). PENNDOT relied on us to provide them with maintenance reports of the environmental controls. Thus, we became more of an inspector rather than an environmental overseer. Since a monitor is responsible for reviewing many items, this has the potential to detract from the time needed elsewhere. This issue, however, was discussed with PENNDOT and rectified early in the construction phase.

Communication and Coordination

Although there has been extensive coordination and communication, a mechanism for future monitoring projects should be developed to ensure that the monitor automatically receives all project plans and reports essential for conducting the information and review aspects of the monitor's role.

Chain of Command Structure and Monitor Enforcement

The structure of the chain of command established for the monitor's role (Figure 2) presents some disadvantages. In order to resolve an issue related to the environmental commitments, the monitor must first try to resolve it at PENNDOT's district office level, then move to their central office level, then to FHWA, and finally to the Secretary of the U.S. Department of Transportation. Within this hierarchy, there are additional levels through which the monitor must pass to reach a resolution. This procedure can result in a lengthy and time-consuming process. The chain of command also applies to resolving issues and problems that arise during design and on the construction site. However, we have found that it is much more efficient to resolve issues through special purpose meetings at the district level and have been successful in doing so (Figures 3 and 4).

Owner Enforcement

Early in the construction phase, the monitor felt that strict enforcement penalties were needed to help PENNDOT ensure that the contractor complied with the environmental provisions of the contract. There have been instances where the contractor has not responded to an environmental issue within the time frame set within the provisions. The monitor recommended to PENNDOT a fine and a penalty system for noncompliance with these provisions. However, PENNDOT felt that their current system was adequate and elected not to follow our recommendations. We still recommend that owners ensure that they have detailed specifications for penalties if the contractor fails to comply with the environmental requirements of the contract.

Regulatory Agency Interaction

Based on the scope of the project, the monitor resolves environmental issues directly through PENNDOT. A mechanism for interacting with the regulatory agencies was never established, nor was protocol for coordination. We do attend agency meetings for the project and participate in field views of the project site. At these meetings, we discuss the project with respect to the environmental commitments and any environmental issues and resolutions. The local conservation district, which inspects the project for compliance with the erosion and sediment control permit from the Pennsylvania Department of Environmental Resources, requested and does receive copies of our Environmental Memorandums to PENNDOT.

The monitor does not always agree with agency recommendations or conclusions reached during their field views of the project site. For example, during a stream relocation, the contractor encountered a small outcrop of bedrock in the new channel. The monitor recommended to PENNDOT that the outcrop remain since it would contribute to the diversity of the channel morphology, enhancing it for the existing warm–water fishery. A regulatory agency questioned whether or not the rock had been placed in the new channel, since the rock could be considered as fill and may cause flooding and erosion problems. The agency felt that it must be removed, blasted out if necessary. The monitor wrote a letter to PENNDOT requesting the outcrop remain because

1. It complied with the commitments for varying the channel contours on relocated stream reaches.
2. Blasting in a live stream would create unnecessary impacts as well as violate existing permits.
3. Equipment in the stream to conduct the blasting and remove the material would violate the commitments and cause unnecessary impacts.
4. The outcrop had created a scour pool that was currently being used by fish.
5. Blasting would endanger the structures of the nearby highway bridge.
6. There was no evidence of flooding or erosion problem since the relocation.

Although we do not always agree with agency decisions, it is our role to ensure that PENNDOT abides by the final requests of the agencies and what PENNDOT commits to. Future monitoring contracts should define more clearly the interaction with regulatory agencies.

Contract Administration

Should there be an independent agency responsible for administering the contract for the monitor? Under the current program, the monitor is 90% federally funded and 10% state funded, with the funds being channeled through PENNDOT. This has occasionally raised the question of whether or not we are truly independent. If monitoring ever becomes a routine process for NEPA projects, then perhaps the contract should be administered by a completely independent agency, with the proponent still responsible for their portion of the funding.

Improvements for Future Monitoring

Of the areas of needed improvement, two are critical for future monitoring programs of this kind: the chain of command structure and proponent enforcement. These two items establish the process for resolving issues and ensuring that such resolutions, as well as the commitments and contract provisions, are carried out during construction. Despite these areas for improvement, the I-476 monitoring process overall has been effective. Just initiating such a program and the daily monitoring has helped to ensure that the EIS has been implemented and that the construction activities have complied with the commitments. This does not mean that there have not been difficulties with the process. A new system has been imposed on an existing one that incorporates standard procedures for design and construction. Thus, the interaction of these two systems has resulted at times in differing viewpoints and approaches, lively discussions, and disagreements on issues and ways to resolve them. This has been more evident during construction than during the design phase. The PENNDOT may also elect not to incorporate the monitor's recommendations after due consideration. This may be due to such reasons as cost, constructability, and so forth. Such actions may or may not create additional concerns and could initiate the hierarchial sequence previously discussed for resolving issues. The interaction has been beneficial, however, to the whole process. It has created an awareness of the issues and fostered environmental problem solving. This awareness is especially important for the contractor's and PENNDOT's field personnel since they were not part of the initial environmental planning and public participation process. The key to a successful approach for this kind of monitoring program, however, relies heavily on a theme we have mentioned before: coordination, communication, and cooperation.

Three final questions that are beyond the scope of this paper but that need to be mentioned and hopefully addressed by policy makers in the near future are as follows. Should environmental monitoring be mandatory? If so, should it be used on all NEPA projects or only a select few? Should such monitoring programs include both monitoring for implementation and monitoring to compare predictions with actual impacts? These questions need to be answered and action taken on this part of the environmental process that is crucial to successful implementation of NEPA documents.

REFERENCES

Council on Environmental Quality (CEQ). 1978. Regulations for Implementing the Procedural Provisions for the National Environmental Policy Act. U.S. Government Printing Office, Washington, D.C.

Larson, T. E. 1984. Overview of Departmental Recommendations on the Blue Route (Interstate 476). Pennsylvania Department of Transportation.

Pennsylvania Department of Transportation. 1984. Mid-County Expressway (I-476): Final Supplemental Environmental Impact Statement/Section 4(f) Evaluation, Volume 1.

Tools for Improving Predictive Capabilities of Environmental Impact Assessments: Structured Hypotheses, Audits, and Monitoring

D. P. Bernard, ESSA Environmental and Social Systems, Analysts Ltd., Vancouver, B.C., Canada; D. B. Hunsaker, Jr., Oak Ridge National Laboratory, Oak Ridge, TN; and D. R. Marmorek, ESSA Environmental and Social Systems, Analysts Ltd., Vancouver, B.C., Canada

ABSTRACT

In this paper, we recommend three tools for improving impact predictions: (1) structured impact hypotheses, (2) environmental audits, and (3) environmental monitoring. Use of these tools in combination, and formally integrating them into the environmental impact assessment (EIA) process could ensure continued evolution in our capability to accurately forecast environmental consequences of project activities. Use of these tools represents an inherently different paradigm, one which is based upon an experimental approach which acknowledges our incomplete knowledge of all ecosystem complexities.

Effects monitoring, when carried out conscientiously and with institutional commitment, leads to an expanding body of information and improved understanding of environmental impact mechanisms; both environmental audits and environmental monitoring contribute to formulating more accurate impact predictions. Structured impact hypotheses also play an important role in this process by clarifying the basis for impact predictions and assisting in impact auditing procedures. We believe that there is no role for *general purpose* environmental effects monitoring conducted without clear scientific focus or purpose. Each portion of a monitoring program should be targeted at either measuring some postulated environmental impact or at analyzing causal relationships between project activities and valued ecosystem components (VECs). Ultimately, incorporating auditing and monitoring as feedback loops in the EIA process will directly lead to significant improvements in the accuracy and usefulness of EIA predictions.

To explore the usefulness of these three tools, a careful examination of EIA documents for 11 hydroelectric projects was performed in Canada. The environmental audit yielded 2073 predictions (61% aquatic, 34% terrestrial, and 5% socioeconomic); only 602 were considered testable. The validity of 87 predictions were tested, using both original analyses of pre- and postdevelopment data and published studies. Although the data set in this case study was insufficient to permit quantitative generalizations on the accuracy of all predictions, conclusions were reached on the relative ease of making accurate predictions in different fields and what factors appeared to influence that accuracy. The detailed inventory of EIA predictions represents an objective and, in some cases, semiquantitative evaluation of the types of predictions found in EIA documents, the methodologies used to generate them, their testability, and (for those that were testable) their accuracy.

INTRODUCTION

Over 10,000 environmental impact assessment (EIA) reports have been composed worldwide in the past two decades.[1] These reports probably contain over 100,000 predictions concerning how ecosystems are expected to respond to some anthropogenic activity. Unfortunately, despite the very large number of EIA predictions that have been formulated, few attempts have been made to rigorously test their accuracy.

Credibility of the EIA process is affected by the accuracy of resulting predictions. Challengers contend that environmental forecasts are frequently imprecise or, worse, incorrect (Hecky et al. 1984; Munro et al. 1986). This problem is compounded by the tendency for some EIA authors to repeat forecasts found in existing documents when the activities and environmental conditions for a proposed project are judged to be similar to those described in an existing document (Rosenberg et al. 1981; Ward 1978). Unless the original forecast is known to be accurate, this approach may propagate erroneous predictions.

There have been frequent calls for a more scientifically sound approach to EIA (Draggan et al. 1987; Orians et al. 1986; Rosenberg et al. 1981; Schindler 1976). Adaptive environmental assessment and management (AEAM) is one such approach (Holling 1978; ESSA 1983; Walters 1986). From an AEAM perspective, there are five main points to consider when examining opportunities for improving the accuracy of impact predictions.

[1] In the period 1970–1980, 10,475 EIAs were written in the United States alone (Culhane 1987).

1. Each project is an experiment from which we can learn.
2. Environmental auditing and monitoring are the two main ways to improve our predictive capability.
3. Environmental auditing is an efficient and systematic means for checking the accuracy of EIA predictions and provides an important feedback loop in the EIA process.
4. Pre- and postproject environmental monitoring provides the basis for both impact prediction and auditing, as well as important feedback on actual responses in natural systems. This environmental monitoring is the foundation on which learning depends.
5. Ongoing analysis of failures and surprises can furnish penetrating insights that can provide the basis for improved impact prediction. These failures and surprises often reveal a vital component of ecosystem behavior that the EIA did not anticipate.

In this paper, three key steps to improving impact predictions are emphasized.

1. using structured impact hypotheses
2. monitoring actual environmental conditions both before and after development
3. adopting an adaptive/iterative paradigm in which prediction accuracy is continually evaluated and improved through environmental auditing procedures

Carefully designed environmental auditing and monitoring programs, carried out conscientiously with institutional commitment, can lead directly to significant improvements in the accuracy and usefulness of EIA predictions. Also, developing structured impact hypotheses can help streamline formulation of EIA forecasts while advancing the ability to design monitoring programs for evaluating the accuracy of resulting forecasts. Emphasis in this report is limited to methods and procedures associated with evaluating environmental consequences of proposed projects. Although social, economic, legal, or policy implications are not explicitly dealt with, it is believed that many of the points raised are relevant.

EIA PREDICTIVE CAPABILITIES

During the time the U.S. National Environmental Policy Act (NEPA) of 1969 became law, there were few tools and no accepted procedures available for making environmental forecasts. Before NEPA, there had been little demand for this type of knowledge and experience; however, during the early 1970s, it became clear that if EIAs were to be routinely prepared, as required by NEPA, then advances would be required in both procedural and methodological domains. A number of individuals and institutions, among them the U.S. Environmental Protection Agency (EPA), the President's Council on Environmental Quality (CEQ), and The Institute for Ecology, soon engaged in developing tools for predicting environmental effects of proposed projects. The CEQ also began formulating operational guidelines to help direct the EIA process. Throughout the past two decades, development and refinement of both EIA tools and procedures has continued in the United States and other countries; as a result, there are many tools available for producing environmental predictions (Everitt et al. 1987b; Sonntag et al. 1987) and several dozen methods for determining the significance of predicted environmental consequences (Thompson 1990).

In the EIA process, there have been few penalties for inaccurate predictions and fewer rewards for accuracy. Project proponents and regulatory agencies often have no interest in continuing environmental monitoring once a decision has been made to proceed with a particular option. Thus, the EIA becomes like an exam or essay that is never marked by the instructor. Forecasts that are wrong are not only useless, but also may be misleading when determining a project's net benefit. This, in turn, interferes with selection of an optimum alternative (Culhane 1987) and brings about criticisms of the EIA process.

There have been a number of outspoken critics of the EIA process and of the capabilities for producing accurate predictions of environmental effects. However, as Kuhn (1970) points out, at about the same time as NEPA became law, identifying flaws and defects in an existing paradigm does not by itself induce a shift to a more acceptable model. Rather, there must be some available superior replacement paradigm. Both Culhane (1987) and Tomlinson and Atkinson (1987a) note that recognizing limitations to current EIA processes and predictive abilities is not synonymous with calling for abandoning the concept of EIA or the NEPA process. Although environmental auditing and monitoring do not represent a new paradigm, both of these tools improve the accuracy of EIA predictions. They represent a different paradigm, inherently reflecting an experimental approach which acknowledges our ignorance rather than the often implicit assumption in EIAs that the experts know the answer.

ENVIRONMENTAL AUDITING

The term *environmental auditing*, as used in this paper, refers to technical activities associated with evaluating the accuracy of scientific forecasts of environmental consequences for a specific project. Tomlinson and Atkinson (1987b) recommend calling this type of activity a predictive techniques audit or an EIA procedure audit. As used here, environmental auditing does not refer to an assessment of how well a particular company or government body performs in the technical or management arena relative to voluntary or legislative environmental standards, sometimes referred to as a *green audit* (Jackson 1990).

A number of authors have already written about environmental auditing (Canter 1985; Clark et al. 1987; Culhane 1987; Cunningham et al. 1977; Gore et al. 1976a, b; Hecky et al. 1984; Marmorek et al. 1986; Munro et al. 1986; Sadler 1987; Tomlinson and Atkinson 1987a, b). Generally, two basic approaches to environmental auditing are available. The first, characterized as a post/preaudit, involves identifying the most important or significant environmental impacts resulting from a project and then inspecting the predevelopment EIA to ascertain whether the impact was adequately predicted in advance. In the second type, known as a pre/postaudit, all predictions in the EIA are inventoried and individually evaluated for accuracy. As illustrated later in the case study, this type of approach lends itself well to development of generic impact hypotheses. These approaches can then serve as a framework to aid in developing impact predictions for prospective projects and for guiding design of monitoring programs to generate data needed to test prediction accuracy.

Several factors can complicate environmental auditing. For example, predictions are sometimes predicated on a prerequisite set of conditions that do not develop as expected. Thus, the prediction is ultimately irrelevant. More commonly, predictions are expressed in an unclear or untestable form, or they are scattered throughout the supporting documentation. In either case, faithfully extracting a concrete, comprehensive, and accurate set of predictions from such an EIA can be an arduous task. Finally, a recurring and often insurmountable problem in auditing the accuracy of EIA predictions is the elementary lack of even rudimentary monitoring data.

For the above reasons and many others, environmental auditing can be a complicated and demanding task and is currently seldom performed. Relative to the number of effects predictions that have been presented in EIAs, there have been only spotty attempts to evaluate our proficiency at developing reliable impact forecasts, partly because neither penalties nor rewards are conferred for inaccurate or accurate predictions. A more immediate and likely cause of the shortage of audited EIAs is lack of necessary data.

Although environmental scientists have suspected that many EIA predictions are quantitatively inaccurate, we are now becoming aware of how frequently impact predictions contained in EIAs are erroneous or faulty. After conducting an extensive review of nearly 240 impacts forecast in EIA documents, Culhane (1987) concluded "environmental assessments are inadequately quantified, only guardedly accurate, and generally unimpressive as rational-scientific exercises." As shown in our case study (see below), predictive capabilities tend to be stronger for first-order impacts (e.g., physiochemical phenomena) and are particularly weak for second- and higher-order biological impacts. For some domains, accurate quantitative predictions are simply not possible with our current understanding and tools. For example, the EIA process is unable to accurately predict even short-term, first-order effects of multiple, interacting air pollutants on forest ecosystems (Torrenueva 1988). Our capability of quantifying intermediate or more long-term effects requires considerable strengthening. Nonetheless, in some cases, accurate but qualitative predictions may be adequate.

Even if environmental auditing were to become a high priority, prediction accuracy will be largely undeterminable until monitoring programs are in routine use. As a result of this prediction inaccuracy, government agencies and corporate sponsors are generally unaware of whether EIA documents prepared on their behalf contain accurate environmental predictions. More importantly, without monitoring data, it is difficult to identify whether mitigation measures are required to lessen effects of project actions or if existing mitigation measures are effective. Consequently, no one can know with reasonable certainty whether the NEPA process is actually protecting natural resources.

When coupled with a well-designed environmental monitoring program, auditing can help provide a clear knowledge of where particular EIA methods can be successfully applied to generate accurate predictions and where their application should be restricted. Through such knowledge, needless errors can be avoided when projecting the environmental consequences of project activities. Applying a combination of auditing and monitoring activities will allow us to efficiently gain this knowledge and experience of applying particular EIA methods in a systematic manner.

ENVIRONMENTAL MONITORING

There are many possible objectives for a monitoring program, and the design varies accordingly.

1. determine compliance of a pollution control program
2. environmental sampling to verify data in a database
3. describe extent of environmental effects and resource losses
4. evaluate effectiveness of alternative mitigation measures
5. evaluate accuracy of effects predictions

In all cases, monitoring must consist of more than simple data collection. A carefully formulated hypothesis is a prerequisite for the design of any monitoring program. This hypothesis is necessary for effective identification of measurement variables, sampling frequency and duration, and spatial extent.

Structured Impact Hypothesis

Our experience in developing monitoring programs, especially for addressing the objectives in Steps 3–5, has shown structured impact hypotheses are an excellent starting point. An example of a structured impact hypothesis is provided in Figure 1. Impact hypotheses embody two main elements joined by a linkage. The first element is an action (Holling 1978), typically one or more anthropogenic activities. In Figure 1, the actions associated with project development are reservoir filling and reservoir construction and operation. The second element is one or more VECs (Beanlands and Duinker 1983). For example, in Figure 1, water quality is the VEC. These two elements (actions and VECs) are linked through some functional connection or coupling. The hypothetical linkage may be expressed in any number of ways depending on the level of understanding of the system (e.g., a numerical relationship, a theoretical conceptual connection). Indicators (Holling 1978) are used to measure the effect of an action on a VEC. For example, in the case of biodiversity (a VEC), the relative abundance of species (indicator) can be monitored. In the case of Figure 1, the indicators could include temperature, heavy metal concentrations, dissolved oxygen content, and so on.

Linkages between actions and VECs are not always straightforward and may involve many intermediate connections among ecosystem components. For example, timber harvesting may affect mammals directly through habitat modifi-

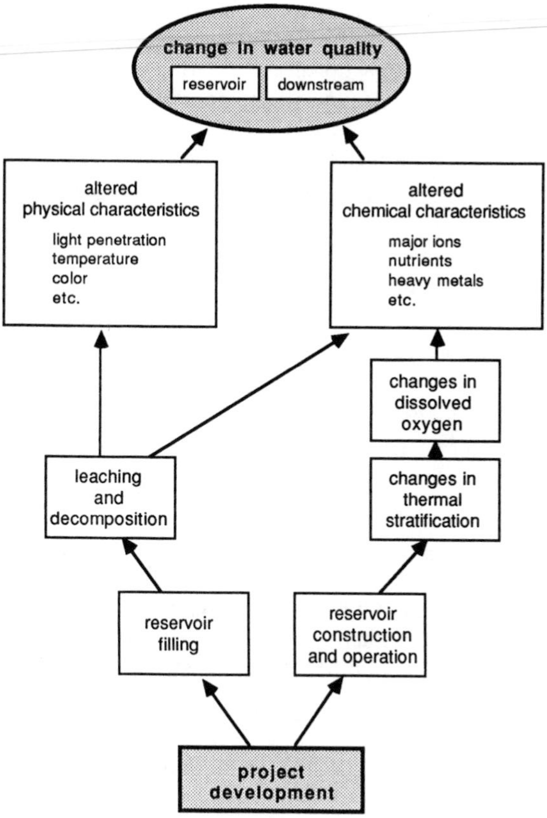

Figure 1. Generic impact hypothesis for water quality impacts from hydroelectric developments.

cation or indirectly through changes in food supply. A conceptual model of an ecosystem may contain many linkages, and when developing an impact hypothesis it is important to identify the key linkages represented in the model. Both primary impacts and secondary (higher-order) effects may be of interest.

Scoping

For the purposes of this paper, scoping is defined as the procedure in which actions, linkages, and VECs are extracted from the conceptual model (Everitt et al. 1987a). With respect to monitoring, scoping helps to select predicted impacts on which scarce resources available for monitoring should be spent. Conover (1987) identifies three key factors that should be considered when identifying impacts for monitoring.

1. significance of the potential impact (significant impacts would receive higher priority)

2. probability of impact (a virtually certain impact would receive lower priority in monitoring)
3. potential for significant environmental improvements that would result from information provided by the monitoring program

Scoping can occur very effectively during a highly focused workshop attended by a diverse group of disciplinary experts. Interdisciplinary workshops can be an important tool in the EIA process and are the core of the AEAM approach (Holling 1978; ESSA 1983; Walters 1986). One major focus of such interdisciplinary workshops is identifying key processes and resource populations most likely to be significantly affected by the proposed actions. In this setting, workshop participants are often called upon to employ not only data and academic knowledge, but also expert judgment and professional intuition. Since available resources normally preclude monitoring all anticipated impacts predicted in the EIA document, it is appropriate for disciplinary experts, working in an interdisciplinary forum, to furnish input to the process of selecting actions, indicators, and spatial and temporal scales for monitoring. In this process, structured impact hypotheses can create an organized and visual framework to guide the workshop discussion. A particularly important set of decisions concerns those impacts that are unlikely to be detected because of their high natural temporal or spatial variation. Are these really worth monitoring? Furthermore, impacts whose magnitudes fall within the existing (i.e., preproject) variability of the impact magnitudes are not likely to be considered significant (and, therefore, potential candidates for monitoring) in the first place.

A number of important technical and scientific issues are involved in developing and operating any monitoring program; these include careful problem definition, detailed design for the experimental or monitoring program, field program implementation, data analysis and management, data interpretation, policy implication of results, integration between groups, and administrative support. Although all these issues are important, in the following paragraphs we focus mainly on the first two.

Design Monitoring During Scoping

Monitoring of actual impacts must be incorporated into the EIA process from the outset. Failure to do so may result in a lack of data measuring key impacts. A number of EIA follow-up studies (Canter 1985; Marmorek et al. 1986; Sadler 1987) have identified lack of monitoring data as an important constraint to the usefulness of follow-up exercise. To be most effective, the monitoring program should be developed during the scoping process when key environmental impacts are identified for analysis. Clark et al. (1987) note that existing monitoring programs often produce data that are not useful in testing predictions in a scientifically acceptable manner. They also note that impact predictions are commonly not phrased in a manner conducive to verification. Frequently, monitoring is carried out in isolation from other EIA activities, such as impact identification and prediction.

Limit Predictions to Monitored Ecosystem Components

A related issue is whether quantitative predictions in EIA documents should be limited to those impacts for which pre- and postdevelopment monitoring of actual effects is planned, based upon criteria such as those proposed by Conover (1987) and discussed previously in this paper (i.e., the probability of the impact, the significance of the potential effect, and the benefits of the information obtained from monitoring vs the cost of conducting the monitoring). Describing such predicted impacts in terms of these criteria would help make the EIA process more efficient and would ensure that we are continually able to improve our predictive capabilities. Impacts whose outcomes are very certain or uncertain would not be monitored unless specific information is needed for mitigation or compensation.

Obtain Key Baseline Data Early On

It is important for monitoring programs conducted in support of the EIA process to establish baseline or reference conditions by measuring key impact parameters before project implementation and to document effects of a proposed action by measuring the same parameters after project implementation (Conover et al. 1987). The statistical ability to detect impacts depends directly on the duration of preproject monitoring (Marmorek et al. 1986). Acquisition of baseline data should be iteratively linked with impact predictions and requirements for impact monitoring (Clark et al. 1987). Baseline conditions need not be measured at the site of the proposed action; conditions outside the zone of influence of a proposed action may constitute an experimental control equal to or greater than site-specific, predisturbance conditions (Conover 1987). Care should be taken to avoid pseudoreplication (Hurlburt 1984; Stewart-Oaten et al. 1986).

Decide on Zones of Influence

Conover (1987) identifies activities with three zones of influence:

1. physical disturbance (i.e., construction activities)
2. normal operations (i.e., the distances to which air emissions, water effluents, and so forth may create effects distinguishable from baseline conditions)
3. major accidental events

Monitoring activities conducted in support of an EIA would normally focus on the zones in Steps 1 and 2, unless the probability of occurrence or significance of accidental environmental damage is unusually large. Lack of adequate baseline monitoring data has been identified as a hindrance to effectively documenting the environmental effects of accidents (Hunsaker and Lee 1987).

Monitor to Detect Surprises

In ecological terms, most pre- and postdevelopment monitoring programs are confined to a relatively short time frame. Thus, over intermediate or longer time periods, we should remain open to the possibility of surprises resulting from our project activities. Consequently, monitoring should be viewed as an activity that is done not only to measure the magnitude of short-term impacts, but also to analyze cause-and-effect relationships. The principal goal in measuring impact magnitudes is to measure the component of the impact caused by the action being evaluated. Understanding mechanisms through which effects are caused by a particular activity may assist in predicting intermediate or longer-term environmental consequences.

Combine Modeling and Monitoring

The fallibility of predictive tools used in forecasting potential impacts makes monitoring essential if we are to improve our capacity for environmental management. Uncertainties in modelling tools, changes in project attributes affecting the environment, and an incomplete understanding of linkages can result in a wide discrepancy between predicted impacts and actual environmental effects. Monitoring and predictions using computer simulation models can be considered as tools designed to perform the same task (i.e., identify magnitude of impacts) but on opposite ends of a continuum of uncertainty and a continuum of cost. Computer simulations allow quantitative expression of an impact at a lower cost than monitoring but usually with an uncertainty or error-bar that can be a large fraction of the prediction. Properly designed and implemented monitoring, on the other hand, provides a quantitative measure of impacts but with greater certainty and higher cost. Combinations of modelling and monitoring can be used to provide quantitative information within budget constraints. For example, in the United States, plans to attain ambient air quality standards were developed using a combination of fixed air monitoring stations and computer modelling predictions to fill gaps in the data of the monitoring network. In terms of a feedback loop, data monitoring helps to refine predictions, and predictions (which can be created with fewer resources) help to define and scope the nature and extent of required monitoring activities.

Direct monitoring of actual effects may not always provide all necessary information to develop mitigation strategies to reduce impact magnitude. For example, in the case of cumulative impacts (e.g., regional air quality), it may be difficult to identify the contribution of a particular source or group of sources to the measured impact (e.g., high sulphur dioxide levels or acid rain). In these cases, additional tools, such as receptor modelling or tracer studies, may be needed to obtain the necessary information.

CASE STUDY

Recognizing the need for a more scientifically sound approach to the EIA, the Canadian Electrical Association sponsored a research project to examine the extent to which environmental impact predictions from EIAs of hydroelectric projects in Canada have been confirmed by postdevelopment experience (Marmorek et al. 1986). Results from this study provide a uniquely detailed and comprehensive set of environmental audits that can be used to illustrate the importance and value of environmental monitoring.

In Canada, more than 370 hydroelectric dams and impoundments are already in place, affecting the hydrology and ecology of over 200 major rivers and streams. On the basis of a number of criteria, a subset of these potential sites was selected for detailed examination. The criteria included requirements for (1) existence of a detailed EIA encompassing aquatic, terrestrial, and socioeconomic effects; and (2) existence of at least some pre- and postdevelopment data. Application of the criteria resulted in elimination of most Canadian hydroelectric reservoirs, primarily because of the absence of an existing EIA (Criterion 1). However, 11 hydroelectric projects did satisfy all the criteria and were selected for detailed investigation. The hydroelectric projects selected were

- Site One Dam, Peace Canyon, British Columbia
- Mica Dam, McNaughton Lake, British Columbia
- Seven Mile Dam, Pend d'Oreille River, British Columbia
- Revelstoke Dam, British Columbia
- Nipawin Hydroelectric Project, Saskatchewan
- Churchill-Nelson, Lake Winnipeg, Manitoba
- LaGrande, James Bay, Eastmain-Opinaca and Caniapiscau Diversions, Quebec
- Wreck Cove Hydroelectric Project, Nova Scotia
- Cat Arm Development, Newfoundland
- Hinds Lake Hydroelectric Project, Newfoundland
- Upper Salmon, Newfoundland

The analysis proceeded through three phases: enumeration of all predictions within the EIA documents, a careful audit of each forecast to determine traits such as testability, and analysis of testable predictions to determine accuracy. During Phases 1 and 2, a number of complications were encountered: (1) some predictions were based upon conditions that never developed; (2) other predictions were expressed in an unclear or untestable form; (3) predictions were occasionally scattered throughout the supporting documentation; and (4) available impact monitoring data were sometimes inadequate to evaluate the predictions. Structured impact hypotheses were developed to help us confront these difficulties and to focus the audit activities.

A careful review of the EIA documents generated an inventory of 2073 impact predictions. Of these, 61% were aquatic, 34% were terrestrial, and 5% were socioeconomic. Of the total, only 602 predictions were judged to be testable after our first pass. Again, most of these impact predictions dealt with aquatic impacts such as morphometry, hydrology, lower trophic levels, fish, and water quality. To simplify this case study, predictions revolving around water quality are the main focus; readers interested in other results may consult Marmorek et al. (1986).

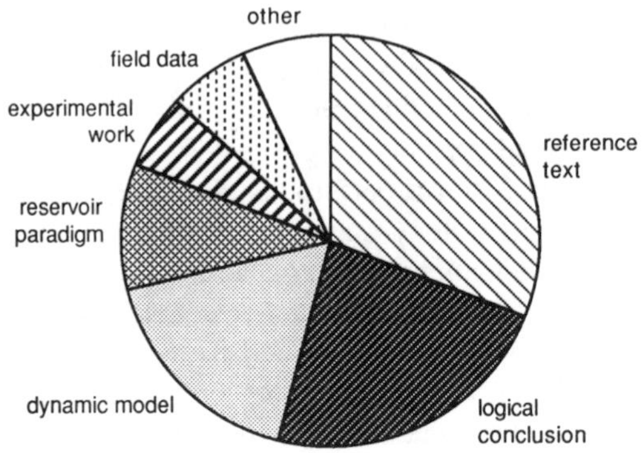

Figure 2. Supporting basis for impact predictions.

In total, 239 water quality predictions were identified from the 11 EIAs. All of the predictions can be represented within a single generic impact hypothesis, illustrated earlier in Figure 1. In this example of a structured impact hypothesis, project development involves two sets of actions: reservoir filling and reservoir construction and operation. The VEC is water quality and there are two main pathways, composed of several linkages each. For example, reservoir filling is hypothesized to result in vegetation decomposition in the flooded watershed. This, in turn, leads to leaching, which can affect both physical characteristics of the water column (e.g., light penetration) and chemical composition (concentration of dissolved nutrients). Finally, modifications to either the physical or chemical characteristics of the reservoir waterbody result in changes in water quality. Using this framework, we were able to examine underlying methods, logic, and assumptions associated with the formation of each water quality prediction in the audited EIA documents.

About two thirds of the 239 water quality predictions were the result of either consulting a reference text, referring to the reservoir paradigm, or simply making a logical conclusion (Figure 2). Only about 30% of the predictions for the 11 hydroelectric projects had been formed using field data, experimental work, or computer modeling. As a result, roughly 70% of all water quality predictions in these 11 EIAs were qualitative. A typical qualitative prediction contains directional information but no numerical forecast (e.g., "nutrient concentrations will increase").

Predicted water quality impacts were further examined to distinguish between testable and nontestable forecasts. Of the 239 predictions, under 30% were testable, but mainly only for directionality, not magnitude of change. For those predictions that could not be tested, the main reason was data unavailability; some predictions were not testable because they were either too trite or too vague.

To test impact predictions, normally both pre- and postdevelopment monitoring data are required. In some of the EIAs examined, testable impact predictions were available along with predevelopment data, but lack of follow-up monitoring during the postdevelopment phase prevented examination of prediction accuracy. Consequently, on the second pass (looking in detail at available data), it was found

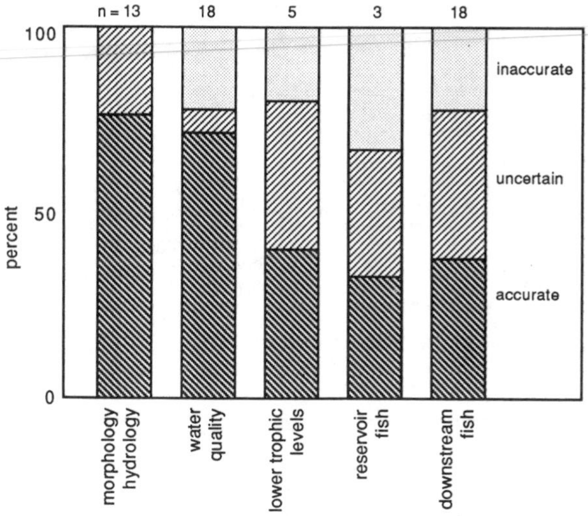

Figure 3. Prediction accuracy for five valued ecosystem components (VECs).

that only 70 of the 239 water quality predictions were actually testable, and only about one third of these could be analyzed statistically.

To further evaluate the general degree of correctness for the water quality predictions, a statistical analysis was performed on 18 predictions from the subset of those that were statistically testable. We found that 13 of the 18 examined predictions were accurate, and only 4 of 18 were inaccurate; for one of the tested predictions, results were uncertain (Marmorek et al. 1986). This outcome is encouraging and suggests that in cases where high quality environmental data exist, predictions of water quality effects can be reasonably accurate.

Prediction accuracy was also evaluated for five general types of VECs: morphology and hydrology, water quality, lower trophic levels, reservoir fish populations, and populations of fish downstream of the reservoir. Water quality and morphology and hydrology impacts were found to be the most accurate of the aquatic resource predictions (Figure 3). In these categories, 70–75% of the impact predictions were judged to be accurate. Impacts involving biological resources, particularly higher-order impacts, tended to be less accurate (Figure 3).

From such analyses, it appears that prediction accuracy depends not only upon availability of data from a monitoring program, but also on the number of conceptual links in the underlying impact hypothesis. The longer the chain of linkages, the greater the probability of generating an inaccurate prediction. Also, the longer the chain, the greater the challenge in determining why a prediction failed. This suggests a relatively high priority for environmental monitoring of biological components of aquatic ecosystems, particularly to help in identifying and verifying higher-level impacts and for revising our conceptual and computer models of how these systems behave.

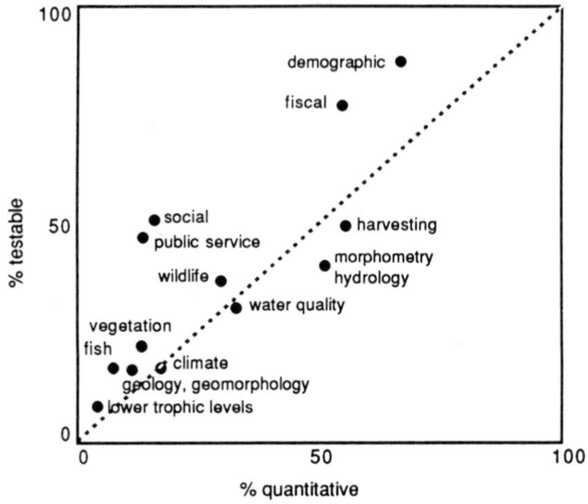

Figure 4. Relationship between percent of predictions that are quantitative and percent that are testable.

It appears that there is a rough relationship between the degree to which predictions are quantitative and their testability (Figure 4). Impacts to climate, for example, tended to be about 20% quantitative and about 20% testable. Impacts dealing with morphometry and hydrology were about 50% quantitative and about 40% testable. Not surprisingly, some of the most testable forecasts are those pertaining to demographics and fiscal impacts (Figure 4), both of which are often expressed as numerical predictions. To ensure that projected changes can be examined for accuracy, it is therefore important that as many predictions as possible be formulated in quantitative terms or at least explicit qualitative terms, since vaguely defined impacts are difficult to classify as either accurate or inaccurate. Risk assessment is playing an increasingly important role as a tool in predicting environmental effects, primarily because of the difficulties in making accurate predictions or monitoring the occurrence of such predictions over long periods of time.

DISCUSSION

In our case study of EIA documents, we found that data generated through environmental monitoring programs are important both in formulating testable impact predictions and for judging the accuracy of effects forecasts. A major finding was that the accuracy of impact predictions depends partly upon the number of links in the hypothesis and upon the completeness of our understanding of the overall system defined by the linkages. The longer the chain of reasoning, the more likely that the overall hypothesis will prove inaccurate, thereby generating unexpected effects. This is consistent with the findings of Hecky et al. (1984) who reported that relative to their experiences at Southern Indian Lake in northern Manitoba, environmental impact predictions pertaining to physical portions of the natural environment were qualitatively correct, while biological responses above the primary trophic level were mostly not predicted or were predicted incorrectly.

A structured approach to formulating impact predictions and monitoring programs can help reduce both the frequency and seriousness of surprises. Perhaps even more important, data from a well-designed monitoring program can help indicate where errors have been made in our reasoning. Such information is essential if we are to learn from past failures to accurately predict environmental impacts.

Globally, many national policies to protect natural resources are approximately 20 years old. As a consequence of both their age and increased social understanding of environmental issues, many of these policies are now under careful scrutiny. In several jurisdictions, new legislation to strengthen the environmental impact assessment process is currently being suggested. In some cases, proposed enhancements include incorporating some of the ideas advanced in this paper. For example, under proposed Canadian legislation (Bill C-78), the EIA process will require in most cases follow-up programs for the dual purpose of verifying EIA prediction accuracy and evaluating efficacy of mitigation measures.

In the United States, calls have recently been made to enhance or extend NEPA to include the capacity to address large-scale international environmental issues such as stratospheric ozone depletion, acidic deposition, tropical deforestation, and global climate change (Caldwell 1989). There are, however, practical limits to environmental monitoring. For some of these global issues, it may take too long to observe the planet's responses through ecological monitoring; by the time we have tested our predictions, we may well have passed the time for changing our actions to lessen impacts. On another front, the United Nations Environment Programme is currently promoting the idea of an international treaty that would require an EIA of any major project or action that could adversely affect the environment of another nation or the global commons (Caldwell 1989).

If the NEPA process is to continue, it will be important to persevere in developing and refining methods capable of producing accurate EIA forecasts. As a first step toward that goal, an increasing amount of effort should be directed toward evaluating accuracy and precision of EIA forecasts generated using existing methods.

CONCLUSIONS AND RECOMMENDATIONS

We recommend three tools for improving impact predictions: structured impact hypotheses, environmental audits, and environmental monitoring. These tools should be used in combination and, if formally integrated into the EIA process, would provide the means to ensure continued evolution in our capability to accurately forecast environmental consequences of project activities. As illustrated in Figure 5, effects monitoring leads to an expanding body of information and

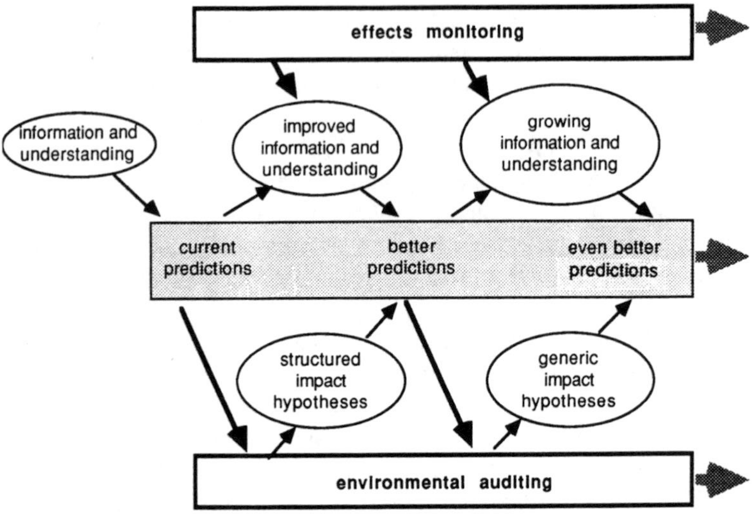

Figure 5. Process by which structured impact hypotheses, environmental audits, and effects monitoring programs contribute to improving impact prediction accuracy.

improved understanding of mechanisms, both of which contribute to formulating more accurate impact predictions. In addition, structured impact hypotheses play an important role in this process by both clarifying the basis for impact predictions and assisting in impact auditing procedures.

Ideally, environmental monitoring programs should be based upon structured impact hypotheses. We believe that there is no role for *general purpose* environmental effects monitoring. Each portion of a monitoring program should be targeted at either measuring some postulated environmental impact or at analyzing causal relationships between project activities and VECs. Ultimately, incorporating auditing and monitoring as feedback loops in the EIA process will reduce the occurrence and magnitude of adverse impacts from proposed actions, which is the ultimate goal of environmental protection legislation and the EIA process.

REFERENCES

Beanlands, G. E., and P. N. Duinker. 1983. An Ecological Framework for Environmental Impact Assessment in Canada. Institute for Resource and Environmental Studies, Dalhousie University and Federal Environmental Assessment Review Office, Ottawa.

Caldwell, L. K. 1989. A constitutional law for the environment: Twenty years with NEPA indicates the need. *Environment* 31(10):6–11, 25–28.

Canter, L. W. 1985. Impact prediction auditing. *Environ. Prof.* 7:255–264.

Clark, B. D., R. Bisset, and P. Tomlinson. 1987. Environmental assessment audits in the U.K.: Scope, results, and lessons for future practice. pp. 519–540. In B. Sadler (ed.), *Audit and Evaluation in Environmental Assessment and Management: Canadian and*

International Experience. Volume 2: Supporting Studies. Beauregard Press, Ltd., Canada.

Conover, S. A. M. 1987. Environmental effects monitoring and environment Canada: A synthesis of the findings of four workshops. pp. 408–434. In B. Sadler (ed.), *Audit and Evaluation in Environmental Assessment and Management: Canadian and International Experience. Volume 2: Supporting Studies.* Beauregard Press, Ltd., Canada.

Conover, S. A. M., L. White, and A. Janz. 1987. Use of Effects Monitoring in Environmental Impact Assessment and Environmental Management. Report prepared for Environment Canada, Federal Environmental Assessment and Review Office, Ottawa.

Culhane, P. J. 1987. The precision and accuracy of U.S. Environmental Impact Statements. *Environ. Monit. Assess.* 8:217–238.

Cunningham, P. A., S. M. Adams, and K. D. Kumar. 1977. Evaluation of Environmental Impact Predictions. CONF-770516-7. Oak Ridge National Laboratory, Oak Ridge, TN.

Draggan, S., J. J. Cohrssen, and R. E. Morrison. 1987. *Environmental Monitoring, Assessment, and Management.* Praeger, New York.

ESSA Environmental and Social Systems Analysts Ltd. 1983. Review and Evaluation of Adaptive Environmental Assessment and Management. En21-36/1983, p. 116. Environment Canada, Ottawa.

Everitt, R. R., D. L. Colnett, and P. Bunnell. 1987a. Methods for Determining the Scope of Environmental Assessments. Federal Environmental Assessment Review Office, Ottawa.

Everitt, R. R., D. L. Colnett, and P. Bunnell. 1987b. Initial Environmental Assessment: Procedures and Practice. Instructor's Manual, Federal Environmental Assessment Review Office, Ottawa.

Gore, K. L., et al. 1976a. Evaluation of Haddam Neck (Connecticut Yankee) Nuclear Power Plant, Environmental Impact Prediction, Based on Monitoring Programs. BNWL-2151. Battelle Northwest Laboratories, Richland, WA.

Gore, K. L., et al. 1976b. Evaluation of Monticello Nuclear Power Plant, Environmental Impact Prediction, Based on Monitoring Programs. BNWL-2150. Battelle Northwest Laboratories, Richland, WA.

Hecky, R. E., et al. 1984. Environmental impact prediction and assessment: The Southern Indian Lake Experience. *Can. J. Fish. Aquat. Sci.* 41(4):720–732.

Holling, C. S. (ed.). 1978. *Adaptive Environmental Assessment and Management.* Wiley, New York.

Hunsaker, D. B., Jr., and D. W. Lee. 1987. Environmental impact analysis of abnormal events: A follow-up study. pp. 379–398. In B. Sadler (ed.), *Audit and Evaluation in Environmental Assessment and Management: Canadian and International Experience. Volume 2: Supporting Studies.* Beauregard Press, Ltd., Canada.

Hurlburt, S. H. 1984. Pseudoreplication and the design of ecological field experiments. *Ecol. Monogr.* 54(2):187–211.

Jackson, S. 1990. The green litmus test: An environmental audit. *Inside Guide* 4(5):45–49.

Kuhn, T. S. 1970. *The Structure of Scientific Revolutions.* University of Chicago Press, Chicago.

Marmorek, D. R., et al. 1986. Predicting Environmental Impacts of Hydroelectric Developments in Canada. CEA 317-G-472. Canadian Electrical Association, Montreal.

Munro, D. A., T. J. Bryant, and A. Matte-Baker. 1986. Learning from Experience: A State-of-the-Art Review and Evaluation of Environmental Impact Assessment Audits. Canadian Environmental Assessment Research Council, Ottawa.

Orians, G. H., et al. 1986. *Ecological Knowledge and Environmental Problem Solving: Concepts and Case Studies.* National Academy Press, Washington, D.C.

Rosenberg, D. M., et al. 1981. Recent trends in environmental impact assessment. *Can. J. Fish. Aquat. Sci.* 38(5):591–624.

Sadler, B. (ed.). 1987. *Audit and Evaluation in Environmental Assessment and Management: Canadian and International Experience.* Vols. 1 and 2. Beauregard Press, Ltd., Canada.

Schindler, D. W. 1976. The impact statement boondoggle. *Science* 192:509.

Sonntag, N. C., et al. 1987. Cumulative Effects Assessment: A Context for Further Research and Development. CEARC Canadian Environmental Assessment Research Council, Ottawa.

Stewart-Oaten, A., W. W. Murdoch, and K. R. Parker. 1986. Environmental impact assessment: Pseudoreplication in time? *Ecology* 67(4):929–940.

Thompson, M. A. 1990. Determining impact significance in EIA: A review of 24 methodologies. *J. Environ. Manage.* 30:235–250.

Tomlinson, P., and S. F. Atkinson. 1987a. Environmental audits: A literature review. *Environ. Monit. Assess.* 8:239–261.

Tomlinson, P., and S. F. Atkinson. 1987b. Environmental audits: Proposed terminology. *Environ. Monit. Assess.* 8:187–198.

Torrenueva, A. L. 1988. Effects of Air Pollution on Forests: Phase II — Comparison of North American and European Decline. CEA-318G-394B. Canadian Electrical Association, Montreal.

Walters, C. J. 1986. *Adaptive Management of Renewable Resources.* MacMillan Publishing Co., New York.

Ward, D. V. 1978. *Biological Environmental Impact Studies: Theory and Methods.* Academic Press, New York.

Implementation of Mitigation at U.S. Army Corps of Engineers Water Projects in the Southern United States

B. M. Young, The Earth Technology Corporation, Alexandria, VA

ABSTRACT

Environmental impact statements (EISs) identify mitigation measures that are to be incorporated into the design and operation of proposed projects. This paper presents an assessment of the extent to which mitigation measures identified in EISs for 22 proposed U.S. Army Corps of Engineers water projects were implemented in the completed projects. There were 212 mitigation measures identified in EISs written between 1972 and 1982. A field survey was conducted during which a compliance score of 0 through 4 was assigned to each of the mitigation measures. The scores indicate the extent to which mitigation measures identified in an EIS were implemented on a project (0 = mitigation not completed with no plans to complete; 4 = mitigation completed beyond that described in an EIS). Five factors that were thought to influence the extent to which mitigation measures are implemented were analyzed: strategy of mitigation, threatened resource, date of EIS preparation, caretaking authority, and project purpose. A chi-square test was used to measure the strength of the relationship between compliance score and the five different factors.

Overall, 65% of the 212 mitigation measures were completed as described or beyond that described in the EIS. The remaining 35% were either inoperable, partially complete, or incomplete. It appears that the U.S. Army Corps of Engineers was moderately successful in implementing the mitigation measures identified in the EISs. Results also show that mitigation compliance was correlated with the date of the EIS; however, compliance was not associated with the other four factors.

INTRODUCTION

The intent of this study was to test whether or not a large number of mitigation measures identified in environmental impact statements (EISs) had, in fact, been completed. A brief discussion of background is provided followed by a discussion of the methodology, the results of the research, and some policy implications. Field observation that provided the basis for this paper took place in the winter and spring of 1986 (Young 1986).

The National Environmental Policy Act (NEPA) requires a federal agency to document the potential environmental consequences of their actions significantly affecting the environment in an EIS. If significant adverse impacts are found, the agency identifies mitigation measures to reduce the magnitudes of the potential effect. Mitigation measures are designed to avoid, minimize, or compensate for the anticipated adverse environmental impact. NEPA regulations require the consideration of mitigation in the planning process; however, compliance is rarely monitored by the responsible agency and environmental groups or actively enforced by NEPA's administering agency, the President's Council on Environmental Quality (CEQ).

The importance of adequate implementation of mitigation measures identified in EISs was alluded to by Lynton Caldwell, the advisor to Congress who is credited with the EIS concept. In his book, *Science and NEPA*, he writes "to make the [environmental impact] statement an end product of the act is plainly contrary not only to the stated purpose of the act, but also to its legislative history. The EIS can neither be understood nor properly evaluated if treated as an end in itself; it was intended and should be regarded as no more than a means to an end" (Caldwell 1982, p. 6).

Methodology

Mitigation measures were compiled from 22 EISs written by the U.S. Army Corps of Engineers (Corps) between 1972 and 1982. These EISs were selected from a large pool of final EISs prepared by the Corps in EPA Region 4 and submitted to EPA Headquarters in Washington, D.C. between January 1, 1970 and December 31, 1983. The criteria applied in the selection process are shown in Table 1.

Each EIS was read to determine mitigation measures identified in them. A total of 212 mitigation measures were identified. Site visits and interviews were conducted at 22 water projects to determine the extent to which mitigation measures were implemented. A minimum of two interviews, lasting 1–3 h, was conducted for each project. The primary interviewee was a Corps' employee or local authority responsible for the day-to-day operations of the project. The second interviewee was either a participant in the EIS commenting process, a person named during the initial interview, or an informed resident in the vicinity of the site. An onsite inspection was then conducted during which a field description and photographs of each mitigation measure were obtained.

Table 1. U.S. Army Corps of Engineers EISs Eliminated in Screening.

Pool:	279 (Final EISs in EPA Region 4 between 1970 and 1983)
	–29 (Generic EISs)
	–129 (Coastal EISs)
Subtotal:	121 (Inland EISs)
	–18 (Operation and maintenance EISs)
	–13 (Permit EISs)
	–12 (Projects deauthorized or never built)
	–32 (Projects not completed by February 1986)
	–15 (EISs filed first two years of NEPA; 1970 and 1971)
	–4 (EISs with no appreciable mitigation)
Subtotal:	27 (EISs suitable for inclusion in the research)
	–5 (EISs eliminated due to logistical problems)
Total:	22 Inland project EISs in sample

A rank of 0–4 was used to measure the extent to which mitigation measures were implemented. The score was based on the degree to which the description of a mitigation measure in the EIS matched the onsite implementation. Each mitigation measure was given one of the following five *compliance* scores:

0 = not completed with no plans to complete
1 = not completed with plans to complete
2 = completed but inoperable or less than described in EIS
3 = completed and operating as described in EIS
4 = completed beyond description in EIS

Five factors that could affect compliance scores were then evaluated for each of the 212 mitigation measures (Figure 1): (1) mitigation strategy, (2) threatened resource, (3) date of EIS preparation, (4) caretaking authority, and (5) project purpose. The components of each factor are described below:

- Three strategies for implementing mitigation measures were adapted from the legal definition of mitigation (U.S. CEQ 1978: 40 CFR 1508.20). These were (1) avoidance strategy (elimination of the impact), (2) minimization strategy (modification), and (3) compensation strategy (restoration or replacement of the resource lost by a project).
- Eight resources potentially threatened by a proposed action were adapted from a previous Corps study (U.S. COE 1984, p. 5). These were (1) soil, (2) surface water, (3) terrestrial habitat and species, (4) aquatic habitat and species, (5) water recreation resources, (6) archaeological and historical resources, (7) aesthetic qualities, and (8) overall environmental quality.
- The four time periods were identified based on the guidelines and regulations in effect at the time of EIS preparation. These were 1972–1973, 1974–1975, 1976–1977, and 1979–1982 (the sample contained no EISs prepared in 1978).
- Two caretaking authorities were identified based on the entity responsible for maintenance of the project. They were the Corps and a local authority.
- Four project purposes were (1) reservoirs (for flood control and hydroelectric power), (2) channelization and levees (for flood control), (3) navigation (lock and dams or ports), and (4) land sale (for real estate transfers).

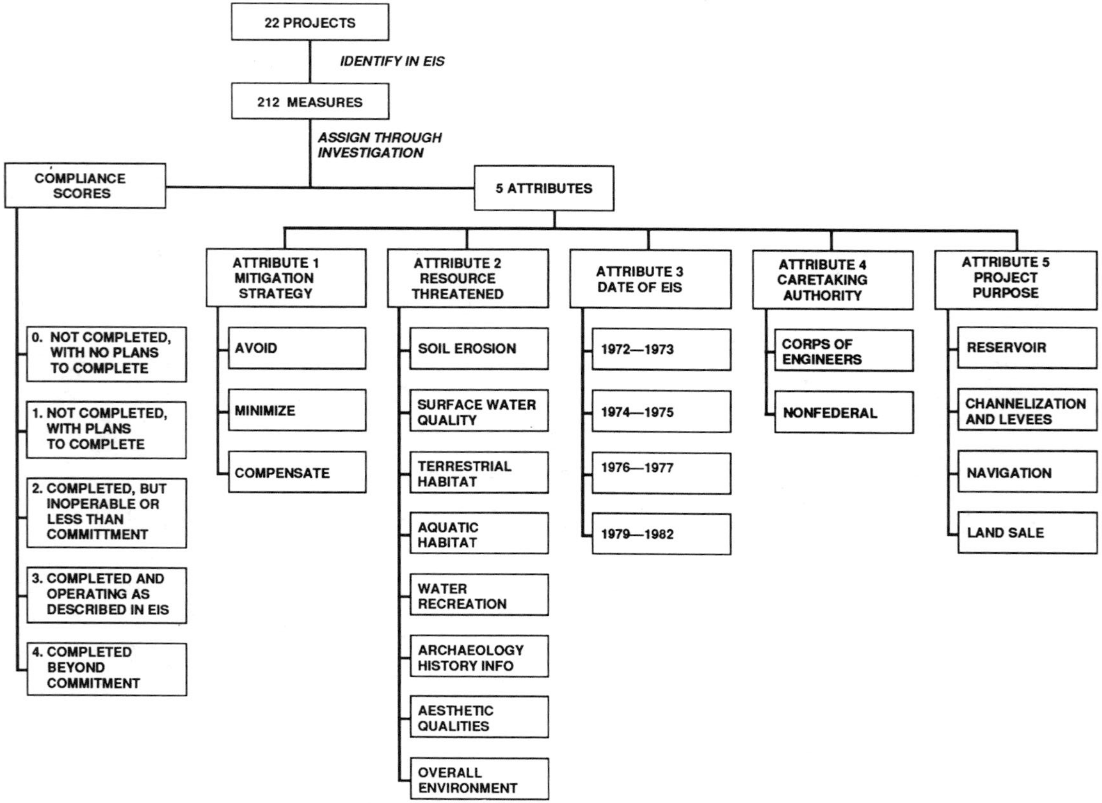

Figure 1. Schematic of research methodology.

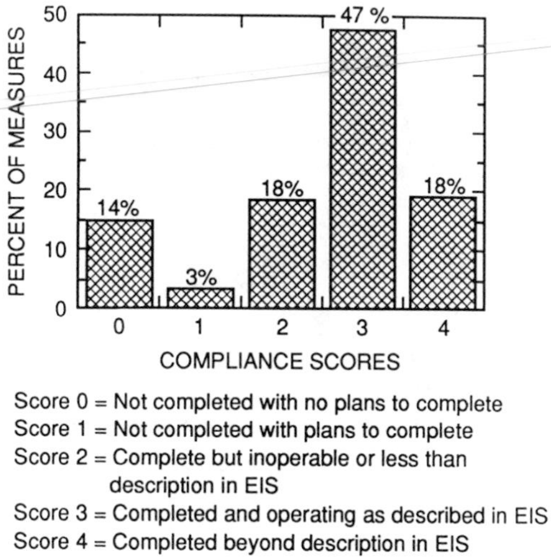

Figure 2. Cumulative mitigation compliance scores.

The data analysis tested the relationship between the five factors and the compliance scores. The following five hypotheses were tested.

- Compliance scores are related to the strategy of mitigation chosen.
- Compliance scores are related to the resource threatened.
- Compliance scores are related to the date that the EIS was prepared.
- Compliance scores are related to the caretaking authority of a project.
- Compliance scores are related to a project's purpose.

The validity of the hypotheses was tested using a chi-square contingency test. The strength of the relationship between the compliance score and each factor was measured at the 0.05 level of confidence. For example, for date of EIS preparation, the null hypothesis — that compliance scores were independent from the EIS preparation date — was rejected based on chi-square testing. Therefore, it was shown that there is a dependence (or a relationship) between the compliance score of a mitigation measure and the date of the EIS where it was identified.

The results that follow use percentage figures rather than actual frequencies so that comparisons can be made between samples that have an unequal number of observations.

RESULTS

The Corps appears to have implemented the majority of mitigation measures identified in the EISs. Figure 2 shows the compliance scores for all 212 mitigation measures. Approximately 65% of the measures were completed as described or beyond the description in the EISs (combination of Scores 3 and 4). Thirty-five percent of the measures were completed below levels described in the EIS

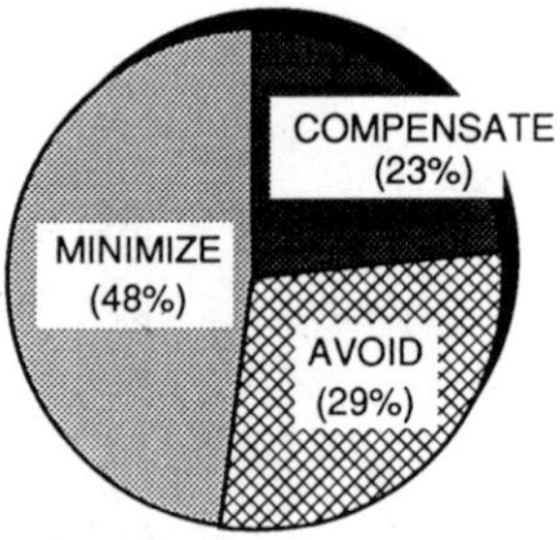

Avoid = Elimination of the Impact
Minimize = Modification to Lessen Impact
Compensation = Restoration or Replacement of the Resource

Figure 3. Distribution by mitigation strategy.

(combination of Scores 0, 1, and 2). Less than 14% of the 212 mitigation measures were never completed (Score 0).

The data contradicted the view that incomplete mitigation is often scheduled for completion in the future. Only 3% of the measures received a score of 1, indicating that a minority of unfinished mitigation may eventually be completed. It was more often the case that mitigation was not completed at the levels described in the EIS (18%) or not completed with no plans to complete (14%).

The 65:35 ratio of compliance/noncompliance was observed regardless of the mitigation strategy (Factor 1), the resource threatened (Factor 2), the caretaking authority (Factor 4), or project purpose (Factor 5). However, the ratio of compliance:noncompliance varied considerably depending on the date of EIS preparation (Factor 3). These factors are discussed in more detail below.

The chi-square test could not identify any statistically significant variations in compliance scores due to Factor 1 (mitigation strategy). However, it is interesting to note the distribution of mitigation for Factor 1 because it shows minimization as the strategy most employed over the 11-year period (Figure 3). Forty-eight percent of the measures minimized impacts by modifying the design of the project. Twenty-three percent of the measures compensated impacts with such plans as in-kind replacement, and 29% avoided impacts by preserving the current state of the environment.

The chi-square test could not identify any statistically significant relationship between compliance score and Factor 2, threatened resource (e.g., mitigation for soil, surface water quality, terrestrial and aquatic habitat, and overall pollution). The distribution of mitigation for Factor 2, however, was a rough indicator of the Corps requirements for mitigation. Surface water quality control, the largest sample, comprised 25% of all the features (Figure 4). Terrestrial habitat loss and soil erosion were each 17% of the 212 mitigation measures tested and loss of aquatic habitat accounted for 12% of the observations.

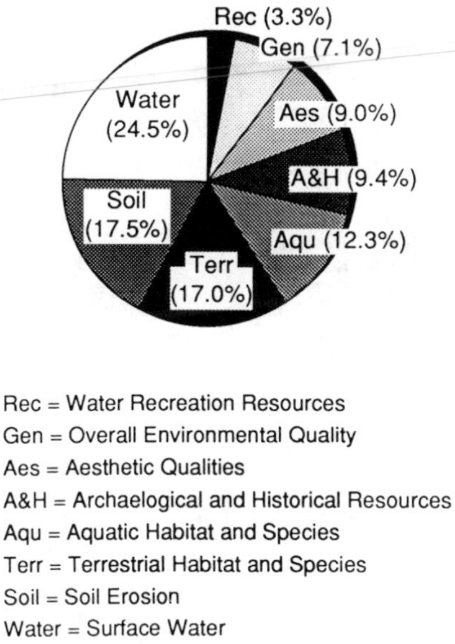

Figure 4. Distribution by threatened resource.

Factor 3, the date of EIS preparation, was the only factor to be significant at a 5% level using the chi-square analysis; therefore, there was a correlation between compliance scores and the date of the EIS.

The analysis indicates that mitigation compliance improved with time; there are higher compliance scores observed for mitigation measures identified in more recent EISs (Figure 5). Regardless of the year in which an EIS was prepared, close to half of the mitigation measures identified were completed as described (Score 3); however, the percentage of mitigation completed beyond expectations increased dramatically in the most recent years. Conversely, the proportion of mitigation completed at levels less than described in the EIS or the percentage of inoperable mitigation measures steadily declined through time. This trend may be attributed to more realistic mitigation measures or better maintenance of newer projects.

The choice of mitigation strategy seems to have changed with time (Figure 6). The early years of the sampling period (1972–1975) were dominated by strategies of minimization, while in the later years (1976–1982), the trend was towards mitigation strategies of avoidance. This shift has occurred because many of the EISs from the early years of NEPA were written as construction of the project proceeded, often too late to avoid impacts but in time to develop mitigation measures that compensate for or minimize impacts. Mitigation strategies appear to have evolved with the environmental movement and the refinement of the EIS process.

The chi-square analysis showed no significant relationship between the compliance score and the caretaking authority. The Corps retained operations on 12

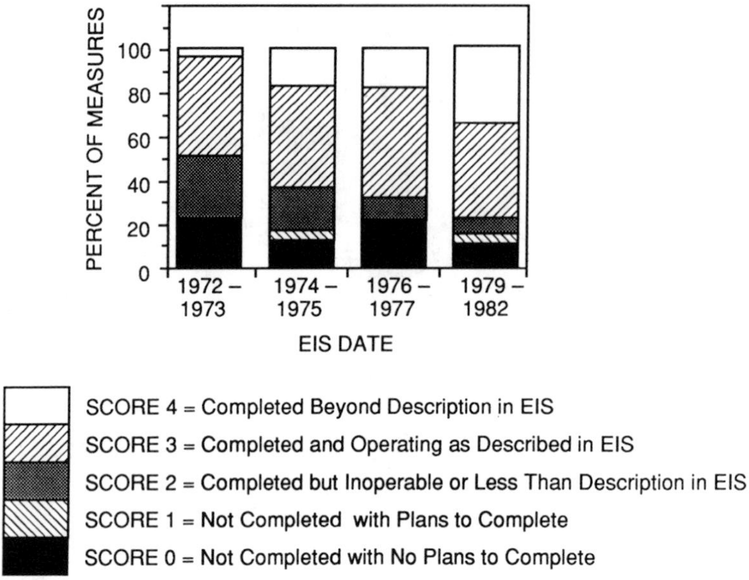

Figure 5. Mitigation compliance by date of EIS preparation.

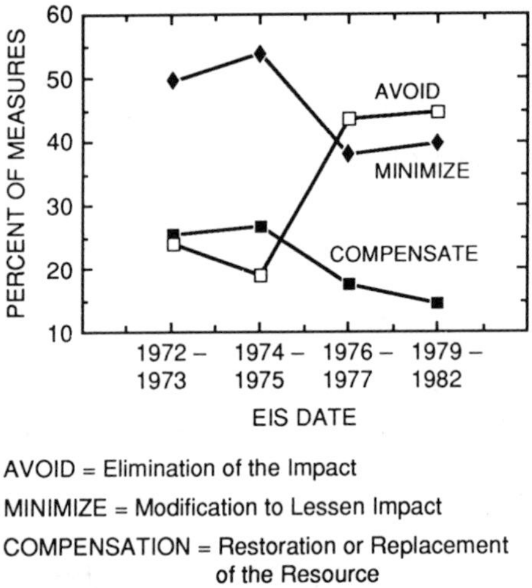

Figure 6. Change of mitigation strategy over time.

of the 22 sites visited, resulting in a total of 130 of the mitigation measures. The remaining 82 mitigation measures were maintained either by local governments or by new landowners.

The chi-square test revealed no significant relationship regarding project purpose and mitigation. Forty-two percent of the measures were at reservoirs, 25% at channels and levees or lock and damns, and 8% on land sold by the Corps.

Policy Implications

Results show that the Corps appears to be moderately successful in mitigation compliance in the southeastern United States, especially considering that no formal system of postconstruction inspection of mitigation was in effect during the time frame covered by this study. Compliance has improved with time and urgent changes are not necessary, but the following recommendations may improve mitigation compliance:

- Because mitigation strategies of minimization and compensation were replaced with avoidance in more recent years of NEPA and because compliance scores have improved with time, avoidance mitigation strategies appear to ensure the best compliance through the life of the project.
- Passage of time decreases the likelihood that mitigation will be completed. There is a need to ensure that mitigation measures are implemented into the design, construction, and operation of projects. Random sampling of mitigation measures or a systematic mitigation compliance audit program at federal facilities may also encourage implementation of the mitigation measures identified in EISs prepared for U.S. Army Corps of Engineers projects.

REFERENCES

Caldwell, L. K. 1982. Science and the National Environmental Policy Act, Redirecting Policy Through Procedural Reform, Tuscaloosa, AL.

U.S. Army Corps of Engineers. 1984. Demonstration of Environmental/Recreation Compliance Inspection, Lessons-Learned, Warm Springs Dam/Lake Sonoma, San Francisco, California. U.S. Army Corps of Engineers, South Pacific Division.

Young, B. M. 1986. Implementation of mitigation features at Corps of Engineers water projects in the southeastern United States. Masters thesis. University of Maryland, College Park.

Postproject Analyses of Fish and Wildlife Mitigation

J. E. Roelle, U.S. Fish and Wildlife Service, Fort Collins, CO; and K. M. Manci, TGS Technology, Inc., Fort Collins, CO

ABSTRACT

Successful mitigation of impacts to fish and wildlife requires that recommended measures be accepted, implemented correctly, and biologically effective. The question of biological effectiveness is examined in this paper by reviewing the results of some recent mitigation evaluation studies. In 61 case studies where all of the mitigation recommendations for a single development action were evaluated, mitigation was considered effective for 54%, partially effective for 30%, and ineffective for 16%. Estimates of overall effectiveness did not vary greatly with the evaluator, the evaluation method, or the type of mitigation. In technique-oriented studies that considered multiple examples of similar measures, mitigation was rarely regarded as successful at more than 60–65% of the sites examined. Most of these technique-oriented studies involved wetland restoration and creation. Many of the studies reviewed were based only on brief site inspections rather than detailed investigations. Results must therefore be interpreted carefully. Additional quantitative evaluations of mitigation success are needed.

INTRODUCTION

Successful mitigation of impacts to fish and wildlife at development projects requires that measures recommended by resource agencies be accepted by the action agency, that accepted measures be correctly implemented by the developer, and that implemented measures be biologically effective. The purpose of this paper is to summarize some recent information concerning effectiveness of fish and wildlife measures applied in a mitigation context. We make no attempt to summarize the vast amount of experimental information that is available concerning some techniques, nor do we address the questions of acceptance and implementation, except insofar as they are directly related to effectiveness. We also provide a brief historical perspective concerning the legal and institutional framework for mitigation, describe some of the most important older mitigation evaluation studies, and make some recommendations regarding future work.

HISTORICAL PERSPECTIVE

The general concept of mitigation to offset losses of fish and wildlife due to development projects is not unique to the National Environmental Policy Act (NEPA). Several federal statutes provide natural resource agencies with the opportunity to comment on the impacts of development proposals, and some of these predate NEPA by a number of years. Public Law 121 of March 10, 1934 stated that the Bureau of Fisheries and the Bureau of Biological Survey would be given the opportunity to make use of waters impounded by the federal government for fish culture stations and migratory bird resting and nesting areas, and that dams constructed by the federal government would be equipped with economically practicable measures to ensure fish passage (Stutzman 1980). Major amendments to this law in 1946 and 1958 resulted in the Fish and Wildlife Coordination Act (FWCA) (16 U.S.C. 661 et seq.), which, among other things, authorized measures for enhancing fish and wildlife, in addition to those for mitigating or compensating for losses. Other federal statutes having a clear relationship to mitigation include the Coastal Zone Management Act of 1972, the Clean Water Act of 1977, and the Surface Mining Control and Reclamation Act of 1977.

In fact, NEPA itself makes no specific mention of mitigation. However, regulations promulgated by the President's Council on Environmental Quality (CEQ) brought the term *mitigation* to the fore and provide the most frequently cited definition (40 CFR 1508.20).

"'Mitigation' includes:

1. avoiding the impact altogether by not taking a certain action or parts of an action
2. minimizing impacts by limiting the degree or magnitude of the action and its implementation
3. rectifying the impact by repairing, rehabilitating, or restoring the affected environment
4. reducing or eliminating the impact over time by preservation and maintenance operations during the life of the action
5. compensating for the impact by replacing or providing substitute resources or environments"

The Fish and Wildlife Service (Service or FWS) subsequently adopted the CEQ's definition and, furthermore, "considers the specific elements to represent the desirable sequence of steps in the mitigation planning process" (U.S. Fish and Wildlife Service Mitigation Policy, *Fed. Regist.* 46(1):7657).

Interest in evaluating the effectiveness of mitigation measures is presumably as old as the concept itself. Indeed, many mitigation measures are nothing more than common habitat management techniques, which have been studied for years, applied in a particular legal or institutional setting. Among the first efforts directed at evaluating mitigation was a series of "follow-up" reports prepared by Service personnel beginning in the late 1950s (e.g., U.S. Fish and Wildlife Service 1962, 1963). Presumably a response to enactment of the FWCA, these reports examined measures recommended for fish and wildlife at dam and reservoir projects primarily from the perspective of acceptance and implementation by the action agencies. Comments or judgments about effectiveness were made in some cases, usually based only on site inspections.

These investigations by Service personnel continued into the 1970s, at which time interest in mitigation seems to have increased, perhaps partially in response to the passage of NEPA and other environmental legislation. For example, Nelson et al. (1978) examined mitigation measures at 90 dam and reservoir projects in 19 western states and made judgments concerning the success of over 200 implemented measures. Judgments concerning effectiveness were made relative to each measure's intended purpose based on available information; no field studies were conducted by the authors. A major symposium concerning mitigation was held in Fort Collins, CO in 1979 (Swanson 1979). Again, however, only a small proportion of the papers presented dealt with the effectiveness of implemented mitigation measures. From 1974 through 1983, the Sport Fishing Institute, under contract to the U.S. Army Corps of Engineers (Corps), conducted a study of the adequacy of fish and wildlife planning at 20 Corps reservoir projects (Martin et al. 1983). This study focused on comparisons of pre- and postproject angling and hunting use as two measures of project impacts. However, implementation of individual mitigation recommendations was examined, and the authors concluded that most of these recommendations were well-conceived. Finally, in 1986, a national symposium focused on mitigation of impacts to and losses of wetlands (Kusler et al. 1988).

Within the Service, interest in assessing the efficacy of mitigation was rekindled in 1985 in the form of a mitigation evaluation project (Roelle and Ellison 1988). Early in the development of the project, we conducted a survey of Service personnel involved in the mitigation process to determine their needs and perceptions (Roelle 1986). One conclusion of this survey was that many of the potential benefits of mitigation evaluation studies are not being realized because the results are not widely communicated. In response to this need, we developed a Mitigation Evaluation Data Base, which contains citations, abstracts, and key words for articles and reports pertaining to mitigation evaluation (Hamilton and Roelle 1987). Entries range from journal articles to single-page memos, with many being unpublished (e.g., the Service follow-up reports mentioned above).

We felt that it was highly desirable to include unpublished documents because part of the purpose of the database was to catalog information not readily available elsewhere. Most of the material presented in the remainder of this paper was taken from entries in the database; where applicable, we also included information from sources not yet included in the database.

APPROACH

The data base presently contains 302 entries. To provide a representative but manageable number for this summary, we limited our consideration to those dated 1984 or later. The choice of 1984 as the starting point was subjective, motivated in part by prior knowledge that Service personnel conducted several evaluations of mitigation in that year. Of course, the date of a report is not an indication of when the mitigation actually took place; in fact, in many cases we were unable to determine when the mitigation occurred. Nevertheless, we believed that information reported in 1984 or later would have the highest probability of being relevant in terms of the current legal and institutional framework, which has changed rapidly at times.

Of the 145 entries dated 1984 or later, about 60 contained at least some information on mitigation effectiveness, with the remainder dealing with acceptance, implementation, or the applicability of mitigation techniques to certain situations. We grouped the reports that discuss effectiveness into two categories: case studies and technique studies. Case studies considered all or most of the mitigation for a single development action, while technique studies considered multiple examples of a single type of recommendation. For example, a case study might involve minimum flows, fish habitat improvements, and land acquisition recommended for a single development, whereas a technique study might involve evaluations of several wetland restoration sites. Obviously, there is overlap between the kinds of recommendations considered in the two types of studies.

We then attempted to summarize results from these studies to obtain an overall picture of what is known about mitigation effectiveness. Rather than imposing our own definition of effectiveness, we simply accepted interpretations of the original authors. In a few cases where the original authors provided relatively detailed information but no summary statement about effectiveness, we made our own evaluation based on the general tone of the report and the extent to which we believed the mitigation goals were achieved. In a few cases, we found it difficult to distinguish between effectiveness and implementation. For example, implementation and effectiveness are essentially synonymous with respect to a recommendation to avoid a particular habitat. Nevertheless, we included these kinds of recommendations because they are an important part of the mitigation process. Not all of the development actions analyzed in these studies have a direct connection to NEPA; rather, they cover the full range of legal mandates under which mitigation recommendations are offered.

RESULTS

Case Studies

The database contained a total of 68 case studies conducted in 27 states. (Citations are not provided for the case studies because most are unpublished and many are unavailable from the authors.) Service personnel were involved in 52 (76%) of these evaluations; the remaining 16 (24%) were conducted by personnel from universities, consulting firms, and other government agencies. Forty-nine (72%) involved only site visits or qualitative observations, and 19 (28%) involved collection of at least some quantitative data for comparison with preproject conditions or control areas. Highway and flood control projects dominated the development activities represented in the case studies (Table 1), largely because personnel in two administrative regions of the Service elected to focus on these types of projects in the evaluations conducted in 1984. Often, highway and bridge projects also involved channelization or channel realignment.

For 61 of the case studies, the authors either provided an overall assessment of mitigation effectiveness or presented sufficient information to allow us to make such an assessment. Of these 61 projects, 33 were judged to be mostly or completely effective, 18 partially effective, and 10 mostly or completely ineffective. Ineffective cases included three where wetland creation or restoration attempts failed due to poor construction practices, four where significant resources were lost despite some attempt at mitigation, one where the recommended measures were not implemented at all, a fish ladder that did not function as planned, and a waterfowl nesting island that was not sufficiently isolated from predators. We found no striking differences in effectiveness between 46 projects evaluated by Service personnel (54% effective, 28% partially effective, 17% ineffective) and 15 evaluated by others (53% effective, 33% partially effective, 13% ineffective). Similarly, there were only minor differences in effectiveness between 44 projects evaluated by site visits (50% effective, 27% partially effective, 23% ineffective) and 17 evaluated using quantitative methods, with the exception that none of the projects evaluated using quantitative methods was rated ineffective (65% effective, 35% partially effective, 0% ineffective).[1] Finally, we grouped the 61 projects into those involving mostly recommendations to avoid or reduce impacts, mostly recommendations to compensate for impacts (e.g., through habitat restoration or creation), and recommendations to both reduce and compensate for impacts. Again, we found no obvious differences in the effectiveness of mitigation of these general types (Table 2).

Slightly over 200 individual mitigation recommendations directed at fish and wildlife were mentioned in the 68 case studies. For 151 of these, the authors either provided an estimate of effectiveness or presented sufficient information for us to make such an estimate (Table 3). It is often difficult to assign an individual

---

[1] Some totals do not add up exactly to 100% due to rounding.

Table 1. Estimated Overall Effectiveness of Mitigation for the Case Studies.

Project Type	Number of Case Studies	Number of Case Studies Evaluated	Number of Evaluated Case Studies (percent)		
			Mostly or Completely Effective	Partially Effective	Mostly or Completely Ineffective
Highway or bridge	22	19	10(53)	7(37)	2(10)
Flood control and bank stabilization	16	15	9(60)	3(20)	3(20)
Dam and reservoir	6	5	3(60)	1(20)	1(20)
Construction and maintenance dredging	3	3	2(67)	1(33)	0(0)
Commercial construction	3	2	1(50)	1(50)	0(0)
Sand and gravel extraction	3	2	2(100)	0(0)	0(0)
Surface mining	2	2	1(50)	0(0)	1(50)
Agriculture	2	2	0(0)	0(0)	2(100)
Other[a]	11	11	5(45)	5(45)	1(10)
Total	68	61	33(54)	18(30)	10(16)

[a] Includes a single project of each of the following types: ditch relocation, bulkhead and boat ramp, pipeline, domestic water intake, housing, airport expansion, sewage treatment, nuclear power plant, hydropower diversion, port development, and stream restoration.

Table 2. Estimated Overall Effectiveness of Mitigation for the 61 Case Studies Grouped According to General Type of Mitigation.

	Number of Evaluated Case Studies (percent)		
Mitigation Type	**Mostly or Completely Effective**	**Partially Effective**	**Mostly or Completely Ineffective**
Mostly impact avoidance or reduction	16(53)	9(30)	5(17)
Mostly compensation	12(60)	5(25)	3(15)
Both reduction and compensation	5(45)	4(36)	2(18)

recommendation to one or another of the specific elements in the CEQ definition of mitigation. However, we believe that the order of the ten categories in Table 3 (top to bottom) generally reflects the sequence in the CEQ definition (avoidance to compensation).

Overall, 66% of the individual recommendations were judged to be mostly or completely effective, 19% partially effective, and 16% mostly or completely ineffective.[1] The proportion of effective recommendations (66%) was thus somewhat higher than the proportion of effective projects (54%, Table 1), perhaps reflecting the idea that several recommendations offered for a project can be individually effective and still not completely offset losses due to the project. The proportion of effective recommendations was lowest for the category water supply and circulation. Most of the partially effective or ineffective cases in this category involved culverts that either were too small to accomplish their intended purpose or became clogged with debris and failed to function. Recommendations involving avoidance, minimization, or rectification (i.e., those in Table 3 down to and including replanting of disturbed areas) were perhaps considered effective more frequently than those involving maintenance or compensation, but sample sizes are small.

Technique Studies

Wetland Restoration and Creation

The majority of the technique-oriented studies contained in the database pertain to wetland restoration and creation (Table 4). California, in particular, has a relatively long history of work in the field of wetland restoration and creation, although we again limited our consideration to information reported in 1984 or later. In fact, many restoration projects in California have been evaluated by a number of different individuals, often with conflicting opinions about results (Race 1985; Zentner 1988a). At times, it is even difficult to tell whether there is overlap between the projects considered by two authors.

In the broadest and most recent of these reviews, Zentner (1988a) concluded that 41 of 63 coastal restoration projects completed in California between 1954 and 1985 could be regarded as successful. Study methods included "monitoring

Table 3. Estimated Effectiveness of Various Types of Recommendations Evaluated in 68 Case Studies.

	Number of Recommendations			
Type of Recommendation	**Mostly or Completely Effective**	**Partially Effective**	**Mostly or Completely Ineffective**	**Total**
Avoid certain areas during construction	16(76)	1(5)	4(19)	21
Minimize vegetation clearing	8(73)	2(18)	1(9)	11
Construction materials and techniques	26(93)	1(4)	1(4)	28
Dredged material management	6(67)	1(11)	2(22)	9
Replant disturbed areas	10(71)	3(21)	1(7)	14
Water supply or circulation	7(39)	9(50)	2(11)	18
Fishery management	4(40)	4(40)	2(20)	10
Wildlife management	5(56)	0(0)	4(44)	9
Wetland restoration or creation	14(54)	6(23)	6(23)	26
Habitat acquisition	3(60)	1(20)	1(20)	5
Total	99(66)	28(19)	24(16)	151

of biotic and abiotic elements," and success was defined by similarity of wetland functions to those of natural sites and by opinions of scientists involved in project planning as to whether project goals were met. Lack of success was most often attributed to improper substrate elevations. In a separate report, Zentner (1988b) concluded that six of ten restoration projects (we deleted one experimental site) sponsored by the California State Coastal Conservancy were successful. Presumably, there is overlap between the projects considered in these two reports, but we could not determine this with certainty. In an earlier study, Baker (1984) reviewed ten projects in the San Francisco Bay area involving loss of wetlands. Mitigation had been completed for five of these projects, and Baker judged that only two sites were successful. Success, in this case, was rated on the basis of a subjective comparison of factors such as primary productivity and flood protection between the impacted wetland and the mitigation site. In one additional study from the West Coast (Fishman Environmental Services 1987), we estimated that four of eight sites might be judged successful based on narrative information provided by the authors.

Three studies contained in the database examined wetland restoration and creation sites on the East Coast. Shisler and Charette (1984) sampled vegetation, sediment, and macroinvertebrates at eight artificial marshes and eight control (natural) marshes in New Jersey. They regarded *Spartina alterniflora* sites as generally successful and *S. patens* sites as generally unsuccessful. Attempts to plant *S. patens* usually resulted in an area dominated by other vegetation. Many sections of the artificial marshes were observed to have substrate elevations inappropriate for the type of marsh desired. CE Maguire, Inc. (1985) judged that 9 of 19 wetland replacement sites in Virginia were successful and that four others were likely to be successful with time. Success was apparently judged largely on the basis of vegetation condition as indicated by estimates of percent cover.

Table 4. Results of Technique-Oriented Mitigation Studies.[a]

Source	Location	Type of Mitigation	Evaluation Method	Evaluator	Number of Sites	Size (acres)	Results
Baker 1984	CA	Wetland restoration or creation	Site inspections	University	5	1–113	2 sites successful 1 site partially successful 2 sites unsuccessful
Zentner 1988a[b]	CA	Wetland restoration or creation	Site inspections, monitoring of biotic and abiotic elements	Consultant	63	>0.1	41 sites successful 16 sites partially successful 6 sites unsuccessful
Zentner 1988b[b]	CA	Wetland restoration or creation	Site inspections	Consultant	10	0.5–80	6 sites successful 4 sites unsuccessful
Fishman Environmental Services 1987	OR	Wetland restoration or creation	Site inspections	Consultant, for state	8	0.01–15	4 sites successful 2 sites partially successful 2 sites unsuccessful[c]
Shisler and Charette 1984	NJ	Wetland restoration or creation	Quantitative comparisons of vegetation, sediment, and macroinvertebrates with natural sites	University	8	0.02–2.5	*Spartina alterniflora* sites generally successful, *Spartina patens* sites generally unsuccessful
CE Maguire, Inc. 1985	VA	Wetland restoration or creation	Site inspections, vegetation cover estimates	Consultant, for Corps	19	0.01–8	9 sites successful 6 sites partially successful 4 sites likely to be successful with time
Reimold and Cobler 1986	CT, MA, NH	Wetland restoration or creation	Site inspections	Consultant, for EPA	4	1.8–13	2 sites marginally successful 2 sites ineffective
Dial and Deis 1986, Quammen 1986	FL	Wetland restoration or creation	Site inspections, vegetation sampling, observations of fish and wildlife	Consultant, for FWS	9	0.01-10	3 sites successful 4 sites partially successful 2 sites unsuccessful[d]

Cobb 1987	TX	Wetland restoration or creation	Quantitative comparisons of vegetation with natural sites	University, for FWS	15	0.01–1	5 sites successful 4 sites partially successful 6 sites unsuccessful
Neill and Turner 1987	LA	Backfilling canals to restore coastal marsh	Quantitative sampling of water depth, vegetation, and soils	University	33		Degree of success varied with several factors, but at least somewhat effective for all canals
Abernathy and Gosselink 1988	LA	Backfilling canals to restore coastal marsh	Quantitative sampling of water depth, marsh vegetation, and submerged aquatic vegetation	University	100 canal segments		Success varied with marsh type, dredging method, soil type, and condition of adjacent marsh
Cobb 1987	TX	Scrapedowns to restore tidal influence or create shallow water or intertidal habitat	Site inspections	University, for FWS	17	0.03–17	5 sites successful 6 sites partially successful 6 sites unsuccessful
Cobb 1987	TX	Seagrass restoration or creation	Quantitative comparison with natural sites	University, for FWS	5	0.55–3.2	1 site successful 1 site partially successful 3 sites unsuccessful
Mangrove Systems, Inc. 1985	FL	Seagrass restoration	Site inspections, quantitative cover sampling	Consultant	20	0.1–17	9 sites successful 11 sites unsuccessful Of 47.54 ac planted, 34.63 (72.8%) either met or will likely meet criterion of 80% cover after 1 year
U.S. Fish and Wildlife Service 1985	ND	Wetland restoration or creation	Site inspections, waterfowl observations	FWS	216	Unknown	Ponds created as mitigation for highway construction generally functioning well as replacement habitat

Table 4. Continued.

Source	Location	Type of Mitigation	Evaluation Method	Evaluator	Number of Sites	Size (acres)	Results
Murphy 1988	AZ	Reseeding of sites disturbed during construction of Central Arizona Project	Not stated	Bureau of Reclamation	29	9–680	18 sites successful or very successful 2 sites partially successful 9 sites unsuccessful
Murphy 1988	AZ, CA	Revegetation, mostly tree and shrub plantings	Site inspections, survival rate	Bureau of Reclamation	17+[e]	0.5–50	Very mixed results, depending on type of stock, planting techniques, irrigation, and predator control
Kinler 1988	LA	Planting desirable bottomland hardwood seedlings on spoil banks	Survival rate of seedlings	FWS	14		0% survival on 6 sites planted under nonspecific guidelines; predation by nutria suspected 57% survival on 6 sites planted under specific guidelines; wire baskets used to discourage predation by nutria Mixed results on 2 sites using chemical repellents to discourage nutria
Cobb 1987	TX	Avoidance[f]	Site inspections	University, for FWS	15		10 sites successful 3 sites partially successful 2 sites unsuccessful[g]
Cobb 1987	TX	Water quality and circulation[h]	Measurements of DO, temperature, and conductivity in canals and adjacent water body	University, for FWS	4	—	4 sites successful

DeHaven and Michny 1987	CA	Rock revetment to protect berm and riparian habitat	Site inspections, wildlife observations	FWS	144	—	7% of sites had high habitat value for wildlife 39% of sites had moderate habitat value for wildlife 54% of sites had low habitat value for wildlife 41% of sites had high potential habitat value for wildlife
DeHaven and Michny 1987	CA	Environmental easements to ensure preservation of riparian habitat	Site inspections, wildlife observations	FWS	28	0.49–49	32% of sites had high habitat value for wildlife 21% of sites had moderate habitat value for wildlife 46% of sites had low habitat value for wildlife 82% of sites had high potential habitat value for wildlife
Dell 1987	LA	High-pressure spray disposal of dredge spoil	Qualitative estimates of vegetation cover and spoil height	FWS, state university	2	—	Marsh vegetation recovered to near preproject densities in 5-18 months; technique considered useful
Slattery and Holland 1987	LA	Plug canals, cut wildlife openings, install weirs, repair levees	Site inspections	FWS	22	—	19 sites effective "structurally" 3 sites partially effective "structurally"

[a] Experimental sites and sites where mitigation was not complete at the time of the study have been omitted where they could be identified.
[b] It is not clear whether there is overlap among the projects discussed in these two reports.
[c] Success estimated by authors of the present paper on the basis of descriptions in the original report.
[d] Assumes that in-kind replacement of vegetation was the criterion for success.
[e] Some locations had more than one planting site; includes some experimental sites.
[f] Mitigation primarily involved avoiding wetland and shallow water habitats with respect to dredging, dredge spoil disposal, and bulkhead and pier replacement.
[g] Data in Tables 15 and 21 of original report are inconsistent.
[h] Mitigation primarily involved maintenance of water quality by specifying canal dimensions and placement of culverts.

Reimold and Cobler (1986) examined five wetland mitigation projects in New England, four of which had been completed. They considered two to be marginally successful and two to be ineffective, apparently based primarily on site elevation, slope, and vegetation condition.

Wetland mitigation evaluations in the South have involved several techniques and habitat types. Dial and Deis (1986) and Quammen (1986) reported on nine sites in the Tampa Bay area. Based on in-kind habitat replacement as the evaluation criterion, Quammen (1986) considered only three of these successful. Cobb (1987) quantitatively compared vegetation at 15 planted and natural sites in Texas. The artificial sites were mostly plantings of *S. alterniflora*. Based on these comparisons, she considered only five of the sites successful.

Two studies have examined the effectiveness of backfilling canals in Louisiana. Neill and Turner (1987) measured canal depth and vegetation recolonization at 33 backfilled canals in fresh, intermediate, and salt marshes. Success depended on factors such as marsh type, canal location and age, soil characteristics, and machinery operator performance, but they concluded that the technique was at least somewhat effective in all cases. Abernethy and Gosselink (1988) measured canal depth, marsh vegetation, and submerged aquatic vegetation at 20 points in each of 100 0.62-km segments of a backfilled pipeline traversing fresh, intermediate, brackish, and salt marshes. They also found differences in success related to marsh type, equipment used in the original dredging, soil type, and condition of the adjacent marsh.

Cobb (1987) also examined scrapedowns to create shallow water or intertidal habitats and seagrass restoration projects. Based on site inspections, she considered only 5 of the 17 sites to be successful. In addition, seven seagrass restoration sites were examined, but two of these were experimental sites. Of the five mitigation sites, only one was considered successful based on quantitative sampling and comparison with adjacent controls. In Florida, Mangrove Systems, Inc. (1985) rated 9 of 20 seagrass restoration sites as successful, but 72.8% of the area planted either met or would likely meet the criterion of 80% coverage after 1 year. In both of these studies, proper site selection was an important determinant of success.

We found only one recent study of wetland mitigation effectiveness in the prairie pothole region (U.S. Fish and Wildlife Service 1985). Service biologists examined 216 wetlands created as mitigation for highway construction, noting water conditions and use by wildlife. They concluded that these wetlands were generally functioning well as replacement habitat. We also noted in the case studies that there was a tendency for these kinds of created depressional wetlands to be regarded as successful.

Other Techniques

Mitigation evaluation studies of techniques other than wetland restoration or creation appear to be relatively scarce, and some of those that are available (e.g., planting riparian vegetation) might well be regarded as wetland restoration or creation, depending on characteristics of the specific site.

Murphy (1988) examined 18 revegetation projects in the lower Colorado River area. One of these projects involved reseeding 29 areas disturbed during construction of the Central Arizona Project. Of these 29 areas, 18 were considered successful based on vegetation survival and growth. The remaining 17 projects mostly involved planting of trees and shrubs. Some experimental sites were involved, and some projects had multiple planting locations. Success was highly variable depending on several factors, including type of stock, planting method, irrigation method (if any), and measures used to control depredations.

Kinler (1988) studied plantings of bottomland hardwood tree seedlings as a means of speeding the revegetation of spoil banks with desirable species. Results differed greatly depending on the specificity of guidelines provided to those doing the planting. At six sites planted under general guidelines, no seedlings survived. Depredation by nutria was suspected as the primary cause of mortality. Seedling survival averaged 57% at six sites planted under specific guidelines, one of which was to place wire baskets around the seedlings to exclude nutria.

As we did in the case studies described earlier, Cobb (1987) grouped 15 projects that primarily involved avoidance of impacts. Ten of these projects were considered successful. Failures and partial successes were largely due to noncompliance with recommendations. Cobb (1987) also evaluated the success of four projects where mitigation was intended to maintain water quality and circulation in canals. Based on measurements of dissolved oxygen, temperature, and conductivity, all of these mitigation measures were considered successful.

DeHaven and Michny (1987) studied two forms of mitigation for a large bank protection project on the Sacramento River. At 144 of their sites, rock revetment was placed to protect a berm and associated riparian vegetation. Based on qualitative observations, they concluded that only 7% of these sites were of high value as wildlife habitat. However, 41% were considered to have high potential value. Mitigation at 28 sites involved easements to ensure long-term preservation of riparian habitat. Of these sites, 32% had high existing value for wildlife and 82% had high potential.

In another study concerning mitigation for canals in Louisiana marshes, Dell (1987) studied the impacts of high-pressure spray disposal of dredged material as an alternative to more conventional methods. Spray disposal deposits material over a larger area than normal sidecasting. At both sites examined, marsh vegetation had recovered to near preproject densities within 5–18 months.

Finally, Slattery and Holland (1987) examined offsite mitigation techniques such as plugging old canals, installing weirs, and cutting openings for wildlife. Most (19 of 22) of these measures were considered *structurally* effective, but the authors did not attempt to assess their biological significance.

DISCUSSION

As defined by the CEQ, mitigation is an extremely broad topic, and this breadth is reflected in the variety of studies discussed above. Nevertheless, there appears to be a fair degree of consistency in the results. Mitigation was regarded as mostly or completely successful in 54% of the 61 case studies, and effectiveness did not seem to be strongly related to the evaluator, the evaluation method, or the type of mitigation. Of 151 individual recommendations, 66% were considered mostly or completely effective. And in technique-oriented studies, mitigation was rarely regarded as successful at more than 60–65% of the sites examined.

Despite this apparent consistency, however, attempts such as this to summarize information on mitigation effectiveness must be interpreted cautiously for a number of reasons. For example, it is difficult to know exactly how representative the studies are. They almost certainly should not be thought of as a representative sample of mitigation recommendations offered by resource agencies; the sample size is simply too small. Service biologists alone make thousands of mitigation recommendations each year (Roelle 1986). Nor can we claim that the database from which these studies were extracted is exhaustive in its coverage of mitigation evaluation studies. Because of the breadth of the topic, it is extremely difficult to survey all of the potentially relevant information, particularly that which is not formally published.

We do believe, however, that the studies discussed above are representative of the mitigation evaluations conducted in the last 5 years, with two or three notable exceptions. We are aware that the literature on surface mine reclamation is underrepresented, as is material on stream habitat improvements. We also suspect that seagrass restoration is not adequately covered. Exclusion of information from experimental sites is also an important limitation; however, we believe that it is justified because success in an experimental setting is no guarantee of effectiveness in a mitigation context.

If the studies described above are representative and even reasonably comprehensive, then it is clear that the amount of effort being devoted to mitigation evaluation is small and that much of the work is relatively subjective. Many of the studies involved only reviews of project files and brief site inspections. Especially in the area of wetland restoration and creation, few studies have involved quantitative sampling of more than vegetation and occasionally wildlife. Hydrologic relationships (e.g., flood desynchronization and groundwater discharge and recharge) have been examined even less frequently.

The reasons for this apparent lack of attention to mitigation evaluation are many. Sound, quantitative studies of complex mitigation efforts can be technically difficult, expensive, and time consuming. Appropriate standards of comparison may be difficult to define. Preproject baseline data are often lacking or insufficient, rarely covering the range of variability inherent in natural systems. Suitable control sites may not exist, especially in heavily developed areas (Race 1985). And finally, there may be a tendency for mitigation evaluation to fall into what we view as an institutional gap, at least within resource management agencies. Operational entities within these agencies rarely have the funding or personnel to undertake detailed evaluations of mitigation. Research and development entities, on the other hand, tend to view mitigation evaluation as routine monitoring rather than true research and thus are reluctant to commit their scarce resources.

The net result of these factors is that much of the information we do have is relatively subjective, and this is another reason for cautious interpretation. Criteria for evaluating success are often not explicitly stated, and they certainly vary from investigator to investigator. For example, at least three different investigators have examined a project known as Bel Marin Keys Unit IV in Marin County, CA. Mitigation for the project consisted of purchasing 105 ac of diked former wetlands and restoring tidal action. Baker (1984) described this mitigation as unsuccessful because it did not replace the habitats lost to development. Eliot (1985) considered it unsuccessful because the result was a tidal *lake* rather than a tidal marsh as stipulated in the authorizing permit. Service biologists, however, regarded it as a success because wintering waterfowl made extensive use of the area (U.S. Fish and Wildlife Service 1984).

Numerous authors have pointed to the need for more explicit mitigation goals and objectives, both as a means of achieving more effective mitigation and as a way of reducing subjectivity in postproject analyses. Although we agree that explicit goals are an important part of any planning and evaluation process, an additional point needs to be made; explicit goals will not eliminate subjectivity from the mitigation process as a whole. While well-stated, quantitative goals provide standards of comparison and bases for postproject evaluation, subjectivity will always be involved in deciding whether the goals are appropriate, particularly in the case of out-of-kind mitigation, when one resource or habitat type is traded for another.

Another limitation of some of the studies described above is that they include results from techniques that are very different. This is especially true in the area of wetland restoration or creation, where mitigation actions may range from simply restoring the natural hydrology to creating a wetland on a former upland site by reducing the elevation and planting wetland vegetation. These kinds of actions clearly differ in terms of technical difficulty and probably in terms of their likelihood of success. Managers seeking guidance in formulating future mitigation must therefore interpret the results of such studies with care.

Finally, caution is warranted because mitigation effectiveness by itself is not an appropriate measure of the overall value of environmental legislation such as NEPA. Studies of mitigation effectiveness do not account for the impact of such legislation on the planning process, particularly in terms of projects that no longer are proposed because they would never pass rigorous scrutiny for environmental acceptability.

REFERENCES

Abernethy, R. K., and J. G. Gosselink. 1988. Environmental conditions of a backfilled pipeline canal four years after construction. *Wetlands* 8:109–121.

Baker, G. F. 1984. An Analysis of Wetland Losses and Compensation Under the Clean Water Act Section 404 Program: Managing Natural Resources Through Mitigation. M.S. thesis. University of San Francisco, CA.

CE Maguire, Inc. 1985. Wetland Replacement Evaluation. Contract No. DACW65-85-D-0068. U.S. Army Corps of Engineers, Norfolk, VA.

Cobb, R. A. 1987. Mitigation Evaluation Study for the South Texas Coast. 1975–1986. Center for Coastal Studies, Corpus Christi State University, TX.

DeHaven, R. W., and F. J. Michny. 1987. Evaluation of Environmental Measures and Wildlife Values of Sacramento River Bank Protection Sites, Units 27–36 of Phase II, Part I. U.S. Fish and Wildlife Service, Sacramento, CA.

Dell, D. A. 1987. High Pressure Spray Disposal of Material Dredged for Oil and Gas Exploration Canals Permitted by New Orleans District, Corps of Engineers, through October 1, 1987. U.S. Fish and Wildlife Service, Lafayette, LA.

Dial, R. S., and D. R. Deis. 1986. Mitigation options for fish and wildlife resources affected by port and other water-dependent developments in Tampa Bay, Florida. *U.S. Fish Wildl. Serv. Biol. Rep.* 86(6).

Eliot, W. 1985. Implementing Mitigation Policies in San Francisco Bay: A Critique. State Coastal Conservancy, Oakland, CA.

Fishman Environmental Services. 1987. Estuarine Mitigation Evaluation Project, Mitigation Site Evaluation Notebook. Prepared for Oregon Department of Land Conservation and Development and Oregon Division of State Lands, Portland.

Hamilton, D. B., and J. E. Roelle. 1987. Mitigation Evaluation Data Base Draft User's Guide. NERC-88/01. U.S. Fish and Wildlife Service, National Ecology Research Center, Fort Collins, CO.

Kinler, Q. J. 1988. Establishment of Desirable Bottomland Hardwood Tree Species on Oil and Gas Exploration Canal Spoil Banks. U.S. Fish and Wildlife Service, Lafayette, LA.

Kusler, J. A., M. L. Quammen, and G. Brooks. 1988. Proceedings of the National Wetland Symposium: Mitigation of Impacts and Losses. Assoc. State Wetland Managers Tech. Rep. 3, Berne, NY.

Mangrove Systems, Inc. 1985. Combined Final Report, Florida Keys Seagrass Restoration Project. Mangrove Systems, Inc., Tampa, FL.

Martin, R. G., N. S. Prosser, and G. C. Radonski. 1983. Adequacy and Predictive Value of Fish and Wildlife Planning Recommendations at Corps of Engineers Reservoir Projects. Contract No. DACW31-79-C-0005. U.S. Army Corps of Engineers, Washington, D.C.

Murphy, S. K. 1988. Documentation of Revegetation Efforts in the Lower Colorado Region: Preliminary Draft. Unpublished Report. U.S. Bureau of Reclamation, Boulder City, NV.

Neill, C., and R. E. Turner. 1987. Backfilling canals to mitigate wetland dredging in Louisiana coastal marshes. *Environ. Manage.* 11(6):823–836.

Nelson, R. W., G. C. Horak, and J. E. Olson. 1978. Western Reservoir and Stream Habitat Improvement Handbook. FWS/OBS-78/56. U.S. Fish and Wildlife Service, Fort Collins, CO.

Quammen, M. L. 1986. Measuring the success of wetlands mitigation. *Natl. Wetlands Newsl.* 8(5):6–8.

Race, M. S. 1985. Critique of present wetlands mitigation policies in the United States based on an analysis of past restoration projects in San Francisco Bay. *Environ. Manage.* 9(1):71–82.

Reimold, R. J., and S. A. Cobler. 1986. Wetland Mitigation Effectiveness. Contract No. 68-04-0015. U.S. Environmental Protection Agency, Boston, MA.

Roelle, J. E. 1986. Mitigation evaluation: Results of a User Needs Survey. NEC-87/01. U.S. Fish and Wildlife Service, National Ecology Research Center, Fort Collins, CO.

Roelle, J. E., and R. A. Ellison. 1988. The U.S. Fish and Wildlife Service's Mitigation Evaluation Project. pp. 252–255. In J. A. Kusler, M. L. Quammen, and G. Brooks (eds.), Proceedings of the National Wetland Symposium: Mitigation of Impacts and Losses, Assoc. State Wetland Managers Tech. Rep. 3, Berne, New York.

Shisler, J. K., and D. J. Charette. 1984. Evaluation of Artificial Salt Marshes in New Jersey. New Jersey Agric. Exper. Sta. Publ. No. P-40502-01-84.

Slattery, T. E., and P. V. Holland. 1987. Off-Site Mitigation for Oil- and Gas-Related Activities Permitted by the New Orleans District, Corps of Engineers, in 1984. U.S. Fish and Wildlife Service, Lafayette, LA.

Stutzman, K. F. 1980. Issues in Fish and Wildlife Planning: Water Resources Planning under the Fish and Wildlife Coordination Act. FWS/OBS-80/44. U.S. Fish and Wildlife Service, Kearneysville, WV.

Swanson, G. A. (tech. coord.). 1979. The Mitigation Symposium: A National Workshop on Mitigating Losses of Fish and Wildlife Habitats. General Technical Report RM-65. Rocky Mountain Forest and Range Experiment Station, U.S. Forest Service, Fort Collins, CO.

U.S. Fish and Wildlife Service. 1962. Supplementary Follow-Up Report for Second Unit, Payette Division, Boise Project, Payette River, Idaho. U.S. Fish and Wildlife Service, Portland, OR.

U.S. Fish and Wildlife Service. 1963. Initial Follow-Up Report for Benbrook Dam and Reservoir Project, Texas. U.S. Fish and Wildlife Service, Albuquerque, NM.

U.S. Fish and Wildlife Service. 1984. Follow-Up Report, Corps of Engineers Permit Mitigation Requirements: A Success and a Failure. U.S. Fish and Wildlife Service, Sacramento, CA.

U.S. Fish and Wildlife Service. 1985. Evaluation of Mitigation Wetlands, U.S. Highway 83 Between Bismarck and Minot. U.S. Fish and Wildlife Service, Bismarck, ND.

Zentner, J. J. 1988a. Wetland restoration success in Coastal California. pp. 216–219. In J. Zelazny, and J. S. Feierabend (eds.), *Increasing our Wetland Resources*. National Wildlife Federation, Washington, D.C.

Zentner, J. 1988b. Wetland projects of the California State Coastal Conservancy: An assessment. *Coastal Manage.* 16:47–67.

Follow-Up of Impact Assessments: International Activities

A. Norton Miller, Special Programs and Analysis Division, U.S. Environmental Protection Agency, Washington, D.C.

ABSTRACT

A Task Force on Environmental Impact Assessment (EIA) Auditing was created in 1988 under the auspices of the United Nations Economic Commission for Europe. Ten countries from East and West Europe and North America were represented on the task force, which examined 11 case studies in order to reach conclusions and recommendations on how to undertake postproject analyses (PPAs).

This paper outlines the study approach and summarizes the conclusions and recommendations of the study. The task force concluded that PPAs can be an effective and necessary means of continuing the EIA process into the implementation phase of projects that undergo environmental reviews. The task force also concluded that PPAs can contribute hard evidence needed for future environmental reviews, as well as allow good decisions to be made in the absence of certainty about project impacts. While recognizing that PPAs are not appropriate for all projects, the task force's general recommendation was that "[p]ost-project analysis should be used to complete the EIA process by providing the necessary feedback in the project implementation phases both for proper and cost effective management and for EIA process development." Specific recommendations are also presented.

INTRODUCTION

In 1981, the senior advisers for Environmental and Water Problems of the United Nations Economic Commission for Europe (UNECE) established a group called the Experts on Environmental Impact Assessment. This group met annually to exchange information on environmental impact assessment (EIA) activities being undertaken by member countries. In 1983, the group established a Task Force on Application of EIA, which was sponsored by the Government of the Netherlands. The results of this task force were published by the United Nations (UN) in the document *Application of Environmental Impact Assessment: Highways and Dams* (ECE/ENV/50) in 1987. One of the conclusions of the task force, which examined EIA case studies on highways and dams from six countries, was that the "provision for, and implementation of, monitoring programs appears to be one of the most neglected areas in EIA."

Following the successful conclusion of the Task Force on Application of EIA, the UNECE Experts discussed a proposal from the Government of Canada for a task force to study approaches to postproject analysis (PPA). After some preliminary work on PPAs by interested countries, the Task Force on EIA Auditing was created by the senior advisers on Environmental and Water Problems in March 1988. Ten countries were represented on the task force, which was headed by Canada: Canada; Finland; Germany, Federal Republic of; Hungary; the Netherlands; Norway; Poland; Sweden; the Union of Soviet Socialist Republics; and the United States of America. This report summarizes the results of that task force, which was published in their entirety by the UN in 1990 as *Post-Project Analysis in Environmental Impact Assessment* (ECE/ENVWA/11).

BACKGROUND

Since the inception of the National Environmental Policy Act (NEPA) of 1969 in the United States, a number of countries have instituted processes for EIA in order to ensure that major projects, generally those undertaken or approved by the national government, receive adequate consideration of their potential environmental consequences in advance of project approval. EIA is implemented in various procedural and institutional contexts and is applied to a wide range of activities. Because EIA is not strictly a scientific or technical activity, the form it takes and the results it accomplishes are largely dependent on the particular national setting where implementation takes place. The Task Force on Application of EIA did conclude, however, that "a successful EIA is one which ensures that all relevant impacts associated with the proposed project are adequately and fully taken into account in the decision-making process."

As countries implemented EIA in their own ways, it is fair to say that they generally focused on examining the potential impacts of the proposed project and developing mitigation measures to bring those impacts into an acceptable range. Rarely did the federal officials responsible for the environmental reviews revisit the projects at a later date to determine if the predicted impacts did, in fact, occur as predicted or if the mitigation measures proposed were implemented and, if implemented, were appropriate. And where such reviews did occur, rarely, if ever, were the results published so that they might be readily available for review by professionals in the field or by the public at large. The Netherlands is the only country with a legal requirement for postproject analysis of projects subjected to the formal EIA system, but this requirement only went into effect in 1986.

The UNECE Experts on EIA believed that PPAs have the potential to substantially strengthen EIA and that there is a need for undertaking more PPAs and making the results known. Consequently, the group believed that there would be value in pulling together the experiences of a number of countries to determine if there were broadly applicable conclusions that might be drawn from those experiences. Recommendations might then be made that would be useful to any country developing or implementing an EIA process.

STUDY APPROACH

The Government of Canada developed a proposal to study PPA and took the lead in reviewing the case studies and preparing both the basic documents used by the task force in its deliberations and the final report.

The original proposal included a description of the benefits that PPAs can provide for improving the effectiveness of EIA, a preliminary typology of PPAs, and an outline of the proposed study to be conducted by the task force. The proposal suggested that the study focus on case studies to determine the PPA methods and approaches that had been successful. The following criteria were developed for the selection of case studies:

1. The project must have been subjected to a prior environmental review.
2. The project must have been implemented following the environmental review — at least to a stage where the major consequences (impacts) of the project could reasonably be determined.
3. A subsequent analysis of the project must have been undertaken in which certain results were reviewed carefully. This study was the postproject analysis or PPA.
4. The PPA should have been designed to provide useful information so that subsequent EIA activities could benefit from the information gained.

It should be noted that the task force subsequently treated the criteria as guidelines; not all cases were required to meet all criteria.

The task force defined PPA as the environmental studies undertaken following the decision to proceed with a given activity. The studies are undertaken in order to facilitate or ensure that the terms imposed by the environmental assessment process are implemented or they may be undertaken in order to learn from the particular activity being studied. The *activities following the decision to proceed* specified in the definition of PPA include preconstruction, construction, operation, and abandonment of the activity. PPAs are also known as follow-up studies or environmental audits.

Two different classifications of PPAs were used to describe the case studies. The first of these classifies PPAs according to their *purpose* or the *use* to which they are put (i.e., environmental management or EIA process de-

Table 1. Summary of Classifications for PPAs.

Classification by Use of PPA	
Project management PPA	Undertaken for the purpose of managing the environmental impacts of the activity
Process development PPA	Undertaken to determine the lessons to be learned from the activity in order that future reviews of similar projects can benefit
Classification by Type of PPA Studies	
Scientific and technical PPA	Deals with the scientific accuracy of impact predictions or the technical suitability of mitigation measures
Procedural and administrative PPA	Deals with EIA process effectiveness; may deal with the project as implemented or with the EIA process

Source: Post-Project Analysis in Environmental Impact Assessment (ECE/ENVWA/II).

velopment). The second refers to the *type* of study undertaken (i.e., scientific and technical or procedural and administrative — these classifications are summarized in Table 1).

The task force examined 11 case studies which are listed in Table 2. The objective was to study existing practices and procedures in the participating countries based on an analysis of case studies, to learn from the collective experience, and to reach conclusions and recommendations on how to undertake PPAs. It was also agreed that the work should concentrate on administrative methods and approaches rather than on scientific methods.

The conclusions and recommendations of the task force are primarily based upon its review of these case studies. It should be noted, however, that the group also relied upon the experiences and opinions of the individual task force participants, as well as the results of interviews that the principal author, Dr. William Ross of The University of Calgary, conducted in the process of gathering information for the report.

GENERAL CONCLUSIONS AND RECOMMENDATIONS

General Conclusions

In examining the case studies, it became clear that the two major purposes listed above can be further broken down into five reasons for undertaking PPA. With regard to environmental management of the activity, there are three reasons:

1. to monitor compliance with the agreed conditions set out in construction permits and operating licenses
2. to review predicted environmental impacts for proper management of risks and uncertainties and
3. to modify the activity or develop mitigation measures in case of unpredicted harmful effects on the environment

Table 2. Case Studies Used by Task Force.

Canada: oil development	Oil field expansion and pipeline construction in northern Canada
Canada: railway	Railway track construction and tunneling in western Canada
Finland: hazardous waste	Integrated hazardous waste treatment plant in southern Finland
Federal Republic of Germany: water extraction	Groundwater extraction for the City of Hamburg
Netherlands: hazardous waste	Secure landfill for middle class hazardous wastes on the North Sea
Netherlands: contaminated silt	Disposal site for contaminated silt from dredging Rotterdam harbor
Norway: oil development	Facility to receive natural gas from the North Sea in western Norway
Poland: copper mine	Copper mine, ore processing plant, and tailings pond in southwestern Poland
Union of Soviet Socialist Republics: pulp mill	Pulp and paper mill on Lake Baikal in south-central U.S.S.R.
United States of America: highway	Relocation of highway route 1A through a salt marsh in Harrington, Maine
United States of America: forestry	Timber sale in Siskiyou National Forest in Oregon

Source: Post-Project Analysis in Environmental Impact Assessment (ECE/ENVWA/II).

EIA process development includes two reasons:

1. to determine the accuracy of past impact predictions and the effectiveness of mitigation measures in order to transfer this experience to future activities of the same type and
2. to review the effectiveness of the environmental management or review process that is in place for the activity in order to improve overall program management in the future

Based upon the case studies, the task force concluded that PPAs are an effective and necessary means of continuing the EIA process into the implementation phase of projects that undergo environmental reviews. The task force also concluded that PPAs can meet the objectives listed earlier as reasons for undertaking PPAs and that they can contribute the hard evidence needed for future environmental reviews, as well as allow good decisions to be made in the absence of certainty about project impacts.

GENERAL RECOMMENDATIONS

The first general recommendation of the task force is the following:

> Postproject analysis should be used to complete the environmental impact assessment process by providing the necessary feedback in the project implementation phase both for proper and cost effective management and for EIA process development.

The task force further recommended that

> ECE member countries should apply the lessons learned here to suitable projects in their jurisdictions and should report back to the Senior Advisers in three years time on the results of those PPAs.

SPECIFIC CONCLUSIONS AND RECOMMENDATIONS

The task force developed a number of specific recommendations that are split into four categories:

1. the *role* of PPAs in the EIA process
2. the *content* of PPAs
3. the *development and design* of PPAs
4. the *management* of PPAs

For each category, the conclusions will be summarized followed by the task force's recommendations.

THE ROLE OF PPAs IN THE EIA PROCESS

Conclusions

The task force concluded that successful PPAs are closely integrated into the overall EIA process. However, the way in which the content of a PPA develops from the environmental review can vary greatly. Public concern about an impact can result in the issue being a very important one with monitoring, evaluation, and management being required as a result. The lack of understanding of an impact can lead to monitoring and evaluation, or the studies undertaken during the environmental review may not address a topic adequately and further study may then become a requirement of the permit or license.

The authority under which the PPA is required is important. Where the authority is sectoral legislation, such as a specific pollution control law, the monitoring is apt to be more limited in scope than if the authority is under a broader mandate for EIA. In either case, it is critical that the information developed during the PPA can be used to modify the project's implementation where such modification is necessary to control adverse environmental impacts.

Finally, PPA need not be undertaken for all projects or all aspects of any one project. Where project impacts and the effectiveness of mitigation measures are well-understood and there are no major public concerns, monitoring and evaluation during implementation may not be necessary. As complexity and/or uncertainty increase, however, the attention that should be given to PPA increases as well.

Recommendations

The task force listed the following recommendations regarding the relationships between environmental impact assessment and postproject analysis:

1. A preliminary plan for the PPA should be prepared during the environmental review of a project; the PPA content requirements should be fully developed when the EIA decision on the project is made.
2. The PPA should focus on important impacts about which there is insufficient information; identification of these impacts and their priorities is undertaken during the environmental review process.
3. The authority to undertake a PPA should be linked to the EIA process so that the concerns identified for inclusion in the PPA during the environmental review can be properly addressed.
4. The conditions of approval for a project should be such that the environmental management for that project will take into account the findings of the PPA.
5. PPAs should be done for all major projects with potentially significant impacts. In addition, for other projects, focussed PPAs may be suitable either for environmental management of the project or to learn from the project.

CONTENT OF PPAs

Conclusions

The case studies included a wide range of material. However, four generic features became evident in reviewing them.

1. Baseline data are required in order to properly interpret the results of subsequent monitoring.
2. Testable hypotheses, generally based on predicted impacts or standards to be met, should be developed for examination.
3. Good documentation throughout the project is very important, especially where there is personnel turnover.
4. In all the case studies, monitoring was an essential component, particularly environmental effects monitoring.

Recommendations

The following list of recommendations follows from the task force's study of PPA content.

1. The development of hypotheses to test should be a part of PPAs. The hypotheses will depend greatly on the nature of the PPA and may involve comparisons of impacts with predictions or with standards or may relate to how well the environmental management system worked.

2. In order to undertake PPAs effectively, relevant baseline data should be collected as fully as is possible.
3. Monitoring and evaluation of data collected are needed to test the hypotheses and should be an essential part of PPA.
4. Documentation of the project and its impacts should be encouraged in order to improve PPAs.

DEVELOPMENT AND DESIGN OF PPA

Conclusions

The task force determined that the development and design of PPAs are particularly important and drew a number of conclusions related to

1. the definition of the purpose of the PPA
2. the definition of the roles and responsibilities of the participants
3. the definition of the environmental management links
4. the treatment of environmental surprises
5. the use of independent expert advice on designing PPAs
6. the relation of the PPA to the different project phases (preconstruction, construction, operation, and abandonment)
7. the need for integration of the various components of the PPA
8. the importance of linking the effects measured to the project
9. the need for good planning of PPAs

The task force concluded that it was important to determine at the very beginning whether the PPA would be management oriented or process development oriented, although the group recognized that the case studies all had elements of both. Environmental management PPAs are those whose results will be used primarily to improve the project being assessed, while process development PPAs are those whose results would be expected to apply primarily to future projects. This becomes especially important where the results of the PPA may be used in litigation to modify the project and thus must meet the requirements for judicial review.

As part of the PPA design, the roles and responsibilities of the proponent, of government agencies, of scientific and technical advisers, and of the public must be defined. The allocation of costs, research roles, analysis roles, and management decision roles must all be a part of this process, and participants should be aware of the financial and other resource implications of participation. These requirements for allocating roles and responsibilities are particularly important for management-oriented PPAs if the results are to feed back usefully into the project.

For management-oriented PPAs, it is also important to determine the management links early in the process and to specify clearly whose responsibility it will be to respond to the results of the PPA. The results will include the standards and other requirements included in the project approval. Unanticipated project impacts, or environmental surprises, should also be considered in the definition of roles and responsibilities, however. These can result from poor impact predictions, from mitigation measures that do not work properly, from poor project management, or from a complete failure to consider the impacts. The task force concluded that environmental surprises can best be handled by ensuring the ability to detect them and a commitment to respond to them, although this may be easier said then done in some cases.

Use of independent advice not only improves the quality of the PPA, but also adds to its credibility. This is of particular value where there is a high degree of public controversy over project approval.

Depending upon the individual case, PPA may vary in relation to different project phases (i.e., more intense monitoring might occur during construction of the project, with a broader plan implemented during operation or abandonment). Also, for certain cases, it may be critical to seek an integrated picture of all consequences of the project, while in others stand-alone research tasks may be identified. The latter are probably more apt to occur in process development PPAs. In all cases, however, effects must be carefully monitored so as to ensure that they, in fact, are being caused by the project under review. Where cumulative impacts are expected, this should be taken into account during the design of the PPA. The careful design of the PPA is also important if it is to be cost-effective, especially since changes during the analysis can slow down the PPA's implementation.

RECOMMENDATIONS

The following list of recommendations follows from the task force's study of PPA development and design:

1. The first and most crucial step in developing a PPA is to define its purpose, including the development of a specific purpose and focus for each of its components.
2. Once the purpose of the PPA is known and its conceptual content is identified (from the environmental review), it is essential to define the roles and responsibilities of the various participants in the PPA — the proponent, the various government agencies, scientific and technical advisers, and the public.
3. Management and participant responses required in response to PPA findings should be, as far as possible, specifically addressed.
4. The need to deal with environmental surprises must be built into PPAs. Monitoring should be done in such a way that unexpected results have a good chance of being detected and those responsible for the PPA should have the power to respond appropriately to unexpected results.
5. The use of independent experts to help design the PPA should be encouraged, as it leads to a better and more credible PPA.
6. The detailed development of the PPA should consider features such as the different phases of the project (preconstruction, construction, operation, and

abandonment), the need for integration of different aspects being studied, and the need to relate the effects being monitored to the project (separating out confounding effects of other activities).

MANAGEMENT OF PPAs

Conclusions

Four conclusions were drawn regarding the management of PPAs. These deal with the value of advisory boards to guide the PPA, public participation in PPAs, adaptive approaches to PPA, and the use of independent researchers for sensitive parts of a PPA.

Joint boards or advisory bodies involving industry (or the proponent), government, technical experts, and the public can be very helpful in managing PPAs. This is separate from the value of seeking independent advice during the design of the PPA. However, the role of these boards must be clearly set forward. They may serve to advise on monitoring programs, help to select researchers, review reports, help to revise monitoring programs or mitigation measures, and explain results to communities. Through their influence with industry or government, they may also serve to help obtain funding for necessary research programs.

The role of the public can also be critical in project implementation and during the PPA. The public may assist in monitoring, may serve on an advisory board, or assist in disseminating information. This is, in part, a continuation of the public participation process in the prior EIA review and can lead to better environmental management.

Some PPAs were designed in phases with opportunities to refine them depending on what results were obtained. More attention was paid to areas that were considered more critical or more problematic, with less effort being spent where impacts were expected to be minimal or unimportant. As the PPA progresses, however, the analysis plan itself may be revised or modified as well as the project it is examining.

Finally, independent researchers can be as helpful in managing the PPA as in ensuring objectivity in its design. This enhances the credibility of the findings, especially where a proposed project has been the subject of public controversy.

Recommendations

The following list of recommendations follows from the task force's study of PPA management:

1. As a tool for managing PPAs, advisory boards consisting of industry, government, contractors, independent experts, and public representatives should be used. Such boards with well-defined terms of reference increase the credibility and quality of the PPA.

2. Public participation in the PPA should be encouraged.
3. PPA reports should be made public.
4. The use of independent researchers should be encouraged for doing particularly sensitive parts of a PPA for which work done by the proponent (or possibly even by a government agency) may not be regarded as credible.
5. PPAs should be managed adaptively with opportunities to refine them depending on the results obtained. More effort should be put into examining those effects that are observed and important, and less effort should be expended for those that the PPA indicates are not resulting in significant impacts.

Concluding Remarks

The task force found that PPAs improve environmental protection and can result in more efficient investment of resources. By improving our ability to predict impacts, we create a more solid base for decision making. By improving our understanding of the effectiveness of mitigation measures, we can assure that approvals require only those measures that can reasonably be expected to be successful. And through ensuring an appropriate follow-through, we can facilitate project approvals and public acceptance where uncertainty still exists at the conclusion of the environmental review. PPAs are no panacea, but better information about project impacts will result in both better projects and wider public acceptance of projects undergoing environmental review.

Much more experience with PPAs is needed and is being gained in a number of countries. The task force hopes that the lessons obtained through the study of the 11 PPA case studies can help others in undertaking PPAs or in developing systems to required PPAs. The task force also hopes that the use of PPAs will become more frequent and that their results will be more broadly disseminated in the future.

REFERENCE

United Nations (UN). 1987. Application of Environmental Impact Assessment, Highways, and Dams. ECE/ENV/50. United Nations, New York.

United Nations (UN). 1990. Post-Project Analysis in Environment Impact Assessment. ECE/ENVWA/11. United Nations, New York.

CHAPTER 8

FEDERAL AND STATE EXPERIENCE

Introduction

R. M. Reed and C. R. Boston, Oak Ridge National Laboratory, Oak Ridge, TN

The implementing regulations for procedural provisions of the National Environmental Policy Act (NEPA) (40 CFR 1500–1508) require each federal agency to adopt the Council on Environmental Quality (CEQ) procedures and supplement them as needed to address specific needs of the agency. Under this requirement, agencies have issued their own regulations or guidelines for complying with NEPA. Some agencies simply adopted the CEQ regulations, while others published more extensive requirements that addressed unique or particularly important aspects of the agency's mission. NEPA itself and the CEQ requirements, however, provide the common basis for all agencies.

Although NEPA is directed specifically at federal government agencies, states are also affected by NEPA requirements for actions involving federal funding or approvals. Coordination with state governments is required for federal NEPA reviews, and state agencies generally are asked to review draft environmental impact statements (EISs) and, in some cases, environmental assessments. Twenty-seven states have enacted their own legislation, *little NEPAs,* that parallel to some degree the requirements of NEPA for state actions.

This chapter describes a sample of federal and state experiences with NEPA and state *little NEPA* requirements. The papers indicate that each federal agency and state government must conduct NEPA or state environmental reviews in the context of their own mission and constraints.

Judith Gottlieb of the Minerals Management Service discusses in her paper the challenges of preparing EISs on oil and gas exploration and development for Alaska's Outer Continental Shelf planning areas. These EISs are particularly difficult because they are prepared before a well-defined proposed action is in place and they must address all foreseeable consequences of development, even when the discovery of oil is unlikely. As the NEPA process has evolved, the agency has had some success in responding to CEQ requirements to reduce the length of these EISs and to focus them on real alternatives and issues. Public review and input during the NEPA process, however, continues to demand an encyclopedic approach.

In his paper, Kenneth Mowll describes the U.S. Department of Transportation's approach to defining alternatives for their EISs. He states a point-of-view that the current NEPA process is used by agencies to justify actions that have already been selected. He believes that a *better management* alternative designed to *maximize the effectiveness of existing natural and manmade systems* should be developed as the base case for comparison, and other alternatives could be described in terms of *minimum performance specifications and maximum environmental impacts.*

Earl Evans and his colleagues at the Department of Energy's Pittsburgh Energy Technology Center describe their experience with developing NEPA documentation for 21 projects funded under the Clean Coal Technology Program. The NEPA strategy for the program evolved during three rounds of solicitations to the private sector to help fund demonstration of innovative technologies. By the third solicitation, a programmatic EIS was developed to look at the environmental implications of deployment of these technologies in the 21st century. Because the projects have few long-term impacts and are expected to result in a net reduction of atmospheric emissions, environmental assessments or internal documentation to support categorical exclusions provide adequate NEPA documentation for the majority of the specific projects.

John Taves of the Department of Energy's Bonneville Power Administration (BPA) discusses the challenges of incorporating impacts on ecological resources as a decision factor. Some of these challenges include the lack of appropriate data, the absence of demonstrated and accepted methodologies, political sensitivities and resistance, and goal conflicts. Of particular interest to BPA activities is their potential impact on salmon and concerns related to the cultural or religious value of salmon to Native Americans. Comparison of these impacts with more quantitative concepts from engineering and economic analysis is particularly difficult. BPA has supported extensive efforts to mitigate impacts to salmon fisheries by coordinating its marketing activities to achieve flow levels that facilitate effective fish spawning, incubation, and migration.

Bruce Leon reviews a random sample of 38 EISs prepared in 1989 to examine the types of analyses that are included currently. He found a lack of rigor in current EISs and a heavy reliance on *expert* judgement, and he believes that *environmental policy is subverted* as a consequence. Leon proposes that impacts and the underlying assumptions of analysis should be stated as testable hypotheses. Follow-up studies should be designed in the EIS so that accuracy of impact predictions can be determined and unanticipated impacts identified.

In his paper, Steve Ugoretz discusses the use of risk assessment in the environmental assessment process. He believes that risk assessments can be an important component of NEPA and state environmental assessments. Public concerns about the meaning of the results of risk assessment and the uncertainties associated with the technique provide a rationale for including them in environmental assessments along with appropriate explanations. The high cost of doing detailed risk assessments may impede widespread adoption of the technique.

Robert Read describes his state's experience with the Wisconsin Environmental Policy Act that was passed in 1972 and patterned closely after NEPA. This act has changed little over the years, and its interpretation has been strongly influenced by court decisions on NEPA. Each state agency has substantial freedom in interpreting compliance requirements because there is no centralized state oversight. Overall, the process has been successful, but critics feel that it is duplicative and wasteful.

Alaska Outer Continental Shelf EIS Evolution Since 1976

J. C. Gottlieb, Minerals Management Service, Anchorage, AK

ABSTRACT

To date, 34 oil and gas environmental impact statements (EISs) have been prepared for Alaska Outer Continental Shelf (OCS) planning areas. The approach to these EISs has changed significantly over the years.

Acquiring oil and gas leases entitles a company to perform preliminary activities. To further explore a lease, a company submits an exploration plan on which the Minerals Management Service (MMS) of the U.S. Department of the Interior (DOI) performs a National Environmental Policy Act (NEPA) review and applies for a permit to drill. The odds are against successful exploration and development; in fact, the most likely outcome of an OCS lease sale is no oil development because the probability of finding oil in most Alaska OCS planning areas is extremely low. If a company does discover hydrocarbons and submits a development plan, MMS evaluates the plan using NEPA criteria.

The OCS EISs are particularly difficult to write because there is no proposed definite action to be analyzed as there is with a proposed construction project. This uncertainty creates difficulties for EIS authors because NEPA requires them to look ahead and analyze all foreseeable consequences of a proposed action. Because the purpose of a lease sale is to discover oil, the EIS authors must assume that oil is found, even if the odds are against it. The MMS uses hypothetical estimates of the amounts of oil and gas that might be found. These estimates lead to further assumptions, such as the number of industry personnel involved, number of helicopter flights, amount of icebreaker support, and total muds and cuttings discharged. There are no absolutes on which to base the analyses. The tenuous relationships among the assumptions have lead some critics to charge MMS with drawing worst-case or "gloom-and-doom" conclusions in its EISs.

INTRODUCTION

The Leasing and Environment Office of the Minerals Management Service (MMS) is responsible for leasing mineral rights to submerged lands extending from 3 mi off the shore to the outer U.S. offshore boundary, which could be over 200 mi out. Our mandates include obtaining a fair and equitable return for these mineral rights as well as protecting the human, marine, and coastal environments. The 70 people in the Leasing and Environment Office have a variety of backgrounds, including oceanography, biology, social sciences, mineral and legal specialists, program analysts, editors, and others. The office contracts for environmental studies, performs categorical exclusion reviews, and prepares environmental assessments (EAs) and environmental impact statements (EISs) on prelease and postlease activities. We also keep the legal records of lease ownership, follow litigation and official notices relating to our program, and actually hold the lease sales that are conducted much like sealed-bid auctions.

The lease sale is the final step in MMS's complex, ten-step prelease process, which begins at least 28–32 months earlier with a call for information and nominations and a Notice of Intent to prepare a lease-sale EIS. This EIS is one of the Leasing and Environment Office's major responsibilities and is the focus of this paper. I will compare three lease-sale EISs as examples to show how we have developed and improved methods and procedures in our attempt to meet the scientific (and other) challenges of the National Environmental Policy Act (NEPA) while also meeting the Alaska Outer Continental Shelf (OCS) Program's mandate to publish almost 40 draft and final oil and gas lease-sale EISs during the past 13 years. I will begin with one of the Alaska OCS Program's earliest documents (a 1976 EIS that analyzed the potential effects of oil and gas leasing at Lower Cook Inlet), move on to one of our first post-CEQ regulation EISs (the 1982 EIS for the Beaufort Sea leasing proposal), and then complete my comparison with our most recent oil and gas lease-sale EIS (the 1987 Chukchi Sea Sale 109 EIS).

LEASING PROGRAM AND PROCESS

The OCS EISs are particularly difficult to write because there is no proposed definite action to be analyzed as there is, for example, with a proposed construction project. This uncertainty creates difficulties for EIS authors because NEPA requires them to look ahead and analyze all foreseeable consequences of a proposed action. Because the purpose of a lease sale is to allow exploration for oil, the EIS authors must assume that oil is found, even if the odds are against it. The MMS uses estimates of the amounts of oil and gas that might be found assuming there is oil in the basin that is explored. These estimates lead to further assumptions about such considerations as scenarios, the number of industry personnel involved, oil-spill risk analyses and trajectories, location of onshore facilities, tanker/pipeline routes, number of helicopter flights, amount of icebreaker support, and total drilling muds and cuttings discharged. There is, therefore, a myriad of assumptions but no absolutes on which to base the analyses. In addition, if new information such as drilling in the area changes any estimates, then all the associated estimates would also change.

Acquiring oil and gas leases entitles a company to perform preliminary activities. The company files an exploration plan and an EA is written. The odds are against successful exploration and development on the leases; in fact, the most likely outcome of an OCS lease sale is no oil development because the probability of finding oil in any individual Alaska OCS lease-sale area is low. The potential oil reserves that are always assumed for the Alaskan OCS are based on all planning areas and will probably take many lease sales to discover. The odds for the success of each individual sale are, therefore, low. If a company does discover hydrocarbons and submits a development plan, MMS evaluates the plan using NEPA criteria.

OCS EISs usually have drawn controversy, public and U.S. Department of the Interior (DOI) attention, and litigation. The DOI's Office of Environmental Project Review and the Office of the Solicitor have always reviewed OCS prelease EISs. All sign-offs and other major decisions relating to OCS lease sales have also been made at the MMS headquarters and departmental levels in Washington. Consequently, little authority or autonomy regarding schedules for publishing EISs is held by regional level offices.

The typical progression of our process is as follows:

1. We query industry for interest and the public for concerns on large planning areas (Figure 1).
2. A smaller amount of acreage (separated into *lease blocks*) is evaluated in the EIS.
3. Often, not all of these blocks are offered for lease (Figure 2).
4. A small percentage of leases are bid upon, and not all bids are accepted as having fair market value (Figure 3).
5. Fewer still result in exploration plans.
6. Even fewer have wells drilled on them (Figure 4).
7. To date, there have been no development plans for the Alaskan OCS.

An Early EIS — Lower Cook Inlet EIS

The Lower Cook Inlet EIS was prepared by the Bureau of Land Management (BLM) in 1976. This EIS, which was about 1400 pages long, evaluated 900,000 ac and used a relatively simple model to assess oil-spill trajectories. The EIS team was small, had a designated coordinator who was also a technical specialist, and did not include an editor. The format followed general NEPA guidance. Definitions of level of impact did not exist. Not all sections had summary/concluding statements. Some examples of the EIS conclusions include the following:

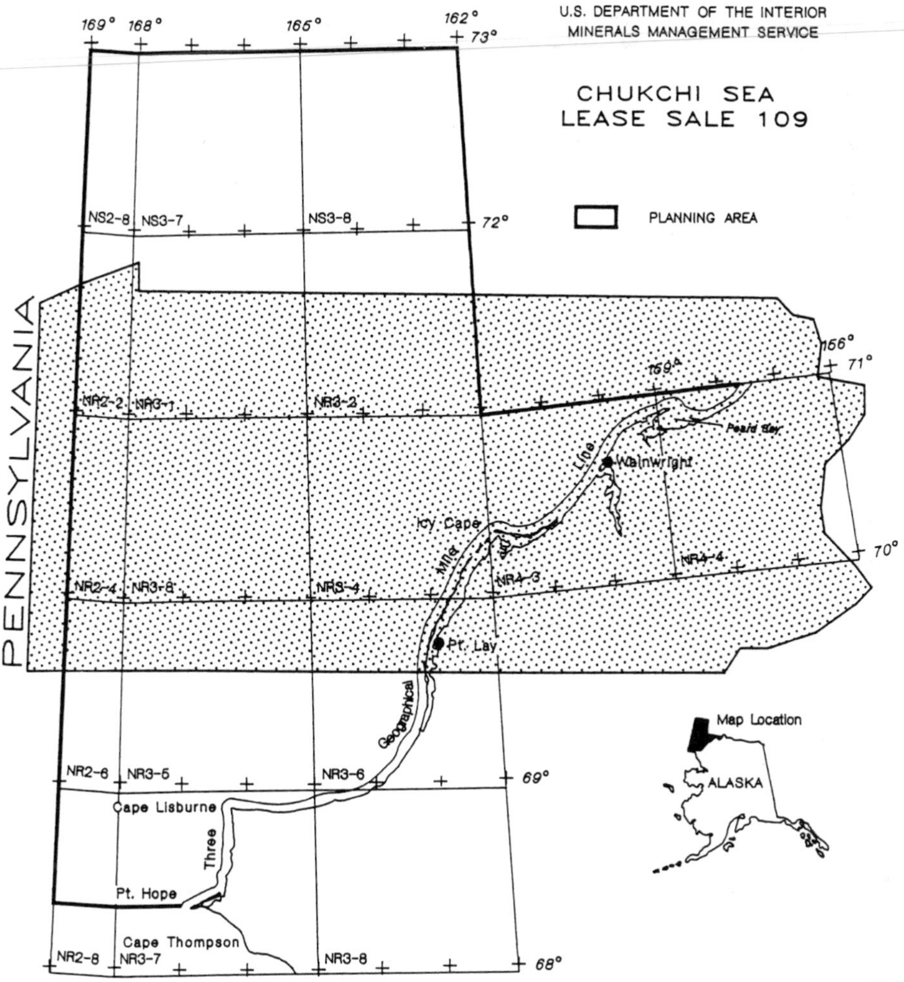

Figure 1. Planning area for the Chukchi Sea Lease Sale 109.

- In summary, oil pollution will contribute to a continuing attrition and, under certain conditions, catastrophic losses of migrant and resident birds.
- Overall, the salmon resources in Cook Inlet will be affected by oil development. Although the impact cannot be well-quantified, it would be marginal only if controls are rigorous and accidents average.
- For the most part, Alaska Native communities are sociologically relatively unknown, even to Alaskan social scientists. Projecting OCS sociological impacts on these coastal communities requires field research.

No overall conclusion was provided.

There was not an extensive amount of information available for describing the environment or for assessing impacts. This lack of data proved to be the catalyst for funding many social, economic, and environmental studies — over $215

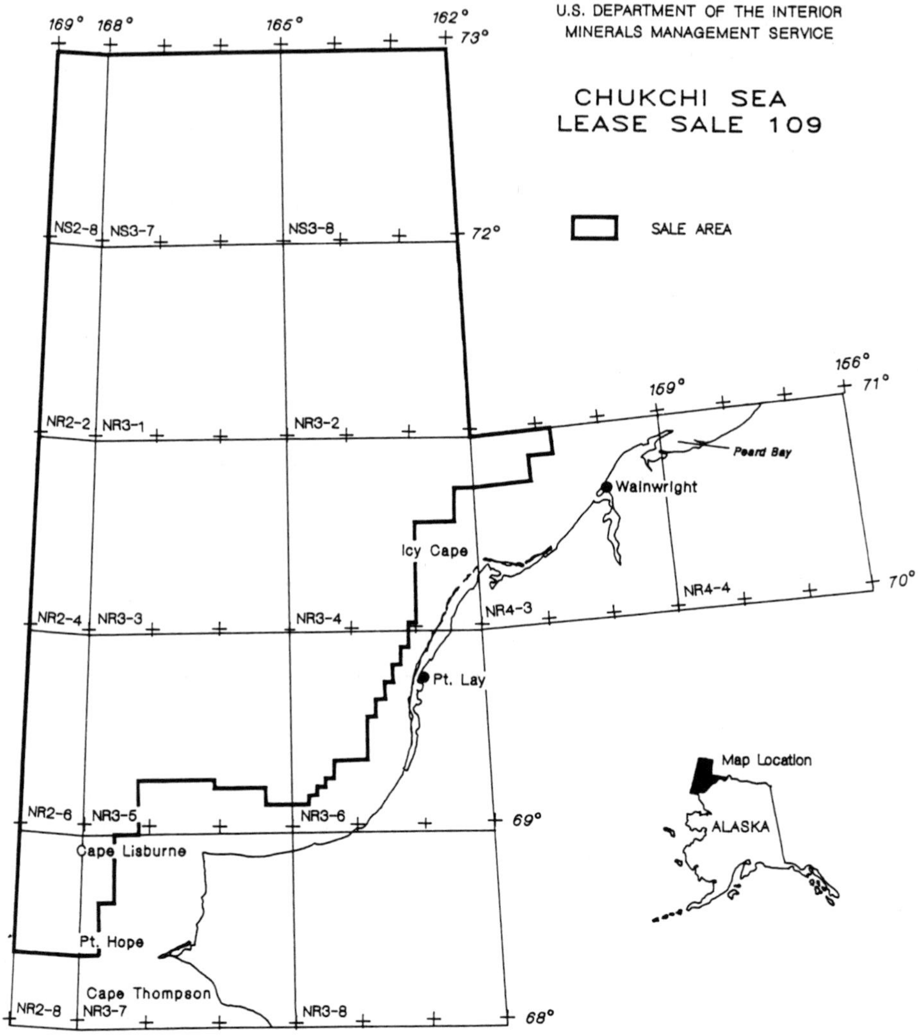

Figure 2. Sale area for the Chukchi Sea Lease Sale 109.

million has been spent on studies in Alaska because of our program. Because the Alaskan population is so small, almost any incoming oil industry workers would have significant effects. Cultural differences contribute to these effects as well; we were the only region in MMS to have a social and economic studies program.

In the Cook Inlet EIS, as in all our early EISs, estimates were provided for minimum and maximum recoverable oil and gas resources within the proposed sale area. The EIS was written, however, on the maximum case to give *maximum consideration to environmental impacts.* A development timetable based on the maximum-case estimate of recoverable resources showed number of wells, miles

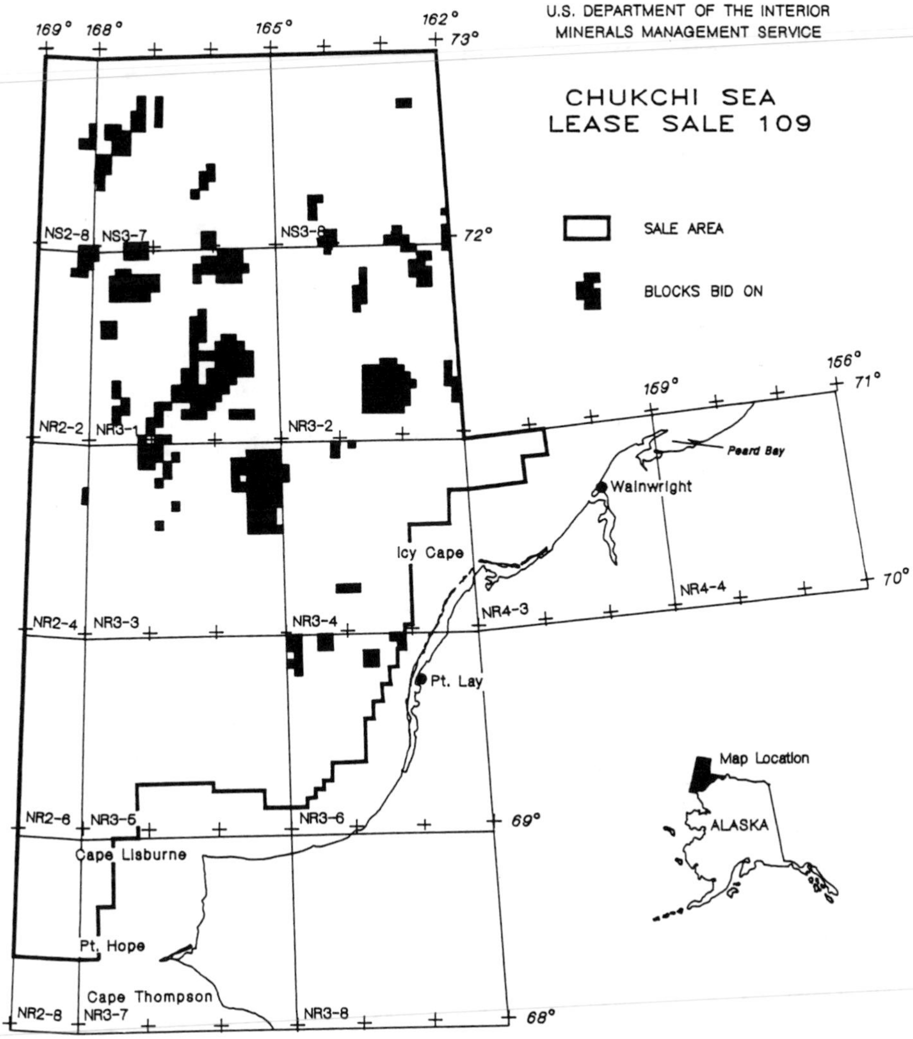

Figure 3. Blocks bid on for the Chukchi Sea Lease Sale Area 109.

of pipeline, and number of terminals. We also provided a map showing possible onshore support and terminal sites. (These assumptions received considerable criticism because readers thought we were predicting and permitting projects.)

The first EIS authors used this information as basic assumptions for assessing impacts; in our EISs, we have always assumed the discovery of recoverable resources. Many people have objected to this assumption; however, we feel the assumption is necessary to fully comply with the Council on Environmental Quality's (CEQ's) requirement to look at all foreseeable effects. Additionally, local communities often have objected to our scenario assumptions (they may not

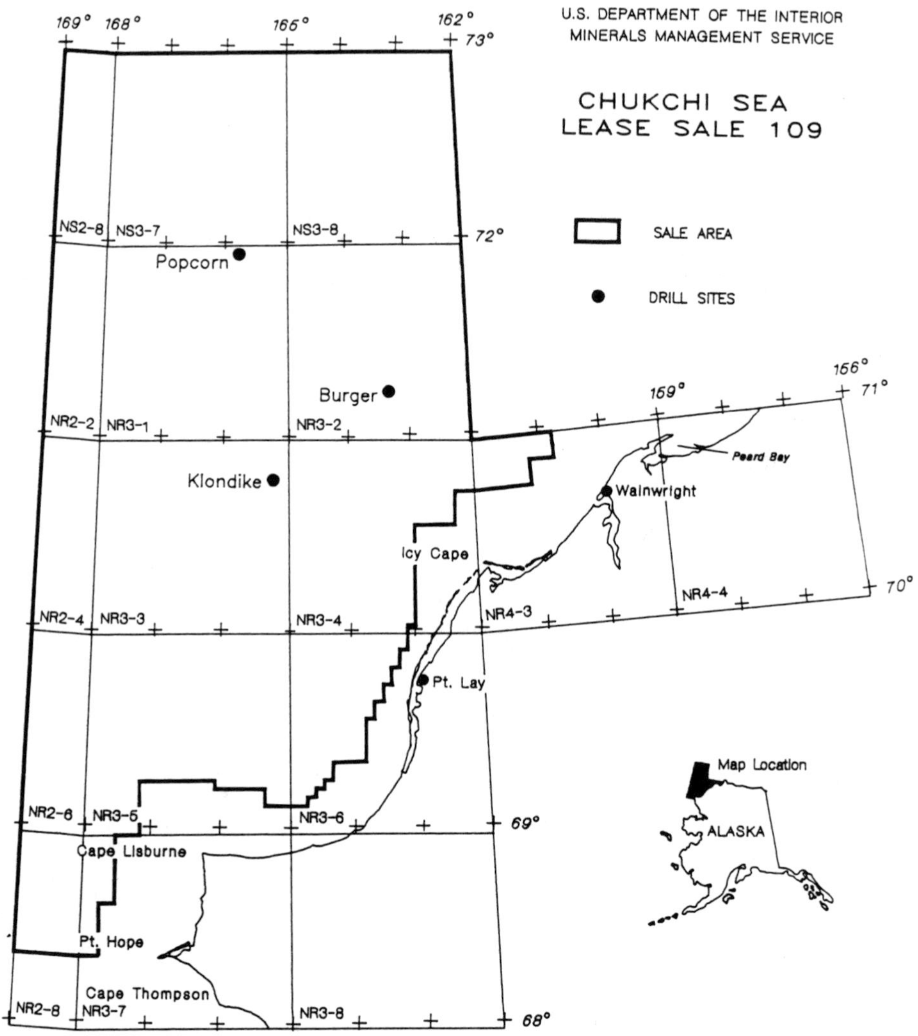

Figure 4. Drill sites for the Chukchi Sea Lease Sale Area 109.

want a hypothetical oil terminal or they wanted to use the EIS as a land use and infrastructure planning document), and our impacts may have been high in certain ways because of these high hydrocarbon-resource-level assumptions. At a later date, we switched to a *most likely* case of the level of resources we expected to be found. Regardless of the case of the level of resources, however, the reliability of the resource-estimate numbers has always been questioned. Because factors such as the number of wells and miles of pipeline depend on these resource estimates, it is not surprising that the early EISs, which used the maximum-case number, were called *gloom and doom* documents.

The Cook Inlet EIS included the following analyses:

- A trajectory evaluation and summary sensitivity analysis (i.e., a matrix) was used to analyze the potential impacts of the lower Cook Inlet sale. A table illustrated the distance from shore of each lease block (3 mi^2, water depth in the area, and the number of days before a spill would impact a cultural or natural resource.
- An oil spill model of surface oil spill behavior was conducted to evaluate surface transport and spreading. Wind and tidal- and surface-current data were included in the model and a grid system was developed.
- Spill sites were selected and results were listed in a table giving times to impact shoreline in hours up to 6 days.
- In addition, probabilities were given for oil exposure. These indicated which portions of the lower Cook Inlet were most likely to be impacted from a spill from the chosen site.
- The cumulative case was covered very generally with a description of federal and state oil and gas lease sales and a couple of possible state construction projects. Cumulative effects were not estimated in the impacts section, but occasionally an author would mention cumulative effects and/or events.

A great deal of public contact and participation occurred during the time of our early EISs (1976–1982). The EIS team usually flew over the proposed area, stopping in key villages or communities to make many personal inquiries and hold numerous meetings. This practice became a way to have a personal orientation and establish good working relationships in the communities, as well as to acquire valuable information.

For the Cook Inlet EIS, there were two public hearings that were well attended; 60 people testified and 47 written comments were received. Within the DOI, there was high-level interest in chairing and attending the hearings. Several officials from various bureaus traveled from Washington, D.C. to participate. The public and state, federal, and local agencies submitted numerous comments. Most people testifying were opposed to the OCS Program and any potential exploration or development that could result from our lease sale.

To date, there has not been any oil developed from any Alaskan offshore federal lands. Disappointing exploration, economic constraints, oil prices, and transportation difficulties have constrained any development. The litigation that followed many of the early EISs did not significantly delay sale timing, but it often caused additional information to be added to later EISs.

A POST-CEQ REGULATION EIS — BEAUFORT SEA EIS

The final EIS for Beaufort Sea, published under MMS in 1982, assessed 1.8 million ac of offshore lands for lease — about twice as much as in the Lower Cook Inlet EIS in almost 1000 fewer pages. Under the CEQ regulations, MMS considered this EIS as *complex* and aimed for the 300-page limit. The EIS was 500 pages long, with 200 pages of appendices. Additionally, several reference documents were prepared to help keep the EIS a manageable size. These included papers on topics such as legal mandates and regulations relating to the OCS program; generic effects from OCS activities on marine mammals, endangered species, seabird populations, and rare plants; federal and state coastal management programs; arctic dredging and artificial island construction; and archaeological resources of the Arctic. By incorporating this information by reference, MMS was able to significantly reduce the size of the description of the environment. Unfortunately, however, in later EISs, much of this information crept back into the documents. Reviewers were not happy with incorporation by reference and asked for more and more extensive summaries of the data. The reviewers expected explanatory and background information to appear in our EISs because they had seen this information in the past. Additionally, commenters expressed frustration at having to obtain separate documents and then cross reference, rather than having complete information in one document. We are, therefore, again attempting to streamline our documents.

By 1982, the OCS sale schedule was very full. Instead of working on one geographic area at a time, MMS often would have three or four frontier areas for which no EIS had previously been written. We hired more people and formed two teams — one to work on arctic EISs (i.e., those from the Norton Basin northward) and one for the subarctic (the Bering Sea area and the Gulf of Alaska). We also established full-time project managers/EIS coordinators for each EIS; whenever possible, the coordinator would have an assistant. We recognized the skills it takes to manage a large project and a large team. We also hired editors to be part of the teams. The project managers and editors greatly improved the organization, readability, and image of the documents. They assisted the analysts so that their work and conclusions were presented more effectively and made the documents more easily understood by the broad audiences we serve. As a result, our EISs became much more readable. There were headers for clear reference, information in tables and graphics was presented more clearly and effectively, and both syntax and substance were improved.

Our conclusions began to read like the following:

- Contacts between fish and the following causal agents are likely ...
- Overall, significant regional impacts to fish are not likely and ...
- Marine mammals are very likely to come in contact with oil pollution and suffer some disturbance ...
- Significant long-term regional population reductions are unlikely ...

Hydrocarbon resource estimates — again, the basis for analyses — were given as three figures: maximum (5% confidence level), mean (50% confidence level), and minimum (95% confidence level) numbers were evaluated. The mean was analyzed as the most likely, as opposed to the maximum used in our early EISs. This change helped the EIS move away from being a predictor of *gloom and doom.* Again, the conditional assumption was that oil and gas existed within the proposed lease-sale area. Resource estimates also were given for each alternative analyzed in the EIS. Our alternatives usually defer acreage from the proposal; therefore, we need to calculate new numbers for the reduced areas. We also always examine *no action* and *delay the sale* alternatives.

By 1982, a very sophisticated oil spill risk analysis (OSRA) had been developed for each sale area. The probability of oil spills occurring was factored into the work, and spill rates were developed for platforms, pipelines, and tankers. We used 35 spill points throughout the planning area and along the transportation routes to evaluate the proposal and alternatives. One hundred spills were *launched* from each point; the trajectories were calculated and applied to land/boundary segments and biological/cultural resources to determine the environmental risk factors. We also took into account seasonality of climatic forces and ice cover, and results included the probability of a spill contacting a specific point. This number in turn was combined with the expected spill rates, as based on the resource estimate of how much oil might be in the area. We now had a wealth of information for the analysts to use in the form of tables and illustrations. They were able to use the OSRA — a very valuable tool — to assess the probabilities of these resources being contacted by oil and being impacted by the proposal.

The cumulative impact conclusions immediately followed the conclusions of the proposal and were just as succinct. Major projects for the cumulative case were illustrated in a color graphic and listed in a table. Complete descriptions were given in an appendix. Twelve projects, in addition to state and federal leasing, were listed — a definite expansion from our earlier efforts.

At the five public hearings for the 1982 Beaufort Sea EIS, 78 people testified. We also received written comments from 30 individuals and/or agencies. Patterns of comments developed over the years, and we found that we could group the comments into issues and answer the concerns. Commenters usually focused on assumptions and methods used in the impact assessment and on the effectiveness of mitigation measures. Litigation centered on Native-ownership claims of off-shore lands and MMS's authority to lease. There were also concerns about boundaries and differential areas. By this time, high-level interest within the agency in chairing and attending the public hearings had waned; in these later hearings, regional staff chaired the sessions. Our headquarters staff sometimes attended, but, in general, we summarized the meetings and sent reports to head-quarters.

A RECENT EIS — CHUKCHI SEA EIS

In 1987, area-wide leasing and years of litigation history around the state resulted in an EIS that addressed 29 million ac in the Chukchi Sea in 1000 pages. At this point, MMS had experienced, seasoned authors, editors, and EIS coordinators; the database in and around the state had increased substantially (although it is still quite small compared with areas outside Alaska); and a great deal of fine tuning had occurred. Definitions were given to impact levels (five categories each), the document was tabbed for easy reference, each section contained a table of contents, and the conclusions were very tight, for example, "the effect of the proposed sale on marine and coastal birds is expected to be minor" (minor as defined in the established definitions).

Additionally, (1) resource estimates were given for the base (expected to be leased, explored, developed and produced), low, high, and cumulative cases (factoring in *exploration-only* in some areas); (2) tables showing assumptions about development were significantly expanded to include such details as number of river crossings of pipeline, miles of seismic data gathered, work boat trips, and amounts of muds used; (3) authors were required to factor in more causal agents in their analyses; and (4) the OSRA and cumulative case were presented similarly to our earlier Beaufort Sea EISs but were more refined.

During the public review period on the Chukchi Sea EIS, we held five hearings, had 34 testifiers, and received 22 written comments. The comments were broken down into 560 specific topics and answered individually. Litigation continued challenging federal ownership and leasing authority. In other areas, we were sued on whether we met NEPA and other requirements. In almost every case, OCS EISs have held under these challenges and have been deemed adequate by the courts.

MMS is currently involved in *focused* leasing in a continued attempt to concentrate the NEPA effort on areas of hydrocarbon potential and to reduce the size of the area to be analyzed in EISs, although the size would still not be as small as in the early years. Theoretically, this should lead to a smaller EIS that takes less time to complete. However, we still face some difficulties:

- Our reviewers (internal to DOI) want EISs to be stand-alone documents, which makes it difficult to scope out issues or incorporate by reference. Neither has tiering really occurred, which makes sale-specific EISs very large. We must include a great deal of internal review time in our schedules.
- We lean toward the environment, perhaps conservatively, leading some to say still — as in the early days — that we project gloom and doom.
- We do not recognize benefits (e.g., economics), which reinforces the gloom and doom image.
- The idea of staged/phased leasing and exploration and development has been argued in the courts and reinforced by the laws.
- Ideally, we could write exploration phase EISs now and developmental EISs after a discovery. Our EISs have always addressed the full suite of possibilities.

QUANTIFYING SOURCES AND FATE OF ENVIRONMENTAL POLLUTANTS

To predict the quantity and source of pollutants, we depend upon the hydrocarbon-resource estimates, statistically established spill rates, and the results of the OSRA. The resource estimates are derived from geological and geophysical information. A probabilistic model (PRESTO, Probabilistic Resource Estimates

Offshore), which takes into account that some prospects may not yield oil or gas, is used to produce numbers, if economical, for oil and gas. These numbers change as new information is obtained from drilling results or geological/geophysical surveys. I mentioned some of the difficulties in basing so much on these tenuous and often changing figures. This model (and others that we use) undergoes numerous peer reviews and a great deal of updating and verification. All of the models are explained in the EIS text and appendices and in reference papers.

Fates of environmental pollutants are illustrated through probabilities derived from the OSRA, land segments, and resource boundaries. Results are factored into conclusions for each resource category.

CONCLUSION

In closing, let me emphasize that our program has become very complex. We begin with a complicated ten-step leasing process. I have discussed in detail only two of these steps, the draft and final EISs. There have been many changes over the years in the level of industry interest in offshore Alaska. The same holds true for the general public who initially opposed our program, then began to welcome the idea of more jobs or land sales, and now oppose the program again because of the Exxon Valdez spill. We used to have many attendees at our public meetings and hearings, but interest and budgets have waned; we now sometimes hold regional-only meetings. And as our program has become more routine, commenters no doubt have become more cynical that they cannot stop the sales or change things. Comments have evolved and become more sophisticated over the years. Governments and organizations have hired people to follow our process and interpret it for their constituents. The oil companies have also done this and have sent representatives to communities to explain us! Some groups have hired professional commenters on our EISs. They have zeroed in on and challenged our models, as well as endangered species and subsistence issues.

We still strive to make our EISs smaller but better documents, and thus make them even more useful to the decision makers and more understandable to the public. In short, we hope to serve our audiences better.

Accommodating Price and Technology Competition within the EIS Process

K. U. Mowll, U.S. Department of Transportation, Washington, D.C.

ABSTRACT

The current environmental impact statement (EIS) process is used to justify already selected major projects and mitigate their impacts. Because the proposed project is usually compared with an unrealistic no-action alternative and no others, the economic efficiency of the proposed action can never be determined. If the alternatives considered in an EIS are changed, more efficient solutions to social and environmental problems can be identified. First, and most important, is the inclusion in the EIS of a *better management* alternative as the base case, consisting of management actions designed to maximize the effectiveness of existing natural and manmade systems and therefore avoid the need to construct a major project and mitigate its impacts. Second, including several possible technological solutions to the problem being addressed in the draft EIS results in implementation costs that can be lowered through the competition of greater numbers of firms representing several technologies in the bidding for the construction contracts. Rather than describe each technological alternative in detail in an unmanageably large document, several technologies could be included in one alternative by describing it in terms of minimum performance specifications and maximum environmental impacts. The result would be either the adoption of a low-cost, low-impact better management alternative which usually does not require an EIS or a cheaper implementation of a major project because of increased competition.

STATEMENT OF THE PROBLEM AND INTRODUCTION

Most environmental impact statements (EISs) prepared today are used to justify the construction of a project whose design, technology, and implementation have already been largely decided. Although the EIS is supposed to be the decisionmaking document for project implementation, in most cases, the only decisions made in an EIS are those related to the mitigation of impacts. Although this approach is a vast improvement over pre-National Environmental Policy Act (NEPA) days and usually results in the mitigation of significant environmental impacts, it tends to discourage the adoption of the most efficient solution to the problem. This drawback results from the severe limitations on the alternatives considered in the EIS.

In the first place, the baseline or no-action alternative is usually developed only to demonstrate how bad things would be without the proposed action or project. In fact, unless the economic system is not functioning at all, self-regulating actions will be taken which tend to mitigate against the catastrophes forecast in the no-action alternatives. For example, driving habits change to prevent gridlock from actually occurring and the price of energy increases, resulting in more efficient use of this resource before it is used up. Second, regulatory and practical limits on the size of the EIS discourage the consideration of competing alternatives in the document. By emphasizing only one alternative or technology, the EIS severely limits the possibilities of reducing the cost of implementing these projects through competition.

Over the past decade, the Federal Transit Administration (FTA) — formerly the Urban Mass Transportation Administration — has found that including a low-cost, low-impact better management alternative tends to lead to more efficient use of federal funds and to more environmentally sensitive projects. Recently, the FTA has tried to further increase the effectiveness of its funds by encouraging the inclusion of many competitive alternatives in the EIS process. This paper summarizes FTA's experience in accommodating price and technology competition in the EIS process.

INCLUDING PRICE AND TECHNOLOGY WITHIN THE NEPA PROCESS

For more than a decade, the demand for federal funds to build fixed guideway mass transportation systems such as subways, light rail lines, busways, monorails, and other automated systems has exceeded FTA's supply of federal funds. In order to manage these scarce funds, the FTA has applied economic principles and cost-effectiveness analysis to decide which projects should be funded. This economic decision process has been combined with the EIS process, resulting in an EIS which includes data not only on environmental impacts but also on the efficiency of proposed projects in solving the problems being addressed. Recently, budgetary constraints, as well as increased interest in privatization and competition in the provision of government services, have led the FTA to explore the possibility of including competing technologies and increased privatization within the EIS process in order to make even more efficient use of these federal funds.

The FTA's use of economic analysis as part of the draft EIS is to determine if any major project is needed at all. This subject is covered in the first subsection and deals primarily with FTA's experience with including a better management alternative in an EIS. If a major project is justified, then the FTA has attempted to maximize the use of technology competition to reduce the cost of implementing the viable projects. The FTA's limited experience in this area is discussed in the second subsection. Finally, the third subsection examines the applicability of this approach to other kinds of projects covered by NEPA.

This section is followed by a brief case study of Houston's experience in attempting to accommodate price and technology competition in the preparation of a draft EIS for a proposed *System Connector* transit project.

INCLUSION OF ECONOMIC CONSIDERATIONS AND BETTER MANAGEMENT ALTERNATIVES IN AN EIS

The FTA has developed a unique approach to the preparation of EISs through the incorporation of the economic and environmental analysis of a wide range of alternatives in the draft EIS. Two factors unique to the FTA have contributed to the development of this approach. First, unlike many federal agencies, the FTA can legally fund a variety of different technologies including light rail transit (trolleys), heavy rail transit, buses on city streets, busways, high occupancy vehicle (HOV) facilities, and automated systems such as monorails and automated guideway transit (AGT). These may be considered *substitutes* in economic terms. Second, the FTA funds up to 80% of the capital costs of these projects from a discretionary program for which demand for funds exceeds supply by several times. Therefore, not only does the FTA have a need to balance demand for funds with its supply, but it also is legally able to fund a wide range of different alternatives (substitutes) with widely varying capital costs.

To attempt to balance the supply of funds for new transit systems with the supply, the FTA has relied on a simplified economic analysis to identify the major projects most worthy of funding. The FTA economic analysis uses only two measures of economic benefits: additional transit patrons and travel time savings for baseline transit patrons. Indirect benefits, such as reduction in air pollution, increased densities of development, and reduced automobile congestion, are directly correlated with the number of additional transit patrons attracted to a new transit system. Thus, economic efficiency is measured in terms of costs per additional transit patron. Because all the data required to calculate this measure of economic efficiency would be developed for each alternative anyway, the inclusion of an economic analysis requires no significant amount of additional work, unless the alternatives normally considered were inadequate.

The FTA quickly learned that the baseline alternative (no action) used to calculate environmental impacts was not usually an appropriate baseline alternative for the calculation of economic impacts. In addition, it was learned that it was usually not the appropriate environmental baseline. For example, the economic benefits (as measured by forecasts of additional transit patrons) of proposed light rail systems in two different cities varied greatly depending upon the quality of the bus service in the no-action alternative. Those cities which had neglected their bus systems claimed many more additional riders would be attracted to their proposed light rail lines than would be attracted to proposed lines in those cities which had maintained good bus service. In other words, by using the no-action alternative as the base case in an economic analysis, those cities which had neglected their transit systems could claim that their light rail lines were more cost-effective. Thus, the cities which had not devoted local resources to transit would be rewarded with federal funds for their light rail proposals because they would be ranked higher in terms of cost-effectiveness.

To correct this inequity, the FTA required that all major projects proposed by urban areas across the country be compared to the best transit system which could be developed without requiring a major capital investment. This better management alternative has been named by the FTA as the *Transportation System Management* alternative or TSM. Management techniques such as traffic signal timing and preemption, reservation of curb lanes for buses, improved coordination, use of larger buses, and express services are included in the better management alternative. These techniques are designed to work within the existing transportation system to make it function more efficiently, rather than forcing a new element on that system whose impacts not only are imperfectly understood but also require significant mitigation efforts. In some cities, the better management alternative was a significant improvement over the no-action alternative, and in other areas with excellent existing transit systems, the improvements were minor. This better management alternative not only allowed the FTA to compare all of the *build* alternatives to similar base cases in all cities, but also (in many cases) showed the local decision makers that low-cost, low-impact alternatives were capable of solving the problem at hand.

Although originally introduced as a baseline for economic analysis, the better management alternative has also pointed out some of the problems with using the no-action alternative as the baseline for environmental analysis. For the past 40 years, transportation gridlock has been forecast as being inevitable if *no action* were taken, yet it has never happened — probably because the self-regulating features of the natural and manmade systems were not considered in the forecasts, nor were efforts to better manage these systems included. In many cases, the *gloom and doom* forecasts of the no-action alternative were used to develop *straw-men* alternatives to justify the construction of unneeded projects with significant environmental impacts.

An interesting side effect of the use of better management alternatives (Mowll 1987) is that, if adopted, they usually do not require an EIS. In other words, better management techniques can be implemented immediately without needing to complete the EIS.

ALLOWING COMPETITION OF TECHNOLOGICAL ALTERNATIVES WITHIN AN EIS

If it has been determined that transit improvements are required, which cannot be provided by a better managed transportation system, then the FTA has attempted to implement the major projects required in the most cost-efficient manner by maximizing competition.

It is an economic truth that the opportunities for the more efficient provision of a product increase with competition. The government has tended to limit the amount of competition by waiting until after a project is designed and the technology selected before initiating the competitive bidding process. However, at this point, many of the firms willing to provide the services or facilities needed with alternative technologies have been eliminated from the competition. Although government procurement practices could accommodate different technologies in the final bidding, the environmental processes seemed to stand in the way.

The two obvious alternative approaches to encouraging more competition in the bidding process seemed unworkable within the traditional NEPA process. The first approach would be to carry all technological alternatives through the whole EIS process and then hold the competitive bidding process, but this approach would produce an unmanageably large document. Furthermore, many of the competing technologies may be proprietary, and their firms would not want many of the technical details of their operations made public. The second approach would be to open the bidding to all technologies before the EIS was prepared and then prepare the EIS on only that most cost-effective technology and design. While this process would open the bidding to all technologies, it would not be allowed under NEPA since an alternative would be selected before the environmental impacts were examined. In addition, the winning contractor could not be expected to endure the uncertainty of a multiyear EIS preparation before work could begin. The compromise approach that the FTA attempted in Houston for the proposed System Connector project is to include all of the technology alternatives in the draft EIS, tentatively select a winning technology and supplier after the draft, and prepare the final EIS on the winning technology. To simplify the draft, the technologies were grouped together in one alternative which consisted of performance standards and limits on the environmental impacts. For example, the alternative could call for the capacity to move so many people per hour at a determined speed and along a specified alignment. Also specified were maximum noise and air pollution levels, visual intrusion, and the environmental impacts common to all of the technologies due to the existence of the structure along the specified vertical and horizontal alignment.

This approach is examined in detail in the case study that follows of the Houston System Connector EIS.

APPLICABILITY TO NONTRANSIT EIS

Nearly all actions requiring EISs have better management or low-cost, low-impact alternatives. Certainly, urban transportation systems have many inefficiencies which could be relatively easily eliminated with better management. At the other end of the spectrum from urban transportation projects, the Northwest Coalition for Alternatives to Pesticides advocates good range and grazing management techniques as an alternative to the use of herbicides (O'Brien 1987). Other better management alternatives to high-cost, high-impact alternatives are (1) substitution of peak hour electricity pricing or purchase of power agreements for the construction of new nuclear power plants and (2) relocation of businesses and residences rather than constructing flood control projects.

HOUSTON SYSTEM CONNECTOR CASE STUDY

For more than a decade, Houston has been planning for an urban rail system. After numerous modifications and local setbacks, Houston recently moved forward to plan and implement a new rail system with heavy private sector participation. This case study summarizes what has happened since then as the Houston Metro (technically the Metropolitan Transit Authority of Harris County) and the FTA struggled to find a cost-effective solution to Houston's transportation problems within the context of the EIS process, while maintaining the possibility of as much competition as possible from private sector suppliers of significantly different transit systems.

Background

Houston's rapid growth, lack of continuous arterial streets, and relatively small expressway network created significant highway congestion in the area before the oil bust. This congestion was alleviated to some extent by the downturn in the local economy and will be further reduced by an ongoing program of toll road construction. However, forecasted economic growth in the core area will probably lead to higher highway congestion levels in the future. There are four major centers of employment in the Houston area: downtown, Post Oak, Greenway Plaza, and the Texas Medical Center (see Figure 1). These activity centers are very large, with the second largest (Post Oak) being larger than Atlanta's Central Business District (CBD) and downtown twice as big. Houston also has five separate transitway facilities providing express bus and car pool services along expressways; however, these transitway facilities are not connected (see Figure 2).

The proposed System Connector project was developed to address the problem of providing transit service between the first three activity centers and within the first two while connecting most of the existing transitways together. Scoping

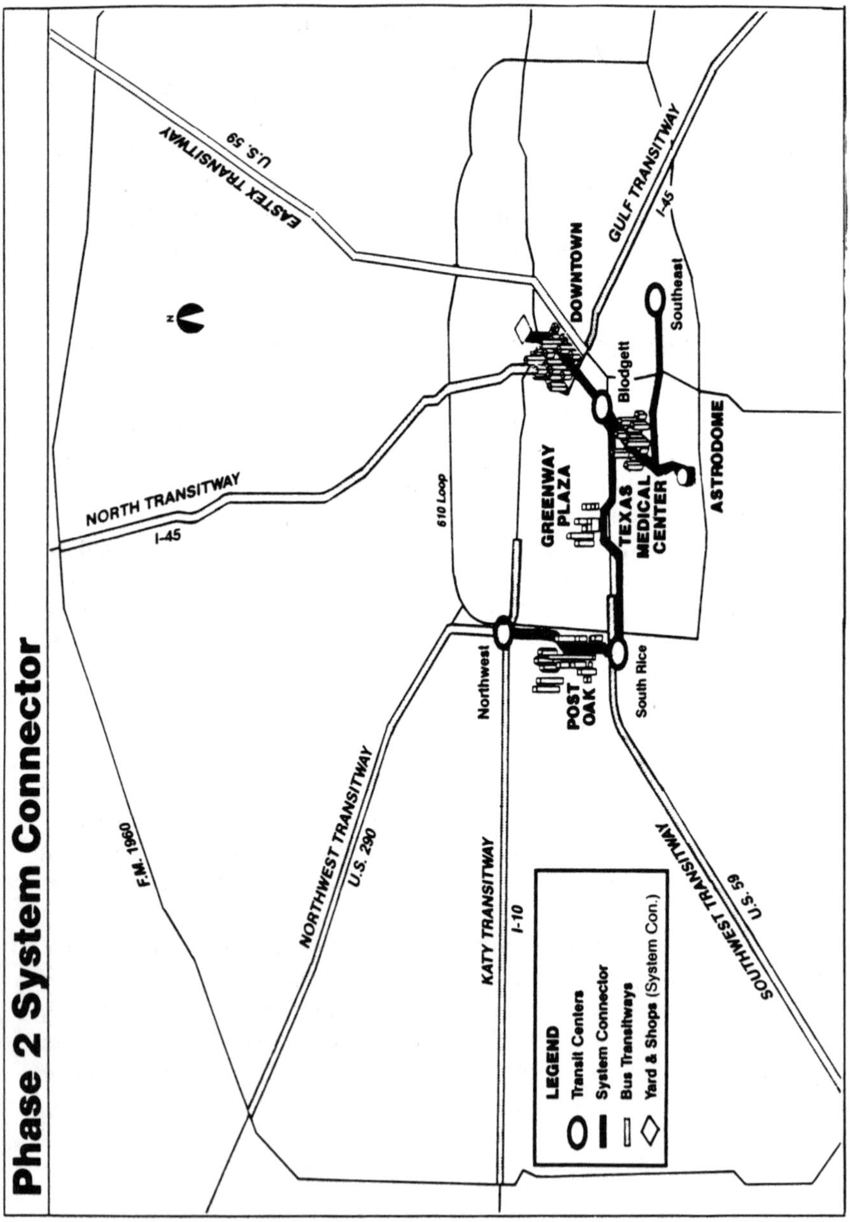

Figure 1. Houston activity center.

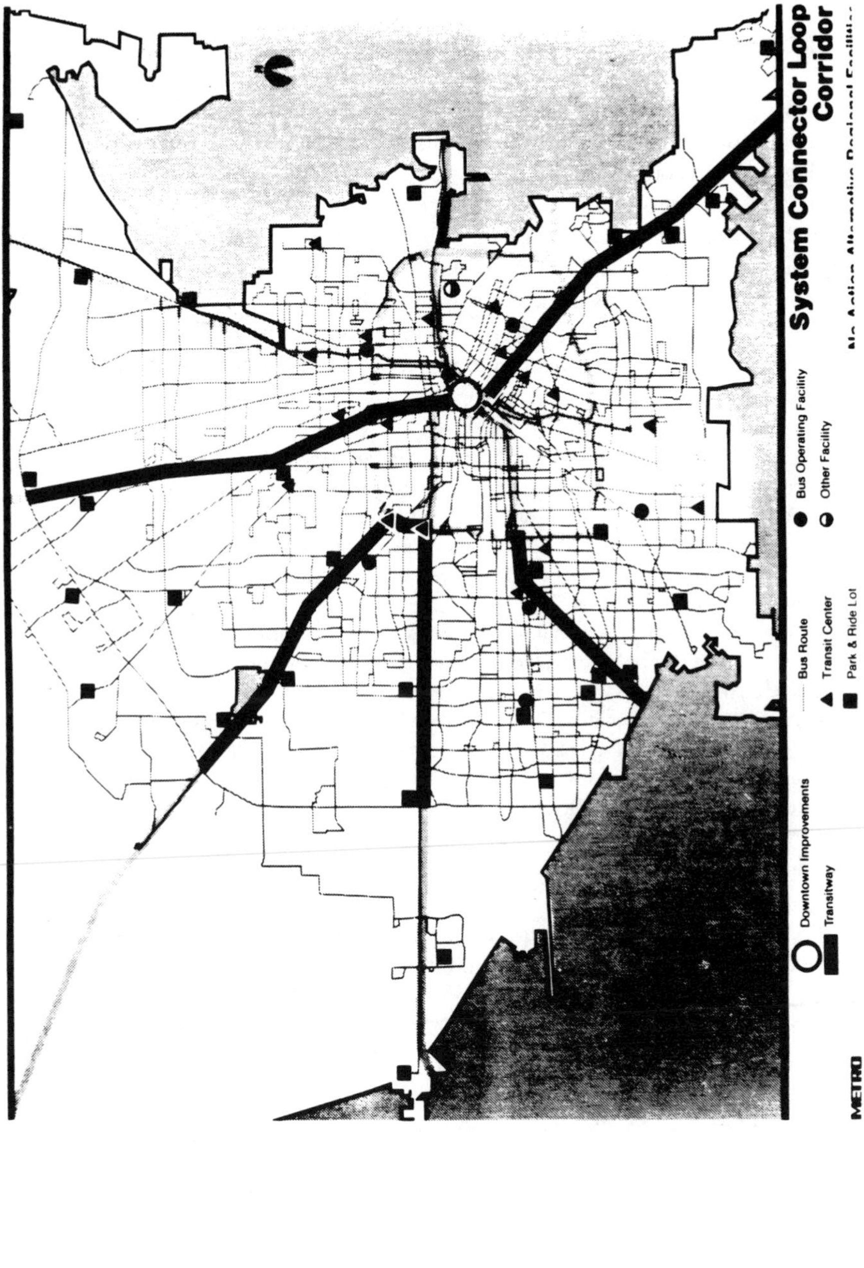

Figure 2. Existing Houston transitways.

meetings for the System Connector EIS were held in August 1986 and presented four alternatives — no action, Transportation System Management (TSM or better bus), bus on busway, and light rail operating partially in streets or in a totally grade separated right-of-way. General alignment alternatives for the last two alternatives were also presented.

Alternatives Considered During EIS Preparation

Following the scoping meetings, promises were made of lower costs for implementation by private sector suppliers representing different transit technologies. These promises coincided with tight financial conditions at both the local and federal levels and resulted in the FTA and Metro being encouraged to explore the possibility of involving multiple, competing private sector suppliers throughout the planning and environmental processes.

In January 1987, the private sector was invited to attend a meeting to express interest in providing rail transit service along the System Connector alignment. Numerous transit system suppliers were represented and the following technologies were offered.

- Standard Light Rail Transit with drivers operating large vehicles on steel tracks in totally grade-separated rights-of-way or in the streets along with automobile traffic with power supplied from overhead catenary
- Automated Light Rail Transit with computer-controlled, driverless operation operating only on an exclusive right-of-way and with either third-rail or overhead power supply
- Straddle Beam Monorail with fully automated, driverless transit vehicles operating on a completely grade-separated guideway by straddling the beam
- Suspended Monorail which is the same as above, but the vehicles operate suspended from a single beam
- Automated Guideway Transit (AGT) operating either on steel wheels on steel tracks or on rubber wheels in a concrete trough guideway without a driver in a completely separate right-of-way

These transit systems were estimated to cost $1 billion to construct, give or take $200 million or so.

The technologies themselves did not appear to be sufficiently different or to have significantly different environmental impacts from the light rail alternatives originally proposed in scoping to require rescoping the EIS, so the introduction of these new technologies was handled in the ongoing citizen participation process. Interest in the development community for significant changes in the alignment raised significant concerns among the environmental professionals regarding the possibility of not only rescoping but also having to document the environmental impacts of a very large number of alignments. Decision makers also expressed concerns about multiple, alternative alignments with different technologies, since it was unclear how they could be compared to one another. Developers and environmental staff worked closely to develop the potentially best alignment. The alignment (see Figure 3) is totally grade separated, includes vertical as well as

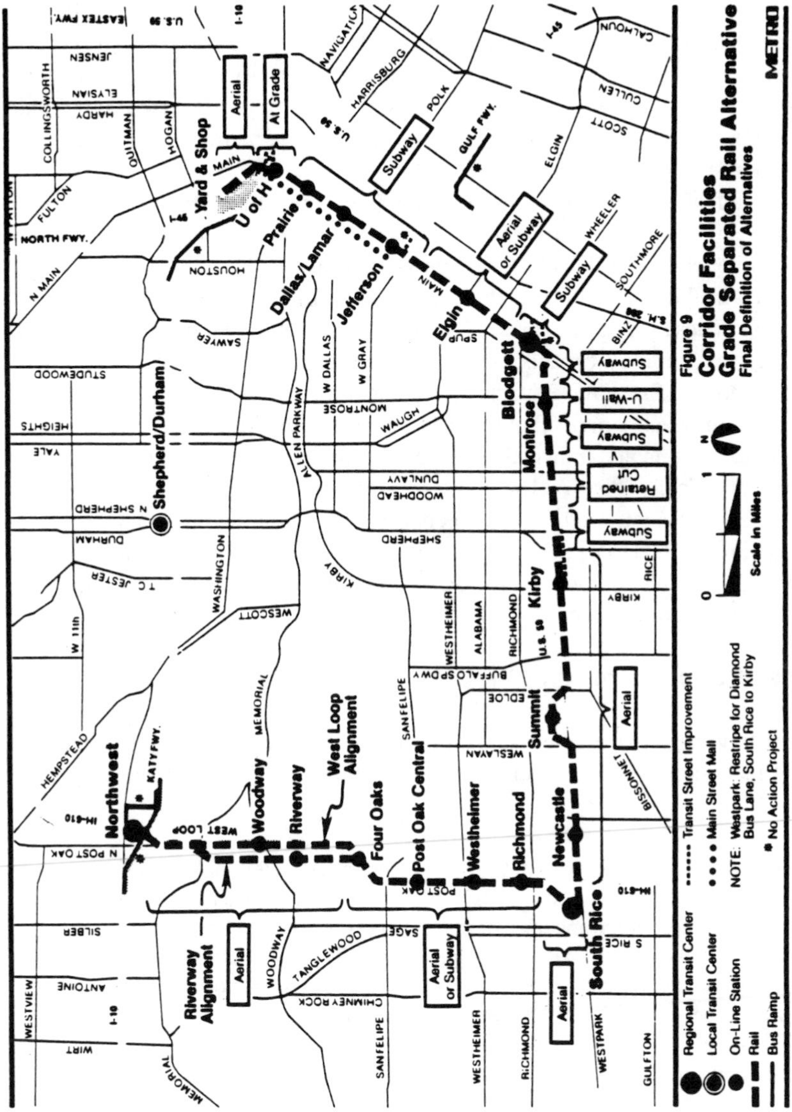

Figure 3. System Connector: automated guideway alternative.

horizontal specifications, and was required for all automated technologies. Both the FTA and Metro recognized that the alignment could change to reflect environmental and other concerns, but decided that for the purposes of simplicity, all automated technologies would be evaluated using one alignment.

The Metro began a detailed analysis of the technologies to determine if each could provide the service levels required and to identify significantly different environmental impacts. All of the automated guideway alternatives were judged capable of meeting the service levels and requirements specified, although questions were raised on the capability of monorail switches to operate fast enough and on the safety of suspended monorails (in an emergency, patrons cannot walk out of a suspended monorail).

Nearly all of the environmental impacts (e.g., air, noise, vibration, historic, parklands, wetlands, and traffic) for all the technologies were amazingly similar. Some differences were detected in impacts such as visual intrusion because of differences in the design of the guideway for the different technologies. Some apprehension has also been expressed about possible unknown impacts generated by the high technology aspects of some of the technologies such as the linear induction motors. Since the impacts of all technologies were very similar, it was decided that the treatment of all of these automated technologies could be included in one generic automated transit alternative. The impacts of this generic alternative could be specifically identified within a very narrow envelope for most impacts. Specific detail was required for each technology in just a few impact areas, such as visual intrusion. By presenting the automated guideway alternative in this manner, all of the technologies could be included in the draft EIS without making the document cumbersome.

Proposed Decisionmaking Process

The following steps, in suggested chronological order, were proposed in order to select a technology in a manner consistent with both NEPA and sound economic practice:

- Publish and distribute the draft EIS with a generic automated guideway alternative specifying performance standards and maximum environmental impact.
- Select the most promising three or four suppliers of automated guideway systems to prepare fixed-cost (capital and operating) bids for the automated alternative (the preparation of these bids and nearly complete construction drawings would have to be subsidized by the Metro and FTA).
- Tentatively select the lowest-cost supplier meeting the performance and environmental requirements included in the draft EIS.
- Prepare and circulate the final EIS on the tentatively selected technology using the data developed by the winning bidder.
- Sign a fixed-price contract with the selected supplier after the environmental process has been completed.

This approach incorporates price and technology competition within the EIS process. It does not conflict with NEPA nor does it produce an overly long document. Furthermore, it seems to satisfy the private sector bidders' need to limit the amount of uncertainty by specifying the acceptable environmental impacts and service requirements before bidding. It also limits the amount of time between bidding and contract signing to the time necessary to assemble and circulate the final EIS.

The cost of this approach is not cheap, since each of the final bidders was expected to receive at least $1 million to help them to prepare data in sufficient detail for them to feel comfortable offering their fixed-price bids. Decision makers assumed that this cost would be more than made up by lower bids due to increased private sector competition. It is very interesting to note that the Metro staff did not necessarily agree and preserved the option to bid on the construction of an automated guideway in competition with the private sector, since they felt they could build a system just as cheaply as the private sector.

Status as of October 1989

The prime force in the development of this process has been the chairman of the Metro Board, Mr. Robert Lanier. He committed the board to making a decision on the kind of transit system needed by Houston based upon sound financial and economic practice. This approach lead to a massive examination of the costs of the alternatives and the benefits in terms of patronage of these alternatives in Houston. The result has been the most careful and thorough examination of the need for transit improvements for any city in the United States. Several trends in the analysis are worth noting. First, the TSM (better management) alternative has been completely updated to offer much better and cost-efficient service. Second, the projected benefits of the expensive guideway alternatives have been reduced by as much as half. Third, a much more realistic and less gloomy forecast of the congestion likely to occur under no-action conditions has been made.

The Metro Board used the information developed in the DEIS to make several very interesting decisions. The Board first chose the monorail technology and then, after intense competition between two suppliers, selected one monorail technology. However, about a year later, after an election in which the mayor was defeated, the Board reversed its decision and selected the "Better Bus" alternative.

CONCLUSION

The experience of Houston in selecting their preferred transit system leads us to conclude that the EIS process can be used to evaluate competing technologies. This case also provides evidence that the EIS process can result in the selection of an alternative which consists of: (1) better management of the existing system, (2) a more cost-effective solution than a high-capital alternative, and (3) few, if any, environmental impacts.

REFERENCES

Mowll, K. U. 1987. The Avoidance of Impacts Through the Use of Low-Cost, Low-Impact Alternatives in EISs, 1987, Environmental Impact Assessment. In N. A. Robinson (ed.), Proceedings of Conference on the Preparation and Review of Environmental Impact Statements. Co-sponsored by the Council on Environmental Quality and The Environmental Law Section of the New York State Bar Association in West Point, New York.

O'Brien, M. 1987. How Does the Scoping Process Affect the Substance of an EIS, Environmental Impact Statements (edited by Nicholas A. Robinson). Co-sponsored by the Council on Environmental Quality and the Environmental Law Section of the New York State Bar Association in West Point, New York.

Activities in the Department of Energy's Clean Coal Technology Program: A New Dimension in NEPA

E. W. Evans, R. A. Hargis, and T. C. Ruppel, U.S. Department of Energy, Pittsburgh, PA

ABSTRACT

The Clean Coal Technology (CCT) program started with the passage of legislation in 1985 and has evolved into the administration's primary initiative to respond to concerns over acid rain. The National Environmental Policy Act (NEPA) aspects of this program differ from those in typical government programs for a number of reasons. First, government participation is through cooperative agreements with industrial partners with the government share limited to 50%. Second, most of these projects involve the retrofit of existing facilities. Finally, the program itself is environmental so that the main environmental impacts of it are positive (i.e., the reduction of emissions of acid rain precursors). By a review of the status of the NEPA documentation for the Clean Coal projects being administered by the Pittsburgh Energy Technology Center (PETC), the unique features of this program and how these features are addressed in the NEPA process are presented.

INTRODUCTION

The Clean Coal Technology (CCT) program is a government industry, co-funded effort to demonstrate new, advanced concepts for using coal more cleanly and efficiently. Begun by Congress in 1986 and expanded by President Reagan in 1987, the program is expected to finance more than $5 billion of commercial-scale demonstration projects when completed in the 1990s. At least half of the funding will be provided by the private sector.

Projects for the CCT program are selected from a series of competitions in which private companies, states, or other nonfederal organizations submit proposals for evaluation by the Department of Energy (DOE). The fact that these projects are administered under cooperative agreements limits the nature and extent of government direction. This, in turn, can limit the range of alternatives under direct government control during project execution. In addition, the nature of cooperative agreements combined with the requirements of the National Environmental Policy Act (NEPA) can result in an exposure of the industrial participant to financial risk that can affect the potential for the success of the project. Ensuring that the objectives of the NEPA process are met without adversely affecting the cost or schedule of the overall project involves close interaction with the industrial participant through all phases of the project.

A unique feature of the CCT program, at least for those projects administered at Pittsburgh Energy Technology Center (PETC), is that most of the projects are retrofits. That is, these projects involve the addition or modification to an existing utility or industrial facility with only minor changes to the existing site and overall plant operations. Construction is usually of limited extent compared to the total plant site and most often takes place on previously disturbed land. The equipment to be added to the facility, in most cases, can be removed at the conclusion of the demonstration period. Because of the limited construction impacts and the temporary nature of the demonstrations, there is less cause for concern over cumulative, secondary, or long-term impacts from these projects. Nevertheless, these projects will be commercial scale, and the data to be obtained are necessary for future commercialization of the technologies.

There is the potential for minor adverse impacts for individual projects, such as activity in a floodplain, increases in the generation of solid wastes, or temporary increases in sedimentation or fugitive dust due to construction. In all cases to date, these minor adverse impacts have been shown not to be significant. As a result, the no-action or delayed-action alternatives have not been chosen.

THE DEPARTMENT OF ENERGY'S CLEAN COAL TECHNOLOGY PROGRAM

Overview

The DOE ran the first clean coal competition (CCT-1) in 1986 and the second (CCT-2) in 1988. The third competition (CCT-3) began in May 1989, with projects to be selected by December 27, 1989. When the program is completed as currently planned, five rounds of competition will have been conducted.

To date, 29 projects have been selected from the first two rounds of competition. The projects have a total value in excess of $2.3 billion with the private sector's share amounting to more than 60%. Project costs have ranged from less than $1 million to as much as $242 million. If CCT-1 and CCT-2 are used as an approximate gage, perhaps as many as 75–100 projects could be financed when the CCT program is completed in the 1990s. However, it must be said that it would be very fortuitous if all projects selected would proceed satisfactorily through negotiations to signing of the cooperative agreement. Financial, and in some cases environmental, impediments may terminate projects.

Objectives

The objectives for each of these three rounds has varied slightly with each solicitation. A reference to the enabling legislation for each solicitation, language in the legislation, and the objectives stated in the Program Opportunity Notice (PON) for each round are provided below:

- CCT-1: Public Law (P.L.) No. 99-190, December 19, 1985, P.L. 99-500, October 18, 1986, P.L. 99-591, October 30, 1986.

"provides funds to conduct cost-shared clean coal technology projects for the construction and operation of facilities that would demonstrate the feasibility of future commercial applications of such technology."

PON objective: "conduct cost-shared clean coal technology projects to demonstrate the feasibility of these [emerging clean coal] technologies for future commercial applications."

- CCT-2: P.L. 100-202, December 22, 1987.

"provides funds to conduct cost-shared innovative clean coal technology (ICCT) projects to demonstrate emerging clean coal technologies that are capable of retrofitting or repowering existing facilities."

PON objective: "conduct cost-shared innovative clean coal technology (ICCT) projects to demonstrate technologies that are capable of being commercialized in the 1990s, that are more cost-effective than current technologies, and that are capable of achieving significant reduction of SO_2 and/or NO_x emissions from existing coal burning facilities, particularly those that contribute to transboundary and interstate pollution."

- CCT-3: P.L. 100-446, September 27, 1988.

"provides funds to conduct cost-shared Clean Coal Technology (CCT) projects for the design, construction, and operation of facilities that would demonstrate the feasibility of future commercial applications of such ... technologies capable of retrofitting or repowering existing facilities and shall be subject to all provisos contained under this head in Public Laws 99-190 and 100-202 as amended by this Act."

PON objective: "conduct cost-shared Clean Coal Technology projects to demonstrate innovative, energy efficient technologies that are capable of being commercialized in the 1990s. These technologies must be capable of (1) achieving significant reductions in the emissions of sulfur dioxide and/or the oxides of nitrogen from existing facilities to minimize environmental impacts such as transboundary and interstate pollution and/or (2) providing for future energy needs in an environmentally acceptable manner."

NEPA STRATEGY

An overall strategy for compliance with NEPA has been developed for the CCT program through the combined efforts of the DOE's Office of Environment, Safety, and Health and Office of General Council. This strategy is consistent with the Council on Environmental Quality (CEQ) regulations (40 CFR Parts 1500–08) and the DOE guidelines for compliance with NEPA (52 FR 47662, December 15, 1987). The strategy, which includes consideration of both programmatic and project-specific environmental impacts during and subsequent to the selection process for each solicitation, has three major elements as outlined below.

Programmatic Environmental Analysis

The first element of the DOE NEPA strategy involves the preparation of a comparative programmatic environmental impact analysis, based on information supplied by the offerors and supplemented by the DOE as necessary. This environmental document analyzes the environmental consequences of the solicitation and the technologies supported by the program compared with the no-action alternative.

The deadlines imposed by the statutes defining the CCT program required the DOE to make selections of projects within 100, 160, and 120 days of receipt of the proposals for the three solicitations to date, respectively. These requirements did not allow adequate time to prepare the required programmatic EIS prior to receipt of the proposals as recommended by CEQ guidelines. The DOE's Offices of Environment, Safety, and Health, and General Council advised the Office of Fossil Energy to follow alternative procedures to ensure that environmental factors were fully evaluated and integrated into its decisionmaking process.

For CCT-1, the DOE required all offerors to provide information in their proposals on the environmental impacts of the potential commercialization of their respective technologies. For the project-specific evaluation, each offeror was required to provide a brief discussion of the purpose of the proposal; the salient characteristics of the proposal site and/or process including reasonable alternatives; the environmental impacts of the proposal; an analysis and summary of the environmental merits of the proposal as compared to environmental criteria provided in the PON sent to offerors; and, to the extent known, a list of permits, licenses, and approvals needed to implement the proposal.

For CCT-2, the DOE's environmental and legal offices recommended a similar but expanded strategy. With 160 days available until project selection, the Office of Fossil Energy was advised to include in a programmatic NEPA document a comprehensive, comparative analysis of the various projects represented in the offerings. However, in light of the confidentiality requirements of the procurement process, this comparative part was not required to be made public.

For CCT-3, there was adequate time to develop a programmatic environmental impact statement (EIS) before selection, using the CCT-2 programmatic document as a basis. A draft programmatic EIS has been issued, public comments have been received, and the final EIS is being written at this time.

With the above procedure, the DOE has been able to integrate the programmatic NEPA requirements into the requirements imposed by the enabling legislation of the CCT program.

Preselection Project-Specific Review

The second element of the DOE's strategy for NEPA compliance involves the preparation of a preselection environmental review based on project-specific environmental data and analyses that offerors supply to the DOE as part of each proposal. This analysis contains a discussion of the site-specific environmental, health, safety, and socioeconomic issues associated with the demonstration project. It includes a discussion of the advantages and disadvantages of the proposed and alternative sites and/or processes reasonably available to the offeror, a discussion of the environmental impacts of the proposed project, and a list of all permits that must be obtained to implement the proposal. The document describes options for controlling discharges and for management of solid and liquid wastes and assesses the risks and impacts of implementing the proposed project. Because this preselection, project-specific environmental review contains proprietary and/or business confidential information provided to the DOE in the proposal, this document is not publicly available.

Site-Specific NEPA Documents

The third element of the DOE's NEPA strategy provides for the preparation of site-specific NEPA documents for each of the projects selected for financial assistance under each solicitation. The remainder of this paper focuses on the status of these individual NEPA documents prepared for each project in the CCT program.

PETC's EXPERIENCE WITH INDIVIDUAL CLEAN COAL PROJECTS

General Procedure

The steps involved in the preparation of a site-specific NEPA document for each project are shown in Figure 1. The starting point for preparation of a NEPA document is the receipt of a completed Environmental Information Volume (EIV)

Task	Action By	Duration (weeks)
EIV Review, Comment and Transmit	Support Contractor	(6) ======
Revise EIV	Participant	(4) ====
Prepare Draft NEPA Document and Transmit	Support Contractor	(12) ============
Review/Comment Draft NEPA	HQ/FE,EH	(4) ====
Prepare Final NEPA Document and Transmit	PETC	(4) ====
Final Approval of NEPA Document ====....=	HQ/FE,EH	(4 to 9)
	TOTAL	33 to 38 weeks

Figure 1. Preparation of site-specific NEPA documents.

from the project participant. The steps shown in Figure 1 do not include this activity, since completion is beyond the DOE control. Experience has shown that the preparation of the EIV can take as little as 2 months or as much as a year, depending on the level of effort required of the project participant. Each of the subsequent steps involves the preparation of documents by PETC through a support contractor. Some overlap of scheduled tasks can occur to reduce the overall schedule. However, if more than one iteration is required for any review cycle, the overall schedule will be lengthened. The need for more than one iteration, especially in the preparation of the draft NEPA document, has been more the rule than the exception. The review cycles at the headquarters (HQ) level can also cause some delays, especially if a number of NEPA documents are transmitted to HQ in the same time frame. If final NEPA approval requires a review at the Secretary of Energy level, an additional 5 weeks may also be required. Therefore, the completion of the NEPA documents for each project will usually take nearly a year and even more in some cases.

A dilemma occurs when the site-specific NEPA document is not completed before the initiation of detailed design. Since the DOE will not provide federal funds for any work beyond preliminary design until final NEPA approval has been obtained, the only options available for the project when this occurs are (1) the participant stops work on the project until NEPA approval is received or (2) the participant proceeds with the project without federal funds and then makes a claim for government reimbursement of these expenses after NEPA has been completed. The first option will undoubtedly lead to an increase in project cost and time required for project execution. The second option leaves the participant at risk for the funds expended if the NEPA documents are not approved. For some projects, the industrial participant may be unwilling or financially unable to continue the project without the availability of DOE funds. To date, no projects have been terminated because of delays caused by the NEPA process. However, delays have occurred that have affected the cost and schedule for at least one of the projects, and the completion of NEPA documents remains on the critical path for several other projects.

Determining the Appropriate Level of NEPA Documentation

The DOE NEPA guidelines provide three general categories which define typical classes of actions: (1) actions which normally do not require environmental assessment (EAs) or EISs; (2) actions which normally require EAs but not necessarily EISs; and (3) actions which normally require EISs. The actions in the first category are referred to as categorical exclusions. Within the DOE NEPA guidelines is a provision for actions for which "neither an EA nor an EIS is required where it is clear that the proposed action is not a major Federal action significantly affecting the quality of the human environment." The NEPA document prepared for such actions is called a Memo-To-File (MTF).

All CCT projects at PETC have required or were expected to require either MTFs or EAs; none of the projects has required an EIS. The decision as to whether an MTF or EA is required hinges on a determination as to whether the impacts are *clearly insignificant*. If extensive data gathering or analysis is required to reach a conclusion as to significance of impacts, then an MTF is not appropriate and an EA is required.

Discussion of the Projects at PETC

There are 21 CCT projects from CCT-1 and CCT-2 that have been assigned to PETC. These projects are listed in Table 1, along with a notation of the solicitation from which the project was selected (CCT-1 or CCT-2) and the status of the NEPA documentation for the project. There are 22 NEPA documents required for 21 projects, since one of the projects involves two sites and two separate NEPA documents were deemed necessary.

The funding levels for the 21 projects range from less than $1 million to over $200 million. Most of the total project costs are in the tens of millions of dollars. The actual project costs and funding levels are available for only those projects for which cooperative agreements have been negotiated. These are, with the corresponding project costs, Coal Tech ($0.8 million), B&W LIMB ($19.4 million), EER Gas Reburning/Sorbent Injection ($30 million, combined for both sites), Ohio Clean Fuels ($226 million), TRW Advanced Slagging Combustor ($49 million), and Bethlehem Steel Coke Oven Gas Cleaning ($35 million). A breakdown of project cost by project phase and participant's share, as well as additional technical and project management information, is contained in the individual

Table 1. Status of NEPA Documents for PETC Clean Coal Projects (as of the date of the conference, October 1989).

CCT Project	CCT No.	NEPA Document	Status
Precombustion Retrofit Projects			
Ohio Clean Fuels	1	EA	Preliminary[a]
OTISCA Fuels	2	MTF	Preliminary
C-E/EPRI Coal Quality Expert	1	MTF	Preliminary
Western Energy Advanced Coal Conversion	1	EA	Draft[b]
United Coal — Microbubble	1	EA	Preliminary
Combustion-Zone Retrofit Projects			
TRW Advanced Slagging Combustor	1	EA	Final[c]
Coal Tech Advanced Combustor	1	MTF	Signed[d]
TransAlta Low NO_x/SO_x Burner Retrofit	2	EA	Draft
B&W LIMB	1	MTF	Signed
EER Gas Reburning/Sorbent Injection			
Hennepin Site	1	MTF	Signed
Lakeside Site	1	EA	Signed
B&W Coal Reburning	2	EA	Draft
SCS Low-NO_x Wall-Fired	2	MTF	Signed
SCS Low-NO_x Tangential-Fired	2	MTF	Signed
Postcombustion Retrofit Projects			
B&W SO_x-NO_x-RO_x Box	2	MTF	Signed
C-E WSA-SNO_x	2	MTF	Preliminary
SCS Selective Catalytic Reduction	2	MTF	Signed
SCS Chiyoda—Advanced FGD	2	EA	Draft
Pure Air—Advanced FGD	2	EA	Draft
C-E Dry Sorbent Injection	2	EA	Preliminary
Industrial Projects			
Passamaquoddy Tribe SO_2 Scrubbing for Cement Kilns	2	EA	Draft
Bethlehem Steel Coke Oven Gas Cleaning	2	EA	Final

[a] Preliminary indicates that a final EIV has not been received.
[b] Draft indicates that a draft of the NEPA document is being prepared or reviewed.
[c] Final indicates that the final NEPA document is being prepared or reviewed.
[d] Signed indicates that the NEPA process has been completed.

Report To Congress prepared for each project for which negotiations have been completed.[1]

Most of the PETC projects involve the retrofit of an existing facility. A retrofit can be defined as the addition of emission control equipment to existing plants that does not result in additional capacity or extension of boiler life. Retrofit technologies can be categorized as precombustion, combustion, and postcombustion. Precombustion retrofit technologies include coal cleaning and coal liquefaction. Combustion-zone retrofit technologies include advanced combustors, low NO_x burners, furnace sorbent injection, and reburning. Postcombustion retrofit technologies include duct injection, advanced flue gas desulfurization, selective catalytic reduction, and combined SO_2/NO_x removal.

[1] These reports are publicly available through the U.S. Department of Energy, Office of Fossil Energy, Forrestal Building, Washington, D.C.

Because retrofits involve the addition of emission control equipment, the net effect of these projects is generally an improvement in air quality impacts, combined with other minor impacts, such as increased solid waste generation. Many of the projects, however, are of limited duration, after which the site will be restored.

Precombustion Retrofit Projects

There are five projects that will demonstrate precombustion retrofit technologies: Otisca Fuels, C-E/EPRI Coal Quality, Western Energy Advanced Coal Conversion Process, Ohio Clean Fuels, and the United Coal Microbubble Project. All of these projects will involve the production of a cleaned fuel from coal and the combustion of the fuel in an existing utility or industrial boiler. Depending on how the project is structured, the primary environmental impacts could be associated with the production of the fuel, with secondary impacts from the combustion of the fuel, or vice versa. For the Otisca Fuels and C-E/EPRI Coal Quality projects, it is anticipated that the primary impacts will be associated with the combustion of the fuels, since the fuels will be produced from existing facilities. For the Western Energy project, the primary impacts will be associated with the construction and operation of the coal cleaning facility; secondary, positive impacts to air quality are anticipated from the combustion of the fuel. In the case of the Ohio Clean Fuels (formerly Ohio-Ontario Clean Fuels) project, both production and combustion of the fuel may be considered primary impacts, since the coal-oil coprocessing demonstration plant is currently envisioned to be closely integrated with an existing utility. The United Coal project is still in preliminary stages, which does not allow any definite statements regarding environmental impacts to be made. For all these projects, there will be impacts associated with the generation and disposal of waste products resulting from the removal of impurities (ash and sulfur) from the feedstock coal.

Combustion-Zone Retrofit Projects

This category can be divided into two groups: advanced slagging combustors and combustion-zone modifications. The Coal Tech, TRW, and TransAlta projects will involve the change of combustors on existing boilers to advanced slagging type combustors (i.e., combustion retrofit technologies). This change takes place inside the existing boiler building. The combustors represent a small part of the fuel handling, boiler, steam turbine, and generator components of the total system at an electric utility plant. The combustors will be largely fabricated offsite and brought in as a unit to be installed. The Coal Tech project will result in minor increases in air pollutants, since the coal-fired combustor will replace a combustor on a small, industrial boiler currently designed to be operated on oil or gas. For the TRW and TransAlta projects, there will be a reduction of SO_2 and NO_x emissions, with a corresponding increase in the amount of solid waste due to the addition of a limestone sorbent for SO_2 control. The solid waste, which results from a mixture of coal ash and limestone, will leave the combustor in the form of a molten slag and is expected to have a low concentration of leachable compounds.

Another group of projects that can be loosely defined as combustion-zone modification projects include the B&W LIMB, the EER Gas Reburning/Sorbent Injection, the B&W Coal Reburning, and the two Southern Company Services (SCS) Low NO_x projects. Each of these projects will be full-scale demonstrations on boiler units ranging in size from 80 to 500 MW(e). There is no construction outside of the boiler building in the two SCS Low NO_x projects, since these will require only minor modification to the combustors, on a smaller scale than that involved in the slagging combustor projects. For the B&W LIMB, EER, and C-E Dry Sorbent Injection projects, construction outside of the boiler building consists of a limestone storage and feeding hopper which occupies less than one fourth of an acre on the existing site. Similarly, the B&W Coal Reburning project will require construction of a small addition to the existing building to house a pulverizer, which will provide pulverized coal to a reburn burner.

The major environmental changes for the two SCS Low NO_x projects and the B&W Coal Reburn project will involve a reduction of up to 65% of the NO_x emissions. There will not be any change in the SO_2 emissions or wastewater discharges. The amount of solid waste will not increase, but there could be a slight change (up to 5%) in the amount of carbon in the fly ash for the two SCS Low NO_x projects and an increase in the ratio of fly ash to bottom ash for the B&W Coal Reburn project. In addition, for the B&W Coal Reburn project, there may be a slight increase in particulate emissions, based on conservative estimates of the performance of the existing precipitator.

The B&W LIMB and EER projects are expected to achieve reductions of both SO_2 and NO_x emissions. The EER project currently involves demonstrations at two separate sites, and two separate NEPA documents have been prepared. One project site (Hennepin) involves an increase in water used for sluicing the fly ash to the ash pond. The other site (Lakeside) uses dry fly ash disposal. The important change at either site is that the solid waste is expected to approximately double because of the limestone added to the flue gas for SO_2 emission control. For the EER projects, the amount of coal will be reduced due to replacement by the natural gas used for reburning (up to 20% of the total energy input to the boiler).

Postcombustion Retrofit Projects

This category of projects can be divided into slipstream projects and full-scale projects. Three of the PETC projects involve the demonstration of a postcombustion retrofit technology on a slipstream at an existing facility. The B&W SO_x-NO_x-RO_x Box, C-E WSA-SNO_x, and SCS Selective Catalytic Reduction projects are examples of such projects. These are proof-of-concept scale (i.e., about 3 to 10 MW(e) equivalent slipstream of flue gas) projects that involve the installation of a temporary flue gas duct that is fastened to the existing ductwork so that a small portion of the flue gas is diverted. This small portion, or slipstream, is treated by the ICCT and the gas is then routed back into the total flue gas duct upstream of the existing electrostatic precipitator. Slipstream testing is conducted to obtain data for establishing full-scale design characteristics of future commercial installations. The environmental impacts of this type of project are limited to construction of a temporary slipstream duct and test equipment which is adjacent to the existing ductwork of the generating station. After a test period of limited duration (about 1–3 years), the slipstream duct and test equipment are removed and the station is restored to original condition. Changes in total station emissions, wastewater discharges, and solid waste are small because the slipstream is only a small portion of the total station output.

The three postcombustion retrofit projects that are full-scale projects are the Combustion-Engineering Dry Sorbent Injection, the SCS Chiyoda, and the Pure Air projects. The Dry Sorbent Injection project is designed to demonstrate three separate processes: in-duct injection, in-duct spray drying, and convective pass injection. However, this project is on hold pending further project definition. The Chiyoda and Pure Air projects will involve the construction and operation of large flue gas scrubbers that will treat the entire flue gas from one 100-MW(e) boiler (Chiyoda) or will treat the total flue gas from a 600-MW(e) station (Pure Air). The major environmental changes for these two projects involve a 90% reduction of the SO_2 in the flue gas and handling and disposition of a gypsum residue produced in the scrubbing process. At the Chiyoda demonstration site, the gypsum will be stacked in a 13-ac lined area for future utilization or permanent disposal in place by covering and revegetating. The Pure Air project will produce a high quality gypsum that will be used in wallboard manufacturing. The water recovered during dewatering the gypsum will be recycled to the Chiyoda scrubber so that there will be no significant change in the wastewater discharges at this site. Part of the Pure Air scrubber water will be evaporated in the flue gas so that there will be little change in wastewater discharged at this site.

Industrial Projects

In addition to the above projects that can be broadly defined as retrofit projects, there are two industrial projects: the Passamaquoddy Tribe project and the Bethlehem Steel project. The Passamaquoddy Tribe project involves the innovative reuse of waste cement kiln dust as a scrubbing reagent for control of emissions of SO_x and NO_x from coal-fired cement plants. The Bethlehem Steel project involves the installation of an innovative coke oven gas cleaning technology at an existing steel plant. Both of these projects will reduce existing emissions of air pollutants as well as the generation of wastes and/or unwanted by-products. In addition, the Bethlehem Steel project has as a specific objective, the elimination of visible emissions associated with the combustion of coke oven gas in order to satisfy requirements of an Administrative Consent Order by the state of Maryland.

CONCLUSION

The CCT program started with the passage of legislation in 1985. It is an environmental program with stated objectives to demonstrate CCTs that are capable of being commercialized in the 1990s, that are more cost-effective than current technologies, and that are capable of achieving significant reduction of SO_2 and/or NO_x emissions from existing coal burning facilities, particularly those that contribute to transboundary and interstate pollution. The CCT program consists of projects with at least 50% cost-sharing by private industry, and the projects are administered through cooperative agreements rather than contracts. Cooperative agreements allow the DOE a monitoring role rather than a strong technical direction role, as is the case with typical government contracts.

There are 21 projects assigned to the PETC. These can be broadly classified as either retrofit projects or industrial projects. The retrofit projects can be further classified as precombustion, combustion, and postcombustion retrofit technology demonstrations. There are five precombustion retrofit projects that involve the production of a cleaned fuel from a coal feedstock and combustion of the fuel in existing utility or industrial boilers. Three projects are combustion-zone retrofits involving the installation of slagging combustors on small boilers; five additional combustion-zone retrofit projects involve the installation of low-NO_x burners with or without a sorbent injection system for control of SO_2. There are six postcombustion retrofit projects: three involve slipstream testing, two are full-scale installations of advanced flue gas scrubbers, and one is a full-scale demonstration of dry sorbent injection. Finally, there are two industrial projects involving a steel plant and a cement plant.

Most of the projects have minor adverse impacts associated with construction on existing sites. Many of the projects are of limited duration, after which the site will be restored. For these projects, there will be no long-term impacts. In almost all cases, there is expected to be a net reduction in emissions of air pollutants during operation and a corresponding improvement in ambient air quality. The most common adverse impact during operation is an increase in the amount of solid waste, especially for those technologies that reduce emissions of SO_2. However, the wastes generated are not expected to be hazardous and, in most cases, can be handled in a dry form, thereby reducing impacts to water quality.

The NEPA process for the CCT program has evolved during the three rounds of solicitations to the point where a draft programmatic EIS has been prepared for the latest solicitation and published for comment. A preselection environmental review is also performed on each round of proposals, although the review is not publicly available since it involves procurement-sensitive information. To date, approximately 1 year after the CCT-2 selections were announced, 7 MTFs and 1 EA have been completed out of the total of 22 NEPA documents to be prepared by PETC. Of the remaining documents, 3 are expected to be MTFs, and 11 are expected to be EAs. The timing of the remaining NEPA documents is on the critical path for some of the projects, since the DOE will not expend federal funds for detailed design, construction, or operation until the NEPA process has been completed. This requirement imposes an additional constraint on project schedule which has to be factored in with other constraints such as a limited construction season in cold weather climates and construction during a scheduled outage of the host utility.

Impacts to Ecological Resources as a Decision Factor in Power System Development and Operation

J. M. Taves, U.S. Department of Energy, Portland, OR

ABSTRACT

The Council on Environmental Quality's (CEQ) regulations for implementing the provisions of the National Environmental Policy Act (NEPA) specify that when undertaking actions requiring preparation of an environmental impact statement (EIS), agencies must prepare a record of decision (ROD). Section 1505.2(b) of the CEQ regulations states the ROD must contain an identification and discussion of all factors that were balanced by the agency in making its decision and "state how those considerations entered into its decision." This paper discusses some of the special challenges encountered by the Bonneville Power Administration (BPA) in undertaking this directive.

INTRODUCTION AND BACKGROUND

The Pacific Northwest region of the United States is rich in natural resources; these include fisheries, wildlife, irrigated agriculture, and hydroelectric power potential. In 1927, the U.S. Army Corps of Engineers began a study of the potential for development of the hydroelectric capability of the Columbia Basin. This resulted in a proposal to construct ten major hydroelectric projects across the Columbia River. The first of these projects, Bonneville Dam, approximately 40 mi upriver from Portland, OR, was begun as a public works project in 1933 by the Corps. Another of the proposed facilities, Grand Coulee Dam, was begun by the Bureau of Reclamation the following year.

In 1937, as construction of Bonneville Dam was nearing completion, Congress passed the Bonneville Project Act. This legislation established the Bonneville Power Administration (BPA) and assigned to it the responsibility for marketing the power produced by the hydroelectric facilities of the Federal Columbia River Power System (FCRPS). In addition, the agency was assigned responsibility for developing transmission facilities necessary to move the power from the power plants to load centers.

Today, the FCRPS is capable of producing ~10,000 average MW of electrical power — roughly ten times the power needed to serve the load of Seattle — under average water conditions. The BPA also operates and maintains the largest high-voltage transmission system in the free world. Most of the facilities of these federal power and transmission systems were constructed before the passage of the NEPA.

In recent years, the BPA has experienced a significant surplus of generation capacity. To economically dispose of this surplus, the BPA has expanded its capability to market surplus power outside the Pacific Northwest. This has involved the construction of additional transmission capability between the Pacific Northwest and California and the establishment of policies for allocating the use of interregional transmission by the BPA and other utilities in the Northwest and California. This paper recounts how impacts to ecological resources have been taken into account in arriving at the decisions the BPA has made concerning surplus power marketing, expansion of interregional transmission facilities, and the allocation of access to federally controlled interregional transmission facilities.

INTERTIE DEVELOPMENT AND USE EIS

On August 31, 1964, Congress authorized the construction of high-voltage transmission facilities between the Pacific Northwest and California. By 1969, two 1600-MW alternating current (AC) lines were in service between the Northwest and California. The following year, the BPA and the Los Angeles Department of Water and Power completed a 2000-MW direct current (DC) *intertie* between the Celilo Converter Station on the Columbia River and The Sylmar Converter Station near Los Angeles. The justification for these facilities included the potential for marketing both federal and nonfederal power generated in the Northwest, but surplus to the needs of the Northwest, to California.

Much of this surplus power is referred to in the utility business as *nonfirm.* Whereas the availability of firm power is guaranteed, the availability of nonfirm power is not; nonfirm power is sold on an *as available* basis. A substantial portion of the power produced by the FCRPS is nonfirm because of annual variations in flows through the hydroelectric system. For planning purposes, the BPA assumes its hydroelectric resources can produce firm power equal to that which current generating facilities could produce under the most adverse set of water conditions (referred to as *critical water*) that have occurred in the Columbia Basin

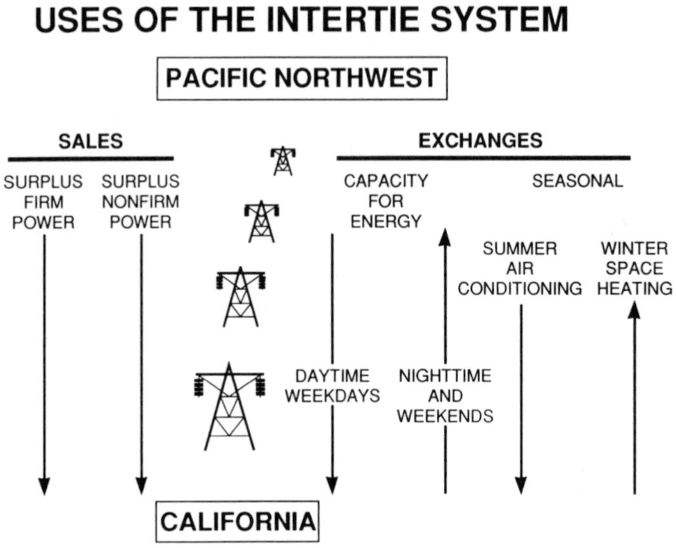

Figure 1. Uses of the intertie system.

from 1929 through 1978. Under more favorable conditions, nonfirm power may be produced. California utilities can substitute purchases of nonfirm power from the Northwest for more costly oil- and gas-fired California generation during years when nonfirm power is available, and the Northwest gains the advantage of recovering a portion of the cost of the FCRPS from California.

In addition to surplus nonfirm power, there is potential for taking advantage of differences between the Northwest and California in the seasonal distribution of demand for power. Whereas the highest power demands in the Northwest occur during the winter as a result of space heating loads, California's peak demand occurs during the summer air conditioning season. This creates the potential for seasonal exchange transactions between the regions, whereby Northwest resources can be used to supplement California resources during the summer, and California resources can, in turn, supplement those of the Northwest during the winter. These seasonal exchanges have the potential to allow deferral of new resource construction in both regions (Figure 1).

Another type of interregional exchange that can occur between the Northwest and California consists of the exchange of Northwest electrical capacity for California energy. The hydroelectric system in the Northwest is capable of producing very large amounts of power at any given moment. However, the amount of energy that can be produced over time is limited by river flows. As a result of capacity/energy exchanges with California, surplus generating capacity on the FCRPS can be used to meet peak electrical demands in California, and, in return, California energy can be returned to the Northwest to supplement the Northwest's power supply to the extent that flows on the FCRPS are not sufficient to fully satisfy energy demand on the system.

Finally, to the extent that Northwest or California resources are constructed ahead of need, due either to economic factors or the inability to precisely forecast load growth, surplus generation can be exchanged between the regions either to defer construction of new resources or to displace existing resources which have higher operating costs.

During the last decade, a significant firm (in contrast to nonfirm) surplus of resources has developed in the Northwest. Although this surplus is now declining substantially, its existence has contributed to an increased interest in expanding the capability for power transfers between the Northwest and California. It has also increased demand for transmission over the BPA's existing intertie facilities to the point where some mechanism became necessary for allocating access to the federally controlled portion of the Intertie. (Two Northwest investor-owned utilities, Portland General Electric and Pacific Power and Light, also own portions of the AC Intertie.) The BPA developed several proposals in response to those needs, which included proposals to (1) increase the capacity of the DC Intertie from 2000 to 3100 MW, (2) participate with California in constructing an additional 1600-MW AC intertie, (3) develop and implement a policy for according access to the federally controlled portions of the interties for both firm and nonfirm power transactions, and (4) embark on an effort to market surplus firm federal power to California.

In late 1984, the BPA announced its intent to prepare an EIS as a part of its effort to develop a policy for according access to the interties. During the scoping process for this document, it became clear the proposed expansions of intertie capacity, as well as the BPA's efforts to market its firm surplus, were related actions and should be included in the EIS. By mid-1985, the *Long Term Intertie Access Policy EIS* had been redesignated the *Intertie Development and Use EIS*.

ENVIRONMENTAL ANALYSES

The analyses for the Intertie Development and Use (IDU) EIS covered a broad range of factors. Although other environmental documents addressed the physical impacts of constructing the proposed expansions of the interties, the IDU EIS considered the impacts associated with the use of the expanded system. The first step in the analysis was to determine how increased power transfers between the Northwest and California might impact the operation and construction of generating resources in each region. Once these effects had been forecasted, this information was used to estimate their potential environmental consequences.

A primary concern of the analysis was to identify potential effects on the operation of the hydroelectric plants of the FCRPS. It was believed that the operation of these facilities had significant potential for affecting the rich anadromous fisheries of the Columbia Basin as well as resident fish located in the system's major storage reservoirs. The anadromous fisheries are composed of stocks that spawn in fresh water, but spend the majority of their lifecycle in the Pacific Ocean. These include various species of salmon and steelhead. These fish have significant commercial and recreational value. Salmon also represent important religious values for Northwest Indian tribes. Nonmigratory, or resident, fish constitute a sport fishery and are eaten by a variety of birds including the bald eagle.

In addition to potentially effecting Northwest fishery resources, the FCRPS operations could impact cultural resource sites located along river or reservoir banks. They could also impact recreational opportunities, including campgrounds and boating facilities.

Interregional power transfers also have the potential to affect thermal generation in both the Northwest and California. This would include primarily coal-fired generation in the Northwest and gas- and oil-fired generation in California. Northwest coal plants are generally located in sparsely populated areas with good air quality. However, many of California's gas- and oil-fired facilities are located in densely populated urban areas with relatively poor air quality.

Two computer models, the System Analysis Model (SAM) and the Fishpass Model (Fishpass), were used to forecast the impact of the proposed actions on anadromous fish. SAM models the operation of the Northwest power system and provides a simulation of the optimum economic operation of the system given specified loads and operating constraints. Operating constraints define certain limits within which operations for power production must occur in order to accommodate other uses of the FCRPS, including flood control, irrigation, recreation, and navigation. The specification of firm loads includes information on the types of firm contracts being served by the system. Various scenarios of firm sales to California were included in the analyses, along with service to forecasted Northwest loads.

SAM provides information on river flow rates, reservoir elevations, and amounts of spill occurring on the FCRPS. Spill is water that, for any of several reasons, may be spilled over the top of dams rather than being run through generation facilities. Information on thermal resource operation is also produced by SAM. Fishpass makes use of the information from SAM on flow rates, reservoir elevations, and spill levels to forecast effects on survival of migrating juvenile anadromous fish.

For purposes of the IDU EIS analyses, the SAM and Fishpass models were linked so that information generated in SAM could be input directly to Fishpass. This technique permitted an integrated analysis of the effects of various electrical load scenarios on anadromous fish. The thermal operations information generated by SAM was combined with emissions data for Northwest coal plants to arrive at estimates of air quality effects. A separate analysis was prepared concerning the operation of thermal resources in California and the consequent impacts on emissions from those resources.

FINDINGS

The provisions proposed for allocating access to the AC Intertie were not found to have a potential for significantly impacting ecological resources. The use of the interties to deliver power to California on a firm basis did appear to have the potential to affect resident fish, due to effects on reservoir levels. The projected contracts would lower Northwest reservoir levels in the fall, reducing reservoir volumes and food supplies during the period of highest growth for these fish. However, given certain limits on the amount of such contracts, the effects could be limited to a single reservoir. Increasing the capacity of the interties resulted in small, discernible increases in coal plant emissions in the Northwest and small reductions of emissions in California. The magnitudes of change were not considered significant, however. Increased capacity also resulted in small (averaging less than 3%) decreases in anadromous fish stocks. These were judged not to be significant, provided other measures to increase anadromous fish survival were implemented.

IMPACTS TO ECOLOGICAL RESOURCES AS DECISION FACTORS

Using impacts to ecological resources as decision factors in power system development and operation presents some special challenges in the decisionmaking process. These include such phenomena as lack of appropriate data, the absence of demonstrated and accepted methodologies, political sensitivities and resistance, and goal conflicts. In many instances, these challenges are not independent of one another.

NEPA has made it necessary for the BPA and other federal agencies to develop new areas of informational and analytical expertise. In the BPA's case, it is no longer sufficient to be able to prepare power system proposals that are based on sound engineering and economic analyses. Information on environmental factors must also be prepared and considered. Because this is a comparatively recent requirement, the data and techniques of analysis have not matured to the point evidenced in the areas of economics and engineering.

We are often confronted with developing cutting edge technology at relatively basic levels in the environmental arena. As a consequence, many of the technologies employed are challenged. They undergo scrutiny reserved for the untried and untested. They generate controversy and, occasionally, suspicion. This situation also provides an opportunity for interest groups to selectively support or challenge information that is difficult to validate. As a result, considerable effort often goes into explaining and defending analytic approaches. In some cases, information may be abandoned simply because the method of its acquisition or interpretation is perceived to be too controversial.

Another problem relates to the ability to provide an effective comparative analysis of decision factors. Whereas engineering and economic concepts generally lend themselves to quantitative analysis, this is often not the case with ecological concepts. It is not evident what value to place on recreational resources. The cultural or religious value of salmon to the Native Americans of the Northwest is difficult to capture in a medium that can be readily compared with the economic value of hydroelectric power production. To attempt to assign a dollar value to such phenomena is often perceived as insensitive and inappropriate, yet it is difficult to make meaningful comparisons between decision factors unless a common medium of valuation can be established. Yet another challenge relates to the temptation to reach beyond the precision of our data or the rigors of our analysis. Much time and effort may be expended attempting to quantify magnitude of effect, when really all that can be reasonably determined is direction of effect. This is an error to which both agencies and audiences are often prone, again complicating the effective achievement of a comparative balancing of decision factors.

In short, it has not been easy to apply the CEQ's directives in determining how to develop and operate the BPA's system. Nevertheless, several features of the decisions arrived at by the BPA reflect consideration of potential effects on ecological resources. The Long-Term Intertie Access Policy (LTIAP) was a primary vehicle for addressing these concerns. Provisions were added to the policy to preclude the use of the interties for delivery of power from new hydroelectric facilities constructed in areas that would be harmful to fishery resources. These areas were termed *protected areas*. Since 1980, the BPA has been making substantial investments in a variety of fish enhancement projects designed to counteract the adverse effects that development of the FCRPS had on fish resources in the Columbia Basin. The protected areas provisions of the LTIAP were intended to preserve the value of these investments.

As previously stated, analyses of firm contracts effects indicated that seasonal exchange transactions between the Northwest and California could adversely affect resident fish by lowering reservoir levels in the fall. The LTIAP limits the amount of nonfederal firm access to the interties to 800 MW. Although this capacity can be used for any type of firm transaction desired by a nonfederal utility, the limit of 800 MW ensures that total transactions, including seasonal exchanges, will not exceed a level found to have relatively limited effects on resident fish. Furthermore, the BPA is implementing programs to mitigate any potential effects by monitoring fish production and by improving habitat and undertaking imprint planting of resident species in several tributary streams.

In addition to these LTIAP provisions, the BPA is also strongly supporting the development of effective bypass systems at various hydroelectric plants along the Columbia River. It is also exploring the feasibility of managing the level of squawfish present in storage reservoirs. These fish prey on juvenile anadromous fish during their migration to the ocean and are believed to account for significant losses of young salmon and steelhead. The BPA also coordinates its own marketing activities to achieve certain flow levels on the FCRPS intended to facilitate effective fish spawning and incubation and, during the spring, successful migration of juvenile anadromous fish.

Survey of Analyses in Environmental Impact Statements

B. F. Leon, Brown & Root, Houston, TX

ABSTRACT

In light of the National Environmental Policy Act's (NEPA's) requirement for objective, scientifically based impact analyses, I examine the state of the analytical art in a random sample of 38 of the 239 environmental impact statements (EISs) filed between January 1, 1989 and August 15, 1989. These 38 reports contain 442 impact analyses, which I categorized as to the extent of original work and supporting evidence. The most common methods of analysis are (1) judgment or declaration (50%), (2) tally or inventory of current conditions (14%), and (3) dismissal of the impact or postponement of its analysis (12%). Preexisting computer algorithms or models were used in 8% of the analyses, and citation of previous studies was the means of analysis in 8% of the cases. Investigations that specifically forecast and provide direct evidence for postulated impacts comprise 7% of the impact analyses.

Beneficial impacts, although accounting for only 10% of the total impacts discussed, are 36% of the specific studies. Most of these analyses concern operational or economic benefits of the project.

Follow-up studies, essential to assess the accuracy of the impact analysis and to determine the actual impacts as they occur, are specified in 10% of the impact analyses. In many cases, no impact analysis is given, and the report contains a statement that the analysis will be done at an unspecified time in the future.

I suggest a minimum standard for an acceptable EIS impact analysis. An analysis of environmental impacts should be specific to that resource in that environment (requiring more tailoring of computer models than is currently being done). In addition, there should be "rules of evidence" to ensure that the projected impacts can be tested against the actual outcome after the project is completed. Finally, follow-up studies of the affected environment should be done after the project is approved to allow the accuracy of the analyses to be checked and to benefit future analyses.

INTRODUCTION

The environmental impact statement (EIS) — the objective analysis of environmental impacts in a public document — is the centerpiece of the federal environmental protection program. It is the EIS that is scrutinized by interest groups, and it is over the EIS that lawsuits are raised by project opponents. It is in anticipation of a not-so-complimentary EIS that planners moderate environmentally damaging features of their projects. It is in the EIS that details of a project (and alternative projects that accomplish the same end) are displayed in public before one plan is set and ground is broken. And it is the EIS that binds agencies to mitigation plans and continuing monitoring of the environment.

The writers of National Environmental Policy Act (NEPA) intended that a diverse group of natural scientists, social scientists, historians, and designers would provide the objective, quantitative analyses necessary for a credible environmental program. These investigators would work together to determine how much a project would affect the human (and nonhuman) environment, publish the results, and let political forces work to determine a project's future. The effectiveness of the program depends on the perception that environmental impacts are predicted accurately in an EIS.

I have conducted a review of the state of environmental impact analysis 20 years after the first EISs were produced. Although many authors have discussed the philosophy of environmental impact studies (e.g., Caldwell 1987), few have analyzed the contents of EISs to determine what is done in practice. Culhane (1987) examined 29 EISs that were produced between 1974 and 1979 to see if predicted impacts did occur once the project was built. His comparison between predicted and real impacts was frustrated by vague and untestable forecasts, and although he did report that few forecasts were clearly inaccurate, only 30% of the predicted impacts were stated clearly enough to be considered accurate. Duinker (1985) makes the case that risking wrong forecasts is how science progresses, and nothing is gained by making safely vague forecasts. The purpose of this study is to see if writers of EISs in 1989 are presenting testable, quantitative, operational forecasts of environmental changes or are supplying less worthy material for the realization of environmental policy.

METHODS

I selected 38 EISs at random from the 239 EISs that were filed with the Environmental Protection Agency (EPA) Office of Federal Activities between January 1, 1989 and August 15, 1989. I used a calculator's random number

Table 1. Distribution of Environmental Impact Statements by Agency.

Agency	Number in Sample (n = 38)	Percentage in Sample (n = 38)	Percentage in All EISs (n = 239)
U.S. Army Corps of Engineers	7	18.4	10.5
Federal Highway Administration	7	18.4	23.0
Forest Service	6	15.8	23.4
Bureau of Land Management	4	10.5	7.9
Urban Mass Transit Administration	2	5.3	1.3
Fish & Wildlife Service	1	2.6	0.8
Housing & Urban Development	1	2.6	2.1
Navy	1	2.6	1.3
Federal Aviation Administration	1	2.6	2.1
Bureau of Reclamation	1	2.6	2.1
National Park Service	1	2.6	2.5
Department of Energy	1	2.6	2.5
Bureau of Prisons	1	2.6	4.2
Army	1	2.6	2.5
National Aeronautics and Space Administration	1	2.6	0.8
National Oceanic and Atmospheric Administration	1	2.6	0.8
Soil Conservation Service	1	2.6	2.5
Other Agencies	0	0.0	9.6
		$\chi^2 = 17.4$ $(p < 0.05)$	

generator (Casio® model FX-115) to produce a list of fractional values that when multiplied by 239 yielded EPA sequence numbers. The distribution of statements by federal agency is listed in Table 1; the sample matches the distribution of all 239 EISs ($p < 0.05$).

I examined the "Environmental Consequences" chapter of each EIS to extract and classify impact analyses. Areas of environmental impact (e.g., air pollution, wildlife) were often labeled by subheadings or were clearly denoted in the text. I aggregated all statements about future impacts in each environmental area and assigned a classification code according to the most specific impact analysis. The categories of impact analyses (Table 2) are arranged such that less specific analyses have higher values than more specific analyses. For example, if a subheading "Water Quality" includes three lists of current water quality conditions (Category 7), two statements about future water quality after project construction (Category 6), and one standard EPA oxygen-sag model run (Category 4), I would record the analysis of water quality impacts as Category 4: "Computer model/algorithm with supplied parameters." I would further note that it is an adverse impact, and that no follow-up studies are stipulated.

A specific impact study is defined here as an investigation of the practical results of a change in the environment, using means that are appropriate to the specific situation, so as to produce a specific forecast that can be tested against a null hypothesis. A computer model or algorithm developed for a particular EIS can be part of a specific study. However, if the model was developed generically and applied to the situation, it is not a specific analysis; if model parameters were

Table 2. Categories of EIS Impact Analyses.

1. Specific study with original data
2. Specific study with previously collected data
3. Computer model/algorithm with original parameters
4. Computer model/algorithm with supplied parameters
5. Reference to previous studies
6. Judgment or declaration
7. Inventory or tally of current conditions
8. Impacts dismissed or analysis postponed
9. Means of analysis unclear

Table 3. EIS Impact Analyses Categorized.

Category	Number of Analyses	Percent of Total
1. Specific study with original data	26	5.9
2. Specific study with previously collected data	6	1.4
3. Computer model/algorithm with original parameters	1	0.2
4. Computer model/algorithm with supplied parameters	36	8.1
5. Reference to previous studies	33	7.5
6. Judgment or declaration	221	50.0
7. Inventory or tally of current conditions	63	14.3
8. Impacts dismissed or analysis postponed	54	12.2
9. Means of analysis unclear	2	0.5
Total	442	100.0

supplied in advance of the impact study such that only operational data were input to the model, then the analysis is a "plug-in" model (Category 4). Categories greater than 3 are not specific studies and do not provide testable predictions about environmental factors of interest, although they may be logically consistent and draw upon previous work.

There was little ambiguity about which category should be assigned to each impact analysis. Fewer than 1% of the analyses were unclear (Category 9).

RESULTS

The 38 EISs presented 442 discernible impact analyses or an average of 11.6 analyses per impact statement. Table 3 is a breakdown of these analyses by category. Of the 442 impact analyses, 33 are specific studies (Categories 1, 2, or 3). Thus, 7% of the impact analyses have measurable predictions about future impacts, and 93% do not. Most of the analyses are not impact studies at all, but are either citations of other studies (Category 5), verbal declarations of possible impacts (Category 6), tallies of current conditions (Category 7), or assertions of no impact or postponement of the analysis (Category 8). Indeed, 47% of the EISs do not have even one specific study.

Computer models/algorithms with supplied parameters (Category 4) — the so-called "plug-in models" that are usually prepared by an agency for all their EISs — were used in 29% of the EISs. These models have their own default parameters for typical sites and require users only to input project data. Such models (and parameters) are based on assumptions about the conditions of the site and the processes being modeled. In no case did an EIS describe the operation of a "plug-in model" that was used, nor did any present a model's assumptions. Thus, errors that are inherent in this approach are not readily traceable, and the results are not subject to scrutiny. Loehle (1987) observes that errors can occur in the evaluation and interpretation of models as well as in the construction of them.

The specific study of beneficial impacts is overrepresented in the EIS. Although only 10% of the impacts are beneficial, 36% of the specific studies are of beneficial impacts. Many of these analyses are of technical improvements that the project is designed to perform, such as greater traffic capacity or flood protection, or of economic benefits such as increased tax base or more jobs. Since most projects involve a feasibility study where these areas are weighed against the results of not taking action, it is understandable that these beneficial impacts would have been thoroughly studied. However, in the absence of analyses of adverse impacts at the same level of detail, the EIS appears to be a document that supports a project instead of objectively considering all its impacts.

Follow-up studies were proposed for 44 (10%) of the environmental impact areas. Many follow-up studies consist of observing the site during construction so as not to destroy resources such as archaeological sites or endangered species habitat. In some cases, a long-term monitoring program is stipulated. Environmental areas that are specified for monitoring include water quality, wildlife, and radioactive substances.

Other characteristics of the sample of EISs were also examined. Impact statements that were prepared by consultants have on average slightly more impacts per EIS, more "plug-in" models (Category 4), and more impact analyses consisting of tallies or inventories of current conditions (Category 7) than agency-prepared EISs. Also, draft EISs are not significantly different from final EISs in their categories of impact analyses.

CONCLUSIONS

The random sample of impact statements examined in this study shows that most EISs are in large part collections of untestable hypotheses about future impacts, supported with tallies of current conditions, ready-made models, and logic. While these hypotheses may ultimately prove true, apparent inaccuracies of prediction can be explained away with the same procedures that were used in the original analyses, and forecasts will rarely be unambiguously wrong. Where a model is used, inaccurate forecasts can be blamed on calibration errors in the model and referred to the agency bureau that produced the model. However, predicted impacts will seldom be checked against actual impacts after construction, since follow-up monitoring is infrequently called for.

The consequences of the prevailing lack of rigor in current EISs are numerous. An untestable prediction has no credibility except for that of its author who must now become an *expert* if the report is to be believed. It is easy to find an *expert* who disagrees, and no one can prove either to be correct. Lawyers can discredit *experts* much more easily than they can discredit a carefully thought-out experimental design and testing program. If an EIS does stand because *experts* say it should, then the *experts* become the true decision makers and are inextricably drawn into the political process of project approval. Environmental policy is subverted, and one suspects projects of having hidden environmental impacts.

RECOMMENDATIONS

An impact analysis should, at a minimum, be specific to that resource in that environment and predict alternative future states of the environment that can be measured and compared. I suggest that the following "rules of evidence" be used to ensure objectivity:

- Impacts are to be stated as hypotheses that are to be tested in the future against the null hypothesis that no impacts occur.
- Testing of predictions is to be done through appropriate experimental design, with original data or previously published studies.
- Data gaps are to be identified and a reasonable range of values substituted for the missing data; appropriate sensitivity analyses are to accompany the use of substituted values.
- All analysis assumptions are to be treated as hypotheses and be subject to testing.
- Computer model assumptions, parameters, and processes are to be itemized and explained in the EIS. Inappropriate assumptions and parameters are to be corrected.
- Follow-up studies are to be designed in the EIS. They will determine the accuracy of the impact predictions and identify unanticipated impacts.

Follow-up studies of environmental areas affected by a project are necessary to gage the accuracy of the impact analysis. Such studies should begin immediately, before the project is in place, to provide a baseline from which to compare the effects of the project. Funding for follow-up studies is not currently provided for many projects; it should be made available for at least each significant adverse impact.

This standard of specificity could be maintained through the present review network, including the sponsoring agency, the public, and the EPA. If accepted, it would do much to restore credibility to the environmental protection process and advance the level of scientific understanding of the environment.

ACKNOWLEDGMENTS

This research was supported in part by Quadrant Consultants, Inc. I gratefully acknowledge the assistance of the Office of Federal Activities, U.S. Environmental Protection Agency, for providing access to environmental impact statements in their files.

REFERENCES

Caldwell, L. K. 1987. The contextual basis for environmental decision-making: Assumptions are predeterminants of choice. *Environ. Prof.* 9(4):302–308.

Culhane, P. J. 1987. The precision and accuracy of U.S. Environmental Impact Statements. *Environ. Monit. Assess.* 8(3):217–238.

Duinker, P. 1985. Forecasting environmental impacts: Better quantitative and wrong than qualitative and untestable! pp. 399–407. In Proceedings of the Environment Canada Conference on Audit and Evaluation in Environmental Assessment and Management, October 13–16, 1985. Vol. 2.

Loehle, C. 1987. Errors of construction, evaluation, and inference: A classification of sources of error in ecological modeling. *Ecol. Model.* 36(3–4):297–314.

Risk Assessment and Environmental Impact Assessment

S. M. Ugoretz, Wisconsin Department of Natural Resources, Madison, WI

ABSTRACT

Risk assessment is a topic of intense interest for regulatory agencies. Increased public concern with the health effects of projects such as incinerators and industrial wastewater discharges is forcing agencies to expand the scope of their review of these projects to include risk assessments. An informal survey of several state agencies with environmental impact assessment (EIA) responsibilities showed that risk assessments are becoming part of the EIA process. Concern was expressed with the public's reactions to the findings of the risk assessments and with the difficulty of computing accurate projections in the face of uncertainties in the information base underlying risk assessment. Risk assessment does appear to have a logical connection to EIA in light of the techniques and analytical and communications skills needed for both.

INTRODUCTION

Our society is becoming increasingly conscious of the risks created by the technologies we employ. The price of an industrial society is being highlighted by the dramatic problems at Love Canal and Times Beach, as well as the more common leaking landfills and freight train derailments. Often, it seems that even a hint that a carcinogenic substance is involved is all that is needed to mobilize a worried public into opposing a project.

Regulatory agencies are bearing the brunt of much of this concern as they evaluate permit applications for facilities associated with human health risks. Risk assessments (RAs) are one tool many agencies are using to cope with *risky* proposals. This paper examines the use of RAs in the environmental review process under several State Environmental Policy Acts (SEPAs).

The RA is most generally defined as "the process of determining the adverse consequences that may result from the use of a technology or some other action" (Conservation Foundation 1985). It is most commonly used to project toxicological effects on human health such as carcinogenesis. One frequent product of an RA is an estimate of lifetime excess cancer risk, such as "one in one million excess cancer cases." RAs may be conducted as stand-alone analyses or as part of a broader process.

A health RA normally consists of four components:

- Hazard Identification: Gathering and evaluating data on the types of adverse effects produced.
- Dose-Response Evaluation: Describing the quantitative relationship between the amount of exposure and the extent of adverse effect.
- Exposure Evaluation: Describing the nature and size of the population exposed to a hazard and the routes by which they are exposed to it.
- Risk Characterization: Integrating the data and analysis of the first three components to determine the likelihood that any of the adverse effects will be experienced by the population (U.S. EPA undated).

Possible uses of RAs include establishing priorities for additional RA or for research, informing the public about risks and helping to decide whether a risk should be regulated, and to what extent (Conservation Foundation 1985). The results of an RA may be applied in risk management to attempt to ameliorate the identified risks. Risk management techniques range from banning the manufacture and use of a substance, such as polychlorinated biphenyls, to issuing advisories urging citizens to avoid consuming certain substances.

The state of Wisconsin, Department of Natural Resources, has used the RA for several purposes, and my program (Environmental Analysis and Review) is currently examining how the RA could be integrated into the EIA process.

To gather some information for that examination, I used a nonrandom survey of several agencies in states with SEPAs to learn whether those agencies were preparing RAs and how they were being conducted and used. I sent questionnaires to state agencies which were on the mailing list for this NEPA symposium and which were likely to have an EIA requirement.

At this writing, I had received responses from the agencies listed in Table 1. In addition, I interviewed persons who had prepared RAs for my agency and the Minnesota Pollution Control Agency. These sources illustrate how the RA is being employed on the front lines of applied environmental analysis.

RESULTS AND DISCUSSION

Reasons for Preparing Risk Assessments

The agencies I contacted gave a variety of reasons for preparing RAs. These mainly fit into the categories listed earlier. Questions or uncertainty about human health implications appeared to be important in determining whether an RA would be prepared.

Table 1. Summary of Agency Responses to the Risk Assessment Survey.

Agency (function)[a]	RA Used in EIA?	Uses of RA	Project Types	Prepared by
Florida Department of Community Affairs (DE)	No			
Maryland DNR Tidewater Administration Power Plant and Environment Review Division (EPA)	Yes	Regulatory decision; risk management	Incinerators; power plant discharges; fish and wildlife advisories	Staff
New Jersey Department of Environmental Protection (EP)	Yes	Regulatory decision; risk management; public information; scoping	Landfills; incinerator; permitting discharges; discharge limits; fish and wildlife advisories	Staff; consultant; permit applicant
New York City Department of Environmental Protection (EP)	Yes	Regulatory decision; risk management; EIS threshold; EIS, land use decisions	Landfills; industrial discharges; contaminated site evaluation	Consultant; Department of Health
New York City Environmental Quality Review (EP)	Yes	Public information; EIS threshold	Incinerator; contaminated site evaluation	Consultant
Newfoundland Department of Environment and Lands (EP, RM)	No			
North Carolina Department of Human Resources (PM)	Yes	Regulatory decision; risk management; public information; scoping, EIS threshold	Landfill; incinerator; permitting discharges; discharge limits; fish and wildlife advisory	Staff
North Carolina Department of Transportation (TR)	No			
Texas Attorney General, Environmental Protection Division (L)	No			
Virginia Department of Transportation (TR)	No			

[a] DE = development, EP = environmental protection, L = legal, PH = public health, RM = resource management, and TR = transportation.

Some RAs were prepared primarily to gather and provide information; these were generally prepared in response to public concerns, to anticipate and be prepared to deal with public concerns in evaluating issues, or to put risks into perspective (for the public as well as the agency). This involves providing a sense of the relative risk posed by the proposed action.

Others were prepared as part of the regulatory decisionmaking process. Minnesota routinely prepares RAs for air quality permits, especially for projects which would emit certain organic compounds. North Carolina Environmental Health will use RAs on request from another agency to determine allowable levels of substances to protect human health. Wisconsin has used RAs to set discharge limits and evaluate certain disposal processes, but not to determine whether to issue or condition a specific permit.

Maryland's power plant review division uses RAs to apply risk management in decisionmaking when there are potential health risks beyond standard permitting criteria. New Jersey uses RAs to set priorities for action and attention by identifying problems with higher actual risks to the public. New York City has used RAs to help reach permit decisions on resource recovery facilities and to evaluate development proposals on previously contaminated sites.

New York City has also used RAs to help make EIS threshold decisions. Wisconsin is planning to use an RA on a resource recovery facility to help reach an EIS threshold decision by determining whether there are risks which are not controlled by normal regulatory authority.

Types of Risk Assessments Prepared

Most of the agencies which returned the survey and used RAs prepared quantitative RAs. Most of these were human health related. Minnesota, North Carolina, and New Jersey indicated that they included ecological risks in their RAs. The documents ranged from less than two pages to "massive documents."

Most agencies did not prepare fault-tree-type assessments. There appeared to be less concern with the modes of catastrophic failures than with the health impacts of routine operations or emissions. It is likely that emergency management agencies would be concerned with failure situations; none of those agencies received the survey.

Projects Evaluated with Risk Assessments

The types of activities evaluated with RAs are most often projects with the potential to generate controversy, such as incinerators and other solid waste disposal facilities, wastewater discharges (industrial and municipal) and sludge disposal plans, power plants and other emitters of hazardous substances (especially organic compounds), and development proposals in areas where there is a preexisting hazard.

RAs are also commonly used for nonproject purposes such as developing fish and wildlife consumption or drinking water advisories. Wisconsin, New Jersey, North Carolina, and Maryland all use RAs in this way.

Public Responses to Risk Assessments

Public reactions to the RAs relates to the subject of risk communications. There is a large and growing body of literature on this subject. Frequently, the methods of communication and their effectiveness will largely determine public responses (Hance et al. 1988).

In the Wisconsin case, the public's concern about the potential health effects of a proposed municipal incinerator led to a lawsuit in which the judge ordered the Wisconsin Department of Natural Resources (DNR) to revise an EA. This convinced the DNR to prepare an RA as part of the revised EA. Even so, the risk assessor felt that the public did not understand the significance of the risk estimates. He felt the public ignored the limitations of the RA and focused on the results, in effect saying: "If the risk is so bad, how can the state permit this?"

One reason for this response may have been the ad hoc decision to add an RA to the EA and its relative lateness in the evaluation process. There was not enough time to lay the groundwork for a complete understanding of the process and the assumptions underlying the RA. To make matters worse, the remand of the original EA probably undermined confidence in the agency.

Here, the level of public concern about a project led to the decision to prepare an RA. The results, coupled with the level of concern, have influenced the agency's decision to place more stringent monitoring requirements on the project and to impose more stringent dioxin limits.

The Minnesota case also involved a proposed municipal incinerator. In this case, the RA decision was made at the same time as the decision to prepare an EIS. Even so, the analyst expressed concern about the public's understanding of the RA. He felt that the project opponents made specious comments about the risks of the project at the EIS hearing and succeeded in confusing the public about the technical issues involved. Two separate RAs were prepared for this project, with different calculated levels of risk. This may have increased the confusion in an already volatile situation. Which estimate was more credible? Which was more accurate? These questions probably plagued the public as they tried to understand the results.

Public responses to the RAs did not change the agencies' evaluations of the risks associated with the projects discussed above. Other states dealt with the public in a similar way. Maryland indicated that the public was involved in the risk communication and management ends of the entire process, not in the actual assessment. New Jersey responded that the state employs an individually tailored feedback mechanism in the risk communication process. They are experimenting with factoring the public into exposure assessments.

Thus, it appears that most of the agencies using RAs tended to carry out the process without direct input from the public. Many authorities on risk communications are urging agencies to involve the public more fully in the entire risk cycle (Hance et al. 1988). Apparently, agencies have not yet put that advice into practice.

An RA and the public's response to it seem to have an influence on the details of the decision the agency makes. The cases discussed in this paper and the survey results show that this influence is more likely to be exerted upon the conditions of approval than on the determination of whether to issue a permit. In addition, public reaction to an RA may persuade the proponent to modify or withdraw a proposal. Local officials may also rethink their support for a project in light of the depth of public response.

Agency Reactions and Lessons Learned

All of the agencies which used the RA expressed some reservations about its limitations. The Wisconsin analyst was concerned with the extreme conservatism of the assumptions and methods used to prepare the RA. He felt that adding risks from different exposure pathways produced a result that exceeded *safe* levels, perhaps artificially. This fostered the public's perception that the project was unacceptable for public health reasons.

Minnesota's analyst concluded that in health RA, the weakest link in the equations could dominate the numbers produced. The more state of the art the methods, the more data required and the greater the costs of the analysis. This, he felt, argues for caution in using state-of-the-art methods. North Carolina also felt that the value of an RA was limited by the strength of the body of knowledge behind it. Even so, it still serves as a useful benchmark for assessing risk and provides the most accurate perspective possible.

Maryland also cited problems due to the lack of good information on dose-response effects, complex mixtures and chemical synergies, and the mechanisms of carcinogenesis. The responder stated that problems were more likely to occur when an incomplete analysis was carried out, especially if it is done to downplay the potential adverse effects of a proposal. New Jersey said an RA was very useful "when there is no alternative."

A response from New York City cited disagreements among the agencies responsible for different aspects of a project. Each agency's experts had different interpretations of the results of the RA. These were only resolved when one agency established primary authority over the project.

High costs were also a problem. Several states mentioned costs as a concern in being able to carry out adequate RAs. Complete health RAs take time to research, and computer models must be run under different exposure scenarios. Staff and computer time can run into tens of thousands of dollars. Several agencies had sources of funds to carry out RAs ranging from putting the burden on the applicant to tapping a special assessment on electric power rates (Maryland).

There is likely to be pressure on the agency to cut corners as much as possible to save internal funds or to save the applicant money in an already expensive permitting and environmental disclosure process. This could lead to "quick and dirty" RAs. Several states expressed the opinion that with RAs you get what you pay for. Poor RAs could cause more problems than they solve.

This survey has shown that the RA is viewed as a mixed blessing by the state agencies which use it in the environmental review process. It is seen as an integral part of the process of evaluating controversial projects when human health effects are in question. On the other hand, there are reservations about how accurate the state of the art is and the potential to exaggerate risks and confuse or alarm the public needlessly.

The newness of the subject is certainly reflected in its limitations, as is the complexity of the processes involved. However, the public is concerned with the health effects of the substances released by many projects, and environmental protection agencies are forced to deal with this concern in the EIA and regulatory processes.

Perhaps the New Jersey comment that the RA is a "very useful when there is no alternative" sums up the apparent ambiguity about the process seen in the responses I received. There was a sense that the RA, as it is used now, is serving, quite appropriately, as a "best guess" to make reasonably informed decisions until more definitive knowledge about the effects of risky projects is available. RAs appear to provide the agencies with some sense of the relative risk of a project.

In some cases, an RA may at least demonstrate to the public that the agency is aware of the potential for adverse health effects. Whether this is enough to assuage public fears in the absence of an obvious effect on the agency's actual decision goes beyond the scope of this paper.

Risk Assessment and Impact Assessment

Since state agencies appear to be using RAs to communicate with the public and decision makers about potential adverse environmental consequences, it is appropriate to consider how RAs could be integrated into the EIA process.

EIAs are the existing mechanisms for explaining technical information to the public and to decision makers. Thus, it is logical, and even economical, to add an RA to the process. People are familiar with the EIA process and know how to use it. Feedback from the public and outside technical experts is already part of the process — from issue identification through notification of the final decision. This fits well into the guidance for risk communications coming from academic studies and the experience of practitioners.

In a way, an EIA is already an analysis of risks — risks to components of the natural and human environment from physical alterations, conventional pollution, or changes in social and economic factors. Adding the human and ecological health risks evaluated in RAs would broaden the perspective of environmental analyses.

The RA has been adopted by the federal government and will probably become a normal part of the evaluation of substances and projects. The states too will have to deal with it. Several state agencies have already recognized this and are using RAs in the environmental review process.

SUMMARY

The results of this brief investigation show that SEPA states are beginning to use RAs in their decisionmaking processes. Some states are finding a broad array of uses for it; others restrict it to nonregulatory purposes. There is enough uncertainty attached to the techniques of risk analysis to give its users some level of discomfort in using it.

Three points raised in this paper can be valuable to other states:

- First, RAs can be an important component of the EIA process. Sometimes RAs are used to inform and influence the decisionmaking process in the fullest sense under NEPA and the SEPAs. Other RAs are attempting to communicate to the public the magnitude of risks associated with an agency action.
- Second, agency analysts are concerned about the uncertainties inherent in the current state of the art of analyzing risks. RAs are only as good as the methods and information used to prepare them. This underscores the importance of risk communications in explaining the strengths and weaknesses of the process to the public and decision makers. The experience and orientation of environmental impact assessors *preadapts* them for this purpose.
- Third, the high costs of thorough RAs may be an impediment to their widespread adoption. This may also tempt agencies to prepare *quick and dirty* analyses. Doing this could lead to additional problems ranging from misinforming the public to creating disagreements among agencies about the meaning and applicability of the results.

Other state agencies may be considering the use of RAs, whether as part of the EIA process or for some other purpose. The field is new and is still developing; important lessons are being learned on the front lines. Those lessons should be evaluated to address some of the difficulties inherent in the state of the art and communicated to others working in the field of EIA. Thus, it would be helpful for the current and prospective users to share information and experiences.

The results and interpretations expressed in this paper are intriguing, but a more scientific evaluation of what is actually occurring in the field would be appropriate. I would like to encourage state agencies to open a forum to implement this suggestion; we all have much to gain.

REFERENCES

The Conservation Foundation. 1985. *Risk Assessment and Risk Control.* The Conservation Foundation, Washington, D.C.

Hance, B. J., C. Chess, and P. S. Sandman. 1988. Improving Dialogue with Communities: A Risk Communication Manual for Government. New Jersey Department of Environmental Protection, Trenton, NJ.

U.S. Environmental Protection Agency. Undated. Principles of Risk Assessment: A Nontechnical Review. U.S. Environmental Protection Agency, Washington, D.C.

Eighteen-Year Trends in Natural Resources Protection Under Wisconsin's Environmental Policy Act

R. H. Read, Wisconsin Department of Natural Resources, Madison, WI

ABSTRACT

Like the National Environmental Policy Act (NEPA), the Wisconsin Environmental Policy Act (WEPA) is nearing the end of its second decade of existence. This paper reviews how one of the nation's first state-level "little NEPAs" has functioned within Wisconsin and especially in the Department of Natural Resources. Trends in WEPA implementation are described, and possible future directions on how WEPA perhaps should be changed to meet the needs of the 1990s, from the perspective of WEPA practitioners, are discussed.

BACKGROUND

The Wisconsin Environmental Policy Act (WEPA) was passed in 1972 and is patterned quite closely after the National Environmental Policy Act (NEPA). WEPA requires all state agencies in Wisconsin to evaluate and disclose the environmental implications of their activities. The environmental focus and regulatory nature of the Wisconsin Department of Natural Resources (DNR) have placed it in a leadership position in the implementation of *Wisconsin's premier environmental law*, as the state Public Intervenor Office[1] often reminds us in its legal briefs and appeals to us.

---

[1] The Public Intervenor in Wisconsin is an assistant attorney general charged specifically with protecting public rights in environmental issues. The Public Intervenor may become involved in administrative proceedings and may file legal actions.

Like NEPA, WEPA has remained essentially unchanged in the statute book. There have been some minor statutory augmentations of WEPA-associated implementation laws, most notably dealing with the ability of the state to require the preparation of environmental impact reports (EIR) by project proponents and the recovery of EIS preparation fees.

WEPA interpretation has evolved over the years in the courts. Since WEPA is modeled after NEPA, the numerous federal court decisions which have molded NEPA have affected our state-specific interpretations as well. The most formative Wisconsin case law has focused on how the threshold for an EIS is determined (i.e., has the state developed a reviewable record and does the state's determination not to do an EIS follow logically from this record), whether the analysis was adequate, and whether the procedural aspects of the process were followed.

There is no centralized oversight in Wisconsin on how each agency complies and interprets WEPA. There is no state counterpart of the Council on Environmental Quality (CEQ), although WEPA directs state agencies to *substantially* follow federal CEQ guidelines. Each state agency has had substantial freedom to determine how it will comply with the law.

From time to time, mainly through the mechanism of executive order, there have been attempts to provide interagency consistency through an Interagency Coordinating Committee (ICC). The effectiveness of these ICCs (the last of which was convened in 1983) has been sporadic because of the administration-specific lifetime of executive orders in Wisconsin. As goes the governor — and his interest in environmental policy — so goes the ICC.

Most of the major *action* agencies in Wisconsin have developed administrative rules to implement their interpretation of WEPA. In the DNR, the heart of the WEPA rules (NR 150, Wisconsin Administrative Code) is a comprehensive list of activities for which the DNR has responsibility, each categorized into one of four types, based upon an activity's potential for environmental impacts:

- Type 1 activities are those major state actions for which an EIS is always necessary (examples are issuing permits for metallic mining or energy production or a significant resource management project such as new state property acquisitions over 1000 ac).
- Type 2 activities are those which require an initial screening document, an environmental assessment (EA), to determine whether an EIS is needed (e.g., new department policies or programs or new municipal sewage treatment plants).
- Type 3 activities are considered environmentally benign under most circumstances and require issuance of only a news release in the vicinity of the proposed activity (examples are establishing public access on lakes and streams, sale of surplus lands, and many activities from the Type 2 list that are below established thresholds of size, extent, or cost).
- Type 4 activities are minor actions which normally do not cause significant environmental impacts and are therefore exempt from environmental screening and public notification (examples are issuance of licenses, state facility operations, and law enforcement activities).

If circumstances warrant, Type 2, 3, or 4 activities can be given a higher level of scrutiny if nontypical impacts are anticipated. Also, any proposed activity or type of activity can be evaluated in a generic EIS or EA if there is reason to believe that there will be recurring actions and/or impacts of a cumulative nature. For example, in cooperation with Minnesota, we are currently evaluating the proliferation of small boat marinas for recreational users along the St. Croix and Mississippi River boundary waters in such a manner.

EXPERIENCE WITH IMPLEMENTATION

The Wisconsin DNR has the most extensive environmental review program of any agency in the state. Our program has been staffed since 1970, and the program has been a DNR bureau since shortly after WEPA was passed. Although staffing levels have fluctuated somewhat over the years, being correlated with such factors as economic conditions and societal attitudes toward the environment, the DNR has maintained a stable staffing core of at least 20 over the last decade.

As part of a decentralized decisionmaking agency, the Environmental Analysis and Review Program has two Environmental Impact Coordinators in each of our six district offices. The central office professional program staff in Madison currently number 14 full-time environmental specialists and supervisors.

The central office administers the program statewide and is also involved with development of EISs, complex project coordination, document reviews, issue analysis, and technical support. The district-based field staff are responsible for ensuring consistency and compliance with WEPA for most DNR actions. They also provide a vital coordinative role with other DNR staff, private applicants, the Wisconsin Department of Transportation (through a legislatively mandated liaison process), and federal agencies such as the Fish and Wildlife Service, Federal Energy Regulatory Commission (hydropower relicensing), the U.S. Army Corps of Engineers, and the Environmental Protection Agency.

Since the inception of WEPA, well over 5700 environmental documents involving DNR proposals or regulatory actions have been written. The vast majority of these have been EAs, screening documents ranging from rather brief to very detailed evaluations of a proposed action. Sixty-eight EISs have been prepared by the DNR for major state actions. Among the most frequent subjects of EISs over the years have been state land acquisition and development (26), solid waste disposal facilities (7), metropolitan sewage district plans (6), major new regulatory programs or budgets (6), and major power plant developments (6). EAs have been written for a much wider array of activities, ranging from minor facility development (e.g., parking lots and toilets) to regulatory permit evaluations of air emissions and enlargements or fills in navigable waterways.

IMPLEMENTATION TRENDS

Cumulative experience and the extensive court history of NEPA/WEPA have produced distinct trends in Wisconsin's environmental analysis procedures over the last 18 years. Advocates of the law have characterized these trends as an evolution or maturation of the law, while critics describe it as stagnation. The most notable trends are discussed below.

- Through periodic revisions of the *type list*, which reflects the level of potential environmental impacts of various actions, the number of EAs and EISs has decreased remarkably. For example, EISs have decreased from a high of 17 in 1973, to an average of 1 or 2 per year. Annually, EAs have stabilized at about 180 per year, down from nearly 1300 per year in the late 1970s.
- The DNR relies much more heavily on the issuance of brief news releases to inform the public of a proposed action in their neighborhood, rather than using EAs as the primary vehicle for public notification. If the news release elicits public concern, a higher level of analysis (EA or EIS) can then be applied.
- More extensive EAs are likely to be done, instead of making an initial decision to prepare an EIS. Our theory, although yet untested, is that more substantially detailed EAs can be utilized equally well as *mini-EISs*, without major content revision, if circumstances eventually indicate an EIS rather than an EA is needed to evaluate a proposed action. Other benefits that we have actually realized from this approach are a more streamlined and cost-effective process that still fulfills the important evaluation and disclosure components of WEPA.
- Rather than go through a draft and final EIS process, which often necessitates a dual public participation process, there has been a trend to a single iterative process — one document, one public input process.
- The DNR has made an effort to integrate WEPA processes with other permit or property master planning document preparation and public participation requirements that are parallel.
- The DNR relies more heavily on applicant-submitted materials in WEPA document preparation, as long as such materials are credible in total and critical elements are verified by the DNR or other experts. Certain parts of WEPA documents are still developed internally by the DNR staff, most notably the environmental consequences and alternatives to the proposed action sections.
- To the extent that it can be done and still adhere to the requirement for a comprehensive analysis of the proposed action, emphasis is given to analysis and alternatives which can materially influence the decision(s) being made under the standards applicable to the proposed action.
- There is an increasing tendency to use the WEPA process to evaluate generically those actions that may be repetitively proposed, that involve unfamiliar new technologies, or that may be individually insignificant but cumulatively significant. Generic analysis also is being used more frequently as a means of evaluating new or proposed policy initiatives.

THE "REALITY" OF ENVIRONMENTAL POLICY IN WISCONSIN NOW AND IN THE FUTURE

As discussed earlier, the statutory language of WEPA (and NEPA) has changed little over the last two decades. Is this because these policy laws are working so well, or is it rather that they are not really working so well but are politically off limits because they reflect good societal values towards meeting environmental goals? At least in our state and in our agency, the debate continues. The truth is found, like in many debates, somewhere in between.

The context of the value and success of WEPA/NEPA lies in the period in which they were passed into law. The late 1960s and early 1970s were the mid years of the *Environmental Decade.* The decisionmaking processes and public participation in those decisions were substantially different then. NEPA and the *little NEPAs* of the various states have fundamentally changed how decisions are made that affect the environment. Most, if not all, of the major environmental laws that followed the environmental policy acts included many of the broad environmental information requirements, alternatives analysis, mitigation objectives, and public notification elements of these foundation laws.

Still, frustration lingers on how far and how well WEPA improves environmental decisions. To some, WEPA is inefficient, duplicative, intrusive into other processes, overly process oriented, and time consuming; most importantly, it does not necessarily cause the most environmentally beneficial decision to be made.

As we approach the 20th anniversary of our state environmental policy act, we in the DNR have been reflecting on how to improve the WEPA process. In evaluating options, we have looked to the federal and other state experiences. Recently, an internal committee has put together a menu of options on key issues of concern to many of us. Some issues have multiple and sometimes opposing alternatives which need further *sifting and winnowing.* Some issues and alternatives are fairly simple to implement within existing legal frameworks; others will definitely need minor to major legislative changes to WEPA. The issues and major alternatives we have considered follow.

- Agency Applicability
  WEPA currently applies to state agencies only. Many decisions on the local government level, land use in particular, have environmental significance. There is also no statewide oversight on WEPA consistency.
  Major Alternatives:

  - Statewide CEQ with local government WEPA responsibilities.
  - No CEQ but local government WEPA responsibilities.
  - Delegate WEPA responsibilities to applicants.

- Project Applicability
  What types of projects should WEPA apply to? Should there be automatic thresholds, or should determinations be case specific?
  Major Alternatives:

  - Establish a checklist system to simplify the screening process.
  - Redefine *major action.* Definition can be set by example or specific threshold criteria.

- Functional Equivalency
  Should certain regulatory program procedures and documents be accepted as

equivalent to the WEPA process and thus used to meet WEPA requirements?
Major Alternative:

- Amend administrative rules to allow equivalent regulatory procedures to functionally substitute for WEPA.

- Substantive Reach
  Should WEPA have a substantive nature whereby results of the WEPA analysis are actively applied in decisionmaking, or, alternatively, should WEPA provide only a process?
  Major Alternatives:

  - Add substantive reach to WEPA.
  - Specify in the statute that WEPA has no substantive reach.

- Public Involvement
  Should public involvement in the WEPA process be expanded or restricted over current levels?
  Major Alternatives:

  - Expand the scope of WEPA requirements to provide for contested case hearings on all EISs.
  - Prohibit contested case hearings on all EISs.

- Litigation Potential
  Lawsuits most often are filed to appeal some WEPA decision the state has made. Most appeals are related to decisions not to do an EIS. Also, parties frequently challenge the adequacy of EAs and sometimes EISs.
  Major Alternatives:

  - Combine the appeal of the EA/EIS and the project action.
  - Consolidate the EA/EIS into a single process.

- Repeal WEPA
  WEPA may have outlived its usefulness in environmental decision making in today's society.
  Major Alternatives:

  - Repeal WEPA completely.
  - Cutback to minimal compliance.

- Scope of Analysis
  Redefine the scope of environmental analysis from all encompassing (often based on public issue scoping) to more narrowly related to the standards used in the decision.
  Major Alternatives:

  - Define *Human Environment* as the physical–biological environment; exclude the economic and social components.

- Limit the scope of EAs/EISs to those issues that can be considered in the agency's decision and to those alternatives that can be implemented.
- Remove the *substantially following CEQ* and the EIS table of contents language from WEPA.

As of late 1989, we have not initiated follow-through on any specific alternative identified. We do intend, however, to evaluate and act on selected options during the development of our program's strategic plan for the 1990s.

WEPA has survived nearly 20 years of application and scrutiny and, in the opinion of some, has even become institutionalized into the way society thinks. Despite the chronic grumbling of those who must comply with the law's procedures in order to achieve their ends, it is difficult to find much administrative or political support for tinkering with the law. One is left to assume, then, that it must be working and is of some value, even in today's environmentally aware society. Changes to WEPA will therefore likely be made only with great caution.

But for a law that found its beginnings in the original Earth Day and evolved through a societal process of natural selection, change itself is natural. Earth Day plus 20 is approaching quickly.

ACKNOWLEDGMENTS

My appreciation is extended to the staff, past and present, of the DNR's Environmental Analysis and Review program (formerly Environmental Impact Program). I also must acknowledge the ideas which are presented here which resulted from the DNR's WEPA Committee (Gary Birch, Jim Morrissey, Bill Clark, Bill Smith, Marcia Penner, George Albright, Roger Fritz, and Robert Ramharter). I appreciate the comments on an earlier draft of this paper from Steve Ugoretz and Ed Jepsen. And lastly, appreciation is extended to the Wisconsin Environmental Decade, and Kathy Falk in particular, for the painful lawsuits during her tenure at Decade that were inflicted on various state agencies, including the DNR, in order that WEPA interpretation could evolve to where it has today.

CHAPTER 9

The International Influence of NEPA

Introduction

S. Rayner, Battelle Pacific Northwest Laboratories, Washington, D.C.

The contributions to this chapter address the impact of the U.S. National Environmental Policy Act (NEPA) on the environmental planning and assessment processes of other jurisdictions. The process of environmental impact assessment (EIA) established by NEPA has been incorporated, at least in principle, into the laws of some 75 national jurisdictions in the 20 years between 1969 and 1989.

Perhaps surprisingly, the adoption of EIA requirements and procedures does not seem to reflect obvious patterns of economic and social development. Although Canada, Australia, and New Zealand implemented EIA laws within 5 years of the passage of NEPA, so did Malaysia. The European Community (EC), however, took 20 years to establish EIA requirements, and, within the community, only The Netherlands has vigorously pursued implementation of a full-blown EIA process.

Nicholas Robinson, in the first paper of this section, identifies seven trends for EIA worldwide:

- EIA works in all legal systems.
- It is spreading rapidly.
- It provides local people with a voice in the decisions of central government.
- It marshals data for decision makers.
- It is usually resisted for contradictory reasons — as restricting development or because it is ineffective.
- There is a tendency to do EIA only for large projects.
- It is not uniformly successful, especially when lacking independent oversight.

All of these observations are useful and valid insights. However, Reinhard Coenen's review of NEPA's impact in the EC indicates that the first and the last trends seem to provide particularly important clues to the effectiveness of EIA around the world. Although the EC introduced broad requirements for EIA, the implementation and effectiveness of these measures have proved highly uneven, largely because of their compatibility with existing legal and administrative processes. Much of the success of EIA in the United States was because it altered previous standard procedures of decision makers. In Europe, however, legislators have preferred to integrate the EC directive into existing planning procedures, especially in the U.K., Ireland, and Denmark. France and The Netherlands have established new procedures; however, the Dutch regulations are the most ambitious in Europe, while the French requirements are very weak.

McCormick's paper highlights the point that most of the stimulus for adoption of EIA in developing countries has been *top down*, at least in the beginning. Often, as in the Indonesian case described by Katili, the requirements of lending agencies have been the catalyst for developing indigenous EIA legislation and rules. This point is particularly significant in the light of Robinson's emphasis on the importance of public participation and independent oversight to the success of EIA. As Wilbanks et al. point out, there is relatively little development of effective indigenous nongovernmental organizations to pressure the authorities in developing countries.

Like the underdevelopment of indigenous environmental constituencies with adequate resources to intervene in EIA proceedings, the main obstacles to effective EIA in developing countries are ones that cannot be overcome by improved procedures or by more rigorous implementation. They are systemic resource problems of Third World poverty. Countries like China and Indonesia simply cannot afford the no-action alternative. The choice is seen as develop or perish. Lack of baseline environmental and socioeconomic data is another resource problem that cannot be addressed legislatively or administratively. For the Third World, EIA remains, at best, a Band-Aid™ [1] to mitigate the worst consequences of rapid industrial development because it is wealth, not legislation, that leads to indigenous demands for clean energy, stable populations, and stewardship of the land and water.

---

[1] Registered Trademark of Johnson and Johnson.

EIA Abroad: The Comparative and Transnational Experience

N. A. Robinson, Pace University School of Law, White Plains, NY

INTRODUCTION

Environmental impact assessment (EIA) is today increasingly being established as a routine decisionmaking technique worldwide. Since Congress conceived the EIA in Section 102(2)(C) of the National Environmental Policy Act (NEPA) of 1969, the EIA has been required by law in more than 75 separate jurisdictions. When the European Community (EC) issued a directive in 1985 requiring that its members adopt EIA procedures, the Dutch and French, in particular, already had considerable experience with EIAs. Indeed, it is remarkable that except for EC member states, each legislature that has followed the lead of Congress in enacting the EIA has done so unilaterally. No duty imposed under a framework treaty or exhortation of a United Nations' resolution produced this result. Rather, the EIA has been embraced on its own merits.

The EIA is a proven technique, used to ensure that governmental actions avoid or minimize otherwise unanticipated adverse effects. The EIA provides systems to institutionalize *foresight*. While its essential structure is substantially the same wherever used throughout the world, the EIA is flexible and has been adapted successfully to operate within the cultural, political, and socioeconomic development conditions prevalent in each jurisdiction which has enacted EIAs.

This Ninth Oak Ridge National Laboratory Life Sciences Symposium examines the analytic techniques by which various scientific disciplines can most effectively conduct the studies needed in an EIA process. A comparable body of scientific experience is growing around the world; nonetheless, the 20 years of decisionmaking under the United States' National Environmental Policy Act has produced the world's most extensive and advanced experience with EIA science and technology. The presentations at this symposium, therefore, will be of substantial interest in all the various jurisdictions with EIA requirements.

By way of introduction, this paper explores the extent of the legislation that creates the EIA mandate. A more comprehensive study of all EIA laws is in preparation by the Commission on Environmental Law of the International Union for The Conservation of Nature and Natural Resources (IUNC, also known as the World Conservation Union), but this study will not be completed before 1992. In the absence of such an exhaustive analysis, this paper sketches the global legislative trends and some of the comparative law variations in EIAs.

ENVIRONMENTAL STEWARDSHIP

The need for EIAs was apparent before the process was conceived. President Theodore Roosevelt, in his 1908 White House Conference on Conservation, called for "foresight."

> We have become great in a material sense because of the lavish use of our resources, and we have just reason to be proud of our growth. But the time has come to inquire seriously what will happen when our forests are gone ... when the soils shall have been further impoverished and washed into streams. ... These questions do not relate only to the next century or to the next generation. One distinguishing characteristic of really civilized men is foresight ... and if we do not exercise that foresight, dark will be the future (Roosevelt 1909).

More recently, U.S. Secretary of State James A. Baker articulated the need for EIAs, both in his prior post as Treasury Secretary and in his initial public address as Secretary of State. In 1987, he described his request that the World Bank institute EIA procedures:

> Growth and development are essential for conservation, and conservation is essential for growth. Despite some assertions to the contrary, these concepts are not mutually exclusive. In fact, they should not necessarily be deemed mutually antagonistic. I am not saying that growth and development do not put new and difficult strains on the natural environment. The lessons of centuries is that they often do — and with tragic results, when men and women are careless. ... I think we have to pursue, both in the United States and abroad, a philosophy of growth combined with conservation. ... What [the United States] wants the World Bank and the other development banks to do is make environmental analysis, systematically and routinely, a central part of every loan proposal. We want the Bank to draw on the expertise of trained environmental analysts — both from its own staff and outside consultants — who know developing countries and can assess just what impacts any new project or policy will have on the ecology of those countries. It should then incorporate that analysis into its lending decisions and assistance from the very beginning of the lending process (Baker 1988).

In response to the urging of representatives of both governments, such as Secretary Baker, and nongovernmental organizations, such as the Environmental Defense Fund (Rich 1985), the World Bank has adopted its initial rules on environmental assessment (EA). These World Bank procedures are modeled on experience with NEPA and the knowledge gleaned from EIAs in Australia, Canada, and elsewhere. Secretary Baker's exhortation is being put into practice also in the other multilateral development banks.

NEPA's procedures anticipate possible environmental problems and identify alternative courses of action to avoid or mitigate adverse impacts. When newly confirmed as Secretary of State, Baker urged NEPA-like, reasoned prudence in his address to the Intergovernmental Panel on Climate Change (IPCC) convened by the United Nations Environmental Programme and the World Meteorological Organization to evaluate global warming. In 1989, Secretary Baker told the IPCC that "the political ecology is now ripe for action" and that, while scientists refine our knowledge about the dangers and dynamics of global climate change caused by pollutants, "we can probably not afford to wait until all of the uncertainties have been resolved before we do act" (Shabecoff 1989). The process of making cautious and informed decisions, with preventative measures to avert unwanted environmental degradation, is the essence of the EIA process. The use of NEPA's EIA process is emerging as an important tool to assist in curbing unnecessary accumulation of greenhouse gases in the atmosphere.

Just as all Canadian provinces and 25 states in the United States have enacted EIA procedures (Robinson 1982, 1989), some of which include innovations improving upon NEPA's techniques, foreign countries adopting EIAs have also found ways to make the EIA increasingly more effective. The transfer of NEPA abroad has not been characterized by rote mitigation of our environmental impact statements, but rather by thoughtful adaptation.

WORLDWIDE EIA TRENDS

Encountering EIA practices around the world, one finds that each process is tailored to match the area's geographic characteristics, environmental needs, and states of socioeconomic development, as well as the cultural and governmental traditions in the jurisdiction. Seven trends in EIA practice can be discerned from its use.

First, the EIA works in all political systems. It can be and has been established alike in common law, civil law, and socialistic law traditions. It is equally useful in developed and in developing countries. It is used by small villages, state agencies, major military divisions, regional authorities, and international agencies, mutatis mutandis. The technique is adaptable to meet the type of governmental decision making involved.

Second, while the EIA is a young (even pioneering) analytic tool for decision makers, its use is spreading fairly rapidly. As it is adopted, different jurisdictions modify and often refine the EIA process. There is a sharing of methodologies. For instance, the EIA works best when some independent authority provides oversight of the process; for NEPA, courts provide this through judicial review. In jurisdictions without a comparable tradition of litigation, this oversight can be provided by analogous administrative arrangements; thus, the Dutch EIA process adapts the concept of an independent commission to judge the sufficiency of an environmental impact statement from Canadian experience in which the tasks of delineating the scope of the EIA and its preparation are done by separate authorities independent of the decision maker. In like fashion, Massachusetts developed the step known as *scoping* as a means to provide better substantive focus for each environmental impact assessment, and the Council on Environmental Quality (CEQ) in turn adopted the scoping process for the revised NEPA regulations. This dynamic system of sharing innovative techniques is likely to continue as the EIA process becomes more widely adopted.

Third, the EIA is effective in providing local people with an opportunity to be heard and to participate in decisionmaking that affects their environment. It facilitates democratic decisionmaking and consensus building regarding new development. The EIA process in the U.S.S.R., known as ecological expertise, has allowed residents in the Altai Alps to review plans for a proposed hydroelectric facility and to call for revision and further review of the plan. It equally gives voice to the often unrepresented interests of indigenous peoples, local residents, and inner-city communities. To be sure, the EIA can be contentious, as countervailing interests use the EIA studies to emphasis their various positions. In a democracy, however, it is better to have the reasoned examination of these contending views in the factually informed context of an EIA than to ignore them or treat them exclusively as political views.

Fourth, the EIA is demonstrably effective in marshaling environmental data for decision makers. Invariably, the EIA encourages interagency communication and consultation. Experience reveals that environmental issues that were unanticipated in the process of project preparation, in fact, are identified before unintended damage occurs.

Fifth, despite its evident value, the EIA is not easy to establish at the outset. The EIA is often, almost always, resisted until decision makers and administrators become educated about its utility. There is, of course, innate institutional resistance to any change; many agency managers have a strong sense of their traditional mission and have not added the coequal duties of environmental stewardship to their *primary* responsibilities. Moreover, in many agencies there is a preference for short-term, business-as-usual procedures. Busy administrators doubt that there is enough time to try new, apparently slower procedures. Some fear a risk to their projects or authority by being subjected to an EIA. These concerns result in politicians and civil servants opposing the use of EIAs; they advance arguments to oppose the establishment of EIAs or to avoid using it once an EIA is in place. In developing countries, opponents of establishing EIAs variously have labeled the process "anti-development, expensive, or a mere paper tiger" (Ahmad and Samny 1985). In developed countries, the canard often has alleged that the environmental impact statement process is the excessively time-consuming generation of too many studies which are never read. Often, inefficiencies in a protracted EIA or occasional mistakes in an EIA process are marshaled as an excuse either to abolish the EIA altogether or, more often, to exempt a project from an EIA. In developed countries, critics of EIAs often generalize from an isolated, notorious instance of an EIA in trouble and assume without verification that the whole system has these flaws. Such critics ignore the thousands of EIA applications successfully completed each year throughout North America and elsewhere.

Foreign assistance agencies, in particular, have resisted the use of EIAs, for instance, for development aid grants, on such diverse grounds as a belief that a donor's EIA would infringe on the recipient's sovereignty or would complicate the administration of aid. However, as with the adverse environmental side effects of the High Aswan Dam in Egypt and in the Mediterranean Sea (George 1972; Kassas 1972), experience demonstrates that it is the failure to study and avoid unintended adverse environmental impacts of overseas aid that harms the recipient nation and often not only wipes out the value of the aid but requires additional expenditures to repair the damage. The EC requires that states routinely examine the environmental impacts of their actions in other states. Canada's courts require that impacts abroad be evaluated.[1] Nonetheless, in the case of NEPA, those who wish to avoid using EIAs suggest that the study of the environmental consequences of U.S. actions abroad would constitute extraterritorial interference in other states' affairs, rather than constituting a stewardship exercise in assuring that the United States does not cause unintended harm abroad. Invariably, whether within a state, across borders, or abroad, the opponents of using EIAs are persons who have rarely or never participated in the process personally. As the EIA is extended to new spheres of decisionmaking, this trend of initial resistance gradually declines.

Sixth, there is a tendency to use EIAs only for large projects. Many nations promulgate a list setting out the types of projects that require EIAs. Other states set a low threshold for EIAs; they recognize that even a small project can cause unintended environmental harm and elect to require use of the analysis widely. In jurisdictions like California or New York, even small villages must follow EIA procedures. Since environmental *significance* is not merely a function of *bigness*, the trends toward using lists and restricting EIAs to large projects do not assure an effective application of the EIA. The tendency to limit EIAs to large projects reflects a desire for administrative convenience on the part of those establishing the EIA process, rather than a mature application of the technique. Experience with EIAs suggests that the use of lists and restrictions based on project size are evidence of an immature phase of the EIA, in which resorting to a clear rule of thumb is preferred to a more sophisticated and initially open analysis based on scientifically based data.

---

[1] See *Canadian Wildlife Federation et al. v. Minister of the Environment and Saskatchewan Water Corp.*, Docket No. T-80-89 (Federal Court of Canada, Trial Division, Ottawa, April 10, 1989), in which the court, per Mr. Justice Cullen, granted *mandamus* compelling a full EIA compliance for a proposed project in Saskatchewan's Souris River Basin (Rafferty/Alemeda Dam Projects) undertaken pursuant to the International River Improvement Act because environmental impacts had not adequately been considered, including the impacts in the state of North Dakota (U.S.) and Province of Manitoba (Canada).

Seventh, the EIA is not uniformly successful. Even in jurisdictions with many years of EIA experience, it is rare to require postproject monitoring to find out whether all adverse impacts were accurately anticipated or whether mitigation plans in fact were successful. Where EIA systems lack an oversight process, they can more easily be subverted by politically or economically persuasive project sponsors. When EIA systems lack automatic public disclosure, as in Thailand, the educational consensus-building and peer-review benefits of the process are lost. When an agency's decision makers are inept in administering an EIA, the adversaries of an action can delay inordinately the start of a project until it loses its essential political or economic sponsors. There is a constant need to evaluate the effectiveness of each jurisdiction's EIA process, to improve, streamline, and weed out dysfunctional aspects. Jurisdictions often engage in a reform process of their EIA procedures after 6–10 years of experience, as was the case in Australia, The Netherlands, the United States, and several states and provinces.

These seven trends are gleaned from examining EIA processes throughout federal and state practices in the United States, in Canada, and abroad. The worldwide experiences with EIAs are too extensive to cover in detail in this short paper. Nonetheless, it may be useful to examine briefly two aspects of EIAs: (1) a brief comparison of EIA functions generally and (2) some thoughts about how EIAs should be used to cope with global, transnational, and transboundary environmental impacts.

COMPARATIVE EIA

The EIA process is best understood by comparing how differing jurisdictions have instituted it. To fully appreciate NEPA's limitations, it is instructive to compare it to the stronger *little NEPAs* of states such as Wisconsin, New York, Washington, or California, and vice versa (Robinson 1989). Those studying EIA procedures established in other nations and in the provinces or states of other nations can and do learn from one another. The states of New South Wales and Victoria regularly compare their practices to those functioning under NEPA. Instead of comparing these state or provincial practices, this paper surveys foreign experience.

After the enactment of NEPA in 1969, the EIA was quickly adopted by the mid-1970s in Australia, Canada, and New Zealand. It has since been instituted in many jurisdictions, including Argentina, Belgium, Brazil, China, Columbia, Costa Rica, Denmark, France, Germany, Greece, Hong Kong, India, Indonesia, Ireland, Israel, Italy, Japan, Korea, Kuwait, Luxembourg, Malaysia, The Netherlands, Norway, Pakistan, Papua New Guinea, Peru, the Philippines, Portugal, Sri Lanka, South Africa, Spain, Taiwan, Thailand, Turkey, United Kingdom, the U.S.S.R., and Uruguay (Table 1). In federal nations, many states and provinces have unilaterally enacted EIAs within their respective jurisdictions (Robinson 1982). The EIA is not yet widely used in large parts of Latin America, the Middle East, or Africa. In the 1970s, the North Atlantic Treaty Organization (NATO)

Table 1. EIA Statutes Cited.

1. **Australia**
Australia's Commonwealth Government adopted the first EIA law in 1974: Environment Protection (Impact of Proposals) Act. Since then, the states of New South Wales, Victoria, South Australia, and Western Australia and the Northern Territory have adopted EIA procedures.
2. **Belgium**
Introduction of the EIA by separate laws (decrees) on the national and regional levels. Integration of EIAs into existing administrative procedures.
3. **Brazil**
Brazil's federal government has some 40 officials conducting EIAs, primarily in Amazonian. Brazilian procedures are described in Braun, R. A. 1976. "Environment Impact Assessment in Brazil." p. 10. In The Legal Procedure Worldletter. The International Newsletter for Environmental Assessment (Sept./Oct. 1976). See "Law Containing Provisions on National Environmental Policy."
4. **Canada**
The EIA began in Canada at the provincial level with Ontario's Environmental Assessment Act (1975) and at the federal level with the establishment of the Environmental Assessment and Review Process (EARP) by Cabinet Decision on Dec. 20, 1973 and revised by Cabinet Decision on Feb. 15, 1977. An independent panel was appointed by the Minister of the Environment to undertake a scope of review defined in Terms of Reference prepared by the Federal EHR Office. EARP Guidelines Order S.O.R./84-467.

   EARP was established by the Cabinet Decision of Dec. 20, 1973 and adjusted by a second Cabinet Decision on Feb., 15, 1977. Responsibility for EARP was assigned to the Minister of the Environment, as reaffirmed in the Government Organization Act (1979). An Order-In-Council entitled the "Environmental Assessment and Review Process Guidelines Order," S.O.R./84-467, was proclaimed under the Act on June 22, 1984. This guideline order replaced prior cabinet decisions.

   For the references regarding the *Province of Quebec,* see the Loi sur la qualite de l'environnement (1972), modifee en 1978, L.R.Q., 1981 c. Q-2, et le Reglement sur l'evaluation et l'examon des impacts sur l'environnement R.R.Q., 1981, c. Q-2, r.1. Environment Quality Act (1972), amended in 1978 and the Environment Impact Assessment and Review — General Regulations. See also Chapter 2, Quebec Environmental Quality Act, R.S.Q. c-2.

   For the reference to *Ontario*, see the Environmental Assessment Act (1975).
5. **China**
   a. Environmental Protection Law 1979. For the People's Republic of China, see §6, Environmental Protection Law, and "Management Guidelines on Environmental Protection of Construction Projects," and supplemental rules on EIA for large and medium construction projects.
   b. For Taiwan, see Executive Yuan (the Cabinet) of Aug. 22, 1987, R.O.C. See R.O.C. Environmental Protection Administration, "Environmental Protection in the Republic of China" (April 1988).
6. **Columbia**
National Code of Natural Resources and Environmental Protection §§28–29, Decree 2811 (1974). See Decree partially regulating Title I of the Act of 9, 1979 and also see Chapter 2, Title IV of Dto. 2811–74 concerning the use of water and liquid waste.
7. **Costa Rica**
Procedures of Environmental Protection Agency.
8. **Denmark**
Implementation of the directive by amending the acts on national and regional planning. Integration of EIAs into the regional planning procedure.
9. **France**
France elected to set a low threshold encompassing most of the European

Table 1. EIA Statutes Cited (continued).

9. **France (continued)**
Community's Directive Annex II actions in its Nature Protection Act of 1976, specified by a decree of 1977. Some 4000–5000 assessments are done annually in France.
10. **Gambia**
National Environmental Management Act of 1987.
11. **Germany**
Introduction of EIAs through a so-called "article law" which determines the basic principles of EIAs in Art. 1 and the necessary amendments to special laws in the following articles. The EIA is integrated into existing procedures.
12. **Greece**
Introduction of the EIA within the framework of the Environmental Protection Act of 1986. Legal EIA regulations for industrial plants since 1981. Integration of EIAs into existing administrative procedures.
13. **Hong Kong**
Town Planning Ordinance 1939 (with amendments). White Paper: Pollution in Hong Kong — A Time to Act.
14. **Ivory Coast (Cote D'Ivoire)**
Decree prescribing the duties of the Minister of the Environment and laying down the Organization of the Ministry. J.O. 19811015, No. 44, pp. 532–533.
15. **India**
Constitution (42nd Amendment) Act 1977. The Environment (Protection) Act 1986. The process in India is limited to major federal projects in only four categories.
16. **Indonesia**
Act of the Republic of Indonesia: No. 4 of 1982. Concerning Basic Provision for the Management of the Living Environment.
17. **Ireland**
Implementation below legislation level by regulations under the "Local Government (Planning and Development) Act" and other relevant laws. Integration of the EIA into existing administrative procedures.
18. **Israel**
Israel adopted EIA regulations for building plans in 1981. See Planning & Building Regulations (Environmental Impact Statements), Kovetz Ha-Takanet of 5742, p. 502.
19. **Italy**
Before a law on the implementation of the EC directive is passed, interim provisions on the basis of the law No. 349 of 1986. Performance of the EIA as a separate procedure preceding the permitting procedure.
20. **Japan**
Environmental Scheme Measures involving Various Public Works Act 1972.
21. **Korea**
Environment Preservation Act 1977 (amended 1979, 1981, 1982, and 1986). Regulations for the Preparation of Environmental Impact Assessment 1981.
22. **Kuwait**
Law No. 62 of 1980, establishing the Environment Protection Department within the Ministry of Health.
23. **Luxembourg**
Project de loi No. 3257, pending since September 1988, for adoption by the Chamber of Deputies of Luxembourg.
24. **Malaysia**
Environmental Quality Act 1974 (amended 1985). Environment Preservation Act 1977 (amended 1979, 1981, 1982, and 1986). Regulations for the Preparation of Environmental Impact Assessment 1981.
25. **Mexico**
Federal: General Act on Ecological Balance and Environmental Protection.
26. **The Netherlands**
Introduction of EIAs by amending the General Environmental Protection Act of 1979. The EIA as a separate procedure; however, legal provisions to coordinate

Table 1. Continued.

26. **The Netherlands (continued)**
EIAs with other procedures. The Netherlands adopted EIAs in 1985, with a special commission of independent experts to review an EIS. A working group of 6–8 specialists is assembled from the same 110 members of the Review Commission. The commission's evaluation is delivered to the competent decisionmaking authority. The Dutch law on EIA supplemented the General Environmental Act (WABM) of 1979 and became effective May 13, 1986.
27. **New Zealand**
New Zealand instituted Environmental Impact Reporting and Assessment practices in 1974 through the Cabinet Decision of Aug. 7, 1972, which established the Commission on the Environment.
28. **Norway**
Experimental systems are described in Tor Lorstang's essay "Challenges for a Proposed EIA System in Norway," Scandinavian Planning & Housing Research, vol. 1, pp. 107–121 (1984).
29. **Pakistan**
Pakistan Environmental Protection Ordinance 1983.
30. **Papua New Guinea**
Environmental Planning Act 1978.
31. **The Philippines**
Presidential Decree No. 1586, 1976: Establishing the Environmental Impact Statement (EIS) System. Presidential Decree No. 1151 1977: Philippine Environmental Policy. Council Resolution No. 4: Revised Rules and Regulations Implementing P. D. 1586, 1986. Presidential Proclamation No. 2146. National Environmental Protection Council: Office Circular No. 3, 1983. Technical definitions and scope of the environmentally critical projects and areas enumerated in Proclamation No. 2146.
32. **Portugal**
Introduction of EIAs within the framework of the Environmental Protection Act of 1987. The EIA procedure precedes the actual permitting procedure.
33. **Democratic Socialist Republic of Sri Lanka**
National Environmental Act, No. 47, 1980.
Coast Conservation Act, No. 57, 1981.
Coast Conservation (Amendment) Act, No. 64, 1988.
34. **South Africa**
No. 100 of 1982. Environment Conservation Act, 1982.
35. **Spain**
Introduction of EIA by separate legislation (Royal Legislative Decree of June 1986). The EIA procedure is integrated into existing administrative procedures.
36. **Thailand**
Improvement and Conservation of National Environmental Quality Act 1975. Last amendment (No. 3), B.E. 2522 (1979). Proclamation for Types and Sizes of Projects Required: Environmental Impact Assessment 1981. EIA in Thailand is based on the National Environmental Quality Act of 1975. Proclamations under §17 of the act cover the activities requiring EIAs. The first such proclamation in 1981 covered major industrial and mining and dam projects, irrigation, commercial airports, and large hotel/resort facilities.
37. **Turkey**
The Environmental Law, and Decree 222/19 and Laws 3301/1986 and 3416/1988. Turkey: Report to UN ECE Seminar on EIA, Warsaw, Poland, Sept. 21–25, 1987 (on file with the CEQ).
38. **United Kingdom**
Implementation below legislation level by regulations under the Town and Country Planning Act and other relevant laws. Integration of EIAs into the existing permitted procedures.
39. **United States**
The National Environmental Policy Act of 1969 (NEPA) is codified at 42 U.S.C. 4321. The EIA provision is in §102(2)(C). Generic regulations governing all

Table 1. EIA Statutes Cited (continued).

39. **United States (continued)**
agencies appear at 40 CFR 1500. Each federal agency can also promulgate its own EIA regulations. For instance, Agency for International Development NEPA procedures appear at 22 CFR 216, having been issued in the *Fed. Regist.* 41:26913 (June 30, 1976).

State *little NEPA* statutes are independently issued by state agencies. See, for instance:

California — Cal. Pub. Res. Code 21000-21174;
Massachusetts — Mass. Gen. L. Ann., ch. 30, §61,62,62II;
New York — 4 N.Y. Envir'l. Conserv. L., Art. 8.;
Wisconsin — Wisc. Stat. §1.11; and
Washington — Wash. Rev. Code §§43.21C.010 to 43.21C.910 (1974).

40. **U.S.S.R.**
See Instructions of Goskompriroda on EIAs, directive of 1990.

41. **Venezuela**
Organic Law on the Environment, Article 21.

Note: For Argentina, Peru, and Uruguay, see Moreira 1988.

was instrumental in explaining the use of EIA practices based on NEPA among NATO member nations, and the Organization for Economic Cooperation and Development (OECD) studied its use extensively. These educational efforts led to early acceptance of EIAs in Western Europe.

In these other countries, EIA procedures rarely rely upon the courts to oversee the accuracy of the environmental impact statement or the procedures, as does the United States under NEPA. However, these EIA procedures do recognize that project proponents often can have a real, conscious, or unconscious bias in favor of their proposals and that the persons preparing an EIA must have some independence in order to ensure objectivity in the evaluation of a project's negative side effects and to ensure identification of alternatives to the project or mitigation measures. Rather simple and inexpensive measures, such as providing for public disclosure of environmental impact reports and providing an opportunity for public comment, guarantee that there will be some measure of objectivity and completeness for an environmental impact analysis. A few countries, such as Thailand, have not yet incorporated a public disclosure process. Most separate the role of EIA preparation from the role of approving the adequacy of an assessment. A variety of institutional measures are employed to divide up the roles of performing the impact assessment and project decisionmaking.

Given Canada's extensive experience with EIAs, it may be useful to outline the federal process in Canada. Since 1973, Canada assiduously has applied and refined the EIA process. It continues to do so, recognizing that sustainable development depends upon the EIA. In 1984, the federal Minister of the Environment established the Canadian Environmental Assessment Research Council (CEARC) as a standing body responsible for advancing research to improve the scientific, technical, and procedural aspects of Canada's EIA. Canada has helped develop EIAs abroad as well. The Canadian International Development Agency (CIDA) and Federal Environmental Assessment and Review Office (FEARO) have prepared an EIA process and sustainable development plan with Indonesia for the marine and coastal resources of the Indonesian archipelago. CIDA incorporates an EIA process into its foreign aid and advice.

Canada's Environmental Assessment and Review Process (EARP) may be revised by incorporating it into new federal legislation in coming years. It was established initially by cabinet decisions, and the EARP guidelines order has been deemed by the courts to be equivalent to a statutory duty, which can be enforced by the judiciary. Given the statutory equivalency of EARP, there is little administrative reason not to embody EIAs into the stronger legislative format. The Canadian process is so well-accepted that any new legislation is likely to confirm and refine the administrative system described here.

EARP applies to all federal proposals. EARP can begin at the planning stage or when the project is advanced as a proposal. The decisionmaking authority is the "initiating department," and the "proponent" is the entity that will undertake the proposal, whether a governmental agency or a private applicant. The current practice is illustrated on the chart set forth in Figure 1.

Each initiating department must have its own screening procedures to decide what approval authorities it has in order to determine when it will be expected to comply with EARP. When a proponent submits a proposal for such an approval, the initiating department at the outset ascertains whether the proposal may have potentially significant adverse effects or is the object of public concern because of its environmental affects. It then refers projects with such effects to the Minister of the Environment for public review by an Environmental Assessment Panel. Initiating department decisions either to refer or not to refer a proposal are reported to FEARO and published.

The FEARO, an independent body somewhat analogous to the CEQ, approves each department's EARP rules. It also provides the secretariats to staff the public review conducted by the environmental assessment panels. The panels are chaired by the FEARO executive chairman or his delegate. Three to seven members of each panel are named by the Minister of the Environment, having been selected based upon their objectivity, public credibility, and special knowledge of factors associated with the proposed action. FEARO then prepares an outline of the scope of the environmental assessment in the form of a draft "Terms of Reference" for the EIA. The Minister of the Environment issues this scoping document after consulting the initiating department.

The panel convenes, establishes its operating procedures within an FEARO framework, and consults the proponent and the public regarding preparation of the environmental impact statement (EIS). The proponent prepares the EIS following the panel's directions. The panel then makes the EIS available to the public and receives comments about it at public meetings and in writing. The panel then prepares a report containing its recommendations, which is given to both the Minister of the Environment and the minister of the initiating department. These ministers release the panel's report to the public, and the initiating department thereafter makes its decision on the proposed action.

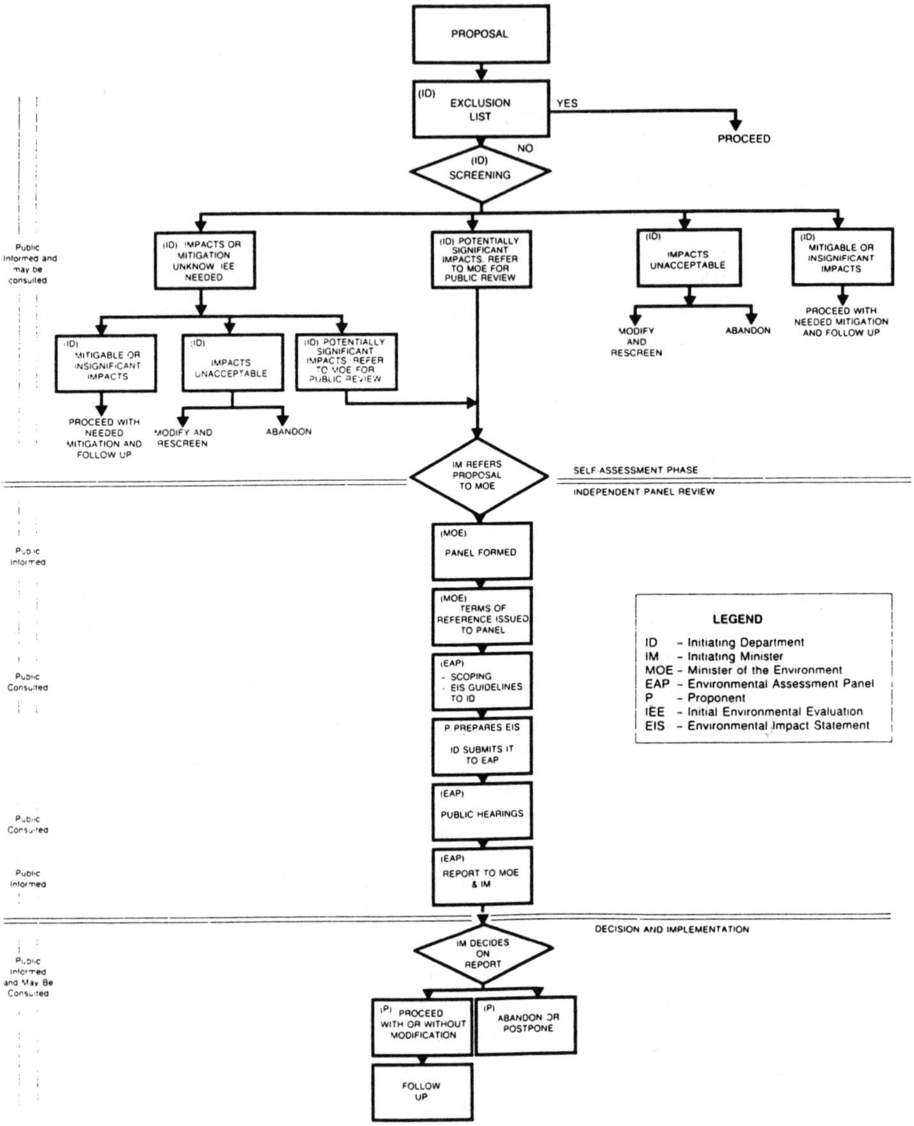

Figure 1. Federal environmental assessment and review process. (From Couch, W. J. 1988. Environmental Assessment in Canada. Can. Council of Res. Environ. Min., Ottawa. With permission.)

This federal process has benefitted from innovations established at the provincial level in provinces such as Ontario and Quebec. The Quebec experience employs the Service Techniques du Ministere Quebecois de l'Environnement (Technical Services of the Quebec Ministry of Environment) to apply the EIA process; consult the proponent and the public; evaluate the EIA; and make recommendations on the environmental aspects of the project, including any necessary surveillance and monitoring. The Public Hearings Bureau, or Bureau d'Audiences Publiques sur l'Environnement (BAPE), assists the public and holds public hearings. Citizens have the right to sue proponents for violations of the EIA procedures.

The federal process follows the Quebec practice in adapting EIA procedures to facilitate participation by the peoples living in the region, the Cree and Inuit and the Naskapi. The indigenous peoples participate in the EIA process through committees to which they name representatives. These committees constitute a systematic outreach, adapting public participation to the cultural traditions of the indigenous community. The procedures in Quebec and Canada for making the EIA effective within traditional communities of indigenous peoples should be studied as a model for improving EIA use in analogous contexts in other jurisdictions.

Canada's experience with EIAs demonstrates how the process can be adapted and evolve. By dividing up the responsibility for the various stages of the EIA, Canada has detached the process from the proproject bias of the department sponsoring the governmental action. In the United States, NEPA relies mostly on judicial review and after-the-fact correction by the courts whenever such a bias would impair the integrity of the EIA; a small office in the U.S. Environmental Protection Agency (EPA) does have a statutory mandate to comment on draft environmental impact statements, but the EPA rarely invokes this authority. Canada has a preventative process with administrative oversight and without much litigation; the United States has a corrective process of oversight, with preferred recourse to litigation and an atrophied administrative role for the EPA or CEQ.

Both Canada and the United States employ scoping, rigorous scientific and technical analysis, preparation of a draft statement, public disclosure of the draft, public comment, and preparation of a revised final statement of environmental impacts, taking into account all comments. Decision makers are then required to consider the statement and make a decision.

At a minimum, NEPA, as construed by the U.S. Supreme Court, merely requires the development and disclosure of the environmental impacts and the means for mitigating them. The court interprets NEPA as a procedural statute. Some early NEPA interpretations were of the view that NEPA imposed substantive duties as well as procedural ones and required the decision maker to select the least impacting alternative or to provide for mitigation measures. Jurisdictions that adopted EIAs on this early model often require mandatory mitigation of impacts disclosed (e.g., the states of California and New York and the Province of Ontario). Canada tends in this latter direction and thus has a substantive environmental protection mandate, whereas NEPA today is a procedural full-disclosure process relying on voluntary measures for environmental protection in light of the disclosure.

Having outlined how EIA procedures have evolved from NEPA in Canada, the closest neighboring nation to the first jurisdiction to establish the EIA in NEPA, it is instructive by contrast to consider how the EIA has been implemented in a very different society. The People's Republic of China has usefully developed its own EIA process. China holds a quarter of the world's population and is actively developing to expand its economy to provide for its growing population. China has been developing strong policies favoring environmental improvements, from afforestation to pollution control, in tandem with its market economic reforms.

China initiated EIAs in 1979, when it required that "for either new construction, extension or expansion projects, the environmental impact statement must be prepared. ... The facilities for pollution control and prevention of other hazards must be designed, constructed, and put into operation simultaneously with the main project." This statutory requirement was based upon studies begun in 1973 at China's first Environmental Protection Conference to gather baseline data on the quality of China's environment. The decision to initiate EIAs was reached by 1978. Following adoption of the law, in 1981, the State Council issued China's EIA rules. Preparation of these rules was a join project of the State Council's Environmental Protection Committee, the State Planning Committee, and the State Economic Committee.

In the Chinese procedure, the lead agency prepares the EIS which the Chinese Environmental Protection Agency is to review and approve. New plans for construction require environmental review; feasibility, choice of location, and preliminary design are all to be part of a larger EIA. There are express provisions for what must be covered in a construction project's EIS, including the surrounding environmental conditions and the technical and economical feasibility of measures to avoid adverse impacts. In its first 5 years of use, some 455 projects in 23 cities had an EIS prepared, for which 287 had received the EPA's approval, reflected in the grant of a "Certificate of Comprehensive Assessment" (Jin and Wen 1987).

Although China's EIA is undertaken only for a relatively few large projects in contrast to the vast country's needs, the process of using an EIA has begun. China adapts the EIA to use techniques appropriate to its circumstance. In one EIA, it was necessary to monitor air quality over a large geographic region where coal was burned; EIA officials were able to assign 10,000 people the tasks of taking simultaneous measurements at prescribed time periods. China's specialists well understand the challenge of introducing EIAs. However functionally different the Chinese EIA process may be from NEPA or EARP, the basic task is the same as that encountered in establishing and refining any EIA process. The words of two Chinese specialists express this similarity well:

> The key problem is: a critical line must be drawn to balance the relation between development and environment. The environmental problem caused by development must be restricted within the limit which human beings and other living things can accept (some people suggest a "bearable limit" principle), so that the economy can develop continuously without degrading environmental quality. A suitable developmental pace should be found to meet environmental requirements and harmonize the environment/economy relationship. In doing so, the economy must be developed in a gradual and sound manner, and the environment must be protected

and improved. We must do our best to integrate the benefits of environment, economy, and society.

In theory, the problem seems easy, but in practice, it is much more complex and difficult. It requires great effort (Jin and Wen 1987).

Independently from the People's Republic of China, the Republic of China on Taiwan also is moving toward use of EIAs. In 1987, it established a cabinet-level Environmental Protection Administration and since has undertaken an EIA process (Chien 1991). Latin America is moving toward greater use of EIAs (Moreira 1988). Other regions are also considering how to institute EIAs and are likely to follow suit; for instance, the Arab League issued a declaration in 1986 urging that EIAs be used for new development projects in the Middle East.

AN INTERNATIONAL LAW OF EIA

All these domestic illustrations of EIAs are evidence of an emerging pattern of state practice. It is becoming a norm of customary international law that nations should engage in effective EIAs before taking action which could adversely affect shared natural resources, another country's environment, or the earth's commons (see Figure 1). The EIA is the way to ensure that no state acts in a manner that harms the environment of another state, a guideline that all states are required to adhere to under international law (Principle 21 of the U.N. Stockholm Declaration on The Human Environment). The European Community's EIA directive to this effect is paralleled by the emergent duty negotiated by the Association of South East Asian Nations (ASEAN). The ASEAN Convention on the Conservation of Nature provides the following: proposals of any activity which may significantly affect the natural environment shall as far as possible be subjected to an assessment of their consequences before they are adopted, and they shall take into consideration the results of their assessment in their decision-making process. As this ASEAN provision is implemented, common EIA procedures can be expected to emerge in Thailand, Indonesia, Singapore, the Philippines, and Malaysia, just as they have in Western Europe through the Common Market.

The United Nations Environmental Programme and the U.N. General Assembly have endorsed the use of EIAs by nations. The World Charter of Nature, adopted by the General Assembly, expressly calls for the use of EIAs. Article 206 of the Law of the Sea Convention provides for the use of EIAs, and agency practice among international organizations increasingly requires it.

As noted above, the World Bank has established procedures for EIAs. The various international organizations borrow from and adapt the state practice to an international organization's needs. The World Bank has adopted a six-step process, not dissimilar to the EIA process of Canada or the United States: (1) screening the proposal, (2) preparing an initial executive project summary, (3) preparing Terms of Reference for an environmental assessment, (4) preparation of the assessment, (5) review of the assessment and incorporation of its findings into the project, and (6) postproject evaluation. The latter provision is an enormously useful and innovative step; only a very few federal Canadian and Dutch EIAs have had monitoring to gage the effectiveness of the analysis and any impact mitigation measures. There is a dearth of postproject evaluation under NEPA. The United States can learn from experience as it develops under these EIA procedures. The World Bank will rely on the nation where the project is planned to implement these EIA procedures.

Thus, international organizations, as well as nations, are moving to reflect EIAs as a basic management tool. Between 1974 and 1986, the Organization for Economic Cooperation and Development (OECD) issued 11 recommendations encouraging the use of EIAs (Table 2). The "soft law" embodied in the resolutions of international agencies is in accord with the state practice reflected in national EIA law and practice.

What does this array of national and international practice tell us? First, there is a growing body of useful experience that deserves more empirical analysis. The United States needs programs such as those of the Canadian Environmental Assessment Research Council to consider how to improve our EIA process; perhaps the President's Council on Environmental Quality (CEQ) can undertake or stimulate this. Prior to 1980, the CEQ had begun to make such valuative studies (CEQ 1981). However, the EIA is too important to leave to unstudied evolution. Second, environmental professionals conducting EIAs can learn from and also derive encouragement from this growing volume of EIA work. Third, this worldwide EIA experience provides guidance on how jurisdictions relying on EIAs should cope with global, transnational, and transboundary environmental impacts. The EIA is emerging as a basic tool for restoring, maintaining, and enhancing environmental quality.

TRANSBOUNDARY, TRANSNATIONAL, AND GLOBAL EIA

Interesting procedural issues arise for EIAs when the impacts studied crossover jurisdictional lines or add incrementally to global environmental trends. In many jurisdictions, transnational impacts are routinely studied (e.g., under the EC Directive or in Canadian EIA federal practice). Global impacts, although harder to define, are also studied (e.g., additions of chlorofluorocarbon emissions impacting on the deterioration of the stratospheric ozone layer). For the reasons mentioned below, these aspects of EIAs are likely to expand in coming years.

NEPA's initial draftsmen anticipated the need for such analysis, although the U.S. federal agencies have done rather little to implement NEPA's mandate. In Section 102(2)(f) of NEPA, Congress directed that, "to the fullest extent possible," all agencies of the federal government shall "recognize the worldwide and long-range character of environmental problems and where consistent with the foreign policy of the United States, lend appropriate support to initiatives, resolutions,

Table 2. International EIA Provisions Cited.

1. **European Community (EC)**
The European Economic Community Directive on EIAs has an Annex I requiring a full assessment and an Annex II for an optional assessment under specified conditions. Council on European Communities Directive "On the Assessment of the Effects of Certain Public and Private Projects on the Environment," 85/337/EEC, O.J.L. 175, July 5, 1985.
2. **Association of South East Asian Nations (ASEAN)**
§14(1), ASEAN Convention on the Conservation of Nature and Natural Resources (1985).
3. **United Nations Economic Commission for Europe (UNECE)**
The Convention on Environmental Impact Assessment in a Transboundary Context was signed at Espoo, Finland, on February 25, 1991, UN Doc. E/ECE/1250.
4. **UNEP Regional Seas Conventions**
The following regional treaties on marine pollution: 1978 Kuwait Regional Convention (Art. 11); 1982 Jeddah Regional Convention (Article 11).
5. **United Nations Environment Programme (UNEP)**
See UNEP Governing Council, "Goals and Principles of EIA," June 17, 1987; Chapter C, 1981 UNEP Conclusions on Off-Shore Mining and Drilling, UNEP Working Group of Experts on Environmental Law, endorsed by the U.N. General Assembly in (Conclusion No. 8), of March 24, 1983.
6. **Organization for Economic Cooperation & Development (OECD)**
OECD (Paris) Recommendations on EIA include the following:
(1) *General EIA Recommendations*
1974 OECD Council Recommendations C(74)216 on Analysis of the Environmental Consequences of Significant Public and Private Projects (Paragraph (1)
(2) *Recommendations on Chemicals*
1974 OECD Council Recommendation C(74)215 on the Assessment of the Potential Chemical Effects of Chemicals (Paragraph 1)
1977 OECD Council Recommendation C(77)97 (Final) on Guidelines in respect of Procedures and Requirements for Anticipating the Effects of Chemicals on Man and on the Environment
(3) *Recommendations on Energy Production*
1976 OECD Council Recommendation C(76)162 (Final) on Reduction of Environmental Impacts for Energy Production and Use [Paragraph 2(5)]
1979 OECD Council Recommendation C(79)117 on Coal and the Environment (Paragraph 5)
(4) *Recommendations on Development Assistance*
1985 OECD Council Recommendation C(85)104 on Environmental Assessment of Development Assistance Projects and Programmes
1986 Draft OECD Council Recommendation C(86)26 on Measures Required to Facilitate the Environmental Assessment of Development Assistance Projects and Programmes
(5) *Recommendations on Exports of Hazardous Wastes*
1984 OECD Council Decision-Recommendation C(83)180 (Final on Transfrontier Movements of Hazardous Waste
1986 OECD Council Decision-Recommendation C(86)64 (Final) on Exports of Hazardous Wastes from the OECD Area
7. **World Bank**
Operational Directive 4.00, Annex A: Environmental Assessment, in Operational Manual for the IBRD, IDA, IFC, and MIGA.
8. **United Nations General Assembly**
World Charter of Nature, UNGA Res. 37.7 of Oct. 28, 1982 (Articles 11 and 16).
9. **UN Law of The Sea**
1982 UN Law of the Sea Convention, Article 206: "When States have reasonable grounds for believing that planned activities, under their jurisdiction or control may cause substantial pollution of or significant and harmful changes to the marine

Table 2. International EIA Provisions Cited (continued).

9. **UN Law of The Sea (continued)** environment, they shall, as far as practicable, assess the potential effects of such activities on the marine environment and shall communicate reports of the results of such assessments [at appropriate intervals to the competent international organizations, which should make them available to all states]."

and programs designed to maximize international cooperation in anticipating and preventing a decline in the quality of mankind's world environment."

Under authority of this provision, the CEQ itself has recognized the worldwide and long-range character of environmental problems in its CEQ annual reports, most especially in "Global 2000 Report to the President," submitted in 1979 (CEQ 1979). Much of the content in the latter report has subsequently been confirmed independently by the U.N. World Commission on Environment and Development in its report *Our Common Future* (UNWCED 1987).

The depletion of stratospheric ozone; gradual warming of the atmosphere; increasing loss of biological diversity; expanding desertification; relative rise of sea levels; and growing pollution, including effects of "acid rain," pose international environmental challenges. None of these trends can be resolved by single nations acting alone, and no country is immune from these problems no matter how good its own environmental protection programs may be. All these problems are caused by the worldwide accumulation of many discrete, isolated acts. The EIA is one of the few environmental management tools fashioned to consider such isolated actions and their cumulative impacts.

Section 102(2)(f) of NEPA expressly states that this worldwide perspective is conditional upon its being "consistent with the foreign policy of the United States." Since 1969, the content of U.S. environmental foreign policy has been modest and imprecise. However, where treaties create express obligations, as in the Convention on the International Trade in Endangered Species (CITES), the policy can be clear. In response to CITES and the U.S. Endangered Species Act, the U.S. Agency for International Development issued rules, for instance, which expressly require that foreign assistance programs consider how to protect endangered species. For most foreign policy questions, however, environmental protection has been subject to countervailing tendencies, and the inertia of past policies in the State Department and other foreign affairs agencies has tended to restrict advancing new environmental protection positions. Most of these past policies were framed with scant attention to trends in environmental degradation.

The most explicit U.S. foreign policy directives for EIAs abroad are provided for in Executive Order 12114. Promulgated in President Carter's Administration, Executive Order 12114 requires the use of EIAs under NEPA for actions on the commons (e.g., the oceans and Antarctica); when an action will affect uninvolved nations ("innocent bystander" situations); when an action is strictly regulated in the United States (e.g., actions involving radioactive materials or toxic substances); and/or when the president or secretary of state designates a natural or ecological resource to be of global importance.

This executive order is being considered for revision in the Bush Administration. It was promulgated following a series of court decisions that had applied NEPA to certain U.S. agency actions abroad. The cases dealt with U.S. aid to build part of the Pan American Highway in Panama, to spray herbicides on marijuana crops in Mexico, and other acts outside the United States. The foreign affairs agencies complained to the CEQ that EIAs would hinder their operations and sought to avoid letting disputes arise which might result in court decisions. The decision to issue the executive order tended to retard new litigation by clarifying foreign policy, but at the same time it embodies a compromise that has stifled further agency innovation under §102(2)(f) of NEPA. Some agency NEPA managers remain ignorant of the existence of Executive Order 12114 because it is not incorporated in their own agency's NEPA regulations. Moreover, the executive order falls short of the full requirements of §102(2)(f) and was at best only a partial step toward implementing the congressional mandate.

Since "the foreign policy of the United States" has traditionally not considered conservation or environmental policies to be a high priority, and Executive Order 12114 appeared to excuse agencies from trying to identify innovative ways to assess environmental impacts of government actions abroad, very little federal agency attention has been devoted to §102(2)(f). Exceptions, of course, exist. The U.S. Army has been distinguished by developing methodologies to comply with NEPA and Executive Order 12114; the Army's EIS for returning weapons, including chemical munitions, from Europe to the United States for dismantling is a useful example.

Notwithstanding the disuse of §102(2)(f) during NEPA's first two decades, NEPA could become an important tool in coping with global trends. As Secretary of State Baker told the IPCC, global climate change in the biosphere seems to present enough serious problems to warrant pursuing preventative measures. Climate modification, like stratospheric ozone depletion, is not the result of any one major act. It follows from many small, apparently innocent emissions of waste gases. To cope with climate change, a wide range of data must be assembled and analyzed. At the same time, the mitigation measures to curb unnecessary gaseous emissions are needed, and alternatives to achieve social objectives without emissions seems desirable. The EIA both assembles data and offers remedial measures. Since climate management has become a foreign policy objective, EIAs would appear to be a low cost and already available tool by which jurisdictions could begin to address climatic change.

For instance, analysis of completed environmental impact statements done pursuant to NEPA would provide substantial baseline data. Data on emissions of greenhouse gases could be discerned from a past EIS, and a future EIS could consider mitigation measures to curb emissions for a wide range of federal agency action (e.g., a review of lignite mining by the EPA, a Bureau of Indian Affairs review of a biomass electric generating facility, the loss of carbon fixing and carbon sequestration in a National Forest Service timber sale, or reduced carbon dioxide emissions achievable from an Urban Mass Transit Agency grant for a suburban railroad replacing automobile commuters). If climate managers existed, they might find that the Federal Energy Regulatory Commission (FERC) could serve useful foreign policy ends through FERC's NEPA reviews.

One particular aspect of NEPA could be especially useful in shaping EIAs as a foreign policy tool. This is the provision for tiering programmatic environmental impact statements with each more action-specific EIS. An agency could undertake a programmatic EIS for the range of possible climate modifications associated with one of its statutory mandates and thereby identify the actions which it might control either to produce data needed for better understanding climate change or to mitigate trends deemed to present adverse impacts on climate. The programmatic EIS would allow each action-specific EIS to focus efficiently on specific points keyed into the prior, generic analysis. Cumulative impacts could thus be examined systematically. Too few federal agencies today take NEPA's cumulative impact process seriously.

The EIA is perhaps the single best management handle for reaching the point of decisionmaking in whatever agency. If, for instance, a nation's foreign policy established goals to maximize tree planting in order to foster the photosynthesis process to fix carbon dioxide as a mitigation measure to stabilize climate, to promote biological diversity, and to avert desertification, each EIA could be directed to examine these discrete issues and shape decisions to advance these goals. State *little NEPA* procedures could be integrated into the same process; since in the U.S. federal system 40% of all energy decisions are implemented by state and local government, the state EIA procedures can be crucial to helping the United States achieve its possible foreign policy ends.

How quickly will foreign affairs agencies come to perceive this positive role for the EIA? In part, the answer depends on how successful we all are as EIA professionals in making the case for more effective use of NEPA and the §102(2)(f) role. In part, the answer may depend on Congress and on whether or how soon it enacts legislation mandating such measures. In part, it may depend on how rapidly consensus builds in the IPCC for a global climate treaty or similar measures.

The need for a more expanded use of the EIA in the case of global trends is already evident because the effects of biospheric change are being detected. The EIA can evaluate these physical manifestations and provide a practical tool for their consideration in concrete decisionmaking. The coastal area environmental assessments and impact statements must consider relative sealevel rise, not because any official orders such analysis but because the sealevel rise is a measurable fact. Acidification of lakes in Canada or New York's Adirondack Mountains has measurable impacts on biota. The EIA exposes such phenomena first to scientific quantification and second to policy scrutiny. The EIA is becoming international because scientific analysis increasingly can and does identify impacts that are transnational and even global.

Consensus has emerged regarding the uses of EIAs in a transboundary context. Through the U.N. Economic Commission for Europe, a negotiation has proceeded to prepare a "Convention on Environmental Impact Assessment Transboundary Context." This treaty's preparation began with a Seminar on Environmental

Impact Assessment in Warsaw, Poland in 1987 and culminated in signing a final agreement at Espoo, Finland on February 25, 1991. The parties agreed to "prevent, reduce, and control significant adverse transboundary environmental impact from proposed activity." To do so, an EIA is required for projects on a "List of Activities" included as Appendix I (e.g., oil refineries, thermal power stations, pipelines, ports, dams, large groundwater abstractions, large deforestations, new highways or long-distance rail lines, and airports). For all other activities that may cause such impacts, any party may request an EIA, and the parties must consult to determine if an EIA should proceed for the unlisted activity. Such other activities are to be examined in terms of their size, location, and effects.

The Convention provides an EIA process which includes (1) notification of the proposed activity to other states with "any available information on its possible transboundary impact;" (2) a response from other states indicating whether or not they will participate in the EIA process; (3) an exchange of sufficient information to evaluate impacts, with a nine-part recitation of the content of the EIA documentation specified in Appendix II to the treaty; (4) where consultations do not reach agreement on the nature of the impacts and their mitigation, guidelines enabling a party to request that an independent three-person inquiry commission be established to conduct its own EIA (the inquiry commission's final report "shall be based on scientific principles," a majority and any dissenting view shall be set forth, and the report shall be sent to all parties to the inquiry); (5) provisions for dispute settlement procedures, including a detailed arbitration process; and (6) provisions for post project analysis ("objectives include ... monitoring compliance with the conditions as set out in the authorization or approval of the activity and the effectiveness of mitigation measures ... review of an impact for proper management and in order to cope with uncertainties ... [and] verification of past predictions in order to transfer experience to future activities of the same type").

Once the state proposing a project completes its transboundary EIA, it must "provide to the affected party the final decision on the proposed activity along with the reasons and considerations on which it was based." Any relevant information arising thereafter must be immediately exchanged and consultations held as to whether the new information requires a change in the EIA.

The Espoo Convention on Transboundary EIA provides a useful model for interstate cooperation. Doubtless, close neighbors will wish to tailor the process with simpler, specific bilateral agreements on EIAs. States with comparable EIA systems may designate a single process, valid in each jurisdiction. As nations begin to ratify and use this treaty, a more routine use of EIAs in the transboundary situation will emerge.

What may retard innovations in the affirmative use of EIA procedures abroad is the legacy of bureaucratic suspicion of the process. In the United States, the State Department has given only relatively modest acknowledgment of NEPA. The Agency for International Development (AID) embraced the EIA process initially only under the pressure of federal court litigation. The Army Corps of Engineers learned the NEPA process after participating in numerous court actions, and now the military is often ahead of civilian agencies in knowing how to use NEPA. Foreign affairs personnel apparently have liked the administrative freedom of ignoring EIAs when it suits them to do so.

The diplomatic community needs education and training in EIAs. Outside forces are unlikely to compel a quick change in attitude. For instance, in the case of NEPA, it is clear that the courts do not lightly intervene in foreign affairs issues. The judiciary properly gives substantial deference to the executive branch in its decisions abroad. The CEQ has urged repeatedly that the foreign affairs agencies do more to evaluate impacts abroad, but the CEQ is advisory and does not direct foreign policy any more than it does domestic policy. However, the CEQ's advice has not been readily embraced. An express presidential directive on EIAs abroad, a stronger Congressional mandate, or both are needed to speed up the process.

The best evidence that EIAs will become a strong foreign policy tool is the fact that its use is increasing. The EIA is a valuable management tool. Its use abroad grows despite the absence of court orders, presidential intervention, or the sympathy of foreign affairs agencies.

EIA AND FORESIGHT

The EIA reflects the well-established duty under international law that each nation act so as not to harm the environment of any other nation. In order to avert harm, one must examine the consequences of one's actions and adjust as necessary. This rule of good neighborly relations is an ancient one.

The EIA can instruct how to protect the environment worldwide just as it has done so locally. The EIA is not a linear process, but a feedback loop. (1) We study and learn; (2) once informed, we strive to employ enlightened self-interest to (3) avoid or mitigate adverse environmental effects (4) so that we can monitor and evaluate our precautions and know better what to study or do in eliminating impacts the next time we take a similar action.

This dynamic system works in small settings, as in protecting a village water supply or preserving the isolated habitats of a migratory species, and also works well on a global scale as with the accumulation of many actions involved in climate stabilization. It is in each nation's best interest to foster the more efficient use of EIAs in all jurisdictions. One of the architects of NEPA, Professor Lynton Keith Caldwell, states the rationale for this national interest as follows:

> NEPA may be seen as a contrived, institutionalized answer to a people's recognition of its deficiencies. Through the impact assessment process written into law we compel ourselves, as participants in self-government, to do what we know should be done in undertaking actions that may have consequences not immediately apparent. The EIS process institutionalizes patience, caution, and looking before leaping. Few if any among the critics of NEPA would act in their personal affairs in the manner that government decision makers formerly acted in relation to the environment (Caldwell 1984).

Congress was inspired 20 years ago when it adopted NEPA. The ready, voluntary adoption of EIAs around the world is testimony to that congressional good sense. The EIA has moved from being an innovative experiment to becoming a staple tool of efficient decisionmaking. It carries on the torch that Roosevelt passed on back in 1908, urging foresight in the care for nature and for the needs of succeeding generations.

REFERENCES

Ahmad, Y., and G. Samny. 1985. *Guidelines for EIA in Developing Countries.* Hodden & Staughtor.

Anderson, F. 1973. *NEPA in the Courts.* Johns Hopkins University Press, Baltimore.

Baker, J. A., III. 1988. Economic growth & conservation: Partners not enemies. In V. Martin (ed.), *For the Conservation of the Earth.* Fulcrum, Golden, CO.

Caldwell, L. K. 1984. *Science & NEPA.* p. 150. University of Alabama Press, Tuscaloosa.

CEQ. 1979. The Global 2000 Report to the President. G. O. Barney, Study Director. Vols. 1 and 2. Government Printing Office, Washington, D.C.

CEQ. 1981. Environmental Trends. Superintendent of Documents. Government Printing Office, Washington, D.C.

Chien, E. 1991. Working Towards a Society with Comprehensive Environmental Protection Legislation in the 21st Century — The Experience of the EPA in the Republic of China. International Conference on Environmental Criminal Law.

Couch, W. J. 1988. *Environmental Assessment in Canada.* Canadian Council of Res. Environ. Min., Ottawa.

George, C. J. 1972. The Role of the Aswan High Dam in changing the fisheries of the Southeastern Mediterranean. and M. Kassas, Impact of river control schemes on the shoreline of the Nile delta. pp. 159–188. In M. T. Farvar, and J. P. Milton (eds.), *The Careless Technology.* Doubleday, New York.

Jin, R. L., and L. Wen. 1987. Environmental policy and legislation in China. p. 163. In D. Macraw et al. (eds.), *Proceedings of Conference on Chinese Environmental Law.* University of Colorado, Bolder.

Moreira, I. V. 1988. EIA in Latin America. pp. 239–254. In P. Wathern (ed.), *Environmental Impact Assessment Theory & Practice.* Unwin Hyman, Ltd., London.

Rich, B. 1985. The multilateral development banks, environmental policy, and the United States. *Ecol. Law Q.* 12:681.

Robinson, N. A. (ed.). 1989. Environmental Impact Assessment — Proceedings of a Conference on the Preparation and Review of Environmental Impact Statements. pp. 71–75. NYS Bar Association and President's Council on Environmental Quality.

Robinson, N. A. 1982. SEQRA's siblings: Precedents from little NEPAs in the sister states. *Albany Law Rev.* 46:1155.

Robinson, N. A. 1981. Introduction: Emerging international environmental law. *Stanford J. Int. Law* 17:229–244.

Roosevelt, T. R. 1909. Opening address by the president. In Proceedings of a Conference of Governors in the White House, Washington, D.C., May 13–15, 1908. Government Printing Office, Washington, D.C.

Seymour, W. N., Jr. 1971. Report of the Legal Advisory Committee to the President's Council on Environmental Quality. CEQ, Washington, D.C.

Shabecoff, P. 1989. Joint Effort Urged to Guard Climate. *New York Times*, Jan. 31, 1989, p. C7 (col. 1).

Warren, C. 1978. A look before we leap: Applying NEPA to U.S. actions abroad. Address at 3rd Annual Conference, National Association of Environmental Professionals, Feb. 8, 1978.

UNWCED (United Nations World Commission on Environment & Development). 1987. *Our Common Future*. Oxford University Press, New York.

NEPA's Impact on Environmental Impact Assessment in European Community Member Countries

R. Coenen, Department for Applied Systems Analysis, Nuclear Research Center, Karlsruhe, FRG

ABSTRACT

In 1985, 16 years after the passing of the National Environmental Policy Act (NEPA) and after many years of deliberations and negotiations, the Council of the European Communities (EC) adopted the so-called "EC directive on the assessment of effects of certain public and private projects on the environment." EC member countries had to incorporate the EC directive into their national legal systems by mid-1988. NEPA can be seen as the godfather of the EC directive. Because of the large difference in time between the passing of NEPA and the adoption of the EC directive, the experiences encountered in the United States could have been taken into consideration. This paper examines the extent to which this has been done by comparing the NEPA process as it has developed over time with the provisions of the EC directive, emphasizing focal aspects such as goals and functions of environmental impact assessment (EIA), areas of application, content, scoping, public consultation and consultation with other authorities, and decisionmaking. The general conclusion is that — with regard to an ideal concept of the EIA process — the EC directive is retrogressive compared with the NEPA process in many respects. However, since the EC directive concedes member countries a certain scope for regulating the EIA process, national differences in implementing the EC directive must be taken into consideration. The comparison of the planned EIA regulations in the various EC member countries shows that some member countries, especially The Netherlands, have by far exceeded the minimum requirements of the EC directive by implementing very ambitious EIA concepts.

INTRODUCTION

It took most member states of the European Community (EC) as many as 20 years after the enactment of the National Environmental Policy Act (NEPA) to introduce comprehensive environmental impact assessment (EIA) provisions. Even then, many countries introduced EIAs only because of obligations imposed by a directive of the European Community (Council of the EC 1985). This directive on EIAs was adopted by the EC Council in 1985, and member states were given a 3-year period to incorporate EIAs into national law. Many member states failed to meet this deadline.

However, in the early 1970s, EIAs had already become an issue on the political agenda in many member states (i.e., demands to introduce EIAs were voiced shortly after the adoption of NEPA). However, France was the only country to follow the example of the United States and in 1976 introduced comprehensive legal EIA provisions within the framework of the Nature Protection Act. It must be said, however, that the French EIA concept can hardly be regarded as ambitious and the efficiency of EIAs in France is viewed relatively critically.

In other countries, attempts to introduce EIAs failed. In 1974, a legislative proposal on EIAs for public projects was presented in the Federal Republic of Germany (FRG). This proposal, however, lacked essential elements of an EIA procedure, such as public consultation. The legislative proposal did not even reach the stage of parliamentary debate. The only remaining result was an administrative order on EIAs for federal projects which produced hardly any effect at all. Corresponding provisions were also implemented in individual states. However, they were as ineffective as those introduced on the federal level.

In 1976, Ireland introduced within the framework of the planning legislation EIA provisions that were, however, very restricted in their fields of application. All public projects were exempted, and most private development projects were not subject to EIAs either because of a financial threshold that was high by Irish standards. Moreover, the legal provisions gave the competent authorities discretion to require an EIA for projects exceeding this threshold. As a result, the legal provision had little effect.

In the United Kingdom (UK), there were also early initiatives to introduce EIAs that failed to reach the stage of a legislative proposal. However, in the UK, substantial EIA experience has been gained since the early 1970s (see, for example, Turnbull 1988). It started with onshore oil development in Scotland. Because of the complexity of these development proposals, the authorities required the developers to perform detailed impact studies for this new type of development. Furthermore, there was a tendency on the part of private enterprises to use the EIA instrument on a voluntary basis in order to avoid delays in planning and implementing projects. By consulting authorities, nature protection organizations, and the public at an early stage, they hoped to identify and clarify critical issues from the beginning and thus to reduce opposition in the permitting phase. The government supported such initiatives (e.g., by giving guidance) but refused to incorporate formal EIA provisions into the planning legislation.

Only The Netherlands has striven to implement an ambitious EIA concept since the mid-1970s. It was implemented by law in 1986. Legal introduction of EIAs was preceded by several pilot EIAs to test individual provisions in practice. In The Netherlands, legal provisions come very close to an ideal EIA concept.

The reasons that EIA introduction failed or produced little effect in most EC countries cannot be discussed in detail here. However, they can be summarized as follows (see, for example, Wathern 1988):

- It was argued that existing provisions already met the requirements of an EIA. This argument may, indeed, be justified to some extent. In the UK, Ireland, and Denmark, for example, planning legislation enables the competent authorities to impose comprehensive information obligations on the developer, which can include the preparation of an EIA. In the FRG, existing licensing procedures stipulate information requirements similar to an EIA study for certain types of projects.
- Further arguments that were advanced against a formal EIA procedure included unnecessary bureaucratic and administrative efforts, delays in licensing procedures, and increased cost.

It was only by the end of the 1970s that EIAs became an issue on the political agenda of the Mediterranean EC member states. The main interest of these economically less developed countries — in contrast to that of most of the northern member countries — was to promote economic development. Because there was no political pressure by environmental movements, economic concerns were given almost absolute priority.

The breakthrough for the introduction of EIAs in the member states was finally brought about by the EC directive on environmental impact assessment mentioned above. However, whether it is a real breakthrough may be called into question. The EC directive originates from the Action Program on Environmental Protection of the EC adopted in 1972, which marked the change from a merely repairing to a preventive environmental policy. To prepare a draft EIA directive, the EC Commission awarded several research contracts, among others to Lee and Wood of the University of Manchester. At that time, experience in the field of EIAs existed only in the United States. Therefore, the recommendations made by Lee and Wood were substantially influenced by the NEPA procedure. Recommendations included the following:

- Member states should establish an institution to elaborate guidelines for the field of application and the performance of EIAs. Obviously this proposal used the President's Council on Environmental Quality (CEQ) guidelines as a model.
- Similar to the American example, a draft and final EIS were recommended, the latter taking into consideration the comments received in the consultation phase.
- In accordance with the American example, the draft and the final EIS were to be elaborated by the competent authority. This provision should apply at least during the phase of introduction.
- The consideration of alternatives was also regarded as an essential requirement.

Contrary to NEPA, the EIA was to be integrated, if possible, into existing procedures according to the recommendations of Lee and Wood. Furthermore, the recommendations included monitoring of sample EISs.

Based on the preparatory work of Lee and Wood, the commission elaborated a first proposal of the directive which was then discussed in detail by a group of experts from member countries. These expert meetings had already condensed the original concept. In 1980, the commission submitted an official draft of the directive (Commission of the European Community 1980). After 5 more years of deliberations within the bodies of the commission, the directive was adopted in 1985. In general, negotiations within the EC bodies are governed by national interests of the individual member states. In most cases, these negotiation result in very weak compromises that dilute the intended effects. This is also true of the EIA directive. However, the meager result of this compromise was partly compensated for by a provision of the directive (EC directive, Art. 13) that allows member states to impose stricter requirements. As will be explained in more detail, some countries have made only very modest use of this possibility.

As a result of the weak compromise, the EC directive provides for three procedural steps only:

1. preparation of the EIS by the developer
2. comment on the EIS by other authorities and the public after the permitting procedure has been initiated
3. consideration of the EIS in the decisionmaking process by the competent authority

Procedural steps such as screening, scoping, revision of the draft EIS after the commenting phase, and monitoring are not required.

LEGAL AND PROCEDURAL IMPLEMENTATION OF THE EIA

In this section, the different ways in which individual member states have implemented the directive are discussed. Because it is not possible to discuss all the details here, only major aspects will be dealt with. (For a more detailed presentation of EIA implementation in the EC member countries, see Coenen and Jörissen 1989.)

EC directives are not directly applicable law but must be incorporated into national law through appropriate measures. Because of different national legal systems, member states are given much discretion with regard to the form of incorporation of directives. It must be ensured, however, that the objectives of the respective directive are fully accomplished. Most member states have introduced the directive on the legislative level either by a separate law or by amending existing general or sectoral laws. Exceptions are the UK and Ireland, which have implemented the directive beyond legislative level by regulations under relevant laws. In general, it can be stated that introducing EIAs by a separate law, as it is envisaged in Belgium, Italy, and Spain, may raise the political status of an EIA as a preventive measure and can have a signal effect on environmental policy.

The introduction into a general environmental law, as provided for in France, Greece, The Netherlands, and Portugal, can also generate such an impetus. However, this may not be true of the approaches of Ireland and the UK as well as the FRG which incorporate EIAs by amending sectoral laws.

Whether an EIA is introduced as a separate procedure or is integrated into existing procedures appears to be more important for the efficiency of the EIA instrument than the form of legal incorporation. According to evaluations of American EIA experts (see Paschen 1989), the undisputed success of the EIA in the United States results partly from the fact that it has altered the standard procedures of decisionmaking of public authorities (e.g., by involving new actors in the procedures or by strengthening the influence of authorities already involved and thus increasing control over the deciding authorities).

According to the directive [EC directive, Art. 2(2)], member states can choose to integrate EIAs into existing administrative procedures or to establish new procedures. The fact that there are only few procedural requirements imposed by the EC directive favors procedural implementation by integrating EIAs into existing procedures. Thus, most member states have opted for integrating EIAs into existing procedures, with the result that compared with those of the United States, the standard procedures of administrative decisionmaking have been changed only to a very limited extent.

In some countries, the EIA procedure factually consists only of preparing an EIS, which then has to accompany the project application in the subsequent phases of the decisionmaking process. There are no new actors, and the position of those already involved is not strengthened at all or only to a limited extent. The EIA becomes a mere supplement to existing procedures. Obvious examples for such a strategy are the FRG and the UK, where a minimum number of procedural changes have been explicitly defined as the guiding criteria for an optimal implementation of an EIA.

The Netherlands, however, introduced the EIA as a separate procedure, despite the fact that new projects are subject to numerous sectoral procedures anyhow. The EIA procedure is used to coordinate the different procedures in order to avoid delays and unnecessary repeating of analyses and procedural steps and thus to streamline the permitting process for projects requiring several licenses. The southern countries, Greece, Italy, and Portugal, also introduce EIAs as a separate procedure preceding the actual licensing procedure.

AREA OF APPLICATION

The EC directive [EC directive, Art. 2(1)] requires that projects which according to type, size, and location may have considerable effects on the environment are made subject to an EIA. Regarding the area of application, there are two main differences to NEPA. First, the EC directive restricts the area of application on the project level. Second, it determines the respective projects in two annexes. Thus, there is no obligation to subject other projects not listed in the two annexes to an EIA. This may be critical in some cases (e.g., in the case of projects which normally have only minor impacts but may produce significant impacts if they are located in sensitive areas).

Whereas projects of Annex I — which includes projects with obviously major environmental impacts, such as large power stations, refineries, large chemical complexes, nuclear installations, waste disposal installations for toxic and dangerous wastes, and motorways — mandatorily require an EIA, member states have some discretion to determine which of the projects listed in Annex II are subject to an EIA or to establish criteria or thresholds for this purpose (EC directive, Art. 4).

The EC member countries have chosen different ways of implementing EIAs for Annex II projects. Belgium follows an approach similar to that of NEPA and will decide on the necessity of an EIA for Annex II projects on the basis of a brief assessment similar to the environmental assessment (EA) in the United States. Other countries, namely Greece, Italy, Portugal, and Spain, will subject some Annex II projects to an EIA, and the FRG will subject the majority of Annex II projects to an EIA.

The Netherlands has established binding thresholds for Annex II projects. But these thresholds are so high that, at present, only large-scale Annex II projects will require an EIA. France, on the other hand, has established very low thresholds, with the result that 4000–5000 impact studies are performed each year. Furthermore, simplified assessments (notice d'impact), that is to say, small EIAs are required for projects with minor environmental impacts. Ireland and the UK decide on the necessity of an EIA for Annex II projects on a case-by-case basis using thresholds or other criteria, however, which are not binding. This means that exceeding these thresholds does not automatically trigger the need for an EIA.

RESPONSIBILITIES IN THE EIA PROCEDURE

As in the NEPA process, the EC directive also requires the competent authority to be responsible for the performance of the EIA procedure. Contrary to what had been envisaged originally, the EIS must be prepared by the developer (EC directive, Art. 5). However, this does not differ conceptually from the provision of the NEPA process in which the EIS must be prepared by the competent authority, since NEPA had been originally designed for public actions on the federal level only. Thus, the competent authority and the developer were identical. A difference from the EC provision exists only because the area of application of the NEPA process was extended by litigation. The courts have made private projects and projects of state and local governments that are financed by federal authorities or that require a federal license or any other federal involvement subject to the NEPA process if potentially significant environmental impacts are expected. The courts have argued that granting a federal license or other federal involvement must be considered a major federal action. Consequently, in accordance with NEPA provisions, the EIS must be prepared by the authority that has to take such major federal actions rather than by the developer.

There has been a controversial discussion of the provision of the EC directive requiring the developer to prepare the EIS. There are arguments in favor of and against this provision. On the one hand, the developer has better access to important parts of the information required (i.e., the project-specific information) because he is in charge of planning the project and, as a rule, has the technical expertise.

A further advantage, generally considered more important, is the fact that the developer is forced to deal intensively and in detail with the environmental effects of his project if he has to prepare the EIS. This may encourage him to give greater consideration to environmental concerns when planning his project. On the other hand, it cannot be ruled out that the EIS will be biased if it is prepared by the developer. The developer's impact statement may give a more positive picture of his project than is justified in order not to jeopardize the planned implementation. Furthermore, as converging interests of the developer and the development-oriented licensing authority cannot be excluded in certain cases, it has been argued that the EC directive does not provide for an independent review of the EIS by a neutral body as is the case in the United States where such functions are fulfilled by the U.S. Environmental Protection Agency (EPA) and by the CEQ to some extent.

Although not required by the directive, several EC countries have provided for a review of the EIA by an independent neutral body or have implemented other provisions to counteract such interest-oriented EISs. The Netherlands and Belgium have established new bodies for an independent review of the EIS and the supervision of the whole EIA process. In addition, Belgium provides for the preparation of the EIS by a neutral state-recognized organization or person, whereas the developer has mainly the function of supplying relevant information.

In other countries, namely Denmark, Greece, Italy, Portugal, and Spain, the function of carrying out an independent review lies with the ministers responsible for environmental affairs. Only the FRG, Ireland, and the UK do not provide for any form of independent review. In France, the Minister of the Environment has the right to review any EIS; however, he rarely makes use of this power.

DEFINING THE SCOPE AND CONTENT OF THE EIA

In general, the EC directive — like the NEPA regulations — defines minimum requirements regarding the content; these regulations are specified in more detail in an Annex III to the directive. However, these specifications still have the character of minimum requirements because it was not considered possible and appropriate to develop a detailed catalog of requirements applicable to very different classes of projects.

The EC directive does not require scoping as a separate formal procedural step to define the scope and content of the EIS, although it has been demonstrated in the United States that a scoping process involving other authorities and the public may improve the efficiency of the EIA procedure by focusing the analysis on the relevant issues and thus avoiding unnecessary analyses and paperwork. It is assumed that by early identification and discussion of critical issues scoping can reduce resistance to planned projects and avoid delays in the permitting procedure. Therefore, scoping was introduced in the United States with the regulations of 1978 (40 CFR).

It is remarkable that one of the main arguments advanced in EC member countries against the introduction of a formal scoping process was the possible delay of already lengthy permitting procedures, whereas in the United States this argument was brought forward to support the implementation of a scoping process. In the FRG, it was also argued that without binding and detailed requirements regarding the content of the EIS the principles of legal security and equal treatment for the developer could be violated. The applicant should know from the beginning the requirements imposed upon him. Furthermore, it would not be tolerated if, as a result of scoping, different requirements regarding the content of the EIS were demanded for comparable projects from the respective developers. Therefore, in the FRG and in some other EC countries, there is a tendency to develop detailed binding catalogs of issues to be dealt with and methods to be applied in the EIS. If negotiations about the content are envisaged at all, they should take place only between the competent authority and the developer. In other EC countries, there is a more positive approach toward scoping. The Netherlands has introduced a scoping process that provides for the participation of other authorities, the public, and the established neutral EIA commission. Other countries recommend scoping with involvement of other authorities and the public on a voluntary basis.

There is another important difference between NEPA and the EC directive with regard to the analysis of alternatives to the proposed project. Whereas in the NEPA process consideration of alternatives is regarded as an essential requirement, the EC directive attributes only minor importance to this issue. According to Annex III of the directive, only an overview of the main alternatives studied by the developer and an indication of the main reasons for his choice, taking into account the environmental effects, should be given when appropriate. As a result, the developer has no obligation.

In the United States, the study of alternatives is done in only a cursory manner in many EISs just to fulfill the legal requirements. The EC directive, on the other hand, has nearly given up the requirement to consider alternatives altogether, although for some types of projects (e.g., infrastructure projects) a thorough analysis of alternatives seems to be appropriate for many reasons. Some EC member countries have recognized this necessity, especially The Netherlands, where the no-action alternative has to be analyzed in any case as well as the alternative that makes optimal use of all opportunities to protect the environment.

INVOLVEMENT OF OTHER AGENCIES IN THE EIA PROCEDURE

The cross-sectoral approach of EIA stringently requires the involvement of all authorities that are affected by the respective project or that have relevant environmental expertise. Uncoordinated planning and decisionmaking were a main motivation for Congress to enact NEPA. Therefore, the NEPA process provides for the involvement of other authorities in all phases of the EIA procedure. This early involvement ensures that other agencies contribute their special knowledge during the planning phase of a project and that the developer and competent authority can make use of it. However, the EC directive does not meet this obvious requirement of early involvement of other authorities. Consultation with other authorities is not required until the application is submitted together with the EIS.

Although it was initially intended, the EC directive does not provide for a draft EIS that after the consultation phase has to be revised and transformed into the final EIS, taking into account comments on the draft. Thus, before the final decisionmaking, the other authorities have no control over the competent agency's use of the comments received. Under these circumstances, agencies are not motivated to comment thoroughly and in detail, and commenting may become an onerous obligation that is not seriously fulfilled. However, some EC member countries will implement consultation requirements that go far beyond those of the EC directive. This is especially true of The Netherlands, which provides for continuous consultation in all phases of the EIA process. Other countries subject the EIS only to the usual consulting processes of the existing administrative procedures, which usually means that consultation does not take place until the application for a project is submitted.

PUBLIC PARTICIPATION

The provisions regarding public participation are also very restrictive compared with those of the NEPA process [EC directive, Art. 6(2)]; it is not mandatory before the application is submitted together with the EIS. Furthermore, the EC directive adopts a funnel-shaped model of public participation; the EIS is made available for the general public, but the opportunity to comment is limited to the public concerned. Thus, the EC directive attributes to public participation primarily the function of preliminary protection of legal rights of those possibly affected. However important, this should not be the only function of public participation in the EIA procedure, especially since legal rights should already be guaranteed by the participation provided for in the existing permitting procedures.

The directive ignores other functions of public participation in the EIA procedure. It can be a valuable source of special site-specific information and of potential social impacts. It can help to reduce delays in the permitting procedures and to promote public consensus on planned developments. However, the fulfillment of both functions presupposes that public participation starts as early as possible and is organized in a flexible manner. Only early and flexible participation opens up opportunities for the developer and the competent authority to make use of the special site-related information of the public in an appropriate manner and ensures that controversial issues can be identified and clarified right at the beginning.

The experience in the FRG, in particular, demonstrates that late and very formal public participation does not reduce public opposition but stiffens conflicts and reinforces positions. Because major alterations of a project design are unlikely at this stage, public participation is perceived mainly as an exercise to fulfill legal obligations. Therefore, the fact that the planned provisions of the FRG regarding public participation meet only the minimum requirements of the EC directive deserves particular criticism.

Nearly all other EC countries, however, implement participation provisions that exceed the minimum requirements of the EC directive by providing for public participation at earlier stages of the EIA procedure (e.g., during scoping). Furthermore, most countries, with the exception of the FRG and Spain, do not restrict commenting on to the affected public.

LINKAGE OF THE EIA AND DECISIONMAKING

The EC directive and NEPA regulations regard the EIA as a procedure to improve the information bases of the competent agency. The EIA procedure should ensure that the permitting agency makes its decision in full knowledge of the potential environmental impacts of the planned project. In other words, the EIA should enable the competent authority to appropriately consider environmental concerns in the balancing process. Like NEPA, the EC directive does not give priority to environmental concerns. However, it must be ensured that the discretionary decision of the competent authority can be reviewed. Therefore, the NEPA process provides for a disclosure of the balancing process by a record of decision. In this respect, the EC directive (EC directive, Art. 8) requires only that the EIS and the comments received be taken into account in decisionmaking and that the decision, as well as the reasons and considerations on which the decision is based, must be made public. The latter is necessary only if the legal provisions of the member states so provide (EC directive, Art. 9).

Thus, the linkage of the EIA with decisionmaking is comparatively weak. As a result, the competent authority may make insufficient use of the results of the EIS or may not take them into consideration at all. Most member states tried to counteract this danger by establishing a closer linkage between EIAs and decisionmaking. As in the NEPA process, this was done primarily by introducing special decision documents in which the authority must explain in detail the influence the results of the EIA had on the decision taken.

The EIA procedure has a prejudicial effect in the southern member states of the EC. This applies especially to Portugal, where the permitting procedure is not initiated if the minister who is responsible for environmental affairs gives a negative judgment on the environmental compatibility of the respective project on the basis of the EIS. In Italy and Spain, the decision is taken by the Council of Ministers in case of disagreement on the environmental compatibility of the project between the Minister of the Environment and the minister in charge of the permitting procedure. In Greece, the ministry responsible for environmental affairs and the ministry that grants permission for the project have to make a joint decision on the environmental mitigation measures that the developer has to implement.

The fact that the EIA procedure has a prejudicial effect in some countries can perhaps be explained by the comparatively less developed environmental legislation.

CONCLUSIONS AND EVALUATION

In summary, the EC directive, compared with the NEPA process, represents only a rudimentary EIA concept, although initial drafts leaned fairly closely upon the NEPA model. However, as mentioned before, the directive is a result of compromise between the member countries. Some countries may be satisfied with this compromise because they can easily implement the EC directive in their standard procedures. However, Henk Brouwer from the Dutch environmental ministry, one of the promoters of an ambitious EIA concept in The Netherlands, has described the compromise as "more the result of the cumulative resistance from the development promoters and bureaucracies in the member countries than a synthesis of the best ideas for the protection of the environment" (Brouwer 1986). However, some member countries have made use of the opportunity offered by the directive (EC directive, Art. 13) to implement stricter requirements. At present, an evaluation of the EIA provisions of the various member countries cannot be based on experiences but only on concepts.

There is no doubt that the implementation strategy pursued by The Netherlands comes closest to the ideal concept of an EIA. It is characterized by introducing the EIS on the legislative level and as a separate procedure, by an early and continuous consultation of other authorities and the public, by a formal scoping process, by an adequate consideration of alternatives, and by an external independent review of the EIA. Furthermore, monitoring is provided for. The introduction of the EIA was preceded by extensive practical tests. Although The Netherlands had a comprehensive sectoral environmental law before, introduction of the EIA was considered necessary in order to counteract the deficits of a sectorally oriented environmental policy and to guarantee a comprehensive cross-sectoral evaluation of the environmental effects of projects. Furthermore, the EIA was regarded as an instrument to streamline and harmonize existing administrative procedures. It must be stated, however, that for the time being the area of application of the EIA is very limited as a result of high thresholds for projects of Annex II to the EC directive. In view of the environmental problems caused by Dutch agriculture, it is especially remarkable that installations for large-scale animal rearing and food production are generally exempted from the EIA procedure, although these types of projects are listed in Annex II to the directive.

The member states of southern Europe — Greece, Italy, Portugal, and Spain — also pursue relatively ambitious strategies of implementation. Their EIA concepts are characterized by giving the ministries responsible for environmental affairs a very powerful role in the EIA process. In such cases, the EIA can influence more than the permitting conditions for environmental protection. It may even determine whether the project proceeds at all.

However, in reality seem to be two deficits in these countries. On the one hand, a deficit of regulation and enforcement in the area of environmental law and, on the other hand, an unsatisfactory consideration of environmental concerns in administrative decisionmaking. These deficits are expected to be offset by ambitious EIA concepts and by strengthening the position of the ministry responsible for environmental affairs. However, the lack of qualified technical and scientific personnel in the field of environmental protection in industry and administration is considered a major bottleneck for an efficient implementation of the EIA provisions. Therefore, EIA-specific training programs are provided for, but it will require 5–10 years for them to produce the necessary qualified manpower.

Furthermore, in view of the low level of industrial and infrastructural development of these countries compared with those of most northern EC member states, one has to raise the question of whether the ministries responsible for environmental affairs will be able to use their formally strong position resulting from the new EIA provisions to ensure an appropriate consideration of environmental concerns in the reality of administrative decisionmaking. Belgium is also planning to implement ambitious EIA concepts. However, an overall evaluation is not yet possible because existing regulations are still fragmentary.

The EIA provisions of Denmark, the FRG, the UK, and Ireland do not exceed the minimum requirements of the EC directive. These countries take the view that the existing permitting procedures for projects with environmental impacts already meet the requirements of an EIA. As a result, they implement EIA provisions that are characterized by a minimum of procedural and institutional change: the EIA is integrated into existing procedures; the competence of authorities is not challenged at all; external review of the EIA process is not provided for; and a scoping process is, if anything, only recommended. Essentially, the EIA process is reduced to a report by the developer, which must be taken into consideration by the competent authority as any other application document would be. Thus, the efficiency of the EIA procedure will depend ultimately on the commitment of the competent authority to the environmental goals and principles of EIA. If they regard the EIA more as an onerous obligation, the results will be correspondingly poor.

France may serve as an example of a country in which the EIA was introduced in 1976 in a similar way (i.e., with a minimum of procedural and institutional change). The efficiency of the EIA procedure in France is rated fairly low. As practical experience shows, the quality of EISs strongly varies from area to area and from authority to authority. This clearly demonstrates that the results of the assessment procedure depend largely on the commitment of the competent authorities to the environmental goals of the EIA.

REFERENCES

Brouwer, H. 1986. Experience in implementing environmental impact assessment in The Netherlands. Paper to the CEMP Conference on the EEC Environmental Assessment Directive, January 30–31, 1986, London. p. 201. In P. Wathern (ed.), *Environmental Impact Assessment — Theory and Practice.* Unwin Hyman, London.

Coenen, R., and J. Jörissen. 1989. Environmental Impact Assessment in the Member Countries of the European Community — Implementing the EC Directive: An Overview. KfK 4507 B, Kernforschungszentrum Karlsruhe, 7500 Karlsruhe FRG.

Commission of the European Communities. 1980. Proposal for a Council Directive Concerning the Assessment of Environmental Effects of Certain Public and Private Projects. Official Journal of the European Communities, c 169, 9.7.1980, 14–17.

Council of the European Communities. 1985. Council Directive of June 27, 1985, on the Assessment of the Effects of Certain Public and Private Projects on the Environment. Official Journal of the European Communities, No. L 175/40.

Lee, N., and C. M. Wood. 1976. Die Einführung von Ummweltverträglichkeitser klärungen in den Europäischen Gemeinschaften. Env/197/76-D. EEC Document.

Paschen, H. (ed.). 1989. The role of environmental impact assessment in the decisionmaking process. pp. 89–102. In *Proceedings of an International Workshop held in Heidelberg, FRG, August 1987.* Erich Schmidt Verlag, Berlin.

Turnbull, R. 1988. Environmental impact assessment. pp. 17–29. In M. Clark, and J. Herington (eds.), *The Role of Environmental Impact Assessment in the Planning Process.* Manselt Publishing, London and New York.

Wathern, P. 1988. The EIA directive of the European community. pp. 192–209. In P. Wathern (ed.), *Environmental Impact Assessment — Theory and Practice.* Unwin Hyman, London.

Implementation of NEPA and Environmental Impact Assessment in Developing Nations

J. F. McCormick, The University of Tennessee, Knoxville, TN

ABSTRACT

The National Environmental Policy Act (NEPA) and the process of environmental impact assessment (EIA) have undergone significant improvements during the past two decades. Implementation of NEPA-type legislation and EIA has spread globally. Environmental assessment is a multibillion-dollar international profession. Currently, EIA is a procedure in economic development planning which makes it possible to save money, save time, avoid significant mistakes, and produce a superior product. In this process, environmental information is obtained by using rigorous scientific methods. This information and fundamental scientific principles are used to predict consequences of alternative policies or projects. A variety of planning methods are used to evaluate consequences of alternatives. Evaluations are presented in a concise and clear form to decision makers.

Nowhere is EIA more important to environmental quality, human health, and survival of civilization than in developing countries. It is in the developing countries that the world's future is being determined. This is where most of the world's population resides, where population is increasing most rapidly, and where natural resources are being devastated at an alarming rate. If we wish to put "ecology into action" (MAB-UNESCO 1981), the action is in the developing countries.

As developing countries undergo a transition from agrarian to industrialized societies, traditional impacts of intensive agriculture persist, while new impacts of industrial and urban expansion are being added. Both agriculture and industry are dependent upon natural resources. Sustained economic productivity is dependent upon sustained ecological productivity. Government leaders in most countries realize that sustained economic productivity is necessary to sustain the support of their people and their government. Unfortunately, too few leaders of governments understand the underlying dependency upon sustained ecological productivity. Good EIAs go beyond superficial economic assessments and reveal fundamental ecological interdependencies that govern the sustainability of economic and political systems. During the design of an agricultural development project in Central Africa, the president of Rwanda described the situation to us most eloquently, 'If in the economic struggle for human survival, we jeopardize the sustained productivity of our natural resources, the struggle will have been in vain.' A good EIA is one means of ensuring the struggle will not have been in vain.

CONSTRAINTS TO IMPLEMENTATION OF A GOOD EIA IN DEVELOPING COUNTRIES

In most developing countries, the greatest constraint to attaining the benefits of a good environmental impact assessment (EIA) is education (Braun 1987). Most urgent is the need to train people to conduct EIAs. Of related importance is environmental education sufficient to ensure the participation of a concerned and informed public. Recognizing education as the factor most limiting environmental quality in the developing countries, The University of Tennessee joined others over the past decade to provide international training in ecological science, resource management, and EIAs. Another constraint is that in some developing countries there is a strong temptation to disregard or delay an EIA in the false belief that the EIA will delay or stop development or increase costs. This idea may be the consequence of misguided efforts to implement the National Environmental Policy Act (NEPA) in the 1970s. Too often, the EIA process was confrontational rather than cooperative. Too often, NEPA and other environmental legislation have been abused by those seeking to stop development for reasons unrelated to environmental quality.

In developing countries, the no-action alternative itself may be a negative impact. In the context of a rapidly growing and poverty stricken population, no action most certainly will have negative physical, biological, and social impacts. In developing countries, the appropriate role of an EIA is to improve development, not to stop or delay development. Disregard for the no-action alternative necessitates acceptance of a greater number of unavoidable impacts. Under these conditions, creative mitigation becomes all the more essential.

In Indonesia, there is a strong emphasis upon mitigation and total project benefit/risk assessment (Soemarwoto 1984). If cumulative benefits exceed cumulative risk, the project is implemented even if there are individual high-risk components of the project. Considerable effort is devoted to mitigation of the high-risk components. China is another nation which, at present, cannot accept the no-action alternative. The Chinese cannot afford to constrain economic development, nor can they afford the pollution-control technologies being implemented in the West. So long as economic growth continues to increase at 9–10% per year, the Chinese have good reason to believe they can implement improved pollution control in the near future when their economy is as strong as that of Western nations. In the meantime, the most common form of mitigation is site selection. Pollution control focuses upon toxic chemicals (Russell 1987) rather than microbes or other contaminants that can be removed by boiling or other household procedures.

It is important to realize that in developing countries major environmental issues are issues of natural resource management rather than issues of pollution, which are predominate in the United States. Gus Speth, President of the World Resources Institute, has pointed out that the United States only recently has begun to acknowledge these differences and adopted a "global environmental agenda." Today, according to Speth (1986), the ten most serious global environmental issues are as follows:

1. loss of agricultural land to desertification, erosion, and nonfarm use
2. depletion of tropical forests and consequential erosion, flooding siltation, and changes in global climate
3. a greatly accelerated rate of species extinction
4. rapid population growth and associated urbanization
5. mismanagement and shortages of fresh water
6. overfishing, habitat destruction, and pollution of marine environments
7. hazardous waste management
8. climate change resulting from burning fossil fuels
9. acid rain and other forms of air pollution
10. mismanagement and shortages of energy fuels

Some of these problems are attributable to carelessness and excessive use of resources and technologies in industrialized countries, but most are problems of resource management arising from poverty and population growth in developing countries. More recently, former U.S. Senator Albert Gore (1989) has called for a "sacred agenda" that assigns environmental issues highest priority for resolution in bilateral and multinational conferences and summit meetings.

An ever-present problem in developing countries is limited availability of capital to finance economic development. As a result, even the relatively low cost of EIAs (0.2–1% of project development) may seem excessive. A significant proportion of development costs is borne by international, often multinational, institutions with historical disregard for EIAs. Recent efforts of a few international institutions to require that EIAs accompany economic assessment are extremely significant. Unfortunately, these efforts to improve development planning are often resentfully interpreted by developing countries as interference in domestic affairs.

A common strategy of developing countries is to remind industrialized nations that during their period of economic development they ravaged forests, mined without reclamation, and dumped hazardous wastes with impunity. Industrialized nations can now afford to rectify environmental problems created in the past. Developing countries are willing to accept this model of delayed mitigation provided by the United States and other industrialized nations. Their message is, "Just be patient! You experienced your industrial and economic revolutions, and later you began to clean up your act. Do not constrain our opportunity for economic growth, and we will follow your example as soon as we can afford to do so." When domestic situations make it difficult for developing countries to implement EIAs, international incentives or constraints may provide the only pressure to implement EIAs. Therefore, such pressure should be maintained rigorously.

Insufficiency of baseline information is another serious problem throughout the world. It is impossible to identify, measure, or evaluate impacts without comparative descriptions of nonimpacted baseline conditions. This is an especially serious problem in developing countries. A baseline condition is not a constant; it is a variable. Ecosystems are dynamic systems. Living organisms and their physical environments constantly interact and modify one another. A period of several years is required to monitor limits of natural variability in the physical environment and in ecosystem performance. In the United States, we only recently have implemented the long-term monitoring and long-term research programs needed to provide baseline information. Few developing countries have the means to develop such programs.

Another scientific constraint is our limited understanding of ecosystem resiliency. It is relatively easy to predict ecosystem damage resulting from most disturbances. It is much more difficult to predict ecosystem resiliency or recovery following disturbances. We must understand both damage and recovery in order to predict or describe impacts upon ecological systems. For example, most contemporary models of atmospheric carbon dioxide and global warming fail to account for natural recovery processes following the cutting of tropical forests (Detwiler and Hall 1988). Consequently, these models contain considerable error. Underestimates of ecosystem resiliency probably contribute to overestimations of environmental impacts.

A related scientific problem in developing countries is our inability to predict impacts upon ecosystems with a long history of occupation. Human occupation of North America has been of such sufficiently short duration, low intensity, or limited extent that some examples of most natural ecosystems persist in a mature state of development in equilibrium with their environment. On the other hand, human occupation of North America has been of sufficient duration, intensity, and extent to result in the disturbance of most types of ecosystems. We have opportunities to observe rates and patterns of recovery of disturbed ecosystems, as well as opportunities to observe their undisturbed counterparts. Consequently, our knowledge of ecosystem development includes an understanding of starting conditions, final conditions, and the way in which site-specific factors influence rates and pathways to equilibrium.

In eastern China, the duration, intensity, and extent of human occupation and land use have obliterated our knowledge of beginning and ending conditions of ecosystem development. Contemporary forest ecosystems lack original species, contain exotic species, and have developed under a variety of management practices. Undisturbed natural forests essentially are nonexistent. Occasional old-growth forests are invariably associated with the remains of Buddhist temples. Old trees owe their longevity to protection by monks. However, these same monks significantly altered species composition of forests by replacing native species with exotics. It is exceedingly difficult to build predictive process models without descriptions of initial or terminal stages of the process.

Implementation of EIA is quite variable in developing countries. Impressive environmental institutions and laws exist as a matter of record. Often, they are supported by funding from a variety of international organizations, but seldom do representatives of those organizations travel to remote areas to observe implementation of EIAs or other environmental programs. I participated in an economic assistance program in central Africa in which environmental assessment and mitigation were significant components. These requirements were inserted by well-intended public servants in Washington, D.C. However, overseas regional officials of the same agency felt they best understood the regional context in which the assistance program was being implemented. As a result, the regional office modified objectives and procedures of EIA and mitigation. Next, the senior public servant residing in the host country expressed his disdain for the wisdom of the regional office and made it quite clear that nothing happens in *his* country without *his* approval. Entirely new EIA procedures and mitigation were proposed. They were based upon personal and professional relationships between the local United States public servant and host country public servants. Regardless of the success of the projects, they scarcely resembled those agreed upon in the contract with the Washington, D.C. office of foreign assistance. At no time did a Washington-based, regionally based, or local U.S. public servant venture to the field to observe EIAs or mitigation projects. To this day, the economic assistance agency probably has three different interpretations of what was accomplished, and, most likely, none has been verified.

Enforcement of environmental laws and policies is also quite variable. In China, all development projects are, of course, government projects. If a project requires an EIA, there is little objection. Public support and participation are impressive. For example, in central China (Wang et al. 1984) it was necessary to monitor air quality over a large geographic area of extensive coal burning. The government simply assigned 10,000 people the task of taking simultaneous measurements at prescribed time intervals. Before the Sandinista revolution, we conducted EIAs of water development projects in Nicaragua. We presented our recommendations for mitigation efforts directly to President Somoza. Within 1 hour, they were accepted; within 1 month, they were initiated. They were rigorously enforced. There were no public objections, unless one considers environmental policy or its mode of implementation to be among the issues that sparked the Sandinista revolution.

In many developing countries, environmental standards are not enforced as they are in the United States. Standards are considered long-term goals.

Improvement from 10% compliance to 40% compliance is greeted with celebration rather than justification for terminating project operation. Perhaps because of questionable law enforcement, some developing countries seek compliance more through incentives than through punitive action. In Brazil, government subsidies are made available at sequential stages of compliance.

Another enforcement issue, more common in developing countries than in the United States, is exemption by decree. A development project may be required to conduct an EIA on Monday, and by Tuesday the governor or president may exempt the project by decree. A procedure very similar to this allowed completion of the Tennessee Valley Authority (TVA)–Tellico Dam project in East Tennessee.

Another significant difference between the United States and developing countries is that in many developing countries there is less effort to conceal corruption than in the United States. This increases frustration and anxiety among those attempting to carry out EIAs or to enforce environmental laws. A particularly blatant example occurred recently during the Balbina hydroelectric project in Brazil. On one occasion, to satisfy one audience, the proponent announced that only seven families were displaced by the project. On another occasion, the proponent announced that citizens had been compensated for the acquisition of 47 parcels of land. When queried about the ownership of the excess 40 parcels of land, the proponent stated that a businessman from Sáo Paulo had purchased 40 parcels of land early in the development planning and sold them later at an exaggerated price. In the United States, we go to greater efforts to conceal such transactions, thereby minimizing public anxiety.

In some developing countries, EIA is more decentralized than in the United States. The lead agency with licensing responsibilities is often a state agency rather than a federal agency. The roles of central government are primarily those of coordination and resolution of appeals. Perhaps the most significant deficiency of EIAs in most developing countries is that these assessments are project oriented. The practice of conducting an EIA of domestic or foreign policy or legislation has not yet been accepted.

Another significant constraint to effective EIAs in developing countries is economic and political instability. Although an EIA is nonetheless important during periods of political revolution or economic depression, other national priorities relegate an EIA to a relatively low status. Vice President Gore's "strategic environmental initiative" seeks to overcome this. In some ways, Nicaragua is an exception to those countries that downplay EIAs during periods of political or economic instability. The new revolutionary government implemented several forms of environmental protection, even under constraints of agrarian land reform and armed invasion.

In some developing countries, individual ministries of government closely guard their information; information is power, and power is not readily shared. Often, an expatriate consultant can obtain multiministry information more easily than a national.

There are four constraints to EIAs that are universal in developing countries. First, the EIA is not included in development planning early enough. It is belatedly and reluctantly accepted as a constraint to the development process. Accordingly, major benefits of an EIA are often lost. Second, procedures for EIAs are mostly categorical rather than procedural. Procedures are similar to those implemented in the United States in 1970. Today, NEPA focuses on procedures. These procedures emphasize brevity, clarity, and rapid response. In developing countries, the focus is upon categories of projects to be assessed and categories of parameters to be monitored. The goal of assisting a decision maker to make the best possible decision is not served well by this approach. Third, the major environmental impact in developing countries, the one which has the most profound influence, is the rate of change. Unprecedented rates of change are visible on the landscape, but the most profound impacts may be social and cultural. These impacts are less visible but no less significant. Traditional social values, lifestyles, and cultures are being replaced within a single generation. We are best prepared to assess impacts on the physical environment; we are less precise in assessing biological impacts; unfortunately, we are less competent in assessing social impacts which may be of greatest significance in developing countries. Last, EIA in the United States is the result of public pressure applied to the political process. The environmental movement was a popular social revolution in the United States. Environmental protection persists only as long as pressure is exerted from the bottom up. In many developing countries, EIA is a top-down process. An EIA is often initiated by international pressure in the form of conditions attached to foreign assistance. As a consequence, public involvement in the EIA process is very limited. At present, many developing countries are attempting to involve the public in environmental decisionmaking (McCormick 1989).

SIMILARITIES IN EIA PROCESSES AMONG DEVELOPING COUNTRIES

Most developing countries have adapted the NEPA process to local needs and realities. Basic components of the NEPA process that appear in EIAs in most countries include the following:

1. preliminary assessment of the need for an EIS
2. emphasis upon alternatives, scoping, and mitigation
3. categorical exclusion
4. notification of the public
5. draft and final EIS
6. public comment
7. multiagency review
8. provisions for appeal

Figures 1 and 2 provide schematic outlines of EIA procedures in Brazil and China. Brazil is not a developing country, but Brazil does include vast areas (Amazonas) that are relatively underdeveloped. In the Chinese system, the activity most deserving comment is site selection. An obvious advantage of site selection early in the EIA process is that EIA can be based on site-specific data. A disadvantage is that subsequent analysis may reveal problems that could have been resolved by selecting alternative sites.

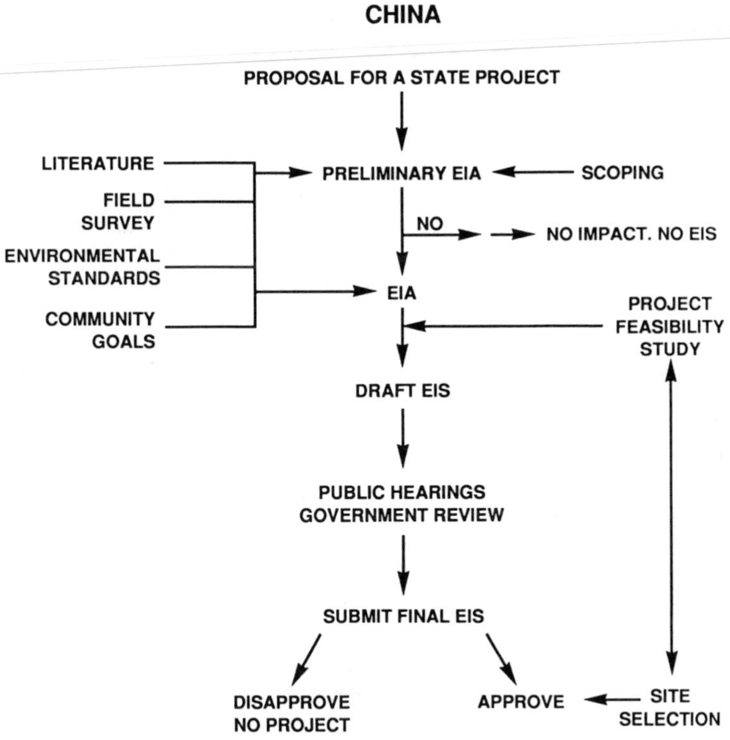

Figure 1. Schematic outline of EIA procedures in China.

One conspicuous feature of the Brazilian system is opportunity for presidential decree to exclude a project from the entire EIA process or to overrule decisions during the process. Another significant feature is the requirement for independent consultants to conduct the EIA and to prepare the EIS with financial support from the proponent. A particularly commendable feature of the system is continued monitoring to ensure compliance with environmental standards.

ADVANTAGES IN DEVELOPING COUNTRIES

An advantage some developing countries have over the United States, in terms of implementing EIAs, is that they often can implement EIAs more expeditiously. Another advantage is that they begin their experience in EIAs with science and technology that is two decades more advanced than that available when the United States initiated the EIA process. Applied ecology was an emerging profession in 1970, remote sensing and geographic information systems were still in research and development stages, and regional and environmental planning were still considered socialist conspiracies. Developing countries have more advanced models of NEPA to simulate than the 1970 version. Also, subsequent environmental legislation, such as the Surface Mining Reclamation Act, provides additional guidelines for environmental assessment not available in 1970.

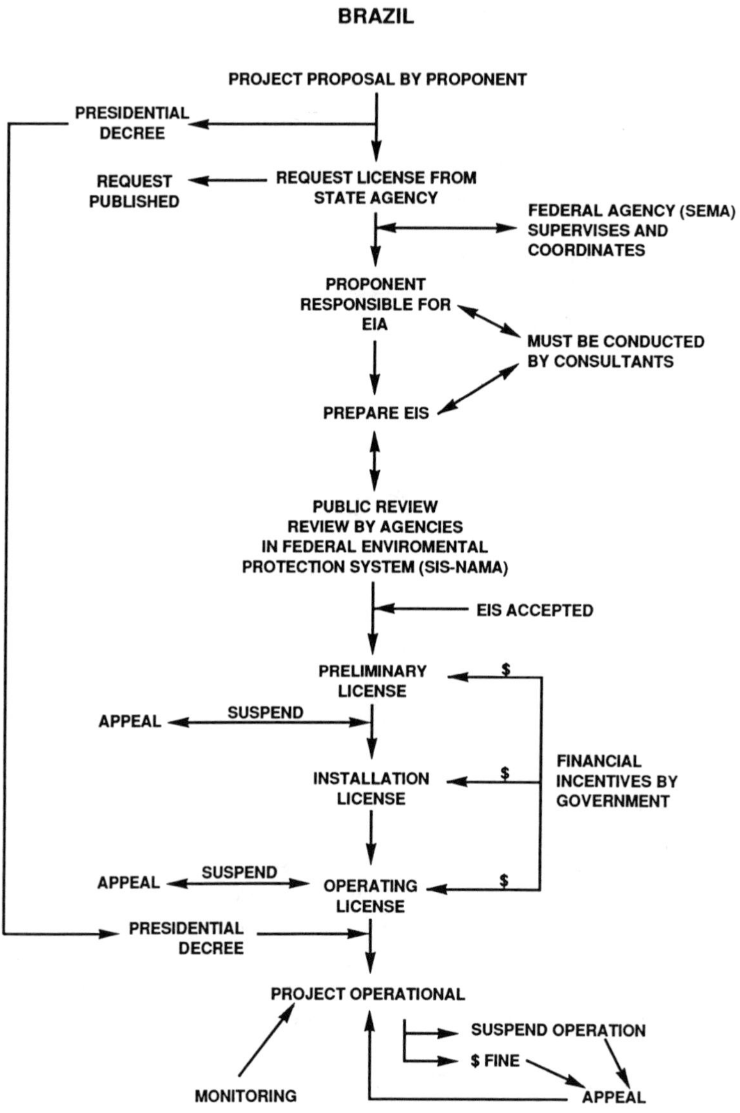

Figure 2. Schematic outline of EIA procedures in Brazil.

SURPRISES

The enthusiastic reception of EIAs by governments has been a pleasant surprise. In China, I was surprised to find EIAs widely implemented. I encountered well-informed Chinese Environmental Protection Agency officers working with development officials in remote regions of the country. Laws requiring EIAs are enforced. Over 1000 EIAs have been conducted in this decade (Russell 1987).

The Chinese Environmental Protection Agency publishes a newsletter that it distributed to over 800,000 individuals.

The most disturbing surprise was in the Amazon of Brazil. Environmental legislation is in place, and conscientious and highly competent people are in positions of responsibility. However, human behavior is out of control. Exponential rates of exploitation exceed even the most diligent protection efforts. Until recently, economic development and environmental degradation in the Amazon were somewhat regulated by the inaccessibility of natural resources. That source of protection has disappeared. Foreign funds and domestic policies are both responsible for the rate and magnitude of environmental degradation. Families being settled in the interior are required by policy to deforest 67% of their land in order to receive title to the land.

While the world has focused its attention upon the cutting and burning of Amazonian forests, a lesser known but more serious problem has emerged. The government has already approved over 60 hydroelectric projects, which collectively are sufficient to put 40% of the Brazilian Amazon underwater. Rivers are being converted to relatively shallow, low-flat, high-temperature reservoirs, whose waters inundate uncut tropical forests. This combination of conditions can be expected to reduce dissolved oxygen to levels insufficient to support desirable life forms and human uses.

Even as foreign investment and domestic policy support forest cutting and inundation, additional significant impacts are resulting from urbanization. Resettlement policies are supported by internationally financed highway development. A duty-free port has been established to encourage international trade in Amazonas. The strategy has been successful in attracting people to the Amazon. In just a few years, the capital city of Manaus has grown from 300,000 to over 1,200,000. It has been impossible for municipal services such as water and waste disposal to keep pace with urbanization and industrialization. Many have found employment and housing; many others have not. They have turned or returned to slash-and-burn agriculture to make a living, but they did not go far — they merely moved from the city to the nearest unoccupied land. As a consequence, slash-and-burn agriculture and intensive fishing have created a zone of resource exploitation that constantly expands from Manaus. This zone increases at a rate approximately equal to the exponential rate of population growth.

PREREQUISITES TO EFFECTIVE EIAs IN DEVELOPING COUNTRIES

1. The most important prerequisite to effective EIAs in developing countries is a commitment from leaders of developing countries to use EIAs in decisionmaking.
2. Host countries are most likely to use EIAs in decisionmaking if the EIA responds to their needs and is designed and implemented by their people.
3. Because of financial and manpower limitations in developing countries, they must pursue the most cost-effective approaches to EIAs. This places a heavy reliance upon scoping. Scoping can be carried out only in the context of a clear national policy for sustained economic and ecological productivity.

4. Ministries of government in developing countries must share environmental information and EIA procedures to a greater degree than in the past. International research programs such as the International Biological Program and Man and the Biosphere have provided generic models of most biomes of the world. Several international nongovernmental organizations cooperated in developing the "World Conservation Strategy," which is a blueprint for environmental decisionmaking at global, national, and local scales. A major part of this global strategy is information sharing and transfer.

   Insufficient information transfer is currently a greater problem than insufficient information. One form of information transfer is environmental education, including public education or intensive training in special skills. At this time, environmental education deserves even higher priority than environmental research. If we fail to acknowledge this, the ecosystems, species, environments, and societies we propose to study or manage may be lost before we commence these noble efforts.
5. The best training in resource management and the EIA is training in the home country. Too often, U.S. institutions or agencies attempt to promote their own image or supplement their own budgets by offering training courses in the United States. Seldom is such training truly relevant to a developing countries' needs. This approach also takes personnel in developing countries away from essential duties for prolonged periods of time. Training in the home country makes it nearly impossible for trainers to ignore cultural differences that influence the effectiveness of the training. International training provided in the United States invariably ignores significant cultural differences, unless trainers themselves have significant overseas experiences in developing countries. To provide training for those from developing countries, go there.

REFERENCES

Braun, R. A. 1987. Environmental Impact Assessment in Brazil. p. 10. In *The Legal Procedure Worldletter. The International Newsletter for Environmental Assessment.* September/October, 1976.

Detwiler, R. P., and C. A. S. Hall. 1988. Tropical forests and the global carbon cycle. *Science* 239:42–47.

Gore, A. 1989. Forum on Global Change and Our Common Future. Address to the National Academy of Sciences, May 1, 1989.

MAB-UNESCO. 1981. Ecology in Action. MAB Poster Production. UNESCO, Rome.

McCormick, J. F. 1989. Seminar on Public Involvement in Environmental Decisionmaking, The University of Tennessee, Knoxville, November 8–9, 1989. Organized for the delegation of Brazilian (Amazon states) Specialists. Support by U.S.A.I.D. and Partners of Americas.

Russell, M. 1987. Environmental Protection under Alternative Economic Systems: A Public Policy Workshop. Department of Economics, The University of Tennessee, Knoxville, November 13, 1987.

Soemarwoto, O. 1984. EIA and energy developments in Indonesia. p. 13. In Proceedings, International Training Course on Environmental Impact Assessment and Land Use Planning, Hong Kong, January 16–21. Padjadjaran University, Bandung, Indonesia.

Speth, G. S. 1986. A new environmental agenda. pp. 22–35. In A. A. Van Vleet (ed.), *Forum for Applied Research and Public Policy*. Tennessee Valley Authority, Knoxville, TN.

Wang, Y. T., L. W. Qi, and Z. Z. Zi. 1984. *The Environmental Impact Assessment System in China*. Environmental Management Institute, Beijing, China.

Potential to Transfer the U.S. NEPA Experience to Developing Countries

T. J. Wilbanks, D. B. Hunsaker, Jr., C. H. Petrich, and S. B. Wright, Oak Ridge National Laboratory, Oak Ridge, TN

ABSTRACT

The U.S. National Environmental Policy Act (NEPA) of 1969 has been important as a learning experience for the United States, but it has also served as a model for the rest of the world. In particular, other countries which benefit from U.S. foreign aid have found NEPA concerns to be of growing importance in the 1980s, and multinational bodies such as The World Bank have increasingly suggested U.S. standards and practices as guidelines to countries without well-defined environmental policies of their own.

With environmental protection becoming a dominant issue in development assistance, it is timely to assess the relevance of the U.S. experience to developing countries. Many lessons are likely to be of substantial importance (e.g., the importance of cumulative impacts, finding a balance between estimating impacts prior to action and monitoring impacts subsequent to action, and potentials to apply modern technologies for impact assessment and mitigation). Other outgrowths of the NEPA process in the United States may be less appropriate to the conditions that exist in many developing countries. Examples might include standards for certain emissions, standards for impact analysis, the size and complexity of impact assessment documents, and mechanisms for public involvement.

In these kinds of connections, the U.S. NEPA experience is already becoming an issue in the expanding dialog about global environmental change. Considered in the light of the experience of other advanced countries, such as West Germany and Great Britain, and in the light of differing circumstances of less-developed countries, such as urgent needs for job creation and shortages of technical capabilities, the U.S. experience has the potential to be transformed from what now appears too often to be a rigid straightjacket into what can be a rich, robust body of lessons learned. In the process, an enhanced transfer of U.S. professional experience and monitoring and control technologies can contribute to our competitiveness in the world economy and an acceleration of environmental management improvements in developing countries.

INTRODUCTION

Driven by concerns about the sustainability of economic growth for developing countries and the world as a whole, environmental management has become one of the most salient issues for development assistance to the Third World. A wide range of motives and messages from all sides of the issue are still tangled, and many of what will eventually be the most important policy directions for U.S. and international organizations are still taking shape. It is clear that systematic assessments of projected environmental impacts are becoming a requirement for major development projects in developing countries, at least where major public sector lenders are involved. In many cases, however, the institutional infrastructures for carrying out these assessments and relating them to the decisionmaking process are poorly developed, and models of appropriate processes for use or adaptation by developing countries are being sought (OECD 1989).

Obviously, the U.S. National Environmental Policy Act (NEPA) experience is one possible model, based on 20 years of well-documented experience and associated with an impressive record in building a national consensus about environmental protection. This paper is a preliminary review of the U.S. NEPA experience as it speaks to developing countries' needs. What have we learned that is relevant? How can programs to assist developing countries be made as effective as possible?

This paper stems from two bodies of experience at the Oak Ridge National Laboratory (ORNL) in Oak Ridge, TN. First, for nearly two decades ORNL has been a major participant in implementing NEPA in the United States and has prepared more than 400 environmental impact statements and assessments, covering every major category of energy technology, as well as programmatic and policy assessments and research and development (R&D) to improve the state of the art (see other ORNL contributions to this volume). Second, since 1982 ORNL has been deeply involved in activities related to the energy and environmental needs of developing countries. Working in 31 countries in Asia, Africa, Latin America, and the Near East (with workshop participation in 10 additional countries), ORNL has addressed issues ranging from energy planning, policymaking, and institution building to power system planning, household fuel options, efficiency improvement potentials, and energy needs of rural areas. For example, ORNL has played a major role in the activities of the Office of Energy and Infrastructure of the U.S.

Agency for International Development (AID). Related to environmental management, these activities have been focused mainly on the power sector (Wilbanks 1990). ORNL is also engaged in related international programs for the U.S. Department of Energy (DOE) and is working in collaboration with AID, the World Bank, and other international agencies on activities such as developing decision support tools for power system lenders and borrowers to use in considering environmental implications of technology and policy options.

ENVIRONMENTAL MANAGEMENT CHALLENGES IN DEVELOPING COUNTRIES

Environmental management is not just a concern of the world's affluent. Developing countries now recognize serious problems with the air quality in many urban-industrial areas. They live with topsoil loss, reservoir siltation, and other effects of deforestation. Impacts on human health of poor sanitation, water pollution, industrial accidents, unregulated pesticide use, and other environmental hazards are a day-to-day reality for many. The concerns of these countries tend to be relatively near term, pragmatic, and operational in a context where environmental impacts are only one part of a much larger development puzzle, but they are real and immediate in many countries.

From their different point of view, industrialized countries tend to care more about the effects of decisions within developing countries on the global environment and on the longer-term sustainability of the development process. Recent attention to global climate change is part of a more general concern about the fragility of the world as an integrated ecosystem, expressed for years in connection with such issues as population growth and the loss of species diversity but given new impetus by evidence of global warming.

Sharing both concerns and, to an increasing degree, bridging between the two perspectives are large nongovernmental environmental interest groups such as the Environmental Defense Fund, the World Resources Institute, the Natural Resources Defense Council, and the International Union for the Conservation of Nature and Natural Resources. From deforestation to energy resource and technology choice, these groups are pushing hard for public attention, policy intervention, and institutional development to improve environmental management in developing countries, and their collective record as a political force in the United States and several European countries gives them influence beyond their membership numbers alone. For example, at least seven far-reaching bills related to environmental management in developing countries were being considered by the U.S. Congress in 1989, and such multilateral lending institutions as the World Bank are moving quickly to respond to criticism that they have not been environmentally responsible enough.

In the meantime, some progress is being made. Several developing countries have enacted environmental legislation and conducted environmental impact assessments. Assistance has been provided with environmental policy and training, either directly through public sector development assistance programs or indirectly through the initiatives of interest groups (e.g., international conferences to promote the exchange of information and to encourage local leaders). In quite a few countries, there are signs that environmental constituencies are developing (Morrison 1989; Durning 1988).

But the challenges that remain are difficult and serious (Wilbanks 1990). Some concern *what* to do in environmental protection. For instance, how much environmental protection makes sense for a particular developing country, given urgent needs to develop resources in order to create jobs and income and heavy demands on investment resources for other priorities? Who should pay for environmental protection in developing countries, both within and between countries? Obviously, where the health of the global environment is the issue, others will have to be prepared to share the costs. Other challenges have to do with *how* to ensure environmental protection. For example, how should decisions be made about environmental protection policies, actions, standards? Who will handle the implementation? How can it be ensured that the desired results will be achieved? How effective are initiatives that are international in origin likely to be without local institutional and political support?

It is in this latter connection — institutions and processes — that the U.S. NEPA experience should be especially instructive, because it has been a living laboratory for institutional development and learning for two decades. The question, of course, is this: To what degree are the lessons we have learned transferable to societies whose cultures, institutions, and environmental contexts are often so different from our own?

THE U.S. NEPA EXPERIENCE AS A MODEL

Central Elements of the Experience

The Essence of NEPA

The U.S. NEPA experience is more than a piece of legislation alone. NEPA provides a statement of national policy (Title I) and assigns to the President's Council on Environmental Quality (CEQ) the responsibility for setting up procedures to implement the policy (Title II). In effect, it establishes a comprehensive structure for environmental protection in the United States: an articulation of national policy (against which particular actions and circumstances can be measured), certain actions and procedures expected of federal agencies when they undertake major activities, and an institution to issue further rules and regulations to assure compliance with federal policy.

More broadly, though, it also includes two decades of experience in implementing NEPA: defining and redefining, shaping and reshaping, a rich body of social learning including components of legislative action, judicial action, administrative action, and social evolution. For example, the effects of NEPA

Table 1. Action-Forcing Provisions of NEPA.

- Preparing for major federal government actions a "detailed statement" on projected environmental impacts
- Using an interdisciplinary approach
- Considering presently unquantified amenities and values
- Studying, developing, and describing alternatives
- Recognizing the wide ranging character of environmental problems
- Making available useful information to individuals and state and local governments
- "Initiating" ecological information in the planning and development of resource-oriented projects
- Assisting the CEQ

have been considerably shaped by a finding by the Supreme Court that the policies and goals set forth in Title I do not constitute judicially enforceable requirements; the finding stated only that federal agencies should consider NEPA's goals when making decisions (Vanderver 1985).

In the absence of intrinsic enforceability, NEPA's importance has derived mainly from eight provisions that can be enforced (Table 1). One provision — requiring an environmental impact statement (EIS) — has dominated the other seven generally considered components of the EIS process, even though this may not have been the original intent of the U.S. Congress. Three other provisions (regarding interdisciplinarity, unquantified values, and alternatives) have received judicial attention beyond the scope of the EIS requirement (Mandelker 1987), although the focus has remained on the NEPA process. Not coincidentally, the action-forcing provisions that have received the least attention are the ones that are least likely to be an explicit part of the EIS process.

This process of implementation has been the catalyst for an evolving dialog about how to incorporate environmental and health concerns in federal governmental decisionmaking. In particular, it has brought the public into the dialog (Bear 1989a) as one of the few ways that individuals and relatively small groups could make their views heard in a world that seemed dominated by large institutions (Wilbanks 1981). At the start of the EIS process, scoping focuses the EIS on key issues, with an opportunity for public involvement. Under NEPA, the subsequent impact assessment is comprehensive and interdisciplinary, extending beyond issues addressed by specific pieces of environmental legislation. For instance, it considers interrelationships among impacts, potential effects from actions (past, present, and future) outside the control of the proponent of the action, and socioeconomic factors that are related to environmental effects. Resulting analyses are presented for public review and comment in a draft EIS, and lengthy hearings are not uncommon. Finally, the EIS process requires that the decision maker explain to the public the rationale for the decision that is the outcome of the process. In essence, then, the U.S. NEPA process has come to focus on *public participation, comprehensiveness, and open communication.*

In the meantime, differences of opinion about how the provisions of NEPA should be interpreted have led to a large body of case law and administrative precedent which is itself a significant part of the U.S. NEPA experience.[1] For example, consider the legislative requirement to "initiate ecological information" (Table 1). When faced with "incomplete or unavailable information," federal agencies preparing EISs are directed by the CEQ regulations (40 CFR 1502.22) to obtain the information if it is relevant to impacts that can reasonably be foreseen and essential to a "reasoned choice among alternatives," as long as the "overall costs of obtaining" the information "are not exorbitant." What constitutes reasonably foreseen, essential, or exorbitant can be a matter of dispute; this requirement has been the focus of a great deal of activity in the lower courts, changes in CEQ regulations,[2] and a Supreme Court decision in 1989.

More broadly, the courts have even looked beyond NEPA itself to shape its reach. For instance, the U.S. Administrative Procedure Act has been applied in a finding that limited substantive reviews to determine if the agency's actions were arbitrary, capricious, or an abuse of discretion (Rivkin 1989). This further illustrates the powerful role of courts in shaping the U.S. NEPA experience, as expressed through uses of EIAs in federal agencies.

In several senses, then, the NEPA process — centered on the EIS process — is more than a piece of legislation. It is a rich collective body of experience with identifying actions, alternatives, and relevant information; comprehensively assessing impacts, with the scope of "comprehensiveness" generally expanding as the state of the art has developed; facilitating public participation, seeking ways to make it better informed and more effective, and learning how to relate it to decisionmaking; sorting out the respective roles of the judicial, legislative, and executive branches of government; and responding to feedback from affected parties, the effects of which have ranged from the redesign of proposed projects to reduce or mitigate impacts to changes in policies and rules.

The NEPA experience is also a body of decisions about federal actions. It is not easy to measure the effect of the process on federal governmental decisionmaking, but it is clear that (after some initial resistance) federal agencies now take the process very seriously indeed (Bear 1989b). Agencies have adopted practices consistent with CEQ regulations, requests for exemptions from NEPA are rare, and requests for deviations from the normal process are also rare. NEPA-related litigation is declining, which suggests that agency practices are falling into line with public expectations (Bear 1989b). And there is a very large stock of anecdotal evidence about project modification (even, in a few cases, cancellation) due to environmental information uncovered through the EIS process. Certainly, one can argue that whole categories of projects that were being proposed before and during the early years of NEPA have disappeared — mainly those with the most serious potentials for adverse environmental impacts (Friesema 1989).

This is not to say that the process is approaching perfection. NEPA has been overtaken by a variety of other environmental legislation and is less dominant

---

[1] In some cases, regulations that pertain to the implementation of NEPA are not in accord with relevant provisions of the act itself.

[2] The only revision since their promulgation in 1978 (Bear 1989a).

in environmental decisionmaking than it once was (Renwick 1988). A recent survey of 75 EISs on coal-fired electric power plants found that 44% of the facilities were already under construction when the final EIS was issued, although the percentage was much smaller for cases in which the EIS process was initiated after the CEQ regulations were promulgated (Schevitz 1987). Caldwell (1989) suggests that many of the "positive opportunities inherent in NEPA have been passed over" as the public has come to view the process as a way to prevent harm rather than to promote good. He cites, for instance, a lack of strong continuing support from the executive and legislative branches of government, leaving too much of the burden of interpretation and implementation to the judicial branch; an overemphasis on procedure compared with substance, perhaps as a result; and a misdirected tendency to criticize NEPA or the EIS process rather than those responsible for environmental policy.

Strengths of NEPA Related to the Developing Context of Countries

For developing countries, the U.S. NEPA experience is by no means an ideal of how to balance economic growth and environmental protection, but some of its components are attractive starting points for designing procedures to minimize environmental degradation. Clearly, the policies and goals articulated in Title 1 of NEPA — especially the goals of (1) attaining the widest range of beneficial resource uses without degradation, risk to health or safety, or other undesirable and unintended consequences and (2) achieving a balance between population and resource use which will permit high standards of living and a wide sharing of life's amenities — are of interest to developing countries. And the eight action-forcing provisions of NEPA are reasonable, well-tested guidelines for impact assessment.

At least as important, though, are the strengths of the EIS process as it has evolved in the United States. These include

- the use of an open scoping process to engender broad participation in defining the focus of impact assessment
- the requirement that alternatives to the proposed action be identified and considered
- a comprehensive approach to assessing impacts, including cumulative effects and interrelationships
- full disclosure of information to interested parties and the public as a basis for interaction and discussion
- mechanisms for objections to be considered openly and fairly, with final recourse to the courts if disputes cannot be resolved otherwise
- the capacity of the approach to evolve with experience

The main strength of the U.S. NEPA process, in fact, is probably its openness. It offers ample opportunities for involvement. It provides strong incentives for federal agencies to develop projects that will be judged environmentally acceptable by a broad cross-section of society. It ensures that all parties to the dialog have access to essentially the same base of information, which has contributed to a much more extensive process of national consensus building about resource and technology choices. As one consequence, it has encouraged parties to decisions about particular facilities (or programs) to consider ways to modify proposed actions to improve their environmental acceptability. It is not a law mandating environmental protection; it is a law creating a process for social dialog about environmental protection.

Limitations of NEPA Related to the Developing Country Context

As indicated previously, however, the NEPA process is not a panacea for developing countries any more than it is for the United States. For example, it is an approach aimed more at information and procedures than implementation and enforcement. It lacks teeth. It does not require follow-up. It does not address *actual* effects as distinguished from *forecasted* effects. It does not apply to private sector actions, in most cases. Perhaps more fundamentally, it has led to a punishment-based system rather than a reward-based system, and it has often led to an emphasis on litigation as the way to resolve disputes. In these senses, at least, it has proved to be a process with weaknesses as well as strengths. Finally, the U.S. NEPA process has evolved out of the U.S. political culture. It is neither culture free nor ideology free. This raises the possibility of limitations in applying it to other cultural and political contexts. Certainly, evidence is already appearing that U.S.-type EIA techniques do not necessarily have the desired effects if they are introduced to an institutional environment unsuited to make use of their results (Wandesforde-Smith et al. 1985).

The Main Lessons We Can Pass on to Developing Countries

After 20 years of living with the evolving NEPA process, the United States has learned some lessons about institutional settings for environmental management. Might some of these lessons be useful to developing countries as they shape their own processes? As a starting point for identifying such lessons, we suggest the following:

- *The likelihood of identifying significant impacts is increased by treating impact assessment (in part) as a social process.* Experience in the United States indicates that such characteristics as comprehensiveness, interdisciplinarity, and broad consultation are essential to ensure that significant impacts of a proposed action are identified. In a world where the most powerful impacts of an action may be indirect, due to interrelationships with other actions, or beyond the scope of standard analytical models, the only way to avoid costly and disruptive mistakes is often to involve a wide range of parties and perspectives in the discussion. The EIA is and should be far more than the mere exercise of a set of quantitative tools.
- *The likelihood of producing decisions and actions that avoid social conflict is increased by ensuring widespread participation.* Not only does broad participation

help in identifying impacts, it also helps in reducing sociopolitical tensions associated with facility, program, and policy decisions. In the very short term, the process of participation may occasionally seem acrimonious — and therefore better avoided — but in the longer run, it (1) informs decision makers about issues and concerns that might be more difficult to address later; (2) reduces the possibility that groups in society will feel alienated by the decisionmaking process; and (3) helps to build a broad consensus about resources, technologies, and institutional arrangements that are acceptable.

- *The environmental management process should be rooted in the indigenous culture for decisionmaking.* Based on experience in the United States and reinforced by experience in other countries, we know that the organizational structure for implementing environmental policies is at least as important as the policies themselves and the tools available for environmental analysis (Horsberry 1983). The particular genius of the NEPA process in the United States has been its close harmony with the way we prefer to make public decisions: in an open, democratic, sometimes confrontational, but eventually consensual manner. Its effectiveness is rooted in our social processes, our communication networks, our interest groups, our longstanding tendency to work toward agreement rather than disagreement (Lakoff 1983), the historic role of our courts as conflict resolvers, our tradition of the accountability of public officials, and the importance in our culture of such conditions as consent and participation (Rayner and Cantor 1987). A process that was structured to work at cross-purposes to our preferences and traditions would be much less effective. (Attempts to implement a nuclear waste disposal process based on federal government primacy, for instance, have been notably ineffective.) Meanwhile, comparative analyses of environmental protection experiences in different countries indicate that no single organizational structure is superior in all cases (Lowry and Carpenter 1985); the structure best suited for a particular country will be related to the ways decisions are most effectively made and implemented in that cultural and political context. In many developing countries, of course, this context is itself evolving, which complicates any attempt to design an ideal approach for a relatively long term.
- *An effective process requires effective nongovernmental environmental interest groups.* We have learned from our experience that a NEPA-like process works effectively only if someone in society "speaks for the environment": participates in the process as an advocate of environmental protection, presses decision makers to give some weight to evidence about adverse environmental impacts, and mobilizes officials and public opinion with regard to environmental issues. Without such a presence, the political balance which determines the eventual decision is much less likely to consider (or at least appropriately value) environmental impacts. Available information goes unused — or, more often, pertinent questions about information requirements are not asked, and inaccurate or incomplete answers are not challenged. Balance requires some degree of organizational and technical strength on every side of an issue (Kash et al. 1976, pp.456–459).
- *Environmental management should be based on operational experience as well as preconstruction estimates.* Clearly, the best information about the impacts of a combination of resources, technology, and institutions is operational experience and not preconstruction estimates — especially for innovative technologies without an extensive track record. Our U.S. experience suggests that relying almost exclusively on impact projections in EIAs is inadequate; in fact, it makes

unrealistic demands on the state of the art for impact assessment. An EIA process should give considerable attention to plans for postconstruction impact monitoring and, based on observed project effects, strategies for mitigation and reassessment, including procedures for sharing information with affected groups and involving them in decisions about responses. Provision should also be made for using information about actual impacts to improve the prediction of similar impacts from future projects (see Bernard, Hunsaker, and Marmorek, in this volume).

Issues in Transferring Our Experience

It would be hubris at its worst, however — and possibly counterproductive to progress with effective environmental management in developing countries — to assume (explicitly or implicitly) that our experience is directly transferable to other cultures and development contexts. The most serious issues related to transfer appear to be the following:

- *The acceptance of impacts and risks in order to accelerate economic development.* In a society where poverty is the most serious form of pollution, some air or water pollution may seem an acceptable price to pay for economic growth, especially when the history of so many of the affluent countries seems to associate industrialization with environmental impacts. Sustained development has historically involved an increased emphasis on some kinds of activities associated with environmental and health impacts, such as industrial development, transportation system development, and pesticide and fertilizer use. For growing populations, progress in these respects will usually be a requirement for ecological balance in some areas, as well as for economic progress. In this sense, increases in environmental impacts of certain types may be necessary in order to reduce environmental impacts of other types. In transferring the U.S. NEPA experience, however, the U.S. process does not appear to be very well-suited for evaluating impacts on a relative basis, either environment vs environment or environment vs economy, except very narrowly for alternatives for a particular function (e.g., generating 100 MW of electric power). Comparing the impacts of, for example, a new hydroelectric power facility with the impacts on sustainable development of a shortage of electricity in the terms relevant to a developing country lies outside the scope of our experience with analyzing the need for power and the associated public dialog — mainly because, in our system, we are not convinced that any significant environmental impacts must be accepted in order to reach economic goals. In addition, it is possible that an EIA process developed mainly to avoid unwanted changes will not necessarily transfer smoothly to a context where the main focus is on wanted changes, such as increased yields and other forms of productivity.
- *Resource requirements for a NEPA-like process.* For many developing countries, the U.S. NEPA process, with its heavy documents and extensive hearings, is too expensive in its requirements for time, expertise, data, and financial support (Luhar and Khanna 1988). For any transfer of the U.S. experience to make sense for these countries, it needs to be accompanied by simplification, especially because it is important that EIAs be conducted by indigenous experts. In some

developing countries, especially in Asia, the local experience with EIAs is already extensive (Roque 1985), but the interest in developing simpler processes remains strong. For instance, the Asian-Pacific Development Centre is currently producing a handbook on "rapid environmental assessment" approaches for its members. Some of this experience with simplification may eventually offer useful lessons to the United States in reevaluating what we ourselves have created.

- *Missing baseline data.* The heart of an EIA is comparing projected future conditions that would occur with a proposed activity with the conditions that would be obtained if that action does not take place. In either case, the starting point is baseline data; that is, a historical time series reporting preexisting levels and trends (including projections without the proposed actions) of air quality, water quality, and other indicators of concern. In most developing countries, certain critical baseline data and other building blocks for characterizing the local environment are missing or unreliable. Frequent problems include lack of appropriate measurements, inadequate specification of data requirements, recording errors, gaps in time series, uncertainty whether data are exact or estimated, and inconsistencies in classification schemes and map scales. But without such a foundation of information, many of the analytical and conflict-resolution approaches characteristic of the U.S. system are difficult to apply. It is encouraging to see that some progress is being made. For example, a *Thailand Natural Resources Profile*, produced in 1987 by the International Union for the Conservation of Nature and Natural Resources (IUCN) and the Thailand Development Research Institute, sets a standard that many countries might emulate; the IUCN is involved in preparing similar profiles with local counterparts in Jordan and Madagascar. Data gaps, however, continue to be a constraint in many areas.
- *The apparent importance of public participation.* It is hard to imagine the U.S. NEPA process without extensive public participation. The entire philosophy imbedded in the process is built on broad, well-informed public involvement as a way to guard against intentional or inadvertent omission or insensitivity. In many developing countries, however, free and open public participation is not a common practice. Decisionmaking is often hierarchical, top-down. To propose a radical departure from this practice is to propose significant sociopolitical change. Moreover, to assume that a broad cross-section of the public can become sufficiently well-informed about technical issues to participate knowledgeably may be unrealistic in some situations, although traditional mechanisms for public dialog have been shown to be effective where they have been tried (Phantumvanit and Nandhabiwat 1989). In any event, wherever public participation cannot reasonably be expected, NEPA-like laws and procedures need to be viewed with great care.
- *Skepticism about U.S. motives.* Finally, since the Stockholm Conference in 1972, if not before, many observers from developing countries have suspected that global environmental protection might be a thinly veiled strategy to maintain the North-South economic gap by discouraging growth. When, for example, influential parties in the United States try to bring pressure on the World Bank to deny loans for coal-fired electric power plants in developing countries in order to reduce the rate of growth in greenhouse gas emissions, while CO_2 emissions from the United States itself rose 7% in 1987–1988 (faster than the global

average) (*World Resources Institute* 1989), such suspicions are reinforced. Obviously, this skepticism can complicate any effort by the United States or others to encourage the use of U.S.-developed tools for environmental management in developing countries.

CONCLUSIONS AND RECOMMENDATIONS

Despite the complexities of transferring U.S. experience to developing countries, we are convinced that the U.S. NEPA experience is relevant to the ongoing dialog about environmental management policies and procedures for developing countries. We suggest that this experience can be converted into several relatively specific guidelines for U.S. and international development assistance agencies in their efforts to improve the art and practice of environmental management in developing countries:

- *Simpler environmental impact assessment processes are needed.* We believe it is possible, given the sizeable experience with many combinations of resources, technologies, and institutions, to translate a U.S. NEPA–like EIA process into a simpler version that will have a relatively high probability of identifying significant adverse impacts and impact issues, focused on potential "showstopper" impacts and postconstruction monitoring and reporting plans. In many cases, microcomputer-based decision-support tools may be applicable. Such tools lead a user through the likely alternatives, major impacts, and principal issues usually associated with each plan and point to ways to make qualitative evaluations, while at the same time documenting the process followed. As mentioned previously, preliminary work on such tools, supported by AID, the World Bank, and others, has already begun.
- *Effective environmental management requires capability and constituency building in developing countries.* Environmental management in developing countries depends mainly on the priorities and capabilities of local institutions and human resources. It cannot be forced from outside, and it cannot be managed from outside, at least for the long term. Helping these institutions and resources to grow in quality, strength, and confidence is a key to making self-sustaining environmental management a reality in most countries. This suggests at least two needs.

    - Indigenous cadres need to be trained to carry out environmental impact assessment and monitoring. Not only do EIA professionals require training and the establishment of institutions, including staff (with equipment) to carry out environmental monitoring, but developing countries need the capability to renew their resources through further in-country training by indigenous institutions. This is a very high priority indeed, even given the fact that these professionals will have to be drawn from a pool of human resources that is already hard pressed in many countries.
    - Indigenous nongovernmental organizations are needed to "speak for the environment" in the national policymaking process. Experience in developing countries is already showing the importance of this dimension of an effective

process. For example, a leading Sri Lankan environmental organization recently won reforms in a controversial tropical forest logging project supported by the World Bank, with the assistance of the Natural Resources Defense Council, a U.S.-based environmental interest group (which funded a trip by a Sri Lankan representative to argue its case with World Bank and U.S. officials). In another case, in this instance within formal governmental EIA processes, local advocates participated in local public hearings along traditional lines in considering the proposed Nam Choan Dam in Thailand, an enormous project which has since been postponed indefinitely, at least partly because of expected impacts on the country's largest natural forest (Phantumvanit and Nandhabiwat 1989). Without such informed local participation, the outcomes might have been different. But in many countries, the local institutions are only now emerging and can benefit from assistance in accelerating their development. It is encouraging to see that such agencies as AID are recognizing this need. In its sector strategy paper for the environment (1983), AID calls for the

> provision of technical assistance to conservation groups in LDCs which have organized in response to the continued degradation of their country's environment and natural resources. These groups can carry out national awareness campaigns, establish local chapters, and sponsor grass roots activities such as tree planting and training in soil conservation measures. By providing assistance to these organizations, A.I.D. has enhanced their effectiveness in promoting public participation in formulating national environmental policies.

More recently, for example, a new AID Regional Environmental and Natural Resources Project in Central America is intended specifically to contribute to regional environmental awareness and related institution building.

- *Rewards should be offered for doing the environmental impact assessment right.* One of the central elements of the U.S. NEPA experience as it has evolved over the last two decades is that NEPA has generally been interpreted as a law that will punish those who fail to comply properly. This punishment, or the threat thereof, usually manifests itself as a potential cancellation or delay of a particular federal action due to failure to properly conduct the NEPA process. As noted by Caldwell (1989), there has been little done to take advantage of the positive opportunities inherent in NEPA. Agencies suffer if they do it wrong, but they are not rewarded for doing it right. As a result, our experience does not have a great deal to offer in the way of a model for implementing a positive reward system for effective EIAs in developing countries. The CEQ is currently working on techniques that will "visibly recognize innovative and vigorous NEPA compliance" in federal agencies (Bear 1989b). Some type of annual award, for example, not for excellent NEPA-like documents but rather for situations in which progressive environmental assessment principles, goals, and actions led to the implementation of more environmentally benign proposals, might be an appropriate starting point for such recognition.
- Decisions about requirements and standards for developing countries should be developed through a dialog with the developing countries concerned. It would

seem inconsistent for us to emphasize information sharing and dialog in our approach to environmental management, at least as defined by NEPA, and then seek to ensure environmental protection in developing countries by emphasizing substantive requirements by external agencies. We have learned from our own experience that dialog can inform and enrich both parties. Rather than focusing narrowly on the question of North-to-South directives about standards, it would seem appropriate for the United States, other affluent countries, and multinational development assistance agencies to devote more attention to structures and procedures for environmentally oriented discussion with developing countries in the spirit of NEPA. At any rate, it is clear that international donors and lenders should be careful not to lay too heavy a hand on the establishment of a formal EIA process (Horsberry 1985). A formal structure can be co-opted by existing power structures for other purposes, unless the local will to manage the environment is strong.

- *Potentials for an EIA process to serve as a catalyst for broader attention to sustainable development should be recognized.* Just as global climate change is serving as a stimulus for discussing broader issues, such as frameworks for international cooperation, an effective EIA process is likely in many countries to be a stimulus for attention to issues such as carrying capacity and national heritage, which are more fundamental than an EIA alone. We believe this potential for an EIA to serve as a trigger for a much broader national and international dialog about development strategies is important and exciting.

These kinds of actions, taken together, offer a real prospect of rapid progress with indigenous environmental management in developing countries, based on institutional mechanisms and resource/technology choices that are suited to each particular national context. If this end can be reached, we will have realized one more significant benefit from our rich experience with NEPA.

REFERENCES

Bear, D. 1989a. NEPA at 19: A primer on an "old" law with solutions to new problems. *Environ. Law Rep.* 19:10060–10069.

Bear D. 1989b. Does NEPA make a difference?. pp. 238–240. In *Environmental Impact Assessment, Proceedings of a Conference on the Preparation and Review of Environmental Impact Statements.* New York State Bar Association, New York.

Bernard, D. P., D. B. Hunsacker, and D. R. Marmorek. 1993. Tools for improving predictive capabilities of environmental impact assessments: structured hypotheses, audits, and monitoring. In S. G. Hildebrand, and J. B. Cannon (eds.), *The Scientific Challenges of NEPA: Future Directions Based on 20 Years of Experinece.* Lewis Publishers, Inc., Chelsea, MI.

Caldwell, L. K. 1989. NEPA and the EIS — What we should have learned. pp. 16–18. In *Environmental Impact Assessment, Proceedings of a Conference on the Preparation and Review of Environmental Impact Statements.* New York State Bar Association, New York.

Durning, A. 1988. Action at the Grassroots: Fighting Poverty and Environmental Decline. Worldwatch Paper No. 88. Worldwatch Institute, Washington, D.C.

Friesema, H. P. 1989. The unclear connection between the scientific quality of environmental impact statements and public policy: Experiences in the U.S.A. pp. 313–322. In H. Paschen (ed.), *The Role of Environmental Impact Assessment in the Decision-Making Process.* Eric Schmidt Verlag, Federal Republic of Germany.

Horsberry, J. 1983. Establishing environmental guidelines for development of AID projects: The institutional factor. *Environ. Impact Assess. Rev.* 4(1):98–102.

Horsberry, J. 1985. International organization and EIA in developing countries. *Environ. Impact Assess. Rev.* 5:207–222.

Kash, D. E., et al. 1976. *Our Energy Future.* University of Oklahoma Press, Norman.

Lakoff, S. 1983. The 'energy crisis' and the liberal tradition: Ideological factors in the policy debate. *Mater. Soc.* 7:453–464.

Lowry, K., and R. A. Carpenter. 1985. Institutionalizing sustainable development: Experiences in five countries. *Environ. Impact Assess. Rev.* 5:239–254.

Luhar, A. K., and P. Khanna. 1988. Computer-aided rapid environmental impact assessment. *Environ. Impact Assess. Rev.* 8:9–25.

Mandelker, D. R. 1987. *NEPA Law and Litigation.* Callaghan and Company, Wilmette, IL.

Morrison, D. 1989. Personal communication.

OECD. 1989. Strengthening Environmental Cooperation with Developing Countries. Organization for Economic Cooperation and Development, Paris.

Phantumvanit, D., and W. Nandhabiwat. 1989. The Nam Choan Dam controversy: An EIA in practice. *Environ. Impact Assess. Rev.* 9:135–147.

Rayner, S., and R. A. Cantor. 1987. How fair is safe enough? The cultural approach of societal technology choice. *Int. J. Risk Anal.* 7:3–9.

Renwick, W. H. 1988. The eclipse of NEPA as environmental policy. *Environ. Manage.* 12(3):267–272.

Rivkin, D. 1989. The environmental impact assessment process, administrative decisionmaking, citizen advocacy, and the courts. In H. Paschen (ed.), *The Role of Environmental Impact Assessment in the Decision-Making Process.* Eric Schmidt Verlag, Federal Republic of Germany.

Roque, C. R. 1985. Environmental impact assessment in the Association of Southeast Asian Nations. *Environ. Impact Assess. Rev.* 5:257–263.

Schevitz, G. 1987. Empirical Analysis of 75 Environmental Impact Statements for Coal-Fired Power Plants in the United States of America. Kernforschungszentrum Karlsruhe, Abteilung für Angewandte System Analyse, Karlsruhe, Federal Republic of Germany.

Vanderver, T. A., Jr. 1985. National Environmental Policy Act. pp. 370–397. In *Environmental Law Handbook.* Government Institute, Rockville, MD.

Wandesforde-Smith, G., R. A. Carpenter, and J. Horsberry. 1985. EIA in developing countries: An introduction. *Environ. Impact Assess. Rev.* 5:201–206.

Wilbanks, T. J. 1981. Building a consensus about energy technologies. ORNL-5784. Oak Ridge National Laboratory, Oak Ridge, TN.

Wilbanks, T. J. 1990. Implementing environmentally sound power sector strategies in developing countries. *Annu. Rev. Energy* 15.

World Resources Institute. 1989. U.S. contribution to global carbon dioxide emissions. Memorandum from J. J. Mackenzie, September 20, 1989.

The Application of Environmental Assessments in Indonesia: Lessons from NEPA

A. N. Katili, Agency for the Assessment and Application of Technology, Hakarta, Indonesia

ABSTRACT

Environmental preservation and protection in Indonesia are stipulated in its environmental law. Formally titled Indonesia Act No. 4 of 1982, the law covers rights, obligations, authorities, protection of the living environment, institutions, compensation and restoration, and penalties.

Enforcing the provisions in the law is the responsibility of the Ministry of Population and the Environment. The ministry also develops air and water quality standards, as well as creates guidelines for environmental impact analysis. The guidelines have become a regulation, effective 1987. The regulation stresses the requirement of environmental impact analysis, environmental management, monitoring, and rehabilitation on activities that will affect people and their environment. The regulation covers planned and established activities.

This paper explores the way in which environmental impact analysis affects the environmental policy in Indonesia. The methods of and the approaches to predicting environmental impacts in Indonesia are based on National Environmental Policy Act (NEPA) experiences.

The environmental impact analysis of a project in Indonesia involves five documents: the description of the existing environment, the terms of reference, the environmental impact analysis, the environmental management plan, and the environmental monitoring plan. This process has its costs and benefits. Resources are needed to develop the document because the project owner has to pay for consultants. In addition, the documents have to be approved by responsible people in the government, which leads to bureaucracy. On the other hand, people's perception of environmental protection has improved. Courses regarding environmental analysis were developed; thus, more and more people have the ability to conduct environmental impact analysis.

INTRODUCTION

A country's development in general covers the utilization and processing of natural resources, as well as the preservation of these resources and the environment. Indonesia, a country with 179 million people, has a land area of 1.9 million km^2 and three times as large an ocean area. Development in the country is essential to increase welfare, reduce unemployment, minimize dependency on agricultural land, and limit slash-and-burn activities.

The large percentage of Indonesia's gross domestic production in 1987 comes from the agricultural sector (25.5%), followed by commerce (16.8%), manufacturing (13.9%), and mining (13.1%) (Indonesia Investment Coordinating Board 1987). Approximately 40% of the country's export earning comes from the mining sector.

Environmental problems associated with development include domestic waste, industrial waste, and social change. The World Environment Center sees Indonesia's serious environmental problems as caused by rapid agricultural and industrial development and by population growth (Baker et al. 1985). These problems are enhanced by the lack of society's participation in waste disposal, minimum safe waste disposal area, and inefficiency of waste collection.

GOVERNMENTAL REGULATIONS

Concerns regarding environmental preservation and protection were stipulated in Indonesia's environmental law, formally titled Indonesia Act No. 4 of 1982, which contains basic provisions for the management of the environment. In general, the law covers rights, obligations, authorities, protection of the living environment, institutions, compensation and restoration, and penalties. Emphasis is on the forests and the freshwater resources. A statement on environmental impact assessment states the requirements for Article 16: "Every plan which is considered likely to have a significant impact on the environment must be accompanied with an analysis of environmental impact carried out according to government regulations."

Thus, Indonesia legislation resembles National Environmental Policy Act (NEPA) Subchapter I, Section 4232(C), which says, "Include in every recommendation or report on proposals for legislation and other major federal actions significantly affecting the quality of the human environment, a detailed statement by the responsible official on (i) the environmental impact of the proposed action." Most Indonesian literature on environmental impact analysis states that NEPA is the prototype for environmental impact analysis.

APPLICATION OF ENVIRONMENTAL IMPACT ASSESSMENT

The Ministry of Population and the Environment is responsible for enforcing the provisions of Indonesia's environmental law. The ministry also develops air and water quality standards, as well as creates guidelines for environmental impact analysis. The guidelines have become a regulation, effective 1987. This regulation stresses the requirement of environmental impact analysis, environmental management, monitoring, and rehabilitation on activities that will affect people and their environment.

Based on this regulation, a committee has to be formed in the government departments responsible for agriculture, commerce, manufacturing, and mining. In the case of mining, the committee has to be formed in the Department of Mines and Energy (Katili 1988). One of the committee's tasks is evaluating environmental impact assessments submitted.

The Department of Mines and Energy coordinates all mining development in Indonesia. There are four principal director generals in the department. Currently, mineral development responsibilities lie with the Director General of General Mines who coordinates development activities in mining industries and the Director General of Geology and Mineral Resources who coordinates the government's geoscience activities. The other two director generals are the Director General of Oil and Natural Gas and the Director General of Electricity and New Energy.

A license for mining is issued if a company meets all requirements. Mining authorization is divided into several categories: survey, exploration, exploitation, beneficiation, transportation, and selling. The regulations from the Department of Mines and Energy start at the exploration stage, where a report is needed on the environmental setting of a prospective mining area. Descriptions are also given of a project's positive and negative impacts on the physical, chemical, biological, sociological, economic, and cultural environment. An authorization for exploitation will not be granted if an environmental impact statement (EIS) is not submitted. In general, an EIS of a mining operation covers the objectives and benefits of a project and its alternatives; the project proposal; the area's natural resources and its environmental, sociological, economic, and cultural situation; and the possible environmental impact and mitigation, as well as the possible follow-up impact. Once authorization is given and the operation is under way, a company has to submit a yearly environmental report on the environmental activities being done in the area. This is a much broader version of an exploration report (Katili 1988).

Other government agencies also ensure that the plan for environmental protection is included in most industrial activities. An example is a regulation issued by

Indonesia's Investment Coordinating Board. Domestic and foreign capital investment in Indonesia has to be controlled by the board, which requires documentation and information if a company is to invest in Indonesia.

Domestic capital investment needs the articles of association, the company's tax registration code number, a bank reference, a description of production process, and a description of pollution-prevention methods. Foreign capital investment (joint venture) requires a power of attorney to sign/submit applications, the foreign companies' Annual General Report, the articles of association of both the foreign and the Indonesian companies, a tax registration code number, a joint venture agreement, the production process, a bank reference, and a description of pollution-prevention methods (Indonesia Investment Coordinating Board 1987).

THE COSTS AND BENEFITS OF ENVIRONMENTAL IMPACT ASSESSMENT (EIA) IN INDONESIA

Environmental laws have been promulgated and governmental agencies have passed internal regulations to integrate the laws in major activities. However, as environmental groups argue, Indonesia's environment will not be improved unless the laws are enforced strictly and more personnel are trained to deal with resource management.

Among the government's efforts to enforce the laws is the development of Environmental Study Centers. Since 1979, a total of 29 such centers have been opened in state universities around the country (Danusaputro 1984). Each center specializes in certain fields depending on its location; these specializations include marine environment, human settlement, environmental health, industrial health, and agriculture.

A few of these centers cooperate with national enterprises or state-owned companies to assess the environmental impact of proposed projects. The centers also provide two levels of environmental impact courses: level A, a 2-week course, and level B, a 3-month course. The first course covers basic environmental science, regulations, and methods of preparing an EIA, as well as visits to projects that conducted EIAs. The latter is an advanced course with practice in the field.

The courses have been attended by people with different educational backgrounds, ranging across science, engineering, social studies, law, and agriculture. Their occupations also vary; participants come from industry, the private sector, central government, local government, nongovernmental organizations and research agencies, and include university lecturers and consultants. Attendance for a single course in an environmental center may reach 20, and each center may have 2 or 3 level A courses per year. Their differing educational backgrounds and occupations ensure that the participants can share their expertise and experiences in dealing with environmental problems. By attending the course, the participants also develop their networking contacts.

An EIA of a proposed project is considered valid if it is organized by personnel who have passed the level B course, who have been trained abroad, or who have experience preparing EIAs. The creation of new jobs in the environmental field also has increased applicants to master's programs in environmental management; thus, more college-educated personnel were trained in the subject of EIA. One benefit of EIA regulations is that more and more people will be trained in conducting environmental impact analysis as well as gaining environmental knowledge. In addition, cooperation has been established between environmental centers and private companies to monitor environmental matters, including pollution and environmental impacts, in the companies' area of operation.

The regulation to conduct EIAs in Indonesia was stipulated in 1987; however, EIAs were carried out years before the law was stipulated even though these early reports have a different format. Most loans from international agencies such as the World Bank and the U.S. Agency For International Development require an EIA; as a result, the EIA became widespread even though it was not fully understood. An EIA was conducted even for projects that were already in operation (Soemarwoto 1988; Soeratmo 1988). Thus, even before 1987, EIAs were conducted for projects such as transmigration, dams, power plants, agribusiness, and other industries such as pulp and fertilizers.

Soemarwoto (1988) indicates that as sometimes happened in developed countries such as the United States, some EIAs in Indonesia were made only to fulfill a requirement. The reports were not used for planning or development of a project since they were finished very late, there is no monitoring rule, and there is a tendency to use an EIS to justify a project.

Opponents of the regulation say that EIAs may create bureaucracy and high-cost economy, while supporters say that it is the way to protect the country's environment. The following section describes the regulation and the methods by which an EIA is applied in Indonesia.

REQUIREMENTS AND METHODS

The EIA of a project in Indonesia involves five documents: (1) the description of environmental setting of a proposed project area, (2) the terms of reference, (3) the EIA, (4) the environmental management plan, and (5) the environmental monitoring plan.

The first document is used for screening to see whether the project will cause a significant impact on the environment. If there will be significant impact, then the terms of reference and the EIA must be prepared. The proposed project will be denied if the negative impact is greater than the positive impact. The fourth and fifth documents are required even if there is no significant impact resulting from a proposed project. One EIA may take about 12–18 months to finish because a committee evaluates the five documents for a proposed project at least five times until the report is approved.

The familiar steps of an EIA applied in Indonesia cover screening, scoping, evaluating impact, proposing the environmental management plan, and report writing. Methods of impact evaluation applied in Indonesia include the Checklist

Method, the Battelle Environmental Evaluation System, the Leopold Matrices, the Network Method, and the Overlay Method.

From the summaries of eight EISs related to coal development in Indonesia prepared between 1982 and 1987, one can conclude that the steps generally taken include survey of the proposed project area; observation; interview; literature study of relevant reports; and review of the area's statistics, which are available in agency or governmental offices. The reports consist of activities proposed, impacts predicted, assessments of environmental quality (water, air, forest area), and conclusions and recommendations (KLH and PDIN-LIPI 1987).

Soemarwoto (1988) implies that the utilization of EIAs in Indonesia is sometimes ineffective and suggests the following ways to increase their effectiveness:

1. Explain to project planners and owners that the EIA is a tool to improve planning for development, as well as to save unnecessary environmental costs in the future.
2. Improve the scoping of the assessment because most EISs contain irrelevant data.
3. Improve the writing so that the reports may be understood by the layperson while remaining scientifically justifiable.
4. Specify the recommendations by eliminating general statements such as "environmental management is needed in the proposed project." The recommendations must explain what steps should be taken.
5. Assign an agency to monitor whether the recommendations in the reports are carried out by project owners.

Project planners' and owners' perception of EIAs may be changed if they are faced with hard facts and cases concerning its benefits. Most cases reviewed, however, were those that occurred in developed countries, especially the United States.

CONCLUSIONS AND RECOMMENDATIONS

The government of Indonesia tries to protect the country's environment by issuing laws and regulations. To enforce the law, all Indonesian agencies and departments whose activities cause environmental impact integrate the law into their operation. The application of environmental impact assessment is the main focus of Indonesia's environmental policy. The methods utilized for conducting EIAs in Indonesia are based on those applied in the United States.

EIA regulation has increased the number of people who are experts as well as those who are interested in environmental and resource management. On the other hand, the necessary regulation may increase bureaucracy and high-cost economy.

The EIA regulation in Indonesia was catalyzed by the requirements from international agencies. EIA reports available indicate that before the regulation was promulgated in 1987, EIAs had been conducted in Indonesia for approximately 10 years. Weaknesses of EIA application should not be permitted to hinder cooperation between departmental and nondepartmental agencies in Indonesia to provide environmental data. EIA regulation has been in effect for only 2 years, and evaluating EIAs based on 2 years of data is misleading. After the EIA regulation in Indonesia has been in effect for 5 years, a study should be conducted. The study should not be limited to reviewing available reports, but should include interviews with personnel who prepare them as well as those who utilize the reports. The personnel can provide input on the applicability of the regulation. The study also should be able to evaluate the extent to which the regulation has created a high-cost economy.

REFERENCES

Baker, M., L. Bassett, and A. Ellington. 1985. The World Environment Handbook: A Directory of Government Natural Resource Management Agencies and Nongovernmental Environment Organizations in 145 Countries. The World Environment, New York.

Danusaputro, M. 1984. Hukum Lingkungan. Buku V: Sektoral. Jilid 2, Binacipta, Bandung, Indonesia.

Indonesia Investment Coordinating Board. 1987. Indonesia: A Brief Guide for Foreign Investors. BKPM-BAI, Jakarta, Indonesia.

Katili, A. N. 1988. Evaluation of environmental management of one nickel mining company in Indonesia with special emphasis on water quality. Ph.D. dissertation. University of Michigan, Ann Arbor.

KLH and PDIN-LIPI. 1987. Sari Penelitian Analisis Dampak Lingkungan. KLH and PDIN-LIPI, Jakarta, Indonesia.

Soemarwoto, O. 1988. Analisis Dampak Lingkungan. Gadjah Mada University Press, Yogyakarta, Indonesia.

Soeratmo, G. F. 1988. Analisis Mengenai Dampak Lingkungan. Gadjah Mada University Press, Yogyakarta, Indonesia.

INDEX